# Lecture Notes in Computer Science    10305

Commenced Publication in 1973
Founding and Former Series Editors:
Gerhard Goos, Juris Hartmanis, and Jan van Leeuwen

Ignacio Rojas · Gonzalo Joya
Andreu Catala (Eds.)

# Advances in Computational Intelligence

14th International Work-Conference
on Artificial Neural Networks, IWANN 2017
Cadiz, Spain, June 14–16, 2017
Proceedings, Part I

 Springer

*Editors*
Ignacio Rojas
Universidad de Granada
Granada
Spain

Gonzalo Joya
University of Malaga
Malaga
Spain

Andreu Catala
Polytechnic University of Catalonia
Vilanova i la Geltrú, Barcelona
Spain

ISSN 0302-9743          ISSN 1611-3349   (electronic)
Lecture Notes in Computer Science
ISBN 978-3-319-59152-0       ISBN 978-3-319-59153-7   (eBook)
DOI 10.1007/978-3-319-59153-7

Library of Congress Control Number: 2017940386

LNCS Sublibrary: SL1 – Theoretical Computer Science and General Issues

Printed on acid-free paper

This Springer imprint is published by Springer Nature
The registered company is Springer International Publishing AG
The registered company address is: Gewerbestrasse 11, 6330 Cham, Switzerland

# Preface

We are proud to present the set of final accepted papers for the 13th edition of IWANN – the International Work-Conference on Artificial Neural Networks – held in Cadiz, Spain, during June 14–16, 2017.

IWANN is a biennial conference that seeks to provide a discussion forum for scientists, engineers, educators, and students about the latest ideas and realizations in the foundations, theory, models, and applications of hybrid systems inspired by nature (neural networks, fuzzy logic, and evolutionary systems) as well as in emerging areas related to these areas. As in previous editions of IWANN, this year's event also aimed to create a friendly environment that could lead to the establishment of scientific collaborations and exchanges among attendees. The proceedings include all the presented communications to the conference. The publication of an extended version of selected papers in a special issue of several specialized journals (such as *Neurocomputing, Soft Computing,* and *Neural Proccesing Letters*) is also foreseen.

Since the first edition in Granada (LNCS 540, 1991), the conference has evolved and matured. The list of topics in the successive call for papers has also evolved, resulting in the following list for the present edition:

1. Mathematical and theoretical methods in computational intelligence. Mathematics for neural networks. RBF structures. Self-organizing networks and methods. Support vector machines and kernel methods. Fuzzy logic. Evolutionary and genetic algorithms.
2. Neurocomputational formulations. Single-neuron modelling. Perceptual modelling. System-level neural modelling. Spiking neurons. Models of biological learning.
3. Learning and adaptation. Adaptive systems. Imitation learning. Reconfigurable systems. Supervised, non-supervised, reinforcement and statistical algorithms.
4. Emulation of cognitive functions. Decision-making. Multi-agent systems. Sensor mesh. Natural language. Pattern recognition. Perceptual and motor functions (visual, auditory, tactile, virtual reality, etc.). Robotics. Planning motor control.
5. Bio-inspired systems and neuro-engineering. Embedded intelligent systems. Evolvable computing. Evolving hardware. Microelectronics for neural, fuzzy and bioinspired systems. Neural prostheses. Retinomorphic systems. Brain–computer interfaces (BCI). Nanosystems. Nanocognitive systems.
6. Advanced topics in computational intelligence. Intelligent networks. Knowledge-intensive problem-solving techniques. Multi-sensor data fusion using computational intelligence. Search and meta-heuristics. Soft computing. Neuro-fuzzy systems. Neuro-evolutionary systems. Neuro-swarm. Hybridization with novel computing paradigms.
7. Applications. Expert systems. Image and signal processing. Ambient intelligence. Biomimetic applications. System identification, process control, and manufacturing. Computational biology and bioinformatics. Parallel and distributed computing. Human–computer interaction, Internet modelling, communication and networking.

Intelligent systems in education. Human–robot interaction. Multi-agent systems. Time series analysis and prediction. Data mining and knowledge discovery.

At the end of the submission process, and after a careful peer review and evaluation process (each submission was reviewed by at least two, and on average 2.8, Program Committee members or additional reviewers), 126 papers were accepted for oral or poster presentation, according to the recommendations of the reviewers and the authors' preferences.

It is important to note, that for the sake of consistency and readability of the book, the presented papers are not organized as they were presented in the IWANN 2017 sessions, but classified under 21 chapters. The organization of the papers is in two volumes, arranged according to the topics list included in the call for papers. The first volume (LNCS 10305), entitled "Advances in Computational Intelligence. IWANN 2017. Part I" is divided into nine main parts and includes the contributions on:

1. Bio-inspired Computing
2. E-Health and Computational Biology
3. Human–Computer Interaction
4. Image and Signal Processing
5. Mathematics for Neural Networks
6. Self-Organizing Networks
7. Spiking Neurons
8. Artificial Neural Networks in Industry, ANNI 2017 (Special Session, organized by: Dr. Ahmed Hafaifa, Dr. Kouzou Abdellah, and Dr. Guemana Mouloud)
9. Machine Learning for Renewable Energy Applications (Dr. Sancho Salcedo Sanz, and Dr. Pedro Antonio Gutiérrez)

In the second volume (LNCS 10306), entitled "Advances in Computational Intelligence. IWANN 2017. Part II" is divided into 12 main parts and includes the contributions on:

1. Computational Intelligence Tools and Techniques for Biomedical Applications (Special Session, organized by: Dr. Miguel Atencia, Dr. Leonardo Franco, and Dr. Ruxandra Stoean)
2. Assistive Rehabilitation Technology (Special Session, organized by: Dr. Oresti Baños and Dr. Jose A. Moral-Muñoz)
3. Computational Intelligence Methods for Time Series (Special Session, organized by: Dr. German Gutierrez and Dr. Héctor Pomares)
4. Machine Learning Applied to Vision and Robotics (Special Session, organized by: Dr. José García-Rodríguez, Dr. Enrique Dominguez, Mauricio Zamora, and Dr. Eldon Caldwel)
5. Human Activity Recognition for Health and Well-Being Applications (Special Session, organized by: Dr. Daniel Rodríguez-Martín and Dr. Albert Samà)
6. Software Testing and Intelligent Systems (Special Session, organized by: Dr. Manuel Núñez and Pablo Cerro Cañizares)
7. Real-World Applications of BCI Systems (Special Session, organized by: Dr. Ricardo Ron and Dr. Ivan Volosyak)

8. Machine Learning in Imbalanced Domains (Special Session, organized by: Dr. Jaime S. Cardoso and Dr. María Pérez Ortíz)
9. Surveillance and Rescue Systems and Algorithms for Unmanned Aerial Vehicles (Special Session, organized by: Dr. Wilbert Aguilar)
10. End-User development for Social Robotics (Special Session, organized by: Igor Zubrycki, Hoang-Long Cao, and Dr. Emilia Barakova)
11. Artificial Intelligence and Games (Special Session, organized by: Dr. Antonio J. Fernández-Leiva, Dr. Antonio Mora-García, and Dr. Pablo García Sánchez)
12. Supervised, Non-supervised, Reinforcement and Statistical Algorithms

In this edition of IWANN 2017, we were honored to have the following invited speakers:

- Dr. Matthias Rauterberg, Technische Universiteit Eindhoven, The Netherlands: "How to Design for the Unconscious"
- Prof. Ulrich Rückert, Bielefeld University, Germany: "Cognitronics: Resource-efficient Architectures for Cognitive Systems"
- Prof. Le Lu, U.S. National Institutes of Health, USA: "Towards Big Data, Weak Label and True Clinical Impact on Medical Image Diagnosis: The Roles of Deep Label Discovery and Open-Ended Recognition"

The 14th edition of the IWANN conference was organized by the University of Granada, University of Malaga, Polytechnical University of Catalonia, together with the Spanish Chapter of the IEEE Computational Intelligence Society. We wish to thank to the University of Cadiz for their support and grants.

We would also like to express our gratitude to the members of the different committees for their support, collaboration, and good work. We especially thank the local Organizing Committee, Program Committee, the reviewers, invited speakers, and special session organizers. Finally, we want to thank Springer, and especially Alfred Hofmann and Anna Kramer for their continuous support and cooperation.

June 2017                                                    Ignacio Rojas
                                                             Gonzalo Joya
                                                             Andreu Catala

# Organization

## Program Committee

| | |
|---|---|
| Leopoldo Acosta | University of La Laguna, Spain |
| Vanessa Aguiar-Pulido | RNASA-IMEDIR, University of A Coruña, Spain |
| Arnulfo Alanis Garza | Instituto Tecnologico de Tijuana, Mexico |
| Ali Fuat Alkaya | Marmara University, Turkey |
| Amparo Alonso-Betanzos | University of A Coruña, Spain |
| Juan Antonio Alvarez-García | University of Seville, Spain |
| Jhon Edgar Amaya | University of Tachira (UNET), Venezuela |
| Gabriela Andrejkova | University of Pavol Jozef Safarik Kosice, Slovakia |
| Cesar Andres | Universidad Complutense de Madrid, Spain |
| Miguel Angel Lopez | Lopez University of Cádiz, Spain |
| Anastassia Angelopoulou | University of Westminster, UK |
| Plamen Angelov | Lancaster University, UK |
| Davide Anguita | University of Genoa, Italy |
| Cecilio Angulo | Universitat Politecnica de Catalunya, Spain |
| Javier Antich | Universitat de les Illes Balears, Spain |
| Angelo Arleo | CNRS, Pierre and Marie Curie University of Paris VI, France |
| Corneliu Arsene | SC IPA SA |
| Miguel Atencia | University of Malaga, Spain |
| Jorge Azorín-López | University of Alicante, Spain |
| Davide Bacciu | University of Pisa, Italy |
| Javier Bajo | Universidad Politécnica de Madrid, Spain |
| Juan Pedro Bandera Rubio | ISIS Group, University of Malaga, Spain |
| Cristian Barrué | Technical University of Catalonia, Spain |
| Andrzej Bartoszewicz | Technical University of Lodz, Poland |
| Bruno Baruque | University of Burgos, Spain |
| David Becerra Alonso | University of the West of Scotland, UK |
| Lluís Belanche | Universitat Politecnica de Catalunya, Spain |
| Sergio Bermejo | Universitat Politecnica de Catalunya, Spain |
| Francesc Bonin | Universitat de les Illes Balears, Spain |
| Francisco Bonnín Pascual | Universitat de les Illes Balears, Spain |
| Julio Brito | University of La Laguna, Spain |
| Antoni Burguera | Universitat de les Illes Balears, Spain |
| Joan Cabestany | Universitat Politecnica de Catalunya, Spain |
| Inma P. Cabrera | University of Malaga, Spain |
| Tomasa Calvo | University of Alcala, Spain |
| Jose Luis Calvo Rolle | University of A Coruna, Spain |

| | |
|---|---|
| Francesco Camastra | University of Naples Parthenope, Italy |
| Carlos Carrascosa | GTI-IA DSIC, Universidad Politecnica de Valencia, Spain |
| Luis Castedo | Universidad de A Coruña, Spain |
| Pedro Castillo | University of Granada, Spain |
| Andreu Catalá | Universitat Politecnica de Catalunya, Spain |
| Ana Cavalli | Institut Mines-Telecom/Telecom SudParis, France |
| Miguel Cazorla | University of Alicante, Spain |
| Wei Chen | Eindhoven University of Technology, The Netherlands |
| Jesus Cid-Sueiro | Universidad Carlos III de Madrid, Spain |
| Maximo Cobos | Universidad de Valencia, Spain |
| Valentina Colla | Scuola Superiore S. Anna, Italy |
| Pablo Cordero | Universidad de Málaga, Spain |
| Francesco Corona | TKK |
| Ulises Cortes | Universitat Politecnica de Catalunya, Spain |
| Marie Cottrell | SAMM, Université Paris 1 Panthéon-Sorbonne, France |
| Raúl Cruz-Barbosa | Universidad Tecnológica de la Mixteca, Mexico |
| Manuel Cruz-Ramírez | University of Córdoba, Spain |
| Erzsébet Csuhaj-Varjú | Eötvös Loránd University, Hungary |
| Daniela Danciu | University of Craiova, Romania |
| Suash Deb | C.V. Raman College of Engineering, India |
| Angel Pascual Del Pobil | Universitat Jaume I, Spain |
| Enrique Dominguez | University of Malaga, Spain |
| Julian Dorado | Universidade da Coruña, Spain |
| Abrahan Duarte | Universidad Rey Juan Carlos, Spain |
| Richard Duro | Universidade da Coruna, Spain |
| Gregorio Díaz | University of Castilla - La Mancha, Spain |
| Emil Eirola | Aalto University, Finland |
| Patrik Eklund | Umea University, Sweden |
| Javier Fernandez De Canete | University of Malaga, Spain |
| Francisco Fernandez De Vega | Universidad de Extremadura, Spain |
| Alberto Fernandez Gil | CETINIA, Rey Juan Carlos University, Spain |
| Enrique Fernandez-Blanco | University of A Coruña, Spain |
| Manuel Fernández Carmona | Universidad de Málaga, Spain |
| Antonio J. Fernández Leiva | Universidad de Málaga, Spain |
| Francisco Fernández Navarro | University of Córdoba, Spain |
| Carlos Fernández-Lozano | Universidade da Coruña, Spain |
| Jose Manuel Ferrandez | Politecnica Cartagena, Spain |
| Ricardo Ferreira | Nove de Julho University |
| Aníbal R. Figueiras-Vidal | Universidad Carlos III de Madrid, Spain |
| Oscar Fontenla-Romero | University of A Coruña, Spain |
| Colin Fyfe | University of the West of Scotland, UK |
| Emilio Garcia | Universitat de les Illes Balears, Spain |

| | |
|---|---|
| Miquel Massot | University of the Balearic Islands, Spain |
| Francesco Masulli | University of Genoa |
| Montserrat Mateos | Universidad Pontificia de Salamanca, Spain |
| Jesús Medina-Moreno | University of Cadiz, Spain |
| Maria Belen Melian Batista | University of La Laguna, Spain |
| Mercedes Merayo | Universidad Complutense de Madrid, Spain |
| Gustavo Meschino | Universidad Nacional de Mar del Plata, Argentina |
| Margaret Miro | University of the Balearic Islands, Spain |
| Jose M. Molina | Universidad Carlos III de Madrid, Spain |
| Augusto Montisci | University of Cagliari, Italy |
| Antonio Mora | University of Granada, Spain |
| Angel Mora Bonilla | University Malaga, Spain |
| Claudio Moraga | European Centre for Soft Computing, Spain |
| Gines Moreno | University of Castilla-La Mancha, Spain |
| Jose Andres Moreno | University of La Laguna, Spain |
| Juan Moreno Garcia | Universidad de Castilla-La Mancha, Spain |
| J. Marcos Moreno Vega | University of La Laguna, Spain |
| Susana Muñoz Hernández | Technical University of Madrid, Spain |
| Pep Lluís Negre Carrasco | University of the Balearic Islands, Spain |
| Alberto Núñez | Universidad de Castilla La Mancha, Spain |
| Manuel Ojeda-Aciego | University of Malaga, Spain |
| Sorin Olaru | Suplec |
| Iván Olier | The University of Manchester, UK |
| Madalina Olteanu | SAMM, Université Paris 1, France |
| Julio Ortega | Universidad de Granada, Spain |
| Alfonso Ortega de La Puente | Universidad Autonoma de Madrid, Spain |
| Alberto Ortiz | University of the Balearic Islands, Spain |
| Emilio Ortiz-García | Universidad de Alcala, Spain |
| Osvaldo Pacheco | Universidade de Aveiro, Portugal |
| Esteban José Palomo | University of Málaga, Spain |
| Diego Pardo | Barcelona Tech, Spain |
| Miguel Angel Patricio | Universidad Carlos III de Madrid, Spain |
| Alejandro Pazos Sierra | University of A Coruña, Spain |
| Jose Manuel Perez Lorenzo | Universidad de Jaen, Spain |
| Vincenzo Piuri | University of Milan, Italy |
| Hector Pomares | University of Granada, Spain |
| Alberto Prieto | Universidad de Granada, Spain |
| Alexandra Psarrou | University of Westminster, UK |
| Francisco A. Pujol | University of Alicante, Spain |
| Pablo Rabanal | Universidad Complutense de Madrid, Spain |
| Juan Rabuñal | University of A Coruña, Spain |
| Vladimir Rasvan | Universitatea din Craiova, Romania |
| Ismael Rodriguez | Universidad Complutense de Madrid, Spain |
| Juan A. Rodriguez | Universidad de Malaga, Spain |
| Sara Rodríguez | University of Salamanca, Spain |

| Fernando Rojas | University of Granada, Spain |
| Ignacio Rojas | University of Granada, Spain |
| Samuel Romero-Garcia | University of Granada, Spain |
| Ricardo Ron-Angevin | University of Málaga, Spain |
| Eduardo Ros | University of Granada, Spain |
| Francesc Rossello | University of the Balearic Islands, Spain |
| Fabrice Rossi | SAMM, Université Paris 1, France |
| Fernando Rubio | Universidad Complutense de Madrid, Spain |
| Ulrich Rueckert | University of Paderborn, Germany |
| Addisson Salazar | Universidad Politecnica Valencia, Spain |
| Sancho Salcedo-Sanz | Universidad de Alcalá, Spain |
| Albert Samà | Universitat Politècnica de Catalunya, Spain |
| Francisco Sandoval | Universidad de Málaga, Spain |
| Jose Santos | University of A Coruña, Spain |
| Jose A. Seoane | University of Bristol, UK |
| Eduardo Serrano | Universidad Autonoma de Madrid, Spain |
| Luis Silva | University of Aveiro, Portugal |
| Olli Simula | Helsinki University of Technology, Finland |
| Jordi Solé-Casals | Universitat de Vic, Spain |
| Carmen Paz Suárez Araujo | Universidad de las Palmas de Gran Canaria, Spain |
| Peter Szolgay | Pazmany Peter Catholic University, Hungary |
| Javier Sánchez-Monedero | University of Cordoba, Spain |
| Ricardo Tellez | Pal Robotics |
| Ana Maria Tome | Universidade Aveiro, Portugal |
| Carme Torras | IRI (CSIC-UPC) |
| Joan Torrens | University of the Balearic Islands, Spain |
| Claude Touzet | University of Provence, France |
| Olga Valenzuela | University of Granada, Spain |
| Oscar Valero | University of the Balearic Islands, Spain |
| Miguel Ángel Veganzones | Universidad del País Vasco (UPV/EHU), Spain |
| Francisco Velasco-Alvarez | Universidad de Málaga, Spain |
| Sergio Velastin | Kingston University, UK |
| Marley Vellasco | PUC-Rio |
| Alfredo Vellido | Universitat Politecnica de Catalunya, Spain |
| Francisco J. Veredas | Universidad de Málaga, Spain |
| Michel Verleysen | Université Catholique de Louvain, Belgium |
| Changjiu Zhou | Singapore Polytechnic, Singapore |
| Ahmed Zobaa | University of Exeter, UK |

## Additional Reviewers

Azorín-López, Jorge
Ballesteros Tolosana, Iris
Benítez Caballero, María José
Bermejo, Sergio

Camacho, Carlos
Camacho, David
Castillo, Pedro
Cazorla, Miguel

# Contents – Part I

## E-Health and Computational Biology

## Human Computer Interaction

## Image and Signal Processing

**Mathematics for Neural Networks**

## Self-organizing Networks

## Machine Learning for Renewable Energy Applications

# Contents – Part II

## Machine Learning Applied to Vision and Robotics

## Human Activity Recognition for Health and Well-being Applications

## Software Testing and Intelligent Systems

## Real World applications of BCI Systems

## Machine Learning in Imbalanced Domains

## Surveillance and Rescue Systems and Algorithms
## for Unmanned Aerial Vehicles

**End-User Development for Social Robotics**

**Artificial Intelligence and Games**

## Supervised, Non-supervised, Reinforcement and Statistical Algorithms

# Bio-inspired Computing

# A Parallel Swarm Library Based on Functional Programming

Fernando Rubio, Alberto de la Encina[✉], Pablo Rabanal,
and Ismael Rodríguez

Facultad Informática, Universidad Complutense de Madrid,
28040 Madrid, Spain
{fernando,albertoe}@sip.ucm.es, prabanal@fdi.ucm.es

**Abstract.** In this paper we present a library of parallel skeletons to deal with swarm intelligence metaheuristics. The library is implemented using the parallel functional language Eden, an extension of the sequential functional language Haskell. Due to the higher-order nature of functional languages, we simplify the task of writing generic code, and also the task of comparing different strategies. The paper illustrates how to develop new skeletons and presents empirical results.

**Keywords:** Metaheuristics · Parallel programming · Skeletons · Functional programming

## 1 Introduction

When dealing with swarm optimization methods (see e.g. [4–6,11]), one of the first problems is deciding which swarm algorithm should be chosen to solve the problem under consideration. The same issue applies for deciding how a swarm method should be parallelized, out of a given set of available parallel strategies. Under these circumstances, it is very useful to provide programmers with several implementations of several swarm intelligence methods – as long as all of them can be easily adapted and used to solve any problem under consideration. In this regard, the reusability and clear separation of concerns of functional programs fits particularly well. In this paper we present a library of parallel functional swarm intelligence algorithms. The chosen parallel functional language is Eden [7], which is a parallel extension of Haskell, a higher-order functional language that guarantees the absence of side effects. The aim of our library is providing programmers with a tool to quickly test the performance of several swarm intelligence algorithms, as well as several parallelizing strategies.

Smaller pieces of the library have been presented in previous works. In [12] we presented our Eden implementation of Particle Swarm Optimization

This work has been partially supported by projects TIN2012-39391-C04-04, TIN2015-67522-C3-3-R, and S2013/ICE-2731.

I. Rojas et al. (Eds.): IWANN 2017, Part I, LNCS 10305, pp. 3–15, 2017.
DOI: 10.1007/978-3-319-59153-7_1

(PSO) [6], whereas an Eden implementation of the Artificial Bee Colony algorithm (ABC) [5] was given in [15]. In this paper we develop an Eden implementation of Differential Evolution [3]. In addition, we glue together these three Eden implementations (and their parallel variants) by constructing a higher abstraction layer. The goal of this tool layer is providing a common unified interface to all supported methods and help the programmer to automatically test the performance of all of these three methods and their variants (as well as others that could be provided in the future) for the target problem.

The rest of the paper is organized as follows. First, we briefly describe the language used. Then, Sect. 3 summarizes the main metaheuristic used in this work. Next, in Sect. 4 we illustrate how to develop generic higher-order functions to deal with a concrete metaheuristic, while in Sect. 5 we show how to provide new parallel skeletons to deal with the same metaheuristic. Afterwards, Sect. 6 presents results obtained with our library. Finally, Sect. 7 presents our conclusions.

## 2    Introduction to Eden

Eden [7] is a parallel extension of Haskell. It introduces parallelism by adding syntactic constructs to define and instantiate processes explicitly. It is possible to define a new *process abstraction* p by applying the predefined function process to any function \x -> e, where variable x will be the input of the process, while the behavior of the process will be given by expression e. Process abstractions are similar to functions – the main difference is that the former, when instantiated, are executed in parallel. From the semantics point of view, there is no difference between process abstractions and function definitions. The differences between processes and functions appear when they are invoked. Processes are invoked by using the predefined operator #. For instance, in case we want to create a *process instantiation* of a given process p with a given input data x, we write (p # x). Note that, from a syntactical point of view, this is similar to the *application* of a function f to an input parameter x, which is written as (f x).

Therefore, when we refer to a *process* we are not referring to a syntactical element but to a new *computational environment*, where the computations are carried out in an autonomous way. Thus, when a *process instantiation* (e1 # e2) is invoked, a new *computational environment* is created. The new process (the child or instantiated process) is fed by its creator by sending the value for $e_2$ via an input channel, and returns the value for $e_1 e_2$ (to its parent) through an output channel.

In order to increase parallelism, Eden employs pushing instead of pulling of information. That is, values are sent to the receiver before it actually demands them. In addition to that, once a process is running, only fully evaluated data objects are communicated. The only exceptions are *streams*, which are transmitted element by element. Each stream element is first evaluated to full normal form and then transmitted. Concurrent threads trying to access not yet available input are temporarily suspended. This is the only way in which Eden processes synchronize. Notice that process creation is explicit, but process communication (and synchronization) is completely implicit.

Process abstractions in Eden are not just annotations, but first class values which can be manipulated by the programmer (passed as parameters, stored in data structures, and so on). This facilitates the definition of skeletons [2,14] as higher order functions. Next we illustrate, by using a simple example, how skeletons can be written in Eden.

The most simple skeleton is map. Given a list of inputs xs and a function f to be applied to each of them, the sequential specification in Haskell is as follows:

```
map f xs  =  [f x | x <- xs]
```

that can be read as *for each element x belonging to the list xs, apply function f to that element*. This can be trivially parallelized in Eden. In order to use a different process for each task, we will use the following approach:

```
map_par f xs =  [pf # x | x <- xs]     where pf = process f
```

The process abstraction pf wraps the function application (f x). It determines that the input parameter x as well as the result value will be transmitted through channels.

Let us note that Eden's compiler has been developed by extending the GHC Haskell compiler. Hence, it reuses GHC's capabilities to interact with other programming languages. Thus, Eden can be used as a coordination language, while the sequential computation language can be, for instance, C.

## 3 Differential Evolution

Differential evolution (DE) [3] is an evolutionary algorithm for optimizing real-valued multi-modal objective functions. Although it is related to Genetic Algorithms, it is a different option in the universe of evolutionary methods. DE maintains a population of candidate solutions and attempts to improve it by combining existing ones. The method uses $NP$ agents as candidate solutions, where each of these agents is represented by an $n$-dimensional vector. The initial population is randomly chosen and uniformly distributed in the search space. DE generates new solutions by adding the weighted difference between two agents to a third one. If the new vector improves the objective function of a predetermined population member, this new vector will replace the one it was compared with, otherwise, the old vector remains unchanged.

Let $f : \mathbb{R}^n \to \mathbb{R}$ be the function to be minimized (or maximized), and let $x_i \in \mathbb{R}^n$ be an agent ($1 \le i \le NP$) in the population with $NP \ge 4$. The basic DE variant implemented in this paper is explained afterwards. First of all, $NP$ agents are randomly created in the search space. Next, a loop is executed as long as an ending condition is not satisfied (typically, the number of iterations performed does not exceed the limit, or the fitness adequation is not reached). Inside the loop and for each agent $x_i$ in the population, three agents a, b, c are chosen. These agents must be distinct from each other and distinct from agent $x_i$. Next, we pick a random integer R in the range $[1, n]$, and an empty vector y with n positions is created. Then, the values of vector y are created as follows: For each y(j) a random real in the range $[0, 1]$ is assigned to variable r, and if the value of r is lower than the *crossover probability* parameter, CR $\in [0, 1]$, or if j=R then the value a(j)+F×(b(j)-c(j)) is set to dimension $j$ of variable y (y(j)); else y(j) = $x_i$(j). When the initialization of y finishes, if f(y)<f($x_i$) the i-th agent $x_i$ is replaced with the new vector y ($x_i$= y). At the end, the agent with the minimum value of $f$ (or the maximum if maximizing) is returned.

Parameter F $\in [0, 2]$ is called the *differential weight*. Parameters F and CR are experimentally chosen.

## 4  Generic Differential Evolution in Haskell

In this section we show how to develop a new (sequential) metaheuristic by using Haskell (Eden parallelizations will be tackled in the next section). In particular, we consider the implementation of *Differential Evolution*, although we could deal with any other metaheuristic in a similar way.

Functional languages allow creating higher-order functions. Thus, we can take advantage of them to define a generic function deSEQ implementing the Differential Evolution metaheuristic. This function will have as input parameter a fitness function, which can be different in each case. It also needs other input parameters, like the number of candidate positions to be used, the number of iterations to be performed, the boundings of the search space, and the concrete parameters F and CR to be used. Moreover, in order to implement it in a pure functional language like Haskell, we need an additional parameter to introduce randomness. Note that Haskell functions cannot produce side-effects, so they need an additional input parameter to be able to obtain different results in different executions. The type of the Haskell function implementing Differential Evolution can be represented as follows:

```
deSEQ::RandomGen a => a        --Random generator
        ->Params               --Adjustment parameters(F,CR)
        ->Int                  --Number of candidates
        ->Int                  --Maximum number iterations
        ->(Position->Double)   --Fitness function
        ->Boundings            --Search space boundaries
        ->(Double,Position)    --Value and position of best candidate
```

Regarding the representation of Position, it must be able to deal with an arbitrarily large number of dimensions. Thus, we can easily represent it by using a list of real numbers. In this case, the length of such list represents the number of dimensions, whereas the concrete elements represent the coordinate values of each of these dimensions. Note that Boundings can be defined in a similar way, although a pair with the lower and upper bound for each dimension is considered in this case. Finally, the type Params only needs to handle the parameters used in Differential Evolution to tune up the algorithm, that is, F and CR, which are real numbers. Thus, the needed auxiliary types are the following:

```
type Position = [Double]
type Boundings = [(Double,Double)]
type Params = (Double,Double)
```

After defining the types and the interface of the main function deSEQ, we have to define its actual body. First, we have to randomly initialize the candidate solutions. This is done by a simple function initializeCandidates (not shown) that distributes the candidates randomly among the search space. After initializing the candidates, function de' performs the real work of the algorithm by iterating the application of the basic step as many times as needed. As in the case of the main function deSEQ, the auxiliary function de' will also need a way to introduce randomness. This is solved by using function split to create new random generators. Let us finally note that function de' needs the same inputs as the main function (number of iterations, fitness function, etc.), as it has to perform the actual work, but now it uses a list of candidate positions instead of only the number of candidates, as we have already created the appropriate list:

```
deSEQ sg p nc it f bo = obtainBestCandidate (de' sg2 p it f bo initCandis)
   where initCandis = initializeCandidates sg1 nc bo f
         (sg1,sg2) = split sg
type Candidate = (Double,Position)    -- Current value, current position
```

In order to define function de' we only need to use a simple recursion on the number of iterations. The base case will be when zero iterations remains. In that case, we return the same list of candidates without modifying it. Otherwise, we use function oneStepDE to perform one iteration of the algorithm, and then we go on performing the rest of iterations by using a recursive call to function de':

```
de' _ _ 0 _ _ cs = cs
de' sg p it f bo cs = de' sg2 p (it-1) f bo (oneStepDE sg1 p f bo cs)
   where (sg1,sg2) = split sg
```

For the sake of simplicity, we assume that the only finishing condition is the number of iterations, but we can easily modify it to include alternative finishing conditions.

Finally, we only need to define how to perform each step. First, we have to generate the list of needed random numbers. For each candidate solution we need three random indexes (corresponding to the candidates a, b, and c described in Sect. 3, which will be used to generate a new candidate), one random dimension to be modified for sure, and one random real number for each dimension. This list of real numbers will be used to decide whether the corresponding dimension is to be modified or not, comparing the real number with the CR parameter. Function genRanIndexDimR generates the list of random numbers for each candidate, while the predefined higher-order function zipWith allows to combine each candidate with the corresponding random numbers generated by function genRanIndexDimR. The source code is as follows:

```
oneStepDE sg (dw,cr) f bo cs = zipWith combineCandidate cs rs
  where rs = genRanIndexDimR sg (length cs -1) (length bo)
```

The definition of `combineCandidate` is trivial. It only has to combine one candidate with the random candidates selected using the random numbers `rs`, using the formula shown in Sect. 3. The complete program is available at http://antares.sip.ucm.es/prabanal/english/heuristics_library.

After implementing the higher-order function dealing with DE metaheuristic, the user only needs to provide the appropriate fitness function corresponding to the concrete problem to be solved. Note that the user does not need to understand the internals of the definition of `deSEQ`, but only its basic interface. That is, the programmer only has to call `deSEQ` providing the fitness function and the concrete parameters to be used (number of iterations and so on).

## 5   Parallel Skeletons

Parallelizing a problem requires detecting time-consuming tasks that can be performed independently. In our case, in each step of the algorithm we can deal independently with each of the candidates. That is, in function `oneStepDE` we could parallelize the evaluation corresponding to each candidate solution. By doing so, we can create a simple skeleton to parallelize DE algorithms. However, in order to increase the granularity of each of the parallel tasks we should avoid creating independent processes for each candidate. It is better to create as many processes as processors available, and to fairly distribute the candidates among the processes. This can be done by substituting `zipWith` by a call to `zipWith_farm`, a parallel version of `zipWith` that implements the idea of distributing a large list of tasks among a reduced number of processes.

By using `zipWith_farm` the speedup improves. However, for each iteration of the algorithm `zipWith_farm` would create a new list of processes, and it would have to receive and return the corresponding lists of candidates. We can improve the parallel performance of the algorithm by parallelizing function `deSEQ` instead of function `oneStepDE`. We start splitting the list of candidates into as many groups as processors available. Then, each group evolves in parallel independently during a given number of iterations. After that, processes communicate among them to redistribute the candidates among processes, and then they go on running again in parallel. This mechanism is repeated as many times as desired until a given number of global iterations is reached.

The implementation of this approach requires using a function `dePAR` instead of `deSEQ`. The new function `dePAR` uses basically the same parameters as `deSEQ`, but instead of using a parameter `it` to define the number of iterations, it uses two parameters `it` and `pit`. Now, the number of iterations will be defined by `it * pit`, where `pit` indicates the number of iterations to be performed independently in each process without communicating with other processes, whereas `it` indicates the number of parallel synchronous steps to be performed among processes. In addition to that, we also include a new parameter `nPE` to define the number of independent processes to be created. In the most common case, this parameter will be equal to the number of processors available. Taking into account these considerations, the type interface of the new function is as follows:

```
dePAR::RandomGen a => a        --Random generator
            ->Params           --Adjustment parameters (F,CR)
            ->Int              --Number of candidates
            ->Int              --Iterations per parallel step
            ->Int              --Number of global steps
            ->Int              --Number of parallel processes
            ->(Position->Double)  --Fitness function
            ->Boundings        --Search space boundaries
            ->(Double,Position)   --Value and position of best candidate
```

The definition of the body of the main function dePAR requires creating as many processes as requested in the corresponding parameter. Thus, before defining this function, we will show how to define a function to deal with the behaviour of each process. Such function will need the corresponding parameter to create random values, and it will also receive the tuning parameters of the metaheuristic (i.e. F and CR), the number of iterations to be performed in each parallel step pit, the fitness function f, and the boundings of the search space bo. Then, the process will receive a list with it lists of candidates through an input channel, and it will produce as output a new list with it lists of candidates. Note that the main function dePAR will perform it global synchronous steps, where each step will perform pit iterations in parallel without synchronization. Thus, dePAR will assign it tasks as input to each process, and each process will return it solutions as output, where those solutions will be used as input of other processes in the next global step. Let us remark that, in Eden, list elements are transmitted through channels in a stream-like fashion. This implies that, in practice, each process will receive a new list of candidates through its input channel right before starting to compute a new parallel step. The complete source code defining a process is as follows:

```
deP sg p pit f bo [] = []
deP sg p pit f bo (bs:bss)=de' sg1 p pit f bo bs : deP sg2 p pit f bo bss
   where (sg1,sg2) = split sg
```

As it can be seen, it is only necessary to define it recursively on the number of tasks. When the input list of lists is empty, the process finishes returning an empty list of results. Otherwise, it uses exactly the same sequential function de' described in the previous section to perform pit iterations, and then it goes on dealing with the rest of the input lists.

Let us now consider how to define the main function dePAR. First, it has to create the initial list of random candidates, exactly in the same way as in the sequential case deSEQ. Then, the main difference with the sequential case appears: we create nPE copies of process deP. Each of them receives the main input parameters of the algorithm (tuning parameters F and CR, fitness function, etc.), and it also receives its own list of tasks (pins!!i). Each element of the list of tasks contains an input list of candidates, that will be processed by deP during pit iterations. The output of each process is a new list of lists of candidates. Each inner list was computed after each parallel step, and they must be redistributed among the rest of processes before starting the next global step. This is done by function redistribute. The final result of function bestPAR is obtained by combining the last results returned by each process. The source code is as follows:

```
dePAR sg p nc pit it nPE f bo = obtainBestCandidate (last poutsFlat)
  where initCandidates = initializeCandidates sg nc bo f
        sgs = tail (generateSGs (nPE+1) sg)
        pouts=[process (deP (sgs!!i) p pit f bo # (take it (pins!!i))
                |i<-[0..nPE-1]]
        poutsFlat = flatXsss pouts
        pins = redistribute nPE (initCandidates:poutsFlat)
```

It is important to note that the user of the library does not need to understand the low level details of the previous definition. In fact, in order to use it, it is only necessary to substitute a call to the sequential function deSEQ by a call to the parallel scheme dePAR, using appropriate values for parameters it, pit, and nPE. The last parameter will be typically equal to the number of available processors. Thus, the only programming effort will be to decide the values of it and pit. In case pit is very small, the granularity of tasks will be reduced, whereas very large values of pit would reduce the possibility to exchange candidates among processes. As a degenerate case, we could use it = 1 and pit being equal to the total number of iterations to be performed. By doing so, we would create groups searching for a solution in a completely independent way.

The previous parallel skeleton can be easily modified to handle different approaches. For instance, when we are using several computers in parallel, it could be the case that each of them is different. Thus, it would be reasonable to assign more candidates to those computers with faster processors, and less candidates to the slower ones. This can be easily done. First, instead of receiving the number of processes, we need to receive as input parameter the speed of each processor. This can be done by using a list of real numbers. Obviously, given the list we can trivially know the number of processes to be created by computing the length of the list. In the implementation, function dePAR has to be modified to split each list of candidates according to their relative speeds. That is, pins is now created by taking into account the speeds parameter:

```
dePARh sg p nc pit it speeds f bo = obtainBestCandidate (last poutsFlat)
  where nPE = length speeds
        initCandidates = initializeCandidates sg nc bo f
        sgs = tail (generateSGs (nPE+1) sg)
        pouts=[process (deP (sgs!!i) p pit f bo)#(take it (pins!!i))
                |i<-[0..nPE-1]]
        poutsFlat = flatXsss pouts
        pins = redistrRelative speeds initCandidates poutsFlat
```

The redistribution considering the relative speed is done by using function shuffleRelative, an auxiliary function that first computes the percentage of tasks to be assigned to each process, and then distributes the tasks by using function `splitWith`. It is worth to comment that we do not need to change anything else in the skeleton. In particular, the definition of the process `deP` itself remains unchanged.

# 6    Experimental Results

In this section we illustrate the usefulness of the library by performing some experiments. Let us remark that the higher-order nature of the language simplifies the development of tools to analyze properties of the different metaheuristic. In particular, we can write new higher-order functions whose parameters are again higher-order functions dealing with different metaheuristics. For instance, we can compare a list of metaheuristics `mths` for the same input problem (given by a concrete `fitness` function and the `bounds` of the search space) by using a higer-order function as follows:

```
compare::[(Position->Double)->Boundings->(Double,Position)]
         -> (Position->Double) -> Boundings -> [Double]
compare mths fitness bounds
  = map (fst . ($ (fitness,bounds)) . uncurry) mths
```

Note that the higher-order function receives as second and third parameters the fitness function and the boundaries of a concrete problem, while the first input is a list of metaheuristics to be compared, where each of them is again a higher-order function that receives a fitness function and the boundaries of the search space. Let us remark that the metaheuristics can have more parameters than those appearing in function `compare`. For instance, Differential Evolution has more parameters: the number of candidates, number of iterations, etc. However, as functions are first class citizens of the language, any metaheuistic can be partially applied. As an example, we can partially apply metaheuristic `deSEQ` to use a concrete random generator, concrete adjustment parameters (F, CR), a concrete number of candidates (75) and a concrete number of iterations (2000) by writing the following expression

```
  deSEQ sg (0.47,0.88) 75 2000
```

Its type is exactly `(Position->Double) -> Boundings -> (Double,Position)`. That is, we can use it as one element of the first input list of function `compare`. For instance, we can compare three different configurations of function `deSEQ` for a single problem `ackley` by writing the following:

```
compare [deSEQ sg (0.47,0.88) 75 2000, deSEQ sg (0.47,0.88) 100 1500,
         deSEQ sg (0.32,0.76) 75 2000]
        ackleyFitness ackleyBounds
```

That is, we are comparing three different configurations. The first and the second one use the same values for F and CR, but the first one uses 75 candidates and 2000 iterations, while the second one uses 100 candidates and 1500 iterations. The third configuration uses different values for F and CR, while the number of candidates and iterations is the same as in the first configuration. We can also generate larger lists of configurations by combining parameters using comprehension lists:

```
compare [deSEQ sg (f,cr) nc ni | f<-[0.47,0.32], cr<-[0.88,0.76],
                        nc<-[75,100], ni<-[1500,2000]]
       ackleyFitness ackleyBounds
```

As it can be expected, we can easily compare the results obtained by both sequential and parallel metaheuristics. For instance

```
compare ([deSEQ sg (0.47,0.88) 75 2000]
       ++[dePAR sq (0.47,0.88) 75 50 40 n|n<-[1..4]])
       ackleyFitness ackleyBounds
```

compares the sequential version with four parallelizations varying the number of processes to be used from 1 to 4, while

```
compare [dePAR sq (0.47,0.88) 75 pit (div 2000 pit) 4 | pit<-[50,100,200]]
       ackleyFitness ackleyBounds
```

compares three parallel implementations, all of them using 4 processes and 2000 iterations, but varying the size of each global step from 50 to 200 iterations. Obviously, the comparison can also include different metaheuristics as follows:

```
compare [deSEQ sg (0.47,0.88) 75 2000, deSEQ sg (0.47,0.88) 100 1500,
       beesSEQ sg 3000 100 1500, psoSEQ sg (-0.16,1.89,2.12) 100 1500]
       ackleyFitness ackleyBounds
```

where we compare two configurations of Differential Evolution, one configuration of Artificial Bee Colony, and another configuration of Particle Swarm Optimization. Our library provides a larger set of functions implementing different kinds of comparisons. For instance, the previous function is extended to execute each metaheuristic $n$ times and to compute average and standard deviation results. We also allow to receive as input not only a problem, but a list of problems, and we analyze the results obtained for all of them, and so on.

In order to show the information we can obtain by using these tools, we compare the results obtained by three different metaheuristics on a given benchmark. In particular, we compare Particle Swarm Optimization, Artificial Bee Colony, and Differential Evolution by using as benchmark a well-known set of functions defined in [16], where we have removed the last six functions of such benchmark because they are simple low-dimensional functions with only a few local minima.

In order to fairly compare the three metaheuristics, for each function we used exactly the same number of fitness evaluations. This number of function evaluations is the same as that defined in [16]. Regarding the tuning parameters of each of the metaheuristics, we use values available in the literature. In particular, the parameters of PSO are taken from [10], in the case of ABC we use [1], and in the case of DE we follow [9]. The results shown in Table 1 were obtained after computing the average of 50 executions for each metaheuristic. Note that for each metaheuristic we can find a concrete problem where it obtains the best results. However, the metaheuristic that obtains more often the best result in this concrete benchmark is ABC. In fact, by using [8] we can perform an statistical analysis to quantify the differences among the metaheuristics. In particular, aligned Friedman test can be used to check whether the hypothesis that all methods behave similarly (the null hypothesis) holds or not. Let us consider $\alpha = 0.05$, a standard significance level. From results given in Table 1, we calculate that the p-value for aligned Friedman is 0.0027, which allows to reject the null hypothesis with a high level of significance (the p-value is much lower than 0.05). So, the test concludes that the results of ABC, PSO, and DE are not considered similar. Ranks assigned by this test to ABC, PSO, and DE are respectively 19.5, 26.31, and 27.69 (smaller ranks denote better methods).

Regarding the speedups, all of them obtain reasonable speedups taking into account that the effort needed to use the skeletons is negligible: the programmer only changes a call to the sequential higher-order function by a call to the parallel skeleton. Anyway, the speedup obtained by PSO is slightly better (around 10%). The reason is that in each global step PSO only communicates the best position found by each island, while in ABC and DE it is communicated the whole set of bees/candidates computed in the last iteration. Thus, larger communications reduces the speedup.

**Table 1.** Average optimality comparison among metaheuristics

| Funct | Name | Dim | PSO | ABC | DE |
|-------|------|-----|-----|-----|-----|
| $f_1(x)$ | Sphere model | 30 | $1.02 \cdot 10^{-4}$ | $3.87 \cdot 10^{-9}$ | $6.17 \cdot 10^{-4}$ |
| $f_2(x)$ | Schwefel's problem 2.22 | 30 | $8.29 \cdot 10^{-3}$ | $1.74 \cdot 10^{-7}$ | $2.84 \cdot 10^{-5}$ |
| $f_3(x)$ | Schwefel's problem 1.2 | 30 | $1.93 \cdot 10^{-5}$ | $3.58 \cdot 10^{3}$ | $3.33 \cdot 10^{4}$ |
| $f_4(x)$ | Schwefel's problem 2.21 | 30 | $1.45 \cdot 10^{-3}$ | 1.39 | $6.42 \cdot 10^{-1}$ |
| $f_5(x)$ | Generalized Rosenbrock's function | 30 | 26.57 | 0.13 | 24.01 |
| $f_6(x)$ | Step function | 30 | 0 | 0 | 0 |
| $f_8(x)$ | Generalized Schwefel's problem 2.26 | 30 | $-9686.99$ | $-12569.49$ | $-12044.01$ |
| $f_9(x)$ | Generalized Rastrigin's function | 30 | $6.97 \cdot 10^{-8}$ | 0 | $9.95 \cdot 10^{-2}$ |
| $f_{10}(x)$ | Ackley's function | 30 | $2.41 \cdot 10^{-3}$ | $9.37 \cdot 10^{-5}$ | 15.53 |
| $f_{11}(x)$ | Generalized Griewank function | 30 | $4.69 \cdot 10^{-3}$ | $1.12 \cdot 10^{-10}$ | $2.53 \cdot 10^{-4}$ |
| $f_{12}(x)$ | Generalized penalized function I | 30 | $6.33 \cdot 10^{-3}$ | $8.96 \cdot 10^{-11}$ | $5.45 \cdot 10^{-5}$ |
| $f_{13}(x)$ | Generalized penalized function II | 30 | $4 \cdot 10^{-2}$ | $8.7 \cdot 10^{-9}$ | $3.99 \cdot 10^{-3}$ |
| $f_{14}(x)$ | Shekel's foxholes function | 2 | 1.65 | 496.58 | 476.85 |
| $f_{15}(x)$ | Kowalik's function | 4 | $1.22 \cdot 10^{-3}$ | $4.69 \cdot 10^{-4}$ | $3.075 \cdot 10^{-4}$ |
| $f_{16}(x)$ | Six-hump camel-back function | 2 | $-1.0316$ | $-1.0316$ | $-1.0316$ |
| $f_{17}(x)$ | Branin function | 2 | 0.398 | 0.398 | 0.398 |

# 7  Conclusions and Future Work

In this paper we have shown the usefulness of the functional programming paradigm to develop generic solutions to deal with swarm intelligence metaheuristics. In particular, we have shown how to develop parallel skeletons for a given metaheuristic, namely Differential Evolution, but the same ideas can be used to deal with any metaheuristic. The higher-order nature of the language simplifies the development of generic functions comparing the results obtained with different configurations.

The results obtained with our library show that the effort needed to use our skeletons is negligible. However, the obtained speedup is good. Anyway, we do not claim to obtain optimal speedup, but *reasonable* speedups at very low programming effort.

As future work, we want to use our library to deal with NP-complete problems appearing in the context of marketing strategies (see e.g. [13]).

# References

1. Akay, B., Karaboga, D.: Parameter tuning for the artificial bee colony algorithm. In: Nguyen, N.T., Kowalczyk, R., Chen, S.-M. (eds.) ICCCI 2009. LNCS, vol. 5796, pp. 608–619. Springer, Heidelberg (2009). doi:10.1007/978-3-642-04441-0_53
2. Cole, M.: Bringing skeletons out of the closet: a pragmatic manifesto for skeletal parallel programming. Parallel Comput. **30**, 389–406 (2004)
3. Das, S., Suganthan, P.N.: Differential evolution: a survey of the state-of-the-art. IEEE Trans. Evol. Comput. **15**(1), 4–31 (2011)
4. Dorigo, M., Birattari, M.: Ant colony optimization. In: Sammut, C., Webb, G.I. (eds.) Encyclopedia of Machine Learning, pp. 36–39. Springer, Heidelberg (2010)
5. Karaboga, D., Görkemli, B., Ozturk, C., Karaboga, N.: A comprehensive survey: artificial bee colony (ABC) algorithm and applications. Artif. Intell. Rev. **42**(1), 21–57 (2014)
6. Kennedy, J., Eberhart, R.C.: Particle swarm optimization. In: IEEE International Conference on Neural Networks, vol. 4, pp. 1942–1948. IEEE Computer Society Press (1995)
7. Loogen, R.: Eden – parallel functional programming with haskell. In: Zsók, V., Horváth, Z., Plasmeijer, R. (eds.) CEFP 2011. LNCS, vol. 7241, pp. 142–206. Springer, Heidelberg (2012). doi:10.1007/978-3-642-32096-5_4
8. Parejo, J.A., García, J., Ruiz-Cortés, A., Riquelme, J.C.: Statservice: herramienta de análisis estadístico como soporte para la investigación con metaheurísticas. In: MAEB 2012 (2012)
9. Pedersen, M.E.H.: Good parameters for differential evolution. Technical report HL1002, Hvass Laboratories (2010)
10. Pedersen, M.E.H.: Tuning & simplifying heuristical optimization. Ph.D. thesis, University of Southampton, School of Engineering Sciences (2010)
11. Rabanal, P., Rodríguez, I., Rubio, F.: Using river formation dynamics to design heuristic algorithms. In: Akl, S.G., Calude, C.S., Dinneen, M.J., Rozenberg, G., Wareham, H.T. (eds.) UC 2007. LNCS, vol. 4618, pp. 163–177. Springer, Heidelberg (2007). doi:10.1007/978-3-540-73554-0_16

12. Rabanal, P., Rodríguez, I., Rubio, F.: Parallelizing particle swarm optimization in a functional programming environment. Algorithms **7**(4), 554–581 (2014)
13. Rodríguez, I., Rabanal, P., Rubio, F.: How to make a best-seller: optimal product design problems. Appl. Soft Comput. **55**, 178–196 (2017)
14. Rubio, F.: Programación funcional paralela eficiente en Eden. Ph.D. thesis, Universidad Complutense de Madrid (2001)
15. Rubio, F., de la Encina, A., Rabanal, P., Rodríguez, I.: Eden's bees: parallelizing artificial bee colony in a functional environment. In: ICCS 2013, pp. 661–670 (2013)
16. Yao, X., Liu, Y., Lin, G.: Evolutionary programming made faster. IEEE Trans. Evol. Comput. **3**(2), 82–102 (1999)

# A Parallel Island Approach to Multiobjective Feature Selection for Brain-Computer Interfaces

Julio Ortega[1]([⊠]), Dragi Kimovski[2], John Q. Gan[3], Andrés Ortiz[4], and Miguel Damas[1]

[1] Department of Computer Architecture and Technology, CITIC,
University of Granada, Granada, Spain
{jortega,mdamas}@ugr.es
[2] University of Innsbruck, Innsbruck, Austria
dragi@dps.uibk.ac.at
[3] School of Computer Science and Electronic Engineering,
University of Essex, Colchester, UK
jqgan@essex.ac.uk
[4] Department of Communications Engineering,
University of Málaga, Málaga, Spain
aortiz@ic.uma.es

**Abstract.** This paper shows that parallel processing is useful for feature selection in brain-computer interfacing (BCI) tasks. The classification problems arising in such application usually involve a relatively small number of high-dimensional patterns and, as curse of dimensionality issues have to be taken into account, feature selection is an important requirement to build suitable classifiers. As the number of features defining the search space is high, the distribution of the searching space among different processors would contribute to find better solutions, requiring similar or even smaller amount of execution time than sequential counterpart procedures. We have implemented a parallel evolutionary multiobjective optimization procedure for feature selection, based on the island model, in which the individuals are distributed among different subpopulations that independently evolve and interchange individuals after a given number of generations. The experimental results show improvements in both computing time and quality of EEG classification with features extracted by multiresolution analysis (MRA), an approach widely used in the BCI field with useful properties for both temporal and spectral signal analysis.

**Keywords:** Brain-computer interfaces (BCI) · Feature selection · Island model based evolutionary algorithms · Multiresolution analysis (MRA) · Parallel multiobjective optimization

## 1 Introduction

Many classification tasks in bioinformatics deal with patterns defined by a large number of features. Moreover, these high-dimensional classification problems have frequently to be solved with the number of patterns smaller than the number of features,

© Springer International Publishing AG 2017
I. Rojas et al. (Eds.): IWANN 2017, Part I, LNCS 10305, pp. 16–27, 2017.
DOI: 10.1007/978-3-319-59153-7_2

thus presenting curse of dimensionality problems [1]. Therefore, feature selection is usually required in bioinformatics [2] to eliminate redundant and noisy features in order to improve the accuracy and interpretability of the classifiers.

Among the three different approaches for feature selection (filter, wrapper and embedded methods) [2], our proposal in this paper corresponds to a wrapper procedure. Although wrapper approaches use the classifier performance to evaluate the utility of a given set of features and thus they are classifier-dependent, they are usually recognized as the preferable approaches whenever they would be feasible [3].

One of the issues to be taken into account in the design of a feasible wrapper-based feature selection procedure is the number of possible features because the size of the searching space depends exponentially on that number. In high-dimensional classification problems, several hundreds or even thousands of features usually define a very huge searching space where efficient metaheuristics are required. This paper proposes a parallel multiobjective evolutionary algorithm, in which the individuals of a population represent different feature selections, and the fitness of a given individual is determined through the evaluation of the classifier performance after training it with the corresponding patterns defined by the set of selected features. Parallel processing has been previously considered to take advantage of high performance computer architectures for feature selection [4–8]. In [7, 8] feature selection is approached using parallel multiobjective and cooperative coevolutionary procedures implemented through a master-worker parallel model. In this paper, we propose an island model to implement parallel multiobjective feature selection applied to BCI.

In what follows, Sect. 2 describes our approach to feature selection based on multiobjective optimization and its parallel implementation through the island model. The application considered in this paper corresponds to BCI tasks related with motor imagery, where the features of the patterns are obtained by using Multiresolution Analysis (MRA). The details of the application and the patterns in the database used are provided in Sect. 3. Finally, Sect. 4 describes the experimental results and the conclusions are given in Sect. 5.

## 2 A Parallel Island Procedure for Multiobjective Feature Selection

As this paper deals with supervised classification problems in which the labels of the training and test patterns are known, it would be possible to evaluate the performance of a classifier from the accuracy obtained after training it (by using the set of training patterns). Nevertheless, other measures that quantify properties such as the generalization capability should be taken into account in order to improve the behavior of the classifier in a real environment, where patterns that have not been used for training have to be processed. To tackle these issues, we propose a multiobjective evolutionary procedure where the selection of features is optimized for both accuracy and generalization capability, both evaluated by using the training patterns.

Figure 1.a shows a scheme of the wrapper method we propose for feature selection based on a multiobjective optimization procedure that searches a vector of decision variables $\mathbf{x} = [x_1, x_2, \ldots, x_n] \in R^n$ to optimize a function vector $\mathbf{f}(\mathbf{x})$, whose scalar

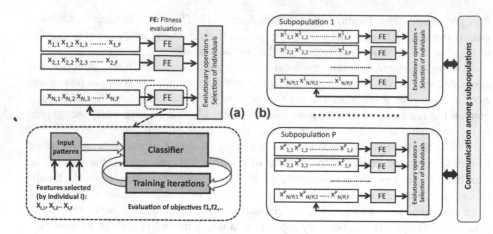

**Fig. 1.** Wrapper approach to feature selection by evolutionary multiobjective optimization: (a) sequential procedure; (b) island parallel procedure proposed in the paper

values $(f_1(\mathbf{x}), f_2(\mathbf{x}), \ldots, f_m(\mathbf{x}))$ represent the m objectives to optimize. These objectives are usually in conflict, and thus multiobjective optimization should obtain a set of *non-dominated* solutions called Pareto optimal solutions that define the *Pareto front* (no solution in the Pareto front is worse than the others when all the objectives are taken into account), from which it is possible to choose the most convenient solution in specific circumstances. To solve the multiobjective optimization problem we have implemented an evolutionary algorithm based on the NSGA-II algorithm [9], with specific individual codification and genetic operators.

**Algorithm 1.** Parallel multi-objective feature selection procedure adopted in Fig. 1.b

```
Parallel_NSGAII_featureselection (N individuals, P threads)
01    Create P Initialization(i,N/P,SP(i)) threads; // i=1,...,P
02    wait(i); // barrier to synchronize P threads;
03    Create P Island_evolution(i,N/P,SP(i),commprof,comm,genpar) threads; // i=1,...,P
04    wait(i); // barrier to synchronize P threads
05    NSGAII_nondomination_sort(SP(1),...,SP(P));// nondomination sort of subpop.
06    save results;
07    end

Island_evolution(i,N/P,SP(i),commprof,comm,genpar)
01        for  j=1 to comm
02            for k=1 to genpar
03                (SP'(i),f(SP'(i)))=NSGAII_tournament_selection (SP(i),f(SP(i)));
04                SP"(i)=Genetic_operators (SP'(i));
05                f(SP"(i)) = Evaluation(SP"(i),DS);
06                (SP*(i))=NSGAII_nondomination_sort ( SP(i);SP"(i));
07                (P,f(P))=NSGAII_replace_chromosome ( SP*(i));
08            end;
09            communication(SP(1),...,SP(P),commprof);
10        end;
```

The main contribution of this paper is a parallel implementation of the procedure summarized in Fig. 1.a based on an island model. This parallel approach, depicted in Fig. 1.b, distributes the N individuals of the population among the available P processors thus defining P subpopulations, each with N/P individuals. The pseudocode description of the parallel procedure is provided in Algorithm 1. The procedure first creates P threads with initialization and evaluate N/P individuals of the population (line 01 of `Parallel_NSGAII_featureselection`). These P threads are synchronized through a barrier in line 02 to perform the evolution of P subpopulations, each including N/P individuals, according to procedure `Island_evolution`. As it can be seen, each thread requires the number of communications (`comm`), the number of generations that the corresponding subpopulation has to complete between communications (`genpar`), and the randomly selected couples of threads that have to communicate after each genpar number of generations (`commprof`). This parallel evolutionary multiobjective procedure, whose behavior is different from the sequential one, allows improvement in the quality of the solutions found by using bigger populations and/or reduction in computing time.

## 3    Feature Selection in BCI with Multiresolution Analysis

The high-dimensional classification problem considered in this paper deals with brain-computer interfacing (BCI) based on the classification of EEG signals corresponding to motor imagery (MI) tasks. This BCI paradigm uses the series of amplifications and attenuations of short duration occasioned by limb movement imagination, the so called event related desynchronization (ERD) and event related synchronization (ERS). In [10] several approaches for multiobjective feature selection in a MRA (Multiresolution Analysis) system for BCI are proposed and evaluated. A MRA system [11] applies a sequence of successive approximation spaces to describe the target signal, thus being useful whenever the target signal presents different characteristics across the approximation spaces. As a specific example of MRA systems, the discrete wavelet transform (DWT) was applied in [10, 12] to characterize EEGs from motor imagery (MI) tasks.

The patterns used in this work are built, from EEG trials, by a feature extraction procedure based on the MRA described in [12]. Each signal obtained from each electrode contains several segments to which a set of wavelets detail and approximation coefficients are assigned. This way, considering S segments, E electrodes, and L levels of wavelets, each EEG pattern is characterized by $2 \times S \times E \times L$ sets of coefficients. The number of coefficients in each level set depends on the level. In the dataset considered here, which was recorded in the BCI Laboratory at the University of Essex, S = 20 segments, E = 15 electrodes, and L = 6 levels. Therefore, 3600 sets of wavelet coefficients in total in each pattern, with from 4 to 128 coefficients in each set, characterize each pattern: a total of 151200 coefficients. Nevertheless, in [12] only one feature is assigned to each electrode and each level of approximation and detail. It is obtained by computing the variance of the coefficient distribution and normalising the obtained values between 0 and 1. This way, $2 \times S \times E \times L = 3600$ features constitute each pattern. Anyway, as the number of training patterns for each subject is approximately 180, it is clear that an efficient procedure for feature selection is required.

In [12] an approach based on the use of several classifiers is considered to reduce the number of features characterizing the patterns applied to each classifier. Figure 2 describes the structure of the classification procedure based on a set of LDA (linear discriminant analysis) classifiers, in which a module for majority voting based on all the LDA outputs provides the final classification output. This way, a set of $2 \times S \times L$ LDA classifiers with the number of inputs equaling the number of electrodes are adopted, as shown in Fig. 2. This procedure is called OPT0 as the baseline method for performance comparison [10].

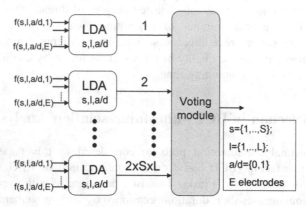

**Fig. 2.** EEG classification with multiple LDA classifiers based on majority voting, with one LDA classifier per segment, per level and, per type (detail and approximation) of wavelet [12]

In [10] two alternatives for feature selection in BCI with MRA, OPT1 and OPT2, were evaluated and compared with the performance of OPT0. The alternative OPT2 selects a set of LDAs among the $2 \times S \times L$ LDAs in the structure of Fig. 2 through the multiobjective optimization procedure described in Sect. 2. OPT1 is a simpler alternative as it uses only one LDA classifier based on a subset of features selected from all the available features. The two cost functions used in the multiobjective feature selection in OPT1 and OPT2 take into account the labels assigned to the training patterns to identify their corresponding classes. Moreover, to characterize the performance of the classifier while it has been trained or adjusted for a given set of features (i.e., an individual of the population), it is important not only to take into account the accuracy obtained for the training set but also its generalization capabilities, i.e., its accuracy for unseen instances. Thus, the first cost function is related with the Kappa index [13] on training dataset, which takes into account the distribution of the per class error as it is computed as $(p_0 - p_c)/(1 - p_c)$, with $p_0$ equal to the proportion of coincidences among the classification outputs and the labels of the patterns and $p_c$ being the proportion of patterns on which the coincidence is expected by chance. The second cost function is the average loss function in a 10-fold cross validation analysis to the training patterns. This paper proposes and evaluates parallel implementation of OPT1 and OPT2.

# 4    Experimental Results

The parallel procedures for OPT1 and OPT2 have been implemented by using the Parallel Computing Toolbox of Matlab® (version 8.3) and executed in a node including two Intel Xeon E5-2620 processors (providing up to 12 threads) at 2.1 GHz and 32 GB DDR3 RAM per node. The experiments were conducted using the dataset recorded in the BCI Laboratory at the University of Essex [12]. For each subject, there is one data file with data recorded for training (training patterns) and another file with data for evaluation (test patterns). Each data file contains about 180 labelled patterns with data from 20 segments (S = 20), six levels (L = 6) of approximation or detail coefficients (a/d = 2), and 15 electrodes (E = 15). The labels correspond to three imagined movements of right hand, left hand, and feet. Each experiment has been repeated ten times in order to complete an analysis to determine the statistical significance of the observed differences among alternatives. The results provided in what follows correspond to the EEG data from subjects 104, 107, and 110 (those achieving the best performance results in [12]).

Table 1 provides the average Kappa index values [13] obtained by the baseline alternative OPT0 [10], which uses the structure of LDA classifiers [12] shown in Fig. 2, and those by the option OPT2 [10], which searches for an optimal selection of

**Table 1.** Comparison of Kappa indexes and execution times of OPT0 and OPT2 with 120 individuals, 50 generations, 5 and 10 generations/communication, 4 and 8 threads on data from subjects 104, 107, and 110 in the BCI dataset of University of Essex

| Sbj. | Procedure | PCnf. | Kappa | Time (s.) | p-val. |
|------|-----------|-------|-------|-----------|--------|
| 104 | OPT0 | | 0.564 ± 0.000 | – | |
| | OPT2 (50/50) | | 0.545 ± 0.035 | 20892 ± 1668 | |
| | OPT2 (120/50) | | 0.547 ± 0.040 | 50135 ± 2222 | |
| | OPT2 (120/50; 4 thr 10 gen/comm.) | PC11 | 0.523 ± 0.014 | 13919 ± 325 | *0.04 |
| | OPT2 (120/50; 8 thr 10 gen/comm.) | PC12 | 0.554 ± 0.019 | 7966 ± 173 | 0.34 |
| | OPT2 (120/50; 4 thr 5 gen/comm.) | PC13 | 0.515 ± 0.023 | 13902 ± 378 | *0.04 |
| | OPT2 (120/50; 8 thr 5 gen/comm.) | PC14 | 0.535 ± 0.034 | 8095 ± 126 | 0.75 |
| 107 | OPT0 | | 0.631 ± 0.000 | – | |
| | OPT2 (50/50) | | 0.634 ± 0.019 | 21167 ± 1134 | |
| | OPT2 (120/50) | | 0.652 ± 0.022 | 50749 ± 1578 | |
| | OPT2 (120/50; 4 thr 10 gen/comm.) | PC11 | 0.657 ± 0.020 | 14214 ± 315 | 1.00 |
| | OPT2 (120/50; 8 thr 10 it/comm.) | PC12 | 0.655 ± 0.017 | 8128 ± 129 | 0.33 |
| | OPT2 (120/50; 4 thr 5 it/comm.) | PC13 | 0.645 ± 0.014 | 14342 ± 134 | 0.20 |
| | OPT2 (120/50; 8 thr 5 it/comm.) | PC14 | 0.642 ± 0.018 | 8225 ± 149 | 0.06 |
| 110 | OPT0 | | 0.648 ± 0.000 | – | |
| | OPT2 (50/50) | | 0.605 ± 0.045 | 18820 ± 1069 | |
| | OPT2 (120/50) | | 0.619 ± 0.021 | 45156 ± 1991 | |
| | OPT2 (120/50; 4 thr 10 it/comm.) | PC11 | 0.628 ± 0.030 | 12986 ± 360 | 0.08 |
| | OPT2 (120/50; 8 thr 10 it/comm.) | PC12 | 0.631 ± 0.020 | 7866 ± 237 | 0.26 |
| | OPT2 (120/50; 4 thr 5 it/comm.) | PC13 | 0.608 ± 0.034 | 13064 ± 501 | 0.15 |
| | OPT2 (120/50; 8 thr 5 it/comm.) | PC14 | 0.629 ± 0.026 | 7863 ± 142 | 0.42 |

LDAs in the structure of Fig. 2. The data in parentheses for the OPT2 alternatives represent the number of individuals in the population and generations of the evolutionary algorithm and, in the case of parallel executions, the number of threads (4 or 8) and generations (5 or 10) executed by each subpopulation. The different parallel configurations are noted as PC11 to PC14 in the column PCnf. of Table 1. The sequential version of OPT2 with 120 individuals and 50 generations for subjects 104, 107 and 110 requires more than twelve hours, and provides solutions with average Kappa indices equal to 0.547, 0.652 and 0.619, respectively. The parallel versions of OPT2 require less computing times to achieve similar levels of performance than the sequential implementation of OPT2 with 120 individuals. The parallel OPT2 implementations also consume less time than the sequential OPT2 with a population of 50 individuals and even improve the performance of this sequential version of OPT2 for some subjects.

**Table 2.** Comparison of Kappa indexes and execution times of OPT0 and OPT1 with 2000 individuals, 50 generations and 100 generations, 5 and 10 generations/communication, 4 and 8 threads on data from subjects 104, 107, and 110 in the BCI dataset of University of Essex

| Sbj. | Procedure | PCnf. | Kappa | Time (s.) | p-val. |
|------|-----------|-------|-------|-----------|--------|
| 104 | OPT0 | | 0.564 ± 0.000 | – | |
| | OPT1 (50/50) | | 0.510 ± 0.056 | 4241 ± 375 | |
| | OPT1 (120/50) | | 0.515 ± 0.047 | 10316 ± 461 | |
| | OPT1 (960/50; 4 thr 10 gen/comm.) | PC21 | 0.653 ± 0.053 | 23279 ± 770 | *0.01 |
| | OPT1 (960/100; 4 thr 10 gen/comm.) | PC22 | 0.606 ± 0.053 | 44787 ± 414 | *0.01 |
| | OPT1 (960/50; 8 thr 10 gen/comm.) | PC23 | 0.634 ± 0.038 | 12104 ± 1261 | *0.01 |
| | OPT1 (960/50; 8 thr 5 gen/comm.) | PC24 | 0.698 ± 0.062 | 12422 ± 1074 | *0.01 |
| | OPT1 (960/100; 8thr 10gen/comm.) | PC25 | 0.673 ± 0.039 | 25751 ± 2000 | *0.01 |
| | OPT1 (1920/100; 8thr 10gen/comm.) | PC26 | 0.665 ± 0.033 | 46383 ± 1343 | *0.01 |
| 107 | OPT0 | | 0.631 ± 0.000 | – | |
| | OPT1 (50/50) | | 0.560 ± 0.041 | 4317 ± 188 | |
| | OPT1 (120/50) | | 0.580 ± 0.052 | 10336 ± 340 | |
| | OPT1 (960/50; 4 thr 10 gen/comm.) | PC21 | 0.613 ± 0.032 | 22653 ± 428 | *0.05 |
| | OPT1 (960/100; 4 thr 10 gen/comm.) | PC22 | 0.636 ± 0.037 | 45039 ± 625 | *0.02 |
| | OPT1 (960/50; 8 thr 10 gen/comm.) | PC23 | 0.603 ± 0.024 | 10746 ± 18 | 0.12 |
| | OPT1 (960/50; 8 thr 5 gen/comm.) | PC24 | 0.609 ± 0.023 | 10778 ± 33 | *0.03 |
| | OPT1 (960/100; 8thr 10gen/comm.) | PC25 | 0.589 ± 0.028 | 22537 ± 1059 | 0.60 |
| | OPT1 (1920/100; 8thr 10gen/comm.) | PC26 | 0.638 ± 0.029 | 45838 ± 975 | *0.02 |
| 110 | OPT0 | | 0.648 ± 0.000 | – | |
| | OPT1 (50/50) | | 0.450 ± 0.045 | 3867 ± 164 | |
| | OPT1 (120/50) | | 0.450 ± 0.021 | 9312 ± 253 | |
| | OPT1 (960/50; 4 thr 10 gen/comm.) | PC21 | 0.569 ± 0.031 | 22640 ± 461 | *0.01 |
| | OPT1 (960/100; 4 thr 10 gen/comm.) | PC22 | 0.564 ± 0.065 | 43796 ± 712 | *0.01 |
| | OPT1 (960/50; 8 thr 10 gen/comm.) | PC23 | 0.566 ± 0.029 | 10850 ± 191 | *0.01 |
| | OPT1 (960/50; 8 thr 5 gen/comm.) | PC24 | 0.569 ± 0.031 | 12037 ± 1407 | *0.01 |
| | OPT1 (960/100; 8thr 10gen/comm.) | PC25 | 0.556 ± 0.074 | 26754 ± 2595 | *0.01 |
| | OPT1 (1920/100; 8thr 10gen/comm.) | PC26 | 0.551 ± 0.028 | 45355 ± 1413 | *0.01 |

Besides the average Kappa index values obtained by the baseline method OPT0, Table 2 gives the results for the option OPT1 [10], which searches for an optima subset of features that constitute the inputs to only one LDA classifier. Similarly, the data in parentheses in Table 2 represent the number of individuals in the population and generations of the evolutionary algorithm and, in the case of the parallel executions, the number of threads (4 or 8) and generations (5 or 10) executed by each subpopulation. The evaluated parallel configurations are noted as PC21 to PC26 in the column PCnf. of Table 2. The sequential version of OPT1 with 120 individuals and 50 generations for subjects 104, 107 and 110 requires more than four and a half hours, and provides solutions with average Kappa indices equal to 0.515, 0.580 and 0.450, respectively. As can be seen, the parallel versions of OPT1 improve the performance achieved by sequential implementations of this approach with less individuals in the population, and requires much less time than the sequential counterpart with the same population size. Moreover, some parallel configurations of OPT1 reduce the performance differences with respect to OPT2 or even overcome it for some subjects and parallel configurations.

Tables 1 and 2 also show results about the comparison tests to identify which alternatives provide statistically different performance, with p-values obtained from statistical tests with a significance level of 5%. The statistical analysis has been done through a Kruskal-Wallis test, which provides the intervals of the Kruskal-Wallis rank for each considered alternative for OPT2 in Table 1 and OPT1 in Table 2. For subjects 104, 107 and 110, the column noted as p-val. in Tables 1 and 2 gives the p-values obtained after comparing each parallel configuration to the corresponding reference sequential alternative. That is, the parallel configurations PC11 to PC14 in Table 1 have been compared to the sequential execution of OPT2 with 120 individuals and 50 generations, and the parallel configurations PC21 to PC26 in Table 2 to the sequential execution of OPT1, also with 120 individuals and 50 generations. A p-value below 0.05 (marked with an asterisk in columns p-val.) means statistically significant difference. In Table 1, except for PC11 and PC13 for subject 104, the differences in the Kappa indices attained by the evaluated parallel configurations are not statistically significant. Therefore, despite the parallel island evolutionary procedure is different from the sequential evolutionary algorithm, the quality of the results for a given configuration of individuals in the population and a given number of generations is similar to the sequential one, while the parallel procedure provides a significant reduction in the execution time. In the case of PC11 and PC13, the quality of the solutions found decreases only by less than 6% (respectively, 4.3% and 5.8%).

The average Kappa indices in Table 2 show improvements in the quality of solutions by all the parallel configurations (PC21 to PC26) for all subjects (104, 107 and 110), with respect to the reference sequential execution of OPT1 with 120 individuals and 50 generations. These improvements are statistically significant in all the cases but two, PC23 and PC25 for subject 107, where improvements are 4% and 1.6% respectively. The sequential OPT2 provides better results than the sequential OPT1. Moreover, OPT0 provides the best performance for subjects 104 and 110 and a performance close to that of OPT2 for subject 107. It should be taken into account that OPT0 and OPT2 are based on the classifier structure of Fig. 2, which uses more features and more complex classifier structure than OPT1 that requires only one LDA classifier. Table 2 shows that the parallel implementation of OPT1 improves its performance compared to the sequential OPT1 and reduces the performance differences from OPT0 and OPT2. In some cases, it even outperformed OPT0 and OPT2. For example, for subject 104 PC24 (the best parallel configuration for subject 104 with OPT1) provides better results than PC12 (the best configuration for subject 104 with OPT2) and OPT0: 0.698 against 0.554 and 0.564 respectively (a difference statistically significant with p-value equal to 0.004). For the subjects 107 and 110, OPT0 and OPT2 still achieved better performances than some parallel configurations of OPT1. Nevertheless, the differences among the best average Kappa indices for the sequential OPT0 and OPT2 with respect to the sequential OPT1 are reduced in the parallel configurations of OPT1: from 0.072 (0.652 − 0.580) to 0.019 (0.657 − 0.638) for subject 107, and from 0.198 (0.648 − 0.450) to 0.079 (0.648 − 0.569) for subject 110.

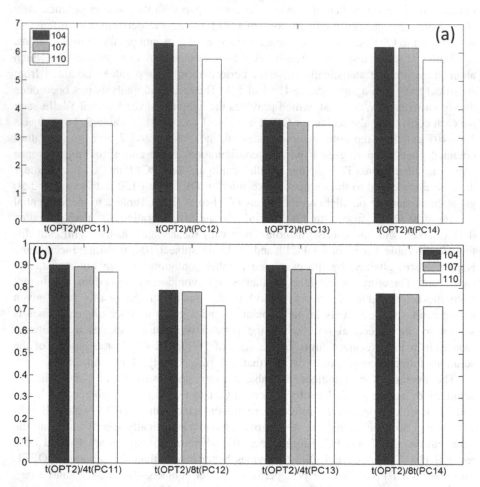

**Fig. 3.** Speedups (a) and efficiencies (b) for different communication profiles and parallel configurations of OPT2 (PC11 to PC14) for subjects 104, 107 and 110

The parallel approach proposed in this paper reduces the execution time with respect to the corresponding sequential version in all the cases, as shown in Table 1 (column Time(s)). Figure 3 shows the speedups (time of the sequential version divided by the time of a given parallel configuration) and efficiencies (speedups divided by the number of processors) provided by the parallel configurations PC11 to PC14 in Table 1, with respect to the sequential execution of OPT2 with 120 individuals and 50 iterations, t (OPT2)/t(PC11) to t(OPT2)/t(PC14). Two different communication profiles have been considered in our experiments. In the first communication alternative, each subpopulation independently implements 10 iterations (generations) before communicating with another subpopulation randomly selected. In the second one, the subpopulations communicate more frequently as it happens after 5 generations of independent evolution in the subpopulations. As shown in Fig. 3, for a given number of threads, the alternative communicating more frequently (5 iterations/communication) provides lower speedups. Consequently, the efficiencies observed in Fig. 3.b are slightly lower in the alternatives communicating more.

The speedups given in Fig. 4 are based on the comparison of execution time for different parallel configurations of OPT1, as shown in Table 2. In this case, we have not compared the parallel time with the corresponding sequential time due to their large values (more than four days in some cases). Figure 4 provides the speedups by using 4 threads with respect to 8 threads, i.e., t(PC21)/t(PC23) and t(PC22)/t(PC25), to compare parallel configurations with the same configuration of individuals in the population, iterations and communication profiles, although t(PC21)/t(PC23) corresponds to 50 generations and t(PC22)/t(PC25) to 100. The speedup t(PC21)/t(PC24) compares parallel configurations with 8 and 4 threads and the same number of generations (50 generations) but different communication characteristics (10 and 5 generations between communications respectively). Finally, t(PC26)/t(PC25) compares parallel configuration with the same number of threads (8 threads), generations (100 generations), and communication profiles (10 generations between communications) but different number of individuals. All those considered speedups should have a value near two, as shown in Fig. 4.

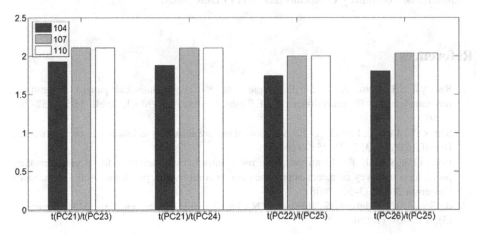

**Fig. 4.** Speedups of different parallel configurations of OPT1 (PC21 to PC22) for subjects 104, 107 and 110

## 5 Conclusions

This paper proposes and analyses parallel island implementations of multiobjective feature selection in BCI tasks with MRA. They are able to improve the quality of the solutions by using greater populations and reduce processing time as well.

On the one hand, we have shown that, despite the parallel algorithm is not exactly the same as the sequential one, it decreases the computing time without a statistically significant reduction in the solution qualities given a number of generations and individuals in the population. Although the speedups shown in Fig. 3.a correspond to efficiencies below one (Fig. 3.b), but the efficiency decrease is relatively small. The scalability behavior of the parallel procedure could be considered adequate for the number of threads used in our parallel executions.

On the other hand, the parallel procedure makes it possible to improve the quality of the solutions by using greater populations in the evolutionary algorithm but similar amount of computing time required by the sequential implementation. It has been shown that by using populations with as many individuals as the reference population multiplied by the number of threads used, OPT1using a simple classifier structure is able to match OPT2 using a complicated classifier structure or even outperform it with similar or even lower computing times.

In [8], a parallel cooperative multiobjective approach was proposed for feature selection in high-dimensional EEG data, which not only evolves independent subpopulations but also assigns different areas of the searching space to different subpopulations. This approach allows superlinear speedups in some cases although with some performance loss. The implementation of such approach to distribute the searching space will be also explored in the present parallel island algorithm for BCI with MRA, along with its use in the feature selection problem for other applications and benchmarks to compare with other previous methods.

**Acknowledgements.** This work was partly funded by grant TIN2015-67020-P (Spanish "Ministerio de Economía y Competitividad" and FEDER funds).

## References

1. Raudys, S.J., Jain, A.K.: Small sample size effects in statistical pattern recognition: recommendations for practitioners. IEEE Trans. Pattern Anal. Mach. Intell. **13**(3), 252–264 (1991)
2. Saeys, Y., Inza, I., Larrañaga, P.: A review of feature selection techniques in bioinformatics. Bioinformatics **23**, 2507–2517 (2007)
3. Pudil, P., Somol, P.: Identifying the most informative variables for decision-making problems - a survey of recent approaches and accompanying problems. Acta Oeconomica Pragensia **2008**, 37–55 (2008)
4. de Souza, J.T., Matwin, S., Japkowitz, N.: Parallelizing feature selection. Algorithmica **45**(3), 433–456 (2006)

5. Sun, Z.: Parallel feature selection based on MapReduce. In: Wong, W.E., Zhu, T. (eds.) Computer Engineering and Networking. LNEE, vol. 277, pp. 299–306. Springer, Cham (2014). doi:10.1007/978-3-319-01766-2_35

6. Zao, Z., Zhang, R., Cox, J., Dulin, D., Sarle, W.: Massively parallel feature selection: an approach based on variance preservation. Mach. Learn. **92**(1), 195–220 (2013)

7. Kimovski, D., Ortega, J., Ortiz, A., Baños, R.: Parallel alternatives for evolutionary multi-objective optimization in unsupervised feature selection. Expert Syst. Appl. **42**(9), 4239–4252 (2015)

8. Kimovski, D., Ortega, J., Baños, A.R.: Leveraging cooperation for parallel multiobjective feature selection in high-dimensional EEG data. Concurr. Comput.: Pract. Exp. **27**, 5476–5499 (2015)

9. Deb, K., Agrawal, S., Pratap, A., Meyarivan, T.: A fast elitist non-dominated sorting genetic algorithm for multi-objective optimization: NSGA-II. In: Schoenauer, M., Deb, K., Rudolph, G., Yao, X., Lutton, E., Merelo, J.J., Schwefel, H.-P. (eds.) PPSN 2000. LNCS, vol. 1917, pp. 849–858. Springer, Heidelberg (2000). doi:10.1007/3-540-45356-3_83

10. Ortega, J., Asensio-Cubero, J., Gan, J.Q., Ortiz, A.: Classification of motor imagery tasks for BCI with multiresolution analysis and multiobjective feature selection. BioMed. Eng. OnLine **15**(Suppl. 1), 73 (2016)

11. Daubechies, I.: Ten Lectures on Wavelets. SIAM, Philadelphia (2006)

12. Asensio-Cubero, J., Gan, J.Q., Palaniappan, R.: Multiresolution analysis over simple graphs for brain computer interfaces. J. Neural Eng. **10**(4) (2013). doi:10.1088/1741-2560/10/4/046014

13. Cohen, J.: A coefficient of agreement for nominal scales. Educ. Psychol. Meas. **20**, 37–46 (1960)

# Deep Belief Networks and Multiobjective Feature Selection for BCI with Multiresolution Analysis

Julio Ortega[1]([⊠]), Andrés Ortiz[2], Pedro Martín-Smith[1], John Q. Gan[3], and Jesús González-Peñalver[1]

[1] Department of Computer Architecture and Technology, CITIC, University of Granada, Granada, Spain
{jortega, pmartin, jesusgonzalez}@ugr.es
[2] Department of Communications Engineering, University of Málaga, Málaga, Spain
aortiz@ic.uma.es
[3] School of Computer Science and Electronic Engineering, University of Essex, Colchester, UK
jqgan@essex.ac.uk

**Abstract.** High-dimensional pattern classification problems with a small number of training patterns are difficult. This paper deals with classification of motor imagery tasks for brain-computer interfacing (BCI), which is a hard problem involving a relatively small number of high-dimensional training patterns where curse of dimensionality issue has to be taken into account and feature selection is an important requirement to build a suitable classifier. Evolutionary meta-heuristics for feature selection are usually more time-consuming than other alternatives, but their high performances in terms of classification accuracy make them desirable approaches. In this paper, feature selection through a wrapper procedure based on multi-objective optimization is compared with the use of deep belief networks (DBN) that constitute powerful classifiers implementing feature selection implicitly. Two different classifiers, LDA (linear discriminant analysis) and DBN, have been used to classify EEG signals with features extracted by multiresolution analysis (MRA) and selected by a multiobjective evolutionary method that also uses LDA to implement the fitness function of the solutions. The experimental results show that DBNs usually provide better or similar classification performances without requiring an explicit feature selection phase. Nevertheless, the DBN's classification performance significantly decreases in problems with a very large number of features. Moreover, to achieve high classification rates, it is necessary to determine a suitable structure for the DBN. Therefore, in this paper we also propose a multiobjective approach to tackle this problem.

**Keywords:** Brain-computer interfaces (BCI) · Deep belief networks (DBN) · Feature selection · Linear discriminant analysis (LDA) · Multiresolution analysis (MRA)

© Springer International Publishing AG 2017
I. Rojas et al. (Eds.): IWANN 2017, Part I, LNCS 10305, pp. 28–39, 2017.
DOI: 10.1007/978-3-319-59153-7_3

# 1  Introduction

Many classification tasks in bioinformatics deal with patterns defined by a large number of features. Moreover, these high-dimensional classification problems have frequently to be solved with the number of training patterns smaller than the number of features, thus presenting curse of dimensionality problems [1]. Feature selection is usually required in bioinformatics [2] to eliminate redundant and noisy features in order to improve the accuracy and interpretability of the classifiers. The brain-computer interfacing (BCI) application considered in this paper is an example of such situation. It implies to classify EEG signals corresponding to motor imagery (MI) tasks, a BCI paradigm that uses the series of amplifications and attenuations of short duration conditioned by limb movement imagination, the so called event related desynchro-nization (ERD) and event related synchronization (ERS). In particular, the BCI considered here uses patterns built by multiresolution analysis (MRA) [3], which applies a sequence of successive approximation spaces to describe the target signal, thus being useful whenever this signal presents different characteristics across the approximation spaces. A specific MRA, the discrete wavelet transform (DWT) applied in [4] to characterize EEGs from MI tasks, has been used here. This way, the patterns are built, from an EEG trial, by a feature extraction procedure, in which each signal obtained from each electrode contains several segments to which a set of wavelets detail and approximation coefficients are assigned. With S segments, E electrodes, and L levels of wavelets, each pattern is characterized by $2 \times S \times E \times L$ sets of coefficients. The number of coefficients in each level set depends on the level. For example, in the case of the dataset recorded in the BCI Laboratory at the University of Essex, S = 20 segments, E = 15 electrodes, and L = 6 levels, therefore there are 3600 sets of coef-ficients, with from 4 to 128 coefficients in each set, and each pattern may contain a total of 151200 coefficients. This huge number of coefficients can be reduced, as in [4], by computing the second moment of the coefficient distribution (variance) in each of the 3600 sets, and normalizing the obtained value between 0 and 1. This way, $2 \times S \times E \times L = 3600$ features constitute each pattern. As the number of training patterns for each subject is approximately 180, it is clear that an efficient procedure for feature selection is still necessary.

One of the issues to be taken into account in the design of a feasible feature selection procedure is the number of possible features because the size of the searching space depends exponentially on that number. As we are involved in high-dimensional classification problems, several hundreds or even thousands of features usually define a very huge searching space where efficient metaheuristics are required. In this paper, we use an evolutionary procedure where the individuals of the population represent dif-ferent feature selections. The fitness of a given individual is determined through a suitable utility function of the selected features. Different alternatives can be considered to define the utility of a given feature selection. If the classes of the training patterns are known (supervised classification problems), it is possible to evaluate feature selection from the performance achieved by the classifier after training with the patterns described by the set of selected features. Nevertheless, it is also possible to assign utilities to the selected features by evaluating the clustering properties of the patterns consisting of these features. This alternative is useful in problems where the pattern labels are not available (unsupervised classification problems) but, although we will not consider it in this paper, could be also used when the labels are known.

This paper investigates the use of Deep Belief Networks (DBN) [5] for EEG classification from two different perspectives. The first one uses a multiobjective evolutionary metaheuristics to select the features for the inputs to the DBN. Nevertheless, in supervised classification problems, the utility function required by the metaheuristics is usually defined by the performance of the classifier once it has been trained by using the training set of patterns with the selected features as components. This approach could prevent the use of powerful classifiers (as DBN) in the feature selection phase, due to the large amount of training time they would require. As the number of classifier trainings depends on the product of the number of individuals (feature subsets) and the number of generations executed by the evolutionary meta-heuristics, the limited extent to which the huge searching space could be explored in a reasonable amount of time could limit the performance of the feature selection. Therefore, the second approach considered in the paper tries to take advantage of the implicit feature selection accomplished by the DBN [5] to avoid the explicit feature selection step, although in this case some optimization procedure should be applied to determine the best DBN structure. In this paper, a multiobjective optimization procedure is devised to optimize the DBN structure.

In what follows, Sect. 2 describes our approach to feature selection based on multiobjective optimization and the issues related with the cost functions proposed to evaluate feature subsets. The main characteristics of DBNs are summarized in Sect. 3, as the purpose of the paper is to analyze the performance of this classifier for BCI applications in comparison with other classifiers such as LDA. The experimental work conducted, the details of the database, and the results obtained are provided in Sect. 4. Finally, Sect. 5 draws conclusions.

## 2 Cost Functions and Multiobjective Feature Selection

Figure 1 shows the feature selection approach adopted in the paper. It is based on a multiobjective optimization procedure that searches a vector of features $\mathbf{x} = [x_1, x_2, \ldots, x_n] \in R^n$ to optimize a function vector $\mathbf{f}(\mathbf{x})$, whose scalar values $(f_1(\mathbf{x}), f_2(\mathbf{x}), \ldots, f_m(\mathbf{x}))$ represent the objectives to optimize. These objectives are usually in conflict, and thus multiobjective optimization should obtain a set of *non-dominated* solutions called Pareto optimal solutions that define the *Pareto front* (no solution in the Pareto front is worse than the others when all the objectives are taken into account), from which it is possible to choose the most convenient solution in specific circumstances. To solve the multiobjective optimization problem an evolutionary algorithm based on the NSGA-II algorithm [6] was implemented, with specific individual codification and genetic operators implemented for the application at hand.

As this paper deals with a supervised classification problem in which the labels of the training and test patterns are known, it would be possible to evaluate the performance of the classifier by the accuracy obtained for the validation patterns after training it (by using the set of training patterns). Nevertheless, other measures that quantify

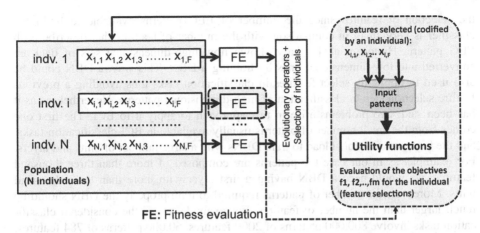

**Fig. 1.** Approach to feature selection by evolutionary multiobjective optimization

properties such as the generalization capability should be taken into account in order to improve the behavior of the classifier in a real environment, where patterns that have not been used for training have to be processed. This is the reason to propose a multiobjective feature selection procedure that optimizes the utility of the selected features evaluated by both the accuracy and generalization capability of the classifier for the available training patterns (with the selected features as components).

As in the considered EEG classification problem the classes corresponding to the training (and test) patterns are known, it is possible to evaluate the performance of a given feature selection by the performance of the classifier for the training patterns described by the selected features. This way, to define this pair of cost functions, we consider an LDA classifier [7] (a brief introduction of LDA is provided in Sect. 3), with f1 and f2 being the accuracy and the generalization capabilities of the LDA. The cost function f1 is related to the Kappa index [8] (f1 = 1-Kappa index) obtained after the learning iterations executed in the feature evaluation step of the evolutionary algorithm. The Kappa index takes into account the distribution of the per class error as it is computed as $(p_0 - p_c)/(1 - p_c)$ with $p_0$ equal to the proportion of coincidences among the classification outputs and the labels of the patterns and $p_c$ being the proportion of patterns on which the coincidence is expected by chance. The cost function f2 is the average loss function in a 10-fold cross validation analysis to the training patterns (mean squared classification error of the validation patterns over all folds).

## 3   The Classifiers: DBN and LDA

Deep belief networks (DBN) [5, 9] are neural network models including several hidden layers of neurons, which have attracted a lot of interest in recent years since the proposal of new algorithms to train deep neural structures [9] and due to their resemblance with the way in which the human brain hierarchizes the information in different layers of abstraction. In this paper, we investigate the performance of DBNs in EEG classification for BCI based on motor imagery. The first problem to cope with is the selection of features since the number of EEG patterns available to build the classifier is very small in comparison with the number of features that describe each EEG pattern. Hinton et al. [5] have shown how high-dimensional sets of data are converted into low-dimensional ones by training a DBN. This way, a DBN could be also used to implicitly select features in classification tasks thus avoiding a previous feature selection step in classification of high-dimensional patterns. Nevertheless, as it has been said, two problems arise in this approach to apply it to BCI. The first one comes from the small number of patterns usually available in BCI classification tasks, and the high computational load to train DBNs with a large number of units and layers. For example, as in our case the patterns are composed of more than three thousand features, we would need a DBN having a first layer with more than three thousand units. Moreover, the number of patterns required to train properly the DBN should be much larger than the number of features. For example, in [5] the considered classification tasks involve 800,000 patterns of 2000 features, 60,000 patterns of 784 features, etc. As we will describe, in our EEG classification problem there are about 180 patterns of over 3600 features.

Therefore, it is worth analyzing whether multiobjective feature selection makes it possible to take advantage of DBNs in hard classification tasks affected by curse of dimensionality problem. It has to be taken into account that the high computational cost of DBNs prevents their use in the feature selection phase, as a large number of training processes are required (the number of individuals in the population multiplied by the number of generations). Thus, it would be useful to analyze the DBN performance after applying different approaches (in particular the faster ones) to evaluating the feature subsets in the multiobjective feature selection procedure. It is also desirable to compare the behavior and performance of DBNs with other classifiers as shown in Fig. 2, which

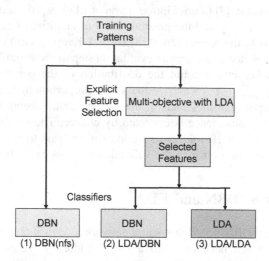

**Fig. 2.** Summary of alternatives evaluated in the paper: (1) DBN classifier with no explicit feature selection (DBN(nfs)); (2) DBN classifier based on multiobjetive feature selection using LDA for fitness evaluation (LDA/DBN); and (3) LDA classifier based on multiobjetive feature selection using LDA for fitness evaluation (LDA/LDA).

summarizes the alternative feature selection approaches and the classifiers compared in the paper. Thus, besides DBNs, Linear Discriminant Analysis (LDA) has been considered for comparison purposes. As detailed descriptions of DBN and LDA can be found elsewhere, in this Section we only provide a brief summary to introduce terms, notations and references to previous works on these classifiers.

*Deep Belief Networks (DBN):* As shown in Fig. 3, DBNs [5, 9] are artificial neural network models built by layers of Restricted Boltzmann Machines (RBMs) [10], which are Boltzmann Machines [11] including stochastic units connected to stochastic binary feature-detecting units through symmetrically weighted connections among different layers but not between units in the same layer. These DBNs work properly only if the initial weights are near to those corresponding to suitable solutions. This is attained by training each RBM through a fast unsupervised learning procedure [12] that changes the weights $w_{ij}^1$ in the first hidden layer by taking into account the error between the training patterns and their reconstruction from the state of the units in the patterns layer. The weights $w_{ij}^2$ in the second hidden layer are changed according to the error between the units in the first hidden layer and their reconstruction from the state of the units in the second layer, and so on. After this unsupervised learning process is finished, the backpropagation algorithm is applied as a supervised learning procedure that takes into account the error between the obtained label and the correct one in the last layer (the classification layer) of the network.

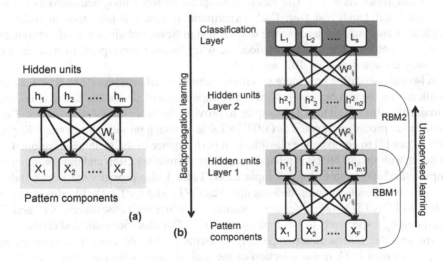

**Fig. 3.** Restricted Boltzmann Machine (RBM) (a) and Deep Belief Network including two RBMs in its hidden layers (b)

*Linear Discriminant Analysis (LDA):* Given a two-class classification problem and a set of training patterns $\mathbf{x} = (x_1, \ldots, x_n)^t$ with their corresponding class labels (for example +1 and −1), the criterion to determine whether a pattern $\mathbf{x}$ belongs to a specific class is given by a linear combination of the features of the pattern. For a constant threshold, T, and a weight vector, $\mathbf{W}$, it should be verified that the dot product $\mathbf{W}^t \mathbf{x} > T$ if $\mathbf{x}$ belongs to one class, and $\mathbf{W}^t \mathbf{x} < T$ if it belongs to the other class. This way, the pattern $\mathbf{x}$ belongs to one class if it is located on one side of a hyperplane perpendicular to $\mathbf{W}$, and to the other class if it located on the other side. LDA supposes that the conditional probability functions for the patterns $\mathbf{x}$ in each class are normally distributed, with means $\mu_{+1}$ and $\mu_{-1}$, and covariances $\Sigma_{+1}$ and $\Sigma_{-1}$. It is also supposed that the class covariances are equal ($\Sigma_{+1} = \Sigma_{-1} = \Sigma$) and have full rank. Under these conditions, $\mathbf{W} = \Sigma^{-1}(\mu_{+1} - \mu_{+1})$. For multiclass problems, there are several alternatives of LDAs that can be applied. For example, it is possible to determine as many LDAs as number of classes, with each classifier adjusted to distinguish a given class against the other classes, and then to combine the results of the classifiers for a given pattern. More details about LDA can be found, for example, in [13].

## 4　Experimental Results

The algorithms have been implemented in Matlab® (version 8.3) and executed in an Intel Xeon E5-2620 processor at 2.1 GHz and 32 GB DDR3 RAM. The experiments use a dataset recorded in the BCI Laboratory at the University of Essex [7]. For each subject, there are a set of 178 patterns for training and another 178 patterns for testing. Each of these EEG patterns is described by components corresponding to 20 segments ($S = 20$), six levels ($L = 6$) of wavelets approximation or detail coefficients (a/d = 2), and 15 electrodes ($E = 15$). The labels correspond to three imagined movements of right hand, left hand, and feet. Each experiment is repeated ten times in order to complete an analysis to determine the statistical significance of the observed differences among alternatives. The results provided in what follows correspond to datasets for three subjects noted as 104, 107, and 110.

In [4] a classification procedure was implemented based on a set of $2 \times S \times L$ LDA classifiers with the number of inputs equaling the number of electrodes, E, and a module for majority voting of all the LDA outputs to provide the final classification output. This classification procedure is noted as OPT0 in Table 1, which provides the average Kappa index values [8] to compare the classification performance of the methods developed in this paper with that of OPT0 [4] that does not use feature selection and thus requires to compute the 3600 features for any input EEG. Table 1 also provides the Kappa index values for the classification procedures noted as OPT1 and OPT2. OPT1, also noted in Table 1 as LDA/LDA, uses a feature selection procedure with cost functions f1 and f2 corresponding respectively to 1-Kappa index of the LDA classifier evaluated on the train patterns and the generalization capability in terms of 10-fold cross validation error. Thus, OPT1 uses LDA in the selection of the features and as the classifier in the final performance evaluation as well, as shown in Fig. 2. OPT2 also uses LDA in the multiobjective feature selection procedure that, in this case, determines the best set of classifiers among the $2 \times S \times L$ LDA classifiers used by OPT0. A more detailed description of OPT0, OPT1, and OPT2 can be found in [14].

Table 1 also provides the results for the alternatives LDA/DBN and DBN(nfs) that use DBN as the classifier. Similar to LDA/LDA, LDA/DBN uses a multiobjective feature selection procedure, but the classification is done by a DBN that uses the features

**Table 1.** Comparison of average Kappa index values for subjects 104, 107, and 110 in the BCI dataset of University of Essex, achieved by different alternatives for feature selection evaluation (with LDA or no feature selection), classification (with LDA or DBN), and number of features (3600 or 300)

| Subj. | OPT0 | OPT2 | 3600 features | | |
|---|---|---|---|---|---|
| | | | LDA/LDA (OPT1) | LDA/DBN | DBN(nfs) |
| 104 | 0.564 | 0.545±0.035 | 0.510±0.056 | 0.584±0.065 | 0.497±0.026 |
| 107 | 0.631 | 0.634±0.019 | 0.560±0.041 | 0.552±0.029 | 0.477±0.010 |
| 110 | 0.648 | 0.605±0.045 | 0.450±0.036 | 0.500±0.049 | 0.476±0.021 |
| Subj. | OPT0 | OPT2 | 300 features | | |
| | | | LDA/LDA | LDA/DBN | DBN(nfs) |
| 104 | 0.564 | 0.545±0.035 | 0.686±0.058 | 0.651±0.052 | 0.720±0.010 |
| 107 | 0.631 | 0.634±0.019 | 0.593±0.069 | 0.599±0.034 | 0.697±0.013 |
| 110 | 0.648 | 0.605±0.045 | 0.614±0.056 | 0.607±0.042 | 0.645±0.027 |

**Table 2.** Number of units in the layers of the DBN configurations used in Table 1 (the averages in columns LDA/DBN and DBN(nfs) are obtained using these DBN configurations)

| Features | DBN1(features) | DBN2(features) | DBN3(features) | DBN4(features) | DBN5(features) |
|---|---|---|---|---|---|
| 3600 | 4000, 4000, 1000, 500, 200, 10 | 4000, 5000, 3000, 1000, 500, 100 | 4000, 6000, 4000, 2000, 500, 100 | 5000, 5000, 5000, 1000, 300, 100 | 5000, 5000, 3000, 1000, 100, 10 |
| 300 | 500,800,1000, 500, 200, 100 | 400, 500, 500, 400, 200, 10 | 400, 500, 500, 500, 200, 10 | 400, 500 600, 300, 100, 10 | 400, 600, 800, 400, 100, 10 |

previously selected by using LDA to determine the corresponding fitness functions f1 and f2. We have used five different configurations (as shown in Table 2) for the DBN with six layers. In DBN(nfs), the DBN classifier does not need feature selection (nfs stand for "no feature selection") as it uses all the features as inputs. In Table 1, the results under 3600 features correspond to the datasets including the 3600 features previously described. We have also performed experiments by considering a subset with 300 of the 3600 features. These 300 features correspond to the 300 values obtained whenever a given level (in our case the first level) of approximation coefficient is considered for the different 20 segments, 15 electrodes ($20 \times 15 = 300$ different features). The multiobjective optimization algorithm used for feature selection uses a population of 2000 individuals and 50 generations in case of 3600 features and 300 individuals and 50 generations in case of 300 features. With respect to the evolutionary operators, we have conducted experiments with mutation probabilities of 0.1, 0.5, 0.9, and 1.0 respectively. The results showed that the best performance was obtained with mutation probability of 1.0.

Table 3 provides information about the multiple comparison tests performed to identify which alternatives obtain significantly different Kappa values on the datasets corresponding to subjects 104, 107, and 110 with a significance level of 5%. These comparison tests have been implemented through the Kruskal-Wallis test that provides the intervals of the Kruskal-Wallis rank for each considered alternative.

**Table 3.** p-values from Kruskal-Wallis tests for the analysis of the statistical significance of differences (values below 0.05 mean statistical significance)

| 104 | LDA/LDA (OPT1) 3600 | LDA/DBN 3600 | DBN(nfs) 3600 | LDA/LDA 300 | LDA/DBN 300 | DBN(nfs) 300 |
|---|---|---|---|---|---|---|
| OPT2 | *p=0.026 | *p=0.0075 | *p=0.001 | *p=0.0086 | *p=0.0080 | *p=0.0086 |
| LDA/LDA (OPT1) 3600 | | *p=0.0088 | *p=0.036 | *p=0.0086 | *p=0.0088 | *p=0.0086 |
| LDA/DBN 3600 | | | *p=0.001 | *p=0.0086 | *p=0.0088 | *p=0.0086 |
| DBN(nfs) 3600 | | | | *p=0.001 | *p=0.001 | *p=0.001 |
| LDA/LDA 300 | | | | | p=0.348 | p=0.5993 |
| LDA/DBN 300 | | | | | | *p=0.0088 |

| 107 | LDA/LDA (OPT1) 3600 | LDA/DBN 3600 | DBN(nfs) 3600 | LDA/LDA 300 | LDA/DBN 300 | DBN(nfs) 300 |
|---|---|---|---|---|---|---|
| OPT2 | *p=0.008 | *p=0.0078 | *p=0.001 | p=0.5959 | *p=0.0078 | *p=0.0078 |
| LDA/LDA (OPT1) 3600 | | p=0.346 | *p=0.001 | p=0.1745 | p=0.1161 | *p=0.0088 |
| LDA/DBN 3600 | | | *p=0.001 | p=0.1732 | p=0.046 | *p=0.0086 |
| DBN(nfs) 3600 | | | | *p=0.001 | *p=0.001 | *p=0.001 |
| LDA/LDA 300 | | | | | p=0.7511 | *p=0.0088 |
| LDA/DBN 300 | | | | | | *p=0.0086 |

| 110 | LDA/LDA (OPT1) 3600 | LDA/DBN 3600 | DBN(nfs) 3600 | LDA/LDA 300 | LDA/DBN 300 | DBN(nfs) 300 |
|---|---|---|---|---|---|---|
| OPT2 | *p=0.008 | *p=0.008 | *p=0.001 | p=0.1850 | p=0.324 | *p=0.0088 |
| LDA/LDA (OPT1) 3600 | | p=0.074 | p=0.410 | *p=0.0088 | *p=0.008 | *p=0.0086 |
| LDA/DBN 3600 | | | p=0.075 | *p=0.0078 | *p=0.024 | *p=0.0086 |
| DBN(nfs) 3600 | | | | *p=0.001 | *p=0.001 | *p=0.001 |
| LDA/LDA 300 | | | | | p=0.184 | p=0.433 |
| LDA/DBN 300 | | | | | | *p=0.036 |

According to the mean values of the Kappa index shown in Table 1 and the statistical significance results of Table 3, some conclusions can be drawn although the relative behavior of different alternatives presents some changes for different subjects 104, 107, and 110. First of all, the relative performances achieved by LDA/LDA and LDA/DBN change with the subject and in general the differences in their obtained mean values are not statistically different for a given number of features (300 or 3600). Therefore, the feature selection by using LDA does not seem to affect the performance achieved by the classifier, LDA or DBN. In some cases, and depending on the subject and number of features in the datasets, LDA/LDA outperforms LDA/DBN, or vice versa. For a given procedure (LDA/LDA, LDA/DBN, or DBN(nfs)) improvements are observed when datasets of 300 features are used instead of datasets of 3600 features. This is expected as, given the huge searching spaces, the efficiency of the evolutionary multiobjective algorithm could be improved by this decrease from 3600 to 300 features.

DBN provides the best performance when the dataset of 300 features is considered. In this case, the inherent feature selection implemented by the DBN is useful to attain good Kappa index values. On the other hand, the use of DBN to classify the datasets with 3600 features produces the worst results among all the considered alternatives. The curse of dimensionality problems arise in this case and the claimed DBN inherent feature selection does not produce a suitable effect.

Nevertheless, it has to be taken into account that only five different DBN structures have been used in our experiments. It should be very useful to have a procedure to search for the optimal DBN structure that provides the best results for a given dataset. Thus, we have implemented a multiobjective evolutionary algorithm that evolves a population of DBNs. The fitness for each DBN is given by two objective functions, f1 and f2. The function f1 is given by the average classification error on ten cross-validation datasets obtained from the given training dataset, and the cost function f2 is the number of units in the DBN structure. This way, the multiobjective procedure searches for the DBN with small classification error and less complexity as well. Table 4 shows the average and best Kappa index values on the test datasets corresponding to subjects 104, 107, and 110, obtained after executing the multiobjective DBN optimization procedure with a population of 50 DBNs with six layers and 20 generations. The average values are obtained considering the solutions in the final Pareto front. The present version of the multiobjective evolutionary procedure explores DBN structures with the same six layers, the number of units in the first layer is between the number of features (300 in Table 4) and the number of features multiplied by 1.75, and the last layer has at least 10 units (the number of DBN outputs is equal to the number of classes, i.e. three outputs in the case considered here). The mutation and crossover operators have been applied. The mutation operator is applied with probability equal to 0.8, changes the units in one layer (in multiples of 20), and resizes the whole DBN to maintain the constraints in its structure. The crossover operator is applied with probability equal to 0.2, selects two individuals and a layer and interchanges the two parts of the corresponding structures. As shown in Table 4, the multiobjective evolutionary procedure (column DBN opt in Table 4) is able to obtain DBN structures (column DBN Structure) that outperforms the results obtained in Table 1 (column DBN(nfs)) with the first set of selected DBN structures.

**Table 4.** Structures obtained by the multiobjective DBN structure optimization procedure (column DBN Structure) and their corresponding Kappa index values (column DBN opt)

| Subject | DBN Structure (L1,L2,L3,L4,L5,L6) | DBN opt | DBN(nfs) |
|---------|-----------------------------------|---------|----------|
| 104 | 450, 560, 1090, 1190, 1020, 700 | 0.750<br>0.733±0.011 | 0.733<br>0.720±0.010 |
| 107 | 360, 700, 740, 620, 560, 180 | 0.733<br>0.723±0.007 | 0.708<br>0.697±0.013 |
| 110 | 510, 510, 750, 680, 550, 420 | 0.683<br>0.672±0.008 | 0.667<br>0.645±0.027 |

## 5  Conclusions

In classification problems with high-dimensional patterns that require feature selection, such as the EEG classification for BCI, it is usual to evaluate a specific feature selection method through the performance of the classifier trained with the patterns consisting of selected features. This approach would require an unacceptable amount of time with powerful classifiers. This is the case for DBNs, which have attracted a lot of interest recently, not only due to their performance capabilities but also for their resemblance of the human brain structure.

This paper considers a multiobjective feature selection procedure using LDA classifier for fitness evaluation and compares the classification performance obtained by LDA and DBN classifiers respectively for a given feature selection. The performance of the DBN classifiers applied to the whole set of features (3600 features in our case) and to a subset of ad hoc selected features (300 features in our case) have been also evaluated. In the experiments, it has been shown that, with the feature selection procedure that uses utility functions based on the performance of classifying the training patterns with LDA, there are no statistically significant differences in the performances of LDA (LDA/LDA alternative) and DBN (LDA/DBN) in most cases. In cases with statistically significant difference between the performances of LDA/LDA and LDA/DBN, as for 300 features, LDA/LDA outperforms LDA/DBN or vice versa (any case implying differences below 5%). This circumstance demonstrates the usefulness of feature selection as a classifier such as LDA, far less complex than DBN, would be still able to produce competitive performances. It would be expected that using DBN to evaluate the utility of the feature selections in the feature selection phase could improve the classification performance of DBN. This DBN/DBN approach has not been considered here due to its high computing time requirements thus constituting an alternative where parallel processing would be useful and will be explored in future. Instead, we have explored the performance of DBN when the whole set of features is used as input to the DBN.

Although DBNs allow an implicit feature selection, they could hardly provide good performance in high-dimensional feature space with a small number of training patterns such as in EEG classification. Nevertheless, good performance can be achieved by DBN structures, without requiring feature selection, on the datasets with a moderate number of features (although still larger than the number of patterns). Any way, it has to be taken into account that feature selection, although usually requires high computing times, provides a reduced set of features to accomplish the classification thus reducing the time required to compute them in a real implementation of the classifier.

The structure of the DBN, along with its other parameters, plays an important role in the classification performance achieved by the DBN, as shown by the multiobjective DBN optimization procedure developed in this paper. There are many alternatives to explore in order to improve this procedure, not only with respect to the parameters to optimize (as in the present version only the number of units in networks changes with a fixed number of layers), but also in the topics related with the acceleration of the multiobjective optimization procedure.

The work presented in this paper opens an interesting approach to using complex DBN classifiers in applications affected by curse of dimensionality. Nevertheless, more experimental work on the tuning of different classifiers is still required to get more definite conclusions. Moreover, recent papers [15–17] have proposed deep neural networks and convolutional deep neural networks to extract features in EEG for BCI tasks. As a comparison with these approaches is difficult due to differences in the datasets and the evaluation indexes used, the implementation of some of these proposals to evaluate them on our datasets, by using the Kappa index, and accomplishing the corresponding statistical analysis of results would be very useful in the future work.

**Acknowledgements.** This work has been funded by grant TIN2015-67020-P (Spanish "Ministerio de Economía y Competitividad" and European Regional Development Fund, ERDF).

# References

1. Raudys, S.J., Jain, A.K.: Small sample size effects in statistical pattern recognition: recommendations for practitioners. IEEE Trans. Pattern Anal. Mach. Intell. **13**(3), 252–264 (1991)
2. Saeys, Y., Inza, I., Larrañaga, P.: A review of feature selection techniques in bioinformatics. Bioinformatics **23**, 2507–2517 (2007)
3. Daubechies, I.: Ten Lectures on Wavelets. SIAM, Philadelphia (2006)
4. Asensio-Cubero, J., Gan, J.Q., Palaniappan, R.: Multiresolution analysis over simple graphs for brain computer interfaces. J. Neural Eng. **10**(4) (2013). doi:10.1088/1741-2560/10/4/046014
5. Hinton, G.E., Salakhutdinov, R.R.: Reducing the dimensionality of data with neural networks. Science **313**, 504–507 (2006)
6. Deb, K., Agrawal, S., Pratap, A., Meyarivan, T.: A fast elitist non-dominated sorting genetic algorithm for multi-objective optimization: NSGA-II. In: Schoenauer, M., Deb, K., Rudolph, G., Yao, X., Lutton, E., Merelo, J.J., Schwefel, H.-P. (eds.) PPSN 2000. LNCS, vol. 1917, pp. 849–858. Springer, Heidelberg (2000). doi:10.1007/3-540-45356-3_83
7. Vapnik, V.N.: Statistical Learning Theory. Wiley-Interscience, Hoboken (1998)
8. Cohen, J.: A coefficient of agreement for nominal scales. Educ. Psychol. Meas. **20**, 37–46 (1960)
9. Schmidhuber, J.: Deep learning in neural networks: an overview. Neural Netw. **61**, 85–117 (2015)
10. Smolensky, P.: Parallel distributed processing: explorations in the microstructure of cognition. In: Information Processing in Dynamical Systems: Foundations of Harmony Theory, vol. 1, pp. 194–281. MIT Press, Cambridge (1986)
11. Hinton, G.E., Sejnowski, T.T.: Learning and relearning in Boltzmann machines. In: Parallel Distributed Processing, vol. 1, pp. 282–317. MIT Press (1986)
12. Ortiz, A., Munilla, J., Górriz, J.M., Ramírez, J.: Ensembles of deep learning architectures for the early diagnosis of the Alzheimer's disease. Int. J. Neural Syst. **26**(7) (2016)
13. Izenman, A.J.: Linear discriminant analysis. In: Izenman, A.J. (ed.) Modern Multivariate Statistical Techniques. Springer Texts in Statistics, pp. 237–280. Springer, Heidelberg (2013)
14. Ortega, J., Asensio-Cubero, J., Gan, J.Q., Ortiz, A.: Classification of motor imagery tasks for BCI with multiresolution analysis and multiobjective feature selection. Biomed. Eng. Online **15**(1), 73 (2016)
15. An, X., Kuang, D., Guo, X., Zhao, Y., He, L.: A deep learning method for classification of EEG data based on motor imagery. In: Huang, D.-S., Han, K., Gromiha, M. (eds.) ICIC 2014. LNCS, vol. 8590, pp. 203–210. Springer, Cham (2014). doi:10.1007/978-3-319-09330-7_25
16. Ren, Y., Wu, Y.: Convolutional deep belief networks for feature extraction of EEG signal. In: International Joint Conference on Neural Networks (IJCNN), 6–11 July 2014
17. Liu, J., Cheng, Y., Zhang, W.: Deep learning EEG response representation for brain-computer interface. In: Proceedings of the 34th Chinese Control Conference, 28–30 July 2015

# IMOGA/SOM: An Intelligent Multi-objective Genetic Algorithm Using Self Organizing Map

Subhradip Aon, Ashis Sau, Prasenjit Dey, and Tandra Pal[✉]

Department of Computer Science and Engineering,
National Institute of Technology Durgapur, Durgapur 713209, West Bengal, India
subhradip.uit02@gmail.com, ashissau001@gmail.com, prasenjitdey13@gmail.com,
tandra.pal@gmail.com

**Abstract.** Multi-objective Genetic Algorithms (MOGAs) are probabilistic search techniques and provide solutions of multi-objective optimization problems. When MOGA reaches near optimal regions, it may face problem in convergence due to its probabilistic nature. MOGA does not pay attention on the neighbourhood of the current population which makes the convergence slow. This scenario may also lead to premature convergence. To overcome this problem, we propose an Intelligent Multi-objective Genetic Algorithm using Self Organizing Map (IMOGA/SOM). The proposed algorithm uses the neighbourhood property of SOM. SOM is trained by the solutions generated by MOGA. SOM performs competition and cooperation among its neurons for better convergence. We have compared the results of the proposed algorithm with two existing algorithms NSGA-II and SOM-Based Multi Objective Genetic Algorithm (SBMOGA). Empirical results demonstrate the superiority of the proposed algorithm IMOGA/SOM.

**Keywords:** Multi-objective genetic algorithms · Self organizing map · NSGA-II · Local search

## 1 Introduction

In last few decades, many researchers have worked on evolutionary computation (EC). Evolutionary computation is a generic population-based meta-heuristic approach which uses the mechanisms inspired by biological evolution such as selection, mutation and recombination. Candidate solutions of the optimization problem play the role of individuals in a population and the fitness function determines the quality of the solutions. Evolution of the population takes place by the repeated application of the above operators. Genetic algorithm (GA) is an evolutionary approach based on natural selection and reproduction.

Real world optimization problems are generally muti-objective in nature having two or more objectives that are conflicting to each other. The main objective of MOGA is to get a diverse set of optimal solutions and converge them to the Pareto front. Diverse solutions are to be distributed uniformly throughout the spectrum of the Pareto front.

© Springer International Publishing AG 2017
I. Rojas et al. (Eds.): IWANN 2017, Part I, LNCS 10305, pp. 40–51, 2017.
DOI: 10.1007/978-3-319-59153-7_4

In MOGA, initially the set of solutions moves towards the Pareto front quickly, but later the rate of convergence towards the Pareto front gradually decreases making the convergence slow. The reason is that the crossover and mutation in GA are random and they do not consider the neighbours of the current population which are topologically similar to them. It makes GA blind about the neighbours of the current solutions. It does not create any problem when the solutions are far away from the Pareto front, but makes the convergence slow when the set of solutions approach near the Pareto front. Some solutions, existing in the current population, which could lead to global optima may get lost in the successive generations. Conventional MOGA also suffers from the problem of genetic drift [1] due to its stochastic nature. The search process is affected by the genetic drift as the algorithm may get stuck into the local optima and hence the global optimal region may remain unexplored. It may lead to premature convergence of MOGA. To avoid these problems, we have used SOM network in the proposed model to make it more intelligent than conventional MOGA. The objective of the proposed algorithm is to use the knowledge incurred from the previous generations, which in turn improves the local search.

The rest of the paper is organised as follows. In Sect. 2, we have presented a review of some works existing in the literature which are similar to the proposed model. The proposed algorithm is presented and explained, in detail, in Sect. 3. The simulation results and comparisons are provided in Sect. 4. In Sect. 5, we have concluded our work and suggested some future directions of research.

## 2   Literature Review

The literature on multi-objective optimization problems is rich. In last two decades, many researchers have worked on MOGAs to achieve better convergence towards the optimal front. Some of the MOGAs are NSGA [2], NSGA-II [3], SPEA2 [4], ABYSS [5], MOEA/D [6], ASMiGA [7], PAES [8], etc.

We observe that a new field of hybrid computing, where knowledge obtained through the learning of neural network has been incorporated in MOGA. It can be viewed as an attempt to introduce local search in the exploration of MOGA. In 2005, Arroyo and Armetano introduced an algorithm IM-MOGLS [9]. They have used multi-objective genetic local search technique for intensifying the search in different local regions. Local search methods have been used in MOGA in many works [10,11] in the literature in order to enhance their performance.

There are some works, where SOM has been used to improve the performance of MOGA. The main reason behind the selection of SOM is its topology and distribution preserving properties. SOM has also the ability to represent high dimensional data into a low dimensional space. In [12] Buche described SOM as a recombination operator for interpolating the parent population. It has increased the possible amount of information to be recombined whereas normal recombination operator uses only two solutions for recombination. An important issue in MOGA is to make a balance between exploration and exploitation. Amour and Rettinger [13] developed an algorithm, Intelligent Exploration for Genetic

Algorithm. SOM has been used in it for mining the information from the evolution process which has been used to enhance the search process. Here SOM also helped to control the problem of genetic drift in [13].

Zhang et al. have introduced regularity model-based multiobjective estimation of distribution algorithm (RM-MEDA) [14]. This approach has used local principal component analysis for building the probability model. Local Principal component analysis, a model for feature extraction, has been hybridized in the model for extracting regularity patterns of the Pareto set from the previous search. Yang et al. developed a hybrid multiobjective estimation of distribution algorithm (HMEDA) [15] that uses local linear embedding, which is a manifold learning algorithm and is used in the optimization process. In order to show better convergence than RM-MEDA [14], Cao et al. devised an algorithm named as manifold-learning-based multiobjective evolutionary algorithm via self organizing maps (ML-MOEA/SOM) [16]. In this algorithm, SOM is used to capture and utilize the manifold structure of the Pareto set. In their algorithm, SOM performs reproduction of new solutions and provide it to MOGA in order to get better convergence. Their method has shown the performance similar to that of RM-MEDA [14].

In [1], the authors have devised an algorithm named SOM-Based Multi Objective Genetic Algoritm (SBMOGA). They have used VNS (Variable Neighbourhood Search) [17] as a local search strategy. SOM uses a multi-objective learning rule for its training based on Pareto Dominance. They have implemented a real-world problem and has found better results than that of NSGA-II. Another SOM-based hybrid algorithm is a self-organizing multi-objective evolutionary algorithm [18], developed by Hu Zhang et al. The authors in [18] used SOM to establish neighbourhood relationship between the current solutions. SOM based hybrid method have also been used in [19,20].

Some of the research works have considered the neighbourhood properties of SOM in the learning algorithm whereas in some other works Pareto dominance property in the learning has been considered. As per our limited survey, we have noticed that no work in the literature has used both of the above mentioned concepts together in the learning of SOM. In the proposed algorithm, we have used both the neighbourhood property and Pareto dominance property for the learning of SOM.

In SBMOGA [1], the SOM neurons get updated when they are dominated by the solutions obtained from MOGA. But, it may not be necessary to update all the dominated neurons, as they all may not be neighbour of any particular solution. In our proposed algorithm, we use a neighbourhood function in SOM that helps to update those neurons that lie within a defined neighbourhood. It is explained later in detail in Sect. 3. In SBMOGA, the SOM neurons remain unaltered when they are not dominated by the solutions obtained from MOGA. Hence, it does not explore the neighbourhood of SOM neurons throughout the execution. As the SOM neurons also provide the optimal solutions, they must explore the search space throughout the execution. Hence, we have incorporated a neighbourhood function in the training of SOM network of the proposed model

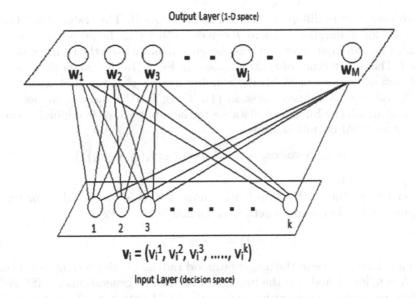

**Fig. 1.** Architecture of SOM

to continue exploring the local regions. Thus, the neighbourhood property of SOM along with the Pareto dominance property are being utilized to overcome the shortcomings of SBMOGA.

## 3   The Proposed Methodology

In the proposed algorithm, multi-objective genetic algorithm has been hybridized with self organizing map (SOM) network [21]. SOM has the capability of preserving the topology and distribution of the input dataset, which are the solutions obtained from MOGA. SOM basically performs a local search in the decision space of multi-objective problems. We have used NSGA-II [3] as MOGA in the proposed model.

Implementation starts with the execution of NSGA-II. NSGA-II initializes its population randomly and then creates its off-springs. The current population and the off-spring population are combined and undergo frontification using fast non-dominated sorting algorithm. NSGA-II uses crowding distance operator to maintain the diversity of the solutions.

The solutions of the first front generated in each alternate generation of NSGA-II are used as input for the training of the SOM. We use one dimensional SOM output neurons in the proposed model. The number of SOM input neurons is same as the number of decision variables in NSGA-II, denoted by $k$. The number of SOM output neurons is denoted by $M$. Weight vectors of SOM, $\mathbf{w}_j$, $(j = 1, 2, ..., M)$ are initialized randomly. The decision variables corresponding to the solutions (in objective space) in the first front, generated by NSGA-II, is used to train the SOM. The number of solutions in the first front is denoted by

$f$, which may vary in different generations of NSGA-II. The training of SOM is carried out for $T$ iterations, where $T$ equals to 50 * $f$ in the proposed model. In each iteration, an input vector $(\mathbf{v}_i)$ is chosen randomly from the training dataset of size $f$. The architecture of SOM is shown in Fig. 1. In the competition among the $M$ output neurons of SOM, the winning neuron $j^*$ is the Best Match Unit (BMU), where $j^*$ is defined below in (1). Thus, BMU is an output neuron of SOM from which the Euclidean distance to the input vector is minimum among all the other SOM output neurons.

$$j^* = \{j|distance_j = min\{distance(\mathbf{v}_i(t), \mathbf{w}_j(t))\}\},\tag{1}$$

where $j = 1, 2, ...M$.

After finding the BMU, its neighbourhood radius is computed according to an exponentially decreasing function as defined below in (2).

$$\sigma(t) = \sigma_0 \exp^{-\frac{t}{\lambda}}\tag{2}$$

Here, $\sigma_0$ = initial value of the neighbourhood radius, $t$ is the current iteration of SOM, $\lambda = n/\log \sigma_0$ and $n$ is the maximum number of generations of NSGA-II.

Then, we perform cooperation among the SOM output neurons. The weight vectors of the SOM neurons, present within the neighbourhood, are updated using (3).

$$\mathbf{w}_j(t+1) = \mathbf{w}_j(t) + L(t) * \sigma_{j^*}(t) * (\mathbf{v}_i(t) - \mathbf{w}_j(t)),\tag{3}$$

where L(t) represents the learning rate, which is a linearly decreasing function as defined below

$$L(t) = 1 - t/T, \ t = 1, 2, ..., T$$

and the neighbourhood function $\sigma_{j^*}(t)$ is defined as

$$\sigma_{j^*}(t) = \exp^{-\frac{d^2}{2\sigma^2(t)}},$$

where $d$ is the Euclidean distance from the BMU $(j^*)$ to the current SOM output neuron.

After updating the SOM neurons, we check the dominance between the newly updated weight vector and the previous weight vector. The dominance is based on the concept of Pareto dominance. If $\mathbf{w}_j(t+1)$ dominates $\mathbf{w}_j(t)$, then the final weight vector for the next iteration is set to $\mathbf{w}_j(t+1)$. Otherwise, it is unaltered. Thus if the solutions of the MOGA deviate from its path towards local optima, it will be restricted by SOM as it always checks Pareto dominance before final updation of weights. If in a generation the SOM neuron(s) reach the global optima, it will not change in the next generation(s). The proposed algorithm IMOGA/SOM is presented in Algorithm 1.

## 4  Results and Discussions

To test the proposed method, we have used nine standard bi-objective uncon-strained test functions, taken from [3]. The summary of the test functions are

---

**Algorithm 1.** IMOGA/SOM

---

1

2 **Hybrid Model:**
3 Initialize the first population
4 Set Generation = 1
5 **/\* NSGA-II Starts its execution \*/**
6 **while** *Generation ≤ Maximum generation(n)* **do**
7    | Perform crossover
8    | Perform mutation
9    | Combine off-springs with their parents
10    | Apply **Fast Non-Dominated Sorting Algorithm** (FNDSA) on the combined population
11    | Save the decision variables of the $f$ solutions of first front obtained from FNDSA
12    | **if** mod $(Generation, 2) == 0$ **then**
13    |   | SOM()
14    | **end**
15    | Generate population for the next generation using Crowding Distance Operator.
16    | Increment Generation by 1.
17 **end**
18 **SOM():**
19 **begin**
20    | Consider the decision variables of $f$ solutions generated by NSGA-II as the training data
21    | Set *size* = $f$
22    | Initialize $t = 1$
23    | Set $T = 50*size$
24    | **while** $t \leq T$ **do**
25    |   | Select an input vector randomly from the training data($\mathbf{v}_i(t)$), where $i = 1, 2, ...f$
26    |   | Find BMU $(j^*)$: $j^* = \{j | distance_j = min\{distance(\mathbf{v}_i(t), \mathbf{w}_j(t))\}\}$, where $j = 1, 2, ...M$
27    |   | Calculate neighbourhood radius $(\sigma(t))$: $\sigma(t) = \sigma_0 \exp^{-\frac{t}{\lambda}}$
28    |   | Find the SOM units present within the neighbourhood radius
29    |   | Learning rate $L(t) = 1 - t/T$
30    |   | Update the weight vectors of neighbourhood neurons:
31    |   | $\mathbf{w}_j(t+1) = \mathbf{w}_j(t) + L(t) * (\sigma_{j*}(t)) * (\mathbf{v}_i(t) - \mathbf{w}_j(t))$
32    |   | Set $j = 1$
33    |   | **while** $j \leq M$ **do**
34    |   |   | **if** $w_j(t+1)$ *dominates* $w_j(t)$ **then**
35    |   |   |   | $\mathbf{w}_j = \mathbf{w}_j(t+1)$
36    |   |   | **end**
37    |   |   | **else**
38    |   |   |   | $\mathbf{w}_j = \mathbf{w}_j(t)$
39    |   |   | **end**
40    |   |   | Increment j by 1.
41    |   | **end**
42    |   | Increment t by 1.
43    | **end**
44 **end**
45 Return the population of GA and the SOM weight vectors as the final solutions

---

presented in Table 1. For performance comparison, we have compared the proposed algorithm with two existing algorithms NSGA-II [3] and SBMOGA [1]. We have used jMetal 4.5.2 [22] and MATLAB 8.1 for its simulation.

## 4.1 Performance Metrics

Three metrics, used to measure the performance of the algorithms, are defined below.

**Table 1.** Summary of the test functions

| Sl. no | Function name | Number of decision variables | Function domain |
|---|---|---|---|
| 1 | SCH | 1 | $[-10^3, 10^3]$ |
| 2 | KUR | 3 | $[-5, 5]$ |
| 3 | POL | 2 | $[-\pi, \pi]$ |
| 4 | FON | 3 | $[-4, 4]$ |
| 5 | ZDT1 | 30 | $[0, 1]$ |
| 6 | ZDT2 | 30 | $[0, 1]$ |
| 7 | ZDT3 | 30 | $[0, 1]$ |
| 8 | ZDT4 | 10 | $x_1 \in [0, 1] x_i \in [-5, 5], i > 1$ |
| 9 | ZDT5 | 10 | $[0, 1]$ |

(i) Convergence Metric ($\Upsilon$) [3] measures how much the obtained set of solutions is close to the Pareto front. Smaller value is considered for better convergence.

(ii) Diversity Metric ($\Delta$) [1] is used to measure how much the solutions are diverse throughout the spectrum of the Pareto front. The range of the values of diversity metric is $[0, 1]$. More close is the value to 1, better is the diversity. Diversity metric, used here, is the revised version of that used in [1]. The diversity metric is defined below in (4).

$$\Delta = \frac{\sum_{Objectives} \sum_{i=1}^{D} count}{No.\ of\ Objectives * D}, \tag{4}$$

where the objective space is divided into $D$ number of divisions for each objective and *count* will be 1 whenever any point from the obtained set of solutions belongs to $i^{th}$ division for the corresponding objectives.

(iii) Inverted Generational Distance (IGD) [16] measures both the convergence and diversity of the obtained set of solutions. It is defined as follows.

$$IGD(F, F^*) = \frac{\sum_{u \in F^*} distance(u, F)}{|F^*|}, \tag{5}$$

where $F^*$ represents the Pareto front, $F$ represents the obtained set of solutions and $u$ represents a point in the Pareto front. Lesser the value better is the IGD.

### 4.2 Parameter Settings

For NSGA-II, population size, maximum number of generations, crossover probability and mutation probability are respectively set to 100, 100, 0.9 and 1/k, where k is the number of decision variables of the test functions. For SOM network, we have considered the number of output neurons $M$ as 100 and initialized the weight vectors randomly in the range of the corresponding decision space. Neighbourhood radius has been initialized to the radius of the decision space. We have used the same parameter values for all of the three algorithms NSGA-II, SBMOGA and proposed IMOGA/SOM. The SOM, used in the proposed algorithm, has been executed for alternate 50 generations, as mentioned earlier in Sect. 3.

### 4.3   Experimental Results

The values of the convergence metric $(\Upsilon)$, diversity metric $(\Delta)$ and the IGD metric for three algorithms: (i) NSGA-II, (ii) SBMOGA and (iii) The proposed IMOGA/SOM are presented in Table 2, where bold entries represent best results.

**Table 2.** Simulation results

| Function | Algorithm | $\Upsilon$ | $\Delta$ | IGD |
|---|---|---|---|---|
| SCH | NSGA-II | 0.0182 | 0.7250 | 0.0183 |
| | SB-MOGA | **0.0135** | **0.8650** | **0.0135** |
| | IMOGA/SOM | 0.0140 | 0.8600 | 0.0140 |
| KUR | NSGA-II | 0.0444 | 0.6750 | 0.0443 |
| | SB-MOGA | 0.0390 | **0.7500** | 0.0389 |
| | IMOGA/SOM | **0.0368** | 0.7350 | **0.0366** |
| POL | NSGA-II | 0.0647 | 0.7050 | 0.0643 |
| | SB-MOGA | 0.0555 | **0.7350** | 0.0553 |
| | IMOGA/SOM | **0.0554** | 0.7300 | **0.0551** |
| FON | NSGA-II | **0.0057** | 0.6750 | **0.0056** |
| | SB-MOGA | **0.0057** | **0.7500** | **0.0056** |
| | IMOGA/SOM | **0.0057** | 0.7350 | **0.0056** |
| ZDT1 | NSGA-II | 0.0228 | 0.7300 | 0.0227 |
| | SB-MOGA | 0.0225 | **0.8300** | 0.0224 |
| | IMOGA/SOM | **0.0221** | 0.7600 | **0.0220** |
| ZDT2 | NSGA-II | 0.0227 | 0.7100 | 0.0227 |
| | SB-MOGA | 0.0224 | 0.7700 | 0.0225 |
| | IMOGA/SOM | **0.0222** | **0.8050** | **0.0222** |
| ZDT3 | NSGA-II | 0.0162 | 0.5650 | 0.0161 |
| | SB-MOGA | **0.0160** | **0.6600** | **0.0160** |
| | IMOGA/SOM | 0.0162 | 0.5850 | 0.0161 |
| ZDT4 | NSGA-II | 0.6012 | 0.0850 | 0.6027 |
| | SB-MOGA | 0.6011 | **0.1400** | 0.6027 |
| | IMOGA/SOM | **0.5942** | 0.1050 | **0.5953** |
| ZDT6 | NSGA-II | **0.2379** | 0.2550 | **0.2378** |
| | SB-MOGA | **0.2379** | **0.2950** | **0.2378** |
| | IMOGA/SOM | **0.2379** | 0.2600 | **0.2378** |

Table 2 shows that for all 9 test functions: SCH, KUR, POL, FON, ZDT1, ZDT2, ZDT3, ZDT4 and ZDT6, the proposed algorithm is better than NSGA-II with respect to all the three metrics: $\Upsilon$, $\Delta$ and IGD. For the convergence and IGD metrics, results show that the proposed algorithm IMOGA/SOM performs better for 5 functions: KUR, POL, ZDT1, ZDT2 and ZDT4, however SBMOGA shows better results for 2 functions: SCH and ZDT3. The remaining functions FON and ZDT6 have shown competitive results for both the convergence and IGD metrics. SBMOGA has greater diversity than the proposed IMOGA/SOM except for the function ZDT2. The simulation graphs are shown in nine figures, Figs. 2, 3, 4, 5, 6, 7, 8, 9 and 10, for nine test functions, where the solutions are plotted in the objective space of the respective function. The figures confirm the convergence of the SOM neurons towards the Pareto front. Though the proposed algorithm IMOGA/SOM gives a diverse set of solutions, SBMOGA

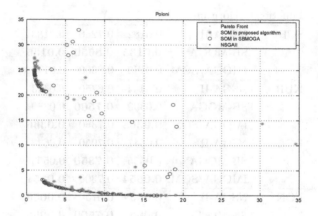

**Fig. 2.** Poloni's function (POL)

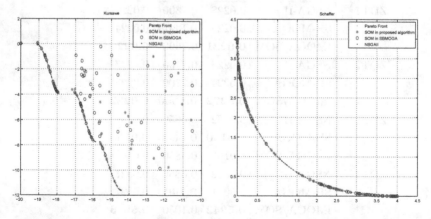

**Fig. 3.** Kursawe's function (KUR)          **Fig. 4.** Schaffer's function (SCH)

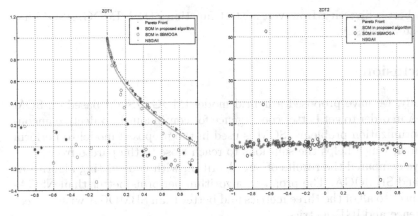

**Fig. 5.** ZDT1 function

**Fig. 6.** ZDT2 function

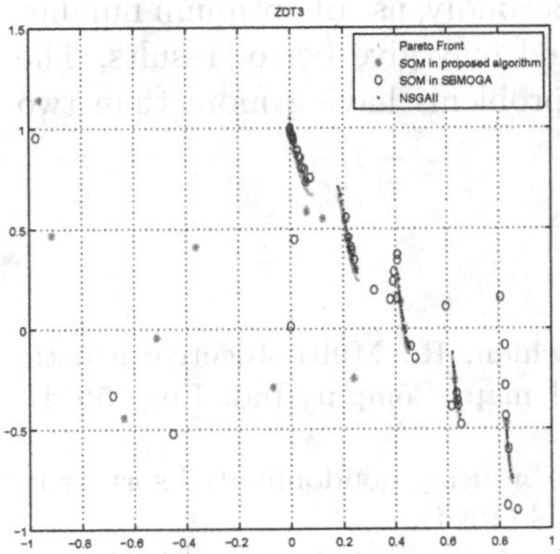

**Fig. 7.** ZDT3 function

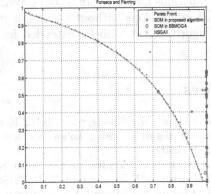

**Fig. 8.** Fonseca and fleming's function

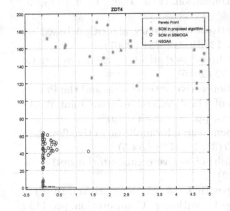

**Fig. 9.** ZDT4 function

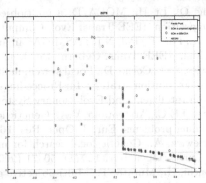

**Fig. 10.** ZDT6 function

has shown better diversity. However, for test function ZDT2, IMOGA/SOM has shown greater diversity.

## 5    Conclusion

In this study, we have proposed a hybrid model IMOGA/SOM based on SOM and MOGA in order to get better convergence for continuous bi-objective unconstrained optimization problems. SOM is used in the proposed model to perform a local search, which helps the solutions to reach close to the pareto front.

We have compared the results of our algorithm with two existing algorithms NSGA-II and SBMOGA. The proposed algorithm performs better than NSGA-II with respect to all of the three metrics and better than SBMOGA with respect to convergence and IGD metrics.

The proposed method may be improved further in two ways. Firstly, neighbourhood functions for SOM can be improved. Secondly, use of optimum number of SOM output neurons without keeping it fixed may give better results. The proposed algorithm can also be extended for problems having more than two objectives.

## References

1. Hakimi-Asiabar, M., Ghodsypour, S.H., Kerachian, R.: Multi-objective genetic local search algorithm using kohonen's neural map. Comput. Ind. Eng. **56**(4), 1566–1576 (2009)
2. Srinivas, N., Deb, K.: Muiltiobjective optimization using nondominated sorting in genetic algorithms. Evol. Comput. **2**(3), 221–248 (1994)
3. Deb, K., Pratap, A., Agarwal, S., Meyarivan, T.: A fast and elitist multiobjective genetic algorithm: NSGA-II. IEEE Trans. Evol. Comput. **6**(2), 182–197 (2002)
4. Zitzler, E., Laumanns, M., Thiele, L., et al.: SPEA2: improving the strength Pareto evolutionary algorithm (2001)
5. Nebro, A.J., Luna, F., Alba, E., Dorronsoro, B., Durillo, J.J., Beham, A.: AbYSS: adapting scatter search to multiobjective optimization. IEEE Trans. Evol. Comput. **12**(4), 439–457 (2008)
6. Zhang, Q., Li, H.: MOEA/D: a multiobjective evolutionary algorithm based on decomposition. IEEE Trans. Evol. Comput. **11**(6), 712–731 (2007)
7. Nag, K., Pal, T., Pal, N.R.: ASMiGA: an archive-based steady-state micro genetic algorithm. IEEE Trans. Cybern. **45**(1), 40–52 (2015)
8. Knowles, J.D., Corne, D.: Local search, multiobjective optimization and the Pareto archived evolution strategy. In: Proceedings of Third Australia-Japan Joint Workshop on Intelligent and Evolutionary Systems, pp. 209–216 (1999)
9. Arroyo, J.E.C., Armentano, V.A.: Genetic local search for multi-objective flowshop scheduling problems. Eur. J. Oper. Res. **167**(3), 717–738 (2005)
10. Jaszkiewicz, A.: Genetic local search for multi-objective combinatorial optimization. Eur. J. Oper. Res. **137**(1), 50–71 (2002)
11. Ishibuchi, H., Murata, T.: Multi-objective genetic local search algorithm. In: Proceedings of IEEE International Conference on Evolutionary Computation, pp. 119–124. IEEE (1996)

12. Büche, D.: Multi-objective evolutionary optimization of gas turbine components. Ph.D. thesis, Universität Stuttgart (2003)
13. Amor, H.B., Rettinger, A.: Intelligent exploration for genetic algorithms: using self-organizing maps in evolutionary computation. In: Proceedings of the 7th Annual Conference on Genetic and Evolutionary Computation, pp. 1531–1538. ACM (2005)
14. Zhang, Q., Zhou, A., Jin, Y.: RM-MEDA: a regularity model-based multiobjective estimation of distribution algorithm. IEEE Trans. Evol. Comput. 12(1), 41–63 (2008)
15. Yang, D., Jiao, L., Gong, M., Feng, H.: Hybrid multiobjective estimation of distribution algorithm by local linear embedding and an immune inspired algorithm. In: IEEE Congress on Evolutionary Computation, CEC 2009, pp. 463–470. IEEE (2009)
16. Cao, W., Zhan, W., Chen, Z.: ML-MOEA/SOM: a manifold-learning-based multiobjective evolutionary algorithm via self-organizing maps. Int. J. Sig. Process. Image Process. Pattern Recognit. 9(7), 391–406 (2016)
17. Hansen, P., Mladenović, N.: Variable neighborhood search: principles and applications. Eur. J. Oper. Res. 130(3), 449–467 (2001)
18. Zhang, H., Zhou, A., Song, S., Zhang, Q., Gao, X.Z., Zhang, J.: A self-organizing multiobjective evolutionary algorithm. IEEE Trans. Evol. Comput. 20(5), 792–806 (2016)
19. Beume, N., Naujoks, B., Emmerich, M.: SMS-EMOA: multiobjective selection based on dominated hypervolume. Eur. J. Oper. Res. 181(3), 1653–1669 (2007)
20. Büche, D., Milano, M., Koumoutsakos, P.: Self-organizing maps for multi-objective optimization. In: GECCO, vol. 2, pp. 152–155 (2002)
21. Kohonen, T.: The self-organizing map. Neurocomputing 21(1), 1–6 (1998)
22. Durillo, J.J., Nebro, A.J.: jMetal: A Java framework for multi-objective optimization. Adv. Eng. Softw. 42(10), 760–771 (2011)

# Solving Scheduling Problems with Genetic Algorithms Using a Priority Encoding Scheme

José L. Subirats[1], Héctor Mesa[1], Francisco Ortega-Zamorano[2],
Gustavo E. Juárez[3], José M. Jerez[1], Ignacio Turias[4],
and Leonardo Franco[1(✉)]

[1] Department of Computer Science, University of Málaga, Málaga, Spain
lfranco@lcc.uma.es
[2] School of Mathematics and Computer Science, Yachay Tech,
San Miguel de Urcuquí, Ecuador
[3] Facultad de Ciencias Exactas y Tecnología,
Universidad Nacional de Tucumán, Tucumán, Argentina
[4] Department of Computer Science, University of Cádiz, Cádiz, Spain

**Abstract.** Scheduling problems are very hard computational tasks with several applications in multitude of domains. In this work we solve a practical problem motivated by a real industry situation, in which we apply a genetic algorithm for finding an acceptable solution in a very short time interval. The main novelty introduced in this work is the use of a priority based chromosome codification that determines the precedence of a task with respect to other ones, permitting to introduce in a very simple way all problem constraints, including setup costs and workforce availability. Results show the suitability of the approach, obtaining real time solutions for tasks with up to 50 products.

**Keywords:** Evolutionary and genetic algorithms · Job shop problems · Priority encoding scheme

## 1 Introduction

Manufacturing companies usually work against clients orders, and unfortunately several times they cannot afford a client order because they had no enough resources to attend it timely. In this sense, a good scheduling plan could be enough to resolve the situation for delivering the orders at the expected time. Thus, an optimized Job planning is essential for manufacturing companies in order to optimize resources, minimize inefficiencies and maximize productivity that usually translates in greater benefits and increased competitiveness [1,10].

There are several levels of organization and planning according to the time horizon of the decisions involved. Flow shop systems are known in the field of production logistics which is called scheduling theory. This theory includes complicated schedules, e.g., production schedules and school schedules, transportation, personal and many others [2]. Even if planning or scheduling are problems

© Springer International Publishing AG 2017
I. Rojas et al. (Eds.): IWANN 2017, Part I, LNCS 10305, pp. 52–61, 2017.
DOI: 10.1007/978-3-319-59153-7_5

affecting most companies there is no systematic solution given the large number of specific variables for each particular case that makes hard to automatize the whole process [3, 4, 7].

Even on simple production scheduling projects, there are multiple inputs, multiple steps, several constraints and limited resources. In general, a resource constrained scheduling problem consists of:

- A set of jobs that must be executed.
- A finite set of resources that can be used to complete each job.
- A set of constraints that must be satisfied.
  - Temporal Constraints: The time window in which the task should be completed.
  - Procedural Constraints: The precedence order in which each task must be executed.
  - Resource Constraints: Are enough resources available when they will be needed?
- A set of objectives to evaluate the scheduling performance.

A typical factory floor setting is a good example of this type of problems where scheduling which jobs need to be completed on which machines, by which employees in what order and at what time. In very complex problems (NP-Hard) such as scheduling, there are no known algorithms for finding an optimal solution in polynomial time, so in the present work we resort to searching for a "good" suboptimal answer. Scheduling problems most often use heuristic algorithms to search for the optimal solution. Heuristic search methods suffer as the inputs become more complex and varied.

Genetic algorithms are well suited for solving production scheduling problems, because unlike several other heuristic methods genetic algorithms operate on a population of solutions rather than on a single solution [6, 8, 11]. In production scheduling this population of solutions consists of several answers that may have different sometimes conflicting objectives. For example, in one solution we may be optimizing a production process to be completed in a minimal amount of time. In another solution we may be optimizing for a minimal amount of defects. By cranking up the speed at which we produce we may run into an increase in defects in our final product.

As the number of jobs are increased, this produces also an increase on the number of constraints and as a consequence an increase on the complexity of the problem. Genetic algorithms are ideal for these types of problems where the search space is large and the number of feasible solutions is small.

To apply a genetic algorithm to a scheduling problem we must first represent it as a chromosome, an ordered set of individual genes. Usually, one way to represent a scheduling genome is to define a sequence of tasks and the start times of those tasks relative to one another. Each task and its corresponding start time represents a chromosome. Nevertheless this kind of approach involves defining the existing constraints as extra conditions, that requires checking for new solutions, slowing down the whole process. A new approach is taken in this work regarding the encoding of the solutions in a chromosome, that contain genes coding the priorities set for each job, permitting from these priorities to construct potential solutions. These type of approach has been applied before on similar type of tasks [5,9] but as far as we know it is the first time to be applied to a scheduling problem with the complexity described in the present work, involving the use of different routes and operations for executing a given job as specified by a real world situation, taking also into account workforce availability and production line setup costs.

## 2   Problem Description

A production order $(J)$ is issued within a factory. The order comprises the production of a number of different products (jobs) $(j_i)$, which can be obtained by using one or more production lines $(L)$. Usually, each job $(j_i)$ has a deadline and a priority value defined. In order to produce the final product $j_i$, a production line is divided into one or more sequential operations $(O)$, each of them needing several kinds of resources (tools, operators, etc.). Moreover, each job $j_i$ can be executed following different production routes $R_{i,k}$ containing different operations.

### 2.1   Factory Description

A factory is modeled as a set of heterogeneous resources:

**Operators $(W)$.** The factory workforce is composed of different specialised operators. The availability of an operator is given by a calendar which reflects shifts and holidays. The operators availability is checked at the time of building a solution from a chromosome.

**Machines or production lines $(L)$.** A set of tools and machines to perform different tasks. The speed of the machine is dependent on the task. Additionally, a machine may be off duty for scheduled periods of time, or unavailable due to being reserved by a previous production order.

$$L = \{l_1, l_2, \ldots, l_l\} \tag{1}$$

### 2.2   Workflow

**Production order $(J)$:** Consists of a set of jobs that have to be scheduled efficiently.

$$J = \{j_1, j_2, \ldots, j_n\} \tag{2}$$

**Jobs $(j_i)$:** A job specifies the quantity of a product $i$ that has to be manufactured according to the production order.

Due to production constraints, customer orders, etc., some jobs have a higher priority than others. A numerical value $p(j_i) \in [0, \ldots, 100]$ is set for each job, and this considered as part of the definition of the problem. Additionally, a job can be constrained by a fixed release date and/or a deadline. In some situations, the completion of certain jobs is required before other jobs can start their task. We define the boolean Jobs Dependency Matrix ($JDM$) such that the element $JDM_{i,j}$ define if the job $J_i$ must conclude before $j_j$ begins.

**Routes ($R^i$):** A job can be done by using a combination of several production lines, and this is specified according to defined routes, that consists in a set of sequential operations needed to complete the job.

$$R^i = \left\{ r_1^i, r_2^i, \ldots, r_m^i \right\} \tag{3}$$

As said before, each possible route $r_m^i$ defines a series of operations ($O_k^{r,i}$):

$$r_r^i = \left\{ O_1^{r,i}, O_2^{r,i}, \ldots, O_k^{r,i} \right\} \tag{4}$$

**Operation ($O_k^{r,i}$):** The operation $O_k^{r,i}$ is the $k^{th}$ non-preemptive action in which a job $j_t$ is divided following a route $r_r^i$.

An operation is characterized by a series of attributes:

- A priority, initially given as the job priority $p_i$.
- A production speed.
- A production line $l \in L$ where the operation will be performed.
- A subset of the available operators $W$:

$$ops_k^{r,i} = \left\{ w_1^{r,i,k}, \ldots, w_o^{r,i,k} | w_m^{r,i,k} \leq w_m, w_m \in W \right\} \tag{5}$$

- A setup time, as the time spent on the preparation of a production line before performing a different type of operation (essentially, specifies the costs of changing the actual operation of a line.

Table 1 shows an example of an order including three jobs, different possible routes in some cases and the operations included in each route.

In conclusion, when a production order arrives at the factory, routes should be chosen for each job, and schedule each operation taking into account the following factors:

- The priority of the operation.
- If all the previous tasks in (JDM) has been finished.
- The set of machines that can execute the operation.
- The needed setup time to prepare the machine.
- The needed time to execute the operation in this machine.
- The operators available that can use this machine in this moment.

The main goal of the scheduling problem is to minimize the production time of the order, finishing each job by its deadline.

**Table 1.** Example of possible routes that can be executed in a hypothetical planning. The production order is composed by three tasks $J = \{j_1, j_2, j_3\}$. The job $j_1$ can be implemented on two routes $R^1 = \{r_1^1, r_2^1\}$, job $j_2$ can be implemented on two different routes too $R^2 = \{r_1^2, r_2^2\}$, however the task $j_3$ should follow just the route $r_1^3$ ($R^3 = \{r_1^3\}$). The 'Operations' column shows the sequential order of operations stipulated by the route, necessary to execute the job.

| Task | Routes | Id | Operations |
|------|--------|----|------------|
| $j_1$ | $r_1^1$ | 1 | $O_1^{1,1}, O_2^{1,1}, O_3^{1,1}$ |
|      | $r_2^1$ | 2 | $O_1^{1,2}, O_2^{1,2}$ |
| $j_2$ | $r_1^2$ | 3 | $O_1^{2,1}, O_2^{2,1}$ |
|      | $r_2^2$ | 4 | $O_1^{2,2}, O_2^{2,2}$ |
| $j_3$ | $r_1^3$ | 5 | $O_1^{3,3}$ |

# 3   Genetic Algorithm Description

In the field of artificial intelligence, a genetic algorithm (GA) generates solutions to optimization problems using operators inspired by natural evolution, such as inheritance, mutation, selection, and crossover. Candidate solutions to the optimization problem play the role of individuals in a population, while a fitness function determines the quality of the solutions. Evolution of the population then takes place after the repeated application of the above operators. Algorithm 1 shows the general scheme of a standard genetic algorithm.

Initialize algorithm;
Evaluate population;
**while** *(not condition end)* **do**
   | Generate new solutions (Elitism, Crossover and Mutation);
   | Evaluate new solutions;
**end**
return leader;
**Algorithm 1.** General scheme of a genetic algorithm

**Initialization:** The initial population is generated randomly. This population is made up of a set of chromosomes from which is possible to create solutions to the problem.

**Evaluation:** A fitness function is set in order to evaluate the goodness of each candidate solution. In the present work, the aim is to minimize the fitness function.

**End condition:** The GA should stop when the optimal solution is reached, but as this is usually unknown, alternative stopping criteria are set. Two criteria are used in this work: (a) setting a maximum number of iterations (generations) related to the maximum amount of time permitted, (b) stopping the evolution of the system when no change in the fitness value is observed for a certain number of generations.

**Selection:** Chromosomes with lower fitness are more likely to be selected, and form part of the next generation.

**Crossover:** Crossover is the main genetic operator. This operator represents sexual recombination, and through its application a new individual is obtained from two parent chromosomes chosen with higher probability for those with lower fitness value.

**Mutation:** The mutation operator, applied to individuals to be included in a new generation, modifies randomly some genes of the chromosome, avoiding the solutions to get caught in local minima.

## 3.1 Chromosome Definition

In a GA, a population of candidate solutions, called chromosomes, is evolved toward better solutions. In the present scheme, each chromosome does not represents directly a solution but contains the necessary information to generate a valid scheduling solution. Each chromosome $c_i$ is composed by a integer vector of genes with length $T + M$. The length of $T$ is the cardinality of the set $J$ (Eq. 2) and $M$ is the number of all the possible operations that can be scheduled. So, the first $T$ genes indicate the route assigned to each planning task using $R^i$ (Eq. 3) for each $j_i$ task. Figure 1 shows the structure of a chromosome for the example task described in Table 1. The first part contains a random chosen route from the possible ones for each scheduled job, and the second part the priorities set for each of the operations needed in order to complete the jobs according to the chosen route. Priorities values in the chromosome are obtained from the priorities assigned to each job (all operations from a route belonging to the same job

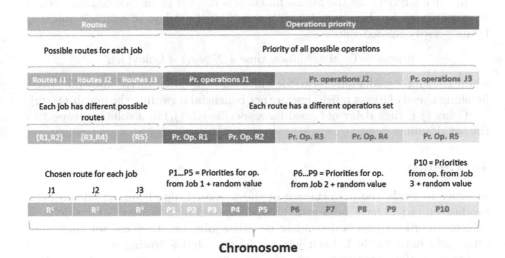

**Fig. 1.** Structure of the chromosome used in the GA, composed by two main parts: the first one containing the information about which route is chosen for each job, and the second one containing the priorities related to the operations.

have the same value) plus a random number in the range $[0, 25]$ that introduce variability in the possible solutions to the problem.

As each chromosome should be a potential solution, we explain now how to build a solution from a given chromosome. The first part of the chromosome indicates for each job which is the chosen route. Now, in order to build a solution, only the priorities of the operations including in these chosen routes are taken into account. Two container sets are created from the priorities, one containing the operations that can be executed at present time, and second container with the rest of operations (those that need to wait for other operations to finish in order to be executed, as stated in the dependence matrix JDM). From the first container, the operation with the largest priority is chosen and assigned to the line that will make that operation to finish earlier, and the same procedure is done with the rest of operations included in the first container (this process is done in a greedy manner, as not all possibilities are analyzed but an order shortcut is taken). In the next step, the operations that can be executed after the one already scheduled are moved from container 2 to container one and the whole cycle is repeated until no operations are left in either container.

### 3.2   Fitness Function Definition

The fitness function computes how 'good' is a potential solution, and thus a proper definition is essential for the success of the GA. Two main aspects should be taken into account for the definition of the fitness function: first, a high correlation between low fitness values and good problem solution, and second that the evaluation of the fitness functions would not be too costly computationally as the algorithm needs to compute it several times during its execution.

In the analyzed case the fitness function is defined as the whole time needed to execute all jobs plus a penalty term that adds the delays of each job weighted by its priority related value:

$$\text{fitness} = \text{Total execution time} + \Sigma \, p(j_i) \times \text{Delay}(j_i),$$

where the sum consider only terms for which the delay is a positive value (i.e., tasks finishing ahead of its set finish time are not beneficial regarding the penalty term).

Delay $(j_i)$: time delay obtained for a specific job $(j_i)$ in a solution respect to the set deadline.

## 4   Results

In order to study the performance of the proposed algorithm, seven synthetic production orders have been generated, composed by sets from 5 to 50 jobs. A 20% of an order's jobs is dependent on other jobs, i.e., they cannot start until other jobs have finished. Each job can be executed following an average of 3 routes, each of one comprising an average of 3 operations, where each operation requires a workforce between 1 and 10 operators.

Each production order has been scheduled 10 times analyzing performance values for the following combination of parameters:

- Population size: $\{25, 50, 100\}$ (*individuals*)
- Elitism (selection): $[0, 10]$ %
- Cross-over rate: 20%, 50%, 80%
- Mutation rate: $[0, 20]$%
- Stagnation at 10000 epochs.

Also as mentioned before, an important aspect for the success of a GA is the correct definition of a fitness function. In order to evaluate the choice followed in this work, we have first computed the correlation (Pearson Correlation value) between the fitness values obtained for problems with different number of products and the total execution time, the total delay and the float times. Total execution time has been defined before, total delay is the sum of all delays produced on individual jobs with respect to the set deadline, and float time is the margin (or flexibility) that every operation has to be delayed without affecting the project completion deadline. The results are shown in Table 2, where high values are obtained in almost all cases, noting that the negative values for the correlation between fitness and float times is expected as for worse solutions (largest fitness value) lower values of float times are expected.

**Table 2.** Correlation between the fitness defined function and the whole problem execution time, delay and float times.

| # of products | Exec. times | Float times | Delay |
|---|---|---|---|
| 5 | 1.0000 | −0.9999 | 0.8936 |
| 10 | 1.0000 | −0.9998 | 0.9272 |
| 15 | 0.8897 | −0.8702 | 0.8943 |
| 20 | 0.9973 | −0.9972 | 0.9990 |
| 30 | 0.9907 | −0.9837 | 0.9917 |
| 40 | 0.9997 | −0.9945 | 0.9929 |
| 50 | 0.9956 | −0.9831 | 0.9767 |

Figure 2 shows the values obtained for the fitness as a function of the size of the GA population for different values of products from 5 to 50 in computer simulations allowed to take a maximum of two minutes for best parameter choices. Fitness values grows approximately linear as a function of the number of products as revealed by analyzing the results across all subplots in Fig. 2, indicating an efficient behavior of the proposal. Further, an increase of the size of the population produces a decrease in fitness values, indicating that it will be possible to improve the current results by increasing further the population size.

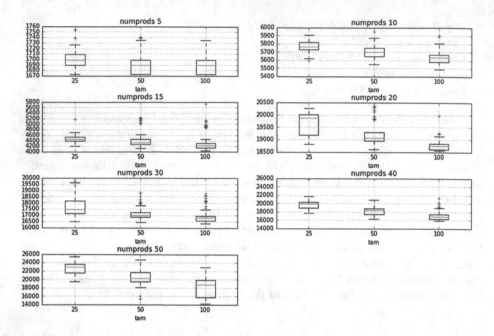

**Fig. 2.** Fitness value as a function of the population size for different number of products.

## 5    Conclusions

A solution for a real world scheduling problem has been proposed using genetic algorithms with a priority encoding scheme. The main novelty in the proposal is the type of chromosome used in the GA that permits to define possible solutions to the problem in a very simple way, taking into account all specified problem variables and restrictions. A fitness function that includes a penalty term related to job production delays seems to be effective, based on the results obtained so far and on a correlation analysis performed. As an overall conclusion, the present proposal has permitted to find acceptable solutions in real time (two minutes time were allowed for obtaining a solution) and further improvements are underway, mainly by introducing an incremental approach that will permit the application of the present proposal to orders with more than 50 products.

**Acknowledgements.** The authors acknowledge support through grants TIN2014-58516-C2-1-R and TIN2014-58516-C2-2-R from MICINN-SPAIN which include FEDER funds.

## References

1. Allahverdi, A., Ng, C., Cheng, T., Kovalyov, M.Y.: A survey of scheduling problems with setup times or costs. Eur. J. Oper. Res. **187**(3), 985–1032 (2008)

2. Cičková, Z., Števo, S.: Flow shop scheduling using differential evolution. Manag. Inf. Syst. **5**(2), 8–13 (2010)
3. Gonzalez, T., Sahni, S.: Flowshop and jobshop schedules: complexity and approximation. Oper. Res. **26**(1), 36–52 (1978)
4. Ham, M., Lee, Y.H., Fowler, J.W.: Integer programming-based real-time scheduler in semiconductor manufacturing. In: Proceedings of the 2009 Winter Simulation Conference (WSC), pp. 1657–1666 (2009)
5. Huang, I., Li, B.: A genetic algorithm using priority-based encoding for routing and spectrum assignment in elastic optical network, pp. 5–11 (2015)
6. Koblasa, F., Sahni, F.M., Vavruška, J.: Evolution algorithm for job shop scheduling problem constrained by the optimization timespan. Appl. Mech. Mater. **309**, 36–52 (2013)
7. Levner, E., Kats, V., Alcaide López De Pablo, D., Cheng, T.: Complexity of cyclic scheduling problems: a state-of-the-art survey. Comput. Ind. Eng. **59**(2), 352–361 (2010)
8. Mesghouni, K., Hammadi, S., Borne, P.: Evolutionary algorithms for job-shop scheduling. Appl. Math. Comput. Sci. **14**(1), 91–103 (2004)
9. Nowling, R., Mauch, H.: Priority encoding scheme for solving permutation and constraint problems with genetic algorithms and simulated annealing, pp. 810–815 (2010)
10. Pinedo, M.L.: Scheduling: Theory, Algorithms, and Systems, 3rd edn. Springer Publishing Company Incorporated, Heidelberg (2008)
11. Ribeiro, F., De Souza, S., Souza, M., Gomes, R.: An adaptive genetic algorithm to solve the single machine scheduling problem with earliness and tardiness penalties (2010)

# Tuning of Clustering Search Based Metaheuristic by Cross-Validated Racing Approach

Thiago Henrique Lemos Fonseca$^{(\boxtimes)}$ and Alexandre Cesar Muniz de Oliveira

Universidade Federal do Maranho, Av. dos Portugueses,
1966 - Vila Bacanga, São Luís, MA 65085-580, Brazil
thiagolemos@lacmor.ufma.br, alexandre.cesar@ufma.br
http://www.ufma.br

**Abstract.** The success of a metaheuristic is directly tied to the good configuration of its free parameters, this process is called Tuning. However, this task is, usually, a tedious and laborious work without scientific robustness for almost all researches. The absence of a formal definition of the tuning and diversity of metaheuristic research contributes to the difficulty in comparing and validating the results, making the progress slower. In this paper, a tuning method named Cross-Validated Racing (CVR) is proposed along with the so named Biased Random-Key Evolutionary Clustering Search and applied to solve instances of the Permutation Flow Shop Problem (PFSP). The proposed approach has reached 99.1% of accuracy in predicting the optimal solution with the parameters found by Irace tuning method. Configurations generated by Irace, even different, have obtained results with the same statistical relevance.

**Keywords:** Metaheuristics · Tuning · Machine learning · Racing algorithms · Irace

## 1 Introduction

Many optimization metaheuristics involve a large number of design choices and performance parameters that need to be properly tuned to reach their best performance. The design and parameter tuning of optimization metaheuristics has been done in a trial-and-error fashion, with poor systematization, extremely designer-expertize dependent. Such approach has a number of disadvantages [1]: time-intensive human effort, highly based on intuition, biased and sometimes not reproducible, limited to a set of instances, poorly explored in designer alternatives, and so on. Instead of manual ad-hoc process, automatized systematics methods has been employed for algorithm tuning. The alternative tuning methods include offline approaches as racing approaches [3,4], and online ones, as adaptive parameter control, hyperheuristic, and so on. Offline approaches are useful especially due to possibility of employ more strict statistical methods to validate hypothesis.

© Springer International Publishing AG 2017
I. Rojas et al. (Eds.): IWANN 2017, Part I, LNCS 10305, pp. 62–72, 2017.
DOI: 10.1007/978-3-319-59153-7_6

Automatic metaheuristic configuration has been described as a machine learning problem [1]. In a primary tuning phase (training), an algorithm configuration is chosen, given a set of training instances representative of a particular problem. In a secondary production (or testing) phase, the chosen algorithm configuration is used to solve unseen instances of the same problem. Automatic algorithm configuration intend to find the parameter set that increases the generalization capability. By this point of view, another negative aspect with respect to manual ad-hoc process is that the same instances are used during tuning phase (training) and final evaluation (test or generalization), probably leading to a biased assessment of performance.

Cross-Validation is a statistical method that allows evaluating and comparing learning algorithms by crossing-over training and test instance groups in successive rounds, computing statistical measures at last. K-fold method is a basic form of cross-validation that involves $K$ rounds, where in each round, $K - 1$ folds are used to training and the remaining one to test [5]. In optimization, an algorithm and its parameters can be understood as the model that solves a given problem. The use of automatic configuration along with machine learning techniques can favor the finding of more accurate and extensible models with great generalization power.

This paper aims to propose Cross-Validated Racing (CVR) as a robust and extensible metaheuristic tuning approach, that is capable of finding the best parameter settings for an optimization metaheuristic independently of the instance set used for training. The so named Biased Random-Key Evolutionary Clustering Search (BRKeCS) is also proposed to solve instances of Permutation Flow Shop Problem (PFSP), making possible to simplify some components of the Clustering Search [2] and to apply the tuning method.

A problem of scheduling operations in *Flow Shop* environment is a problem of scheduling production in which $n$ jobs should be processed by $m$ distinct machines, having the same processing flow in the machines. Usually, the solution to the problem is to determine a sequence of jobs among the $n!$ possibles sequences that minimize the time interval between the beginning of execution of the first job on the first machine and the execution time of the last job on the last machine (*Makespan*). Permutation Flow Shop problem (PFSP) has been investigated and a sort of different optimization methods has been proposed [6–8,10].

This paper is organized as follows. In Sect. 2 previous approaches are considered. Section 3 details the proposed Genetic Algorithm method, considering method particularities, solution encoding, genetic operators and individual reconstruction. In Sect. 4, foundations of racing tuning approaches are presented. Section 5 is devoted to describe the Cross-Validated Racing based on Iterative Race. In Sect. 6 computational results are presented for problem instances found in the Literature. At last, the findings and conclusions are summarized in Sect. 7.

## 2    Previous Approaches

The new resulting hybrid optimization algorithm has been recently employed to solve problems related to minimization of tool switches [17].

Biased Random-Key Genetic Algorithms represent a class of algorithms to solve optimization problems where the solutions are random-keys vectors [15], i.e., an array of real number in continuous interval $[0, 1]$ that need to be decoded before fitness evaluation. A decode function is responsible for converting the array of keys in a feasible solution in the problem domain. Afterwards, the objective function can be computed. An important advantage of the use of a key space is the strong independence of the optimization algorithm in relation to the problem domain, since the solutions are be converted and evaluated, leaving only to the designer the implementation of the *decoder/evaluator* procedure.

The algorithm involves a population of $p$ random keys, generated in the range $[0, 1]$, over a number of iterations called *generations*. The population is partitioned into two groups: a small $p_e$ of elite solutions, composed of the individuals with the best ratings, and the remainders $p - p_e$ denominated as non-elite group. The elite individuals are kept non mutated during next generations. From crossover operations between $p_e$ and the mutants $p_m$ new individuals are produced.

Clustering Search (CS) is a generic way of combining search metaheuristics with clustering in order to detect promising regions so that these regions are subsequently exploited by problem-specific heuristics [20]. CS dynamically divides the search space into *clusters* employing a metaheuristic based methods, responsible to generate candidate solutions continuously during the search process.

Evolutionary Clustering Search (ECS) is a hybrid evolutionary algorithm that employs the CS framework to locate promising search areas, represented by *cluster centers*, $c$. The number of clusters $NumCl$ can be fixed a priori or dynamically determined in running time. The cluster coverage is determined by a *distance metric* that computes the similarity between a given solution and the cluster center. Hamming and Euclidean distance metrics are popular ones [2].

In ECS, a selected individual $s_k$ is presented to the clustering process, yielding in activated cluster centers $c_j$ depending on distance metric $\wp(s_k, c_j)$ and predefined radius $r_j$. Activated clusters receive proportionally votes and can be better exploited posteriorly by local search procedures. A Non-activated cluster can be removed and the respective search area, framed by it, is forgotten. Considering $\mathcal{G}_j$ $_{(j=1,2,\dots)}$ as all current detected clusters, the following rule defines when a new cluster must be created:

$$c_{new} = s_k \text{ if } \wp(s_k, c_j) > r_j, \forall \mathcal{G}_j, \text{ or} \tag{1}$$

When an individual is quite similar to an existent cluster, the assimilation rule is applied to cluster center, $c_j$, and the similar individual, $s_k$, yielding in a new positioning of the cluster, $c_j'$:

$$c_j' = c_j \oplus \beta(s_k \ominus c_j), \text{ otherwise.} \tag{2}$$

where $\oplus$ and $\ominus$ are abstract operations over $c_j$ and $s_k$ meaning, respectively, addition and subtraction of solutions. The operation $(s_k \ominus c_j)$ means how distant are the solutions $s_k$ and $c_j$, considering the distance metric. A learning rate, $\beta$, of this difference is used to update $c_j$, giving $c'_j$. Assimilation plays an important role in clustering process, since clusters must to frame and represent a search area where exists an oversampling of candidate solutions, identifying probably a promising search area.

The number of votes (density) received by each cluster are checked up at regular generation intervals, indicating which are promising. Finally, the local search procedure provides an exploitation mechanism in alleged promising areas.

BRKGA has been employed to generate candidate solutions for the clustering process [17]. BRKGA makes possible to simplify some components of the CS, allowing the need to implement only the decoder and local search heuristics as represented below by Fig. 1.

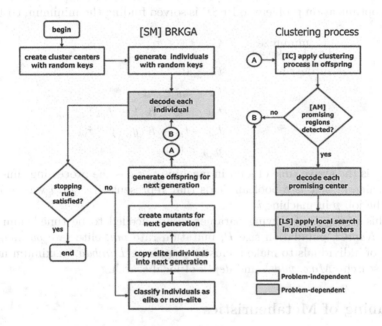

**Fig. 1.** BRKGA+CS conceptual design [17]

Despite the number of applications involving CS based algorithms [12,13], a certain difficulty rests on the need for specific procedures for distance metric, assimilation and local search, beyond native metaheuristic operators. Besides, a lot of performance parameters need to be tuned as well.

## 3 BRKoCS Applied to Permutation Flow Shop Problems

In this section, specific procedures used in this application are presented a well as their performance parameters that are needed to be tunned. Unlike regular

BRKGA, this BRKeCS implementation employs the Blender Crossover (BLX-$\alpha$) [16]. Given two chromosomes $p_1$ and $p_2$, the chromosome $q$ is produced by $q = p_1 + \alpha(p_2 - p_1)$ where $\alpha \simeq U(-\alpha, 1 + \alpha)$, with $U$ represents a uniform distribution. Typically, the alpha values are 0.5 or 0.25.

Local Search procedure is an essential part to effectiveness of BRKeCS, since the promising areas of search space should be explored as soon as they are discovered. BRKeCS use the 2-opt heuristics to intensify this exploration [18]. A complete 2-opt local search compares every possible valid combination of the swapping mechanism, evaluating the candidate solution's neighborhood. This technique can be applied to many permutation related problems [19].

The *Decoder* procedure is designed to guarantee that feasible solution in the problem domain:

$$\Phi(\mathbb{R}^n) \to \mathbb{N} \tag{3}$$

where $n$ is a number of jobs (size of the chromosome) was implemented.

The optimization problem to PFSP is solved finding the minimum cost given by:

$$\underset{i,p = 1..n; \ k,h = 1..m}{\text{minimize}} \quad C = t_{ik} + \tau_{ki}$$

$$\text{subject to} \quad \begin{aligned} & t_{ik} \succeq 0 \\ & t_{ik} - t_{ih} \succeq \tau_{ik}. \\ & t_{pk} - t_{ik} + K(1 - y_{ipk}) \succeq \tau_{ik} \\ & t_{pk} - t_{ik} + K(y_{ipk}) \succeq \tau_{pk} \\ & y_{ipk} \in 0, 1 \end{aligned} \tag{4}$$

where $t_{ij}$ is the arrive time of job $i$ in machine $j$, $\tau_{ik}$ is the processing time of job $i$ in machine $k$, $y_{ipk}$ is a boolean fields that represents if o job $i$ was processed before the job $p$ in machine $k$.

In this work, the following parameters are needed to be tuned: number of clusters *NumCl*, population size $P$, mutation rate *pm*, elite rate *pe*, maximum number of individuals to make a cluster promising *Lambda*, maximum number of local search *rMax*, *width* and *depth* of local search.

## 4   Tuning of Metaheuristics

Following [1], Tuning Problem can be described by the following components:

- $\theta$: set of candidates configurations
- $I$: set of instances
- $P_I$: probability of a instance $i$ to be selected to be solve
- $t : I \to R$: function linking the computing time for each instance
- $c$: random variable that represents the cost of the best solution found by running the $\theta$ setting on the instance $i$ for $t(i)$ seconds
- $C \subset \mathbb{R}$: range of $c$ representing its possible values
- $P_C$: probability that $c$ is the cost of the best solution found by running the $\theta$ setting for $t(i)$ seconds on the $i$ instance. Its notation is given by $P_C(c|, i)$

- $C(\theta) = C(\theta|\Theta, I, P_i, P_C, t)$: criterion that needs to be optimized, measuring the desirability of $\theta$
- $T$: amount of time feasible for experimentation given a set of candidate configurations on a set of instances.

Based on these concepts, the Tuning Problem can be formally described as the 7-Tuple $\langle \Theta, I, PI, PC, t, C, T \rangle$ where the goal is given by:

$$\overline{\theta} = arg\min_{\theta} C(\theta). \tag{5}$$

In a tuning problem, a finite set of candidate configurations $\Theta$ is given along with a class of instances $I$. The cost of the best $i$ solution found with a candidate $\theta$ at a time $t(i)$ is a stochastic quantity described by a conditional measure $P_C$. The problem is to find, within a time T, the best setting according to a criterion $C$, when the measures $P_I$ e $P_C$ are unknown, but a sample of instances can be obtained to test the candidate configurations. One expects to find the cost $\mu$ expressed by the integral

$$\mu = \int cdP_C(c|\theta, i)dP_I(i). \tag{6}$$

The above expression cannot be computed analytically since the values of $P_C$ e $P_I$, are not known. However, the samples can be analyzed and, according to these measurements, the quantities $\mu(\theta)$ can be estimated [11].

## 4.1   Racing Approach

Racing algorithms are inspired by the *Hoeffding Race Algorithm* [14], used to solve selection models on Machine Learning. The idea behind tuning with racing algorithms is that the performance evolution of a candidate configuration can be performed incrementally. In fact, the empirical average $\widehat{\mu_k}(\theta) = \sum_{j=1}^{k} c_j^{\theta}$, of the results obtained for any experiment $k$ is an estimate of the criterion given for Eq. (6).

Since the instances $i_i, i_2, ..., i_k$ are sampled according to the measure $P_I$ and observed the costs $c_1^{\theta}, c_2^{\theta}, ..., c_k^{\theta}$, the best solutions found in a time execution $t$ for a $\theta$ configuration of the metaheuristic. The Algorithm 1 exemplifies the operation of a generic Racing algorithm.

Based on these elements, one can conclude that the racing algorithm therefore generates a sequence of sets of candidate configurations: $\Theta_0 \supseteq \Theta_1 \supseteq \Theta_2 \supseteq ....$ In that the step of a set $\Theta_k$ to $\Theta_{k+1}$ is done by discarding configurations that appear to be suboptimal to the information base of step $k$. If in the step $k$, the set of candidate solutions is still $\Theta_{k-1}$, in other words, there was no change, a new instance is considered.

Each instance $\theta \in \Theta_{k-1}$ is performed over $i_k$ and each cost $c_k^{\theta}$ is added to the vector $c^{k-1}(\theta)$, forming different vectors $c^k(\theta)$, one for each $\theta \in \Theta_{k-1}$. The step $k$ ends with the definition of a set $\Theta_k$ derived from the deletion of $\Theta_{k-1}$ candidates by means of a statistical test that compares the vectors $c^k(\theta)$ for all

---

**Algorithm 1.** Generic Racing algorithm

---

1: **procedure** RACING($M$,Test)    ▷ Total number of experiments and Statistical Test
2:     $numExp \leftarrow 0$                    ▷ Number of experiments
3:     $numInst \leftarrow 0$                   ▷ Number of instances
4:     $C \leftarrow$ ALLOCATEMATRIX($maxInstances, \Theta$)
5:     $survivors \leftarrow \Theta$
6:     **while** $numExp + |survivors| > M$   &   $numInstances + 1 > maxInst$ **do**
7:         $i \leftarrow$ CHOOSEINST()                    ▷ Random selection of instances
8:         $numInst \leftarrow numInst + 1$
9:         **for all** $\theta \in survivors$ **do**
10:            $s \leftarrow$ RUNEXP($\theta, i$)                        ▷ run $BRK_eCS$
11:            $numInst \leftarrow numInst + 1$
12:            $C[numIns, \theta] \leftarrow$ EVALUATE($s$)
13:        $survivors \leftarrow$ DELETECANDIDATES($survivors, C, Test$)
14:     $\bar{\theta} \leftarrow$ BEST($survivors, C$)
15:     **return** $\bar{\theta}$                               ▷ best configuration

---

$\theta \in \Theta_{k-1}$. The described process ends when there remains only one candidate for the surviving configuration, when the maximum number of instances is reached or when the number of defined experiments are executed.

The advantage of the racing approach is the better allocation of computational resources among the candidate configurations. Instead of spending computational time to estimate the performance of lower candidates, the racing algorithms focuses on the most promising candidates and gets the lowest variance estimate for them.

# 5 Tuning by Cross-Validated Racing

## 5.1 Cross-Validation Methodology

The validation task, in the scope of Machine Learning, is a process of determining the degree of reliability of the model built in relation to the data presented. The test methodologies adopted in the project were cross-validation with K-folds used in Machine learning. In the K-folds Cross-Validation method, the training set original is divided into $K$ subsets. Of these K subsets, a subset is retained to be used in the validation of the model and the remaining $K - 1$ subsets are used in training. The cross-validation process is, then, repeated $K$ times, so that each of the $K$ subsets are used exactly once as a test data for model validation.

For this experiments, it will used a iterated racing procedure, which is an extension of racing algorithm proposed by Balaprakash [4] named Irace.

The main purpose of Irace is to automatically configure optimization algorithms by finding the most appropriate settings given a set of instances of an optimization problem. The advantage of using Irace and cross-validation (CVR) is that Irace generates a set of tuning configurations for each step of the validation, but since each fold is a representative set, the configurations generated by Irace, even different, have the same statistical relevance. Allowing that researchers who use the same procedure can be able of to compare their metaheuristics.

The final result of this process is the average performance of the Irace in the K tests. The purpose of repeating multiple test times is to increase the Accuracy of the Irace.

## 6    Computational Results

In all experiments, each run of the metaheuristics takes a limits if 2.000.000 objective function calls and 1000 trials are considered for statistical significance. In each trial, the configuration selected is tested on 10 test instance of the chosen set. For the tests, the Taillard's Benchmark [9] was divided in 5 folds mutually exclusive (no repetitions of instances), each set of instances within each fold are a representative set of Benchmark. The largest group of instances (500 jobs and 20 machines) was removed to a final accuracy test.

Accuracy is given by:

$$Ac = \frac{1}{v} \sum_{i=1}^{v} \epsilon_{y_i, \overline{y}_i} \qquad (7)$$

where v is the number of validation data and $\epsilon_{y_i, \overline{y}_i}$ is the residual error represented by:

$$\epsilon_{y_i, \overline{y}_i} = \frac{y_i - \overline{y}_i}{y_i} \qquad (8)$$

in that $y_i$ and $\overline{y}_i$ are known optimum and found optimum, respectively.

Table 1 represents the $BRKeCS$ results to Cross Validated racing algorithm, the values of each Fold is the average residual error of Irace over the chosen group of instances, in other words, the process of training over each Fold results in a specific parameter configuration that is applied over a random set of instances $(20 \times 5, 20 \times 10...)$ generating a residual error $\epsilon_{y_i, \overline{y}_i} Fold$.

**Table 1.** Cross-validation results to Irace

| Set | $\epsilon_{y_i, \overline{y}_i} Fold1$ | $\epsilon_{y_i, \overline{y}_i} Fold2$ | $\epsilon_{y_i, \overline{y}_i} Fold3$ | $\epsilon_{y_i, \overline{y}_i} Fold4$ | $\epsilon_{y_i, \overline{y}_i} Fold5$ |
|---|---|---|---|---|---|
| $20 \times 5$ | 0 | 0 | 0 | 0 | **−0.0008***  |
| $20 \times 10$ | 0.0006 | 0 | 0 | 0 | 0 |
| $20 \times 20$ | 0 | 0.0004 | 0.0025 | 0 | 0.0021 |
| $50 \times 5$ | 0 | 0.0014 | 0.0003 | 0.0029 | 0.0003 |
| $50 \times 10$ | 0.0085 | 0.0107 | 0.0069 | 0.0022 | 0.01741 |
| $50 \times 20$ | 0.0147 | 0.0188 | 0.0199 | 0.0079 | 0.0140 |
| $100 \times 5$ | 0 | 0.0041 | 0.0034 | 0.0017 | 0.0009 |
| $100 \times 10$ | 0.0046 | 0.0132 | 0.0107 | 0.0208 | 0.0182 |
| $100 \times 20$ | 0.0171 | 0.0283 | 0.0316 | 0.0204 | 0.0177 |
| $200 \times 10$ | 0.0091 | 0.0116 | 0.0112 | 0.0165 | 0.0056 |
| $200 \times 20$ | 0.0222 | 0.0193 | 0.0256 | 0.0253 | 0.0179 |
| $\epsilon_{y_i, \overline{y}_i}$ | 0.0070 | 0.0098 | 0.0102 | 0.0089 | 0.0085 |
| | | | $Ac = 0.0089 == 99.1\%$ | | |

**Table 2.** Comparison between BRKeCS tuned by CVR and different algorithms

| Set | $\epsilon_{y_i,\overline{y}_i}CDS$ | $\epsilon_{y_i,\overline{y}_i}SA$ | $\epsilon_{y_i,\overline{y}_i}PAL$ | $\epsilon_{y_i,\overline{y}_i}CVR$ |
|---|---|---|---|---|
| $20 \times 5$ | 0.0949 | 0.0940 | 0.01082 | $-0.00016$ |
| $20 \times 10$ | 0.1213 | 0.1860 | 0.01528 | 0.00012 |
| $20 \times 20$ | 0.0964 | 0.3260 | 0.1634 | 0.001 |
| $50 \times 5$ | 0.0610 | 0.03 | 0.0534 | 0.00098 |
| $50 \times 10$ | 0.1298 | 0.17 | 0.1403 | 0.00914 |
| $50 \times 20$ | 0.1577 | 0.28 | 0.1794 | 0.1506 |
| $100 \times 5$ | 0.0513 | 0.04 | 0.0251 | 0.00202 |
| $100 \times 10$ | 0.0915 | 0.11 | 0.0913 | 0.0135 |
| $100 \times 20$ | 0.1419 | 0.15 | 0.1555 | 0.02302 |
| $\epsilon_{y_i,\overline{y}_i}$ | 0.1050 | 0.1540 | 0.09272 | 0.0222 |

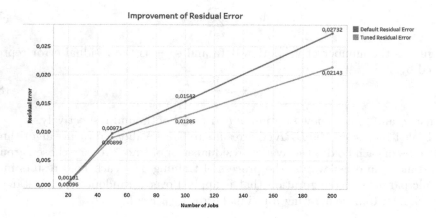

**Fig. 2.** Comparison between tuned residual error and default residual error

Based on Table 1, the average residual error of CVR over Taillard's Benchmark [9] is 0.0089 or 0.89%, in other words, the accuracy of the model in predict the optimum with the parameters found by Irace is 99.1%. In validation test, the set of instances $20 \times 5$ presented a average residual error lower than known optimum of the literature.

Exact solution methods for the problem of PFSP are still limited to small instances, $n \leq 20$ and even to them the running time continues to be large. In Table 2, residual errors of heuristics that are cited in [21]: PAL [22], CDS [23] and the Simulated Annealing (SA) [24] are compared with the $BRKeCS$ tuned by CVR. The efficiency of CVR tuning is representative and wins the others in all groups of instances.

Figure 2 represents the degree of improvement in residual error according to the number of jobs for BRKeCS. The tuning caused a general improvement of 26.44% in the optimum quality. For small instances, the gain was 5%, followed

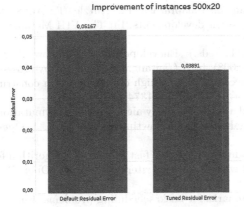

**Fig. 3.** Average residual error of big instances $500 \times 20$

by medium instances with gain of 20%. Using the validated model, a final experiment was performed over the biggest set of instances ($500 \times 20$) with gain of 32.78% as showed by Fig. 3.

## 7  Conclusion

This paper provided some evidence of the effectiveness of Cross-Validated Racing Approach for tuning metaheuristics and more in general for tuning stochastic algorithms making it accessible for future researchers a standard tuning model that allow the comparison of different algorithms. CVR has improved by almost 27% of the default results, showing what is possible compare metaheuristics using the same metodology of test, tune in and still improve residual error.

Other innovative element that can to be considered as original contribution was the developing of the Hybrid Metaheuristic $BRKeCS$ that presented good results for solving the problem PFSP.

**Acknowledgements.** The authors would like to thank CNPq (grant 481845/2013-5) for partial funding of this research.

## References

1. Birattari, M.: Tuning Metaheuristics. A Machine Learning Perspective. Springer, Heidelberg (2009)
2. Oliveira, A.C.M., Lorena, L.A.N.: Detecting promising areas by evolutionary clustering search. In: Bazzan, A.L.C., Labidi, S. (eds.) SBIA 2004. LNCS, vol. 3171, pp. 385–394. Springer, Heidelberg (2004). doi:10.1007/978-3-540-28645-5_39
3. Universit Libre de Bruxelles. http://iridia.ulb.ac.be/IridiaTrSeries/rev/IridiaTr2011-004r001.pdf

4. Birattari, M., Yuan, Z., Balaprakash, P., Stutzle, T.: Automated algorithm tuning using F-races: recent developments. In: The VIII Metaheuristics International Conference (2009)
5. Liu, L., Özsu, M.T. (eds.): Encyclopedia of Database Systems, pp. 123–134. Springer, Naples (2008). http://leitang.net/papers/ency-cross-validation.pdf
6. Nagano, M.S., Moccellin, J.V.: A high quality solution constructive for flow shop sequencing. J. Oper. Res. Soc. **53**, 1374–1379 (2002)
7. Rajedran, C., Ziegler, H.: Ant-colony algorithms for permutation flowshop scheduling to minimize makespan/total flowtime of jobs. Eur. J. Oper. Res. Amst. **155**, 426–436 (2004)
8. Grabowski, J., Wodecki, M.: A very fast tabu search algorithm for the permutation flow shop problem with makespan criterion. Comput. Oper. Res. Amst. **31**, 1891–1909 (2004)
9. Taillard, E.: Benchmarks for basic scheduling problems. Eur. J. Oper. Res. Amst. **64**(2), 278–285 (1993)
10. Taillard, E.: Some efficient heuristic methods for the flow shop sequencing problem. Eur. J. Oper. Res. Amst. **47**, 65–74 (1990)
11. Reuven, Y.R., Dirk, P.K.: Simulation and the Monte Carlo Method. Wiley, Hoboken (2007)
12. Costa, T.S., Oliveira, A.C.M.: Artificial bee and differential evolution improved by clustering search on continuous domain optimization. Soft. Comput. **19**(9), 1–12 (2014)
13. Oliveira, R.M., Mauri, G.R., Lorena, L.A.N.: Clustering search for the berth allocation problem. Expert Syst. Appl. **39**, 5499–5505 (2012)
14. Maron, O.: Dissertation: Hoeffding Races: Model Selection for MRI Classification. Massachusetts Institute of Technology (1992)
15. Bean, J.C.: Genetic algorithms and random keys for sequencing and optimization. ORSA J. Comput. **2**, 154–160 (1994)
16. Eshelman, L., Shaffer, J.: Real-coded genetic algorithms and interval schemata. In: Whitley, D.L. (ed.) Foundations of Genetic Algorithms 2. Morgan Kaufmann Publishers, San Mateo (1993)
17. Lorena, L.A.N., Chaves, A.A., Senne, E.L.F., Resende, M.G.C.: Hybrid method with CS and BRKGA applied to the minimization of tool switches problem. J. Comput. Oper. Res. **67**, 174–183 (2016)
18. Croes, G.A.: A method for solving traveling salesman problems. Oper. Res. **6**, 791–812 (1958)
19. Jackson, J., Girard, A., Rasmussen, S.: A combined Tabu search and 2-opt heuristic for multiple vehicle routing. In: American Control Conference (2010)
20. Oliveira, A.C.M., Lorena, L.A.N.: Hybrid evolutionary algorithms and clustering search. In: Hybrid Evolutionary Systems. Studies in Computational Intelligence, vol. 75, pp. 81–102 (2007)
21. Ruiz, R., Moroto, C., Alcaraz, J.: Two newrobust genetic algorithms for the flowshop scheduling problem. Int. J. Manag. Sci. **34**, 461–476 (2006)
22. Palmer, D.S.: Sequencing jobs through a multistage process in the minimum total time - a quick method of obtaining a near optimum. Oper. Res. Q. **16**, 101–107 (1965)
23. Campbell, H.G., Dudek, R.A., Smith, M.L.: A heuristic algorithm for the n-job, m- machine sequencing problem. Manage. Sci. **16**, B630–B637 (1970)
24. Daya, M.B., Al-Fawzan, M.: A Tabu search approach for the flow shop scheduling problem. Eur. J. Oper. Res. **09**, 88–95 (1998)

# A Transformation Approach Towards Big Data Multilabel Decision Trees

Antonio Jesús Rivera Rivas[(✉)], Francisco Charte Ojeda,
Francisco Javier Pulgar, and Maria Jose del Jesus

Department of Computer Science, University of Jaén, Jaén, Spain
{arivera,fcharte,fpulgar,mjjesus}@ujaen.es

**Abstract.** A large amount of the data processed nowadays is multilabel in nature. This means that every pattern usually belongs to several categories at once. Multilabel data are abundant, and most multilabel datasets are quite large. This causes that many multilabel classification methods struggle with their processing. Tackling this task by means of big data methods seems a logical choice. However, this approach has been scarcely explored by now. The present work introduces several big data multilabel classifiers, all of them based on decision trees. After detailing how they have been designed, their predictive performance, as well as the execution time, are analyzed.

**Keywords:** Multilabel classification · Big data · Decision trees

## 1  Introduction

Pattern classification is among the most popular machine learning (ML) tasks. Usually, each data pattern is associated to one category (the class label). Starting from a set of previously labeled samples, classification algorithms train a model (the classifier). Once trained, the classifier can be shown new unlabeled patterns and it produces the predicted label as output. Decision trees (DT) are well-known classifiers [1], quick to build and easily interpretable. DT ensembles, such as Random Forest (RF) [2], are also quite effective, producing good predictive performance.

A large part of the data generated nowadays is made of patterns linked to several categories at once, instead of only one. Music clips can produce a subset of the existing emotions [3], images can be categorized into several groups [4], blog posts and questions in forums are assigned a set of tags [5], etc. The task of learning from data pattern which are assigned several labels is known as multilabel classification (MLC) [6]. The use of DTs in MLC, multilabel DTs (MDTs), is also a common option.

The amount of new images, videos, music clips, blog posts and other multilabel contents uploaded everyday to the Internet is impressive. As a consequence, MLC algorithms have to be able to process large datasets, a work that

© Springer International Publishing AG 2017
I. Rojas et al. (Eds.): IWANN 2017, Part I, LNCS 10305, pp. 73–84, 2017.
DOI: 10.1007/978-3-319-59153-7_7

usually takes a long time. Facing this problem through Big Data (BD) techniques, notably by distributing the workload among a group of machines, seems a natural choice. Nonetheless, it is a barely explored alternative.

This work aims to propose five different MDT implementations based on BD techniques. Three of them are BD multilabel versions of the well-known ID3 [7], CART [8] and C4.5 [9] algorithms. The other two are ensembles of MDTs, as it is known that ensembles tend to improve classification results. In addition, ensembles are easier to parallelize in a BD environment than other techniques. The five proposals will be experimentally tested with a double goal. Firstly, the predictive performance of each alternative will be compared. Secondly, the improvement in running time as the number of parallel nodes is increased is analyzed.

The remainder of this paper is structured as follows. Section 2 introduces the foundations of MLC and some concepts related to DTs designed to work with BD infrastructure. The five proposed MDTs for BD are presented in Sect. 3. In Sect. 4 the experimental framework is detailed and results are discussed. Finally, some conclusions are drawn in Sect. 5.

## 2 Preliminaries

The methods presented in Sect. 3 are MDTs designed for BD environments. Therefore, it is essential to know the foundations of MLC, introduced in Subsect. 2.1, as well as some notions about BD infrastructures such as Hadoop and Spark, brought in Subsect. 2.2.

### 2.1 Multilabel Classification

Multilabel datasets (MLDs) emerge naturally in certain fields, such as music and video categorization [3,10], image tagging [4], document classification [11] or gene function identification [12]. An MLD can be defined as a subset of $A_1 \times A_2 \times ... \times A_f \times \mathcal{P}(\mathcal{L})$, being $A_i$ the $f$ input features and $\mathcal{P}(\mathcal{L})$ the powerset of $\mathcal{L}$, the full set of labels appearing in the data. There is no difference between the input space of an MLD and a traditional dataset. By contrast, the output space of the former is made of a set of 0s and 1s (labelset), stating which of the labels in $\mathcal{L}$ are relevant for each pattern. Therefore, a classifier have to be able to predict several outputs simultaneously.

Aside of the number of labels, which can be seen as the length of the labelset, several other metrics can be extracted from the samples' labelsets [13]. The two most common are label cardinality (Card) and label density (Dens). $Y_i$ being the labelset of the ith-instance in the MLD $\mathcal{D}$, Card (1) is simply the average number of relevant labels in the MLD. Dens (2) is the normalized label cardinality[1].

---

[1] Card, Dens and many other multilabel characterization metrics can be easily obtained with the mldr package [14].

$$Card = \frac{1}{|\mathcal{D}|}\sum_{i=1}^{|\mathcal{D}|}|Y_i| \qquad (1) \qquad\qquad Dens = \frac{1}{|\mathcal{L}|}\frac{1}{|\mathcal{D}|}\sum_{i=1}^{|\mathcal{D}|}|Y_i| \qquad (2)$$

MLC has been faced mainly through two different approaches [6], data transformation and method adaptation. The former aims to produce binary or multiclass datasets from the MLD, so that they can be processed with traditional classification algorithms. The latter, on the contrary, advocates for rewriting these traditional algorithms, making them able to work with multilabel data natively.

Although several transformation methods have been proposed in the literature, two of them stand out because are frequently used as foundation of many other algorithms. They are Binary Relevance (BR) and Label Powerset (LP).

- BR consists in producing as many binary datasets as labels there are in the MLD, training an independent classifier for each label. The predictions provided by these classifiers are joined to obtain the final labelset. Obviously, the number of models to build (and the time needed to do it) increases linearly with the number of labels.
- Whereas BR relies in binary classifiers, LP do it in multiclass ones. The trick lies in considering each distinct labelset as class identifier. The major drawback of this approach is that theoretically $2^{|\mathcal{L}|}$ different classes could exist.

Regarding the method adaptation approach, multilabel algorithms based on the best known models, such as trees [15], neural networks [16], support vector machines [12] or nearest neighbors [17], can be found in the literature.

## 2.2 Decision Trees for Big Data

The recent advances in communications and storage technologies have led to the emergence of big databases, in a context where "big" has to be understood as beyond the processing capacity of current personal computers. The answer to this scenario was the use of clusters of computers. For facing complex ML tasks different BD frameworks have been developed over time, such as Hadoop and Spark [18], both from the Apache Foundation.

Hadoop relies on an own distributed file system [19], named HDFS, and the approach to distribute the workload is the popular Map-Reduce [20]. Unlike Hadoop, Spark supports in-memory data sharing. This technique produces a noticeable improvement in running time, notably when multiple-pass computations over the data are needed. Depending on specific conditions, Spark runs as 100 times faster than Hadoop. In addition, a basic library of ML methods running over Spark, named MLlib [21], is available. Among the provided ML algorithms, a generic ID3/CART DT can be found.

The cornerstone of Spark is the Resilient Distributed Dataset (RDD). It represents a data collection that is distributed among a set of machines (cluster nodes). Spark is able to cache RDDs in memory, reusing them between successive parallel operations. In the following, we refer to the number of nodes used to process the data as number of RDD partitions. The goal of an Spark cluster is to reduce the total running time by distributing the workload among its nodes. Therefore, the more nodes there are in the cluster, the less time will be spent in processing the data. However, partitioning and distributing the data is also a time-consuming task. Depending on the amount of data, this preparatory work could take longer than the reduction obtained by the parallelization. So the number of RDD partitions is a parameter that could need some adjustment.

## 3    Multilabel Decision Trees for Big Datasets

Taking as foundation the generic tree implementation provided by MLlib [21], the data mining library for Spark, three MDT algorithms were designed, ID3, CART and C4.5. All of them are based on the LP transformation, previously defined, so the labelsets are taken as class identifiers. Then, two ensembles of MDTs are also proposed, BR and RF. The details about these proposals are provided in the following subsections.

### 3.1    Classifiers Based on Single MDT

To learn a single MDT from an MLD, instead of a collection of binary trees, the LP transformation has been used. Thus, each labelset in an MLD is taken as the class identifier. In addition, multilabel versions of entropy and the Gini index, the metrics used to decide the variable used in each split of the tree, are needed.

Based on [15], and being $\mathcal{L}$ the full set of labels in the MLD, $p(l)$ the probability of $l$ being relevant and $q(l) = 1 - p(l)$, the entropy measurement is defined as shown in (3). Analogously, (4) corresponds to the Gini index computation.

$$Entropy = -\sum_{l \in \mathcal{L}} p(l) \log p(l) + q(l) \log q(l) \tag{3}$$

$$Gini = 1 - \sum_{l \in \mathcal{L}} p(l)^2 + q(l)^2 \tag{4}$$

Based on the MLlib implementation of ID3 [7], a multilabel version using (3) and the LP transformation was implemented. In the same way, the MLlib's version of CART [8] was adapted to work with MLDs, using (4) and the same transformation. Since C4.5 is not available in MLlib, it was written as extension of the existing ID3 algorithm following [9]. This implied essentially implementing the pruning procedure of C4.5, producing smaller trees with a better ability of generalization.

The main difference between the classical implementation of these methods and the one made here, based on Spark, is that the latter parallelizes the task of evaluating the goodness of the attributes to be used in each split.

## 3.2 Classifiers Based on Ensembles

Tree ensembles, such as RF, are among the most popular and best performing classifiers. Since they train several independent models, ensembles are specially suitable for work distribution in a cluster. Each tree will be built independently, once the data partitions have been sent to each node.

The first ensemble is based on the BR transformation, using C4.5 as underlying classifier. Therefore, there will be as many trees as labels in the MLD. Each one will learn to differentiate patterns for which one label is relevant against all the others, as explained in Sect. 2. The predictions provided by the individual trees, at test time, are later combined to get the full labelset.

RF is proposed as the second MDT ensemble. As in BR, this approach also generates multiple trees. However each one is a MDT processing all labels, not a binary tree. A random subset of the input features is chosen to train the trees, as usual in RF. The trees are built with the multilabel C4.5 version described in the previous subsection. The maximum tree depth is set to 5 and the ensemble uses 100 trees, as recommended in [22].

# 4 Experimentation

In this section how the previously described methods have been empirically tested is explained. Section 4.1 outlines the experimental framework. The conducted experimentation has two main goals. Firstly, the classification performance produced by each one of the MDT implementations will be assessed in Sect. 4.2. Later, in Sect. 4.3, the execution time of each method, as well as the influence of the number of partitions in running time, will be analyzed. This way, the best algorithm could be chosen according to time restrictions and classification performance demands, as discussed in Sect. 4.4.

## 4.1 Experimental Framework

The MDTs have been tested using six MLDs[2] having disparate traits, as shown in Table 1. Three of them, medical [24], slashdot [25] and tmc2007 [11] come from the text domain, emotions [3] and scene [4] have their origin in the multimedia sources, while yeast [12] was produced from genetic data. The number of instances and attributes will mainly impact the execution time of the algorithms. On the other hand, the number of labels and cardinality are attributes that influence the predictive behavior of the models. Depending on the transformation applied to the data, the number of labels can also increase the running time.

Each MLD was partitioned following a 5 folds cross validation, thus each run used 80% of data to train the model and the remainder 20% as test patterns. Reported results are mean values over these 5 runs per MLD/method.

---

[2] All of them can be found in the RUMDR [23] repository.

**Table 1.** Main characteristics of the MLDs

| Dataset | Instances | Attributes | Labels | Cardinality |
|---------|-----------|------------|--------|-------------|
| emotions | 593 | 72 | 6 | 1.869 |
| medical | 978 | 1 449 | 45 | 1.245 |
| scene | 2 407 | 294 | 6 | 1.074 |
| slashdot | 3 782 | 1 079 | 22 | 1.181 |
| tmc2007 | 28 596 | 500 | 22 | 2.158 |
| yeast | 2 417 | 198 | 14 | 4.237 |

Aiming to analyze how the number of RDD partitions affected execution time, the training and testing was repeated six times using a different configuration. The used values are 1, 2, 4, 8, 16, 32 and 64. Theoretically, as the number of RDD partitions grows execution time should decrease, since the work is distributed among a larger amount of machines. The cluster used has 14 nodes and each node disposes of 2 x Intel Xeon E5-2670v2 and 64 GB of RAM.

The predictions made by each classifier were assessed by means of five performance metrics. Let $Y_i$ be the ground truth labelset of the ith-instance, $Z_i$ the predicted one, $\Delta$ the symmetrical difference, and $[\![\ ]\!]$ the Iverson operator (returns 1 if the expression is true, 0 otherwise). Hamming Loss (5), Accuracy (6), F-Measure (8), and Subset Accuracy (9) are defined as follows. Hamming Loss is a loss metric, so the lower the value the better is performing the classifier. For the other four metrics, higher values are better.

$$Hamming\ Loss = \frac{1}{n}\frac{1}{k}\sum_{i=1}^{n}|Y_i \Delta Z_i| \tag{5}$$

$$Accuracy = \frac{1}{n}\sum_{i=1}^{n}\frac{|Y_i \cap Z_i|}{|Y_i \cup Z_i|} \tag{6}$$

$$Precision = \frac{1}{n}\sum_{i=1}^{n}\frac{|Y_i \cap Z_i|}{|Z_i|}, \quad Recall = \frac{1}{n}\sum_{i=1}^{n}\frac{|Y_i \cap Z_i|}{|Y_i|} \tag{7}$$

$$F\text{-}Measure = 2 * \frac{Precision * Recall}{Precision + Recall} = \frac{2TP}{2TP + FP + FN}. \tag{8}$$

$$Subset\ Accuracy = \frac{1}{n}\sum_{i=1}^{n}[\![Y_i = Z_i]\!] \tag{9}$$

Micro F-Measure (10) differs from F-Measure in the way it is averaged. Instead of computing the metric for each instance, the true positives (TP), true negatives (TN), false positives (FP) and false negatives (FN) for all the instances are aggregated, then the measure is computed. $\mathcal{L}$ denotes the full set of labels appearing in the MLD. Additional details about all these metrics can be found in [6].

$$Micro\ F\text{-}Measure = F\text{-}Measure(\sum_{l \in \mathcal{L}} TP_l, \sum_{l \in \mathcal{L}} FP_l, \sum_{l \in \mathcal{L}} TN_l, \sum_{l \in \mathcal{L}} FN_l) \qquad (10)$$

In addition to the previous metrics, running times were also gathered to compare the influence of the number of RDD partitions in the total execution time.

## 4.2 Classification Performance

Firstly, the interest is in determining which one of the MDTs produces better classification results. The values corresponding to each evaluation metric are depicted in Fig. 1. Each bar plot shows results for the six MLDs processed with the five algorithms. As can be observed, the bar associated to RF is noticeable

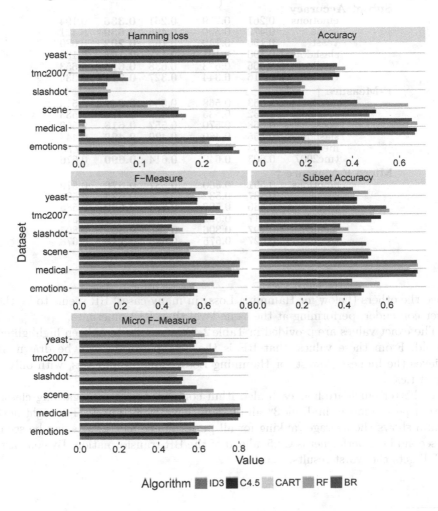

**Fig. 1.** Classification results

**Table 2.** Classification results

| Metric | Dataset | ID3 | C4.5 | CART | RF | BR |
|---|---|---|---|---|---|---|
| **Hamming Loss** ↓ | | | | | | |
| | emotions | 0.260 | 0.250 | 0.254 | **0.211** | 0.246 |
| | scene | 0.174 | 0.162 | 0.163 | **0.113** | 0.140 |
| | yeast | 0.242 | 0.240 | 0.243 | **0.218** | 0.228 |
| | slashdot | 0.048 | 0.048 | 0.053 | **0.045** | 0.047 |
| | medical | **0.010** | **0.010** | 0.011 | **0.010** | 0.011 |
| | tmc2007 | 0.081 | 0.068 | 0.072 | **0.056** | 0.061 |
| **Accuracy** ↑ | | | | | | |
| | emotions | 0.481 | 0.492 | 0.479 | **0.567** | 0.436 |
| | scene | 0.512 | 0.526 | 0.522 | **0.668** | 0.504 |
| | yeast | 0.461 | 0.461 | 0.451 | **0.513** | 0.442 |
| | slashdot | 0.388 | 0.392 | 0.365 | **0.420** | 0.381 |
| | medical | 0.738 | 0.743 | 0.707 | **0.745** | 0.741 |
| | tmc2007 | 0.528 | 0.572 | 0.541 | **0.612** | 0.598 |
| **Subset Accuracy** ↑ | | | | | | |
| | emotions | 0.261 | 0.269 | 0.261 | **0.336** | 0.194 |
| | scene | 0.472 | 0.500 | 0.496 | **0.639** | 0.421 |
| | yeast | 0.160 | 0.149 | 0.155 | **0.202** | 0.080 |
| | slashdot | 0.190 | 0.191 | 0.177 | **0.198** | 0.183 |
| | medical | 0.666 | 0.673 | 0.638 | **0.675** | 0.653 |
| | tmc2007 | 0.316 | 0.344 | 0.327 | **0.371** | 0.334 |
| **F-Measure** ↑ | | | | | | |
| | emotions | 0.553 | 0.568 | 0.554 | **0.644** | 0.519 |
| | scene | 0.532 | 0.535 | 0.531 | **0.678** | 0.533 |
| | yeast | 0.568 | 0.570 | 0.557 | **0.618** | 0.560 |
| | slashdot | 0.455 | 0.460 | 0.429 | **0.499** | 0.447 |
| | medical | 0.763 | 0.767 | 0.731 | **0.770** | **0.770** |
| | tmc2007 | 0.603 | 0.649 | 0.614 | **0.690** | 0.676 |
| **Micro F-Measure** ↑ | | | | | | |
| | emotions | 0.591 | 0.600 | 0.587 | **0.670** | 0.579 |
| | scene | 0.537 | 0.532 | 0.528 | **0.673** | 0.590 |
| | yeast | 0.580 | 0.587 | 0.573 | **0.630** | 0.585 |
| | slashdot | 0.517 | 0.522 | 0.492 | **0.574** | 0.514 |
| | medical | 0.797 | 0.806 | 0.785 | **0.808** | 0.805 |
| | tmc2007 | 0.627 | 0.675 | 0.642 | **0.718** | 0.678 |

above the others (below for Hamming Loss) in many cases. BR seems to be the closer contender, performing at the same level than RF sometimes.

The exact values are provided in Table 2. Best results have been highlighted in bold. From these values, that RF is the best performer can be drawn. It achieves the highest (lowest for Hamming Loss) value in all cases, with only a pair of ties.

To better elucidate how each algorithm compare to others regarding classification performance, in Table 3[3] all of them have been ranked. The rightmost column shows the average ranking for all performance metrics. As can be seen, the second best performer is C4.5, ahead of the BR transformation. By contrast, CART gets the worst results.

---

[3] Names of metrics have been abbreviated to better fit them as column captions.

**Table 3.** Average ranking by metric

| Algorithm | HL | Acc | F-M | SA | MF-M | Avg. rank |
|---|---|---|---|---|---|---|
| RF | 1.000 | 1.000 | 1.083 | 1.000 | 1.000 | 1.017 |
| C4.5 | 2.833 | 2.250 | 2.333 | 2.333 | 2.500 | 2.450 |
| BR | 2.333 | 4.000 | 3.250 | 4.333 | 3.167 | 3.417 |
| ID3 | 4.333 | 3.583 | 3.833 | 3.500 | 3.667 | 3.783 |
| CART | 4.500 | 4.167 | 4.500 | 3.833 | 4.667 | 4.333 |

## 4.3 Execution Time Analysis

The main goal of distributing the workload among a group of machines is to reduce the total execution time taken by the process. In this case, the process is the training of each classifier. The number of RDD partitions have been set to different values, aiming to analyze at which extent increasing the parallelization level decreases running time.

Since it has been already proven that ID3 and CART produce poor classification performance, time analysis will be focused in the other three MDT implementations. Figure 2 shows execution times in seconds for each MLD and method. The X axis is common to all plots, indicating the number of RDD partitions. Y axes are independent, stating the running time in seconds. Experiments taking longer than 10 h were discarded, this is the reason to the lacking of data points in tmc2007 for 1 and 2 partitions.

As would be expected, in general running time decreases as the number of RDD partitions grows. However, there are a few exceptions such as RF while trained with emotions, medical and yeast. In these cases increasing the number of partitions from 32 to 64 implies a deterioration instead of an improvement, taking significantly longer. This could be explained by the fact that the process of dividing the problem and distributing it among the machines in the cluster, takes longer than the savings obtained by sharing the workload.

## 4.4 Discussion

From the observation of the previous results, choosing the best MDT alternative is a matter of deciding what is most important in each case, predictive performance or running time. To obtain the best possible classification of new patterns RF is the correct choice, with a large advantage over the other algorithms. RF is an ensemble, a collection of C4.5 trees each of them generated from a random subset of the features. Therefore, obtaining better results than a single C4.5 classifier is not strange. BR is also an ensemble, but each one of the trees is focused in predicting one label only, working independently of the other trees. The approach of creating several trees taking the relationship among labels into account, through the LP transformation, proves to be superior.

As would be expected, the running times for the ensembles, BR and RF, are longer than for the single C4.5 MDT. However, increasing the number of RDD

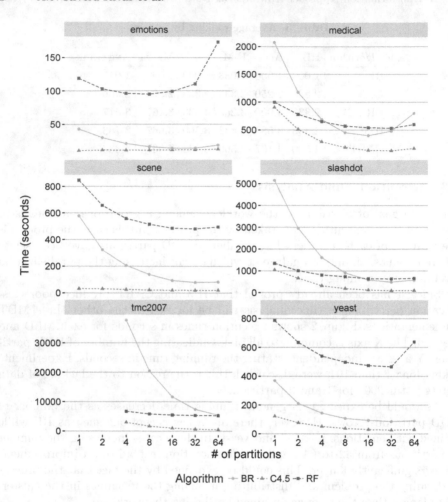

**Fig. 2.** Execution time vs number of partitions

partitions reduces these times until they get quite close in some cases. As can be observed in Fig. 2, BR is slower with the MDLs which have more labels, such as tmc2007 and slashdot, since it has to produce a larger set of classifiers. On the contrary, RF is more affected by the number of samples and attributes, as it has to produce bigger trees as these numbers grow. In general, for MLDs with many labels RF will produce the best classification results in less time than BR, although depending on the number of RDD partitions some surpassing could exists (as with the medical MLD). C4.5 running times are always the lowest, but they only benefit from distributing the work among machines with the larger MLDs. As can be stated from Fig. 2, for MLDs such as emotions, scene and yeast the line for C4.5 is almost flat.

Overall, given its predictive performance and for being able to reduce running time as the number of RDD partitions is increased, RF seems to be the best decision when it comes to choose a multilabel decision tree algorithm for big data environments.

# 5    Conclusions

The amount of new data patterns produced every day is huge, mainly in form of images, videos, sounds and texts due to the emergence of services such as Flickr, Instagram, YouTube! and personal blogs. These patterns are multilabel in nature, and could be grouped/categorized/tagged into several classes. Hence the interest in having methods able to perform multilabel classification with big databases.

In this work five different decision tree based methods have been proposed. Three of them are multilabel versions of well-known ID3, CART and C4.5 algorithms, using an adapted entropy/Gini metric and based on the LP transformation. The other two, RF and BR, are ensembles of classifiers, following two distinct approaches. The former trains several trees with a subset of the input features and all labels, while the latter trains an individual tree for each label with all the features.

A two-way experimental study has been conducted. The first part has led as result that RF is the best choice when only predictive performance matters. The second part analyzed how the total running time could be reduced by increasing the number of RDD partitions. The behavior of RF and BR was dependent of the MLD characteristics, noticeably the number of labels. As ensemble methods, their running time was always higher than that of single-MDT classifiers such as C4.5.

**Acknowledgments.** This work is partially supported by the Spanish Ministry of Science and Technology under project TIN2015-68454-R.

# References

1. Kotsiantis, S.: Supervised machine learning: a review of classification techniques. In: Proceedings of Conference on Emerging Artificial Intelligence Applications in Computer Engineering: Real World AI Systems with Applications in eHealth, HCI, Information Retrieval and Pervasive Technologies, pp. 3–24. IOS Press (2007)
2. Svetnik, V., Liaw, A., Tong, C., Culberson, J.C., Sheridan, R.P., Feuston, B.P.: Random forest: a classification and regression tool for compound classification and qsar modeling. J. Chem. Inf. Comput. Sci. **43**(6), 1947–1958 (2003)
3. Wieczorkowska, A., Synak, P., Raś, Z.: Multi-label classification of emotions in music. In: Kłopotek, M.A., Wierzchoń, S.T., Trojanowski, K. (eds.) Intelligent Information Processing and Web Mining. AISC, vol. 35, pp. 307–315. Springer, Heidelberg (2006)
4. Boutell, M., Luo, J., Shen, X., Brown, C.: Learning multi-label scene classification. Pattern Recogn. **37**(9), 1757–1771 (2004)
5. Charte, F., Rivera, A.J., del Jesus, M.J., Herrera, F.: QUINTA: a question tagging assistant to improve the answering ratio in electronic forums. In: Proceedings of IEEE International Conference on Computer as a Tool, EUROCON 2015, pp. 1–6. IEEE (2015)
6. Herrera, F., Charte, F., Rivera, A.J., Del Jesus, M.J.: Multilabel Classification: Problem Analysis, Metrics and Techniques. Springer, Heidelberg (2016)

7. Quinlan, J.R.: Induction of decision trees. Mach. Learn. **1**(1), 81–106 (1986)
8. Steinberg, D., Colla, P.: CART: Tree-Structured Non-Parametric Data Analysis. Salford Systems, San Diego (1995)
9. Quinlan, J.R.: C4.5: Programs for Machine Learning. Morgan Kaufmann Publishers Inc., San Francisco (1993). ISBN 1-55860-238-0
10. Snoek, C.G.M., Worring, M., van Gemert, J.C., Geusebroek, J.M., Smeulders, A.W.M.: The challenge problem for automated detection of 101 semantic concepts in multimedia. In: Proceedings of 14th ACM International Conference on Multimedia, MULTIMEDIA 2006, pp. 421–430 (2006)
11. Srivastava, A.N., Zane-Ulman, B.: Discovering recurring anomalies in text reports regarding complex space systems. In: Aerospace Conference, pp. 3853–3862. IEEE (2005)
12. Elisseeff, A., Weston, J.: A kernel method for multi-labelled classification. In: Advances in Neural Information Processing Systems, vol. 14, pp. 681–687. MIT Press (2001)
13. Herrera, F., Charte, F., Rivera, A.J., del Jesus, M.J.: Case studies and metrics. Multilabel Classification, pp. 33–63. Springer, Cham (2016). doi:10.1007/978-3-319-41111-8_3
14. Charte, F., Charte, D.: Working with multilabel datasets in R: the mldr package. R. J. **7**(2), 149–162 (2015)
15. Clare, A., King, R.D.: Knowledge discovery in multi-label phenotype data. In: Raedt, L., Siebes, A. (eds.) PKDD 2001. LNCS (LNAI), vol. 2168, pp. 42–53. Springer, Heidelberg (2001). doi:10.1007/3-540-44794-6_4
16. Zhang, M.: Ml-rbf: RBF neural networks for multi-label learning. Neural Process. Lett. **29**, 61–74 (2009)
17. Zhang, M., Zhou, Z.: ML-KNN: a lazy learning approach to multi-label learning. Pattern Recogn. **40**(7), 2038–2048 (2007)
18. Zaharia, M., Chowdhury, M., Franklin, M.J., Shenker, S., Stoica, I.: Spark: cluster computing with working sets. HotCloud **10**(10–10), 95 (2010)
19. Shvachko, K., Kuang, H., Radia, S., Chansler, R.: The hadoop distributed file system. In: 2010 IEEE 26th Symposium on Mass Storage Systems and Technologies (MSST), pp. 1–10. IEEE (2010)
20. Gillick, D., Faria, A., DeNero, J.: Mapreduce: distributed computing for machine learning, Berkley, 18 December 2006
21. Meng, X., Bradley, J., Yavuz, B., Sparks, E., Venkataraman, S., Liu, D., Freeman, J., Tsai, D., Amde, M., Owen, S., et al.: Mllib: machine learning in apache spark. J. Mach. Learn. Res. **17**(34), 1–7 (2016)
22. del Río, S., López, V., Benítez, J.M., Herrera, F.: On the use of mapreduce for imbalanced big data using random forest. Inf. Sci. **285**, 112–137 (2014)
23. Charte, F., Charte, D., Rivera, A., de Jesus, M.J., Herrera, F.: R ultimate multilabel dataset repository. In: Martínez-Álvarez, F., Troncoso, A., Quintián, H., Corchado, E. (eds.) HAIS 2016. LNCS (LNAI), vol. 9648, pp. 487–499. Springer, Cham (2016). doi:10.1007/978-3-319-32034-2_41
24. Crammer, K., Dredze, M., Ganchev, K., Talukdar, P.P., Carroll, S.: Automatic code assignment to medical text. In: Proceedings of Workshop on Biological, Translational, and Clinical Language Processing, BioNLP 2007, pp. 129–136. Association for Computational Linguistics (2007)
25. Read, J., Pfahringer, B., Holmes, G., Frank, E.: Classifier chains for multi-label classification. Mach. Learn. **85**, 333–359 (2011)

# Evolutionary Support Vector Regression via Genetic Algorithms: A Dual Approach

Shara S.A. Alves[(✉)], Madson L.D. Dias, Ajalmar R. da Rocha Neto, and Ananda L. Freire

Department of Teleinformatics, Federal Institute of Ceará, Fortaleza, CE, Brazil
shara.alves@ppget.ifce.edu.br, ajalmar@ifce.edu.br,
{madson.dias,ananda.freire}@ppgcc.ifce.edu.br

**Abstract.** Evolutionary machine learning is an emerging research area that covers any combination of evolutionary strategies and machine learning. In support vector machines, metaheuristics have been widely employed to tune parameters, select features or obtain a reduced subset of support vectors. However, there are only a few works that aim at embedding evolutionary strategies into the support vector regressors training process, i.e., to apply evolutionary methods to solve the quadratic optimization problem. In this paper, we intend to solve the quadratic optimization problem for support vector regression in its dual formulation by employing genetic algorithms. Our proposal was validated in real-world datasets against state-of-the-art methods, such as sequential minimal optimization, iterative single data algorithm, and a classical mathematical method. The results revealed that our proposal is a competitive alternative, which often reduced the generalization error and achieved sparse solutions.

**Keywords:** Support vector regression · Genetic algorithms · Evolutionary machine learning

## 1 Introduction

The SVR training process requires the solution of a quadratic optimization problem, which can be described in its primal or dual formulation. The dual formulation is presented in terms of a set of Lagrange multipliers and the bias [21]. Methods such as Sequential Minimal Optimization (SMO) [9], Iterative Single Data Algorithm (ISDA) [12], and Quadratic Programming (QP) [11] have been applied to obtain the aforementioned set of Lagrange multipliers, as well as the bias. Unfortunately, the classical mathematical methods, such as QP, require large matrices manipulation and may lead to more numerical precision errors, since QP solves quadratic programming problem by numerical optimization.

As known, support vector classifiers and regressors are achieved by solving different quadratic optimization problems. In a nutshell, in their dual formulation, the difference is mainly in terms of the equation to be maximized and

© Springer International Publishing AG 2017
I. Rojas et al. (Eds.): IWANN 2017, Part I, LNCS 10305, pp. 85–97, 2017.
DOI: 10.1007/978-3-319-59153-7_8

also in terms of the number of Lagrange multipliers associated with a certain pattern. Actually, SVC and SVR dual formulations are described as having one and two Lagrange multipliers for each pattern, respectively. Along with these differences, as a constraint, the couple of Lagrange multipliers for a certain pattern should not assume zero values at the same time [25]. Thus, any algorithm proposed to solve this quadratic optimization problem must successfully handle these challenging requirements.

Evolutionary Machine Learning (EML) is an emerging research area concerning any combination of evolutionary strategies and machine learning. Metaheuristics, such as Simulated Annealing (SA) [24], Genetic Programming (GP) [2], Particle Swarm Optimization (PSO) [14], and Genetic Algorithms (GAs) [27] are alternatives to numerical optimization based methods. Due to GAs inherent features, some optimization problems can be solved without supposing linearity, differentiability, continuity or convexity of the objective function. In general, GAs have been widely employed to deal with parameters tuning [3,10], features selection [15] and also to obtain a reduced-set SVM [19,20,26]. The few related works aimed at solving SVC in its primal formulation are presented in [23] and, in its dual formulation in [7,17]. Only recently, a solution we proposed fulfilled each and every constraint for SVCs [7]. Nevertheless, approaches that solve the SVR quadratic optimization problem by GAs are quite scarce in the literature. An approach related to the quadratic optimization problem for regression (SVR) was found in [22], however it is restricted to the primal formulation. In this paper, we aim at solving the SVR quadratic optimization problem in its dual formulation. Our approach employs GAs to handle the quadratic optimization problem and its constraints towards obtaining the dual set of Lagrange multipliers, as well as, the bias. The proposal is validated with real-world datasets and compared with state-of-the-art methods, such as SMO, QP and ISDA. The remaining content of this paper is organized as follows. In Sect. 2 we review the fundamentals of SVR and briefly introduce GAs in Sect. 3. Then, our proposal, the Evolutionary Support Vector Regression, is presented in Sect. 4. After that, we present the simulations in Sect. 5, and finally, our conclusion remarks are presented in Sect. 6.

## 2    Support Vector Machines for Regression

Consider a training set $X = \{\mathbf{x}_i, y_i\}_{i=1}^l$, so that $\mathbf{x}_i \in \mathbb{R}^d$ is an input sample and $y_i \in \mathbb{R}$ is the corresponding dependent variable. The goal is to find a function

$$f(\mathbf{x}) = \mathbf{w}^T \mathbf{x} + b, \tag{1}$$

that has at most $\varepsilon$ deviation from the dependent variable $y_i$ and also it is as flat as possible. In SVR, we measure the error of approximation and it corresponds to dealing with the Vapnik $\varepsilon$-insensitive loss function $|\xi_i|_\varepsilon = \begin{cases} 0, & \text{if } |\xi_i|_\varepsilon \leq \varepsilon, \\ |\xi_i| - \varepsilon & \text{otherwise.} \end{cases}$
The loss is equal to zero if the difference between the predicted and the measured value $y_i$ is less than $\varepsilon$. If the difference is larger than $\varepsilon$, this difference is used as

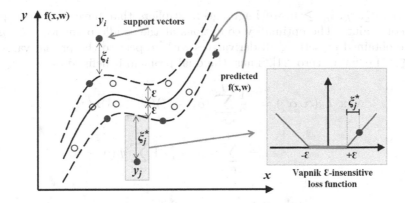

**Fig. 1.** Dashed lines: Vapnik $\varepsilon$-insensitive loss function. Solid line: predicted function. Adapted from [12].

the error. The equation defines an $\varepsilon$ tube as shown in Fig. 1. The predicted values inside the tube mean the loss (error or cost) is zero. For all other predicted points outside the tube, the loss is equals to the magnitude of the difference between the predicted value and the radius $\varepsilon$ of the tube [12].

In order to ensure that $f(\mathbf{x})$ has at most $\varepsilon$ deviation from the dependent variable $y_i$, one has to minimize the norm $\|\mathbf{w}\|^2 = \mathbf{w}^T\mathbf{w}$. The SVR primal problem [25] is defined as

$$\min \ P(\mathbf{w}, \xi_i, \xi_i^*, b) = \frac{1}{2}\mathbf{w}^T\mathbf{w} + C\sum_{i=1}^{l}(\xi_i + \xi_i^*), \tag{2}$$

$$\text{subject to } \begin{cases} y_i - \mathbf{w}^T\mathbf{x}_i - b & \leq \varepsilon + \xi_i \\ \mathbf{w}^T\mathbf{x}_i + b - y_i & \leq \varepsilon + \xi_i^* \,, \\ \xi_i, \xi_i^* & > 0 \end{cases} \tag{3}$$

where $b$ is the bias, $\{\xi_i\}_{i=1}^{l}$ and $\{\xi_i^*\}_{i=1}^{l}$ are the slack variables that allow some errors, the constant $C > 0$ determines the trade-off between the flatness in Eq. (1) and the amount up to which deviations larger than $\varepsilon$ are tolerated.

In most of the cases, the optimization problem in Eq. (2) is more easily to solve in its dual formulation. Therefore, a Lagrange function is built from the primal problem and its constraints by introducing a dual set of variables,

$$L_p(\mathbf{w}, b, \xi, \xi^*, \alpha, \alpha^*) = \frac{1}{2}\mathbf{w}^T\mathbf{w} + C\sum_{i=1}^{l}(\xi_i + \xi_i^*) - \sum_{i=1}^{l}\alpha_i(\varepsilon + \xi_i - y_i + \mathbf{w}^T\mathbf{x}_i + b)$$

$$- \sum_{i=1}^{l}\alpha_i^*(\varepsilon + \xi_i^* + y_i - \mathbf{w}^T\mathbf{x}_i - b) - \sum_{i=1}^{l}(\eta_i\xi_i + \eta_i^*\xi_i^*), \tag{4}$$

where $\alpha_i$, $\alpha_i^*$, $\eta_i$, $\eta_i^* \geq 0$ are Lagrange multipliers that have to satisfy positivity constraints. The optimality conditions at the saddle point for this problem are obtained by setting all derivatives with respect to the primal variables $(\mathbf{w}, b, \boldsymbol{\xi}, \boldsymbol{\xi}^*)$ equal to zero [21]. Thus, the dual problem is defined as

$$\max\ L_d(\boldsymbol{\alpha}, \boldsymbol{\alpha}^*) = -\frac{1}{2} \sum_{i,j=1}^{l} (\alpha_i - \alpha_i^*)(\alpha_j - \alpha_j^*)\mathbf{x}_i^T\mathbf{x}_j$$

$$- \varepsilon \sum_{i=1}^{l}(\alpha_i + \alpha_i^*) + \sum_{i=1}^{l} y_i(\alpha_i - \alpha_i^*), \tag{5}$$

$$\text{subject to}\quad \alpha_i, \alpha_i^* \in [0, C]\ \text{and} \tag{6}$$

$$\sum_{i=1}^{l}(\alpha_i - \alpha_i^*) = 0, \tag{7}$$

so that the Eq. (1) is rewritten as

$$f(\mathbf{x}) = \sum_{i=1}^{l}(\alpha_i - \alpha_i^*)\mathbf{x}_i^T\mathbf{x} + b. \tag{8}$$

There can never be a set of Lagrange multipliers $\alpha_i, \alpha_i^*$ which are both nonzero, since $\alpha_i\alpha_i^* = 0$. We suggest the reader to obtain more details in [21]. The nonlinear versions is achieved by preprocessing the training set $\mathbf{X}$ by using the *kernel trick* replacing the dot product $\mathbf{x_i}^T\mathbf{x}$ with a kernel function $k(\mathbf{x}, \mathbf{x}_i)$.

## 3   Genetic Algorithms

Genetic Algorithms [27] are metaheuristics inspired by natural evolution processes, such as inheritance, mutation, natural selection and reproduction. Such metaheuristics are used to solve optimization problems by adopting a population of candidate solutions, named individuals, which evolve towards better solutions. In this population, each individual has a set of genes, named chromosome, that is changed through mutation or combined with another one by crossover processes. A typical genetic algorithm requires: (i) a genetic representation related to the solution domain, and (ii) a fitness function to evaluate the individuals.

## 4   Proposal: Evolutionary Support Vector Regression

Our proposal, henceforth called Evolutionary Support Vector Regression (ESVR), relies on finding the Lagrange multipliers for SVR by a single objective genetic algorithm. Similarly to SMO, QP, and ISDA, the ESVR inputs are the training set, the regularization parameter C and the deviation $\varepsilon$. Firstly, the initial population is built at random so that each individual represents one possible solution, i.e., the Lagrange multipliers values from Eq. (5). The initial population already satisfies that $\alpha_i\alpha_i* = 0$ and the first constraint in Eq. (6), but needs to be adjusted to comply with the second constraint in Eq. (7). Such adjustment algorithm is presented later. Then, for each next generation and until ESVR reaches a defined stop criteria: ($i$) the elitism is applied, which involves preserving a small proportion of the best individuals into the next generation; ($ii$) a fraction of new individuals are generated by the crossover process, and ($iii$) the remaining individuals are obtained through the mutation operator. The adjustment algorithm is applied when necessary during the aforementioned steps since crossover and mutation operators violate the second constraint from Eq. (7). Finally, when ESVR stops and the Lagrange multipliers were found, the bias value is computed from Eq. (8) and the Lagrange multipliers and bias are returned. For sake of simplicity, the ESVR flowchart is depicted in Fig. 2.

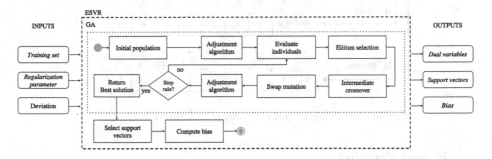

**Fig. 2.** ESVR flowchart: the dotted rectangle comprehends GA steps which are first executed and return the set of Lagrange Multipliers (Dual variables); then the support vectors are selected and the bias computed.

The ESVR stops when ($i$) it reaches a certain generation limit number or ($ii$) the average relative change in the best fitness function value is less than or equal to $1.0e - 4$. Once we find the Lagrange multipliers, we compute the bias $b$ from Eq. (8) as $b = \frac{1}{l}\sum_{i=1}^{l}\left(y_i - \sum_{i=1}^{l}(\alpha_i - \alpha_i^*)\mathbf{x}_i^T\mathbf{x}\right)$.

## 4.1 Genetic Representation

In SVR, each sample is related to the Lagrange multipliers $\alpha_i$ and $\alpha_i^*$. As stated in Sect. 2, $\alpha_i\alpha_i^* = 0$, i.e., $\alpha_i$ or $\alpha_i^*$ is equals to 0 since a set of Lagrange multipliers can not be both simultaneously nonzero. Therefore, considering $\alpha_i^{diff} = (\alpha_i - \alpha_i^*)$, it is straightforward to one realizes, through Eq. 5, that $\alpha_i^{diff} \in [-C, C]$ and $(\alpha_i - \alpha_i^*) \in [-C, C]$, which means that

$$\alpha_i, \alpha_i^* = 0 \quad \text{if } \alpha_i^{diff} = 0,$$
$$\alpha_i = 0 \text{ and } \alpha_i^* = |\alpha_i^{diff}| \quad \text{if } \alpha_i^{diff} < 0, \tag{9}$$
$$\alpha_i = |\alpha_i^{diff}| \text{ and } \alpha_i^* = 0 \quad \text{if } \alpha_i^{diff} > 0.$$

As for the genetic representation, the individual was defined as a real-valued genes vector $\boldsymbol{\alpha}^{diff} = [(\alpha_1 - \alpha_1^*), (\alpha_2 - \alpha_2^*), \ldots, (\alpha_i - \alpha_i^*), (\alpha_l - \alpha_l^*)]$, where $\alpha_i^{diff} \in [-C, C]$. The individual must have $l$ genes, since the training dataset has $l$ samples. The generated individual must handle the two SVR constraints $\alpha_i, \alpha_i^* \in [0, C]$ and $\sum_{i=1}^{l}(\alpha_i - \alpha_i^*) = 0$ from Eqs. (6) and (7), respectively. The Lagrange multipliers $\alpha_i$ and $\alpha_i^*$ are easily retrieved from $\alpha_i^{diff}$, one can see an example in Fig. 3

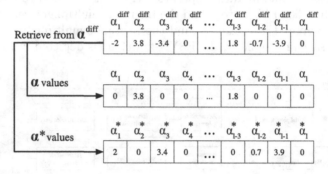

**Fig. 3.** Retrieving $\boldsymbol{\alpha}$ and $\boldsymbol{\alpha}^*$ from $\boldsymbol{\alpha}^{diff}$ through Eq. (9).

## 4.2 Fitness Function

The fitness function is the same as Eq. (5) with a slight modification since our proposal is in terms of minimization $F(\boldsymbol{\alpha}^{diff}) = \frac{1}{2} \sum_{i=1}^{l} \sum_{j=1}^{l} (\alpha_i - \alpha_i^*)$ $(\alpha_j - \alpha_j^*) \mathbf{x}_i^T \mathbf{x}_j + \varepsilon \sum_{i=1}^{l}(\alpha_i + \alpha_i^*) - \sum_{i=1}^{l} y_i(\alpha_i - \alpha_i^*)$, where $\alpha_i$ and $\alpha_i^*$ are retrieved from $\boldsymbol{\alpha}^{diff}$ according to Eq. (9).

## 4.3 Genetic Operators

Since our genetic representation deals with the Lagrange multipliers, the Intermediate Recombination [18] crossover operator was defined. In such arithmetic operator, the offspring is chosen somewhere around and between the variable values of the parents. Let $\mathbf{p} = [p_i, \ldots, p_n]$ and $\mathbf{q} = [q_i, \ldots, q_n]$ be the parents. The offspring $\mathbf{z}$ is computed by $z_i = \beta_i p_i + (1 - \beta_i)q_i$, $i = 1, \ldots, n$, where $\beta_i \in [-g, 1+g]$ is a scaling factor chosen uniformly at random and $g$ defines the area size for possible offspring. When $g = 0$, the $\mathbf{z}$ area is the same size as the area spanned by the parents $\mathbf{p}$ and $\mathbf{q}$. The generated offspring needs to be fixed through the adjustment algorithm (Sect. 4.4) since $\mathbf{z} \in [-C, C]$ and it is feasible just with respect to the first constraint in Eq. (6). The $g$ value is set to 0 in our crossover operator. The swap mutation [5] is defined as the mutation operator. It selects two genes and swaps their values. Since such operator can violate the second constraint in Eq. (7), these ones are fixed by the adjustment algorithm.

## 4.4 Adjustment Algorithm

Our genetic representation already complies with the first constraint in Eq. (6) because it has bounded genes. On the other hand, those individuals may violate the second constraint in Eq. (7). To overcome such situation, the individuals are fixed through the adjustment algorithm. This is a common strategy in the GAs literature [16]. The adjustment algorithm is presented as follows.

```
    ADJUSTMENT(α^diff, C)
1.  α, α* ← RETRIEVE(α^diff)
2.  s ← ∑_{i=1}^{l}(α_i − α_i*)
3.  while |s| ≠ 0 do
4.      if s > 0 then
5.          k ← RAND[1, l]
6.          if α_k > |s| then
7.              α_k ← α_k − |s|
8.          else
9.              α_k ← 0
10.             if RAND[0, 1] < 0.5
11.                 α_k* ← C
12.             end if
13.         end if
14.         α_k^diff ← (α_k − α_k*)
15.     else
16.         k ← RAND[1, l]
17.         if α_k* > |s| then
18.             α_k* ← α_k* − |s|
19.         else
20.             α_k* ← 0
21.             if RAND[0, 1] < 0.5 then
22.                 α_k ← C
23.             end if
24.         end if
25.         α_k^diff ← (α_k − α_k*)
26.     end if
27.     s ← ∑_{i=1}^{l}(α_i − α_i*)
28. end while
29. return α^diff
```

The main idea of our adjustment algorithm is to iteratively reduce an amount from $\alpha_i$ or $\alpha_i^*$ selected at random until $\boldsymbol{\alpha}^{diff}$ complies with the second constraint. In order to explain it, consider $s \leftarrow \sum_{i=1}^{l}(\alpha_i - \alpha_i^*)$, if $s > 0$, that means we need to reduce from $\alpha_i$, or reduce from $\alpha_i^*$ otherwise. If it is not possible to reduce $s$ from $\alpha_i$, there are two (selected at random) alternatives in this case: $\alpha_i = 0$ or $\alpha_i^* = C$. We highlight that our adjustment algorithm reduces an amount from $\alpha_i$ or $\alpha_i^*$ instead of increasing them and it also avoids changing those Lagrange multipliers already equal to zero. This behavior reduces the number of support vectors.

## 5  Simulations and Discussion

We carried out simulations on six real-world datasets from the UCI repository of machine learning databases [1]: (*i*) Servo (SER) with 4 variables and 167 patterns; (*ii*) CPU with 6 variables and 209 patterns; (*iii*) Boston (BOS) with 13 variables and 506 patterns; (*iv*) Ailerons (AIL) with 5 variables and 7129 patterns, and (*v*) Elevators (ELE) with 6 variables and 9517 patterns and also on (*vi*) Motorcycle (MOT) with 1 variable and 103 patterns from [8].

## 5.1   Experimental Setup

First, for each dataset, the data normalization was performed following $v_i' = \frac{v_i - \eta_V}{\sigma_V}$, where $\eta_V$ and $\sigma_V$ are the mean and the standard deviation of $V$, respectively. After this normalization, the data is normally distributed with mean zero and standard deviation of one.

Then, we conducted 30 independent runs on each dataset, 80% of the data samples were randomly selected for training purposes and so the remaining 20% of the samples were used for assessing the regressors' generalization performance. The same selected at random training and test sets were used by SMO, QP, ISDA and ESVR. The ESVR works on reduced training set through k-medoids algorithm, $k = 500$ [13] when $l > 500$. As one will see, such approach have not an effect on ESVR performance. All experiments were performed with Gaussian $\left( k(\mathbf{x}, \mathbf{x}_i) = \exp\left(-\sigma^{-2} \|\mathbf{x} - \mathbf{x}_i\|^2\right) \right)$ and linear kernel. The Gaussian kernel parameter $\sigma$ was tuned by applying a grid search with 5-fold cross-validation over the training dataset, where $\sigma \in [2^{-15}, 2^{-14}, \dots, 2^3]$. The SVR hyperparameter $\varepsilon$ was fixed $\varepsilon = 0.001$ and the regularization parameter $C$ was selected by the following prescription $C = \max(|\bar{\mathbf{y}} + 3\sigma_{\mathbf{y}}|, |\bar{\mathbf{y}} - 3\sigma_{\mathbf{y}}|)$, where $\bar{\mathbf{y}}$ and $\sigma$ are the mean and the standard deviation of the $\mathbf{y}$ values of the training set. For more details see [4].

After all 30 independent runs we present the results and discussion in terms of mean and standard deviation of the MSE and number of support vectors (#SV). These results were validated through the Friedman non-parametric statistical test [6] along with Tukey-Kramer test to compare SMO, QP and ISDA to ESVR.

## 5.2   Results and Discussion

The results for SVR trained by SMO, QP, ISDA, and ESVR are presented in Tables 1 and 2, for linear and Gaussian kernel, respectively. Those results which outperformed the other methods related to MSE value are in bold face. We also present the results of applying the Friedman statistic test [6] in column ST, in which ✓ is a relation of equivalence from the others methods related to ESVR, while ✗ indicates a considerable difference among them.

By analyzing the Table 1 one can conclude that the performances of the ESVR were equivalent or even superior to those achieved by the SMO, QP and ISDA for each dataset and it also achieved the lowest MSE value for 4/6 of the datasets. Moreover, our proposal produced sparse solutions and it was sparser than the others methods for CPU, SER, MOT and ELE. On the ELE dataset, the largest set used in our experiments, the proposal ESVR was equivalent to the ISDA, and outperformed ISDA on AIL dataset, the second largest set used in our experiments. As one can see, there are no results for QP on AIL and ELE datasets due to the computational cost the method demanded.

Related to the results with the Gaussian kernel (see Table 2), one can see that the performance results among the methods were similar and a few better than with linear kernel. The proposal ESVR it is still equivalent to the others methods. On AIL dataset, ISDA achieved the lowest MSE value, and ESVR lower than SMO.

**Table 1.** Results of 30 independent runs with linear kernel.

| Dataset | Method | MSE $(\mu)$ | $(\sigma)$ | ST | #SV $(\mu)$ | $(\sigma)$ |
|---------|--------|-------------|------------|-----|-------------|------------|
| CPU | SMO | 6.27e+03 | 6.54e+03 | ✓ | 1.67e+02 | 0.00e+00 |
|  | QP | 6.24e+03 | 6.49e+03 | ✓ | 1.67e+02 | 1.83e−01 |
|  | ISDA | 6.37e+03 | 6.70e+03 | ✓ | 1.50e+02 | 5.91e+00 |
|  | **ESVR** | **5.21e+03** | **5.14e+03** | - | **1.33e+02** | **2.19e+01** |
| SER | SMO | 1.90e+00 | 7.68e−01 | ✗ | 1.34e+02 | 3.05e−01 |
|  | QP | 1.90e+00 | 7.77e−01 | ✗ | 1.34e+02 | 3.05e−01 |
|  | ISDA | 1.91e+00 | 7.67e−01 | ✗ | 1.33e+02 | 9.10e−01 |
|  | **ESVR** | **1.27e+00** | **4.58e−01** | - | **1.03e+02** | **1.80e+01** |
| BOS | SMO | 2.55e+01 | 8.10e+00 | ✗ | 4.05e+02 | 3.05e−01 |
|  | **QP** | **2.54e+01** | **8.09e+00** | ✗ | **4.05e+02** | **0.00e+00** |
|  | ISDA | 2.64e+01 | 8.52e+00 | ✗ | 3.17e+02 | 1.90e+01 |
|  | ESVR | 3.82e+01 | 1.29e+01 | - | 3.33e+02 | 5.50e+01 |
| MOT | SMO | 2.28e+03 | 5.52e+02 | ✓ | 1.06e+02 | 1.83e−01 |
|  | QP | 2.29e+03 | 5.64e+02 | ✓ | 1.06e+02 | 0.00e+00 |
|  | ISDA | 2.31e+03 | 5.68e+02 | ✓ | 1.04e+02 | 1.27e+00 |
|  | **ESVR** | **2.17e+03** | **4.44e+02** | - | **6.59e+01** | **7.24e+00** |
| AIL | SMO | 4.68e−08 | 8.24e−09 | ✗ | 1.39e+01 | 1.87e+00 |
|  | QP |  |  |  |  |  |
|  | ISDA | 4.37e−08 | 4.26e−09 | ✓ | 1.37e+01 | 1.88e+00 |
|  | **ESVR** | **3.37e−08** | **4.10e−09** | - | **4.07e+02** | **8.07e+01** |
| ELE | SMO | 2.51e−06 | 1.32e−07 | ✗ | 1.44e+03 | 1.90e+02 |
|  | QP |  |  |  |  |  |
|  | **ISDA** | **2.11e−06** | **7.81e−08** | ✓ | **5.57e+02** | **8.42e+01** |
|  | ESVR | 2.16e−06 | 8.03e−08 | - | 3.10e+02 | 1.15e+02 |

The Fig. 4 present the 30 executions of SMO, QP, ISDA and ESVR as a relation between the normalized MSE value and which percentage from training set was used as support vectors. We present in the Fig. 4(e, f) the best performance achieved by the ISDA and ESVR on MOT dataset with the Gaussian kernel. The SV up stands for those samples that $\alpha_i > 0$ and SV low if $\alpha_i^* > 0$. Since the behavior of SMO, QP and ISDA methods are similar, one can see the results in Table 2, we opt to present only the ISDA solution. On MOT dataset, the proposal achieved the lowest MSE value.

**Table 2.** Results of 30 independent runs with Gaussian kernel.

| Dataset | Method | MSE | | ST | #SV | |
|---|---|---|---|---|---|---|
| | | $(\mu)$ | $(\sigma)$ | | $(\mu)$ | $(\sigma)$ |
| CPU | **SMO** | **6.15e+03** | **6.75e+03** | ✓ | **5.12e+01** | **7.90e+00** |
| | QP | 6.15e+03 | 6.76e+03 | ✓ | 5.11e+01 | 7.66e+00 |
| | ISDA | 6.49e+03 | 7.69e+03 | ✓ | 4.47e+01 | 5.98e+00 |
| | ESVR | 6.77e+03 | 8.42e+03 | - | 1.31e+02 | 2.17e+01 |
| SER | SMO | 4.54e−01 | 3.75e−01 | ✓ | 5.83e+01 | 1.66e+01 |
| | **QP** | **4.54e−01** | **3.75e−01** | ✓ | **6.02e+01** | **1.78e+01** |
| | ISDA | 4.60e−01 | 3.91e−01 | ✓ | 5.63e+01 | 1.71e+01 |
| | ESVR | 5.06e−01 | 3.87e−01 | - | 9.00e+01 | 1.59e+01 |
| BOS | SMO | 1.02e+01 | 4.15e+00 | ✗ | 2.79e+02 | 1.35e+01 |
| | QP | 1.02e+01 | 4.16e+00 | ✗ | 2.79e+02 | 1.35e+01 |
| | **ISDA** | **9.86e+00** | **3.63e+00** | ✗ | **2.75e+02** | **1.13e+01** |
| | ESVR | 1.57e+01 | 5.98e+00 | - | 3.28e+02 | 6.01e+01 |
| MOT | SMO | 5.81e+02 | 1.88e+02 | ✓ | 1.06e+02 | 0.00e+00 |
| | QP | 5.81e+02 | 1.88e+02 | ✓ | 1.06e+02 | 0.00e+00 |
| | ISDA | 5.81e+02 | 1.88e+02 | ✓ | 1.06e+02 | 2.54e−01 |
| | **ESVR** | **5.74e+02** | **1.71e+02** | - | **6.20e+01** | **2.26e+01** |
| AIL | SMO | 6.63e−08 | 4.57e−09 | ✗ | 2.12e+02 | 0.00e+00 |
| | QP | | | | | |
| | **ISDA** | **3.80e−08** | **2.99e−09** | ✓ | **1.06e+02** | **0.00e+00** |
| | ESVR | 4.86e−08 | 6.46e−09 | - | 4.35e+02 | 7.09e+01 |
| ELE | SMO | 2.29e−06 | 1.10e−07 | ✗ | 3.74e+03 | 1.48e+02 |
| | QP | | | | | |
| | **ISDA** | **2.09e−06** | **9.78e−08** | ✓ | **2.93e+03** | **1.83e+02** |
| | ESVR | 2.48e−06 | 1.38e−07 | - | 4.53e+02 | 6.42e+01 |

One can see that the behavior of SMO, QP and ISDA are very similar, and it was necessary more support vectors with linear kernel. ESVR tries to keep the same behavior with both linear and Gaussian kernel, and also achieves sparser solutions in both of them. On AIL and ELE datasets, which are the largest datasets used in our experiments, ESVR achieved the lowest MSE with linear kernel and, with the Gaussian kernel the ESVR performance was equivalent to ISDA. Overall, the results indicate that ESVR is very competitive in terms of MSE since it achieved similar or even the lowest MSE values. Furthermore, ESVR is also competitive with respect to the number of support vectors. The performance on larger datasets was superior in some cases and equivalent to ISDA performance. As shown ESVR can be applied successfully in huge datasets

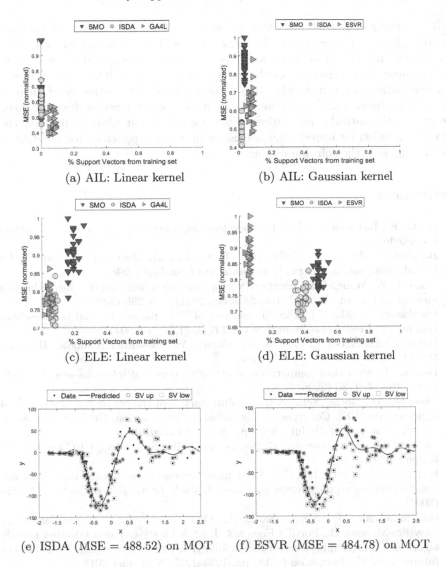

(a) AIL: Linear kernel

(b) AIL: Gaussian kernel

(c) ELE: Linear kernel

(d) ELE: Gaussian kernel

(e) ISDA (MSE = 488.52) on MOT     (f) ESVR (MSE = 484.78) on MOT

**Fig. 4.** MSE (normalized) in test and %SV from training set on AIL and ELE.

since even on a reduced training set, the performance achieved was equivalent or even superior to the compared methods.

# 6   Conclusion

We propose the Evolutionary Support Vector Regression to solve the quadratic optimization problem from Support Vector Regression (SVR) in its dual formulation through Genetic Algorithms. Our proposal successfully embedded the

SVR constraints, so that it is able to handle the dual optimization problem and still obtain the Lagrange multipliers and the bias in a seamlessly way. The simulations and statistical tests results indicate that our proposal is competitive when compared with the state-of-the-art methods such as SMO, ISDA, and some classical mathematical methods while reducing the generalization error, achieving sparse solutions and also maintaining its simplicity since it does not need complex mathematical computations. Future works aim at adding improving the initial population by introducing heuristics to select support vector candidates in order improve the solution quality.

# References

1. Bache, K., Lichman, M.: UCI machine learning repository (2013). http://archive.ics.uci.edu/ml
2. Banzhaf, W., Nordin, P., Keller, R.E., Francone, F.D.: Genetic Programming: An Introduction, vol. 1. Morgan Kaufmann, San Francisco (1998)
3. Chen, K.Y., Wang, C.H.: Support vector regression with genetic algorithms in forecasting tourism demand. Tour. Manag. **28**(1), 215–226 (2007)
4. Cherkassky, V., Ma, Y.: Practical selection of SVM parameters and noise estimation for SVM regression. Neural Netw. **17**(1), 113–126 (2004)
5. Davis, L.: Handbook of Genetic Algorithms. Van Nostrand Reinhold, Hoboken (1991)
6. Demšar, J.: Statistical comparisons of classifiers over multiple data sets. J. Mach. Learn. Res. **7**, 1–30 (2006)
7. Dias, M.L.D., Rocha Neto, A.R.: Evolutionary support vector machines: a dual approach. In: IEEE Congress on Evolutionary Computation, CEC 2016, Vancouver, BC, Canada, 24–29 July 2016, pp. 2185–2192 (2016)
8. Eubank, R.L.: Nonparametric Regression and Spline Smoothing. CRC Press, Boca Raton (1999)
9. Fan, R.E., Chen, P.H., Lin, C.J.: Working set selection using second order information for training support vector machines. J. Mach. Learn. Res. **6**(Dec), 1889–1918 (2005)
10. Gascón-Moreno, J., Salcedo-Sanz, S., Ortiz-García, E.G., Carro-Calvo, L., Saavedra-Moreno, B., Portilla-Figueras, J.A.: A binary-encoded tabu-list genetic algorithm for fast support vector regression hyper-parameters tuning. In: 2011 11th International Conference on ISDA, pp. 1253–1257, November 2011
11. Nocedal, J., Wright, S.: Numerical Optimization. Springer Science & Business Media, New York (2006). doi:10.1007/978-0-387-40065-5
12. Huang, T.M., Kecman, V., Kopriva, I.: Kernel Based Algorithms for Mining Huge Data Sets: Supervised, Semi-Supervised, and Unsupervised Learning, vol. 1. Springer, Heidelberg (2006)
13. Kaufman, L., Rousseeuw, P.J.: Finding Groups in Data: An Introduction to Cluster Analysis. Wiley, Hoboken (1990)
14. Kennedy, J.: Particle swarm optimization. In: Sammut, C., Webb, G.I. (eds.) Encyclopedia of Machine Learning, pp. 760–766. Springer, New York (2011)
15. Kharrat, A., Halima, M.B., Ayed, M.B.: MRI brain tumor classification using support vector machines and meta-heuristic method. In: 15th International Conference on ISDA, pp. 446–451, December 2015

16. Michalewicz, Z., Janikow, C.Z.: Handling constraints in genetic algorithms. In: ICGA, pp. 151–157 (1991)
17. Mierswa, I.: Evolutionary learning with kernels: a generic solution for large margin problems. In: Proceedings of the 8th Annual Conference on Genetic and Evolutionary Computation, pp. 1553–1560. ACM (2006)
18. Mühlenbein, H., Schlierkamp-Voosen, D.: Predictive models for the breeder genetic algorithm i. continuous parameter optimization. Evol. Comput. 1(1), 25–49 (1993)
19. Neto, A.R.R., Barreto, G.A.: Opposite maps: vector quantization algorithms for building reduced-set SVM and LSSVM classifiers. Neural Process. Lett. 37(1), 3–19 (2013)
20. Silva, D.A., Silva, J.P., Neto, A.R.R.: Novel approaches using evolutionary computation for sparse least square support vector machines. Neurocomputing 168, 908–916 (2015)
21. Smola, A.J., Schölkopf, B.: A tutorial on support vector regression. Stat. Comput. 14(3), 199–222 (2004)
22. Stoean, R., Dumitrescu, D., Preuss, M., Stoean, C.: Evolutionary support vector regression machines. In: 2006 Eighth International Symposium on Symbolic and Numeric Algorithms for Scientific Computing, pp. 330–335, September 2006
23. Stoean, R., Preuss, M., Stoean, C., El-Darzi, E., Dumitrescu, D.: Support vector machine learning with an evolutionary engine. J. Oper. Res. Soc. 60(8), 1116–1122 (2009)
24. Van Laarhoven, P.J., Aarts, E.H.: Simulated annealing. In: Van Laarhoven, P.J., Aarts, E.H. (eds.) Simulated Annealing: Theory and Applications, pp. 7–15. Springer, Dordrecht (1987)
25. Vapnik, V.: The Nature of Statistical Learning Theory. Springer, New York (2013)
26. Wang, W., Xu, Z.: A heuristic training for support vector regression. Neurocomputing 61, 259–275 (2004)
27. Whitley, D.: A genetic algorithm tutorial. Stat. Comput. 4(2), 65–85 (1994)

# E-Health and Computational Biology

# Analysis of Electroreception with Temporal Code-Driven Stimulation

Ángel Lareo[1](✉), Caroline Garcia Forlim[2], Reynaldo D. Pinto[3],
Pablo Varona[1], and Francisco B. Rodríguez[1](✉)

[1] Grupo de Neurocomputación Biológica, Departamento de Ingeniería Informática,
Escuela Politécnica Superior, Universidad Autónoma de Madrid, Madrid, Spain
{angel.lareo,f.rodriguez}@uam.es
[2] Clinic and Policlinic for Psychiatry and Psychotherapy,
University Medical Center Hamburg-Eppendorf, Hamburg, Germany
[3] Laboratory of Neurodynamics/Neurobiophysics - Department of Physics and
Interdisciplinary Sciences - Institute of Physics of São Carlos,
Universidade de São Paulo, São Paulo, Brazil

**Abstract.** Temporal code-driven stimulation is a new closed-loop stimulation method for information processing research in biological systems. The biological signal is processed and an event-based binary digitization is performed in real time. Patterns of temporal activity in the system are matched with binary codes and stimulation is triggered after the detection of a predetermined code. This paper presents the characteristics of this closed-loop methodology together with novel analytical possibilities derived from using an information-theoretic approach. The implementation of this method for its application to the study of coding schemes in fish electroreception is presented. Finally, our preliminary results showed that code-driven stimulation decreases the discharge frequency of the electric fish and increases the probability of sparser codes. The relation between those two measures can be used to assess the analysis of factors involved in the information processing in the system.

## 1 Introduction

From a single neuron to entire neural systems, interaction-response loops have a relevant role at all levels of the nervous systems. These closed-loop interactions are also relevant in sensory perception from the environment. In the last few years, there is an increasing advance in closed-loop methodologies applied to the study of biological systems (Chamorro et al. 2012; Schiff 2012; El Hady 2016; Lareo et al. 2016) and a growing need to use real-time protocols to establish such closed loops.

Closed-loop methods allow to dynamically stimulate a biological system depending on its own activity. These activity-dependent methods are a complementary approach to the traditional stimulus-response approaches. Different closed-loop stimulation protocols have been employed in a wide variety of neuroscience related domains (Chamorro et al. 2012; Muñiz et al. 2011; Madhav et al. 2013; Roth et al. 2014; Potter et al. 2014; Forlim et al. 2015; Biró and Giugliano 2015).

© Springer International Publishing AG 2017
I. Rojas et al. (Eds.): IWANN 2017, Part I, LNCS 10305, pp. 101–111, 2017.
DOI: 10.1007/978-3-319-59153-7_9

Most closed-loop techniques only discriminate between the presence or absence of a certain event or use the instantaneous value of a monitored signal to trigger the stimulus (Muñiz et al. 2011; Berényi et al. 2012; Ehrens et al. 2015; Forlim et al. 2015). The coding scheme of a neural system can also be addressed with modern closed-loop protocols. Adaptive sampling is a relevant example where an information-theoretic measure is used for selecting the stimulus presented to the system (Benda et al. 2007).

Taking these characteristics into account, we have recently defined a new closed-loop stimulation method to investigate the presence and role of different coding schemes in biological systems: temporal code-driven stimulation (Lareo et al. 2016). First, a binary digitization of the analog biological signal is performed in real time. Event trains are represented as binary codes and a code is selected to be used as a trigger for stimulation. The response to this code-driven stimulation can be used to assess sequential processing in the system and to characterize it.

As a proof of concept, we have applied this method to the weakly electric fish *Gnathonemus Petersii*. These fish, from the pulse mormyrids family, possess a remarkable electrosensory system and are a well-known example of a biological system with temporal coding (Bullock et al. 2006; Baker et al. 2013). Weakly electric fish generates electric fields using an electric organ (Caputi et al. 2002; Bullock et al. 2006; Von der Emde et al. 2010). Electric organ discharges (EODs) are then detected as distortions in the electric field around the fish body using electroreceptors. The EODs can be easily detected non-invasively in alive and active freely-behaving animals using the appropriate hardware (Jun et al. 2012; Forlim and Pinto 2014; Forlim et al. 2015). The pulse waveform of this fish is stereotyped, but the time between pulses vary considerably (Baker et al. 2013; Carlson and Gallant 2013; Forlim and Pinto 2014). The inter-pulse intervals (IPIs) carry information about the behavioral state of the fish (Carlson 2002; Carlson and Gallant 2013; Forlim and Pinto 2014), i.e., information is encoded in the temporal structure of the IPI patterns. These IPIs also change depending on external electrical stimulation (Kramer 1979; Forlim et al. 2015).

In this paper we present new analyses using this methodology, temporal code-driven stimulation (Lareo et al. 2016), to stablish stimulation-response loops in real time. We have conducted some validation experiments in electroreception. We compare the fish response to closed-loop stimulation to control sessions without stimulation and to open-loop stimulation. In the methods section we describe the real time setup, the stimulation protocols and the validation experiments conducted. Then, the results of these experiments are analyzed and discussed. In particular, here we complement previous analyses based in the IPI distribution with ones based in the distribution of binary codes in the response signal. A possible link between both is discussed.

## 2  Materials and Methods

### 2.1  Hardware and Software Setup

The acquisition platform consisted of a computer running the software implementing the temporal code-driven stimulation algorithm (Fig. 1-A), hardware to process and acquire the fish' signal (Fig. 1-B) and an aquarium with the living system (Fig. 1-C). The aquarium has 4 differential dipoles to measure the fish signal, displayed forming an asterisk (Fig. 1-C) at medium depth in the tank. The signal received by the dipoles was amplified (TL082 JFET-Input Dual Operational Amplifier with a gain of approximately: $91\,k\Omega/2.2\,k\Omega \approx 42$), summed (LM741 Operational Amplifiers) and squared (AD633 Analog Multiplier). Signal was acquired at 17 kHz by a data acquisition (DAQ) board (NI PCI-6251, National Instruments Corporation) in a PC-compatible computer. Stimulation was generated in real time by software and delivered using the same DAQ board by a silver tip dipole, placed at the bottom-middle of the tank (Forlim et al. 2015; Lareo et al. 2016). The fast electric signalling of the fish requires the mentioned high acquisition rate and precise real-time software technology to build closed-loop interactions.

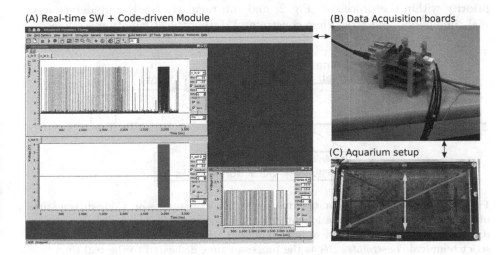

**Fig. 1.** (A) Graphic User Interface of RTBiomanager software running a temporal code-driven stimulation session and acquiring signal from the real system. (B) Acquisition system, the signal received by the dipoles was amplified, summed, squared and acquired at 17 kHz by a data acquisition (DAQ) board in a PC-compatible computer. (C) Dipole setup used for temporal code-driven stimulation in electroreception experiments.

For better re-usability and portability we used general purpose computers with an open-source real time operating system. We chose Ubuntu Linux with a real time kernel patch named Real Time Application Interface, RTAI[1]. Real

---

[1] https://www.rtai.org/.

time technology assures preemption in the tasks related to interacting with the biological system. RTAI provides an application programming interface (API) which permits the execution of periodic tasks in hard real time.

Temporal code-driven stimulation protocols were implemented in a toolbox which can be used in conjunction with real time software responsible for managing the closed-loop (Lareo et al. 2016). We selected RTBiomanager, an open-source real time software for this purpose (Muñiz et al. 2005, 2008, 2009), aside with a code-driven extension (Lareo et al. 2016).

## 2.2  Stimulation Protocols

For comparison purposes, we defined two stimulation protocols in our methodology: a code-driven closed-loop protocol and an open-loop one.

The experimental protocol consists of four ordered sessions (Fig. 3-A): 10 min control session without stimulation (C1); 10 min code-driven stimulation session (CL); 10 min control session (C2); 10 min open-loop stimulation session (OL).

The open-loop protocol does not take into account the activity of the fish. It is designed to compare changes in the system with those that occur under closed-loop stimulation. Both protocols worked as real time periodic tasks with high priority within the processor (Fig. 2) and run regularly (with a predetermined real-time period and strict time constraints) during the stimulation sessions.

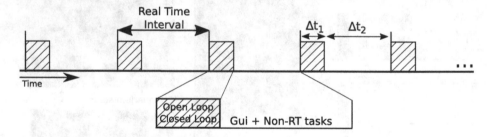

**Fig. 2.** Illustration of real time task management at the processor. Code-driven stimulation and open-loop stimulation work as real time periodic tasks with high priority within the processor. Vertical lines represent real time interruptions, which have strict temporal constraints. $\Delta t_1$ is the processor time dedicated to the real-time task (operations of the protocol being runned). $\Delta t_2$ is the remaining time between real-time interruptions, dedicated to lower priority (non real-time) operations, such as GUI updates.

**Code-Driven Protocol.** The code-driven protocol acquired the signal from the biological system. The protocol processed this signal to binary digitize it in real time and if a triggering code was detected it delivered a stimulus (Fig. 3-A).

To digitize the monitored signal to a binary sequence, we divided it into $N$ time windows of size $\Delta t$. Then we assigned a bit value depending if an event is detected in the time window: 1 when an event is present and 0 otherwise. This results in a discrete temporal sequence $\{e_t; t = 1 \ldots N\}$ where $e_t$ could be 0 or 1 (Fig. 3-B).

We defined a code $C_t^L$ of size L (number of bits) at time $t$ as a sequence of symbols $C_t^L = \{e_{t-L}; \ldots; e_{t-1}; e_t\}$, where $t$ is the time when the sequence is detected. Bits were superimposed between codes as we used a shift of 1 bit. $C_t^L$ was the sequence that happened at time $t$ formed by the events between $t$-$L$ and $t$. Accordingly, there were $N$-$L$+$1$ words of $L$ bits in each time series.

For selecting an appropriate $\Delta t$, the system must be first characterized using data from control sessions without stimulation. We selected its value using a maximum entropy criterion. For that, the signal was digitized using several values of $\Delta t$, obtaining different set of codes $C^L = \{c_1^L, c_2^L, \ldots, c_n^L\}$ and the probability of each code.

Regarding the codes, here we used $L = 4$ and ended by 1, 8 different words to trigger the stimulus ($2^4 = 16$ possible codes, 8 ended by 1). We selected as trigger a code with mean probability of occurrence.

The entropy was then calculated as:

$$H(C^4) = -\sum_i P(c_i^4) log P(c_i^4)$$

where $P(c_i^4)$ is the probability of occurrence of $c_i^4$.

The entropy $H(C^4)$ is related to the variability of the set and it represents the signal capability of encoding information. After calculating the entropy of the signal for different values of $\Delta t$, we selected the appropriate parameter value which maximized entropy.

Once $\Delta t$ and the triggering code were selected, we conducted a code-driven session, the biological signal was acquired and then digitized to a binary sequence, using the same binarization technique explained before. Temporal sequences of events were detected in real time as binary codes. The stimulation was delivered after the detection of the triggering code. We chose a 500 Hz sinusoidal stimulus, 2.5 V in amplitude, lasting between 200 ms and 300 ms.

**Open-Loop Protocol.** We defined an open-loop stimulation protocol (i.e. stimulation that does not take into account the activity of the fish). It aims to compare changes in the system due to code driven stimulation (i.e. stimulation triggered by the activity of the fish) with changes due to open-loop stimulation. This open-loop protocol delivered stimulation at the same average frequency than in code-driven sessions.

The temporal sequence were divided into regular time windows. Only one stimulus was triggered per window. The stimulus is randomly sent any time inside this window. For selecting an appropriate window time, we divide the number of stimuli delivered during a previous code-driven session by the total time of that session.

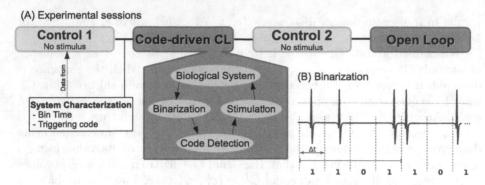

**Fig. 3.** Experimental schema (A) and binarization example (B). Consists of four ordered sessions: First control session, code-driven closed-loop session, second control session and open-loop stimulation session. Characterization of the system is done after every first control session to select an appropriate $\Delta t$ and trigger $C_t^4$ to stablish the closed-loop.

## 2.3   Description of the Experimental Protocols and Analysis

We have conducted 4 experiments using temporal-code driven and open-loop stimulation. Two fish of the species *Gnathonemus Petersii* were used, 10–13 cm long, acquired from local dealers of Madrid, Spain. They were housed in a 30 L ($40 \times 30 \times 25$) cm tank, water temperature was kept at $25\,^{\circ}$C, exposed to natural illumination. All experiments were noninvasive and all animals behaved normally after the experiments.

We analyzed the electrical activity during different sessions: control sessions without stimulation, stimulation triggered by code 0101 and open-loop stimulation sessions. The experimental protocol (Fig. 3) was described in Sect. 2.2.

Histograms of the codes detected in each session were used to study changes in the code distribution under different stimulation conditions Data was recorded in real time and also stored for an offline analysis, where code distributions between control and stimulation sessions were compared.

This analysis was complemented with an offline processing of the original non-binarized signal, detecting time between events (IPIs). We used histograms of the IPIs, quantile-quantile plots (qqplots) and tukey mean-difference plots in order to compare the distribution of IPIs between closed-loop and open-loop sessions (Cleveland 1993). Changes in these distributions due to stimulation are related to changes in the information processing of the system.

Finally, to establish a relation between IPIs analysis and codes, we grouped the codes according to the number of events in it and calculate the probability of each group as the sum of the independent probabilities of occurrence of each code in a group. Depending on the activity of the system represented in each code, we defined three groups *low*, *medium* and *high*: low = $\{0000, 0001, 0010, 0100, 1000\}$; medium = $\{0011, 0101, 0110, 1001, 1010, 1100\}$; high = $\{0111, 1011, 1101, 1110, 1111\}$.

# 3    Results

We compared the response of the system to code-driven stimulation with the control sessions and with the response obtained due to open-loop stimulation to detect a characteristic distinct response. We calculated codes histogram, IPI histogram, qqplots comparing IPI distribution between sessions. We selected code 0101 as the triggering code due to its capacity to produce changes in the system, observed during previous studies (Lareo et al. 2016).

The IPI histogram comparing IPIs during control sessions for all experiments with those during code-driven stimulation using code 0101 (Fig. 4-A) showed an increment of the probability of firing larger IPIs (between 250 and 350 ms). In the qqplot, dots above the reference line confirms that the fish increased the probability of firing longer IPIs for all ranges under code-driven stimulation. This behavior was also observed for each experiment independently. In the tukey-mean difference plot one can see that mean difference and standard deviation for all 4 experiments. It also showed that mean IPIs are longer during closed-loop stimulation when compared to those during the control session. For IPIs around 200 ms, IPIs discharged during stimulation sessions were from 50 to 120 ms longer than those during the control sessions.

Oppositely, when stimulating using open-loop protocol, fish increased the probability of firing shorter IPIs (Fig. 4-B). In the qqplot the dots are below the reference line within from 160 to 260 ms.

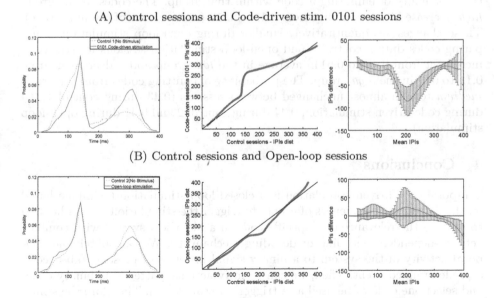

(A) Control sessions and Code-driven stim. 0101 sessions

(B) Control sessions and Open-loop sessions

**Fig. 4.** IPIs histogram, qqplot and tukey mean-difference plot comparing IPIs between sessions. Results from 4 experiments using the same triggering code (0101).

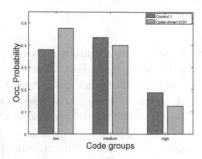 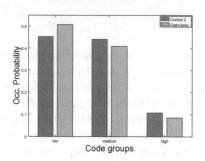

**Fig. 5.** Mean probabilities of each code group for the experiments depicted in Fig. 4. Group *low* represents codes with 3 or 4 zeros. Group *medium* represents codes with 2 zeros. Group *high* represents codes with 0 or 1 zeros. The bar chart showed the mean probability of each group in each session for all experiments. When comparing temporal-code driven stimulation sessions to the control 1 sessions, it showed an increment in the sparser codes (low group, increment of 9%) and a decrement of the high activity codes (high group, decrement 6%). It also showed a minor increment (6%) in the low codes and a minor decrement (3%) in high activity codes when comparing control 2 sessions and open-loop stimulation.

Regarding the codes emmited by the system (Fig. 5), it is shown that the sparser codes in the group *low* increased their probability during code-driven stimulation from 0.38 to 0.47, which means an increment of 9% in average for the probability of emmiting a code within that group. The codes in the group *high* decreased its probability from 0.19 to 0.13 (a decrement of 6% in average). These changes are quantitatively smaller during open-loop stimulation. Comparing codes during control 2 and open-loop stimulation sessions it showed an increment from 0.45 to 0.51 in average in the *low* group and a decrement from 0.11 to 0.08 in the *high* group. The probability of emitting codes from the group *medium* remain almost unchanged between sessions (0.43 during control 1, 0.4 during code-driven stimulation, 0.44 during control 2 and 0.41 during open-loop stimulation).

## 4    Conclusions

Temporal code-driven stimulation is a closed-loop stimulation technique based on the detection of sequences of events to trigger the stimulation. It can be used to address the relevance of a specific code in a biological system with complex activity-dependent encoding or decoding mechanisms. We can match the temporal activity of the system to a binary signal in terms of presence/absence of a predetermined event. Then, we can identify activity patterns as binary codes and select one code to be used as a trigger for stimulation. Therefore the stimulation depends on the system's activity in a closed-loop way. These codes, due to their structure, carry information about the pulse timing. The interaction with the biological system serves to test the relevance of predetermined sequences of events in the coding scheme.

Using a simple code-driven stimulation we have validated this technique addressing the presence of binary codes in electric fish signaling. Previous analysis done in terms of changes in the IPI distribution among sessions (Lareo et al. 2016) were complemented here with an analysis of the binary codes emitted by the system. The application of information-theoretic cost functions has been shown effective in previous experiments using closed-loop approaches (Benda et al. 2007; DiMattina and Zhang 2011, 2014). Accordingly, it is an important step to consider the distribution of binary codes in the system response, and link it to how the information is processed by the system.

A decrease in the IPI frequency (longer IPIs) is observed when the system is under closed-loop stimulation. Thus, an increment in the emission of sparser codes (codes with more zeros) is expected. In average, for all the experiments conducted, we observed minor changes in the code-emission pattern under open-loop stimulation than those under closed-loop (Fig. 5). This is in agreement with the expected behavior and stablishes a relation between IPI distribution and code emission probabilities. IPI distribution is frequently used to detect changes in the information processing of the system (Carlson 2002; Forlim and Pinto 2014; Forlim et al. 2015; Lareo et al. 2016), so this relation suggests that considering code probabilities in the response can improve the analysis. Nevertheless, this is only a validation for the methods and more experiments are needed to quantify this behavior.

We concluded that temporal code-driven stimulation methods can be applied to study the electroreceptive system. The existence of sequential IPI patterns related with the fish behavior expose that sequences of pulses have relevance for the system. We showed that temporal code-driven stimulation can modify the fish electrical behavior in a distinctive way. This can be observed attending to changes in the IPIs distribution, but also to changes in the code emission pattern. Thus, we can use this methodology to test the relevance of different sequential patterns for a given system in a closed-loop way, exposing relevant sequential dynamics hidden until now.

**Acknowledgments.** This work was funded by Spanish projects of Ministerio de Economía y Competitividad/FEDER TIN-2010-19607, TIN2014-54580-R, TIN-2012-30883, DPI2015-65833-P (http://www.mineco.gob.es/), ONRG grant N62909-14-1-N279, Brazilian Agency of Conselho Nacional de Desenvolvimento Científico e Tecnológico (http://www.cnpq.br/) and Fundação de Amparo à Pesquisa do Estado de São Paulo (www.fapesp.br). The funders had no role in study design, data collection and analysis, decision to publish, or preparation of the manuscript.

# References

Baker, C.A., Kohashi, T., Lyons-Warren, A.M., Ma, X., Carlson, B.A.: Multiplexed temporal coding of electric communication signals in mormyrid fishes. J. Exp. Biol. **216**(13), 2365–2379 (2013)

Benda, J., Gollisch, T., Machens, C.K., Herz, A.V.: From response to stimulus: adaptive sampling in sensory physiology. Curr. Opin. Neurobiol. **17**(4), 430–436 (2007)

Berényi, A., Belluscio, M., Mao, D., Buzsáki, G.: Closed-loop control of epilepsy by transcranial electrical stimulation. Science **337**(6095), 735–737 (2012)

Biró, I., Giugliano, M.A.: Reconfigurable visual-programming library for real-time closed-loop cellular electrophysiology. Front. Neuroinform. **9**, 14095–14106 (2015)

Bullock, T.H., Hopkins, C.D., Fay, R.R.: Electroreception, vol. 21. Springer Science & Business Media, Berlin (2006)

Caputi, A.A., Castelló, M.E., Aguilera, P., Trujillo-Cenóz, O.: Electrolocation and electrocommunication in pulse gymnotids: signal carriers, pre-receptor mechanisms and the electrosensory mosaic. J. Physiol.-Paris **96**(5), 493–505 (2002)

Carlson, B.A.: Electric signaling behavior and the mechanisms of electric organ discharge production in mormyrid fish. J. Physiol.-Paris **96**(5), 405–419 (2002)

Carlson, B.A., Gallant, J.R.: From sequence to spike to spark: evo-devo-neuroethology of electric communication in mormyrid fishes. J. Neurogenet. **27**(3), 106–129 (2013)

Chamorro, P., Muñiz, C., Levi, R., Arroyo, D., Rodríguez, F.B., Varona, P.: Generalization of the dynamic clamp concept in neurophysiology and behavior. PLoS ONE **7**(7), e40887 (2012)

Cleveland, W.S.: Visualizing Data. Hobart Press, Summit (1993)

DiMattina, C., Zhang, K.: Active data collection for efficient estimation and comparison of nonlinear neural models. Neural Comput. **23**(9), 2242–2288 (2011)

DiMattina, C., Zhang, K.: Adaptive stimulus optimization for sensory systems neuroscience. In: Closing the Loop Around Neural Systems, p. 258 (2014)

Ehrens, D., Sritharan, D., Sarma, S.V.: Closed-loop control of a fragile network: application to seizure-like dynamics of an epilepsy model. Front. Neurosci. **9**, 58 (2015)

El Hady, A.: Closed Loop Neuroscience. Academic Press, Cambridge (2016)

Forlim, C.G., Pinto, R.D.: Automatic realistic real time stimulation/recording in weakly electric fish: long time behavior characterization in freely swimming fish and stimuli discrimination. PLoS ONE **9**(1), e84885 (2014)

Forlim, C.G., Pinto, R.D., Varona, P., Rodríguez, F.B.: Delay-dependent response in weakly electric fish under closed-loop pulse stimulation. PLoS ONE **10**(10), e0141007 (2015)

Jun, J.J., Longtin, A., Maler, L.: Precision measurement of electric organ discharge timing from freely moving weakly electric fish. J. Neurophysiol. **107**(7), 1996–2007 (2012)

Kramer, B.: Electric and motor responses of the weakly electric fish, gnathonemus petersii (mormyridae), to play-back of social signals. Behav. Ecol. Sociobiol. **6**(1), 67–79 (1979)

Lareo, A., Forlim, C.G., Pinto, R.D., Varona, P., Rodríguez, F.B.: Temporal code-driven stimulation: definition and application to electric fish signaling. Front. Neuroinform. **10** (2016)

Madhav, M.S., Stamper, S.A., Fortune, E.S., Cowan, N.J.: Closed-loop stabilization of the jamming avoidance response reveals its locally unstable and globally nonlinear dynamics. J. Exp. Biol. **216**(22), 4272–4284 (2013)

Muñiz, C., Arganda, S., Borja Rodríguez, F., Polavieja, G.G.: Realistic stimulation through advanced dynamic-clamp protocols. In: Mira, J., Álvarez, J.R. (eds.) IWINAC 2005. LNCS, vol. 3561, pp. 95–105. Springer, Heidelberg (2005). doi:10.1007/11499220_10

Muñiz, C., de Borja Rodríguez, F., Varona, P.: Rtbiomanager: a software platform to expand the applications of real-time technology in neuroscience. BMC Neurosci. **10**(1), 1 (2009)

Muñiz, C., Forlim, C.G., Guariento, R.T., Pinto, R.D., Rodríguez, F.B., Varona, P.: Online video tracking for activity-dependent stimulation in neuroethology. BMC Neurosci. **12**(1), 1 (2011)

Muñiz, C., Levi, R., Benkrid, M., Rodríguez, F.B., Varona, P.: Real-time control of stepper motors for mechano-sensory stimulation. J. Neurosci. Methods **172**(1), 105–111 (2008)

Potter, S.M., El Hady, A., Fetz, E.E.: Closed-loop neuroscience and neuroengineering. In: Closing the Loop Around Neural Systems, p. 7 (2014)

Roth, E., Sponberg, S., Cowan, N.: A comparative approach to closed-loop computation. Curr. Opin. Neurobiol. **25**, 54–62 (2014)

Schiff, S.J.: Neural Control Engineering: The Emerging Intersection Between Control Theory and Neuroscience. MIT Press, Cambridge (2012)

Von der Emde, G., Behr, K., Bouton, B., Engelmann, J., Fetz, S., Folde, C.: 3-dimensional scene perception during active electrolocation in a weakly electric pulse fish. Front. Behav. Neurosci. **4**, 26 (2010)

# A Novel Technique to Estimate Biological Parameters in an Epidemiology Problem

Antone dos Santos Benedito[1] and Fernando Luiz Pio dos Santos[2]([✉])

[1] Institute of Biosciences of Botucatu, São Paulo State University,
Botucatu, Brazil
[2] Department of Biostatistics, Institute of Biosciences of Botucatu,
São Paulo State University, Distrito Rubião Júnior,
Botucatu, SP 18618-689, Brazil
{antone,flpio}@ibb.unesp.br
http://www.ibb.unesp.br

**Abstract.** In this paper, we describe a study of a parameter estimation technique to estimate a set of unknown biological parameters of a nonlinear dynamic model of dengue. We also explore a *Levenberg-Marquardt* (*LM*) algorithm to minimize the cost function. A classical mathematical model describes the dynamics of mosquitoes in water and winged phases, where the data are available. The main interest is to fit the model to the data taking into account the parameters estimated. Numerical simulations were performed and results showed the robustness of *LM* in estimating the important parameters in the dengue disease problem.

**Keywords:** Computational population dynamics · Ordinary differential system · Aedes · Dengue

## 1 Introduction

Dengue is a subject of intense research and it has been a major public health problem worldwide, especially in tropical and subtropical countries such as Brazil, where its incidence has increased in recent years. The World Health Organization (WHO) [1] states that about half of the world's population is now at risk. About 100 million people are infected in more than 100 countries from all continents and many people die as a consequence of dengue. It is a viral infection febrile disease caused by a virus of the family Flaviridae, transmitted by female mosquito bites, usually of the genus Aedes aegypti (A. a.) and living in urban habitats [2,3]. A. a. also transmits Chikungunya and Zika virus infections. There are four different serotypes of the dengue virus (DEN-1 - DEN-4). Despite great improvements in hygiene, sanitation and vector control, disease containment remains one of the biggest challenges for the modern world. Studies addressing the combat of dengue using different methodologies can be found in the literature

The original version of this chapter was revised: An acknowledgement has been added. The erratum to this chapter is available at 10.1007/978-3-319-59153-7_65

I. Rojas et al. (Eds.): IWANN 2017, Part I, LNCS 10305, pp. 112–122, 2017.
DOI: 10.1007/978-3-319-59153-7_10

[4–6]. The dengue vaccine has been licensed for use, but only to people living in endemic areas (WHO). Nevertheless, despite all efforts, knowledge leading to improvement and progress in the development of new tools and strategies for dengue prevention is very important, necessary and still far from ideal [7]. The main focus of this paper is the mathematical and computational investigation of the parameter estimate technique coupled to the *Levenberg-Marquardt* (*LM*) algorithm to solve the problem of estimating a set of biological parameters in the dengue disease.

## 2 The Dynamic Epidemiological Model

Here we describe the mathematical model involving the mosquito population dynamics [16]. Tables 1 and 2 show the state variables and biological parameters, respectively.

**Table 1.** State variables for the mosquito population at time $t$.

| | |
|---|---|
| $A(t)$ | Aquatic phase (immature forms) |
| $I(t)$ | Non-fertilized females |
| $F(t)$ | Fertilized females (after mating) |
| $M(t)$ | Male insects (natural male) |

**Table 2.** Biological parameters for the model (1).

| Parameter | Description | Value [17] | Unit |
|---|---|---|---|
| $\gamma$ | Ratio of transition to winged form | 0.121 | $days^{-1}$ |
| $\beta$ | The effective mating rate | 0.7 | $days^{-1}$ |
| $r$ | The proportions of females | 0.5 | – |
| $(1-r)$ | The proportions of males | 0.5 | – |
| $\mu_A$ | Aquatic phase mortality rate | 0.0583 | $days^{-1}$ |
| $\mu_I$ | Unmating female mortality rate | 0.0337 | $days^{-1}$ |
| $\mu_F$ | Mating fertilized mortality rate | 0.0337 | $days^{-1}$ |
| $\mu_M$ | Male mortality rate | 0.06 | $days^{-1}$ |
| $\phi$ | Intrinsic oviposition rate | 6,353 | $days^{-1}$ |
| $C$ | Carrying capacity | 3 | $mosquito^{-1}$ |

$$
\begin{cases}
\frac{dA}{dt} = \phi \left(1 - \frac{A}{C}\right) F - (\gamma + \mu_A)A \\
\frac{dI}{dt} = \quad r\gamma A \quad - (\beta + \mu_I)I \\
\frac{dF}{dt} = \quad \beta I \quad - \quad \mu_F F \\
\frac{dM}{dt} = (1-r)\gamma A \quad - \quad \mu_M M
\end{cases}
\tag{1}
$$

Let us define $\mathbf{X}(t) = [A(t), I(t), F(t), M(t)]$ representing the continuous solution vector of the system (1) at time $t$. In this study, the system is solved numerically. Therefore, in the next section we describe the discrete-time model and numerical method used to obtain the respective numerical solution $X_i$ at time discretization $i$.

## 2.1  The Discretized Model

To obtain the numerical solutions, the classical backward Euler integration method is used for the time discretization of the ODE system (1). The main advantage of using this numerical scheme in a particular case to estimate a set of unknown parameters is because it is easy to correspond all discrete parameters known available in time $t$ with each time discrete $i$ to be considered within the scheme. More details can be found in [8].

# 3  Parameter Estimation Technique

The parameter estimation technique is often presented in the literature as an optimization problem concerning an objective function in the form of least squares [9–12]. In such problems, having a dataset available, a mathematical model is chosen and it is sought to minimize the sum of the squared distances called residues, between each of the given points and the adjusted curve.

## 3.1  The Problem

Let us introduce the notations to present the problem of parameter estimation below:

- $i \in \{1, \ldots, m\}$, where $m$ is the last day of the data observation;
- $E = \{\mathbf{X}_{i_{obs}}\}$: the set of the observed data, $\forall i = 1, \cdots, m$;
- $\mathbf{X}_{i_{obs}}$: the observed state variables;
- $\mathbf{X}_{i_{obs}}^l$: the $l$-th component of the $\mathbf{X}_{i_{obs}}$, $l = 1, \cdots, k$, $k$ is the number of equations;
- $\mathbf{X}_i$: the solution of a dynamic population model, $\forall i = 1, \cdots, m$;
- $\mathbf{X}_i^l$: the $l$-th component of the solution $\mathbf{X}_i$, $l = 1, \cdots, k$;
- $\mathbf{b} = [b_{i_{obs}}]$: the vector of known parameters (observed data), $\forall i = 1, \cdots, m$;
- $\mathbf{p} = (p_1, \cdots, p_n)$, where $n$ is the number of parameters to be estimated.

**Problem 1 (Find the local minimal).** *Let us consider the function* $w : R^n \rightarrow R^m$, $m \geq n$. *We need to find the minimal of* $\| w(\mathbf{p}) \|$, *i.e., find the local minimal* $\mathbf{p}^*$ *of* $W(\mathbf{p})$, *given by*

$$W(\mathbf{p}) = \frac{1}{2} w^{\mathbf{T}}(\mathbf{p}) w(\mathbf{p}) = \frac{1}{2} \| w(\mathbf{p}) \|^2, \tag{2}$$

*where* $\| \, . \, \|$ *is the euclidian norm and* $w(\mathbf{p})^{\mathbf{T}}$ *is the transpose of the* $w(\mathbf{p})$.

**Definition 1 (Cost function).** *Let the cost function $W$ be defined as:*

$$W(\mathbf{p}) = \frac{1}{2} \parallel w(\mathbf{p}) \parallel^2 = \frac{1}{2} \sum_{i=1}^{m} w_i(\mathbf{p})^2 = \frac{1}{2} \sum_{i=1}^{m} \sum_{l=1}^{k} (\mathbf{X}_i^l(\mathbf{p}, \mathbf{b}) - \mathbf{X}_{i\,obs}^l)^2, \quad (3)$$

In this study, the main problem involving parameter estimation consists of:

**Problem 2 (Minimize $W$).** *Determine $\mathbf{p}$ that minimizes the cost function $W$, subject to $p_k \geq 0, \forall k = 1, \cdots, n$.*

Then, to solve Problem (2), we applied the algorithm described as follows.

### Algorithm 1 (Parameter Estimation)

> **Step 1 - Input** *the observed data;*
> **Step 2 - Set** *initial parameter guess* $\mathbf{p}^0 = [p_1^0, \cdots, p_n^0]$ *for* $\mathbf{p}$;
> **Step 3 - Build** *a function to calculate* $\mathbf{X}$ *from a mathematical model;*
> **Step 4 - Calculate** *the cost function $W$ from Eq. (3);*
> **Step 5 - Build** *a routine to minimize $W$;*
> **Step 6 - Return** $\mathbf{p}$.

The next section describes the *LM* method as an alternative method to use in the parameter estimation problem.

## 3.2 The *LM* Technique to Estimation

The *LM* method is a computational optimization technique used to solve nonlinear square problems [10–13]. It is the result of an improvement of the *Gauss-Newton* method which, in turn, consists of a modification of *Newton's* method. It was proposed by [14], related to a suggestion published by [15]. As *Newton* and *Gauss-Newton's* methods, *LM* is an iterative method inherent to non-linear optimization methods. This means that given a starting point $p_0$ the method produces a sequence of vectors $\mathbf{p}_1, \mathbf{p}_2, \mathbf{p}_3...$ which is expected to converge to the local minimum $\mathbf{p}^*$ of the function to be adjusted. The short form of the *LM* algorithm can be described as following.

### Algorithm 2 (The *Levenberg-Marquardt* short form)

> **Step 1 -** *Input* $\mathbf{p} := \mathbf{p}_0$; $\lambda := \lambda_0$
> *Repeat until it STOPS:*
> **Step 2 -** *Solve* $(J^t J + \lambda I)\mathbf{h}_{lm} = -\mathbf{g}$
> **Step 3 -** $\mathbf{p}_{\mathbf{new}} := \mathbf{p} + \mathbf{h}_{lm}$
> **Step 4 -** *Update* $\lambda$

where $\mathbf{g} = J(\mathbf{p})^{\mathbf{T}} w(\mathbf{p})$ is the gradient of the *LM* method; $J(\mathbf{p})|_{ik} = \frac{\partial w_i}{\partial \mathbf{p}_k}(\mathbf{p})$ the Jacobian Matrix, $\mathbf{h}_{lm}$ is the *LM* descent-direction and $\lambda > 0$ is the *damping parameter* given by *Madsen* [10]. Then, considering the Algorithm 2, the optimization form of the *LM* algorithm applied in this study can be finally summarised as below:

**Algorithm 3 (The $LM$ Technique for Parameter Estimation)**

**Step 1** - *Define* $\mathbf{p} = \mathbf{p}_0$ *and* $\nu = 2$;

**Step 2** - *Calculate* $B = J(\mathbf{p})^{\mathbf{T}} J(\mathbf{p})$ *and* $\mathbf{g} = J(\mathbf{p})^{\mathbf{T}} w(\mathbf{p})$;

**Step 3** - *Do* $\lambda = \tau * max\{b_{ii}\}$;

**Step 4** - *If* $\| \mathbf{g} \|_{\infty} \leq \varepsilon_1$, *go to* **Step 8**;

**Step 5** - *Solve* $(J^{\mathbf{T}} J + \lambda I)\mathbf{h}_{lm} = -\mathbf{g}$, *in* $\mathbf{h}_{lm}$, *with* $\lambda \geq 0$;

**Step 6** - *If* $\| \mathbf{h}_{lm} \| \leq \varepsilon_2(\| \mathbf{p} \| +\varepsilon_2)$, *go to* **Step 8**;

   *Otherwise, Do* $\mathbf{p}_{new} = \mathbf{p} + \mathbf{h}_{lm}$;

   *Calculate* $\varrho = (W(\mathbf{p}) - W(\mathbf{p}_{new}))/(L(0) - L(\mathbf{h}_{lm}))$;

**Step 7** - *If* $\varrho > 0$, *do* $\mathbf{p} = \mathbf{p}_{new}$, *up to date* $B$ *and* $\mathbf{g}$, *and Do*

   $\lambda = \lambda * max\{\frac{1}{3}, 1 - (2\varrho - 1)^3\}$ *and* $\nu = 2$;

   *Otherwise, Do* $\lambda = \lambda * \nu$ *and* $\nu = 2 * \nu$;

   $it = it + 1$; *Back to* **Step 4**;

**Step 8** - *Return* $\mathbf{p}$.

where $\varrho = \frac{W(\mathbf{p}) - W(\mathbf{p}_{new})}{L(0) - L(\mathbf{h}_{lm})}$ defined as the *gain ratio* of the method. The stop-criteria used were $\| \mathbf{g} \|_{\infty} \leq \epsilon_1$, $\| \mathbf{p}_{new} - \mathbf{p} \| \leq \epsilon_2(\| \mathbf{p} \| +\epsilon_2)$ and $it \geq it_{max}$, where $\epsilon_1$ and $\epsilon_2$ are the precisions, $it$ and $it_{max}$ are the number of interations and maximal interation, respectively.

# 4 Results

Here, we present the numerical results obtained by adopting the novel technique for estimation parameters, Algorithm 3, as described in this paper. To perform the numerical simulations and tests, the following information 1–5 should be considered.

1. Parameters to be estimated: $\mathbf{p} = (\mu_F, C, \mu_M)$;
2. Available data: $\mathbf{X}_{iobs} = [A_{iobs}, F_{iobs}, M_{iobs}]$;
3. Observed parameters: $\mathbf{b} = [\phi_{iobs}]$;
4. Initial conditions $(A_0, F_0, M_0) = (0, 100, 30)$;
5. Parameters of simulations: Time-step $dt = 1$, $\tau = 10^{-3}$ and $\varepsilon_1 = \varepsilon_2 = 10^{-8}$.

The data used to test the parameter estimation technique described in this paper refer to the variables $A$, $F$ and $M$ for the population of winged and aquatic mosquitoes, with an oviposition rate per capita $\phi$, at different temperatures $-15\,^{\circ}$C, $20\,^{\circ}$C, $25\,^{\circ}$C and $30\,^{\circ}$C, and from the experiments carried out in two different cities (A and B) from the São Paulo State. The state variable $I$ is neglected in this work, due to its rapid transition to state $F$. In these experiments, a total of 100 newly emerged female and 30 male mosquitoes were placed in a cage containing amber glass with filter paper for egg laying. Inside the cage, the necessary food (water with honey) was available *ad libitum*. Once a day, the mosquitoes received a blood meal from an immobilized rat in order to allow the fertilized eggs to develop. Each day the eggs were counted on the filter paper and then replaced. In order to obtain the ratios and oviposition, the new individuals were removed, decreasing the population at the end of the experiment. The number of male and female survivors was also recorded daily. Table 3 presents all parameter estimations by using the $LM$ algorithm.

**Table 3.** Parameters estimated for cities A and B.

|  | $\mu_F$ | | $C$ | | $\mu_M$ | |
|---|---|---|---|---|---|---|
|  | A | B | A | B | A | B |
| 15 °C | 0.0388 | 0.0392 | 727.0709 | 82.1313 | 0.0624 | 0.0421 |
| 20 °C | 0.0174 | 0.0258 | 132086.0575 | 89881.9725 | 0.0398 | 0.0666 |
| 25 °C | 0.0249 | 0.0311 | 1363006.7101 | 1104267.7859 | 0.0666 | 0.0416 |
| 30 °C | 0.0318 | 0.0282 | 469855.7423 | 639707.7831 | 0.1008 | 0.0795 |

## 4.1   Comparison of the Estimations for $\mu_F$ and $\mu_M$ (25 °C)

Table 4 shows the estimated values for $\mu_F$ and $\mu_M$ at 25 °C compared to those presented by Thomé et al. [16]. Note that the values of $\mu_F$ - city B and $\mu_M$ - city A are very close to those shown by [16]. Concerning parameter $C$, the relatively high values obtained can be justified by the abundant availability of nutrients to the mosquitoes during the experiments, considering the data as described.

**Table 4.** Comparisons $\mu_F$ and $\mu_M$ estimations (25 °C).

| Thomé et al. [16] | | Present work | | | |
|---|---|---|---|---|---|
| $\mu_F$ | $\mu_M$ | $\mu_F$ - City A | $\mu_F$ - City B | $\mu_M$ - City A | $\mu_M$ - City B |
| 0.0337 | 0.06 | 0.0249 | 0.0311 | 0.0666 | 0.0416 |

## 4.2   Comparison of the Estimations for $\mu_F$

In Table 5, it can be observed that for approximately 15 °C and 25 °C, in city B, our estimations for $\mu_F$ are relatively close to those of [17]. This relationship is also observed at 15 °C for city A.

**Table 5.** Comparison $\mu_F$ estimations for different temperatures and cities.

| Yang et al. [17] | | Present work | | |
|---|---|---|---|---|
| Temperature (°C) | $\mu_F$ | Temperature (°C) | $\mu_F$ - City A | $\mu_F$ - City B |
| 15.30 | 0.03608 | 15 | 0.0388 | 0.0392 |
| 20.05 | 0.04216 | 20 | 0.0174 | 0.0258 |
| 25.64 | 0.03043 | 25 | 0.0249 | 0.0311 |
| 31.33 | 0.04391 | 30 | 0.0318 | 0.0282 |

In the next section, the sequences of figures show the curves from numerical test predictions of the model compared to data of $A$, $F$ and $M$, where it is possible to observe the model fit these data.

## 4.3   Estimation Tests

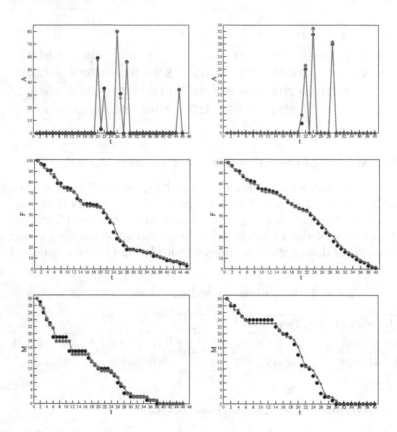

**Fig. 1. Test 1:** Dynamic of dengue mosquito populations $A(t)$, $F(t)$ and $M(t)$ (from top to bottom) for the cities A (left) and B (right) in $15\,^{\circ}\mathrm{C}$. Red line means predictions of the model and black dots depict the observed data of the aquatic phase, fertilized females and males of dengue mosquitoes. (Color figure online)

In **Test 4**, the *Pearson correlation coefficient* $\rho$ [12] values were calculated, indicating the strong linear correlation between the parameters estimated and the real data. Table 6 shows the $\rho$ values for cities A and B at the temperature of $30\,^{\circ}\mathrm{C}$ (Fig. 4).

**Table 6.** $\rho$ values ($30\,^{\circ}\mathrm{C}$).

|        | $A$ | $F$ | $M$ |
|--------|-----|-----|-----|
| City A | 0.999989726 | 0.998431825 | 0.996482096 |
| City B | 0.999976721 | 0.999048742 | 0.995447591 |

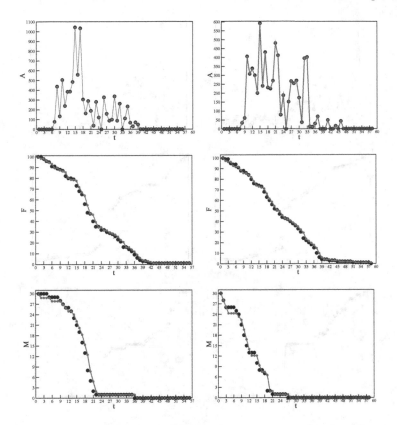

**Fig. 2. Test 2:** Dynamic of dengue mosquito populations $A(t)$, $F(t)$ and $M(t)$ (from top to bottom) for the cities A (left) and B (right) in 20 °C. Red line means predictions of the model and dots depict the observed data of the aquatic phase, fertilized females and males of dengue mosquitoes. (Color figure online)

In the next section, we draw some conclusions about the novel technique to estimate parameters in the dynamic population applied in the dengue disease problem.

## 5    Conclusions

In this paper, we described a novel technique to investigate important biological parameters involving the dengue mosquito population. The parameter estimation technique was used to estimate a set of unknown parameters of a nonlinear model of dengue that describes the dynamics of mosquitoes in water and winged phases. The $LM$ algorithm was explored to minimize the cost function to fit the model to the dengue data available, taking into account the parameters estimated. The results obtained show that this field of parameter estimation can be an important data analysis tool to apply in dynamic population systems, in

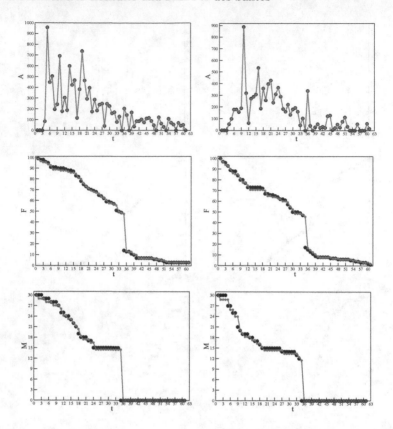

**Fig. 3. Test 3:** Dynamic of dengue mosquito populations $A(t)$, $F(t)$ and $M(t)$ (from top to bottom) for the cities A (left) and B (right) in 25 °C. Red line means predictions of the model and black dots depict the observed data of the aquatic phase, fertilized females and males of dengue mosquitoes. (Color figure online)

particular to analyze the dengue disease problem involving real data. Thus, the main conclusions are:

1. In all tests performed, the model was able to fit the dengue data;
2. The comparisons from the numerical solutions with data show the robustness of the code to fit the dengue data available.
3. There is closeness between the estimated parameters and values from the literature;
4. The results show that this technique can be an important data analysis tool to be applied in dynamic population systems;

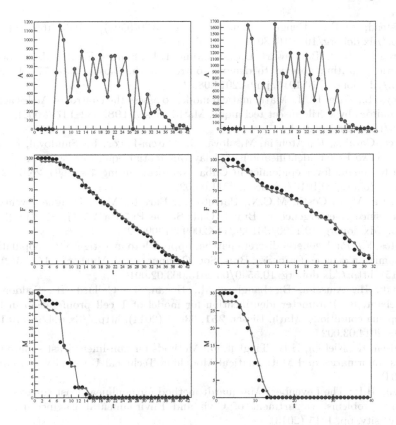

**Fig. 4. Test 4:** Dynamic of dengue mosquito populations $A(t)$, $F(t)$ and $M(t)$ (from top to bottom) for the cities A (left) and B (right) in $30\,^\circ$C. Red line means predictions of the model and black dots depict the observed data of the aquatic phase, fertilized females and males of dengue mosquitoes. (Color figure online)

5. The methodology presented here improved the application of the *LM* algorithm as an optimization alternative to analyze the dengue disease problem involving real data.

**Acknowledgments.** The authors would like to thank the Brazilian agencies CAPES for the master's scholarship provided and FAPESP for the financial support received.

# References

1. World Health Organization. Dengue: Guidelines for Diagnosis, Treatment, prevention and control. WHO, Geneva (2009). http://www.who.int/tdr/publications/documents/dengue-diagnosis.pdf?ua=1
2. McBridea, W.J.H., Bielefeldt-Ohmannb, H.: Dengue viral infections; pathogenesis and epidemiology. Microbes Infect. **2**, 1041–1050 (2000). http://dx.doi.org/10.1016/S1286-4579(00)01258-2

3. Halstead, S.B.: Dengue. The Lancet **370**(9599), 1644–1652 (2007). http://dx.doi.org/10.1016/S0140-6736(07)61687-0

4. Florentino, H.O., Bannawart, B.F., Cantane, D.R., Santos, F.L.P.: Multiobjective genetic algorithm applied to dengue control. Math. Biosci. **256**, 77–84 (2014). http://dx.doi.org/10.1016/j.mbs.2014.08.013

5. Esteva, L., Yang, H.M.: Mathematical model to assess the control of Aedes aegypti mosquitoes by sterile insect technique. Math. Biosci. **198**, 132–147 (2005). doi:10. 1016/j.mbs.2005.06.004

6. García-Garaluz, E., Atencia, M., Joya, G., García-Lagos, F., Sandoval, F.: Hopfield networks for identification of delay differential equations with an application to dengue fever epidemics in Cuba. Neurocomputing **74**, 2691–2697 (2011). http://doi.org/10.1016/j.neucom.2011.03.022

7. Teixeira, M.G., Costa, M.C.N., Barreto, F., Barreto, M.L.: Dengue: twenty-five years since re-emergence in Brazil. Cad. Saúde Pública **25**(1), S7–S18 (2009). http://dx.doi.org/10.1590/S0102-311X2009001300002

8. Santos, F.L.P.: A general discrete patches approach to investigate the populational dynamics in dengue. Proc. Ser. Braz. Soc. Comput. Appl. Math. SBMAC **3**(2), 1–7 (2015). http://dx.doi.org/10.5540/03.2015.003.02.0016

9. Ayoub, H., Ainseba, B., Langlais, M., Hogan, T., Callard, R., Seddon, B., Thiébaut, R.: Parameter identification for model of T cell proliferation in Lymphopenia conditions. Math. Biosci. **251**, 63–71 (2014). http://dx.doi.org/10.1016/j.mbs.2014.03.002

10. Madsen, K., Nielsen, H.B., Tingleff, O.: Methods for non-linear least square problems. Informatics and Mathematical Modelling Technical University of Denmark (2004)

11. Gavin, H.P.: The Levenberg-Marquardt method for nonlinear least squares curve-fitting problems. Department of Civil and Environmental Engineering, Duke University, pp. 1–15 (2013)

12. Press, W.H., Teukolsky, S.A., Vetterling, W.T., Flannery, B.P.: Numerical Recipes in C: The Art of Scientific Computing. Cambridge University Press, Cambridge (1992)

13. Cho, C.-K., Kwon, Y.H.: Parameter estimation in nonlinear age-dependent population dynamics. IMA J. Appl. Math. **62**(3), 227–244 (1999). https://doi.org/10.1093/imamat/62.3.227

14. Marquardt, D.W.: An algorithm for least-squares estimation of nonlinear parameters. J. Soc. Ind. Appl. Math. **11**, 431–441 (1963). http://dx.doi.org/10.1137/0111030

15. Levenberg, K.: A method for the solution of certain non-linear problem in least squares. Q. J. Appl. Math. **2**, 164–168 (1944). http://www.jstor.org/stable/43633451

16. Thomé, R.C.A., Yang, H.M., Esteva, L.: Optimal control of Aedes aegypti mosquitoes by the sterile insect technique and insecticide. Math. Biosci. **223**(1), 12–23 (2010). http://dx.doi.org/10.1016/j.mbs.2009.08.009

17. Yang, H.M., Macoris, M.L.G., Galvani, K.C., Andrighetti, M.T.M., Wanderley, D.M.V.: Assessing the effects of temperature on the population of Aedes aegypti, the vector of dengue. Epidemiol. Infect. **137**, 1188–1202 (2009). https://doi.org/10.1017/S0950268809002040

# Breast Cancer Microarray and RNASeq Data Integration Applied to Classification

Daniel Castillo[(✉)], Juan Manuel Galvez, Luis Javier Herrera,
and Ignacio Rojas

Department of Computer Architecture and Computer Technology,
University of Granada, Granada, Spain
cased@ugr.es

**Abstract.** Although Next-Generation Sequencing (NGS) has more impact nowadays than microarray sequencing, there is a huge volume of microarray data that has not still been processed. The last represents the most important source of biological information nowadays due largely to its use over many years, and a very important potential source of genetic knowledge deserving appropriate analysis. Thanks to the two techniques, there is now a huge amount of data that allows us to obtain robust results from its integration. This paper deals with the integration of RNASeq data with microarrays data in order to find breast cancer biomarkers as expressed genes. These integrated data has been used to create a classifier for an early diagnosis of breast cancer.

**Keywords:** Computational biology · RNASeq · Bioinformatics · Microarray · Classification · SVM · Feature selection · Integration

## 1 Introduction

Cancer is the second leading cause of death worldwide, just behind cardiovascular disease. Specifically, breast cancer is one of the five most dangerous cancers in the world, showing a high mortality rate according to World Health Organization (WHO) and being the cancer with the highest impact among the female population [12]. Today, many breast cancer diagnosis are done when a patient presents several related symptoms, thus increasing the mortality risk. If the cancer has spread in the organism, treatment becomes more difficult, and generally the chances of surviving are much lower. However, cancers that are diagnosed at an early stage are more likely to be treated successfully. Therefore, it is primordial to find biomarkers that allow an early diagnosis of breast cancer. Two sequencing technologies has been used to compute the genes expression, which are explained and compared below:

### 1.1 Microarray Technology

Microarray is a method that allows the measurement of the value expressions of a large number of genes simultaneously from a collection of microscopic DNA spots

© Springer International Publishing AG 2017
I. Rojas et al. (Eds.): IWANN 2017, Part I, LNCS 10305, pp. 123–131, 2017.
DOI: 10.1007/978-3-319-59153-7_11

attached to a solid surface. This technology is based on the DNA hybridization process, so that DNA is hybridized for each of the spots that represent one gene value expression. Once the step is finished with a laser, the expression values are read and written in a file with the extension .CEL.

A summarized microarray workflow is presented at the Fig. 1. As can be seen, once microarray data are available, all of them are processed and filtered from a quality analysis to be later normalized. Once this was done, the last step is the integration of all microarrays. VirtualArray tool [5] has been used for the integration process.

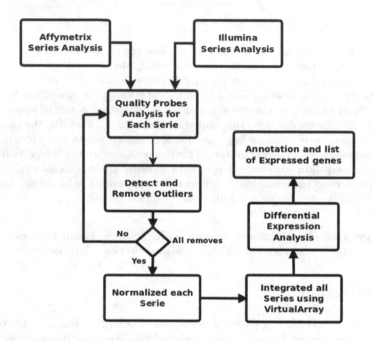

**Fig. 1.** Microarray gene expression integration pipeline

## 1.2   RNASeq Technology

This technology appeared as a revolutionary tool for transcriptome and as a natural evolutionary step in the study of the genome after the massive use of microarray technology. In this sense, one of the most advantageous aspects is that although RNASeq can be used only for transcriptome profiling, it also can be combined with other functional genomics methods to enhance the analysis of gene expression. Expression is quantified by counting the number of reads mapped to each locus in the transcriptome assembly step. This expression level can be calculated for exons or genes using contigs or reference transcript annotations. These observed RNASeq read counts have been robustly validated against previous technologies such as microarrays or quantitative polymerase chain reaction (qPCR) [13].

## 1.3  Comparison Between both Technologies

RNASeq is being widely used within the scientific community for gene expression studies and it is replacing DNA microarray technology. RNASeq offers an important number of advantages over microarrays [17], although the computational cost of RNASeq experiments are also higher than microarray technology:

- RNASeq allows to detect the variation of a single nucleotide.
- Genomic sequence knowledge is not necessary in RNASeq.
- RNASeq provides quantitative expression levels.
- RNASeq provides isoform-level expression measures.
- RNASeq offers a broader dynamic range than microarrays.

In spite of these advantages, microarrays are still used because of their lower costs. Besides, as microarrays have been used for a longer period, there exist many robust statistical method to analyzed them. There are many significant microarray experiments already available and even there are also a high number of microarray datasets that have not been analyzed so far. These datasets might have information that could reveal important facts and candidate biomarkers. In any case, the transition to RNASeq keeps going on. Both technologies can be applied together to the extent that microarray data could be used to create a classifier for RNASeq samples, and like Nookaew et al. explained [11], a high consistency between RNASeq and microarray data exists, which encourages to follow using microarray as a versatile tool for gene expression analysis. This experiment is a first approach that only intends to show the potential in the integration of both technologies, obtaining a significant improvement in classification level with the introduction of some samples of RNASeq together with a huge quantity of samples of microarray.

## 2  Materials and Methods

All analyzed RNA samples were obtained from NCBI GEO web platform [2]. 108 samples from microarray series and 6 samples from RNASeq samples.

Table 1 shows a summary about the series used and their origin. As it can be seen, there are series from different countries, and thus there are samples from different ethnic groups. Furthermore, there are different sequencing technologies in the experiment including samples from Affymetrix [4] and Illumina [6] Moreover, there are data from different generation sequencing. In summary, samples have been integrated from different generation sequencing, technologies, platforms and countries, bringing all of them heterogeneity to the study.

Both microarray and RNASeq data have passed a strict pipeline. Microarray samples require restrictive quality analysis to discard non-representative samples which took place due to incorrect acquisition, as well as normalization during pre-processing in order to adapt the range of quantification variability of the samples considered. Once this is done, from an available set of high quality samples, reliable biomarkers have been obtained from the application of a feature

**Table 1.** Input series, technology, quantification and number of samples/outliers. RNASeq samples are highlighted in darker gray.

| Series | Technology | Quantification | # Quality Samples | # Excluded Outliers | Samples Origin |
|---|---|---|---|---|---|
| GSE52712 | Affymetrix | Gene Expression | 19 | 1 | Manchester (UK) |
| GSE40987 | Affymetrix | Gene Expression | 10 | 0 | Boston (USA) |
| GSE52262 | Affymetrix | Gene Expression | 16 | 0 | Houston (USA) |
| GSE12790 | Affymetrix | Gene Expression | 20 | 1 | San Francisco (USA) |
| GSE46834 | Illumina | Gene Expression | 8 | 0 | New York (USA) |
| GSE68651 | Illumina | Gene Expression | 35 | 1 | Southampton (UK) |
| GSE78011 | Illumina | Counts | 3 | 0 | Louisville (USA) |
| GSE81593 | Illumina | Counts | 3 | 0 | New York (USA) |
| TOTAL | Integrated | | 114 | 3 | |

selection algorithm. Thus, a gene ranking has been obtained to be applied in the classification stage and to train the classifier with an increasing number of genes. 47 genes comply the statistical restrictions of logarithmic fold change ($| logFC | \geq 2$) and p-value $\leq 0.001$ to form the final ranking of relevant genes considered as potential biomarkers of the disease. logFC represents the difference between breast cancer and control expressed values, whilst p-value represents the probability of obtaining a result equal or higher than what it was observed when the null hypothesis is true.

Differently, for the extraction of the RNASeq data from the fastq original files, tools like tophat2 [7], bowtie2 [8], samtools [9] and htseq [1] have been used to obtain the count files. Once the count files have been obtained, the expression values have been calculated using the counts and the NOISeq R package [16].

Later, the normalized expression values have been integrated directly with the normalized values belonging to microarrays. Same statistic restrictions were considered on RNASeq respect to logFC and p-value. For expressions genes in both of them limma R package [14] has been used. This tools allows the expressions values to calculate the expressed genes independently of the technology used. All RNASeq pipeline from SRA data to gene expression values can be seen in Fig. 2.

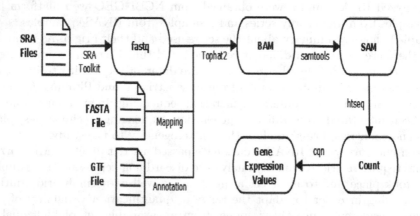

**Fig. 2.** RNASeq gene expression integration pipeline

The genes considered as relevant during the microarray samples processing have been used as a filter, as if they were a mask, to directly select such genes in the quantification obtained with RNASeq samples. From the gene ranking previously obtained, the RNASeq samples have been classified.

The pipeline followed in this integration and classification process of all the samples can be seen in Fig. 3.

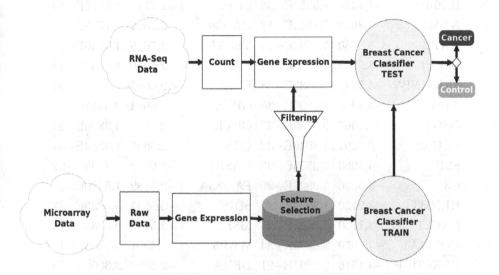

**Fig. 3.** Microarray and RNASeq integration pipeline

For classification, Support Vector Machine (SVM) [10] algorithm had been used using also minimum Redundancy Maximum Relevance (mRMR) [3] for features selection. SVM algorithm is based on the idea of separating the different categories in a problem through a hyperplane. The algorithm calculates the maximum-margin hyperplane that maximizes the distance between different classes. On the other hand, mRMR algorithm will rank in first position the gene that contains the maximum relevance information but minimum redundancy information with respect to the rest of the genes, and so forth, it will proceed with the whole ranking.

## 3  Results

According to the methodology described in the previous section, results obtained in the development of this research are next presented. After obtaining the 46 expressed genes from microarray (as can be seen at the Table 2), two different validations for checking the correct integration of the data have been proposed: on the one hand, validation with exclusively microarray data and, on the other hand, Validation from the integration of RNASeq with the above.

**Table 2.** 46 expressed genes calculated from 108 microarray samples

| Gene name | logFC | P. Value | Gene name | logFC | P. Value |
|---|---|---|---|---|---|
| KRT6A | −5,6245 | 1,2425E−36 | GNA15 | −3,0134 | 6,4947E−39 |
| S100A2 | −5,0766 | 2,8493E−35 | KRT6B | −2,9109 | 1,2327E−27 |
| KRT14 | −4,5092 | 2,3764E−23 | BNC1 | −2,8127 | 5,2979E−41 |
| IL20RB | −4,1428 | 8,5258E−35 | FBP1 | 2,7717 | 2,8626E−12 |
| NNMT | −3,8639 | 2,1313E−18 | RAB38 | −2,7692 | 3,1577E−30 |
| KRT19 | 3,5780 | 1,1504E−14 | TSPYL5 | 2,7625 | 1,4045E−13 |
| SFRP1 | −3,5533 | 1,3229E−41 | NMU | −2,6986 | 1,0537E−24 |
| SERPINB5 | −3,5242 | 1,6435E−38 | EVA1C | −2,6693 | 1,5478E−31 |
| ADRB2 | −3,4898 | 2,6899E−36 | GPR87 | −2,6684 | 2,3311E−31 |
| DSG3 | −3,4567 | 9,8795E−37 | CPVL | −2,6536 | 4,3830E−22 |
| CLCA2 | −3,4201 | 1,8646E−32 | CBS | 2,6525 | 1,5263E−31 |
| SLPI | −3,4063 | 3,3555E−25 | CASP1 | −2,6212 | 8,8105E−22 |
| C3 | −3,3636 | 5,9826E−29 | FAM83A | −2,5882 | 1,5410E−24 |
| HENMT1 | 3,2120 | 1,3924E−26 | SDPR | −2,5647 | 5,0039E−27 |
| CXCL1 | −3,1760 | 5,0443E−25 | MSLN | −2,5447 | 9,6283E−26 |
| COL17A1 | −3,1650 | 8,3370E−34 | WBP5 | −2,5391 | 3,5403E−15 |
| PRKCDBP | −3,1516 | 2,2807E−21 | DFNA5 | −2,5290 | 3,3008E−20 |
| UCP2 | 3,1334 | 4,8393E−18 | IRX4 | −2,4620 | 5,6860E−24 |
| EFHD1 | 3,1332 | 2,4777E−30 | BEX2 | 2,4220 | 1,0773E−18 |
| RGS2 | −3,0960 | 1,3639E−27 | BIRC3 | −2,4075 | 9,3570E−30 |
| IFI16 | −3,0862 | 4,3263E−17 | SLC26A2 | −2,3175 | 4,0183E−33 |
| ZBTB16 | −3,0713 | 3,8496E−33 | C3orf14 | 2,2990 | 9,8396E−22 |
| DNER | −3,0677 | 7,3521E−25 | ACOT4 | 2,2539 | 3,0642E−19 |

For the experiment, 108 microarray samples have been used to create the classifier for breast cancer, with the aim of achieving early diagnosis when unlabeled samples are presented. This classification tool makes use of SVM and the mRMR feature selection algorithm. This algorithm selects the most relevant genes to perform the classification. Leave One Out Cross Validation (LOO) [15] was used to assess the classification process.

Results obtained are shown in the Fig. 4. This figure includes two simulations. Red line represents the accuracy using only microarray data for training and validation through the genes ranking calculated previously by feature selection algorithm. On the other hand, the blue line shows the same results but adding the 6 RNASeq samples to the dataset. As can be seen, RNASeq samples bring to the classifier more accuracy thanks to the existing differentiation in level of quantification between cancer samples and healthy samples. With microarray data the classifier reaches 96% of accuracy with only 6 genes of the ranking. However, when RNASeq data is added, accuracy rises to 99% with the same 6 genes, and to 100% of accuracy when 41 genes from RNASeq are considered.

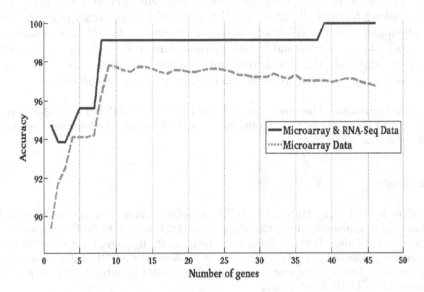

**Fig. 4.** Ranking of variables applied to breast cancer classification using Microarray and RNASeq data integrated

## 4    Conclusions

An heterogeneous data integration from different technologies (microarray and RNASeq) that quantify the quantity of RNA in human biological samples is carried out in this work.

An appropriate merging has been achieved from the combination of both biological samples. Firstly, a mapping has been necessary to convert RNASeq raw data (in counts) towards their equivalent gene expression values. Later, an aggregation was possible because gene expression values from different technologies are comparable.

This work shows how the integration of heterogeneous data from different microarrays platforms and even from different sequencing technologies as RNASeq can be achieved in order to improve statistical significance as well as to obtain results that are independent of the applied sequencing technology. Specifically, this work has satisfactorily integrated data from series of Affymetrix and Illumina microarrays technologies and RNASeq from Illumina HiSeq. Additionally, 46 possible breast cancer biomarkers genes have been found from microarray gene expression. Extensively, these genes were checked using RNASeq samples in the classification step. This fact showed that these genes were independent of the sequencing technologies used.

The most significant novelty of this work is the development of a high-accuracy classification method combining microarray and RNASeq data, thus mixing different sequencing technologies with different operation pipelines. This classifier allows to take advantage of all the available microarray existing data, more abundant than the RNASeq, moreover integrating both types of data. As future work, the authors intend to add a larger number of RNASeq samples with the aim of using only the RNASeq samples for the validation step, using a set of expressed genes selected from the microarray gene expression data.

**Acknowledgements.** This work was supported by Project TIN2015-71873-R (Spanish Ministry of Economy and Competitiveness -MINECO- and the European Regional Development Fund -ERDF).

# References

1. Anders, S., Pyl, P.T., Huber, W.: HTSeq-a python framework to work with high-throughput sequencing data. Bioinformatics **31**(2), 166–169 (2014). btu638
2. Barrett, T., Troup, D.B., Wilhite, S.E., Ledoux, P., Rudnev, D., Evangelista, C., Kim, I.F., Soboleva, A., Tomashevsky, M., Edgar, R.: NCBI GEO: mining tens of millions of expression profiles—database and tools update. Nucleic Acids Res. **35**(suppl 1), D760–D765 (2007)
3. Ding, C., Peng, H.: Minimum redundancy feature selection from microarray gene expression data. In: Computer System Bioinformatics, CSB 2003, Proceedings of 2003 IEEE Bioinformatics (2003)
4. Gohlmann, H., Talloen, W.: Gene Expression Studies Using Affymetrix Microarrays. CRC Press, Boca Raton (2009)
5. Heider, A., Alt, R.: virtualArray: a R/bioconductor package to merge raw data from different microarray platforms. BMC Bioinform. **14**(1), 75 (2013)
6. Illumina: Illumina genes expression arrays (2009). http://www.illumina.com/techniques/microarrays/gene-expression-arrays.html
7. Kim, D., Pertea, G., Trapnell, C., Pimentel, H., Kelley, R., Salzberg, S.L.: TopHat2: accurate alignment of transcriptomes in the presence of insertions, deletions and gene fusions. Genome Biol. **14**(4), R36 (2013)
8. Langmead, B., Salzberg, S.L.: Fast gapped-read alignment with Bowtie 2. Nat. Methods **9**(4), 357–359 (2012)
9. Li, H., Handsaker, B., Wysoker, A., Fennell, T., Ruan, J., Homer, N., Marth, G., Abecasis, G., Durbin, R., et al.: The sequence alignment/map format and samtools. Bioinformatics **25**(16), 2078–2079 (2009)
10. Noble, W.S.: What is a support vector machine? Nat. Biotechnol. **24**, 1565–1567 (2006)
11. Nookaew, I., Papini, M., Pornputtpong, N., Scalcinati, G., Fagerberg, L., Uhlén, M., Nielsen, J.: A comprehensive comparison of RNA-seq-based transcriptome analysis from reads to differential gene expression and cross-comparison with microarrays: a case study in saccharomyces cerevisiae. Nucleic Acids Res. **40**(20), 10084–10097 (2012). gks804

12. OMS: Women's Health (2013). http://www.who.int/mediacentre/factsheets/fs334/en/
13. Peirson, S.N., Butler, J.N.: Quantitative polymerase chain reaction. In: Rosato, E. (ed.) Circadian Rhythms: Methods and Protocols, pp. 349–362. Springer, Heidelberg (2007)
14. Ritchie, M.E., Phipson, B., Wu, D., Hu, Y., Law, C.W., Shi, W., Smyth, G.K.: limma powers differential expression analyses for RNA-sequencing and microarray studies. Nucleic Acids Res. **43**(7), e47 (2015). gkv007
15. Shao, J.: Linear model selection by cross-validation. J. Am. Stat. Assoc. **88**(422), 486–494 (1993)
16. Tarazona, S., García, F., Ferrer, A., Dopazo, J., Conesa, A.: NOIseq: a RNA-seq differential expression method robust for sequencing depth biases. EMBnet. J. **17**(B), 18 (2012)
17. Wang, Z., Gerstein, M., Snyder, M.: RNA-seq: a revolutionary tool for transcriptomics. Nat. Rev. Genet. **10**(1), 57–63 (2009)

# Deep Learning Using EEG Data in Time and Frequency Domains for Sleep Stage Classification

Martí Manzano, Alberto Guillén, Ignacio Rojas, and Luis Javier Herrera[✉]

Department of Computer Architecture and Computer Technology,
University of Granada, Granada, Spain
jherrera@ugr.es

**Abstract.** Polysomnography analysis for sleeping disorders is a discipline that is showing interest in the development of reliable classifiers to determine the sleep stage. The most common methods shown in the literature bet for classical learning techniques and statistics that are applied to a reduced number of features in order to tackle the computational load. Nowadays, the application of deep learning to the sleep stage classification problem seems very interesting and novel, therefore, this paper presents a first approximation using a single channel and information from the current epoch to perform the classification. The complete Physionet database has been used in the experiments. Deep learning has been applied to the time and frequency domains from the EEG signal obtaining a good performance and promising further work.

**Keywords:** Deep learning · Sleep stage classification · Time and frequency domains

## 1 Introduction

Among the various disciplines where deep learning is gaining popularity, Biomedical Engineering is one of the most promising due to the large amount of data available from the different devices recording biosignals, historic records, genetic profiles, etc. Several of the consequences of successful applications are better treatments, alert automatization, preventive monitoring, just to mention a few.

In the field of neuroscience and, more concretely, in the analysis of the sleep process, the number of sensors and signals is large as it involves the brain itself (by Electroencephalography (EEG)) and the rest of the body (Electrooculography (EOG) for the eye and other muscles). When the information is combined the term Polysomnography analysis is used. The classification of sleep stages is useful to study Alzheimer, sleep disorders, epilepsy.

Rechtschaffen and Kales where the ones who, in 1968, defined some criteria to classify the sleep stages that were defined as: wakefulness, REM stage (Rapid Eye Movement) and the NREM stage (no Rapid Eye Movements stage), which is composed of four differentiated stages (NREM1, NREM2, NREM3 and NREM4)

I. Rojas et al. (Eds.): IWANN 2017, Part I, LNCS 10305, pp. 132–141, 2017.
DOI: 10.1007/978-3-319-59153-7_12

[1]. Not too far away in time, the NREM stages were reduced to three taking the name of Slow Wave Sleep (SWS) [2].

The process to obtain the hypnogram that contains all the sleep stages from a polysomnogram requires the division of the original signals in epochs that are periods of 30 s. Each epoch is mapped into a stage (according to the recommendations in [2]. During the epoch, a change in the sleep stage might occur, in these cases, the experts must determine which is the most relevant and assign it, not existing an standard criterion.

Previous approaches of applying machine learning to the task of classifying sleep stages perform feature extraction: in [3] Spectral Relative Power coefficients (SRP) are extracted from the Fast Fourier frequency transformation of the EEG signal, from the spectrum previously divided in the relative bands associated to brain activity during sleep (*Alpha, Theta, Delta, Sigma, Beta*). Once the variables are obtained, a single hidden-layer neural network is designed to perform classification. Other works include combination of different features extraction systems to enhance the classification accuracy [4].

In this paper, all the problem is addressed directly with deep learning so the first phase of feature extraction is performed automatically in the network. One of the benefits of doing so is that all the redundant information can be squeezed improving the accuracy of the classifier.

Previous work was made in [5], in which a set of experiments with a dataset of 25 patients using Deep Learning techniques were carried out. Hand-extracted features are compared with Deep Belief Network extracted features from EEG, EOG and EMG. The deep learning approach attains a 67.4% of accuracy in patient cross-validation (CV), needing a data balance process. However this work did not follow the R&K rules as epochs were 1 s long and transition epochs were eliminated.Another approach apply Convolutional Neural Networks on the time domain together with data from previous and following epochs [6] obtaining an overall accuracy of 74%. However, they use cross-validation on only 20 patients. In a different work [7], a group of 10 patients is used to train a deep belief network to extract features from the polysomnogram signals (2 EEG channels, EOG and EMG) to attain a 91% of patient CV accuracy.

The present work was motivated by the aim of using a single EEG channel as source of information in order to allow the use of simpler devices, and the consideration of complete databases to attain a more realistic performance estimation on unseen data. Two different deep learning approaches were compared. Patient cross-validation was utilized as performance measure. The rest of the paper is organized as follows: Sect. 2 introduces the deep learning architecture and the methodology used. Section 3 presents the results obtained in the Physionet database [8], in Sect. 4 conclusions and further work are drawn.

## 2 Materials and Methods

For the sake of completeness, all patients from the Physionet database [5] and the updated R&K rules have been used in order to optimize a Deep Learning (with a Convolutional Network architecture) classification model and using only information from the current epoch in each classification. Time and frequency domains will be evaluated using deep learning techniques; temporal connectivity in the time domain will be operated using a Convolutional Network architecture.

## 2.1 Deep Learning

Deep Learning can be defined as a set of algorithms and statistical models which form part of Statistical Learning and Artificial Intelligence. These algorithms are based on deep neural networks (classical neural networks with many layers and neurons, see Fig. 1) and usually require larger databases to be trained than traditional neural networks.

Thanks to the higher number of hidden layers and neurons, deep learning allows the designers to obtain transformations and relationships that would be hidden to the expert's eye at first and not requiring a preprocessing stage (like feature selection).

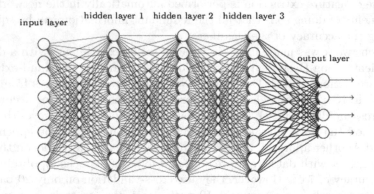

**Fig. 1.** Example of fully-connected deep neural network with four output neurons.

As the architecture becomes bigger, the possibilities regarding its structure increase as well. Apart from fully-connected networks, the two other most well-known deep learning architecture are convolutional neural networks, which perform convolution operations over the inputs, and recurrent neural networks, which present neuron connections that can form loops. For further reading about deep learning, [9] presents a very complete view of the subject.

## 2.2 Convolutional Neural Networks

Convolutional Neural Networks are composed by a set of connected convolutional layers; the convolution operation can be interpreted as a filter/kernel which is applied over a region of the input space. The size of the convolutions is specified during the design phase, but the content of the filter (what is being looked for) is learned during the training from the regional characteristics that the input data presents. Opposite to fully-connected feed forward neural networks, convolutional networks share the synaptic weights. Therefore this network is appropriate to data with local connectivity such as images, sound or time series, as their spatial structure can provide pattern and abstraction extraction of different sizes and levels.

The three main components of a convolutional neural network are [9] (Fig. 2):

- Convolutional layer: It represents the main difference with multilayer perceptron and applies a filter over the inputs space: in a one dimensional problem, along that one; in a 2-D image, along the width and height. Filters have lower dimension than the input space size, in order to detect the presence or absence of patterns. Filters can be applied several times in a deep set of layers.
- Pooling layer: The pooling layers intend to reduce the spacial dimensions of the convolutional layer outputs. They replace the output of the convolutional layer in a specific point by a combination of its neighborhood. This way, although loss of information can occur, it is generally convenient to reduce over-fitting and diminishing the dimensional complexity of the data for the next network layers.
- Fully-connected layer: After the successive operation over a set of convolutional and pooling layers, and in order to extract the relevant information from the high-dimensional data, a final higher level processing is carried out through one or several fully-connected layers. These layers will be directly connected to the output layer to perform the classification itself.

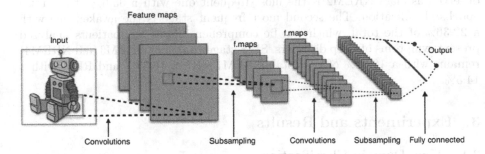

**Fig. 2.** Example of convolutional network [10] used in image classification. Spacial information on different sub-regions of the image are used to perform the classification.

## 2.3 Evaluation of the Models

In order to evaluate the performance obtained by each model, the well-known cross-validation [11] has been used. Since the number of patients is small, a

Leave One Out (LOO) approach was feasible to be applied, giving the chance to be more accurate. This way, any expectable patient variability in EEG signals recorded or in their manual classification, is reflected in the expected performance of the model, making it a more realistic performance measure. This performance measure has been used in most of the latest revised works [4,6,7].

### 2.4   Dataset

Dataset used for the experiments was extracted from the public repository *PhysioNet* [8]. A total of 25 overnight polysomnogram (21 male and 4 female, with ages between 28 and 68 -mean age of 50 years old) from the *St. Vicent's University Hospital*, from whom it was suspected that they suffered from any sleep disturbance, were considered. Recorded polysomnogram signals included: two EEG channels (C3-A2 and C4-A1) at 128 Hz, two EOG -two eyes- at 64 Hz, EMG also at 64 Hz, and other signals related to patient movement, posture and breathing.

Data was provided under EDF format and single EEG channel (C3-A2) was used in this work, which agrees with recommendations in the manual of sleep stage classification of Rechtschaffen and Kales. Before epoch segmentation, preprocessing using a Notch filter at 50 Hz and a high pass filter at 0.3 Hz [5,12] was performed. Also signal was down-sampled to 64 Hz to reduce the input data size and the temporal correlation among the variables. Each epoch's corresponding sleep stage was tagged by a single expert.

Average sleep data duration from the patients is 6.9 h. In total, 20075 epochs are available from the 25 patients, with a total operation time of 173.2 h. The histogram of the sleep stages is given in Fig. 3. Certain data imbalance can be observed as class NREM2 is the most frequent one with a 33.6% of the total epochs classification. The second most frequent stage is the awaken one with a 22.36% of the total, which can be comprehensible as the patients analyzed present symptoms of sleep disorders. SWS stage (union of NREM3 and NREM4) remains with a 12.81% of the total, NREM1 with a 16.37% and REM with a 14.5%.

## 3   Experiments and Results

### 3.1   Time Domain Classification

Local connectivity among the variables in the time domain was the objective aimed to take advantage of by using a Convolutional Neural Network. This type of architecture has been previously used with success on different types of signals [13–16]. Thus taking these as reference, temporal structures from the data were aimed to be extracted automatically using this deep learning technique. Hyperparameters of the network were obtained by trial and error. The filter sizes were established in order to allow the extraction of time patterns from lower frequency bands (Delta waves from 0.5–3 Hz) until higher frequency bands

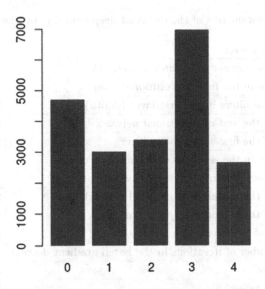

**Fig. 3.** Histogram over the 5 sleep stages in the complete set of data from the 25 patients. 0 awaken, 1 REM, 2 NREM1, 3 NREM2, 4 SWS

(*sleep spindle* or sigma patterns until 16 Hz). Two convolutional layers with pooling were used in order to reduce the input space maps; next a single fully-connected layer to the output was used. An initial subdivision of training and test from a subset of the data was used for hyperparameter obtaining. Later, patient cross-validation was performed for their final tunning.

The resulting convolutional neural network architecture designed can be seen in Fig. 4. Most relevant hyper-parameters of the model are detailed in Table 1.

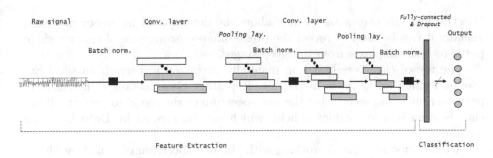

**Fig. 4.** Neural network architecture operating in the time domain

This convolutional network when validated using patient-cross-validation, attained a 68.6% of precision and a 54.6% of mean F-measure over the five sleep stages.

**Table 1.** Hyperparameters of the designed deep neural networks in the work.

| Hyperparameter | Value |
|---|---|
| *EEG time domain convolutional network* | |
| Filter size in the first convolutional layer | 8 |
| Number of filters in the first layer (depth) | 9 |
| *Stride* of the first convolutional network | 2 |
| *Stride* of the first *max-pooling* layer | 2 |
| Filter size in the second convolutional layer | 16 |
| Number of filters in the second layer (depth) | 18 |
| *Stride* of the second convolutional network | 2 |
| *Stride* of the second *max-pooling* network | 2 |
| Number of neurons in the *fully-connected* layer | 2500 |
| Max. number of iterations in the batch gradient descent | 2000 |
| Batch size | 512 |
| Learning rate | 0.01 |
| *Dropout* regularization *keep-rate* | 0.5 |
| *EEG frequency domain fully-connected network* | |
| Number of neurons in the *fully-connected* layer | 1600 |
| Max. number of iterations in the batch gradient descent | 2000 |
| Batch size | 512 |
| Learning rate | 0.01 |
| *Dropout* regularization *keep-rate* | 0.5 |

## 3.2 Frequency Domain Classification

Spectral analysis over a signal may allow the detection of frequency patterns; visually it can be used to reveal the predominance or absence of characteristic patterns and complexes from the sleep stages.

The use of classifiers based on manually extracted features from the EEG was performed for instance in [17] for sleep stage classification with relatively positive results. However using the raw spectrum of the signal to perform direct classification is a less explored field, which can be tackled by Deep Learning techniques.

A direct neural network working with the raw spectrum obtained by Short-Time Fourier Transform STFT (0.5 Hz to 32 Hz, corresponding to the variation of physiological waves [17]) on 30 s epoch was designed. In order to reduce the width of the limit discontinuities (leakage), a Hanning window was used together with the FFT (which according to our tests provided the best results).

According to [18], spectral phenomena happening in different frequency regions are different, thus weight sharing in the network only have sense on a limited bandwidth. A convolutional network wouldn't therefore be that effective in this problem. So, a fully-connected network with a hidden layer of 1600 neurons using Dropout as regularization to reduce the over-fitting was designed and trained. Most relevant hyper-parameters for this network can be seen in Table 1 (Fig. 5).

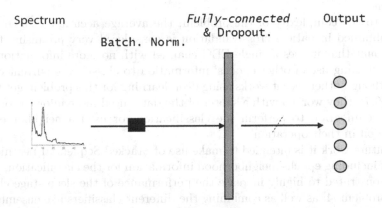

**Fig. 5.** Neural network architecture operating in the frequency domain

Patient leave-one-out cross-validation led to a 68.9% of accuracy and 57.5% of mean F1 measure. These results are similar to that of the previous convolutional network working in the time domain (Table 2).

**Table 2.** Summary of the results of the two proposed deep networks

| Model | Accuracy | Std Acc. | F1-measure |
|---|---|---|---|
| Time domain | 68.6% | 7.53 | 54% |
| Frequency domain | 68.9% | 7.52 | 57.5% |

In relation to the patient-cross-validation results and the confusion matrix associated, two interesting facts are observed, that also were seen in [5]. In most patients, NREM1 stage was the worst classified, which can be due to the little physiological differences among this one and NREM2 stage [19] as well as the special consideration of the manual classification of the two stages. Second, there is a high variability in the classification accuracies per patient, being the two worst ones 51.3% and 65.3%, and the best ones 85.3% and 83.6%. These differences are expectable due to the different interferences in the polysomnogram recordings, as well as the inter individual physiological and brain operation differences.

# 4    Conclusions and Further Work

This paper has presented the adaptation of deep learning to the problem of sleep stage classification in such a way that the problem has been decomposed into two sub-domains and the dataset tackled is complete in comparison with previous approaches.

Due to the complexity of the problem, the average accuracy of 68.6% and 68.9% obtained in patient-cross-validation is considered very promising taking into account that it uses a single EEG channel with no more information, and it is not making use of other epochs' information to classify the current epoch. Thus although other recent works using deep learning for this problem get better results [6,7], they worked with a subset of the data, used more information from the polysomnogram to perform the classification, or used a neighbor epochs information in their operation.

As future work it is intended to make use of Stacked Sequential Learning, as a way of including epochs' neighborhood information for the classification, which has demonstrated to highly increase the performance of the sleep stage classification problem [4] as well as combining the different classifiers like ensembles or integrated in a more complex deep learning architecture.

**Acknowledgements.** This work was supported by Project TIN2015-71873-R (Spanish Ministry of Economy and Competitiveness -MINECO- and the European Regional Development Fund -ERDF).

# References

1. Rechtschaffen, A., Kales, A. (eds.): A Manual of Standardized Terminology, Techniques and Scoring System for Sleep Stages of Human Subjects, 58 p. Public Health Service, Washington, D.C. (1968)
2. American Academy of Sleep Medicine. The AASM Manual for the Scoring of Sleep and Associated Events: Rules, Terminology and Technical Specification (2007)
3. Kerkeni, N., Alexandre, F., Bedoui, M.H., Bougrain, L., Dogui, M.: Automatic classification of sleep stages on a EEG signal by artificial neural networks. In: 5th WSEAS International Conference on SIGNAL, SPEECH and IMAGE PROCESSING - WSEAS SSIP 2005, Corfu, Island/Greece, August 2005
4. Herrera, L.J., Fernandes, C.M., Mora, A.M., Migotina, D., Largo, R., Guillén, A., Rosa, A.C.: Combination of heterogeneous EEG feature extraction methods and stacked sequential learning for sleep stage classification. Int. J. Neural Syst. **23**(3), Article no. 1350012 (2013)
5. Längkvist, M., Karlsson, L., Loutfi, A.: Sleep stage classification using unsupervised feature learning. Adv. Artif. Neural Syst. **2012**, Article No. 5 (2012)
6. Tsinalis, O., Matthews, P.M., Guo, Y., Zafeiriou, S.: Automatic sleep stage scoring with single-channel EEG using convolutional neural networks. Biomed. Eng./Biomed. Tech. (2016). arXiv:1610.01683
7. Zhang, J., Yan, W., Bai, J., Chen, F.: Automatic sleep stage classification based on sparse deep belief net and combination of multiple classifiers. Trans. Inst. Measur. Control **38**(4), 435–451 (2015)

8. Goldberger, A.L., Amaral, L.A.N., Glass, L., Hausdorff, J.M., Ivanov, P.C., Mark, R.G., Mietus, J.E., Moody, G.B., Peng, C.-K., Stanley, H.E.: PhysioBank, PhysioToolkit, and PhysioNet: components of a new research resource for complex physiologic signals. Circulation **101**(23), e215–e220 (2000). Circulation Electronic Pages: http://circ.ahajournals.org/content/101/23/e215.full PMID:1085218; doi:10.1161/01.CIR.101.23.e215

9. Goodfellow, I., Bengio, Y., Courville, A.: Deep Learning. MIT Press, Cambridge (2016)

10. Aphex34. Typical CNN (2015)

11. Kohavi, R.: A study of cross-validation and bootstrap for accuracy estimation and model selection. In: Proceedings of the 14th International Joint Conference on Artificial Intelligence - Volume 2 (IJCAI 1995), vol. 2, pp. 1137–1143. Morgan Kaufmann Publishers Inc., San Francisco (1995)

12. Reddy, A.G., Narava, S.: Artifact removal from EEG signals. Int. J. Comput. Appl. **77**(13), 17–19 (2013)

13. Palaz, D., Magimai-Doss, M., Collobert, R.: Analysis of CNN-based speech recognition system using raw speech as input. In: Proceedings of Interspeech, number Idiap-RR-23-2015, pp. 11–15. ISCA, September 2015

14. Yang, J.B., Nguyen, M.N., San, P.P., Li, X.L., Krishnaswamy, S.: Deep convolutional neural networks on multichannel time series for human activity recognition. In: Proceedings of the 24th International Conference on Artificial Intelligence, IJCAI 2015, pp. 3995–4001. AAAI Press (2015)

15. Abdel-Hamid, O., Mohamed, A., Jiang, H., Penn, G.: Applying convolutional neural networks concepts to hybrid NN-HMM model for speech recognition. In: 2012 IEEE International Conference on Acoustics, Speech and Signal Processing (ICASSP), pp. 4277–4280, March 2012

16. Tóth, L.: Combining time- and frequency-domain convolution in convolutional neural network-based phone recognition. In: 2014 IEEE International Conference on Acoustics, Speech and Signal Processing (ICASSP), pp. 190–194, May 2014

17. Kerkeni, N., Alexandre, F., Bedoui, M.H., Bougrain, L., Dogui, M.: Automatic classification of sleep stages on a EEG signal by artificial neural networks. In: Proceedings of the 5th WSEAS International Conference on Signal, Speech and Image Processing, SSIP 2005, pp. 128–131, Stevens Point, Wisconsin, USA. World Scientific and Engineering Academy and Society (WSEAS) (2005)

18. Yu, D., Abdel-Hamid, O., Deng, L.: Exploring convolutional neural network structures and optimization techniques for speech recognition. In: Interspeech 2013. ISCA, August 2013

19. Bresler, M., Sheffy, K., Pillar, G., Preiszler, M., Herscovici, S.: Differentiating between light and deep sleep stages using an ambulatory device based on peripheral arterial tonometry. Physiol. Meas. **29**(5), 571 (2008)

# Human Computer Interaction

# Application of an Eye Tracker Over Facility Layout Problem to Minimize User Fatigue

Juan García-Saravia[1]([⊠]), Lorenzo Salas-Morera[1],
Laura García-Hernández[1], and Adoración Antolí Cabrera[2]

[1] Area of Project Engineering, University of Córdoba, Córdoba, Spain
{i62gasaj,lsalas,irlgahel}@uco.es
[2] Department of Psychology, University of Córdoba, Córdoba, Spain
aantoli@uco.es

**Abstract.** With interactive evolutionary computation it is possible to introduce the subjective preferences of the decision maker within the general algorithm evolution criteria. The problem that generates this is user fatigue, since it has to evaluate a considerable number of plants designs in each generation. To avoid user fatigue it is proposed to substitute the direct evaluation through the mouse by means of a numerical scale by an eye tracking system in which the system "captures" the evaluation that the user assigns to the plants through the gaze behavior. This article presents a first approximation to this solution. The results obtained in the experiments are promising and a clear relationship between the parameters that define the gaze behavior of the user with the score assigned to the designs can be seen.

**Keywords:** Unequal area facility layout problem · Eye tracking · Interactive genetic algorithm · User fatigue · Gaze behavior

## 1 Introduction

Facility Layout Design (FLD) determines the placement of facilities (sometimes called departments) in a manufacturing plant. The goal is to achieve the most effective arrangement in order to meet one or more objectives and optimize plant efficiency.

Plant layout design is important to achieve production efficiency, since it directly influences manufacturing costs, leading times, work in process and productivity [1]. A good distribution of facilities contributes to the overall efficiency of operations and could reduce between 20% and 50% of the total operating cost [2].

There are many types of facility layout problem (FLP) [3] but all of them depend on the specific features of manufacturing systems (e.g., production variety and volume, material handling system chosen, different possible flows allowed for parts, number of floors, facility shapes and the pick-up and drop-off locations) and different techniques to solve them (e.g., exact approaches, approximated approaches, heuristics, meta-heuristics [4] as for example.

One of the most studied approaches to the FLP is unequal area facility layout problem (UA-FLP) [5]. UA-FLP considers a rectangular plant layout that is made up by unequal rectangular facilities that have to be placed effectively over the plant layout.

© Springer International Publishing AG 2017
I. Rojas et al. (Eds.): IWANN 2017, Part I, LNCS 10305, pp. 145–156, 2017.
DOI: 10.1007/978-3-319-59153-7_13

Initially, different methods were used to approach UA-FLP, only considering quantitative criteria (e.g. material handling cost, closeness or distance relationships, adjacency requirements and/or aspect ratio). These approaches, however, may not adequately represent all of the relevant non quantitative information that affect a human expert involved in design (e.g. engineers, architects, and designers in general) [6]. The qualitative features are complicated to be considered with a classical heuristic or meta-heuristic optimization [7, 8]. Examples of qualitative features can be: facility location preferences, distribution of the remaining spaces, relative placement preferences, or any other subjective preference that can be considered as important by the decision maker (DM). A correct effective facility layout evaluation procedure needs the consideration of qualitative and quantitative criteria [9].

An interactive genetic algorithm (IGA) was proposed to consider qualitative and/or subjective criteria through the interaction of the algorithm with a DM.IGA considers the knowledge of the DM in the search process, adjusting it to their preferences in each generation of the algorithm [8], in this way, the interactive evolutionary computation (IEC) gives the advantage of considering qualitative criteria against the classical heuristic or a meta-heuristic optimization. In addition, these qualitative features can be subjective, not known at the beginning or changed during the process. Therefore, they can not be formulated as an objective function of a classical optimization problem. IEC can greatly contribute to improve optimized design by involving users in searching for a satisfactory solution [10]. In this IEC the fitness function is replaced by a human's user evaluation [11]. Thus, intuition and domain knowledge can be involved in the identification of good designs [12].

Including the knowledge of an expert in the algorithm is essential to be able to consider qualitative features. Also, it provides other advantages: finding a solution that satisfies the DM, it does not have to be an optimal solution (it is not safe to find it) [13]; select the best trade-off solution when a conflict among objectives or constraints exists [14]; help the algorithm in the process of searching, to consider the user's preferences [15–17]; do not have to specify all preferences previously; DM can learn about the progress of their own choices or preferences [14]; stimulate the user creativity [18]; obtain original, innovated and practicable solutions.

In the last years, new approaches have been implemented to improve the operation and performance of the IGA. The goal is normally to reach better solutions, in less time and reducing the fatigue of the DM, and considering user preferences. For example, using a neural network to discharge the DM evaluation task [19], includingniching techniques into the approach in order to preserve population diversity which avoids presenting similar solutions to the designer in the same iteration of the algorithm [20].

According with Takagi [11], one of the major problems of the IEC is user fatigue. For this reason, one of the priorities of IEC application should be to reduce user fatigue. Some factors that directly affect user fatigue are: number of generations of the algorithm and number of representative individuals shown to the user. One problem of reducing user fatigue is that the IEC converges too fast because population diversity is lost too soon. In this way, the algorithm does not explore enough solutions and will be unable to find a satisfactory solution. There has to be a balance between user fatigue and convergence of the solution.

There are different methods for trying to reduce user fatigue: visualizing individuals in a multi-dimensional searching space in a 2-D space [21]; using genetic programming to learn subjective fitness functions from human subjects, using historical data from an interactive evolutionary system for producing pleasing drum patterns [22]; augmenting user evaluations with a synthetic fitness function combining partial ordering concepts, notion of non-domination from multi objective optimization, and support vector machines to synthesize a fitness model based on user evaluation [23, 24] and using a system of interactive differential Evolution (IDE) [25] among others.

The methods commented previously are focused on improving the algorithm performance, but they do not guarantee that the user fatigue could always be reduced. Searching different techniques to reduce DM's fatigue, reducing the active participation of DM could be interesting [26]. In this way, some works analyzed the use of an eye tracking systems to capture the DMpreferences in other fields [27]. Eye trackers measure gaze behavior during task execution, visualize what areas on a screen are inspected, and thus provide clues on what information was included in the decision making process. They are used, for example, in educational or cognitive psychology to understand expert performance [28]. Eye tracking provides a potential for new insight into the reasoning process [29].

Traditionally, in each one of the generations of the IEC the user is asked to evaluate each solution by assigning a rating or selecting with the mouse. Some studies propose a framework that uses in real time gaze information to predict which parts of a screen are more significant for a user [30]. Eye movement based analysis can improve traditional performance, protocol, and walk through evaluations of computer interfaces. Overall, data obtained from eye movements can significantly enhance the observation of users' strategies while using computer interfaces, which can subsequently improve the precision of computer interface evaluations [31]. User fatigue can be reduced following users eyes movement and without sacrificing quality of fitness evaluation [32]. There is a close relationship between the behavior gaze and the final choice or decision on the part of the user. [33].

According to several authors some of the most important parameters to consider in the gazing behavior are fixations and visit. Michalski and Grobelny [34, 35] analyzed the main parameters to be taken into account to be able to obtain useful conclusions from user's eyes movement; Gere et al. [36] in this experiment a close relation was found between gazing behavior and choice by the applied models and where the workflow is well-suitable to similar practical eye-tracking problems and [37] in his conclusion explain that exist significant positive correlations between two eye tracking parameters (fixation count and visit duration) and the choice rate.

The main goal is to find a technique that allows to reduce the fatigue of an expert in the evaluation process of an IGA using eye-tracking, and maintaining the evaluation effectiveness previously obtained. In this paper a new approach combining an IGA with the use of an eye tracker is proposed.

The rest of the paper is organized as follows. In Sect. 2, the materials used and the approach and development of the experiment are presented. In Sect. 3, the results of the experiment are shown and commented. Finally, conclusions and suggested future work are given in Sect. 4.

## 2  Method

### 2.1  Materials and Experiment Design

An eye tracker model Tobii X2-30 and software Tobii Studio v3.3.0.567. have been used for the experiments. Data generated in Tobii Studio were exported to Microsoft Excel 2013 for analysis.

In order to evaluate the performance of eye tracking to evaluate the designs presented to the DM, a previous evaluation, scoring each plant from 1 to 5, has been used. The objective of the experiment is to analyze whether is possible to relate the evaluation made by the DM with the parameters of the eye-tracking.

The experiment has been developed using a problem with 12 facilities to be arranged in an area of 35 × 55 [38]. In this case we have defined a single interest group with the following restrictions, a solution that has the plant layout divided into three bays, where facility 'B' touches any side of the layout, and facility 'A' is located in the bottom right corner of the plant layout adjacent to facility 'F'.

The experiment was designed as follows: to begin with, a complete test of the IGA has been carried out as it has been done so far, in the Fig. 1 shows the execution of the

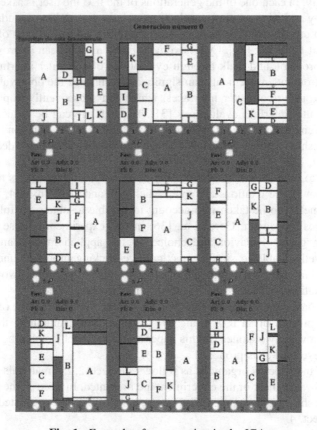

**Fig. 1.**  Example of a generation in the IGA.

IGA. In each of these iterations, screenshots have been made (one for each generation) and the test has been completed with a certain solution.

The next step was to export the images obtained by the IGA to Tobii Studio in order to be evaluated by an expert.

Tobii Studio allows to define areas of interest (AOIs) on the images that are displayed to the user. An AOI is a defined section on an image and could be define more than one AOI on an image. At the end of the experiment Tobii Studio allows to analyze the information of each AOI separately. In this way, an individual analysis of each one of the areas defined on the image can be obtain.

Then, the different areas of interest on each one of the images showed to the expert during the evaluation process have been defined. In each image nine AOIs have been defined corresponding with each one of the plant to be evaluated. This step has been done for each of the images (one for each generation).

Tobii Studio can manage a large number of parameters, metrics and descriptive statistics. It also provides different means to see the results of the experiment: a recording of the experiment, images of the heat maps, duration of the fixations, etc.

## 2.2 Experimental Procedure

At first, expert was informed about the general objective of the study. Then placed in front of the screen with the Tobii located on the lower edge of the screen. Before starting the test Tobii Studio always performs a calibration test. The task of the experts was to observe each of the images presented to him (each corresponding to a generation) and to value them based on the requirements specified at the beginning. The images were presented in the same order in which they were generated by the application by order of generation. During the whole process, subjects' visual activities were registered by the eye tracking system.

## 3 Results and Discussions

In this section an analysis of the results obtained in the experiment was performed. The 9 generations used in the experiment have been analyzed, but the emphasis should be on generations 0, 1, 2, 7 and 8. In the first three it is possible to see how the problem begins to converge towards a concrete solution and the last two show how the algorithm reaches a concrete solution.

The scores indicated by the expert in the execution of the IGA [8] are shown so they can be compared with the data obtained using the Tobii: generation 0 (2,2,2,2,2, 1,1,1,2), generation 1 (2,4,1,2,3,2,2,3,1), generation 2 (1,3,2,2,1,1,3,3,1), generation 7 (4,4,3,4,4,4,4,4,4) and generation 8 (4,4,4,4,4,4,4,4,4).

Analysis has been focused on the data obtained in each of the generations discussed. Each generation includes: the heat map generated by the Tobii Studio, the graphs of the parameters exported from Tobii Studio to Microsoft Excel and the ratings assigned by the expert in the execution of the IGA. To obtain the conclusions the following parameters have been considered as the most relevant: fixation count, total fixation duration, visit count and total visit duration [34, 35]. Now the generations are going to be analyzed and discussed.

Figure 2 shows the results obtained in generation 0. The first generation is totally random so the expert does not know a priori what he is going to see. For this reason many red areas are obtained in the heat map, since the expert has distributed his attention among all the plants. As can be seen in the graphs, there are no significant differences in the values obtained in the different parameters.

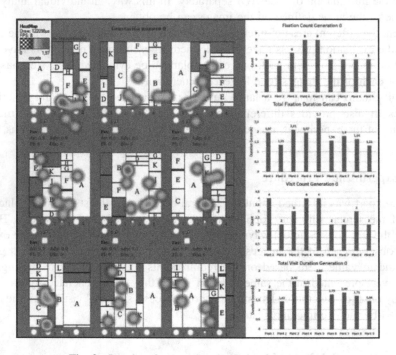

**Fig. 2.** Results of generation 0. (Color figure online)

The data obtained from the IGA experiment showed that the expert assigned a score of 1 to 2 to all the plants, which means that none of the proposed solutions met the requirements and, at the same time, they were all equally interesting. With the eye tracker equivalent results are obtained, no plant has captured the attention of the expert over the others. Moreover, analyzing the images of the plants you can see that none meets the specified requirements.

Figure 3 illustrates the results obtained in generation 1. The heat map shows that the expert focused his attention on plant 2. Graphs show how the parameters of plant 2 stand out over the others. The expert evaluated plant 2 with a higher rating than the others (4), so that the data obtained with the eye tracker and the IGA are consistent. Analyzing the image can be seen that this plant is the one that best suits the requirements. In generation 1 we see that the algorithm converges to a fairly good solution and these are the results obtained. It should be noted the heat zones obtained in the rest of the plants. Green areas are clearer and even blurred, since the expert gives them less

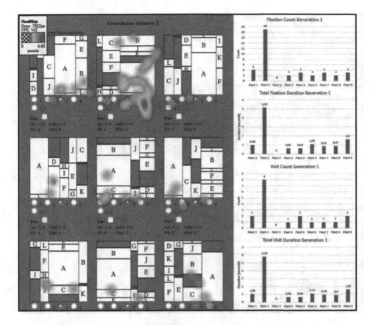

**Fig. 3.** Results of generation 1. (Color figure online)

attention compared to plant 2. The data obtained in the graphs confirm these details, there is a big difference in all parameters between plant 1 and the rest of the plants [30].

Figure 4 describes the results obtained in generation 2. The heat map shows that the expert focused his attention on plant 8. Graphs show how the parameters of plant 8 stand out from each other. The expert evaluated plant 8 with 3 points, as well as plants 2 and 7. In the heat maps and graphs it is shown that the expert does not pay much attention to plants 2 and 7. The plant 2 does not approach a good solution and the plant 7 is worse solution than plant 8. The data obtained with the eye tracker are better than those obtained in the IGA experiment. In this generation, the algorithm converge to a good solution too.

Figure 5 shows the results obtained in generation 7. The images of the plants show that the algorithm converge to a solution (the plants are quite similar due to the loss of the population diversity). The heat map shows that several plants catch the attention of the expert. Graphs show that plants 1 and 3 have less attention of the expert (in this case the plant 3 would not be an acceptable solution), but among the rest of plants there are not significant differences. In the evaluation made by the expert in IGA can see that all plants are rated 3 or 4. This means that there are no plants for which the expert has a particular preference over the others.

Figure 6 illustrates the results obtained in generation 8. Both the heat map and the graphs show that the expert has distributed his attention among most plants. A higher count of fixations in plants 1, 3 and 4 can be seen. Such plants are perfectly valid within the requirements. As in the previous generation the algorithm converge a solution and the population diversity has disappeared. All plants are evaluated with 4 points for the expert in IGA. These scores are perfectly compatible with the results obtained.

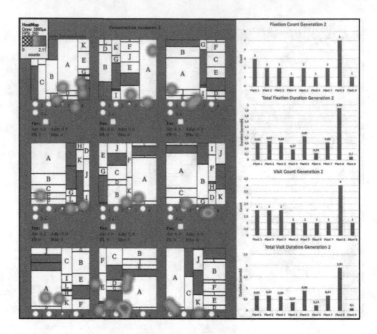

**Fig. 4.** Results of generation 2.

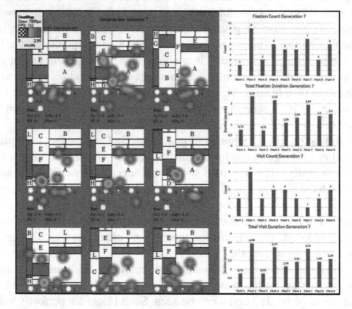

**Fig. 5.** Results of generation 7.

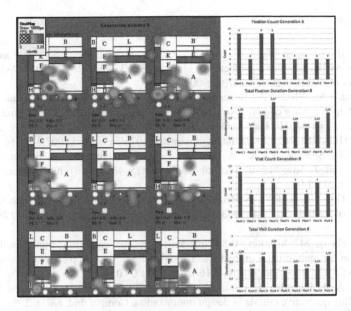

**Fig. 6.** Results of generation 8.

The next link shows the complete execution of the experiment generated with the Tobii Studio. (https://drive.google.com/open?id=0Bw91P0kj7BnUQ2pNLWcySkhWMTQ).

## 4 Conclusions

The objective of this experiment is minimizing the fatigue of the user in each one of the iterations in the plant evaluation process through the eye tracker system. The results obtained are quite positive. In most cases, the scores obtained by the eye tracker system are enough close to the scores assigned by the expert in the process of numerical scale evaluation through the mouse. Even, in some cases the results obtained by the eye tracker have been better. This evaluation method allows a considerable reduction in the evaluation time and fatigue of the user, besides, it reduces the need to have to evaluate plant to plant using a mouse. Only by gaze behavior the same solutions have been obtained that the evaluation by hand of an expert, even in some cases, these solutions have been better.

The fulfillment of the initial approaches of this work open ample possibilities to improve the mentioned evaluation processes, therefore we believe it is convenient to continue working in this area of research. Thus, future lines of work could be other types of problems with different grades of difficulty or different number of facility. Additionally, it could be analyzed by a statistical analyzing.

# References

1. Kouvelis, P., Kurawarwala, A.A., Gutiérrez, G.J.: Algorithms for robust single and multiple period layout planning for manufacturing systems. Eur. J. Oper. Res. **63**(2), 287–303 (1992). https://doi.org/10.1016/0377-2217(92)90032-5
2. Tompkins, J.A., White, J.A., Bozer, Y.A., Tanchoco, J.M.A.: Facilities Planning, 4th edn. (2010). http://eu.wiley.com/WileyCDA/WileyTitle/productCd-EHEP000315.html
3. Drira, A., Pierreval, H., Hajri-Gabouj, S.: Facility layout problems: a survey. Ann. Rev. Control **31**(2), 255–267 (2007). https://doi.org/10.1016/j.arcontrol.2007.04.001
4. Singh, S.P., Sharma, R.R.K.: A review of different approaches to the facility layout problems. Int. J. Adv. Manuf. Technol. **30**(5–6), 425–433 (2006). https://doi.org/10.1007/s00170-005-0087-9
5. Armour, G.C., Buffa, E.S.: A Heuristic Algorithm and Simulation Approach to Relative Location of Facilities, p. 294 (1963). http://pubsonline.informs.org/doi/abs/10.1287/mnsc.9.2.294
6. Babbar-Sebens, M., Minsker, B.S.: Interactive genetic algorithm with mixed initiative interaction for multi-criteria ground water monitoring design. Appl. Soft Comput. J. **12**(1), 182–195 (2012). https://doi.org/10.1016/j.asoc.2011.08.054
7. Brintrup, A.M., Ramsden, J., Tiwari, A.: An interactive genetic algorithm-based framework for handling qualitative criteria in design optimization. Comput. Ind. **58**(3), 279–291 (2007). https://doi.org/10.1016/j.compind.2006.06.004
8. García-Hernández, L., Pierreval, H., Salas-Morera, L., Arauzo-Azofra, A.: Handling qualitative aspects in unequal area facility layout problem: an interactive genetic algorithm. Appl. Soft Comput. J. **13**(4), 1718–1727 (2013). https://doi.org/10.1016/j.asoc.2013.01.003
9. Ertay, T., Ruan, D., Tuzkaya, U.R.: Integrating data envelopment analysis and analytic hierarchy for the facility layout design in manufacturing systems. Inf. Sci. **176**(3), 237–262 (2006). https://doi.org/10.1016/j.ins.2004.12.001
10. Brintrup, A.M., Takagi, H., Ramsden, J.: Evaluation of sequential, multi-objective, and parallel interactive genetic algorithms for multi-objective floor plan optimisation. In: Rothlauf, F., et al. (eds.) EvoWorkshops 2006. LNCS, vol. 3907, pp. 586–598. Springer, Heidelberg (2006). doi:10.1007/11732242_56
11. Takagi, H.: Interactive evolutionary computation: fusion of the capabilities of EC optimization and human evaluation. Proc. IEEE **89**(9), 1275–1296 (2001). https://doi.org/10.1109/5.949485
12. Quiroz, J.C., Louis, S.J., Shankar, A., Dascalu, S.M.: Interactive genetic algorithms for user interface design. In: 2007 IEEE Congress on Evolutionary Computation, pp. 1366–1373 (2007). https://doi.org/10.1109/CEC.2007.4424630
13. Avigad, G., Moshaiov, A.: Interactive evolutionary multiobjective search and optimization of set-based concepts. IEEE Trans. Syst. Man Cybern. Part B (Cybern.), **9**(4), 1013–1027 (2009). https://doi.org/10.1109/TSMCB.2008.2011565
14. Jeong, I.J., Kim, K.J.: An interactive desirability function method to multiresponse optimization. Eur. J. Oper. Res. **195**(2), 412–426 (2009). https://doi.org/10.1016/j.ejor.2008.02.018
15. Quiroz, J.C., Banerjee, A., Louis, S.J.: IGAP: interactive genetic algorithm peer to peer, pp. 1719–1720 (2008). https://doi.org/10.1145/1389095.1389426
16. Luque, M., Miettinen, K., Eskelinen, P., Ruiz, F.: Incorporating preference information in interactive reference point methods for multiobjective optimization. Omega **37**(2), 450–462 (2009). https://doi.org/10.1016/j.omega.2007.06.001

17. Chaudhuri, S., Deb, K.: An interactive evolutionary multi-objective optimization and decision making procedure. Appl. Soft Comput. **10**(2), 496–511 (2010). https://doi.org/10.1016/j.asoc.2009.08.019
18. Sato, T., Hagiwara, M.: IDSET: interactive design system using evolutionary techniques. Comput.-Aided Des. **33**, 367–377 (2001). https://doi.org/10.1016/S0010-4485(00)00128-7
19. García-Hernández, L., Pérez-Ortiz, M., Arauzo-Azofra, A., Salas-Morera, L., Hervás-Martínez, C.: An evolutionary neural system for incorporating expert knowledge into the UA-FLP. Neurocomputing **135**, 69–78 (2014). https://doi.org/10.1016/j.neucom.2013.01.068
20. García-Hernández, L., Palomo-Romero, J.M., Salas-Morera, L., Arauzo-Azofra, A., Pierreval, H.: A novel hybrid evolutionary approach for capturing decision maker knowledge into the unequal area facility layout problem. Expert Syst. Appl. **42**(10), 4697–4708 (2015). https://doi.org/10.1016/j.eswa.2015.01.037
21. Hayashida, N., Takagi, H.: Acceleration of EC convergence with landscape visualization and human intervention. Appl. Soft Comput. **1**(4), 245–256 (2002). https://doi.org/10.1016/S1568-4946(01)00023-0
22. Costelloe, D., Ryan, C.: Genetic programming for subjective fitness function identification. In: Keijzer, M., O'Reilly, U.-M., Lucas, S., Costa, E., Soule, T. (eds.) EuroGP 2004. LNCS, vol. 3003, pp. 259–268. Springer, Heidelberg (2004). doi:10.1007/978-3-540-24650-3_24
23. Llor, X., Sastry, K., Goldberg, D.E.: Combating user fatigue in iGAs : partial ordering, support vector machines, and synthetic fitness. In: Gecco 2005, pp. 1363–1370, February (2005). https://doi.org/10.1145/1068009.1068228
24. Llorà, X., Sastry, K., Alías, F.: Analyzing active interactive genetic algorithms using visual analytics. In: Proceedings of the Annual Conference on Genetic and Evolutionary Computation (GECCO), vol. 8, no. 217, pp. 1417–1418 (2006). https://doi.org/10.1145/1143997.1144223
25. Takagi, H., Pallez, D.: Paired comparisons-based interactive differential evolution, pp. 475–480 (2009)
26. Pallez, D., Collard, P., Baccino, T., Dumercy, L.: Eye-tracking evolutionary algorithm to minimize user fatigue in iec applied to interactive one-max problem. In: Proceedings of the 2007 GECCO Conference Companion on Genetic and Evolutionary Computation, pp. 2883–2886 (2007). https://doi.org/10.1145/1274000.1274098
27. Cheng, S., Liu, Y.: Eye-tracking based adaptive user interface: implicit human-computer interaction for preference indication. J. Multimodal User Interfaces **5**(1–2), 77–84 (2012). https://doi.org/10.1007/s12193-011-0064-6
28. Gegenfurtner, A., Lehtinen, E., Säljö, R.: Expertise differences in the comprehension of visualizations: a meta-analysis of eye-tracking research in professional domains. Educ. Psychol. Rev. **23**(4), 523–552 (2011). https://doi.org/10.1007/s10648-011-9174-7
29. Blondon, K., Wipfli, R., Lovis, C.: Use of eye-tracking technology in clinical reasoning: a systematic review. Stud. Health Technol. Inf. **210**, 90–94 (2015). https://doi.org/10.3233/978-1-61499-512-8-90
30. Pallez, D., Cremene, M., Baccino, T., Sabou, O.: Analyzing human gaze path during an interactive optimization task. In: Proceedings of the 2010 Workshop on Eye Gaze in Intelligent Human Machine Interaction - EGIHMI 2010, pp. 12–19 (2010). https://doi.org/10.1145/2002333.2002336
31. Goldberg, J.H., Kotval, X.P.: Computer interface evaluation using eye movements: methods and constructs. Int. J. Ind. Ergon. **24**(6), 631–645 (1999). https://doi.org/10.1016/S0169-8141(98)00068-7

32. Holmes, T., Zanker, J.: Eye on the prize: using overt visual attention to drive fitness for interactive evolutionary computation. In: Proceedings of the 10th Annual Conference on Genetic and Evolutionary Computation - GECCO 2008, pp. 1531–1538 (2008). https://doi.org/10.1145/1389095.1389390

33. Orquin, J.L., Mueller Loose, S.: Attention and choice: a review on eye movements in decision making. Acta Psychol. **144**(1), 190–206 (2013). https://doi.org/10.1016/j.actpsy.2013.06.003

34. Michalski, R., Grobelny, J.: An eye tracking based examination of visual attention during pairwise comparisons of a digital product's package. In: Antona, M., Stephanidis, C. (eds.) UAHCI 2016. LNCS, vol. 9737, pp. 430–441. Springer, Cham (2016). doi:10.1007/978-3-319-40250-5_41

35. Michalski, R., Grobelny, J.: The effects of background color, shape and dimensionality on purchase intentions in a digital product presentation. In: Antona, M., Stephanidis, C. (eds.) UAHCI 2016. LNCS, vol. 9739, pp. 468–479. Springer, Cham (2016). doi:10.1007/978-3-319-40238-3_45

36. Gere, A., Danner, L., de Antoni, N., Kovács, S., Dürrschmid, K., Sipos, L.: Visual attention accompanying food decision process: an alternative approach to choose the best models. Food Qual. Prefer. **51**, 1–7 (2016). https://doi.org/10.1016/j.foodqual.2016.01.009

37. Jantathai, S., Danner, L., Joechl, M., Dürrschmid, K.: Gazing behavior, choice and color of food: does gazing behavior predict choice? Food Res. Int. **54**(2), 1621–1626 (2013). https://doi.org/10.1016/j.foodres.2013.09.050

38. Salas-Morera, L., Cubero-Atienza, A., Ayuso-Munoz, R.: Computer-aidedplant layout | Distribucion en planta asistida por ordenador. Inf. Tecnol. 7(4) (1996)

# Active Sensing in Human Activity Recognition

Alfredo Nazábal[✉] and Antonio Artés[✉]

Department of Signal Theory and Communications and Gregorio Marañón Health Research Institute, Universidad Carlos III de Madrid, Leganés, Spain
{anazabal,antonio}@tsc.uc3m.es

**Abstract.** This work studies the problem of reducing the energy consumption of wearable sensors in a Human Activity Recognition (HAR) system. A HAR system is implemented using Hidden Markov Models, where decisions over the acquisition of new data are made based on the entropy of the posterior distribution of the activities. This problem is intractable in general, so three different active sensing algorithms are implemented to find numerically the data acquisition events. The performance of these algorithms is evaluated using a HAR database, resulting in a significant reduction on the number of observations acquired, thus reducing the energy consumption, while maintaining the performance of the system.

## 1 Introduction

Human Activity Recognition (HAR) is a field that has grown considerably in the last years, illustrating its huge influence in many modern social applications including ambulatory monitoring of elderly patients, human behaviour characterization or camera surveillance applications [1]. For example, in a monitoring application the knowledge of the activities being performed is critical to evaluate the context in which the patients are being monitored. The target activities of these applications range from simple activities such as walking, sitting or standing to more complex activities such as eating, sleeping or tooth brushing.

Wearable sensors are a common approach in long term monitoring HAR systems [2,3]. The most popular ones are inertial sensors, either used by themselves or combined with other multiple sensors. Some examples include a Parkinson symptoms detection system [4] and a gesture recognition system [5]. Smartphones, in particular, are gaining popularity in HAR systems as they include several built-in sensors that provide information of the daily activities of its user [6].

Alfredo Nazábal is supported by a FPI grant by the Ministerio de Economía y Competitividad of Spain (BES-2013-064825).

This work has been partly supported by MINECO/FEDER ('ADVENTURE', id. TEC2015-69868-C2-1-R), MINECO (AID, id. TEC2014-62194-EXP, grant 'DAMA', id. TIN2015-70308-REDT), and Comunidad de Madrid (project 'CASI-CAM-CM', id. S2013/ICE-2845).

© Springer International Publishing AG 2017
I. Rojas et al. (Eds.): IWANN 2017, Part I, LNCS 10305, pp. 157–166, 2017.
DOI: 10.1007/978-3-319-59153-7_14

Sensor-based activity recognition systems present several inherent problems, like the position of the sensors in the body, the noisy nature of the input observations or the energy consumption of the devices. The energy consumption problem in smart-phones is an important topic [7], since the HAR system shares its resources with the rest of the applications in the device and the embedded sensors are primary sources of power consumption. During short periods of time, it is reasonable to use all the information provided by the built-in sensors, however, when continuous monitoring is considered, the optimization of the battery and memory of the devices employed is critical. The main purpose of this paper is to develop a framework where the joint optimization of the energy consumption of the sensors and the performance of the HAR system is achieved.

The main approach to solve this problem consists of minimizing the data acquisition of the sensors. In [8] the authors propose a framework that uses a hierarchical sensor management strategy to recognize user activities as well as to detect activity transitions and decide which sensors to use at any given time. Instead of the ad hoc approximation considered in [8], we follow a systematic approach to decide when to perform the data acquisition. In this work, a HAR system based on Hidden Markov Models (HMMs) is considered. At a sampling rate of tens of Hertz, the mass of the posterior probability distribution of the activities given the observations is located on a single activity during most of the time. Accordingly, the uncertainty of the performed activity is low in general, as well as the entropy of the posterior distribution. When no observations are available, the posterior probability distribution corresponds asymptotically with the stationary distribution of the HMM, and the entropy of this posterior increases. A novel active sensing strategy is proposed to exploit this property, i.e. to stop acquiring observations when the entropy is low, and to estimate the time instant when the entropy reaches a certain threshold and new samples need to be acquired again. This is a reasonable assumption in a long term activity recognition system, where some activities are performed constantly during extended periods of time and only a few data samples are needed to recognize these activities.

The rest of the paper is organized as follows. Section 2 introduces the problem and the notation used in this work. In Sect. 3 three algorithms are developed to solve the active sensing problem numerically. Section 4 shows the results of the algorithms in a HAR database and Sect. 5 concludes the paper.

## 2   Problem Statement

Data acquired from the sensors, $X = \{\mathbf{x}_1, \mathbf{x}_2, \cdots, \mathbf{x}_N\}$, is defined as a sequence of $N$ vector observations of dimension $D$, $\mathbf{x}_n \in \mathbb{R}^D$. This data is modelled using a HMM (Fig. 1), where $S = \{s_1, s_2, \cdots, s_N\}$ is the sequence of hidden states explaining the data, with $s_n \in \{1, \cdots, M\}$ and $M$ being the number of possible states. A HMM is characterized by three parameters [9], the initial probabilities distribution $\boldsymbol{\pi} = p(s_1)$, the transition matrix $A \in \mathbb{R}^{M \times M}$, with elements $a_{ij} = p(s_{n+1} = j | s_n = i)$ and $\sum_i a_{ij} = 1$, and the observations probability distribution $p(\mathbf{x}_n | s_n)$.

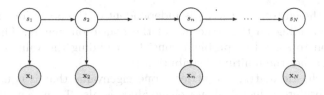

Fig. 1. Graphical representation of the Hidden Markov Model

The posterior of the state $s_n$ given all the observations until time instant $n$, $\mathbf{x}_{1:n}$, is defined by the forward step of the Forward-Backward algorithm. This posterior is denoted as $\boldsymbol{\alpha}_n \triangleq p(s_n|\mathbf{x}_{1:n})$ with elements $\alpha_n(i) = p(s_n = i|\mathbf{x}_{1:n})$.

To compute the probability of any state in the future, $n + n_0$, given the current known data, all the states between the current time instant $n$ and the future time instant $n + n_0$ are marginalized. In a Markov Chain this probability is obtained multiplying $n_0$ times the transition matrix by the posterior of $s_n$

$$\boldsymbol{\alpha}_{n+n_0} = \sum_{s_{n:n+n_0-1}} p(s_{n+n_0}, s_{n:n+n_0-1}|\mathbf{x}_{1:n}) = A^{n_0}\boldsymbol{\alpha}_n \tag{1}$$

Equation (1) represents the knowledge over the posterior of $s_n$ from time instant $n$ to the future. The entropy of $\boldsymbol{\alpha}_{n+n_0}$ provides a measure of the uncertainty of the future activities in terms of $n_0$. The entropy of a random variable $X$ with probabilities $p(\mathbf{x})$ is defined as $H(X) \triangleq -\sum_{i=1}^{|x|} p(x_i) \log p(x_i)$, where $|x|$ is the dimensionality of the vector $\mathbf{x}$.

The active sensing problem can be defined as the set of techniques required to find the values of the future time instant $n_0$ where the entropy of (1) exceeds a certain threshold $H_0$. At this time instant, the posterior distribution becomes unreliable, hence new data needs to be acquired to reduce the entropy. The solution of $n_0$ is obtained from

$$H(\boldsymbol{\alpha}_{n+n_0}) = -(A^{n_0}\boldsymbol{\alpha}_n)^T \log(A^{n_0}\boldsymbol{\alpha}_n) < H_0 \tag{2}$$

where $(\cdot)^T$ denotes the transpose operation. Unfortunately, (2) is intractable in general, so a numerical approximation is needed to obtain $n_0$.

## 3   Active Sensing Strategies

### 3.1   Activity Independent Approximation

The transition matrix of a HMM is a stochastic matrix with a limiting distribution or stationary state $\mathbf{p}$, i.e. as $n_0 \to \infty$, $A^{n_0}\boldsymbol{\alpha}_n = \mathbf{p}$ for every $\boldsymbol{\alpha}_n$. The existence of this limiting distribution implies that the entropy of $\boldsymbol{\alpha}_{n+n_0}$ must converge

$$\lim_{n_0 \to \infty} H(\boldsymbol{\alpha}_{n+n_0}) = -\mathbf{p}^T \log(\mathbf{p}) = H(\mathbf{p})$$

In the limit, the entropy of $\alpha_{n+n_0}$ is independent of the posterior of the activities, only depending on the structure of the transition matrix. Thus, a naive approximation to solve this problem consists of finding the value of $n_0$ where the system reaches the limiting distribution $\mathbf{p}$.

Every stochastic matrix has at least one eigenvalue that is equal to 1 and the largest absolute value of all its eigenvalues is also 1. In fact, the limiting distribution of the stochastic matrix is the eigenvector corresponding to the eigenvalue equal to 1. The n-power of the transition matrix can be expressed as $A^n = U \Lambda^n U^{-1}$, where $U$ is the matrix of eigenvectors of $A$ and $\Lambda$ is a diagonal matrix containing its eigenvalues $\{\lambda_1, \cdots, \lambda_M\}$, where $|\lambda_1| > |\lambda_2| > \cdots > |\lambda_M|$. This means that the eigenvectors of the transition matrix remain unaltered and only the $n$-power of the eigenvalues is needed.

The activity independent method consists of finding the minimum value of $n_0$ that satisfies $|\lambda_2|^{n_0} < \epsilon$, where $\epsilon$ controls the precision of the approximation,

$$n_0^p = \left\lfloor \left| \frac{\log(\epsilon)}{\log(|\lambda_2|)} \right| \right\rfloor \tag{3}$$

With this condition, $A^{n_0} \alpha_n \simeq \mathbf{p}$, as the only contribution of the transition matrix is related to the first eigenvalue $\lambda_1 = 1$ and all the other values of $\Lambda$ decrease to zero faster.

Figure 2 shows an example of $H(\alpha_{n+n_0})$ as a function of $n_0$ for a transition matrix with $M = 10$ states. For any transition matrix, this function is not monotonically increasing in general, with several local maxima in the interval $[0, n_0^p]$ that are greater than the value of the entropy in the limiting distribution. When $n_0 > n_0^p$, $H(\alpha_{n+n_0})$ converges to the entropy of the limiting distribution $H(\mathbf{p})$. When $n_0 = 0$, the value is just the entropy of the posterior of $s_n$, $H(\alpha_n)$. Choosing the value $n_0^p$ as the time instant when new data is sampled is a naive approximation, since there exists in general an interval of values in $[0, n_0^p]$ where the uncertainty is greater than $H(\mathbf{p})$. We need to choose $n_0$ in the first interval where (2) holds.

## 3.2   Threshold Method

A direct approach to solve the data acquisition problem consists of finding numerically the value of $n_0$ that satisfies (2), given that a suitable value of $H_0$ is chosen. If a small value of $H_0$ is considered the uncertainty will be too small and more data than needed would be acquired. If $H_0$ is too large, this value is not reached since the entropy is bounded above and data is not acquired again. The value of $H_0$ is chosen in terms of the entropy of the limiting distribution, $H_0 = cH(\mathbf{p})$, where $c \in [0, 1]$ is a parameter that controls the distance to the entropy of the limiting distribution $H(\mathbf{p})$. Thus, the threshold method consists of finding the maximum value of $n_0$ such as (2) holds,

$$H(\alpha_{n+n_0}) < cH(\mathbf{p})$$

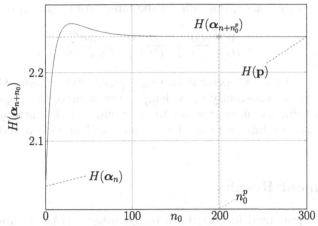

**Fig. 2.** Toy example of the entropy $H(\alpha_{n+n_0})$ for a transition matrix with 10 states

When $c = 1$, the threshold is equal to the entropy of the limiting distribution. Depending on the structure of the transition matrix, this could be problematic. If $H(\alpha_{n+n_0})$ is monotonically increasing in terms of $n_0$, this solution reduces to the naive approximation, i.e., $n_0 = n_0^p$. By choosing $c < 1$ data is acquired sooner, and the posterior distribution of the activities can be updated properly. For small values of $c$, the data is acquired too quickly, so the reduction in the number of observations is negligible. As the value of $c$ increases, the separation of the intervals when new data is acquired becomes larger, reducing the number of observations employed and thus the energy consumption.

### 3.3   Line Intersection Method

Fixing a threshold on the entropy is not always the best approach. An alternative method, where $H_0$ is not fixed in advance, consists of selecting $n_0$ as the intersection between two lines, a constant line defined by the entropy of the limiting distribution $y_1(n) = H(\mathbf{p})$ and the tangent line of the entropy at some $n_k \in [0, n_0^p]$. This line intersection method reaches different values of $H_0$ depending on the activities being performed.

The equation of the tangent line to $H(\alpha_{n+n_k})$ at $n_k$ is defined as

$$y_2(n) = y_k + m(n - n_k)$$

where $y_k$ is the entropy of the posterior at $n_k$, $H(\alpha_{n+n_k})$, and $m$ is the slope of the tangent line, that corresponds to the derivative of the entropy at $n_k$, $H'(\alpha_{n+n_k})$:

$$y_k = -(A^{n_k}\alpha_n)^T \log(A^{n_k}\alpha_n)$$
$$m = -(U \log(\Lambda)\Lambda^{n_k} U^{-1}\alpha_n)^T \log(A^{n_k}\alpha_n)$$

Computing the intersection between both lines $y_1(n) = y_2(n)$ and replacing $y_k$ and $m$ by its expressions, the value of the intersection point is obtained

$$n = n_k + \frac{\mathbf{p}^T \log(\mathbf{p}) - (A^{n_k} \boldsymbol{\alpha}_n) \log(A^{n_k} \boldsymbol{\alpha}_n)}{(U \log(\Lambda) \Lambda^{n_k} U^{-1} \boldsymbol{\alpha}_n)^T \log(A^{n_k} \boldsymbol{\alpha}_n)}$$

The problem of choosing a representative $n_k \in [0, n_0^p]$ still remains. One method consists of increasing constantly $n_k$ as long as the value of the slope is greater than a certain value. As it approaches the maximum of the entropy, the slope will decrease and the intersections of both lines will provide a good estimate of $n_0$.

## 4   Experiment Results

The algorithms presented in Sect. 3 are tested using a HAR database created using APDM Opal [10] wearable inertial sensors.[1] This database contains the measurements from eight different people with two sensors placed on the waist and the ankle. Only one of the sensors is assumed to be available during each of the experiments, evaluating the active sensing algorithms in two different settings. All the sequences contain a combination of five different activities: running, walking, standing, sitting and lying (in no particular order) under semi-naturalistic conditions in an indoor environment during a minimum of 20 min.

A HMM classifier with a Gaussian mixture observation model is trained for each sensor using the standard Baum-Welch algorithm [11]. The HMMs employ the structure described in [12], assigning three states per activity. A leave-one-person-out methodology is used in training, where one sequence is left out to test the entropy algorithms using the model parameters obtained during the training with the rest of the sequences.

The duration of the window where the sensors are acquiring new observations is set as a parameter for all the algorithms. Three different window sizes are used in the experiments, $W = \{5, 10, 20\}$ s. When the time window is over, the sensors stop acquiring data, and the active sensing algorithms decide the next time instant $n_0$ where the sensors need to acquire a new window of observations.

In the activity independent algorithm, three different precision values are used, $\epsilon = \{0.1, 0.01, 0.001\}$. In the threshold algorithm, three different values of $c = \{0.7, 0.8, 0.9\}$ are used. In the line intersection algorithm, the algorithm stops when the value of the slope of the line $y_2$, is less than 0.1, 0.01 and 0.001.

Table 1 shows the average reduction of the number of observations in each of the settings and the average precision loss of the system due to this sample reduction. In particular, in Fig. 3 the precision loss of all of the methods is compared in terms of the number of observations used in the waist sensor experiment.

Decreasing the number of samples acquired reduces the performance of the system. However, under the same conditions, the loss in precision is not heavily

---

[1] The dataset is available at http://www.tsc.uc3m.es/dataproy/har/databases.zip.

**Table 1.** Average percentage of sensor observations employed and precision loss of all the algorithms using both sensors. The first part corresponds to the activity independent algorithm, the second part to the threshold algorithm and the last part to the line intersection algorithm.

| W(s) | Algorithm parameter | Waist precision loss | Ankle precision loss | % of Waist sensor data used | % of Ankle sensor data used |
|------|------|------|------|------|------|
| 5  | 0.1   | −0.0705 | −0.1233 | 27.18% | 24.59% |
| 10 | 0.1   | −0.0827 | −0.1247 | 36.12% | 32.80% |
| 20 | 0.1   | −0.0724 | −0.1048 | 48.23% | 42.36% |
| 5  | 0.01  | −0.1947 | −0.1919 | 16.40% | 15.45% |
| 10 | 0.01  | −0.1742 | −0.1802 | 23.52% | 20.17% |
| 20 | 0.01  | −0.1749 | −0.1311 | 32.58% | 34.01% |
| 5  | 0.001 | −0.2902 | −0.2801 | 11.73% | 11.25% |
| 10 | 0.001 | −0.2546 | −0.2350 | 17.88% | 16.69% |
| 20 | 0.001 | −0.2780 | −0.1819 | 25.31% | 26.81% |
| 5  | 0.7   | −0.0473 | −0.0545 | 36.86% | 36.98% |
| 10 | 0.7   | −0.0317 | −0.0507 | 50.54% | 48.42% |
| 20 | 0.7   | −0.0315 | −0.0296 | 61.16% | 59.57% |
| 5  | 0.8   | −0.0659 | −0.0853 | 29.29% | 29.10% |
| 10 | 0.8   | −0.0655 | −0.0617 | 39.90% | 38.51% |
| 20 | 0.8   | −0.0582 | −0.0628 | 51.37% | 48.45% |
| 5  | 0.9   | −0.1098 | −0.1878 | 20.47% | 20.61% |
| 10 | 0.9   | −0.1127 | −0.1530 | 28.46% | 27.69% |
| 20 | 0.9   | −0.1019 | −0.1369 | 40.34% | 36.52% |
| 5  | 0.1   | −0.0070 | −0.0032 | 54.82% | 55.77% |
| 10 | 0.1   | −0.0053 | 0.0010  | 67.52% | 67.59% |
| 20 | 0.1   | −0.0166 | 0.0045  | 76.25% | 76.83% |
| 5  | 0.01  | −0.0390 | −0.0417 | 31.49% | 34.89% |
| 10 | 0.01  | −0.0212 | −0.0358 | 45.81% | 49.30% |
| 20 | 0.01  | −0.0180 | 0.0174  | 60.99% | 63.00% |
| 5  | 0.001 | −0.1513 | −0.2008 | 16.39% | 13.42% |
| 10 | 0.001 | −0.1242 | −0.1564 | 26.03% | 24.53% |
| 20 | 0.001 | −0.1067 | −0.1348 | 37.79% | 33.30% |

influenced by the reduction in window size. It is more important to update the model with new observations when the entropy increases than to acquire large windows of observations, since the entropy is practically zero during these windows. The best model in terms of precision loss is the line intersection algorithm, though the number of samples used is in general larger than in the other models. The threshold algorithm is the second in terms of performance. The activity independent algorithm performs worse than the others, since the posterior of the activities is not considered while computing the next time instant when new data must be acquired.

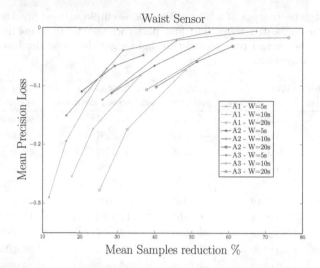

**Fig. 3.** Comparison of all the algorithms in Sect. 3 using three different window sizes for the waist sensor. A1 is the activity independent algorithm, A2 is the threshold algorithm and A3 is the line intersection algorithm.

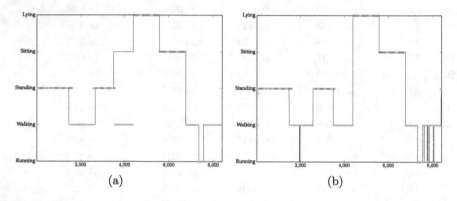

**Fig. 4.** Accuracy of the threshold model with parameter $c = 0.7$ and a window size of 5 s. The activity estimation when observations are available is represented in green, and in red the activity estimation when no observations are available. In (a) the true activity is represented in blue while in (b) it represents the estimated activity when all observations are available. (Color figure online)

Figures 4a and b show an example of the effects of the threshold algorithm over the acquired samples for each activity using one of the sequences. The number of samples required when a static activity like sitting is performed is much less than in the case of a dynamic activity like walking, and also the distance between two consecutive data acquisitions is much larger. Furthermore, Fig. 4b shows that the performance of the algorithm in terms of the estimated activities while using all the observations is not affected by the data reduction.

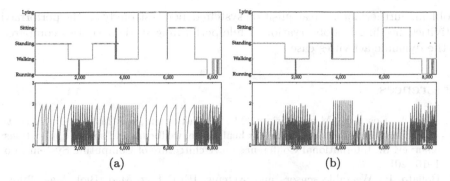

(a)                                              (b)

**Fig. 5.** Comparison between the estimated activities in the upper figure and the entropy of the forward step in the bottom figure using the threshold algorithm (a) and the line intersection algorithm (b). In the threshold algorithm $c = 0.7$ and in the line intersection algorithm the limit of the slope is set to 0.1. The window size is set to 5 s.

There exist several differences in the entropy evolution between the different algorithms. In general, when a static activity is performed, the entropy of the forward step increases more slowly compared to the dynamic activities. This effect can be observed in more detail in Figs. 5a and b. In the threshold algorithm the entropy increases until a certain fixed threshold, where the algorithm acquires a new window of data. The number of windows acquired in each activity differs considerably. In the line intersection algorithm, the next time window is chosen when the derivative of the entropy, i.e. the slope, is less than 0.1. Depending on the shape of the entropy function, we stop at different values. This effect is reduced when the parameter of the model decreases, since the entropy of the limiting distribution is constant for a specific transition matrix. Also, the number of observation windows used in this model is larger in general, leading to shorter periods of time where there is no data acquisition and consequently to reduce the loss in precision of the system.

## 5   Conclusions

A systematic approach to solve the problem of energy consumption reduction in HAR systems with inertial sensors have been proposed. This work shows that using the maximum entropy of the posterior of the activities, the number of observations can be reduced while maintaining the performance of the recognition system. Three different methods to deal with the data acquisition problem have been implemented and evaluated, emphasizing the importance of updating the model with new observations when the entropy increases rather than acquiring large windows of observations. The number of observations acquired can be reduced to 31% while only decreasing the precision by 0.039, although other operation points are also possible depending on the specifications of the problem. Furthermore, the acquisition system depends strongly on the performed activities, needing less observations while performing static activities compared to the dynamic activities case.

# References

1. Avci, A., Bosch, S., Marin-Perianu, M., Marin-Perianu, R., Havinga, P.: Activity recognition using inertial sensing for healthcare, wellbeing and sports applications: a survey. In: International Conference on Architecture of Computing Systems, pp. 1–10 (2010)
2. Bonato, P.: Wearable sensors and systems. IEEE Eng. Med. Biol. Mag. **29**(3), 25–36 (2010)
3. Lara, Ó.D., Labrador, M.A.: A survey on human activity recognition using wearable sensors. IEEE Commun. Surv. Tutor. **15**(3), 1192–1209 (2013)
4. Patel, S., Lorincz, K., Hughes, R., Huggins, N., Growdon, J., Standaert, D., Akay, M., Dy, J., Welsh, M., Bonato, P.: Monitoring motor fluctuations in patients with Parkinson's disease using wearable sensors. IEEE Trans. Inf. Technol. Biomed. **13**(6), 864–873 (2009)
5. Zappi, P., Lombriser, C., Stiefmeier, T., Farella, E., Roggen, D., Benini, L., Tröster, G.: Activity recognition from on-body sensors: accuracy-power trade-off by dynamic sensor selection. In: Verdone, R. (ed.) EWSN 2008. LNCS, vol. 4913, pp. 17–33. Springer, Heidelberg (2008). doi:10.1007/978-3-540-77690-1_2
6. Lee, Y.S., Cho, S.B.: Layered hidden Markov models to recognize activity with built-in sensors on Android smartphone. Pattern Anal. Appl. **19**, 1–13 (2016)
7. Perrucci, G.P., Fitzek, F.H., Widmer, J.: Survey on energy consumption entities on the smartphone platform. In: 2011 IEEE 73rd Vehicular Technology Conference (VTC Spring), pp. 1–6. IEEE (2011)
8. Wang, Y., Lin, J., Annavaram, M., Jacobson, Q.A., Hong, J., Krishnamachari, B., Sadeh, N.: A framework of energy efficient mobile sensing for automatic user state recognition. In: Proceedings of the 7th International Conference on Mobile Systems, Applications, and Services, pp. 179–192. ACM (2009)
9. Rabiner, L.: A tutorial on hidden Markov models and selected applications in speech recognition. Proc. IEEE **77**, 257–286 (1989)
10. APDM, Inc. Opal technical specification. http://www.apdm.com/
11. Murphy, K.P.: Machine Learning. MIT Press, Cambridge (2012)
12. Florentino-Liano, B., O'Mahony, N., Artés-Rodríguez, A.: Hierarchical dynamic model for human daily activity recognition. In: BIOSIGNALS, pp. 61–68 (2012)

# Searching the Sky for Neural Networks

Erich Schikuta[1](✉), Abdelkader Magdy[1], Irfan Ul Haq[2], A. Baith Mohamed[1],
Benedikt Pittl[1], and Werner Mach[1]

[1] University of Vienna, Vienna, Austria
{erich.schikuta,shaabana52,abdel.baes.mohamed,beneditk.pittl,
werner.mach}@univie.ac.at
[2] Department of CIS, Pakistan Institute of Engineering and Applied Sciences,
Islamabad, Pakistan
irfanulhaq@pieas.edu.pk

**Abstract.** Sky computing is a new computing paradigm leveraging resources of multiple Cloud providers to create a large-scale distributed infrastructure. N2Sky is a research initiative promising a framework for the utilization of Neural Networks as services across many Clouds integrating into a Sky. This involves a number of challenges ranging from the provision, discovery and utilization of N2Sky services to the management, monitoring, metering and accounting of the N2Sky infrastructure. This paper focuses on the semantic discovery of N2Sky services through a human-centered querying mechanism termed as N2Query. N2Query allows N2Sky users to specify their problem statement as natural language queries. In response to the natural language queries, it delivers a list of ranked neural network services to the user as a solution to their stated problem. The search algorithm of N2Query is based on the semantic mapping of ontologies referring to problem and solution domains.

**Keywords:** Neural network as a service · Virtual organization · Semantic description · Cloud computing

## 1 Introduction

Sky providers aggregate the services scattered across various Cloud-based infrastructures to provide the concept of sky computing. The sky computing in this way copes with the problem of vendor lock-in and extends the flexibility, transparency and elasticity of the integrated infrastructure as compared to that of a single Cloud. Sky computing has taken another step forward towards the realization of virtual collaborations, where solutions are virtual and resources are logical. The exchanging data among researchers is the main stimulus point for the development. This is just as valid for the neural information processing community as for any other research community [13]. As described by the UK e-Science initiative [16] many goals can be achieved by using new stimulation techniques, such as enabling more effective and seamless collaboration for scientific and commercial communities.

© Springer International Publishing AG 2017
I. Rojas et al. (Eds.): IWANN 2017, Part I, LNCS 10305, pp. 167–178, 2017.
DOI: 10.1007/978-3-319-59153-7_15

In the context of our work, we designed N2Sky a virtual neural network (NN) simulator for the computational intelligence community [14]. It provides access to neural network resources and enables infrastructures fostering multi Cloud resources. On the one hand, neural network resources can be generic neural networks trained by a specific learning paradigm and training data for given problems whereas on the other hand they can represent already trained networks, which can be used for given application problems. The vision of N2Sky is the provisioning of neural networks where any member of the community can access or contribute neural networks all over the Internet.

The number of neural networks is expected to be very large and continuously growing. These neural network objects are distributed on a worldwide scale on the Internet administratively under the umbrella of the N2Sky virtual organisation on participating resource nodes. Searching for specific neural network resources providing solutions to given problems can be a time consuming and difficult task. We developed an ontology-based approach for searching semantically the resource pool of N2Sky, where the generic idea was presented in [12]. In this paper, we present N2Query, an implementation of our generic semantic query approach as a component of the N2Sky infrastructure. It allows N2Sky users to search for neural networks by using natural language. The N2Query component is depicted in Fig. 1. The N2Query architecture consists of:

- A semantic querying interface that allows the user to specify his/her problem description in natural language form (query).
- An ontology mapping mechanism that allows N2Query to recognise the semantic of the natural language query in form of an ontology (called problem ontology). Then a mapping algorithm is applied to match N2Query's problem ontology against an already constructed solution ontology resulting in a list of adequate neural networks.
- A ranking mechanism to deliver a list of links to neural network resources of the N2Sky virtual organisation for solving the problem.
- An XML based Neural Network resource representation language to maintain and search the problem and solution ontologies.

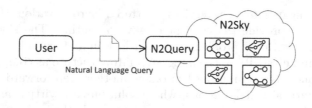

**Fig. 1.** N2Query in context with N2Sky

The structure of the paper, is as follows: The state of the art of neural network simulators and the baseline research are given in Sect. 2. Section 3 highlights the activities of the overall N2Query process. Section 3 discusses the architecture of all services and components of the N2Query engine. A use case and the execution process of the N2Query are presented in Sect. 4. Finally, the paper concludes our findings and presents our plans for future work.

## 2 Related Work

Over the last few years, a lot of simulation environments have been developed to mimic the behaviour of artificial and biological neural networks [9]. IQR [1] is a neurone simulator which allows neuronal models to control the behaviour of real devices in real-time. NeuroSpace [3] aims to integrate neural networks into relational database. NEUVISION [8] is a simulation environment used to simulate large-scale neuronal networks. NeuroWeb [11] lets users exchange information (neural network objects, neural network paradigms) and exploits available computing resources for neural network specific tasks (specifically training of neural networks).

Actually, there is no simulator or environment cover all simulation approaches, being able to solve a different kind of problems in the Cloud. In the course of our research, we designed and developed N2Sky [14], a virtual organization (VO) for the community of computational intelligence (CI), providing access to neural networks and enabling infrastructures to foster federated Cloud resources [12].

N2Sky supports qualified users to easily run their simulations by accessing data related neural network resources that has been published by the N2Sky service manager and the N2Sky data service [5]. Moreover, N2Sky provides a facility to end users to solve their problems by using predefined objects and paradigms. For the purpose of thin clients a simple Web browser, which can execute on a PC or a smartphone, can be used to access the front-end, the N2Sky (Mobile) Web Portal. It is relying on the N2Sky user management service which grants access to the system [14].

N2Sky aroused strong interest even beyond the CI community[1]. This endeavour of providing an environment for the access to practically unlimited resources faces one specific challenge. We propose a centralised registry approach collecting all semantic knowledge of neural network objects by semantic web technologies [9].

N2Sky offers neural network resources as a service which dynamically uses the available computing environment to reduce the execution time. Summing up, the N2Sky environment provides [14]:

- Sharing of neural network paradigms, objects and related information between the researchers and end user world wide.
- Reduction of training time of neural network by automatically selecting appropriate parallel implementations of the neural network services exploiting suitable Cloud resources.

---

[1] http://cacm.acm.org/news/171642-neural-nets-now-available-in-the-Cloud/.

- Transparent access to High-end neural network resources stored in Cloud environment.
- Uniform Look and feel for location independence of computational, storage and Network resources.

N2Sky uses ViNNSL description language for describing neural networks paradigm to allow for easier sharing of resources between the paradigm provider and the customers [6].

# 3    N2Query Architecture

Using the N2Query users can submit his/her query formulated in natural language through the N2Query interface, which further interacts with the N2Sky infrastructure to look for the plausible solutions for the user and responds through the same interface. Figure 2 shows the high level process of N2Query. The detail of every stage through the process will be covered in the subsequent sections however a rather general description is aimed here to gradually build the understanding of the reader. After the user submits his/her query in the natural language form, the Stanford Language Processor is applied for the syntax and semantic analysis of the query. The processed problem statement is then sought out through the already built problem ontology. The nodes of the problem ontology that are marked relevant with the user problem are then linked with the relevant nodes of already available solution ontology producing the ontology mapping for the specific problem. The ontology mapping results into the identification of the neural networks required to solve the submitted problem. The retrieved neural networks are afterwards ranked by applying ElasticSearch and are conveyed to the user as the solution of his problem.

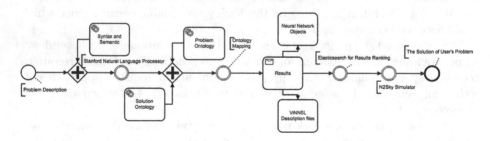

**Fig. 2.** N2Query components control flow

The architecture of N2Query and the system services are depicted in Fig. 3.

*The N2Query (Mobile) Web Portal:* It provides the access point to the N2Query system by a web browser interface which can be used on PCs, tablets or even a smartphones. N2Querry provides two different interfaces to the user, a free text

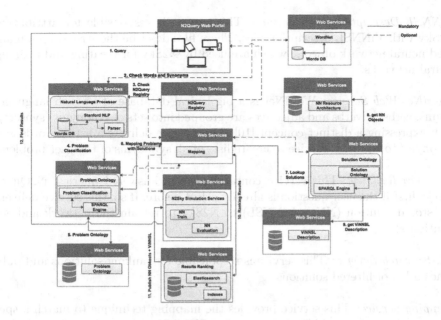

**Fig. 3.** N2Query architecture and components

and a directory search. Figure 5 shows, the N2Query free text interface. The directory interface depicted in Fig. 7 allows a category search and offers a brief description of the mechanism of N2Query interface by providing some examples to show how the user can interact with N2Query interface.

*Natural Language Processor Service:* This service analyzes the semantic definition behind user query. The NLP service applies five steps (Lexical Analysis, Syntactic Analysis (Parsing), Semantic Analysis, Discourse Integration, Pragmatic Analysis) to understand the meaning of statements.

*Problem Ontology Service:* This service is based on a hierarchy of known neural network problems. This service aims to classify the user's problem under one or more categories of typical neural network problems, as approximation, optimisation, searching etc.

*Solution Ontology Service:* This service stores all solutions provided by managed neural network resources in the N2Sky virtual organisation in a hierarchical form. The solution ontology can specify one or more solutions for a specific problem.

*NN Resource Architecture Web Service:* This service administrates N2Sky neural network objects which are considered as solution(s) for a specific problem and can be used under the umbrella of the N2Sky simulator.

*ViNNSL Description Web Service:* This service is responsible for attributing N2Sky with ViNNSL description files. These files describe the structure of managed neural network objects, which are used by N2Sky for creating and training neural networks.

*WordNet Web Service:* WordNet is a big lexical database of English language. Nouns, verbs, adverbs and adjectives are grouped into sets of synonyms (synsets), each expressing a distinct concept [10]. This service delivers a list of words and synonyms to the NLP service to create lists of all synonyms of the user problem.

*N2Query Registry:* This service contains all well answered queries. N2Query checks first if the user question is already asked before. If so, this service delivers the stored solution (NNs + ViNNSL) to N2Sky to retrain the network and get solution(s).

*Elasticsearch Service:* This service is responsible for ranking solutions and publishes a list of filtered solutions.

*Mapping Service:* This service provides the mapping technique to match a specific problem to possible solutions using SPARQL query language and then deliver list of solutions of that problem.

Three ontology combination paradigms can be distinguished, ontology linking, ontology mapping, and ontology importing [4]. For our problem, we apply ontology linking, where individuals from distinct ontologies are connected with links.

The concept is as follows: We administer basically two ontologies, a problem ontology and a solution ontology [6]:

- The *problem ontology* consists of a hierarchical organisation of typical neural network application problems, as classification, optimization, approximation, storage, pattern restoration, cluster analysis, feature extraction etc. In the ontology hierarchy these main domains are finer distinguished till the single problem specifications show up in the leave nodes.
- The *solution ontology* stores all known N2Sky neural networks organized according to their paradigm, as perceptron, multi-layer backpropagation, self-organizing maps (Kohonen cards), recurrent networks (Elman, Jordan, etc.), cellular neural networks, etc. Here the ontology delivers a fine grained structure finally giving the neural network objects (trained neural networks for a specific problem) as leaves.

Figure 4 depicts the two ontologies for solving the problem of "Face Recognition" (see case study section). Our N2Query component has already connected them (the nodes are connected with links). The links between problem-solution ontology simulate the mapping process between the defined problem with solution(s). We generate a mapping of problem ontology nodes, describing a specific problem, to solution ontology nodes, denoting network objects, which deliver a

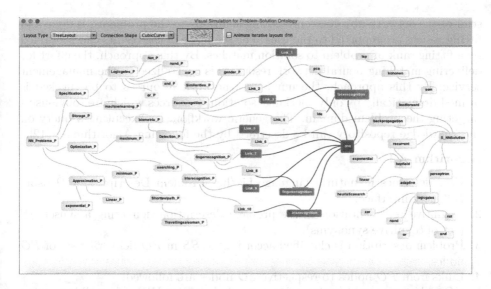

**Fig. 4.** Mapping between solution ontology (right) and the problem ontology (left) for the "face recognition" problem

solution for this problem. Links can be defined not only between leaves of the hierarchies but also between internal nodes.

Based on the ViNNSL semantic information of an N2Sky resource the mapping between problems to solutions can be done by N2Sky administrators manually, and by an automatic mapping during insertion of the new network objects.

**Integrating with the Ontologies.** This workflow for integrating new neural network resources into the knowledge repository of N2Sky can be described by the following Algorithm A1 [12].

**Algorithm A1**

1. N2Sky resource provider ($RP$) attributes its Neural Network Resource ($NNR$) with a ViNNSL Description ($VD$) specifying structural and semantic information.
2. $RP$ sends $VD$ together with $NNR$ or URI of $NNR$ to N2SKy knowledge repository.
3. $VD$ is integrated into Problem Ontology ($PO$) according to problem domain.
4. $NNR$ or its URI is integrated into Solution Ontology ($SO$) according to network paradigm.
5. Link between $VD$ insertion node in $PO$ and $NNR$ insertion node in $SO$ is created.

**Querying the Ontologies.** The search algorithm is as follows: Based on the natural language keywords of the user query a scan over the problem ontology is performed. Hits, patterns matching the scan, resembled by nodes in the hierarchy, are collected and the links to the solution ontology are followed. There, a scan of the network objects, representing solutions to the problems, is done and fitting results are reported to the user. The sequence of the results can be guided by a fitting rank of problem to solution matches. By this approach, the effort for delivering matching neural network resources is centralized in the management service. By this approach, the number of network resources to be checked is pruned dramatically by only checking (solution) resources which are (obviously) targeting the problem domain. The generic workflow for executing a query on the knowledge repository can be described by the following Algorithm A2 [12].

**Algorithm A2**

1. User describes in natural language his/her Problem Description $PD$ using N2Query interface.
2. Cognitive representation of the problem description delivering a synset $SS$ (set of cognitive synonyms).
3. Problem description is classified according to $SS$ in $PO$ delivering set of $PO$ nodes.
4. Links from $PO$ nodes to respective $SO$ nodes are followed.
5. $SO$ nodes are starting points of tree search delivering URIs of possible solution candidates.
6. $VD$ of solution candidates are analyzed and ranked according to match with $PD$.
7. Ranked list is reported to user.

## 4   N2Query Case Study

In this work, we choose the problem of Face-Recognition as a case study example [7].

In our use case, the user submits a query "How to solve face recognition problem" as shown in Fig. 5. The query is then processed and the results are displayed in the same interface.

Our approach to solve this problem consists basically of two phase, the ontology integration phase and the ontology query phase.

### 4.1   Ontology Integration Phase

The prerequisite for the user query is that the required neural network object must be present in the ontology architecture as well as the required ontology mapping should already be available within the system. The search process is performed on this ontology architecture. In the integration phase a neural network resource, e.g. a trained neural network object, which is provided by a member of the N2Sky virtual organization, is entered into the solution ontology. We use as running example the face recognition problem [7]. It is assumed the respective Backpropagation network was realized and contributed to N2Sky. Hereby the Algorithm A1, presented in Sect. 3, has to be executed:

The provider of the NN resource, the Backpropagation network, uses the ViNNSL language for the description of the problem and paradigm domain in step A1/1. Hereby, the `paradigm` and `problem domain` tags of ViNNSL are used.

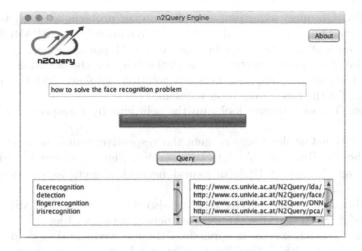

**Fig. 5.** N2Query free text user interface

In step A1/3 the description of the NN resource, Classification → Machine Learning → Biometric → FaceDetection → Gender, is integrated into the problem Ontology.

The NN resource is integrated into the Solution Ontology accordingly to its paradigm family in step A1/4, Backpropagation → DNN → FaceRecognition → PCA.

In step A1/5 an appropriate link from problem to Solution Ontology is created pointing from problem description to the respective physical NN resource, see Fig. 4.

## 4.2    Ontology Query Phase

In the following we show how a natural language free text user query is analyzed by the N2Query system.

The following query analysis steps refer to the workflow from Fig. 3, marked as ordered numbers integrating various components of the N2Query architecture.

1. The user query is sent to the Natural Language Processing (Stanford Parser) web service to analyze the semantic of that query.
2. The NLP web service connects with WordNet web service which delivers a list of words and synonyms of the user problem.
3. If the query has been asked before, the N2Query Registry service sends all details to the Mapping service. This service is responsible for gathering neural network paradigms as solutions of that problem. Afterwards, the Mapping web service publishes solution(s) to N2Sky for retraining networks.
4. The Problem Ontology service receives the recognised statement of the user query in form of tokens. That service classifies the user problem under the hierarchical structure of the most known neural network problems.

5. The SPARQL query algorithm is applied on the problem ontology to match the user problem with stored problems. Conclusively the SPARQL query engine sends all matched classifications to the Mapping service.
6. The Mapping service matches the problem(s) to the solution ontology. Figure 4 shows the mapping between solution ontology and the problem ontology for the face recognition problem.
7. Solution Ontology service looks up the solutions by a respective SPARQL query.
8. The Solution Ontology service gets the respective neural network objects from the NN Resources Architecture service. Figure 6 shows the SPARQL query and the list of URIs of neural network objects as solutions using Protégé ontology tool.
9. The solution Ontology service receives also the respective ViNNSL description file(s) which describe the received neural network objects.
10. The Elasticsearch service is applied on the received solution(s) for ranking and filtering results and publishes a list of solutions of the user problem to the N2Sky simulator.
11. N2Sky receives the published solution(s) (NN objects and ViNNSL) and starts creating, training, retraining and evaluating neural networks for the user problem.
12. N2Sky send the final result(s) to the user as shown in the bottom right pane in Fig. 5. Figure 7 represents another possibility to present results in a structured directory interface.

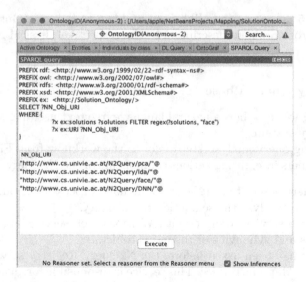

**Fig. 6.** SPARQL query and list of URI's of NN objects as solutions using Protégé

In this process, we use the Stanford parser for Natural Language Processing. The problem-solution ontologies are implemented by RDF and processed by the

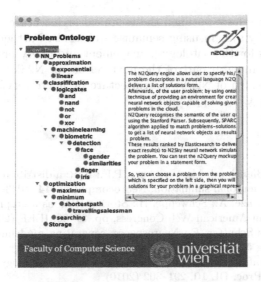

**Fig. 7.** N2Query directory and guide interface

OWL semantic web languages [15]. Huge storage repositories for RDF data have been developed, which store the RDF triples in a relational database (RDB) [2]. So, we use the Eclipse RDF4J Framework for ontology storage. According to the steps in the last section N2Query starts to map from the solution to the problem ontology to retrieve the existing solutions to that problem. In this process we use SPARQL language for the matching process. Thus, specifically the Algorithm A2, see Sect. 3, is executed by applying sematic web techniques:

In step A2/1, querying the N2Query tool by a natural language phrase like, "How to Solve the Face Recognition Problem", the N2Query system tries to recognise the semantic representation of this problem using Stanford natural language processor (step A2/2), and receives in the synnet set a phrase like "Face Recognition". In step A2/3 the SPARQL Query algorithm classifies the user query into the problem domain, and, following the link to the Solution Ontology (step A2/4), delivers by a subtree traversal (A2/5) the possible URIs of the existing NN objects. Based on the ViNNSL descriptions in step A2/6 the system produces and reports a ranked list of qualified solution URIs to the user's problem (A2/7).

## 5   Conclusion and Future Work

The N2Sky project manages neural network resources on a worldwide basis. Provisioning of adequate networks to given problems is a critical success factor of the N2Sky project. Hence we developed the N2Query component which is able to process natural language queries for finding matching networks.

In this paper, we have outlined the technical architecture and implementation of the N2Query system using semantic web tools. Further, we defined the different workflows for the ontology management and query processing.

N2Query is an integral part of the next release of N2Sky which delivers a comprehensive redesign of N2Sky's architecture by micro-services, docker technique and dynamic Cloud deployment.

# References

1. Bernardet, U., Blanchard, M., Verschure, P.F.: IQR: a distributed system for real-time real-world neuronal simulation. Neurocomputing **44**, 1043–1048 (2002)
2. Bönström, V., Hinze, A., Schweppe, H.: Storing RDF as a graph. In: Proceedings of the First Latin American Web Congress, pp. 27–36. IEEE (2003)
3. Cornelis, H., De Schutter, E.: Neurospaces: separating modeling and simulation. Neurocomputing **52**, 227–231 (2003)
4. Homola, M., Serafini, L.: Towards formal comparison of ontology linking, mapping and importing. Proc. DL **10**, 291–302 (2010)
5. Huqqani, A.A., Li, X., Beran, P.P., Schikuta, E.: N2Cloud: cloud based neural network simulation application. In: The 2010 International Joint Conference on Neural Networks (IJCNN), pp. 1–5. IEEE (2010)
6. Huqqani, A.A., Mann, E., Schikuta, E.: Novel concepts for realizing neural networks as services in the sky. Procedia Comput. Sci. **29**, 2315–2324 (2014)
7. Huqqani, A.A., Schikuta, E., Ye, S., Chen, P.: Multicore and GPU parallelization of neural networks for face recognition. Procedia Comput. Sci. **18**, 349–358 (2013)
8. Mulas, M., Massobrio, P.: Neuvision: a novel simulation environment to model spontaneous and stimulus-evoked activity of large-scale neuronal networks. Neurocomputing **122**, 441–457 (2013)
9. Prieto, A., Prieto, B., Ortigosa, E.M., Ros, E., Pelayo, F., Ortega, J., Rojas, I.: Neural networks: an overview of early research, current frameworks and new challenges. Neurocomputing **214**, 242–268 (2016)
10. Princeton University: WordNet: a lexical database for English. https://wordnet.princeton.edu
11. Schikuta, E.: NeuroWeb: an internet-based neural network simulator. In: ICTAI, pp. 407–412 (2002)
12. Schikuta, E., Magdy, A., Mohamed, A.B.: A framework for ontology based management of neural network as a service. In: Hirose, A., Ozawa, S., Doya, K., Ikeda, K., Lee, M., Liu, D. (eds.) ICONIP 2016. LNCS, vol. 9950, pp. 236–243. Springer, Cham (2016). doi:10.1007/978-3-319-46681-1_29
13. Schikuta, E., Mann, E.: A cloud-based neural network simulation environment. In: Rojas, I., Joya, G., Gabestany, J. (eds.) IWANN 2013. LNCS, vol. 7902, pp. 133–143. Springer, Heidelberg (2013). doi:10.1007/978-3-642-38679-4_12
14. Schikuta, E., Mann, E.: N2Sky - neural networks as services in the clouds. In: The 2013 International Joint Conference on Neural Networks (IJCNN), pp. 1–8. IEEE (2013)
15. Staab, S., Studer, R.: Handbook on Ontologies. Springer Science & Business Media, Heidelberg (2013)
16. UK e-Science: UK e-Science programme (2016). http://www.escience-grid.org.uk

# Image and Signal Processing

# Non-linear Least Mean Squares Prediction Based on Non-Gaussian Mixtures

Gonzalo Safont[1(✉)], Addisson Salazar[1], Alberto Rodríguez[2], and Luis Vergara[1]

[1] Institute of Telecommunications and Multimedia Applications, Universitat Politècnica de València, València, Spain
gonsaar@upvnet.upv.es,
{asalazar,lvergara}@dcom.upv.es
[2] Department of Communications Engineering, University Miguel Hernández de Elche, Elche, Spain
arodriguezm@umh.es

**Abstract.** Independent Component Analyzers Mixture Models (ICAMM) are versatile and general models for a large variety of probability density functions. In this paper, we assume ICAMM to derive a closed-form solution to the optimal Least Mean Squared Error predictor, which we have named E-ICAMM. The new predictor is compared with four classical alternatives (Kriging, Wiener, Matrix Completion, and Splines) which are representative of the large amount of existing approaches. The prediction performance of the considered methods was estimated using four performance indicators on simulated and real data. The experiment on real data consisted in the recovering of missing seismic traces in a real seismology survey. E-ICAMM outperformed the other methods in all cases, displaying the potential of the derived predictor.

**Keywords:** Prediction · ICA · Non-linear · Non-Gaussian · Interpolation

## 1 Introduction

Prediction is one of the fundamental problems in statistical signal processing [1], in which some unknown values in a given domain (e.g., time, space, time-space) are to be estimated from some known values of the same nature and domain. It is part of other fields like spectral analysis, coding, time series analysis, interpolation and smoothing, and appears in many areas of application. Many methods exist to implement linear and nonlinear predictors, broadly classified into statistical and deterministic methods. Statistical approaches rely on searching for solutions to the implicit estimation problem, while deterministic methods define some objective function to be minimized under some constraint imposed by the training sample set. In spite of the huge amount of previous work in statistical prediction methods, new developments are still possible if new statistical models appear, so that new solutions can be found. In this paper, it is assumed that the joint probability density of the data can be modeled with an Independent Component Analyzers Mixture Model (ICAMM) [2–4]. ICAMM is a versatile non-Gaussian mixture model which encompasses most of the usual statistical models,

© Springer International Publishing AG 2017
I. Rojas et al. (Eds.): IWANN 2017, Part I, LNCS 10305, pp. 181–189, 2017.
DOI: 10.1007/978-3-319-59153-7_16

including Gaussian Mixture Models, as particular cases. This generality implies that ICAMM can be used in a large variety of scenarios, for instance in the change detection field [5]. Therefore, we have derived a closed-form non-linear Least Mean Squared Error (LMSE) predictor for cases in which ICAMM is an appropriate model for the observed data. The corresponding predictor will be called E-ICAMM. The new method has been compared with already existing predictors, both by simulation and by real data analysis. We have selected four methods, which may be considered representative of many of them: Kriging, Wiener structures, Matrix Completion, and Splines. Kriging [6] is a linear predictor widely-used in geostatistics and other applications where the topographical distribution of the signal is important. It is also a 2D implementation of a linear LMSE predictor, and thus, an appropriate reference with which to compare E-ICAMM. The Wiener predictor [7] is composed by a linear predictor followed by a nonlinear scalar correction term. The Wiener predictor was chosen as representative of methods that try to keep the simplicity of the linear solution while being closer to the non-linear LMSE solution. Matrix Completion [8] is a non-linear method that estimates missing data in a matrix from a few revealed entries. This method was chosen because it does not rely on statistical concepts but rather in structural assumptions about the data. Finally, Splines [9] are representative of deterministic methods which approximate complex function stepwise by local polynomials.

## 2    Independent Component Analysis Mixture Model

The proposed method is based on Independent Component Analysis (ICA) [10]. ICA assumes that the observation at time instant $n$, $\mathbf{x}(n)$, can be modeled as an instantaneous linear transformation of a set of independent sources $\mathbf{s}(n)$, $\mathbf{x}(n) = \mathbf{A} \cdot \mathbf{s}(n)$. Here, $\mathbf{A}$ is the mixing matrix of size $[R \times M]$, where $R$ is the number of variables of each observation; $M$ is the number of sources; and $n$ is the current time instant. We will assume for simplicity that there are as many observed variables as sources ($R = M$) and that $\mathbf{A}$ can be inverted to find $\mathbf{W}$, the demixing matrix. Due to the independence consideration of ICA, the multivariate probability density function of the observations can be obtained as a product of one-dimensional marginal densities, $p_x(\mathbf{x}(n)) = |\det \mathbf{W}| \cdot \prod_{m=1}^{M} p_{s_m}(s_m(n))$. There is a wide range of multivariate non-Gaussian probability densities from real applications that can be modeled by adapting the marginal distributions and the mixing matrix, such as electroencephalographic (EEG) data (e.g., [10, 11]). The standard ICA model was recently extended to an ICA Mixture Model [2, 3]. In ICAMM, it is assumed that the data are separated in $K$ mutually-exclusive classes and each class is modeled using a different ICA. Therefore, the data are modeled as $\mathbf{x}(n) = \mathbf{A}_k \cdot \mathbf{s}_k(n) + \mathbf{b}_k$, where $k$ is the class at time $n$, denoted by $C_k(n)$, $k \in [1, K]$; $\mathbf{A}_k$ and $\mathbf{s}_k(n)$ are respectively the mixing matrix and the sources of the ICA model of class $k$; and $\mathbf{b}_k$ are the corresponding bias vectors. Essentially, $\mathbf{b}_k$ determines the location of the $k$-th cluster and $\mathbf{A}_k$, $\mathbf{s}_k$ determine its shape. Again, we will assume that the mixing matrices can be inverted to find the demixing matrices, $\mathbf{W}_k$.

Given the mixture model and the independence of the sources, the probability density of the observations can be expressed as

$$p(\mathbf{x}(n)) = \sum_{k=1}^{K} P(C_k)p(\mathbf{x}|C_k) = \sum_{k=1}^{K} P(C_k)|\det(\mathbf{W}_k)| \prod_{m=1}^{M} p(s_{k,m}(n)) \qquad (1)$$

where $P(C_k)$ is the prior probability of class $k$; $\mathbf{w}_{K,m}^T$ is the $m$-th row of the demixing matrix of class $k$, $\mathbf{W}_k = \mathbf{A}_k^{-1}$; and $s_{k,m}(n) = \mathbf{w}_{k,m}^T(\mathbf{x}(n) - \mathbf{b}_k)$ is the estimated value of the $m$-th source of class $k$ at time $n$. Mixture models emanate in a natural manner in the field of classification/segmentation methods. Since the data are categorized into several mutually exclusive classes, the class can be estimated by maximizing the posterior probability of the observation, $P(C_k(n) \,|\, \mathbf{x}(n))$.

## 3  Non-linear Least Mean Squares Predictor Based on ICAMM (E-ICAMM)

The Least Mean Squared Error (LMSE) predictor is the conditional mean of unknown data with respect to known data, $\hat{\mathbf{z}}_{\text{LMSE}} = E[\mathbf{z}|\mathbf{y}] = \int z \cdot p(\mathbf{z}|\mathbf{y}) \cdot d\mathbf{z}$. In this work, we propose a novel non-linear LMSE method that implements the conditional expectation by assuming an Independent Component Analysis Mixture Model.

Let us assume that we want to recover $M_{unk}$ unknown values from observation vector $\mathbf{x}(n)$ using its remaining $M_k = M - M_{unk}$ known values and a known ICA Mixture Model. Without any loss of generality, one can split known and unknown values into vectors $\mathbf{y}(n)$ and $\mathbf{z}(n)$, respectively, so that $\mathbf{x}(n) = [\mathbf{y}(n)^T, \mathbf{z}(n)^T]^T$. This split can be different for each data point but we will not denote this explicitly. For brevity, we drop the $(n)$ notation in the following. Given the mixture model, the conditional expectation $E[\mathbf{z}|\mathbf{y}]$ can be obtained as

$$E[\mathbf{z}|\mathbf{y}] = \sum_{k=1}^{K} E[\mathbf{z}|\mathbf{y}, C_k] \, P(C_k|\mathbf{y}) \qquad (2)$$

where $E[\mathbf{z}|\mathbf{y}, C_k]$ is the conditional expectation for class $k$, and it could be interpreted as the solution to the prediction problem if the current observation belonged to that class. Assuming that the observation belongs to class $k$, the sources of that class can be estimated using the ICAMM parameters:

$$\mathbf{s}_k = \mathbf{W}_k(\mathbf{x} - \mathbf{b}_k) = \left[\mathbf{W}_{\langle \mathbf{y}\rangle,k} \ \mathbf{W}_{\langle \mathbf{z}\rangle,k}\right]\begin{bmatrix} \mathbf{y} \\ \mathbf{z} \end{bmatrix} - \mathbf{W}_k\mathbf{b}_k = \mathbf{W}_{\langle \mathbf{y}\rangle,k}\mathbf{y} + \mathbf{W}_{\langle \mathbf{z}\rangle,k}\mathbf{z} - \mathbf{W}_k\mathbf{b}_k \qquad (3)$$

where $\mathbf{W}_{\langle \mathbf{y}\rangle,k}$ and $\mathbf{W}_{\langle \mathbf{z}\rangle,k}$ denote the columns of the demixing matrix $\mathbf{W}_k$ that multiply $\mathbf{y}$ and $\mathbf{z}$, respectively. $\mathbf{W}_{\langle \mathbf{y}\rangle,k}$ is of size $[M \times M_k]$ and $\mathbf{W}_{\langle \mathbf{z}\rangle,k}$ is size $[M \times M_{unk}]$. By taking the conditional expectation with respect to $(\mathbf{y}, C_k)$ on (3) and moving the terms around, we arrive to

$$\mathbf{W}_{\langle\mathbf{z}\rangle,k}\, E[\mathbf{z}|\mathbf{y}, C_k] = E[\mathbf{s}_k|\mathbf{y}, C_k] + \mathbf{W}_k\mathbf{b}_k - \mathbf{W}_{\langle\mathbf{y}\rangle,k}\mathbf{y} \qquad (4)$$

Equation (4) is an overdetermined system of equations with $M$ equations and $M_{unk}$ unknowns (the values $E[\mathbf{z}|\mathbf{y}, C_k]$). There are a number of ways to solve this system, such as using the pseudo-inverse of $\mathbf{W}_{\langle\mathbf{z}\rangle,k}$:

$$E[\mathbf{z}|\mathbf{y}, C_k] = \mathbf{W}_{\langle\mathbf{z}\rangle,k}{}^{+}\left(E[\mathbf{s}_k|\mathbf{y}, C_k] + \mathbf{W}_k\mathbf{b}_k - \mathbf{W}_{\langle\mathbf{y}\rangle,k}\mathbf{y}\right) \qquad (5)$$

where superindex + denotes the pseudo-inverse. Solving the system requires knowledge of conditional expectation of the sources, $E[\mathbf{s}_k|\mathbf{y}, C_k]$, which is generally unavailable. In order to obtain a closed-form solution, we assume that $E[\mathbf{s}_k|\mathbf{y}, C_k] \sim E[\mathbf{s}_k|C_k]$, where $E[\mathbf{s}_k|C_k]$ is estimated for each class directly from the original training set used to estimate the ICAMM parameters. If the sources are centered, $E[\mathbf{s}_k|C_k]$ becomes a vector of zeros.

The E-ICAMM algorithm is proposed in Table 1. The conditional probability $P(C_k|\mathbf{y})$ is computed using Bayes' rule from $p(\mathbf{y}|C_k)$, $k = 1\ldots K$, which may be learned using any statistical modeling from training data (a dimension-reduced ICAMM would be an option). We have named the above method E-ICAMM (Expectation using ICAMM). Table 1 also includes a procedure to compute the covariance of the prediction error $\mathbf{e} = (E[\mathbf{z}|\mathbf{y}] - \mathbf{z})$, $E[\mathbf{e}\mathbf{e}^T|\mathbf{y}] = \sum_{k=1}^{K} E[\mathbf{e}\mathbf{e}^T|\mathbf{y}, C_k]P(C_k|\mathbf{y})$, which can be used to estimate prediction error.

**Table 1.** E-ICAMM algorithm, including the estimation of the prediction error.

| |
|---|
| Initialize for $k = 1\ldots K$ |
| $E[\mathbf{s}_k|\mathbf{y}, C_k] = E[\mathbf{s}_k]$, $\Sigma_{\mathbf{s}_k|\mathbf{y}, C_k} = \mathbf{I}_M$ (an identity matrix of size $[M \times M]$) |
| Calculate the E-ICAMM solution |
| $E[\mathbf{z}|\mathbf{y}, C_k] = \mathbf{W}_{\langle\mathbf{z}\rangle,k}{}^{+}\left(E[\mathbf{s}_k|\mathbf{y}, C_k] + \mathbf{W}_k\mathbf{b}_k - \mathbf{W}_{\langle\mathbf{y}\rangle,k}\mathbf{y}\right)$, $k = 1\ldots K$ |
| $E[\mathbf{z}|\mathbf{y}] = \sum_{k=1}^{K} E[\mathbf{z}|\mathbf{y}, C_k]P(C_k|\mathbf{y})$ |
| Estimate prediction error |
| $E[\mathbf{z}\mathbf{z}^T|\mathbf{y}, C_k] = E[\mathbf{z}|\mathbf{y}, C_k]E[\mathbf{z}|\mathbf{y}, C_k]^T + \mathbf{W}_{\langle\mathbf{z}\rangle,k}{}^{+}\,\Sigma_{\mathbf{s}_k|\mathbf{y}, C_k}\left(\mathbf{W}_{\langle\mathbf{z}\rangle,k}{}^{+}\right)^T$ |
| $E[\mathbf{e}\mathbf{e}^T|\mathbf{y}, C_k] = E[\mathbf{z}\mathbf{z}^T|\mathbf{y}, C_k] - E[\mathbf{z}|\mathbf{y}, C_k]E[\mathbf{z}|\mathbf{y}]^T$ |
| $\quad - E[\mathbf{z}|\mathbf{y}]E[\mathbf{z}|\mathbf{y}, C_k]^T + E[\mathbf{z}|\mathbf{y}]E[\mathbf{z}|\mathbf{y}]^T$ |
| $E[\mathbf{e}\mathbf{e}^T|\mathbf{y}] = \sum_{k=1}^{K} E[\mathbf{e}\mathbf{e}^T|\mathbf{y}, C_k]\, P(C_k|\mathbf{y})$ |

In practice, the ICAMM parameters are not known beforehand, and instead they have to be estimated from training data. The estimation algorithm depends on the kind of data available for training. For supervised data, the parameters of each chain can be calculated using any of the traditional ICAMM estimation algorithms (e.g., [2, 3]).

## 4  Experiments on Simulated Data

The classification performance of E-ICAMM was measured by Monte Carlo experiments on several sets of simulated data. E-ICAMM was compared with the following methods: Ordinary Kriging [6]; a Wiener structure that implemented a nonlinear stage after the Ordinary Kriging prediction [12]; matrix completion via the smoothed rank function method (SRF) [13]; and Splines [9]. The reasons for selecting each of these methods were explained in the Introduction. During each iteration of the simulations, the data were drawn from a randomly-initialized ICA Mixture Model. We considered 8 different sets of parameters for the ICAMM, with different number of variables $(M, M_k, M_{unk})$, number of classes $(K)$, and type of independent sources: uniform (U), Laplacian (L), K distributed with shape parameter $v = 1$ $(K1)$, and K distributed with $v = 10$ (K10). Details on each set are shown in Table 2. Regardless of type, all sources were normalized to zero mean and unit variance. During each iteration of the Monte Carlo experiments, $N = 1000$ data were generated from each set of ICAMM parameters. Then, the first half of the data was used to train the considered methods (e.g., obtaining an estimated ICAMM for E-ICAMM) and the performance of the proposed methods was tested on the second half of the data. Prediction performance was

**Table 2.** Details of all sets of parametes for ICAMM used in the simulations.

| Set # | 1 | 2 | 3 | 4 | 5 | 6 | 7 | 8 |
|---|---|---|---|---|---|---|---|---|
| $K$ | 2 | 2 | 2 | 2 | 2 | 3 | 3 | 3 |
| $M$ | 3 | 4 | 4 | 4 | 3 | 4 | 3 | 4 |
| $M_k$ | 2 | 2 | 2 | 2 | 2 | 2 | 2 | 2 |
| $M_{unk}$ | 1 | 2 | 2 | 2 | 1 | 2 | 1 | 2 |
| Sources | U, L | U, L | L, K1 | K1, K10 | U, L, K1 | U, L, K1 | U, L, K1, K10 | U, L, K1, K10 |

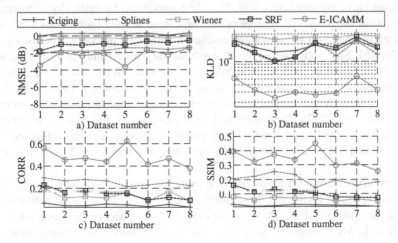

**Fig. 1.** Performance indicators for the 8 datasets.

estimated by four error indicators: the Normalized Mean Square Error (NMSE); the Kullback-Leibler Divergence (KLD); the correlation between the true data and the estimated data (CORR); and the Structural Similarity Index (SSIM, [14]). This process was repeated 500 times per set of ICAMM parameters, for a grand total of 6,500 iterations, and results were averaged for each set. The results for the eight sets of ICAMM parameters are shown in Fig. 1. E-ICAMM yielded the best performance in all cases.

## 5   Application on Seismic Signals for Underground Surveying

The basic technique of seismic exploration [15] consists of generating seismic waves and measuring the response in a series of geophones, usually disposed along a straight line directed toward the source. Typically, the recorded data are subject to signal processing techniques to enhance their quality and to improve their interpretation by the user (see [16] and references therein). There are multiple applications for seismic exploration, the most important of which is underground exploration, for example, to locate mineral deposits [17]. In this work, the proposed prediction methods were tested on a public dataset from BP Amoco [18] (available at http://ahay.org/data/bppublic/ PUBLIC_2D_DATASETS/2.5d/). Out of the whole data set, a single 2-D slice of data (also known as seismogram) was chosen for this experiment. The selected seismogram comprises 240 traces, each one 352 samples long. The vertical sampling period is 9.9 ms and the horizontal sampling period is 25 m. For the experiment, it was assumed that several seismic traces were completely or partially corrupted and the proposed prediction methods were used to interpolate the missing information in these traces. This kind of scenario is relatively common in seismic studies, since they involve large amounts of devices over a large area of terrain, in hostile environments or under unfavorable meteorological conditions.

The selected seismogram was treated as an image. In order to apply E-ICAMM, the image was split into squares or "patches" of fixed size, $[L \times L]$. Then, each patch is transformed into a column vector with $M = L^2$ components by vertically concatenating the columns of the patch. These vectors are considered as the observations $\mathbf{x}(n)$, which are then partitioned into in known and unknown data. This patching process was used for E-ICAMM, SRF and Wiener, but it is unnecessary for Kriging and Splines, which are already designed to work with two-dimensional data. In this work, we considered patches of size $[8 \times 8]$ samples, which were converted to column vectors of $M = 64$ variables. This patch size is typical in image processing applications, owing in part to the importance of JPEG compression and other methods related to the discrete cosine transform (DCT, [19]), and patch size is usually determined using rather empirical methods. Given the size of the seismogram, this resulted in 1320 patches of size $[8 \times 8]$.

The experiment was set up as follows. First, the seismogram was split in $[8 \times 8]$ patches as explained above and these patches were used to train the parameters of the proposed methods. For E-ICAMM, a one-class ICAMM was trained, thus $M = L^2 = 64$ and $K = 1$. After training, a number of traces were removed completely (marked as missing) from the seismogram and the proposed methods were used to reconstruct these missing traces. To determine performance across the whole seismogram, missing traces were positioned so that every patch was missing $\kappa$ traces, that is, missing $\kappa L$ values. For instance, if 6 traces were removed from a $[8 \times 8]$ patch, this means that 48 out of 64 values were missing. In order to test the behavior of the proposed methods with respect to the number of missing data, the number of missing traces per patch waschanged from 1 to 6.

Figure 2 shows the performance of the proposed methods. The results obtained by SRF for this application were considerably lower than those yielded by the other methods, and thus SRF was omitted from the figure for clarity. Results varied depending on the chosen performance indicator; while CORR and KLD show Kriging as the worst predictor, its values of NMSE and SSIM were sometimes close to those of E-ICAMM. In general, Wiener structures achieve a middle ground between Kriging and the proposed non-linear predictor. However, the value of SSIM obtained with the Wiener predictor is much lower than that of the other methods, which indicates that the Wiener method did not properly reconstruct the local structures in the data. Finally, E-ICAMM yielded the best results for all considered error indicators.

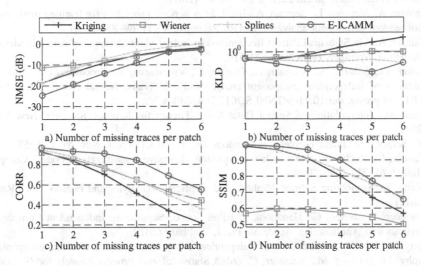

**Fig. 2.** Performance indicators for the prediction of seismic data for patches of size $[8 \times 8]$.

## 6 Conclusions

This paper has presented a new closed-form non-linear LMSE: E-ICAMM. The predictor is derived assuming a non-Gaussian mixture model (ICAMM) for the joint probability density of the observations. The versatility and generality of ICAMM allows a wide scope of applications were the new predictor, which is optimal for the assumed conditions, can outperform other methods. This was verified by comparison with four other well-known prediction approaches which cover a wide range of alternative predictors. The proposed method outperformed the other considered methods in a wide array of simulated conditions, as well as for the prediction of missing reflection seismology data from a real application. There are several lines of research for future development of the method. An important issue that can be developed is the incorporation of prior knowledge in ICAMM as it has been done in semi-blind source separation for single ICA models [20].

**Acknowledgment.** This work was supported by Spanish Administration and European Union under grant TEC2014-58438-R, and *Generalitat Valenciana* under grant PROMETEO II/2014/032.

# References

1. Einicke, G.: Smoothing, Filtering and Prediction - Estimating The Past, Present and Future. InTech, Rijeka (2012)
2. Lee, T., Lewicki, S., Sejnowski, T.: ICA mixture models for unsupervised classification of non-gaussian classes and automatic context switching in blind signal separation. IEEE Trans. Pattern Anal. Mach. Intell. **22**(10), 1078–1089 (2000)
3. Salazar, A., Vergara, L., Serrano, A., Igual, J.: A general procedure for learning mixtures of independent component analyzers. Pattern Recognit. **43**(1), 69–85 (2010)
4. Salazar, A.: On Statistical Pattern Recognition in Independent Component Analysis Mixture Modelling. Springer, Heidelberg (2013)
5. Safont, G., Salazar, A., Vergara, L., Gomez, E., Villanueva, V.: Probabilistic distance for mixtures of independent component analyzers. IEEE Trans. Neural Netw. Learn. Syst. (2017, in press). doi:10.1109/TNNLS.2017.2663843
6. Stein, M.: Interpolation of Spatial Data: Some Theory for Kriging. Springer, New York (2012)
7. Wiener, N.: Nonlinear Problems in Random Theory. M.I.T. Press, New York (1958)
8. Candès, E., Recht, B.: Exact matrix completion via convex optimization. Found. Comput. Math. **9**(6), 717–722 (2009)
9. Wang, Y.: Smoothing Splines: Methods and Applications. Taylor and Francis, Boca Raton (2011)
10. Common, P., Jutten, C.: Handbook of Blind Source Separation: Independent Component Analysis and Applications. Academic Press, Waltham (2010)
11. Jung, T., Lee, T.: Applications of independent component analysis to electroencephalography. In: Wenger, M., Schuster, C. (eds.) Statistical and Process Models for Cognitive Neuroscience and Aging. Psychology Press, Hove (2012)
12. Vergara, L., Bernabeu, P.: Simple approach to nonlinear prediction. Electron. Lett. **37**, 926–928 (2001)
13. Ghasemi, H., Malek-Mohammadi, M., Babaie-Zadeh, M., Jutten, C.: SRF: matrix completion based on smoothed rank function, In: IEEE International Conference on Acoustics, Speech and Signal Processing (ICASSP) (2011)
14. Wang, Z., Bovik, A., Sheikh, H., Simoncelli, E.: Image quality assessment: from error visibility to structural similarity. IEEE Trans. Image Process. **13**(4), 600–612 (2004)
15. Gadallah, M., Fisher, R.: Applied Seismology: A Comprehensive Guide to Seismic Theory and Application. PennWell Books, Tulsa (2005)
16. Zhou, H.: Practical Seismic Data Analysis. Cambridge University Press, New York (2014)

17. Cheraghi, S., Craven, J., Bellefleur, G.: Feasibility of virtual source reflection seismology using interferometry for mineral exploration: a test study in the lalor lake volcanogenic massive sulphide mining area, Manitoba, Canada. Geophys. Prospect. **63**(4), 833–848 (2015)
18. Etgen, J., Regone, C.: Strike shooting, dip shooting, widepatch shooting-does prestack depth migration care? A model study, In: 58th Annual SEG International Meeting and Exposition, pp. 66–69 (1998)
19. Li, X.: Patch-based image processing: from dictionary learning to structural clustering. In: Lukac, R. (ed.) Perceptual Digital Imaging: Methods, and Applications, pp. 223–250. CRC Press, Boca Raton (2012)
20. Llinares, R., Igual, J., Salazar, A., Camacho, A.: Semi-blind source extraction of atrial activity by combining statistical and spectral features. Digital Signal Process.: Rev. J. **21**(2), 391–403 (2011)

# Synchronized Multi-chain Mixture of Independent Component Analyzers

Gonzalo Safont$^{(\boxtimes)}$, Addisson Salazar, Ahmed Bouziane,
and Luis Vergara

Institute of Telecommunications and Multimedia Applications (iTEAM),
Universitat Politècnica de València,
c/Camino de Vera s/n, 46022 Valencia, Spain
gonsaar@upvnet.upv.es
{asalazar,lvergara}@dcom.upv.es

**Abstract.** This paper presents a novel method for modeling the joint behavior of a number of synchronized Independent Component Analysis Mixture Models (ICAMM), which we have named Multi-chain ICAMM (MCICAMM). This allows flexible estimation of complex densities of data, subspace classification, blind source separation, accurate local dynamic learning, and global dynamic interaction. Furthermore, the proposed method can also be used for classification following the maximum a posteriori, forward-backward, or Viterbi procedures. MCICAMM outperformed competitive methods such as ICAMM, SICAMM, and Dynamic Bayesian Networks for the classification of simulated data and the automatic staging of electroencephalographic (EEG) data from epileptic patients performing a neuropsychological test for short-term memory. Therefore, the potential of the method to suit different kind of data densities and to deal with the changing non-stationarity and non-linearity of brain dynamics was demonstrated. MCICAMM parameters provide a structured result that might be interpreted in several applications.

**Keywords:** Dynamic modeling · ICA · HMM · Non-Gaussian · Non-parametric estimation · EEG

## 1 Introduction

Statistical modeling methods pursue an approximated mathematical description of the underlying data-generating process from a certain phenomenon under analysis. In order to simplify this process, most methods assume that the data are stationary. This assumption, however, is not valid in many real-world applications. This difficulty is solved by introducing non-stationarity in the model using e.g. Hidden Markov Models (HMM, [1]), which are usually based on simple linear models. In cases where linear models are not enough to reproduce the dynamics of the data, non-stationarity is treated by considering non-linearities in the probability modeling, e.g., by variational learning of non-linear state-space models [2] or Extended Kalman Filters augmented with local searches [3]. Statistical dynamic models have been used in EEG (electroencephalographic) signal analysis, e.g., brain oscillation analysis [4] and decoding upper limb

I. Rojas et al. (Eds.): IWANN 2017, Part I, LNCS 10305, pp. 190–198, 2017.
DOI: 10.1007/978-3-319-59153-7_17

movement [5]. In general, these analyses assume methods based on Gaussian Mixture Models (GMM), which provide adequate statistical modeling capabilities but it is difficult to associate the model with real physical phenomena.

We propose in this paper a dynamic modeling method based on mixtures of independent component analyzers (ICA). Briefly, ICA is a blind source separation technique that models the observations as a linear mixture of a set of statistically-independent non-Gaussian sources [6]. ICA has been used in many applications on real data (e.g., [7, 8]). The parameters of ICA have been shown to be related with physical processes such as brain sources or similarities between ICA and image processing in the visual cortex [7]. ICA was recently extended to ICA Mixture Models (ICAMM, [9]), where sources from the same class are still independent, but dependencies (non-linearities) between sources from different classes is considered. ICAMM has been successfully applied to different fields such as signal reconstruction [10] and change detection [11].

The method proposed here defines a general framework to characterize the joint dynamic behavior of a number of synchronized ICAMM models, which we have named multi-chain ICAMM (MCICAMM). MCICAMM jointly considers the degrees of freedom provided by multiple ICAMM to model complex masses of densities, subspace classification, and blind source separation, besides accurate local dynamic learning, and global dynamic interaction. Therefore, this method should be suitable to deal with a broad range of real problems where higher flexible density estimation, accurate detection and characterization of changes in the data dynamics are required. Besides, mixing matrices, bias vectors, and transition matrices estimated by MCICAMM can be also used for further analyses. In this work, we applied MCICAMM to EEG signal analysis.

## 2  Independent Component Analysis Mixture Models

Independent Component Analysis [6] assumes that the observation at time $n$, $\mathbf{x}(n)$, can be modeled as an instantaneous linear transformation of a set of independent sources, $\mathbf{s}(n)$, $\mathbf{x}(n) = \mathbf{A} \cdot \mathbf{s}(n)$. This model was extended to an ICA Mixture Model in [9], where it is assumed that the data are separated in $K$ mutually-exclusive classes and each class is modeled using a different ICA. Therefore, $\mathbf{x}_k(n) = \mathbf{A}_k \cdot \mathbf{s}_k(n) + \mathbf{b}_k$, $k = 1, \ldots, K$, where $k$ is the class at time $n$, denoted by $C_k(n)$; $\mathbf{A}_k$ and $\mathbf{s}_k(n)$ are respectively the mixing matrix and the sources of the ICA of class $k$; and $\mathbf{b}_k$ are the corresponding bias vectors. Essentially, $\mathbf{b}_k$ determines the location of the $k$-th cluster and $\mathbf{A}_k$, $\mathbf{s}_k$ determine its shape. For simplicity, we will assume that the mixing matrices can be inverted to find the demixing matrices, $\mathbf{W}_k$.

ICA and ICAMM are usually estimated under the assumption that the data are time independent. This assumption is relaxed in Sequential ICA Mixture Models (SICAMM) [12, 13]. SICAMM is a non-linear hidden dynamic model where each state is associated with a class of an ICA Mixture Model. In essence, the class change is modeled as a first-order Markov process (or Markov chain). Thus, the class at the current time instant is independent from the class at past time instants, given the class at time $n - 1$. The observations during each hidden state (class) are modeled by an ICA, jointly considering ICAMM and Hidden Markov Models.

## 3 Multi-chain ICAMM (MCICAMM)

We propose a general framework to characterize the joint behavior of $L$ synchronized ICA Mixture Models (or "chains"), which we have called Multi-chain ICAMM (MCICAMM). MCICAMM jointly considers the degrees of freedom provided by ICAMM to model complex masses of densities, subspace classification, and hidden source separation, providing accurate local dynamic learning and global dynamic interaction. The temporal dependences between the $L$ groups of data, each modeled by a different mixture of ICA, are characterized by conditional probabilities that model the temporal influences between classes. As a matter of fact, in order to model asymmetrical and complex dependences, the class of each chain at time $n$ is dependent on the classes of all chains at time $n - 1$.

The degrees of freedom of MCICAMM allow it to accurately model complex local non-Gaussian probability densities without losing global modeling capabilities, while also considering time dependences. Furthermore, it is known that ICA can produce not only a valid statistical model of the data, but also sources with physical or physiological meaning, such as the extraction of physiologically significant sources for EEG data (e.g., [7]). This capability is inherited by the proposed method. Thus, unlike models that are purely statistical in nature, MCICAMM can obtain a representation of the data that reflects both their probability density function and their underlying generating models that can be related with the analyzed physical phenomenon.

Before tackling the model itself, we will define several notations which will ease the theoretical development of MCICAMM. Let us assume that there are $L$ data chains, and that the parameters for the $l$-th chain will be denoted by an $^{(l)}$ superscript. The observation from the $l$-th chain at time $n$ is denoted by $\mathbf{x}^{(l)}(n)$, and it is a vector of size $M^{(l)}$. Each chain is assumed to have been modeled by an ICAMM with $K^{(l)}$ classes whose parameters are: $\mathbf{W}_{k_l}^{(l)}$, $\mathbf{s}_{k_l}^{(l)}(n)$, $\mathbf{b}_{k_l}^{(l)}$, $k_l = 1, \ldots, K^{(l)}$. The set of observations from all the chains at a given time is $\mathbf{x}(n) = [\mathbf{x}^{(1)}(n)^T, \mathbf{x}^{(2)}(n)^T, \ldots, \mathbf{x}^{(L)}(n)^T]^T$, and the history of all these sets of observations up to time $n$ is $\mathbf{X}(n) = [\mathbf{x}(0), \mathbf{x}(1), \ldots, \mathbf{x}(n)]$. Finally, we define the class vector $\mathbf{k} = [k_1, k_2, \ldots, k_L]^T$ as a particular combination of classes from each one of the $L$ chains, with $1 \leq k_l \leq K^{(l)}$. The classes from all models at time $n$ are denoted by $\mathbf{c_k}(n) = [C_{k_1}(n), \ldots, C_{k_L}(n)]^T$.

In MCICAMM, the Markov assumption is extended from each individual chain to the joint sets by defining $p(\mathbf{X}(n)|\mathbf{c_k}(n)) = p(\mathbf{x}(n)|\mathbf{c_k}(n)) \cdot p(\mathbf{X}(n-1)|\mathbf{c_k}(n))$. Thus, $p(\mathbf{X}(n)|\mathbf{c_k}(n))$ is a function of the current set of observations for all chains and the probability density of the previous set of observations, $\mathbf{X}(n-1)$. In order to simplify the model, the conditional independence of the data is extended from single chains to the whole MCICAMM chain structure, therefore:

$$p(\mathbf{x}(n)|\mathbf{c_k}(n)) = \prod_{l=1}^{L} |\det(\mathbf{W}_{k_l}^{(l)})| \cdot \prod_{m=1}^{M^{(l)}} p_s\left(\left(\mathbf{w}_{k,m}^{(l)}\right)^T \cdot \left(\mathbf{x}^{(l)}(n) - \mathbf{b}_{k_l}^{(l)}\right)\right) \tag{1}$$

MCICAMM can perform classification by MAP (maximum a posteriori) estimation, i.e., maximizing $P(\mathbf{c_k}(n)|\mathbf{X}(n))$. This probability can be estimated considering the multiple chains in MCICAMM as follows:

$$P(\mathbf{c_k}(n)|\mathbf{X}(n)) = \frac{p(\mathbf{x}(n)|\mathbf{c_k}(n)) \cdot p(\mathbf{c_k}(n)|\mathbf{X}(n-1))}{\sum\limits_{k_1'=1}^{K^{(1)}} \sum\limits_{k_2'=1}^{K^{(2)}} \cdots \sum\limits_{k_L'=1}^{K^{(L)}} p(\mathbf{x}(n)|\mathbf{c_{k'}}(n)) \cdot p(\mathbf{c_{k'}}(n)|\mathbf{X}(n-1))} \tag{2}$$

where $p(\mathbf{x}(n)|\mathbf{c_k}(n))$ is obtained using (1) and $P(\mathbf{c_k}(n)|\mathbf{X}(n-1))$ can be estimated from the posterior probabilities of the previous time instant:

$$P(\mathbf{c_k}(n)|\mathbf{X}(n-1)) = \sum\limits_{k_1'=1}^{K^{(1)}} \sum\limits_{k_2'=1}^{K^{(2)}} \cdots \sum\limits_{k_L'=1}^{K^{(L)}} \pi_{\mathbf{kk'}} \cdot P(\mathbf{c_{k'}}(n-1)|\mathbf{X}(n-1)) \tag{3}$$

where $\pi_{\mathbf{kk'}}$ is the transition probability from combinations of classes $\mathbf{k'} = [k_1', k_2', \ldots, k_L']^T$ to combination $\mathbf{k} = [k_1, k_2, \ldots, k_L]^T$. The initial values for (3) can be estimated as the prior probabilities of each combination of classes, $P(\mathbf{c_k}(0)|\mathbf{X}(-1)) = P(\mathbf{c_k}(0)) = \prod_{l=1}^{L} P(C_{k_l}(0))$. Other classification procedures that exploit the temporal dependences in the data are also possible, such as the forward-backward procedure and Viterbi decoding ([14, 15], respectively).

## 4  Simulations

The performance of MCICAMM was compared with that of the following methods: ICAMM; SICAMM; a Bayesian Network using GMM, named BN1; a Dynamic Bayesian Network that implemented a continuous HMM; and a Dynamic Bayesian Network that implemented a continuous Coupled HMM (CHMM). The Dynamic Bayesian Networks use GMM and were named DBN1 and DBN2, respectively. Classification uses MAP criterion (MCICAMM+MAP), Baum-Welch (MCICAMM+ BW), and Viterbi (MCICAMM+VI) procedures.

The data were randomly drawn from a MCICAMM model: two chains ($L = 2$), two classes ($K^{(1)} = K^{(2)} = 2$), and observations of dimension two in both chains ($M^{(1)} = M^{(2)} = 2$). These values were selected to obtain a model that was simple, yet informative enough to show the behavior of the proposed method. The mixing matrix and centroid for each class were randomly initialized. The sources followed a Laplacian distribution with zero mean and unit standard deviation. Transition probabilities of both chains were set using variables, $\alpha$ and $\beta$:

If the other chain was in class 1 at time $n - 1$ :

$$\pi_{11}^{(l)}(n) = \pi_{22}^{(l)}(n) = \alpha; \ \pi_{12}^{(l)}(n) = \pi_{21}^{(l)}(n) = 1 - \alpha$$

If the other chain was in class 2 at time $n - 1$ :

$$\pi_{11}^{(l)}(n) = \pi_{22}^{(l)}(n) = \alpha - \beta; \ \pi_{12}^{(l)}(n) = \pi_{21}^{(l)}(n) = 1 - (\alpha - \beta)$$

$$(4)$$

where $0 \leq \alpha \leq 1$ and $\alpha - 1 \leq \beta \leq \alpha, l = 1, 2$. $\alpha$ is the intra-chain dependence parameter (sets the time dependence of each chain with respect to past values of the same chain). Conversely, $\beta$ is the inter-chain dependence parameter (sets the time dependence of each chain with respect to past values of the other chain). Transition probabilities $\pi_{kk'}$ were calculated from these values.

For each iteration of the Monte Carlo experiment, a random MCICAMM model was set as indicated above and $N = 1024$ observations, $\mathbf{x}(n), n = 1, \ldots, N$, along with their respective classes, $\mathbf{c}_k(n)$, were randomly drawn from the model. Half the data was used for training and the other half was used for classification testing.

MCICAMM and DBN2 were configured as the multiple-chain model used for data generation ($L = 2; K^{(1)} = K^{(2)} = 2; M^{(1)} = M^{(2)} = 2$). For the other methods that use only one chain (ICAMM, SICAMM, BN1, DBN1), a single model was used for the data of both chains, $\mathbf{x}(n)$. These models were set to $K' = K^{(1)} \cdot K^{(2)}$ classes. Thus, ecah class of single chain methods was considered equivalent to one combination of classes of MCICAMM and DBN2. For instance, class 1 of SICAMM was considered as equivalent to the combination of classes $\mathbf{k}_1 = [1, 1]^T$ of MCICAMM, and different from any other combination of classes. Parameter estimation for the ICA-based methods was performed using supervised training; the ICAMM parameters were estimated using the MIXCA procedure [16] and transition probabilities were estimated by counting. The number of Gaussian components of GMM for each node was selected from 3 to 20 by minimizing the Akaike information criterion [17]. Similar number of components has been used in several applications on EEG data (e.g., [18]).

Two separate experiments were performed to consider the variations of intra-chain and inter-chain dependence. For the first experiment, $\beta = 0.1$ and $\alpha$ was changed from 0.6 to 0.99 (quasi-complete dependence) in steps of 0.025. For the second experiment, $\alpha = 0.8$ and $\beta$ was changed from 0.3 to 0.8 in steps of 0.06. The experiments were repeated 300 times for each $\alpha$, $\beta$ value performing 7500 iterations.

Figure 1 shows the average classification error rate with respect to variations of $\alpha$ (Fig. 1.a) and $\beta$ (Fig. 1.b). Results were consistent for both experiments. The non-dynamic methods (ICAMM and BN1) maintained their performance when sequential dependence increased, while most of the dynamic methods (DBN1, SICAMM and MCICAMM) increased their performance as $\alpha$ and $\beta$ increased. This improvement increased consistently with both $\alpha$ and $\beta$. The low performance of DBN2 was due to problems with estimating the model from the training data. SICAMM and MCICAMM performed better than DBN1, and MCICAMM achieved the best classification performance due to exploitation of time cross-dependences between chains. The best result was always yielded by MCICAMM+BW, which consistently obtained a classification error rate 0.1 lower than that DBN1.

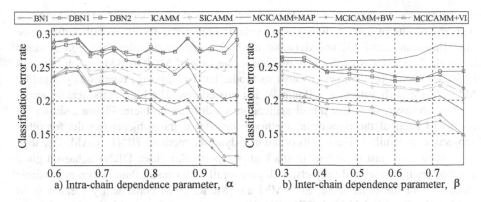

**Fig. 1.** Average classification error rate of the considered methods on data drawn from a MCICAMM with respect to: (a) intra-chain dependence, $\alpha$; (b) inter-chain dependence, $\beta$.

## 5   Experiments on Real Data

EEG is the record of electrical brain activity at the scalp level in real time. It has been used in many applications for medical diagnosis and recently in novel fields such as biometric identification/authentication [19]. MCICAMM was tested on a set of EEG signals from four epileptic patients performing a learning and memory neuropsychological test (Barcelona Neuropsychological Test - TB, [20]). In each trial, the participant is shown an abstract line figure during 10 s. There is a 1-second retention interval and afterward the participant is told to select the shown figure out of a group of four similar figures. The test consists of ten trials and scoring is given depending on the number of correct responses. TB was split in two classes, [stimulus+retention] vs. [response], and the methods were used to classify brain activity.

MCICAMM and DBN2 used two chains, one for each brain hemisphere, while the rest of the methods used all 19 channels at once. Both chains considered the same classes: in this case, chain structure is used to isolate the contributions of each brain hemisphere. Such division could be used, for instance, to determine hemispheric dominance during certain tasks and measure spatial neglect [21]. In terms of the MCICAMM parameters, we have: $L = 2$ models (one for each brain hemisphere); $K^{(1)} = K^{(2)} = 2$ classes ([stimulus+retention] vs. [response]); and $M^{(1)} = M^{(2)} = 9$ EEG channels (channels on the left side of the head vs. channels on the right side of the head).

The parameters of each method were estimated using supervised training on the first half of the data and classification performance was tested on the second half of the data. Given the vastly-different prior probabilities of each class, classification performance was measured using the balanced error rate (BER) and Cohen's kappa coefficient ($\kappa$, [22]). BER is in interval [0, 1], and a low value is better than a high value. Cohen's kappa is in interval [−1, 1] and a high value is better than a low value.

Table 1 shows the average BER and kappa across the four subjects. In concordance with the results in Sect. 4, ICA-based methods obtained a better overall performance than Bayesian networks, and dynamic methods performed better than non-dynamic methods. The latter is particularly marked in the case of MCICAMM, which obtained noticeably better results than the other methods. The best average result was yielded by MCICAMM with Viterbi and Baum-Welch procedures. The high classification errors overall, however, show that this classification is a difficult problem. Figure 2 shows the results of each of the considered methods for one of the subjects (results for other subjects were similar). The results of the non-dynamic methods (BN1, ICAMM) tended to oscillate very fast or to remain stuck at one particular class. DBN1 achieved good and bad results in several subjects, with an overall better result than both non-dynamic methods. On the other hand, SICAMM and MCICAMM consistently yielded good results. Finally, MCICAMM+BW and +VI showed very few rapid changes, yielding very smooth classifications.

**Table 1.** Average results of the automatic staging of EEG data for the four subjects.

| Method | ICAMM | SICAMM | MCICAMM | | | BN1 | DBN1 | DBN2 |
|--------|-------|--------|------|-----|-----|-----|------|------|
| | | | +MAP | +BW | +VI | | | |
| BER | .374 | .266 | .220 | .172 | .189 | .434 | .373 | .324 |
| Kappa | .235 | .417 | .489 | .591 | .574 | .109 | .218 | .323 |

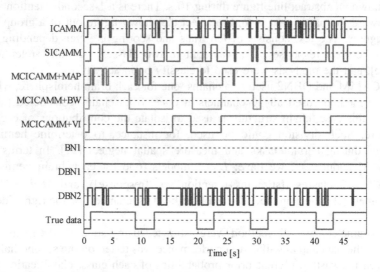

**Fig. 2.** Results for the automatic staging of EEG data for one of the subjects. In all cases, "high" represents the [response] class and "low" represents [stimulus+retention] class.

# 6   Conclusions

We have introduced a novel method to characterize the joint behavior of synchronized ICA Mixture Models (ICAMM), which we have called Multi-chain ICAMM (MCICAMM). In MCICAMM, subspace classification, learning of interaction dynamics, and hidden source separation are jointly considered. The capabilities of MCICAMM to model local and global dependencies were tested by measuring its classification performance on a large number of simulations and automatic staging of EEG signals from epileptic patients performing a neuropsychological memory test. MCICAMM outperformed the other competitive methods: ICAMM, SICAMM, Bayesian (BN) and Dynamic Bayesian Networks (DBN) that implemented a continuous CHMM. Furthermore, this improvement increased with intra- and inter-chain dependence. The configuration of MCICAMM allowed brain hemisphere analysis and accurate detection of small dynamic changes in each hemisphere. The improvement obtained by MCICAMM with respect to the best-performing Dynamic Bayesian Network was 0.15 for the balanced error rate and 0.27 for Cohen's kappa. Furthermore, the MCICAMM parameters provided a structured result that might be interpreted on its own in future works.

**Acknowledgments.** This work was funded by Spanish Administration and EU (TEC2014-58 438-R) and Generalitat Valenciana (PROMETEO II/2014/032).

# References

1. Cappe, O., Moulines, E., Ryden, T.: Inference in Hidden Markov Models. Springer, New York (2005)
2. Frigola, R., Chen, Y., Rasmussen, C.: Variational gaussian process state-space models. In: Advances Neural Information Processing Systems (NIPS), pp. 3680–3688 (2014)
3. Xu, K., Hero, A.: Dynamic stochastic blockmodels for time-evolving social networks. IEEE J. Sel. Top. Sig. Process. **8**(4), 552–562 (2014)
4. Neuper, C., Klimesch, W.: Event-Related Dynamics of Brain Oscillations. Elsevier, Amsterdam (2006)
5. Antelis, J., Montesano, L., Ramos, A., Birbaumer, N.: Decoding upper limb movement attempt from EEG measurements of the contralesional motor cortex in chronic stroke patients. IEEE Trans. Biomed. Eng. **64**(1), 99–111 (2017)
6. Common, P., Jutten, C.: Handbook of Blind Source Separation: Independent Component Analysis and Applications. Academic Press, Cambridge (2010)
7. Jung, T., Lee, T.: Applications of independent component analysis to electroencephalography. In: Wenger, M., Schuster, C. (eds.) Statistical and Process Models for Cognitive Neuroscience and Aging. Psychology Press (2012)
8. Llinares, R., Igual, J., Salazar, A., Camacho, A.: Semi-blind source extraction of atrial activity by combining statistical and spectral features. Digit. Sig. Process.: Rev. J. **21**(2), 391–403 (2011)
9. Lee, T., Lewicki, M., Sejnowski, T.: ICA mixture models for unsupervised classification of non-gaussian classes and automatic context switching in blind signal separation. IEEE Trans. Pattern Anal. Mach. Intell. **22**(10), 1078–1089 (2000)

10. Safont, G., Salazar, A., Rodriguez, A., Vergara, L.: On recovering missing ground penetrating radar traces by statistical interpolation methods. Remote Sens. **6**(8), 7546–7565 (2014)
11. Safont, G., Salazar, A., Vergara, L., Gomez, E., Villanueva, V.: Probabilistic distance for mixtures of independent component analyzers. IEEE Trans. Neural Netw. Learn. Syst., in press. doi:10.1109/TNNLS.2017.2663843
12. Salazar, A., Vergara, L., Miralles, R.: On including sequential dependence in ICA mixture models. Sig. Process. **90**, 2314–2318 (2010)
13. Safont, G., Salazar, A., Vergara, L., Rodriguez, A.: New applications of sequential ICA mixture models compared with dynamic bayesian networks for EEG signal processing. In: 5th International Conference Computational Intelligence, Communication Systems and Networks (2013)
14. Baum, L., Petrie, T., Soules, G., Weiss, N.: A maximization technique occurring in the statistical analysis of probabilistic functions of markov chains. Ann. Math. Stat. **41**(1), 164–171 (1970)
15. Viterbi, A.: Error bounds for convolutional codes and an asymptotically optimum decoding algorithm. IEEE Trans. Inf. Theory **13**(2), 260–269 (1967)
16. Salazar, A.: On Statistical Pattern Recognition in Independent Component Analysis Mixture Modelling. Springer, Heidelberg (2013)
17. Burnjam, K., Anderson, D.: Model Selection and Inference: A Practical Information-Theoretic Approach. Springer, Heidelberg (2013)
18. Thomas, E., Temko, A., Marnane, W., Boylan, G., Lightbody, G.: Discriminative and generative classification techniques applied to automated neonatal seizure detection. IEEE J. Biomed. Health Inform. **17**(2), 297–304 (2013)
19. Safont, G., Salazar, A., Soriano, A., Vergara, L.: Combination of multiple detectors for EEG based biometric identification/authentication. In: International Carnahan Conference on Security Technology (ICCST 2012), Article no. 6393564, pp. 230–236 (2012)
20. Quintana, M., et al.: Spanish multicenter normative studies (neuronorma project): norms for the abbreviated barcelona test. Arch. Clin. Neuropsychol. **26**(2), 144–157 (2011)
21. Dietz, M., Friston, K., Mattingley, J., Roepstorff, A., Garrido, M.: Effective connectivity reveals right hemisphere dominance in audiospatial perception: implications for models of spatial neglect. J. Neurosci. **34**(14), 5003–5011 (2014)
22. Gwet, K.: Handbook of Inter-Rater Reliability. Advanced Analytics LLC, Gaithersburg (2014)

# Pooling Spike Neural Network for Acceleration of Global Illumination Rendering

Joseph Constantin[1,3]([✉]), Andre Bigand[2], and Ibtissam Constantin[1,3]

[1] Laboratoire de Physique Appliquée, Université Libanaise, Faculté des Sciences 2, Campus Fanar, BP 90656, Jdeideh, Lebanon
{cjoseph,ibtissamconstantin}@ul.edu.lb
[2] LISIC, ULCO, 50 rue F. Buisson, BP 719, 62228 Calais Cedex, France
bigand@lisic.univ-littoral.fr
[3] Applied Mathematics Department, Lebanese University, Faculty of Sciences 2, Campus Fanar, BP 90656, Jdeideh, Lebanon

**Abstract.** The generation of photo-realistic images is a major topic in computer graphics. By using the principles of physical light propagation, images that are indistinguishable from real photographs can be generated. However, this computation is a very time-consuming task. When simulating the real behavior of light, individual images can take hours to be of sufficient quality. This paper proposes a bio-inspired architecture with spiking neurons for acceleration of global illumination rendering. This architecture with functional parts of sparse encoding, learning and decoding consists of a robust convergence measure on blocks. Feature, concatenation and prediction pooling coupled with three pooling operators: convolution, average and standard deviation are used in order to separate noise from signal. The pooling spike neural network (PSNN) represents a non-linear mapping from stochastic noise features of rendering images to their quality visual scores. The system dynamic, that computes a learning parameter for each image based on its level of noise, is a consistent temporal framework where the precise timing of spikes is employed for information processing. The experiments are conducted on a global illumination set which contains diverse image distortions and large number of images with different noise levels. The result of this study is a system composed from only two spike pattern association neurons (SPANs) suitably adopted to the quality assessment task that accurately predict the quality of images with a high agreement with respect to human psycho-visual scores. The proposed spike neural network has also been compared with support vector machine (SVM). The obtained results show that the proposed method gives promising efficiency.

**Keywords:** Dynamic learning · Global illumination · Pooling strategies · Sparse coding · Pooling spike neural network · Support vector machine

## 1 Introduction

Generating photo-realistic pictures is a very ambitious goal and it has been one of the major driving forces in computer graphics. Visual realism has always been

© Springer International Publishing AG 2017
I. Rojas et al. (Eds.): IWANN 2017, Part I, LNCS 10305, pp. 199–211, 2017.
DOI: 10.1007/978-3-319-59153-7_18

a strong motivation for research in the field and it is a selling point for many graphics-related commercially available product [1]. Formulating the global illumination problem as the rendering equation allows for a unified approach when computing images. One of the most influential consequences of the rendering equation was the development of Monte Carlo ray tracing [2]. Thus, it became possible to compute photo-realistic images assuming that the algorithm ran long enough in order to reduce noise to an acceptable level. Different strategies have been proposed to quickly compute the global illumination of a scene. Parker et al. described a programmable ray tracing engine designed for the GPU [3]. Another method for obtaining real-time frame rates is to sacrifice accuracy for speed. One such solution is to approximate indirect lighting reflected from surfaces as a set of virtual point lights [4].

Despite the different strategies used for speeding up computation, the processing costs of these methods remain proportional to the number of rays, which effectively limits the inter reflection effects they can simulate. Various perceptual models have been proposed to detect stochastic noise in global illumination algorithms. They are based on perceptual quality metrics [5] and visual attention [6]. However these models which require long computation time have been simplified but their simplifications have not been validated. They require some features and modifications in order to obtain accurate response.

In this paper, we investigate the use of machine learning in order to detect automatically the presence of noise in synthetic images. Building quality metrics based on machine learning for global illuminations have additional challenges over metrics for natural images. The metrics are often full-reference, namely they rely on a non-distorted copy of the image for evaluating the distorted one [7]. However in rendering, such a non-distorted image is not available and a blind quality assessment approach is desirable. In contrast non-reference image quality metrics are inferior in performance to full-reference metrics [8]. In a third type of methods, the reference image is only partially available, in the form of a set of extracted features made available as side information to help evaluate the quality of the distorted image. This is referred to as reduced-reference quality assessment [9]. These methods are limited to small sizes of images, due to the required number of kernels that grows with the size of the training set. In addition, they require large amounts of carefully chosen labeled images to tune the parameters of the perception model in order to give a good precision on the testing scenes. In the objective to solve these drawbacks, the authors realized a spike neural network (SNN) for detecting stochastic noise [10]. However this model, which is based on input vectors with 182 spikes, was unable to learn a complete global illumination scene with different levels of noisy images. In addition, the adopted receptive fields coding technique was not based on bio-inspired approaches in order to meet real time processing.

The main contribution of this paper is to design a simple biological architecture of pooling spike neural network (PSNN) with a small input vector of 26 spikes and two learning SPANs. The system is integrated in a consistent scheme by processing precise-timing spikes where sparse coding, learning and decoding are involved. Through learning, the neurons adapt their synaptic weights and detect surrounding environment by using a dynamic learning parameter for each sub-image on the learning scene based on its level of noise. We experiment the use of different strategies for feature, concatenation and prediction pooling. A comparative analysis of results between the PSNN and the SVM models is realized on different rendering images. We show the advantage of the proposed PSNN in terms of small number of parameters and accurate testing. The paper is structured as follows: Sect. 2 describes how to build the global illumination scenes set and the pooling layers structure, Sect. 3 explains the architecture of the PSNN perception model, whereas Sect. 4 shows the experimental results. Finally Sect. 5 states some conclusions and future research.

## 2  Global Illumination Set and Pooling Layers Structure

The scenes Bar, Class and Cube consist of diffuse surfaces with complex geometry (Fig. 1). In the Deskroom1, DeskRoom2, and Sponza scenes, objects with different shapes and material properties are placed together presenting rich and complicated shading variations under different lighting conditions (Fig. 2). These recent scenes start to be used for visualization with a good rendering and they are challenging for the new technique we propose. The images are cut into sixteen non-overlapping blocks of sub-images of size $128 \times 128$ pixels for the scenes with $512 \times 512$ resolution. We set the maximum number of rays per pixel to 10100 in order to obtain non-distorted copies of the images. The labeling process selects images computed using diffuse and specular rendering and ask the observers for their qualities. The observers which are from different locations have faced the same display conditions [9]. The learning and the testing sets contain images with different percentages of homogeneous regions, edges, and exhibit different light effects. The average number of rays that are required for each block to be perceived as identical to the reference one is shown in Fig. 3.

In order to separate the noise from signal, a sub-image is computed by applying to the luminance component different pooling strategies. The feature pooling is performed element by element on each version of the sub-image in a deep learning process using thirteen layers. We use Averaging (A1–A2), Gaussian (G1–G6) [11], Median (M1–M2) [12] and Wiener (W1–W2) [13] convolutions of depth

(a)            (b)            (c)

**Fig. 1.** (a) Blocks of the scene Bar, (b) blocks of the scene Class, (c) blocks of the scene Cube (Color figure online)

**Fig. 2.** (a) Blocks of the scene DeskRoom1, (b) blocks of the scene DeskRoom2, (c) blocks of the scene Sponza (Color figure online)

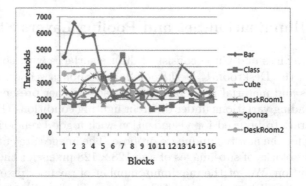

**Fig. 3.** Thresholds variation for the blocks of the scenes Bar, Class, Cube, DeskRoom1, Sponza and DeskRoom2

and spread equal one. Next, the image is denoised using Wavelet decomposition (Wav) [14] (see Table 1). The sub-image noise is estimated as a pixel subtraction between the current sub-image and the sub-image computed by each layer. The mean and the standard deviation pooling are applied to the thirteen activation layers. Next, the feature concatenation pooling is performed by concatenating the versions features vectors in a single longer feature vector in order to obtain a total of 26 noise features vector used as input to the PSNN model. The noise features are obtained by computing the difference between the noise features of quick ray traced sub-image of scene and the current one. Finally, the prediction pooling is performed for each sub-image of the scene.

**Table 1.** Feature pooling architecture

| Layers | L1 | L2 | L3 | L4 | L5 | L6 | L7 | L8 | L9 | L10 | L11 | L12 | L13 |
|---|---|---|---|---|---|---|---|---|---|---|---|---|---|
| Type | A1 | A2 | G1 | G2 | G3 | G4 | G5 | G6 | M1 | M2 | W1 | W2 | Wav |
| Size | 3 × 3 | 5 × 5 | 3 × 3 | 5 × 5 | 3 × 3 | 5 × 5 | 3 × 3 | 5 × 5 | 3 × 3 | 5 × 5 | 3 × 3 | 5 × 5 | |
| Standard deviation | | | 0.5 | 0.5 | 1 | 1 | 1.5 | 1.5 | | | | | |
| Padding | 1 | 2 | 1 | 2 | 1 | 2 | 1 | 2 | 1 | 2 | 1 | 2 | |

# 3    Architecture of the Pooling Spike Neural Network

The system is composed from three layers. The encoding layer generates a set of activity patterns that represents the pooling stochastic noise features, the learning layer tunes the SPANs parameters making sure that they respond to noise features correctly and the decoding layer extracts information about the global illumination images qualities [16] (Fig. 4).

## 3.1    Encoding Neurons

We provide a biologically coding method about how spikes could be generated from noise features. Such method which is based on the sub-threshold membrane oscillations $(SMOs)$ [17] can encode information using sparse coding with a good temporal selectivity. Each encoding neuron unit contains a positive, a negative and an output neurons. The encoding unit is connected to a noise feature signal $I$ and the $SMO$. The potential of the positive neuron is the summation of $I$ and the $SMO$. The potential of the negative neuron is the subtraction of $I$ and the $SMO$. The neuron will fire a spike if the membrane potential crosses the threshold $\theta_e$. The firing of either the positive neuron or the negative neuron will cause a spike from the output neuron (Fig. 4). The equation of the encoding unit is described as:

$$SMO_i = Gcos(wt + \vartheta_i) \tag{1}$$

where $G$ is the magnitude of the $SMO$, $w = 2\pi/T$ is the angular velocity and $\vartheta_i$ is the initial term defined as:

$$\vartheta_i = \vartheta_r + (i-1)\Delta\vartheta \quad \forall i = 1, 2 \ldots N \tag{2}$$

where $\vartheta_r$ is the reference term, $\Delta\vartheta$ is the difference between nearby encoding neurons and $N$ is the number of encoding neurons. We set $\vartheta_r = 0$, $\Delta\vartheta = 2\pi/N$, $G = 0.5$ and the threshold value $\theta_e = 0.5$ in order to obtain sparse spikes. Figure 5 shows the spikes of the scene Bar distributed sparsely for the 26 SPANs.

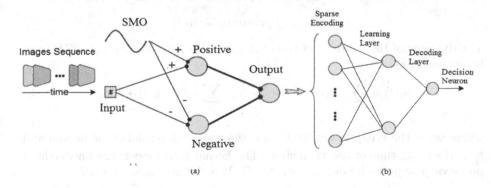

**Fig. 4.** (a) Encoding unit composed from positive, negative and output neurons. (b) Architecture of the PSNN from three layers

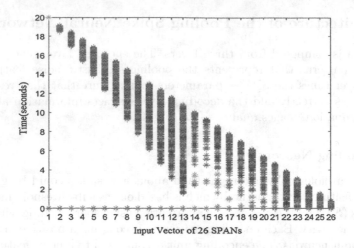

**Fig. 5.** Spikes of the scene Bar distributed sparsely for the 26 SPANS

## 3.2 Learning Neurons

The neurons are Leaky Integrate and Fire (LIF) described by the following equation [15]:

$$\tau_m \frac{du_i}{dt} = -u_i(t) + R_m I_i^{syn}(t) \tag{3}$$

where $I_i^{syn}$ is the input signal current, $\tau_m = R_m C_m$ is the membrane time constant, $R_m = 1M\Omega$ and $C_m = 15nF$ are the membrane resistance and capacitance respectively. The neuron fires a spike if its state variable $u_i$ crosses a predefined threshold $\theta$. The set of all firing times of a neuron $i$ is denoted by:

$$F_i = \left\{ t_i^f, 1 \leq f \leq n \right\} \equiv \{t, u_i(t) \geq \theta\} \tag{4}$$

After firing an output spike at time $t_i^f$, the state variable $u_i$ is reduced and then the neuron receives input from a set of pre-synaptic neurons $j \in \Gamma_i$, where:

$$\Gamma_i = \{j : j \ presynaptic \ to \ i\} \tag{5}$$

The dynamic of the neuron $i$ at time $t$ in the Spike Response Model (SRM) can be expressed as:

$$u_i(t) = \sum_{t_i^f \in F_i} \eta(t - t_i^f) + \sum_{j \in \Gamma_i} w_{ij} (\sum_{t_j^f} K(t - t_j^f)) + V_{rest} \tag{6}$$

where $w_{ij}$ is the synaptic weight, $V_{rest}$ is the resting potential of the neuron and $t_j^f$ is the firing time of the $j^{th}$ neuron. The kernel $K(.)$ models the un-weighted post-synaptic potential of a neuron $j \in \Gamma_i$. It is defined as follows [16]:

$$K(t - t_j^f) = V_0(exp(\frac{-(t - t_j^f)}{\tau_s}) - exp(\frac{-(t - t_j^f)}{\tau_f}))H(t - tj^f) \tag{7}$$

where $\tau_s$ and $\tau_f$ are the slow and fast decay constants respectively. $V_0$ is a normalization factor such that the maximum value of the kernel is 1. $H(t - t_j^f)$ is the heavy-side function which vanishes for $t_j^f \leq t$. The kernel $\eta(.)$ which models the membrane potential during a spike can hence be described by a certain standard time course defined by [18]:

$$\eta(t - t_i^f) = -\eta_0 exp(\frac{-(t - t_i^f)}{\tau})H(t - t_i^f) \qquad (8)$$

where $\eta_0$ is the amplitude of the relative refractoriness and $\tau$ is a decay time constant. The evaluation of neural dynamics is performed on a time step $\Delta t = 0.1\,ms$ which assures a stable PSNN. We find that the final time $t_f = 20\,ms$ is sufficient for the PSNN, so it can fire for the different sub-images [10]. The learning rule for the $nth$ image is defined as follows:

$$\Delta w_{ij}(n) = \begin{cases} \lambda\delta(n)U_t \ if \ P_m \\ -\lambda\delta(n)U_t \ if \ N_m \\ 0 \ otherwise \end{cases} \qquad (9)$$

where $U_t = \sum_{t_j^f < t_{max}} K(t_{max} - t_j^f)$, $Pm$ (Positive misclassified) denotes that the neuron should fire but its state is silent, $N_m$ (Negative misclassified) denotes that the neuron should keep silent but its state is firing and $t_{max}$ denotes the time at which the neuron reaches its maximum potential. The training is stopped either when the neuron successfully separates all training samples or when the maximum iterations is reached.

Employing only one fixed learning parameter $\lambda$ is not an optimal solution because it cannot detect surrounding environment in order to complete learning efficiently. Consequently, weights changes drastically or slowly in every learning period which leads to low learning efficiency. To tackle this problem, we propose a learning technique inspired from the perceptron based spiking learning rule defined in [19]. A dynamic learning parameter $\delta(n)$ is computed for the $nth$ noisy image in the learning base which is not classified correctly depending on its noise level. This parameter is computed as follows:

- In case of positive misclassified neurons, the dynamic learning parameter $\delta(n)$ is computed as the distance between the threshold $\theta$ and the maximum decoding neuron voltage. This method can avoid excessive weights modification if the neuron decoding voltage is close to the threshold.
- In case of negative misclassified neurons, the dynamic learning parameter $\delta(n)$ is expressed as the distance between its firing time and the nearest firing time for the next noisy images classified correctly. Using this technique, weights change slowly if the next noisy image firing time is close to its nearest classified image and conversely weights vary drastically.

The steps of the learning process are explained in Algorithm 1.

---
**Algorithm 1.** *Learning algorithm for the PSNN model*
___
**Require:** *a set of scenes where each scene is divided into* 16 *blocks of sub-images*
 1: Initialize the weights, the parameter $\lambda$ and dynamic the learning parameters $\delta$ for all sub-images extracted from the 16 blocks.
 2: Read randomly a sub-image $I$ from the set of scenes.
 3: Extract the luminance vector and apply to it the pooling strategies in order to obtain the 26 features.
 4: Provide the difference between the features of the test sub-image and a quick ray traced sub-image of the scene.
 5: Convert these features into spikes using *SMOs* encoding units.
 6: Apply the learning rule defined in Eq. (9) for each of the noisy sub-image.
 7: Repeat steps 2–4 for all the sub-images.
 8: Compute the dynamic learning parameters $\delta$ for all sub-images in case of positive or negative misclassified neurons.
 9: Repeat from step 2 until the maximum iteration is reached.
___

### 3.3 Decoding Layer

The decoding layer consists of two SPANS. The first SPAN fires a spike only for noisy images. The second SPAN fires a spike only for denoised images. The decision neuron selects the SPAN with the maximum voltage in order to decide about the quality of the image.

## 4 Experimental Results

In order make a comparative analysis between the PSNN and the SVM models, the sub-images of the scene Bar are used for learning and the sub-images of the scenes Class, Cube, DeskRoom1, DeskRoom2 and Sponza are used for testing. We apply the pooling strategies in order to select a vector of size 26 noise features as input to the SVM. The output of SVM is negative for noisy images and strictly positive otherwise. We use V-times cross validation techniques with different RBF kernels to determine the standard deviation $\sigma$ and the margin trade-off parameter $C$ [9]. The learning set is decomposed into 101 groups of size 16 records each. We find that the optimal values of $C$ and $\sigma$ are equals respectively to 64 and 4. The optimal number of support vectors is equal 258.25. The maximum precision is equal to 93.25% (Figs. 6 and 7).

**Table 2.** Parameters values of the SPAN

| Parameter | $\tau_s$ | $\tau_f$ | $\tau$ | $\theta$ | $V_0$ | $V_{rest}$ | $\eta_0$ |
|---|---|---|---|---|---|---|---|
| Value | 15 ms | 3.75 ms | 1 ms | 1 mV | 1 mV | 0 mV | 1 mV |

In case of PSNN, the parameters are chosen as shown in Table 2 and the input of the network is a vector of 26 sparse spikes. The initial weights are selected randomly in the interval $[0, 1]$ and the initial learning rate is selected equal to 0.01 as it assures a stable PSNN with a good convergence. During learning,

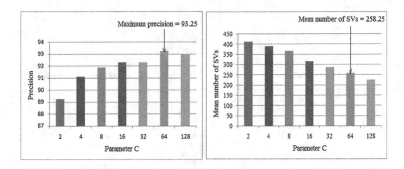

**Fig. 6.** Precision and mean number of SVs for different values of $C$

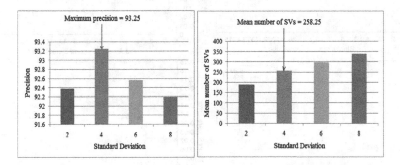

**Fig. 7.** Precision and mean number of SVs for different values of $\sigma$

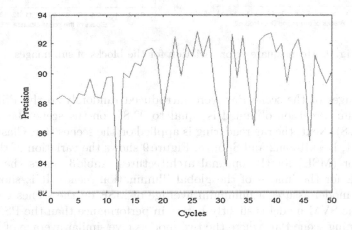

**Fig. 8.** Variation of precision during the learning process

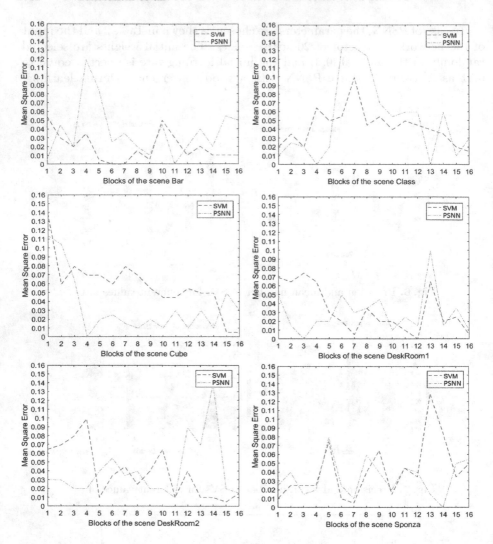

**Fig. 9.** Mean square error variation for the blocks of sub-images

the sub-images of the scene Bar were introduced randomly to the PSNN and the maximum precision obtained is equal to 92.8% on the scene Bar after 50 cycles (Fig. 8). Next, the ray rendering is applied on the scenes Bar, Class, Cube, DeskRoom1, DeskRoom2 and Sponza. Figure 9 shows the variation of the mean square error (MSE) for the optimal architectures. Table 3 shows the average of the MSE for the images of the global illumination scenes. It is shown that the PSNN model assures a minimum average of MSE on the scenes Cube and Sponza. The SVM model is slightly better in performance than the PSNN only on the learning scene Bar, where the two models give similar average of MSE on the scenes Class, DeskRoom1 and DeskRoom2. Figure 10 shows the variation of

the actual thresholds of the perception models and the desired human psycho-
visual scores. We find that the PSNN outperforms the SVM model on the scenes
Cube, DeskRoom1, DeskRoom2 and Sponza, whereas the two models give similar
scores on the scene Class. The SVM assures better convergence than the PSNN
only on the leaning scene Bar. However, the PSNN model needs less number of
parameters than SVM. The number of parameters is equal to $26 \times 258 = 6708$
for SVM, while the number of parameters is equal to $26 \times 2 = 52$ for PSNN.

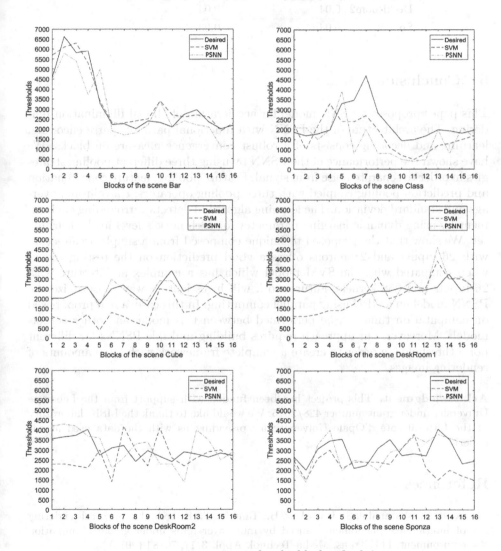

**Fig. 10.** Thresholds variation for the blocks of sub-images

**Table 3.** Average mean square error of the global illumination scenes

| Scene | SVM (average MSE) | PSNN (average MSE) |
|---|---|---|
| Bar | 0.02 | 0.03 |
| Class | 0.04 | 0.04 |
| Cube | 0.06 | 0.04 |
| DeskRoom1 | 0.03 | 0.03 |
| DeskRoom2 | 0.04 | 0.04 |
| Sponza | 0.04 | 0.03 |

## 5   Conclusion

This paper proposes a PSNN model for acceleration of Global Illumination rendering. The architecture of the PSNN with functional parts of sparse encoding, learning and decoding consists of a robust convergence measure on blocks. We have shown the performance of the PSNN by using three different pooling strategies in order to separate noise from signal. Feature pooling, feature concatenation and prediction pooling coupled with three pooling operators: convolution, average and standard deviation. The learning algorithm detects surrounding environment by using dynamic learning parameter for each noise's level in the learning set. We show that the proposed technique composed from a simple architecture with 26 inputs and 2 neurons offers a good prediction on the testing scenes when compared with the SVM model which has a complex architecture with 26 inputs and 258 kernels. Future work will investigate a way to optimize the PSNN model on a GPU using parallel computing. In this case, a real proof based on computation time can be performed between this model and other kernels models. Moreover, our approach requires building multiple PSNNs for different noise thresholds in order to create a complete framework with large amounts of rendering images.

**Acknowledgment.** This project has been funded with support from the Lebanese University under grant number 428/2015. We would like to thank the LISIC laboratory at the Littoral cote d'Opale University for providing us with the data used in our experiments.

## References

1. Ikeda, S., Watanabe, S., Raytchev, B., Tamaki, T., Kaneda, K.: Spectral rendering of interference phenomena caused by multilayer films under global illumination environment. ITE Trans. Media Technol. Appl. **3**(1), 76–84 (2015)
2. Hedman, P., Karras, T., Lehtinen, J.: Sequential Monte Carlo instant radiosity. In: Proceedings of the 20th ACM SIGGRAPH Symposium on Interactive 3D Graphics and Games, pp. 121–128 (2016)

3. Parker, S.G., Bigler, J., Dietrich, A., Friedrich, H., Hoberock, J., Luebke, D., McAllister, D., McGuire, M., Morley, K., Robinson, A., Stich, M.: OptiX: a general purpose ray tracing engine. ACM Trans. Graph. **29** (2010)

4. Thiedemann, S., Henrich, N., Grosch, T., Muller, S.: Voxel-based global illumination. In: I3D, pp. 103–110 (2011)

5. Volevich, V., Myszkowski, K., Khodulev, A., Kopylov, A.: Using the visual differences predictor to improve performance of progressive global illumination computation. ACM Trans. Graph. **19**(1), 122–161 (2000)

6. Shi, J., Yan, Q., Xu, L., Jia, J.: Hierarchical image saliency detection on extended CSSD. IEEE Trans. Pattern Anal. Mach. Intell. **38**(4), 717–729 (2015)

7. Demirtas, A., Reibman, A., Jafarkhani, H.: Full-reference quality estimation for images with different spatial resolutions. IEEE Trans. Image Process. **23**(5), 2069–2080 (2014)

8. Delepoulle, S., Bigand, A., Renaud, C.: A no-reference computer generated images quality metrics and its application to denoising. In: IEEE Intelligent Systems IS12 Conference, vol. 1, pp. 67–73 (2012)

9. Constantin, J., Bigand, A., Constantin, I., Hamad, D.: Image noise detection in global illumination methods based on FRVM. Neurocomputing **64**, 82–95 (2015)

10. Constantin, J., Constantin, I., Rammouz, R., Bigand, A., Hamad, D.: Perception of noise in global illumination algorithms based on spiking neural network. In: The IEEE Third International Conference on Technological Advances in Electrical, Electronics and Computer Engineering, pp. 68–73 (2015)

11. Makandar, A., Halalli, B.: Image enhancement techniques using highpass and lowpass filters. Int. J. Comput. Appl. **109**(14), 21–27 (2015)

12. Dawood, F., Rahmat, R., Kadiman, S., Abdullah, L., Zamrin, M.: Effect comparison of speckle noise reduction filters on 2D echocardiographic. World Acad. Sci. Eng. Technol. **6**(9), 425–430 (2012)

13. Biswas, P., Sarkar, A., Mynuddin, M.: Deblurring images using a Wiener filter. Int. J. Comput. Appl. **109**(7), 36–38 (2015)

14. Gao, D., Liao, Z., Lv, Z., Lu, Y.: Multi-scale statistical signal processing of cutting force in cutting tool condition monitoring. Int. J. Adv. Manuf. Technol. **90**(9), 1843–1853 (2015)

15. Mohemmed, A., Lu, G., Kasabov, N.: Evaluating SPAN incremental learning for handwritten digit recognition. In: Huang, T., Zeng, Z., Li, C., Leung, C.S. (eds.) ICONIP 2012. LNCS, vol. 7665, pp. 670–677. Springer, Heidelberg (2012). doi:10.1007/978-3-642-34487-9_81

16. Qiang, Y., Huajin, T., Kay, C.T., Haoyong, Y.: A brain inspired spiking neural network model with temporal encoding and learning. Neurocomputing **138**, 3–13 (2014)

17. Hu, J., Tang, H., Tan, K.C., Li, H., Shi, L.: A spike-timing-based integrated model for pattern recognition. Neural Comput. **251**(2), 450–472 (2013)

18. Pavlidis, N., Tasoulis, D., Plagianakos, V.P., Vrahatis, M.: Spiking neural network training using evolutionary algorithms. IEEE Int. Joint Conf. Neural Netw. **4**, 2190–2194 (2005)

19. Qu, H., Xie, X., Liu, Y., Zhang, M., Lu, L.: Improved perception based spiking neuron learning rule for real-time user authentication. Neurocomputing **151**, 310–318 (2015)

# Automatic Recognition of Daily Physical Activities for an Intelligent-Portable Oxygen Concentrator (iPOC)

Daniel Sanchez-Morillo[1(✉)], Osama Olaby[2],
Miguel Angel Fernandez-Granero[1], and Antonio Leon-Jimenez[3]

[1] Biomedical Engineering and Telemedicine Research Group,
University of Cadiz, Puerto Real, Cadiz, Spain
{daniel.morillo,ma.fernandez}@uca.es
[2] Department of Control Engineering and Automation,
University of Aleppo, Aleppo, Syria
osa_olaby@yahoo.com
[3] Pulmonology, Allergy and Thoracic Surgery Unit,
Puerta Del Mar University Hospital, Cadiz, Spain
antonio.leon.sspa@juntadeandalucia.es

**Abstract.** In recent years, new autonomous physiological close-loop controlled (PCLC) medical devices for oxygen delivery are being researched. Most of this PCLC devices are based on the feedback of arterial oxygen saturation, measured using a pulse oximeter. However, pulse oximeters may provide spuriously low or high $SpO_2$ values. In this work, a different approach to adjust automatically oxygen dosing in portable oxygen concentrators (POC) according to the physical activity performed by patients with COPD is presented. To that purpose, the ability of various machine-learning algorithms to recognize four human daily activities from sensor signals collected from a single waist-worn tri-axial accelerometer is evaluated. A set of 56 features was considered and recognition accuracy of up to 91.15% on the four activities of daily living was obtained using a SVM classifier. The associated activity recognition error rate was lower than 5%, ensuring a low percentage of time wrongly assigned to a certain activity. The underlying idea is the hardware implementation of the SVM classifier to control the oxygen flow in intelligent portable oxygen concentrators.

**Keywords:** COPD · Human daily activity recognition · Oxygen therapy · Long term oxygen therapy · Pervasive healthcare · Portable oxygen concentrator

## 1 Introduction

Oxygen is essential for human beings and other living organisms. Our cells need a constant supply of oxygen and lungs extract it from air and transfer it to blood, that supplies oxygen to body cells. Many diseases affect the lung's ability to transfer oxygen to blood what can cause hypoxemia, respiratory failure and death. Chronic Obstructive Pulmonary Disease (COPD) is the most frequent respiratory illness what lead to respiratory insufficiency. The global prevalence of COPD in people aged 40 and

© Springer International Publishing AG 2017
I. Rojas et al. (Eds.): IWANN 2017, Part I, LNCS 10305, pp. 212–221, 2017.
DOI: 10.1007/978-3-319-59153-7_19

over is 11.7% [1] and in 2013 is estimated as the fifth leading cause of reduced Disability-Adjusted life years across the world [2]. In addition, the economic burden of COPD is very high in developed countries. In the European Union, direct costs was estimated at over 38.6 billion euros a year [3].

While lung damage because of COPD is irreversible, there are treatments that can increases life expectancy in patients with severe resting hypoxemia, such as long-term oxygen therapy (LTOT) [4]. Oxygen therapy has long become a cornerstone in the treatment of patients with COPD and other hypoxemic and hypercapnic chronic respiratory diseases. Actually, clinical application for the use of oxygen has extended beyond the hospital setting to homes. Home oxygen therapy in COPD patients is usually coupled with physical rehabilitation what improves memory, physical performance, dyspnea and quality of life [5]. Therefore, the need for adequate and light portable oxygen devices to suit patient daily activities remains clear.

However, despite the variety of portable oxygen sources for COPD therapy available currently, existing devices cannot respond to all the requirements of patients. In fact, oxygen, used during COPD patient's therapy, is a drug and just like any other drug it has to be cautiously prescribed and monitored [6]. Oxygen needs for patients with COPD vary during sleep, due to nocturnal oxygen desaturations [7], rest and physical activity [8]. Therefore, oxygen should be delivered to avoid both hypoxia and hyperoxia along the changing daily activities.

In traditional flow oxygen delivery, the amount of administered oxygen is generally adjusted manually by selecting the prescribed level of oxygen flow (in l/min in continuous flow devices and in discrete levels in pulse-based devices). In LTOT, the titration of flow rate depends on the patient profile, him/her mobility, the adequate correction of arterial oxygen saturation ($SpO_2$) during stand and exercise, and other related factors. Flow rate settings can be adjusted by the patient either mechanically (by directly regulating the flow valve using a knob) or electronically, by using a keyboard to select an option among a short discrete range, usually from 1 through 5.

However, manually adjustment of the flow rate is a time-consuming task that requires experienced and trained patients [9]. These limitations convert systems for the automatic flow rate control in a desirable achievement.

In recent years, new autonomous physiological close-loop controlled (PCLC) medical devices for oxygen delivery are being researched. These PCLC devices may be able to automate manual adjustments during oxygenation. The potential benefits of automated oxygen therapy affect:

- patients, by improving control of oxygenation and monitoring
- health care systems by reducing workload, improving monitoring and compliance with recommendations, reducing oxygen use and promoting early hospital discharge following exacerbations of COPD) [10].

In most of these approaches, oxygen saturation ($SpO_2$), measured using a pulse-oximeter, is the process variable for the control algorithm [11]. Oxygen dosing is adjusted automatically to maintain a target arterial saturation, customized for the specific patients' needs. However, robustness of control algorithms, fail-safe mechanisms and limited sensors reliability are among the reasons for the lack of practical application of this new generation of oxygen devices [12].

In this work, we propose a different approach to adjust automatically oxygen dosing in patients with COPD. Human activities recognition from wearable devices is being applied for long-term recording and clinical access to patient's activity information [13, 14]. In this paper, we present and evaluate the ability of various machine-learning algorithms to recognize four human daily activities (walking, standing, walking upstairs and walking downstairs) from sensor signals collected from a single waist-worn tri-axial accelerometer. The underlying idea is that this algorithm would be implemented in a sensor device to control oxygen flow in portable oxygen concentrators. The oxygen flow would be automatically set according to the physical activity performed by the patient and real-time detected.

The rest of the paper is organised as follows. Section 2 presents an overview of the domiciliary oxygen devices. Section 3 details the proposed control scheme, the used dataset, features, and the evaluated machine learning algorithms. Section 4 presents the results of the evaluation of the validated algorithms. Finally, Sect. 5 captures the conclusions and the future work.

## 2    Background on Domiciliary Oxygen Devices

Selection of oxygen device is essential and must match the patient's profile [6]. In this regards, multiple modalities are becoming available for portable oxygen therapy: portable compressed gas cylinders (CGC), portable liquid oxygen systems (LOX), and portable oxygen concentrators (POCs).

CGC is a low cost and widely available source of LTOT. However, CGCs are heavy devices with some safety issues what make them inadequate for active patients.

Some studies showed benefits of LOX over gaseous oxygen in terms of patients' acceptability, duration and hours of therapy received [15]. POCs can work on both DC and AC voltages, and are ideal for home use since there is no need for regular refilling. In addition, the operating cost of a POC is minor [6].

A typical POC comprises an air compressor, two cylinders filled with zeolite pellets, a pressure-equalizing reservoir, some valves, tubing and a cooling system that keeps the POC from overheating. The control algorithm and performance differ among manufacturers and models [16].

While portable CGC and LOX devices are not adequate for COPD patients who need to develop his mobility daily, POCs have allowed many patients with chronic lung disease to travel and maintain active lifestyles [17]. The results from this study will ultimately applied to the automatic adjustment of the flow valve in POCs.

## 3    Methods

### 3.1    Proposed Intelligent Portable Oxygen Concentrator (iPOC)

Figure 1 details schematically the proposed intelligent POC system (iPOC). Feedback control can be implemented with a single input-single output (SISO) table look-up controller. The feedback control law T can be tabulated by a table look-up function.

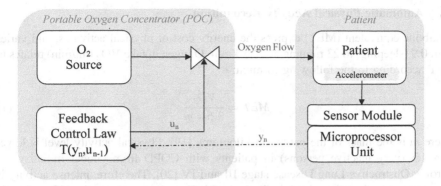

**Fig. 1.** Block diagram of the control scheme for the proposed intelligent portable oxygen concentrator (iPOC).

This look-up function is provided by the clinicians and customized for the specific needs of the patients following well-established titration procedures [18].

A microprocessor receives measurements from a triaxial accelerometer sensor and computes the physical activity in real-time ($y$). A finite state machine (Mealy machine) whose output value is determined by both its current state and the current input is used to determine the physical activity (PA) state that will input the controller.

Figure 2 illustrates the designed Mealy machine considering the four different PA states used in this study. An smartphone, with inbuilt accelerometer sensor, could be used to perform the PA detection tasks [14]. The new control variable $u$ is obtained from the static table taking into account previous state $u_{n-1}$ to avoid errors in the transition between different degrees of PA. The controller then outputs the corresponding control signal to the flow regulator managing the delivery of a suitable oxygen dose to the patient. Accordingly, the oxygen flow target (l/min) updates continuously using the automatic classification of PA.

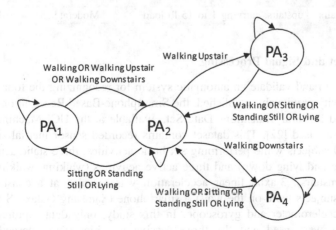

**Fig. 2.** State diagram to limit the evolution between different physical activity degrees. $PA_n$: state corresponding to the detected physical activity $n$.

## 3.2 Automatic Physical Activity Recognition

Metabolic equivalent (MET) express the energy cost of physical activities, and varies from 0.9 (sleeping) to 23 running at 22.5 km/h. Oxygen uptake $VO_2$ (ml/min) relates to MET according to the following formula:

$$MET = \frac{VO_2}{3.5 * m} \tag{1}$$

where $m$ is the mass of the patient in kilograms [19]. Physical activity level is lower than 1.5 (very inactive persons) in patients with COPD from Global Initiative for Chronic Obstructive Lung Disease stage III and IV [20]. Therefore, intense activity is not expected in the targeted users. Since the MET, and accordingly the $VO_2$, of static postures such as sitting, standing still and lying in the bed shows low variations, only four common activities with moderate MET values cover the spectrum of PA performed by users. These activities were selected for automatic recognition because they pose a need for updating the oxygen flow delivered to the patient. Table 1 illustrates the selected activities and their corresponding METs adapted from [21].

**Table 1.** Physical activities for automatic recognition and the corresponding MET values and $VO_2$. The mass of the user in kilograms is referred as $m$.

| Activity | Description | Intensity | MET (W/kg) | $VO_2$ (l/min) |
|---|---|---|---|---|
| Sitting, standing still or lying in bed | Lying down awake, sitting (watching TV, typing, desk work) or standing still | Light | 1.3 | 0.00455*$m$ |
| Walking | Walking (up to 2.5 mph) around home, store or office | Light | 3 | 0.01050*$m$ |
| Walking downstairs | Descending stairs, carrying 15 lb load | Moderate | 3.5 | 0.01225*$m$ |
| Walking upstairs | upstairs, carrying 1 to 15 lb load | Moderate | 5 | 0.01750*$m$ |

## 3.3 Dataset and Signal Processing

In order to train and validate an automatic system for recognizing the four degrees of physical activity previously established, the Smartphone-Based Recognition of Human Activities and Postural Transitions Data Set, available at the UCI Machine Learning Repository was used [22]. This dataset contains recorded sensor inertial signals of a sample of 30 subjects while performing six basic activities: three static activities (sitting, standing and lying down) and three active activities (walking, walking-upstairs, walking-downstairs). 3-axial linear acceleration was sampled at a constant rate of 50 Hz. The subjects wore on the waist a smartphone (Samsung Galaxy S II) with an embedded accelerometer and gyroscope. In this study, only data captured with the accelerometer were used and the three passive activities were merged into one according to the abovementioned rationale.

The dataset contains 400 activity instances and 13.182 s of recording. The obtained dataset was randomly partitioned into two sets, where 70% of the volunteers was selected for generating the training data (7415 samples) and 30% the test data (2996 samples). The accelerometer sensor signal was preprocessed by applying noise filters and then sampled in activity windows of 2.56 s and 50% overlap. The body component of the sensor acceleration signal was separated from the gravitational component using a Butterworth high-pass filter with 0.3 Hz cutoff frequency.

### 3.4    Features, Machine Learning Algorithms and Evaluation Metrics

In order to reduce the computational burden, only time-domain features were considered. From each window, a total of 56 time domain features were estimated: mean, standard deviation, median absolute deviation, largest value, smallest value, skewness, kurtosis, signal magnitude area, average sum of the squares, interquartile range, signal entropy, 4th order Burg autoregression coefficients and Pearson correlation coefficient were computed for each axis and for the accelerometer magnitude signal. Angle between tri-axial signal mean and vector was also computed for each axis. Finally, the calculated features were normalized and bounded within [−1,1].

Features were tested by using various supervised machine-learning algorithms such as decision tree (DT), linear discriminant analysis (LDA), radial basis function network (RBF), feedforward multilayer perceptron (MLP), and support vector machine (SVM).

C4.5 generation algorithm was applied to create a single-tree model. A pruning process was adopted to minimize the cross-validated error. MLP was trained with the conjugate gradient algorithm. One hidden layer with 13 neurons was used. This number was automatically optimized in the range from 2 to 20. Logistic activation function was selected for the hidden and the output layer neurons. Concerning the SVM classifier, C-SVC (regularized support vector classification) model with the RBF kernel function was chosen. The model parameter values are 40.78 and 1.35 for C and Gamma, respectively. Matlab software was used for training and validating the different models.

Regarding the evaluation metrics, we considered the overall accuracy, sensitivity, specificity and the activity recognition error rate (ARER). ARER was estimated as the percentage of time that has been wrongly assigned to an activity [23]:

$$ARER(\%) = 100\frac{Time(sec)\,wrongly\,classified}{Session\,duration(sec)} \qquad (2)$$

## 4    Results

Figure 3 shows the accelerometer data along the x-axis for each of the four activities under recognizing (standing/sitting/laying down, walking, walking upstairs and down-stairs).

All the features described in the previous section were extracted from each recording within 50% overlapped consecutive time windows of 2.56 s. The calculated

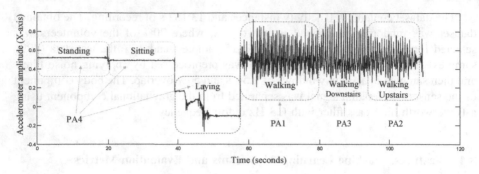

**Fig. 3.** Accelerometer data along the x-axis for each of the four activities of daily living.

feature set constituted the input for the automatic physical activity classification through the abovementioned machine learning algorithms.

Tables 2 and 3 report the confusion matrices obtained while discerning among the four proposed physical activities. In particular, Table 3 shows the classification accuracy and the corresponding ARER values for each classifier. Overall, SVM offered the highest performance, yielding 95.19% accuracy and 4.85% ARER.

**Table 2.** Confusion matrixes and performance parameters for different validated classifiers. DT: decision tree C4.5, LDA: linear discriminant analysys, RBF: radial basis function, MLP: multilayer preceptron, SVM: support vector machine. PA1: walking, PA2: walking upstairs, PA3: walking downstairs, PA4: sitting, standing still or lying.

|  |  | PA1 | PA2 | PA3 | PA4 | Sensitivity (%) | Specificity (%) |
|---|---|---|---|---|---|---|---|
| DT | PA1 | 412 | 70 | 14 | 0 | 83.06% | 94.84% |
|  | PA2 | 107 | 342 | 22 | 0 | 72.61% | 94.50% |
|  | PA3 | 21 | 61 | 338 | 0 | 80.48% | 98.56% |
|  | PA4 | 1 | 8 | 1 | 1599 | 99.38% | 100.00% |
| LDA | PA1 | 472 | 21 | 3 | 0 | 97.18% | 94.84% |
|  | PA2 | 120 | 347 | 4 | 0 | 75.58% | 98.30% |
|  | PA3 | 33 | 19 | 368 | 0 | 89.05% | 99.84% |
|  | PA4 | 1 | 1 | 1 | 1606 | 99.94% | 100.00% |
| RBF | PA1 | 482 | 13 | 1 | 0 | 97.18% | 94.84% |
|  | PA2 | 112 | 356 | 3 | 0 | 75.58% | 98.30% |
|  | PA3 | 17 | 29 | 374 | 0 | 89.05% | 99.84% |
|  | PA4 | 0 | 1 | 0 | 1608 | 99.94% | 100.00% |
| MLP | PA1 | 473 | 1 | 22 | 0 | 95.36% | 95.28% |
|  | PA2 | 95 | 374 | 2 | 0 | 79.41% | 98.65% |
|  | PA3 | 21 | 32 | 367 | 0 | 87.38% | 99.07% |
|  | PA4 | 2 | 1 | 0 | 1606 | 99.81% | 100.00% |
| SVM | PA1 | 480 | 3 | 11 | 2 | 96.37% | 96.52% |
|  | PA2 | 75 | 382 | 14 | 0 | 80.25% | 98.53% |
|  | PA3 | 7 | 32 | 381 | 0 | 90.48% | 99.03% |
|  | PA4 | 0 | 0 | 0 | 1609 | 100.00% | 99.86% |

**Table 3.** Comparison of the accuracy of the evaluated classifiers.

| Classifier | Accuracy (%) | ARER (%) |
|---|---|---|
| Decision tree C4.5 (DT) | 89.82% | 10.18% |
| Linear discriminant Analysis (LDA) | 93.22% | 6.78% |
| Radial Basis function Network (RBF) | 94.13% | 5.87% |
| Multilayer Perceptron (MLP) | 94.13% | 5.87% |
| Support vector machine (SVM) | 95.19% | 4.81% |

## 5  Discussion and Conclusion

Intelligent POC technology could offer a very important benefit to the patient. Close loop oxygen concentrators described in scientific literature are mostly based on the feedback of $SpO_2$, measured by using a pulse oximeter sensor. However, various causes has been documented to provide spuriously low or high SpO2, such as excessive movement, poor finger perfusion resulting from vasoconstriction and/or hypotension, carbon monoxide poisoning, light influence, etc. [24]. The search for more robust and reliable sensors able to operate against patient variability and environmentally changing scenarios is a challenge.

The main cause for what a patient with COPD needs to update the oxygen flow obtained using a POC is related to the changing activities of daily living. The automated recognition of physical activities has been researched intensively during the last decade, and its application to the respiratory field is promising. Accelerometer sensors could support the automatic recognition of the simple physical activities performed by patients with severe COPD degree. Simultaneously, they could support patients in identifying patterns during their daily physical activities, promoting a more active lifestyle.

In this regards, this study poses a first step to devise a new iPOC, to manage automatically the need for oxygen. A selected set of 56 features and different machine learning algorithms were used for evaluating recognition performance using a public database of physical activity signals.

Recognition accuracy of up to 91.15% on four everyday activities using a single tri-axial accelerometer was obtained using a SVM classifier. The associated ARER was lower than 5%, ensuring a low percentage of time wrongly assigned to a certain activity.

Future works include combining classifiers to outperform individual classifiers, collecting of data from users with COPD and implementing the algorithm into a commercial POC.

## References

1. Adeloye, D., Chua, S., Lee, C., Basquill, C., Papana, A., Theodoratou, E., Nair, H., Gasevic, D., Sridhar, D., Campbell, H., Chan, K.Y., Sheikh, A., Rudan, I., GHERG: Global and regional estimates of COPD prevalence: systematic review and meta–analysis. J. Glob Health. **5**, 20415 (2015)

2. Murray, C.J.L., Barber, R.M., Foreman, K.J., et al.: Global, regional, and national disability-adjusted life years (DALYs) for 306 diseases and injuries and healthy life expectancy (HALE) for 188 countries, 1990–2013: quantifying the epidemiological transition. Lancet **386**, 2145–2191 (2015)

3. Global strategy for the diagnosis, management, and prevention of chronic obstructive pulmonary disease (2017 report). http://goldcopd.org/

4. McDonald, C.F.: Oxygen therapy for COPD. J. Thorac. Dis. **6**, 1632–1639 (2014)

5. Nonoyama, M., Brooks, D., Lacasse, Y., Guyatt, G.H., Goldstein, R.: Oxygen therapy during exercise training in chronic obstructive pulmonary disease. In: Nonoyama, M. (ed.) Cochrane Database of Systematic Reviews, p. CD005372. Wiley, Chichester (2007)

6. Meena, M., Dixit, R., Kewlani, J., Kumar, S., Harish, S., Sharma, D.: Home-based long-term oxygen therapy and oxygen conservation devices: an updated review. Natl. J. Physiol. Pharm. Pharmacol. **5**, 1 (2015)

7. Lacasse, Y., Sériès, F., Vujovic-Zotovic, N., Goldstein, R., Bourbeau, J., Lecours, R., Aaron, S.D., Maltais, F.: Evaluating nocturnal oxygen desaturation in COPD – revised. Respir. Med. **105**, 1331–1337 (2011)

8. Bye, P.T., Esau, S.A., Levy, R.D., Shiner, R.J., Macklem, P.T., Martin, J.G., Pardy, R.L.: Ventilatory muscle function during exercise in air and oxygen in patients with chronic air-flow limitation. Am. Rev. Respir. Dis. **132**, 236–240 (1985)

9. Dunne, P.J.: Long-term oxygen therapy (LTOT) revisited: in defense of non-delivery LTOT technology. Rev. Port. Pneumol. **18**, 155–157 (2012)

10. Lellouche, F., Lipes, J., L'Her, E.: Optimal oxygen titration in patients with chronic obstructive pulmonary disease: a role for automated oxygen delivery? Can. Respir. J. **20**, 259–261 (2013)

11. Claure, N., Bancalari, E.: Automated closed loop control of inspired oxygen concentration. Respir. Care. **58**, 151–161 (2013)

12. Physiological Closed-Loop Controlled (PCLC) Medical Devices. In: Physiological Closed-Loop Controlled Devices Workshop, pp. 1–20. Food and Drug Administration (FDA), White Oak, Maryland (2015)

13. Nazabal, A., Garcia-Moreno, P., Artes-Rodriguez, A., Ghahramani, Z.: Human activity recognition by combining a small number of classifiers. IEEE J. Biomed. Heal. Inform. **20**, 1342–1351 (2016)

14. Damaševičius, R., Vasiljevas, M., Šalkevičius, J., Woźniak, M.: Human activity recognition in AAL environments using random projections. Comput. Math. Methods Med. 4073584 (2016)

15. Lock, S., Blower, G., Prynne, M., Wedzicha, J.: Comparison of liquid and gaseous oxygen for domiciliary portable use. Thorax **47**, 98–100 (1992)

16. Chatburn, R.L., Williams, T.J.: Performance comparison of 4 portable oxygen concentrators. Respir. Care. **55**, 433–442 (2010)

17. Maskey, D., Agarwal, R.: Oxygen Therapy in Chronic Obstructive Pulmonary Disease. JP Medical Ltd, Westminster (2013)

18. Egton Medical Information Systems Limited: Use of Oxygen Therapy in COPD, pp. 4–7 (2015)

19. Mortazavi, B., Alsharufa, N., Lee, S.I., Lan, M., Sarrafzadeh, M., Chronley, M., Roberts, C. K.: MET calculations from on-body accelerometers for exergaming movements. In: 2013 IEEE International Conference on Body Sensor Networks (BSN) (2013)

20. Waschki, B., Kirsten, A., Holz, O., Müller, K.-C., Meyer, T., Watz, H., Magnussen, H.: Physical activity is the strongest predictor of all-cause mortality in patients with COPD. Chest **140**, 331–342 (2011)

21. Ainsworth, B.E., Haskell, W.I.L., Whitt, M.C., Irwin, M.L., Swartz, A.M., Strath, S.J., O'Brien, W.I.L., Bassett Jr., D.R., Schmitz, K.H., Emplaincourt, P.O., Jacobs Jr., D.R., Leon, A.S.: Compendium of physical activities: an update of activity codes and MET intensities. Med. Sci. Sports Exerc. **32**, S498–S504 (2000)
22. Reyes-Ortiz, J., Oneto, L., Sama, A., Parra, X., Anguita, D.: Transition-aware human activity recognition using smartphones. Neurocomputing **171**, 754–767 (2016)
23. San-Segundo, R., Montero, J., Moreno-Pimentel, J., Pardo, J.: HMM adaptation for improving a human activity recognition system. Algorithms **9**, 60 (2016)
24. Chan, E.D., Chan, M.M., Chan, M.M.: Pulse oximetry: understanding its basic principles facilitates appreciation of its limitations. Respir. Med. **107**, 789–799 (2013)

# Automatic Detection of Epiretinal Membrane in OCT Images by Means of Local Luminosity Patterns

Sergio Baamonde$^{(\boxtimes)}$, Joaquim de Moura, Jorge Novo, and Marcos Ortega

VARPA Group, Department of Computer Science,
University of A Coruña, A Coruña, Spain
{sergio.baamonde,joaquim.demoura,jnovo,mortega}@udc.es

**Abstract.** This work presents a novel approach for automatic detection of the epiretinal membrane in Optical Coherence Tomography (OCT) images. A tool able to detect this pathology is very valued since it can prevent further ocular damage by doing an early detection. This approach is based in the location of the inner limiting membrane (ILM) layers of the retina. Then, the detected locations are classified using a local-feature based vector in order to determine presence of the membrane. Different tests are run and compared to establish the appropriateness of the approach as well as its practical validity.

**Keywords:** Epiretinal membrane · Retinal layers · Medical imaging · Optical coherence tomography

## 1  Introduction

Epiretinal membrane (ERM), also called macular pucker, is a macular pathology that can cause minor damage to the retina, like central vision decrease and metamorphopsia [11]. This disease can be caused by changes in the vitreous humor [4] and, consequently, the response of immune system to protect the retina can sometimes provoke that a number of cells converge on the macular area. This situation produces a transparent layer (Fig. 1) that, like every scar tissue, contracts causing tension on the retina, specifically on the inner limiting membrane (ILM). This phenomenon contributes to the appearance of ERM.

Since this pathology is frequently asymptomatic, it is imperative to develop a reliable system of detection to avoid further complications caused by its increasing severity.

In order to detect the ERM, ophtalmologists can work with the patient clinical history, looking for diabetes and ocular diseases or surgeries. Also, specialists

This work is supported by the Instituto de Salud Carlos III, Government of Spain and FEDER funds of the European Union throug the PI14/02161 and the DTS15/00153 research projects and by the Ministerio de Economía y Competitividad, Government of Spain through the DPI2015-69948-R research project.

© Springer International Publishing AG 2017
I. Rojas et al. (Eds.): IWANN 2017, Part I, LNCS 10305, pp. 222–235, 2017.
DOI: 10.1007/978-3-319-59153-7_20

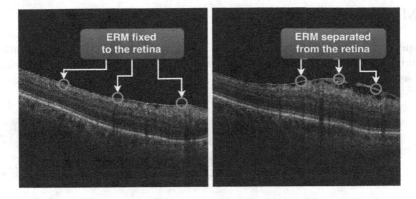

**Fig. 1.** Different appearances of ERM

can perform a complete ophthalmological evaluation to check for ERM, but at an additional cost and work hours.

The most precise way to evaluate the retinal morphology is doing an optical coherence tomography (OCT) scan [2], since the ERM appears as a bright layer on the retina [1]. Also, detecting irregularities on the retinal surface and/or retinal thickening, between others, can also mean that ERM is present on the patient.

Surgery may be needed when facing symptomatic ERM, e.g. vision loss, diplopia or debilitating metamorphopsia. When indicated, pars plana vitrectomy is performed [9]. However, ERM can recur and require further surgery. This recurrence rate can be reduced by undergoing ILM peeling [7].

The detection of the ERM is a manual process done by a specialist, but some tools have been developed to help with this task. Wilkins *et al.* [13] work with OCT pictures in real time, correcting patient's eye movement with image processing algorithms. Once the images are obtained, the specialist manually places computer cursors on the superficial and deep retinal boundaries. These boundaries are based on reflectivity and thickness differences between different areas of the retina.

Comparatively, other studies [3,6,8] work with spectral-domain OCT (SD-OCT). Its main advantage in comparison to time-domain OCT (TD-OCT) is the easier visualization of intraretinal layers (as the photoreceptor layer) through higher resolution pictures and the possibility of obtaining 3D images. This technique allows the specialists to obtain accurate surface maps and capture tension lines caused by the ERM on the ILM.

With this work we aim to create an automatic tool to detect epiretinal membrane presence on OCT pictures. The methodology consists on the processing of the OCT picture to locate the ILM layer of the retina, continuing with the extraction of relevant features of this layer. Finally, we will classify these data using classifiers trained beforehand to identify presence of ERM in the vicinity or adherent to the retina.

## 2    Methodology

Our methodology is based on the classification of the ILM located points to determine the presence of the membrane. To reach this goal, several stages are proposed, as shown on Fig. 2.

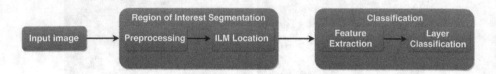

**Fig. 2.** Methodology used for the developing process

The first step is a preprocessing stage in order to remove undesired structures in the OCT input image as well as enhancing relevant ones.

Afterwards, the goal is to locate the ILM as it represents our region of interest (ROI) given the fact that it is the location where the membrane appears. To this end, an active contour model (Snake) [5] is used to get the location of the topmost layer of the retina in the picture. This model will try to adapt its shape to the shape of the inner limiting membrane. Consequently, a fair amount of information is available about the split between background and eye zones.

In the next step, once the ILM is located, a feature extraction procedure takes place in each ILM location point in order to establish the presence of the membrane by using a trained classifier on these features, which would be the last step of our method. The feature vector for a ILM point would be defined in a small local window of the image surrounding that particular point. The idea behind this is to check for the ERM also in the zones where it is not adhered to the retina.

In order to be able to classify the existence of membrane, the vector will be based on local histograms of intensity, as this is the main characteristic of the membrane to be recognized. Following sections explain each step in more detail.

### 2.1    Region of Interest Segmentation

**Preprocessing.** In order to be able to correctly fit the shape of the Snake on top of the ILM, some preprocessing operations are performed on the image to avoid unnecessary elements. Figure 3 shows an example of the different steps at this stage. We remove every black border surrounding the OCT image and then we apply a Gaussian filter with $\sigma = 1.5$. This value was found to be good at preserving relevant features while filtering significantly. Finally, we apply a morphological operator (opening) to finish the cleaning of the picture and ease up the execution of the geometrical model.

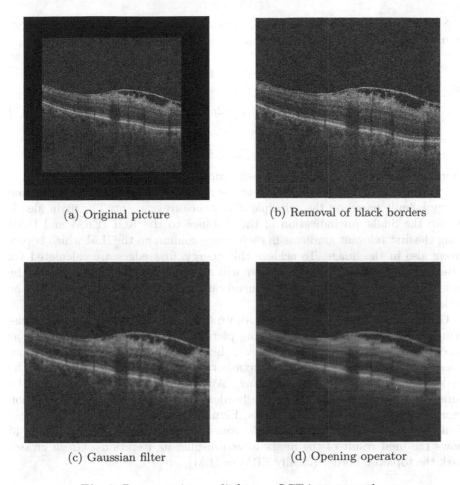

(a) Original picture        (b) Removal of black borders

(c) Gaussian filter        (d) Opening operator

**Fig. 3.** Preprocessing applied to an OCT image sample

**Layer Segmentation.** After preparing the image, since we want to approximate the shape of the inner limiting membrane, we use the active contour model mentioned beforehand. This model is initialized above all the layers of the retina, near the top border of the picture. When executed, it will try to converge on the topmost layer and adapt to its shape.

In this particular case we use a different approach of the Snake. We only allow downwards movement and, if the energy does not decrease, for a particular node this is stopped. This way, we ensure every point remains on their respective start columns. Also, with this approach, the Snake does not converge around an object, but instead lands on top of the upper layer, behaving as intended.

The energy of a Snake is defined in (1).

$$E = \int (\alpha(s)E_{\text{cont}} + \beta(s)E_{\text{curv}} + \gamma(s)E_{\text{img}})ds \tag{1}$$

Snake is defined as $N$ points $p_1, p_2, ..., p_N$, so the formulations for each energy term are explained on (2), (3) and (4).

$$E_{\text{cont}} = \|p_i - p_{i-1}\|^2 \tag{2}$$

$$E_{\text{curv}} = \|p_{i-1} - 2p_i + p_{i+1}\|^2 \tag{3}$$

$$E_{\text{img}} = -\|\nabla I\| \tag{4}$$

where $\nabla I$ is the gradient of the intensity computed at each Snake point.

In order to get the Snake to adapt to the region of interest (ROI), an external energy is built based on the principle of distance to gradient. The main idea is to give the Snake an indication of the distance to the ROI (ERM and ILM) being the first relevant gradients in each image column on the ILM which is very strong also in the image. To achieve this energy, first edges are calculated via Sobel and Feldman [12]. In general, we will aim to detect the limit between the background and the inner limiting membrane as a border, since we want the Snake to position above this sector.

Once we have this region segmented, we apply the Euclidean distance transform for the edge image. The resulting picture will be passed to the Snake as the external energy parameter. This way, the Snake will try to stick to the zones of less energy, that is, the zones where a border exist (the closest one being the border on the inner limiting membrane). An example of this procedure can be found on Fig. 4. Figure 4b shows the borders detected by the edge extraction algorithm, symbolized as white pixels. Figure 4c represent the external energy of the Snake, where dark areas are the zones of minimal energy. Lastly, Fig. 4d shows the final result of the Snake after finishing its iterations. Green crosses mark the topmost border (ideally ERM or ILM).

## 2.2    Feature Extraction

Once the ROI is located in the image it is needed to establish the presence or absence of epiretinal membrane along the retina surface of a particular image. The hypothesis to achieve this is that luminosity of membrane differs sufficiently from ILM, the retina and image background. Thus, local features based on intensity can be defined on a vicinity of each Snake node to determine the existence of membrane in it by analyzing luminosity patterns. This is, a location with darker values above and under the central point should be a floating ERM, while if it only has dark values above and bright points under, can be a ERM next to the ILM or ERM nonexistence, depending on the intensity of the central window (brighter values are associated with ERM presence).

Following this hypothesis, after the Snake finishes its execution in the previous stage, local intensity features are computed for each node. These features need to contain information of the surroundings of the obtained points, more precisely from the vertical area around the point. Having this information allows

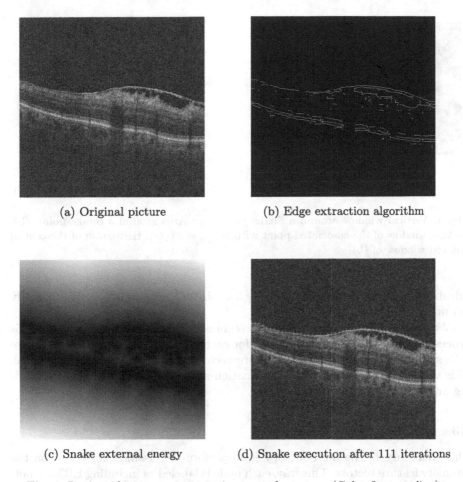

(a) Original picture

(b) Edge extraction algorithm

(c) Snake external energy

(d) Snake execution after 111 iterations

**Fig. 4.** Region of interest segmentation procedure steps (Color figure online)

us to differentiate between points situated on the background from the ones situated on the ILM.

To this end, we will be using a series of vertical areas centered on the points of the Snake (Fig. 5). This area is divided in a series of $W$ squared windows. For each of these windows, we will calculate afterwards the intensity histogram with $N$ bins for the area. By appending all the bin values the feature vector is built. Lastly, all $W$ feature vectors are combined in one full feature vector containing $N \times W$ elements. This vector represents the intensity values of the entire vertical area of the point.

For this work five regions are considered centered around the node. As Fig. 5 shows, data located above or under the limits on the defined windows do not contribute with any meaningful data for ERM location as it only adds redundant information. In the result section, several studies are conducted to establish a

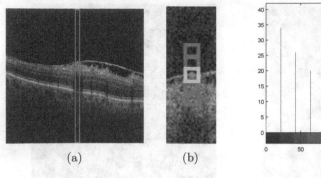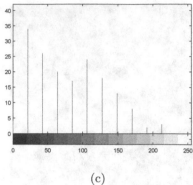

(a)                          (b)

(c)

**Fig. 5.** Vertical window around a Snake point. (a): Area around a Snake point. (b): Feature window of the associated point with $W_{size} = 13$. (c): Histogram of the central squared window of (b).

suitable value for window size. Nevertheless, size of the regions is matter of study in our experiment section.

We calculate the histogram with $N$ bins for each squared window. This process will give us $N$ discrete values for each window. Afterwards, all of these values will be converted to a full feature vector containing $N \times W$ values total. This vector contains all relevant information about the point and the surrounding area, more precisely its luminosity.

### 2.3    Layer Classification

The final stage of the methodology is to perform a classification based on the intensity feature vectors. This way, each node is labeled as including ERM or not.

We will classify the points extracted from the image using a series of classifiers trained previously by using a 10-fold Cross-validation method with a set of samples manually labeled by a clinician. Each fold will use 90% of the samples as training samples and 10% as test samples. The models being used on this section are a Naive Bayes classifier, a Multilayer Perceptron and a Random Forest. We will generate different classifiers for each class with different parameters: number of bins and size of the squared windows. An example of the ERM recognition is shown on Fig. 6.

## 3    Experimental Results

OCT scans were obtained with a tomograph CIRRUS™HD-OCT Zeiss, with Spectral Domain Technology. The resolution of the images is 490×500 pixels.

Our working set is comprised of 129 images showing different sections of the eye. ERM presence can be found in some of the pictures. Training samples have

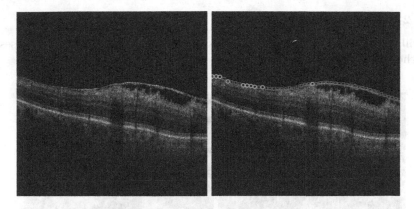

**Fig. 6.** Result of classification with Random Forest, $W_{size} = 13$ and $N_{bins} = 15$. Circles symbolize absence of ERM and squares presence of ERM

been randomly selected from all the pictures. In different experiments, separation between training and test is done accordingly as explained in following sections.

The energy terms used by the Snake in our experiment are shown on Table 1. These values have been selected because with the picture set we are using they give the Snake enough traction to provide a good approximation of the shape of the ILM. The high $\gamma$ value allows the Snake to adapt to the ILM or ERM shape (zones of high energy), while $\alpha$ and $\beta$ are less relevant because we only allow downwards movement so keeping the points clustered is not a relevant problematic.

**Table 1.** Energy terms used for the Snake

| Energy type | Parameter | Value |
|---|---|---|
| $E_{cont}$ | $\alpha$ | 0.8 |
| $E_{curv}$ | $\beta$ | 0.4 |
| $E_{img}$ | $\gamma$ | 2.0 |

Our study of the methodology is done by performing two different experiments. First, we aim to separate the samples between 2 classes (*membrane* and *no membrane*) to get a first approximation about the presence or absence of ERM. Lastly, those samples will be divided instead on 4 classes, subdividing *membrane* class on *membrane* and *floating membrane* (ERM separated from the retina). Similarly, *no membrane* class is split on *no membrane* (points of the ILM with ERM absence) and *background*.

Our goal is to check what is the most accurate approximation (2 or 4 classes) while improving the behavior of the classifiers used by refining the parameters passed as input.

## 3.1    2-Class Classification

We will test first the behavior of the classifiers when using 2 different classes to split the data, as seen on Fig. 7:

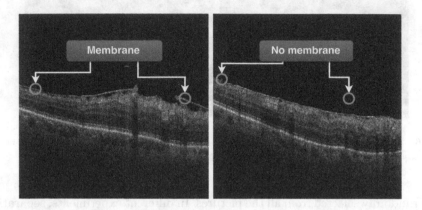

**Fig. 7.** Structure types used on classification

- 60 samples from class *membrane*. These points are the ones belonging to any point where ERM exists.
- 60 samples from class *no membrane*. This class contains any point not possessing ERM, either background or ILM without membrane.

These samples were used to train different classifiers. Each training iteration was repeated 10 times to obtain more accurate metrics. The results appearing here are the average of every iteration. Accuracy is defined in (5).

$$Acc = (TP + TN)/(P + N) \tag{5}$$

where $TP$ and $TN$ are True Positive and Negative values, while $P$ and $N$ are Positive and Negative values.

To evaluate the results, we use a k-fold cross-validation with $k = 10$ [10].

The different types of classifiers we used are a Multilayer Perceptron, a Naive Bayes classifier and a Random Forest classifier. These three approximations provide us a vast array of behaviors, allowing us to conclude what is the best approach for this problem.

The following tables (Table 2a, b and c) show the results of obtaining the accuracy of each classifier for different values of window size ($W_{size}$) and number of bins ($N_{bins}$). The most accurate classifier in each series is bolded for clarity.

A more in-depth comparison between the best approximation of each class is done by comparing side to side the ROC curves of each classifier (Fig. 8). We can conclude that with the results we have obtained (Fig. 9), the best classifier is the one based in Random Forest method, accuracy-wise and with better ROC values.

**Table 2.** Accuracy of different classifiers

(a) Multilayer Perceptron accuracy

| $W_{size}$ | $N_{bins}$ 5 | 10 | 15 | 20 | 25 |
|---|---|---|---|---|---|
| 5 | 75.75% | 73.50% | 72.67% | 72.00% | 69.83% |
| 9 | 83.17% | 80.92% | 77.75% | 76.75% | 76.17% |
| 13 | **86.92%** | 83.33% | 84.92% | 79.25% | 77.75% |
| 17 | 84.08% | 79.50% | 81.92% | 78.08% | 77.50% |
| 21 | 79.75% | 79.25% | 81.83% | 79.83% | 84.17% |
| 25 | 82.25% | 82.33% | 79.25% | 79.50% | 82.25% |

(b) Naive Bayes classifier accuracy

| $W_{size}$ | $N_{bins}$ 5 | 10 | 15 | 20 | 25 |
|---|---|---|---|---|---|
| 5 | 68.92% | 70.17% | 64.58% | 68.17% | 63.33% |
| 9 | 70.92% | 76.08% | 73.58% | 73.92% | 74.42% |
| 13 | 72.33% | **80.83%** | 79.08% | 73.33% | 74.42% |
| 17 | 80.50% | 78.75% | 75.08% | 70.33% | 70.83% |
| 21 | 71.83% | 76.83% | 75.33% | 76.75% | 70.08% |
| 25 | 79.92% | 73.92% | 73.53% | 75.83% | 69.75% |

(c) Random Forest classifier accuracy

| $W_{size}$ | $N_{bins}$ 5 | 10 | 15 | 20 | 25 |
|---|---|---|---|---|---|
| 5 | 79.67% | 78.00% | 78.00% | 75.17% | 75.25% |
| 9 | 84.83% | 85.83% | 85.58% | 83.92% | 82.08% |
| 13 | 87.33% | 91.08% | **91.25%** | 88.08% | 86.75% |
| 17 | 86.33% | 85.92% | 85.92% | 85.67% | 85.83% |
| 21 | 83.83% | 84.42% | 83.58% | 85.75% | 83.75% |
| 25 | 84.67% | 81.25% | 84.75% | 84.58% | 81.83% |

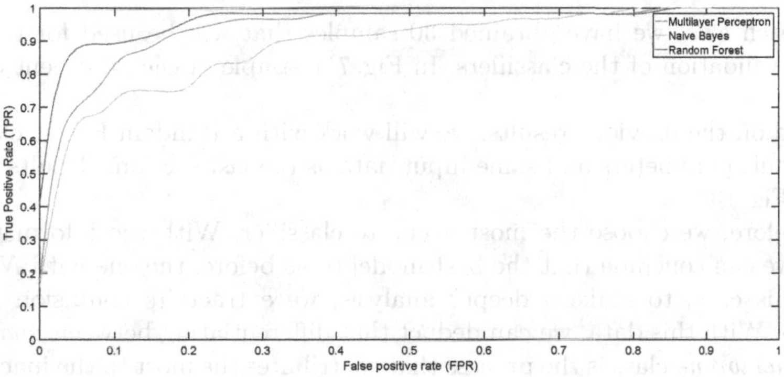


**Fig. 8.** ROC Curves for each best classifier. Random Forest scores above the other 2 classifiers

### 3.2 4-Class Classification

In this experiment the goal was to assess the performance of a classifier able to distinguish four scenarios:

- Class *membrane.* These points are the ones belonging to zones where the ERM is fixed to ILM layer.
- Class *floating membrane.* Here we group ERM points situated on the background.

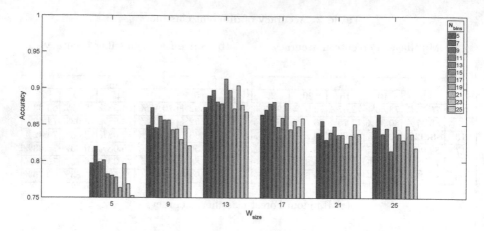

**Fig. 9.** Accuracy for Random Forest classifiers with 2 classes. Best results are found with $W_{size} = 13$ and $N_{bins} = 15$

– Class *no membrane*. This class contains points from the retina not containing ERM.
– Class *background*. Background points not belonging to any of the other classes are classified here.

For each class we have obtained 30 samples that will be used for training, test and validation of the classifiers. In Fig. 7 a sample of each different class is shown.

Based on the previous results, we will work with a Random Forest classifier with default parameters and same input data as the last section. Results can be seen on Fig. 10.

As before, we choose the most accurate classifier. With the information of Table 3 we can conclude that the best model is, as before, the one with $N_{bins} = 15$. In this case, to make a deeper analysis, we extract its confusion matrix (Table 4). With this data, we can deduct that differentiating between *membrane* and *no membrane* class is the process that contributes the most to the inaccuracy of the classifier. This is coherent with our last approach (splitting between those both classes only) being the main focus of this work.

**Table 3.** Accuracy for Random Forest with 4 classes and $W_{size} = 13$

| Number of bins ($N_{bins}$) | | | | | | | | | | |
|---|---|---|---|---|---|---|---|---|---|---|
| 5 | 7 | 9 | 11 | 13 | 15 | 17 | 19 | 21 | 23 | 25 |
| 84.17% | 85.75% | 86.50% | 86.42% | 85.83% | **88.34%** | 86.25% | 86.00% | 88.05% | 86.50% | 84.92% |

We can see that the behavior of each approximation is similar: a low value of $W_{size}$ or too high causes inaccuracy in the classifiers since we are introducing

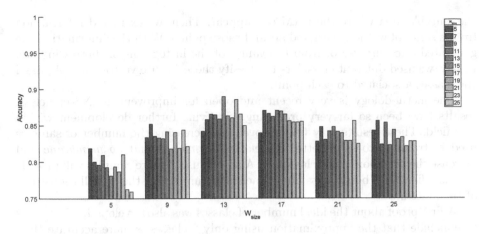

**Fig. 10.** Accuracy for Random Forest classifiers with 4 classes. Best results are found with $W_{size} = 13$

**Table 4.** Confusion matrix for Random Forest with $W_{size} = 13$ and $N_{bins} = 15$

|  | Membrane | Floating membrane | No membrane | Background |  |
|---|---|---|---|---|---|
| Membrane | 24 | 1 | 5 | 0 | 80.0% / 20.0% |
| Floating membrane | 0 | 30 | 0 | 0 | 100.0% / 0.0% |
| No membrane | 7 | 0 | 23 | 0 | 76.7% / 23.3% |
| Background | 0 | 1 | 0 | 29 | 96.7% / 3.3% |
|  | 77.4% / 22.6% | 93.8% / 6.2% | 79.3% / 20.7% | 100.0% / 0.0% | 88.3% / 11.7% |

Ground Truth (left side label)

Membrane   Floating   No membrane   Background
                membrane
**Obtained results**

noise instead of useful information. Also, we see better accuracy with $N_{bins}$ values in the range of 13 to 17. As with window size, values too low or too high provoke inconsistencies in the classification. Better values for 2 classes and 4 classes are situated on $W_{size} = 13$, giving us a good approximation about what is a good value to get the most information but avoiding unnecessary noise.

## 4   Conclusions

Identifying the appearance of epiretinal membrane is an important process in the opthalmologic field, since it can improve the results of ERM extraction surgery.

In this paper, we have developed an automatic process to detect the ERM on OCT pictures with deformable models. First, we situated a number of points on the suitable area where the ERM can appear. Then, we extracted information from a series of windows situated around those points. With this information, we generated a feature vector from the values of the histograms of those windows. Lastly, we used different classifiers to classify those feature vectors and obtained the classes associated to each point.

The methodology is very recent and open for improvement. Nevertheless, results have been so far very promising, justifying further development within this field. These results may be improved by increasing the number of samples used in training, to split better the classes *membrane* and *no membrane* and increase the precision of the classifiers. Also, the use of more samples will provide the classifiers with better data about each class, improving the overall robustness of the system.

A first proof about the ideal number of classes was also developed, allowing us to conclude that the approximation using only 2 classes is more accurate than introducing another 2 classes, giving us 4 in total. In future works, a tool to separate the points of ERM fixed on the retina from the ones on the background will need to be developed to provide that information to specialists.

# References

1. Brancato, R.: Optical coherence tomography (OCT) in macular edema. Doc. Ophthalmol. **97**, 337–339 (1999)
2. Do, D.V., Cho, M., Nguyen, Q.D., Shah, S.M., Handa, J.T., Campochiaro, P.A., Zimmer-Galler, I., Sung, J.U., Haller, J.A.: Impact of optical coherence tomography on surgical decision making for epiretinal membranes and vitreomacular traction. Retina **27**, 552–556 (2007)
3. Falkner-Radler, C.I., Glittenberg, C., Hagen, S., Benesch, T., Binder, S.: Spectral-domain optical coherence tomography for monitoring epiretinal membrane surgery. Ophthalmology **117**, 798–805 (2010)
4. Foos, R.Y.: Vitreoretinal juncture; epiretinal membranes and vitreous. Invest. Ophthalmol. Vis. Sci. **16**, 416–422 (1977)
5. Kass, M., Witkin, A., Terzopoulos, D.: Snakes: active contour models. Int. J. Comput. Vis. **1**, 321–331 (1988)
6. Koizumi, H., Spaide, R.F., Fisher, Y.L., Freund, K.B., Klancnik Jr., J.M., Yannuzzi, L.A.: Three-dimensional evaluation of vitreomacular traction and epiretinal membrane using spectral-domain optical coherence tomography. Am. J. Ophthalmol. **145**, 509–517 (2008)
7. Kwok, A.K., Lai, T.Y., Yuen, K.S.: Epiretinal membrane surgery with or without internal limiting membrane peeling. Clin. Exp. Ophthalmol. **33**, 379–385 (2005)
8. Legarreta, J.E., Gregori, G., Knighton, R.W., Punjabi, O.S., Lalwani, G.A., Puliafito, C.A.: Three-dimensional spectral-domain optical coherence tomography images of the retina in the presence of epiretinal membranes. Am. J. Ophthalmol. **145**, 1023–1030 (2008)
9. Machemer, R.: A new concept for vitreous surgery. 7. Two instrument techniques in pars plana vitrectomy. Arch Ophthalmol. **92**(5), 407–412 (1974)
10. McLachlan, G.J., Do, K.-A., Ambroise, C.: Analyzing Microarray Gene Expression Data. Wiley, Hoboken (2004)

11. Medina, C.A., Townsend, J.H., Singh, A.D. (eds.): Manual of Retinal Diseases. Springer International Publishing, Heidelberg (2016)
12. Sobel, I., Feldman, G.: A 3x3 isotropic gradient operator for image processing. Talk Stanford Artif. Intell. Proj. (SAIL) 271–272 (1968)
13. Wilkins, J.R., Puliafito, C.A., Hee, M.R., Duker, J.S., Reichel, E., Coker, J.G., Schuman, J.S., Swanson, E.A., Fujimoto, J.G.: Characterization of epiretinal membranes using optical coherence tomography. Ophthalmology **103**, 2142–2151 (1996)

# An Expert System Based on Using Artificial Neural Network and Region-Based Image Processing to Recognition Substantia Nigra and Atherosclerotic Plaques in B-Images: A Prospective Study

Jiří Blahuta[⊠], Tomáš Soukup, and Jiri Martinu

The Institute of Computer Science, Silesian University in Opava,
Bezruc Sq. 13, 74601 Opava, Czech Republic
jiri.blahuta@fpf.slu.cz
http://www.slu.cz/fpf/en/institutes/the-institute-of-computer-science

**Abstract.** The presented paper is focused on ways of digital image analysis of ultrasound B-images based on echogenicity investigation in determined Region of Interest (ROI). An expert system has been developed in the course of the research. The goal of the paper is to demonstrate how to interconnect automatic finding of the position of the substantia nigra using Artificial Neural Network (ANN) with supervised learning and ROI-based image analysis. For substantia nigra is able to detect the position using ANN from B-image in transverse thalamic plane. From this is computed echogenicity index grade inside the ROI as parkinsonism feature. The methodology is well applicable for a set of images with the same resolution. The results have shown practical application of ANN learning in this case. The second part of the paper is focused on detection of atherosclerotic plaques. An experimental prospective study shown the using ANN can be highly time-consuming problem due to complexity of B-images. The plaques have no standardized shape and size in comparison with SN. To objective appraisal of using ANN to automatic finding atherosclerotic plaque in B-image we need a large set of images of normal and pathological state. Although it is very important using ANN, automatic detection in highly time-consuming problem for ANN training.

**Keywords:** Ultrasound · Substantia nigra · B-MODE · B-images · Stroke ultrasound · Parkinson's Disease · Neural networks ultrasound

## 1 Introduction to B-MODE Imaging

B-MODE imaging is one of the most important modes for diagnostic ultrasound. A B-image is represented as a two-dimensional image in grayscale. Each pixel $P[x;y]$ of digitized B-image is displayed as a point with a brightness degree $H$

© Springer International Publishing AG 2017
I. Rojas et al. (Eds.): IWANN 2017, Part I, LNCS 10305, pp. 236–245, 2017.
DOI: 10.1007/978-3-319-59153-7_21

corresponding to tissue echogenicity. The echogenicity is defined as the ability to reflect or transmit ultrasound waves in the context of surrounding tissues [1]. There are anechoic, hypoechoic and hyperechoic structures. In the case of the reflection is zero, the structure is anechoic. Hyperechoic structures in B-image are displayed as white, hypoechoic as gray and anechoic as black, i.e. each echogenicity grade is expressed by gray level in B-image. Some structures have variable echogenicity according to many factors. The Fig. 1 demonstrates different echogenicity degrees in B-image.

In this study are used images with 256° of $H$, in other words from $H = 0$ to $H = 255$.

## 2   A Set of Used Images

This study is primarily focused on analysis of transcranial B-images in transverse thalamic plane [2]. The first investigated structure is the substantia nigra in the midbrain followed by different brain structures commonly examined in neurosonology. Within a pilot study are also used B-images of atherosclerotic plaques. The main goal is to demonstrate using the approach for different B-images. No image pre-processing steps have been used.

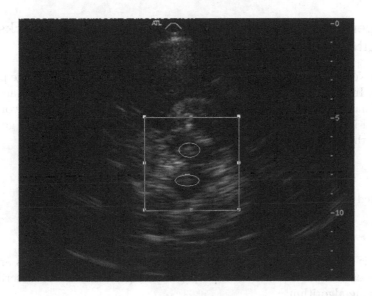

**Fig. 1.** A B-image in transverse thalamic plane with marked window $5 \times 5$ cm consists ipsilateral and contralateral SN

## 3   Principle of the ROI-Based Image Analysis in B-MODE

There are many ways to automatic recognition of structures in the field of digitized image processing. Our study is focused on neurosonology. The substantia nigra is an area in the midbrain. In B-images is characterized by different echogenicity grade depending of death of dopaminergic cells. In normal case the SN is anechogenic, increased echogenicity is caused by the death of the cells which is the main feature of Parkinson's Disease. Thanks to early diagnostics is possible to help prevent the development of symptoms. ROI-based principle is based on fact that is not required to analyze the whole image but only a ROI which is a subset of the image matrix. For example, ultrasound images used in this study have the resolution of $768 \times 576$ pixels. In practice, not only in medical image analysis, is needed to observe only examined parts of the image. Formally, ROI is represented as a submatrix of the image matrix. In neurology we need to define ROI for different brain structures for which the software is adaptable. The adaptability is one of the key features to define a new ROI to extend the functionality for different structure in B-image. In the developed application is able to automatically locate the ROI according to coordinate preset. In case of using images with the equal resolution, is able to set location of the ROI with option of manual shift to correct position. Also has been inspected how to recognize the shape and position of substantia nigra. For this purpose has been used a set of 100 images with identical resolution from the same ultrasound machine. The automatic positioning based on artificial neural network with supervised learning have already published [3,4].

## 4    Supervised Learning of Artificial Neural Networks in Digital Image Processing

Artificial Neural Networks represent one of phenomena in various image processing fields. In this study, the ANN are used to find position of the ROI according to coordinates. To this processing is used a feedforward MLP network with supervised learning based on Error-Backpropagation Algorithm. The goal is to minimize the network error

$$E_t = \sum_{j=1}^{n} E_p \tag{1}$$

represented as the summation of partial errors

$$E_p = \frac{1}{2}(y_j - d_j)^2 \tag{2}$$

where $y_j$ is $j$-th output and $d_j$ is a desired output. The partial errors are computed for each input $x_j$. The backpropagation algorithm can be described by the following algorithm:

1. Weights initialization to random small values
2. Set the input vector and the vector of desired outputs
3. Compute the real output $y_j$ for input $x_j$
4. Compute the network error (1) from all partial errors (2)
5. Weights adaptation

The algorithm ends if $E_t < E_d$ or determined number of cycles or maximal time of the learning is exceeded.

## 4.1   Training of the ANN and Experimental Results

The goal of the training is to learn correct position of the SN depending on the training set. Because is used the set of images with equal resolution and each image is cut in the same size, the correct position of the SN can be predefined. Used ANN is composed of 2 hidden layers. Thanks to equal resolution and window size, the default correct position of the center of the ROI is given by

$$S = [2.00; 2.00] \tag{3}$$

as the initial position of desired output (5). If the output position is incorrect, is necessary to retrain the network by manual positioning. The coordinates of a new position are added into the training set. The goal is to achieve successful detection of the substantia nigra area by minimization of the error of the correct position. See Fig. 2 with experimental results of the ROI position. Totally has been used a set of 100 images to training the ANN with correct position. To train of the ANN have been used images with normal, e.g. low echogenicity grade of SN. It was observed if the results are reliable used also for hyperechoic SN (Fig. 2). The finding of the position can be successfully used for images with equal resolution and in which SN. The ANN should be better adapted with many more samples with significant differences between images. The adaptation of the position finding is the one of crucial goals of the ANN training. In other words, the goal is to train the ANN to reliable recognition of the position of the SN for any B-image in transverse thalamic plane (Fig. 1). There are two following aspects:

- minimal total error $E_t$
- to find a position which is applicable for almost all B-images in thalamic transverse plane

There are two crucial requirements to successful application. In experimental study with a set of 100 images which have equal resolution and size, the ANN evinced acceptable results. Correlation coefficient between desired and observed results was $r > 0.8$ and also have been judged intra- and inter- $\kappa$ coefficients in which were achieved $\kappa_{intra} = 0.75$ and $\kappa_{inter} = 0.82$.

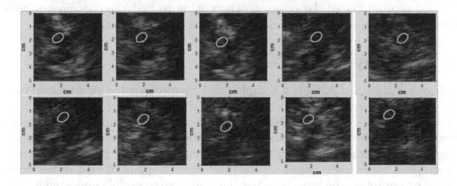

**Fig. 2.** Experimental results of SN positioning by the ANN learning

## 4.2   Brief Description of Used ANN

In this case the following ANN is used:

- multilayer feedforward network
- 1 input layer, two hidden layers, 1 output layer
- supervised learning based on Error Back-Propagation algorithm
- logic sigmoid activation function

To input layer is get an input vector and a weight vector of $n$ inputs. Based on multiplication is computed the inner potential

$$\zeta = x_j \cdot w_j \tag{4}$$

where $x_i$ are inputs of coordinates and $w_i$ are randomly set weights from 0 to 1. Each input is multiplicated by the weight. In this case the inputs represents the coordinates $S$ (3). During the computation in hidden layers is computed the partial error $E_p$ (2) for each input. Log-sigmoid transfer function

$$y = \frac{1}{1 + e^{-\lambda x}} \tag{5}$$

and

$$y = \frac{1}{1 + e^{-x}} \tag{6}$$

if the $\lambda = 1$ (sigmoid steepness). The sigmoid steepness influences learning speed. The goal is to learn ANN to compute optimal coordinates of the position of the SN which can be applicable in general. To global error (2) minimization is used Mean Squared Error (MSE) given by

$$\frac{1}{n} \sum_{j=1}^{n} (y_j - d_j)^2 \tag{7}$$

In general, MSE is considered as the error between predictions and observed values. In this case, MSE describes the error between desired coordinates and observed coordinates of the position. The goal is to find the coordinates with minimal error and which can be used in general for any B-image.

## 4.3   Complexity of Intelligent Learning

Intelligent learning of ANN is a complex and time-consuming problem. In case of B-images, the learning is very complicated due to B-image structure. There are many aspects which make the image processing very complicated, e.g. non-linear speckle noise and ultrasound artifacts (shadowing, reverberation, mirror imaging), etc. The accuracy >95% of the learning is required by an experienced sonographer. From one's own experience and from the point of view of the experienced sonographer, the learning is not the main goal to useful employment. The goal of this research is to use many different ANN approaches. In other words, we can find the optimal activation function, number of hidden layers, weights adaptation and another criteria for an optimal ANN learning. The learning phase is the first phase of the process. Accuracy of the learning phase must be validated by evaluation phase; the comparison of the global error (2) of the ANN. There is highly time-consuming phase with repeating of the learning phase due to complexity of the B-image structure. There is necessity to use a large set of images with different resolution, SN echogenicity grade and position of the SN to evaluate the accuracy.

## 5   Using in Developed Software

We have developed the software tool designed for evaluation of echogenicity grade of the SN [5,6]. Five echogenicity grades are distinguished in neurosonology, i.e. from hypoechoic to hyperechoic. This software is based on binary thresholding algorithm inside the selected ROI and computing echogenic area inside the SN to Parkinson's Disease diagnosis [6]. Greater echogenic area represents higher probability of Parkinson's Disease. Let $H$ represents the intensity level of a pixel and $T$ is a threshold, then

$$H > T \tag{8}$$

For each $T$ is computed how many pixels are $>T$ and subsequently is converted into real mm$^2$. The echogenicity grade of the SN is evaluated on the basis of number white pixels depending on the condition (4) for thresholds $T \in \langle 0; 255 \rangle$. See Fig. 3 with an example of decreasing number of pixels depending on threshold $T$.

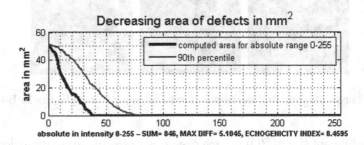

**Fig. 3.** Thresholding results for SN measurement.

Output values, their context and statistical analysis of measured descriptors is discussed in [7–9].

## 6   Prospective Study on Atherosclerosis Recognition

The methodology has been developed with respect to universal using not only for transcranial images. The software could be used for various ultrasound B-images because the principle is designed with respect to universal application. We want to investigate risk evaluation of atherosclerotic plaques in B-images. Firstly, we must find some features in B-images for reliable detection of atherosclerosis. The goal of this prospective study is to inspect how atherosclerotic plaques are distinguishable. Atherosclerotic plaques can be distinguished by size, homogeneity, shape, composition, etc. This prospective study is focused on finding these features to distinguish normal and pathological cases. In comparison with examined brain structures, in case of atherosclerotic plaque is not applicable to use a predefined ROI with equal size and shape. There are two main known crucial limitations. The first limitation is about displaying of atherosclerotic plaques in B-image; how to reliably distinguish plaques. The second limitation is closely related with the principle of the developed software. The predefined ROI is used in case of analysis of SN and some other structures which substantially do not change size and position. The output from the software represents measured values of the echogenic area. Due to using the equal ROI, echogenicity grade can be compared depending on area inside the ROI. To investigation of SN echogenicity grade is used an elliptical ROI with default area $A = 50\,\text{mm}^2$, see Fig. 3. In case of investigation of atherosclerotic plaques is not applicable to use ROI with constant area. Each plaque appears differently and must be defined a special ROI and we cannot compare decreasing area (Fig. 3). Figure 4 shows three different ROIs corresponding to atherosclerotic plaques defined by an experienced sonographer.

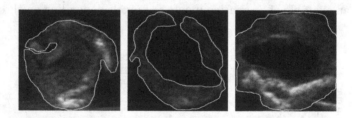

**Fig. 4.** Three different ROI corresponding to atherosclerotic plaques in B-image

Due to different size of ROI in each case, is not possible to compare decreasing area, see Fig. 5. In case of SN and another brain structures is used equal size of the ROI for all cases.

## 6.1 Automatic Recognition of the Plaque Using ANN

For atherosclerotic plaques using ANN is much more complicated in comparison with SN. There is a way how to recognize shape of the plaque using ANN. The automatic recognition of the shape could be based on Region-Based training or training based on echogenicity grade. There are two crucial barriers. Firstly, the learning of the ANN to successful recognition of the plaque shape could be time-consuming problem because the shape has no standardization. So, to train

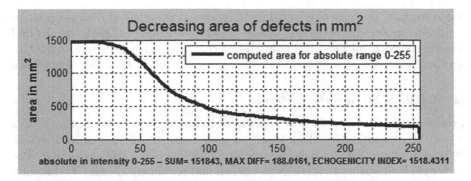

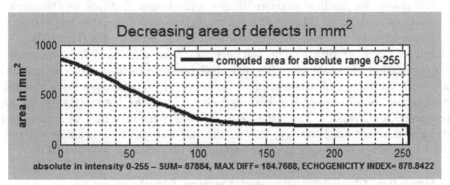

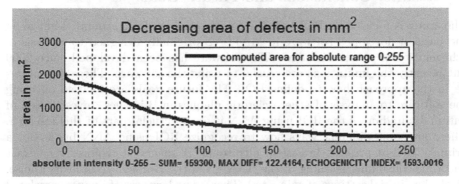

**Fig. 5.** The results of descending area are incomparable

the ANN requires a huge sample of different normal and pathological findings, e.g. based on correlation with histology. Secondly, how to select some features as inputs into ANN input layer to computing? Simply, what we get from one image as reliable feature could not be reliable from another image. The recognition of the plaque by ANN is much more complicated in comparison with finding position of the SN. In B-image each plaque is displayed differently (different size, shape, echogenicity, etc.). On the ground of the differences the recognition of the plaque is highly time-consuming. First of all, very large set of images is needed to start. We have only small set of images at present but a large database is created concurrently with analysis of preserved histological patterns. To find features, the patterns must be analyzed and compared with B-images. After this step could be analyzed if findings in B-image correspond to histological patterns.

## 6.2   Ratio of Hyperechoic Pixels

In connection with these facts is possible to compute ratio instead of area. The core of the computing could be based on ratio between number of hyperechoic pixels to total number of pixels independently on shape and size of the ROI. The ratio could be a feature to distinguish normal and pathological cases. Is this solution acceptable for medical practice? Exist some better features to atherosclerotic plaques analysis in B-images? Even though is one of possible ways to analyze atherosclerotic plaques in B-images with using developed software. It is to research in the near future. The software has been also used for experimental investigation of insula [10]. Recently has been performed an experimental study with 23 images of atherosclerotic plaques [11]. The images were visually distinguished (homogeneity, low heterogeneity and strong heterogeneity) by an experienced sonographer. The aim was to observe reproducibility using selected statistical descriptors. However, the results did not produce reliable markers caused by number of images and the ambiguity of manually selected ROI.

## 7   Results, Conclusions and Future Goals

The study is focused on ROI-based image processing of ultrasound B-image on the basis of developed software in MATLAB originally designed to analysis of echogenicity of substantia nigra. The software has been developed since 2009 but recently the limitations are crucial to using the software. These limitations lie in the approach how to define ROI. The future goals are focused on using the software to analysis of atherosclerotic plaques in B-images. Investigation of automatic finding of the ROI using ANN is one of crucial part of this research. There are some different ways how to recognize ROI using ANN but is not a primary goal at present. In case of ultrasound B-images, it requires very exact and time-consuming learning of the ANN given the B-image complexity (noise, resolution, initial settings of gain, etc.). There is no large set of images correlated with histological analysis. Nevertheless, the results demonstrated that finding the position of the SN is partially applicable, primarily for a set of the images with equal size and resolution. The achieved correlation was >0.8 and inter-/intra- $\kappa$ coefficients were 0.75 and 0.82. To find an optimal learning and evaluating phase of ANN is needed larger set of images and try to set different input parameters of ANN. Using ANN to automatic detection of the atherosclerotic plaque is considered as very progressive way but also highly time-consuming and much more complicated. The performed experimental study with the set of 23 images is not sufficient to unbiased results. First of all, there is necessity to obtain a large set of B-images compared with histological patterns to find corresponding features depending on progression of atherosclerosis.

**Acknowledgments.** This work was supported by The Ministry of Education, Youth and Sports from the National Programme of Sustainability (NPU II) project IT4Innovations excellence in science - LQ1602.

# References

1. Edelman, S.K.: Understanding Ultrasound Physics, 4th edn. E.S.P. Ultrasound (2012)
2. Skoloudik, D., Jelinkova, M., Bartova, P., Soukup, T., Blahuta, J., Cermak, P., Langova, K., Herzig, R.: Transcranial sonography of the substantia nigra: digital image analysis. Am. J. Neuroradiol. **35**(9), 2273–2278 (2014)
3. Blahuta, J., Soukup, T., Cermak, P., Vecerek, M., Jakel, M., Novak, D.: ROC and reproducibility analysis of designed algorithm for potential diagnosis of Parkinson's disease in ultrasound images. In: Mathematical Models and Methods in Modern Science, 14th WSEAS International Conference on Mathematical Methods, Computational Techniques and Intelligent Systems (MAMECTIS 2012) (2011)
4. Blahuta, J., Soukup, T., Cermak, P., Rozsypal, J., Vecerek, M.: Ultrasound medical image recognition with artificial intelligence for Parkinson's disease classification. In: Proceedings of the 35th International Convention, MIPRO 2012 (2012)
5. Blahuta, J., Soukup, T., Cermak, P., Novak, D., Vecerek, M.: Semi-automatic ultrasound medical image recognition for diseases classification in neurology. In: Kountchev, R., Iantovics, B. (eds.) MedDecSup 2012. Studies in Computational Intelligence, vol. 473, pp. 125–133. Springer, Switzerland (2013)
6. Blahuta, J., Soukup, Jelinkova, M., Bartova, P., Cermak, P., Herzig, R., Skoloudik, D.: A new program for highly reproducible automatic evaluation of the substantia nigra from transcranial sonographic images. Biomed. Pap. **158**(4), 621–627 (2014)
7. Blahuta, J., Cermak, P., Soukup, T., Vecerek, M.: A reproducible application to B-MODE transcranial ultrasound based on echogenicity evaluation analysis in defined area of interest. In: Soft Computing and Pattern Recognition, 6th International Conference on Soft Computing and Pattern Recognition (2014)
8. Skoloudik, D., Fadrna, T., Bartova, P., Langova, K., Ressner, P., Zapletalova, O., Hlustik, P., Herzig, R., Kanovsky, P.: Reproducibility of sonographic measurement of the substantia nigra. Ultrasound Med. Biol. **9**, 1347–1352 (2007)
9. Riffenburgh, R.H.: Statistics in Medicine, 3rd edn. Academic Press, Cambridge (2012)
10. Skoloudik, D., Bartova, P., Maskova, P., Dusek, P., Blahuta, J., Langova, K., Walter, U., Herzig, R.: Transcranial Sonography of the Insula: Digitized Image Analysis of Fusion Images with Magnetic Resonance. Georg Thieme Verlag KG Stuttgart, Ultraschall in der Medizin (2016)
11. Blahuta, J., Soukup, T., Cermak, P.: How to detect and analyze atherosclerotic plaques in B-MODE ultrasound images: a pilot study of reproducibility of computer analysis. In: Dichev, C., Agre, G. (eds.) AIMSA 2016. LNCS (LNAI), vol. 9883, pp. 360–363. Springer, Cham (2016). doi:10.1007/978-3-319-44748-3_37. ISSN: 0302-9743

# Automatic Tool for Optic Disc and Cup Detection on Retinal Fundus Images

Miguel Angel Fernandez-Granero[1](✉), Auxiliadora Sarmiento Vega[2],
Anabel Isabel García[2], Daniel Sanchez-Morillo[1], Soledad Jiménez[3],
Pedro Alemany[3], and Irene Fondón[2](✉)

[1] Biomedical Engineering and Telemedicine Research Group,
University of Cádiz, Puerto Real, Cádiz, Spain
{ma.fernandez,daniel.morillo}@uca.es
[2] Signal Theory and Communication Department,
University of Seville, Seville, Spain
{sarmiento,anagarnog,irenef}@us.es
[3] Surgery Department, University of Cádiz, Cádiz, Spain
{soledad.jimenez,pedromaria.alemany}@uca.es

**Abstract.** The aging of the population is a matter of concern due to its association with various diseases in humans that limit their quality of life. Among them, glaucoma is one of the leading causes of blindness in the world. To its early diagnose, retinal fundus images are visually inspected by experts. In recent years, image-based computer aided diagnosis systems have been proposed. Automatic segmentation of Optic Disc (OD) and cup areas are their first and most difficult tasks. In this paper, a computerized technique aimed to their extraction from the original images is presented. The tool is related to human perception due to the use of an advanced color metric, CIE94 within a uniform color space, CIE $L*a*b*$ to compute pixels' color gradients [1]. Based on this information, a classifier assigns a probability value to each of the pixels, meaning its suitability for being part of the Optic Disc and Cup border. The tool has been tested on 200 images from different public databases achieving an accuracy value of 96.63%. This quality level makes the proposed color-based image processing system capable to assist the physicians in glaucoma screening programs.

**Keywords:** Glaucoma · Optic disc · Cup · Cup to disc ratio · Machine learning · Retinal images · Diabetic retinopathy

## 1 Introduction

There is a widespread need of the medical community of tools for the detection and management of diseases involving the retina in a more cheap and efficient way [2]. The reason is clear: with the increase of the average world population age, the number of patients suffering from eye diseases has increased. This growth has led to a relative proliferation of ophthalmologic services, especially in the rural areas of developed countries. Recent studies suggest the existence of 37 million of blind people. The VISION 2020 initiative is having a considerable impact in reducing blindness

© Springer International Publishing AG 2017
I. Rojas et al. (Eds.): IWANN 2017, Part I, LNCS 10305, pp. 246–256, 2017.
DOI: 10.1007/978-3-319-59153-7_22

caused by eye infections, but a greater effort is still needed to solve problems such as cataracts, glaucoma and diabetic retinopathy [3], especially when it is known that three quarters of all blindness can be prevented or treated when are early diagnosed [4].

Ocular screening programs are effective if the disease is identified at an early stage. These ocular procedures consist on the acquisition and study of color images of the retina, currently acquired non-invasively by an expert ophthalmologist. With the large number of patients performing eye examinations in an ordinary way, this added work represents a time consuming task for the medical expert, who must analyze and diagnose each one of them. It becomes evident the need to automate the task of analyzing the large number of retinal images leaving the medical expert with those dubious or problematic. This automation would have a positive effect on the patient since he would receive his results in a shorter time interval. Several eye-screening programs have been developed in the past with an acceptable sensitivity and specificity in relation to the associated costs [5, 6]. However, there is still great potential for cost and time reduction, increased effectiveness and extension to remote areas [2].

Among all eye diseases, the present article proposes a useful tool for the diagnosis and treatment of glaucoma, the second leading cause of blindness in developed countries [2]. This ocular disease is characterized by the progressive loss of nerve fibers in the retina that causes changes in the appearance of the optic disc (OD) and cup. The majority of affected individuals do not present symptoms in the early stages of the disease, suffering visual field defects and progressive loss of vision in later stages. The changes in the appearance of the OD and the cup make this disease suitable for diagnosis using advanced techniques of image processing. These techniques should extract information from the image, especially col-or and form, to translate it into a likelihood of suffering the disease.

The state of the art regarding image processing for the diagnosis of glaucoma is presented below. Given the relevance of the color information for this application, the methods will be presented according to the color model that they use from the most basic that use the image in gray scale to the more advanced ones that use spaces adapted to human perception.

Although OD has well defined characteristics, its automatic segmentation is not straightforward since there are numerous variations on its appearance due to the pathology itself or due to the presence of other anomalies in the patient. The edge detection of the OD has received a great deal of attention from image processing specialists [4].

A large group of works are based on the grayscale image. The reason is that these methods use well-known algorithms that have been developed in the past for other applications. In addition, until relatively recently, invasive gray-scale angiographies were the ones the researchers took as a starting point for their algorithms. In this way, most techniques use these algorithms taking advantage of the work done for years by researchers in the field of image processing. They are, therefore, less innovative not taking into account color information [5-8].

The use of color in retinographies is quite limited. Most of the existing algorithms do not exploit the color image in its three-plane representation (RGB, HSV, etc.) but make use of a single color plane treating it as a grayscale image. Most of the methods employ a plane of the RGB space since this is the simplest representation because the image is usually stored in this format [9–11]. Another frequently used color space is some of the HSV family given the intensity or hue separation. These color spaces are not uniform, that is, the distance between colors measured in them does not correspond to perceived color differences [12].

Finally, there are methods that are based on the luminance coordinate (L*) of the uniform space CIE L*a*b*, [4, 13].

Sometimes, two planes are used but independently, that is, part of the processing is done in one plane and another part is performed in another [14]. Only a few articles mention the use of the three planes (usually in the RGB space in combination with the grayscale image) but, again, processing them separately [15]. The use of color spaces related to human perception is generally forgotten, with a minority being the techniques designed in the uniform color space CIE L*a*b*, [16].

As we can see, the work focused on the detection of OD is numerous. In the case of cup, however, the number of articles is not so high, perhaps because of the greater difficulty in precisely finding the edge of this area, which in many cases is diffuse or presents occluded regions by the strong presence of blood vessels. In some OD, this area is almost imperceptible being very difficult its location by traditional methods [4].

The selection of a color space for the detection of cup is usually based on the search for a greater contrast in the brightness of this zone with respect to the background of the retinography. The most common selection is the use of a single color plane with the G of the RGB space being mostly adopted [17, 18]. There are also some techniques based on the R and B planes [19, 20], being of minor use the color plane a* of the space of color adapted to human perception CIE L*a*b* [21].

As for the use of combined color planes, the processing is carried out in separate planes and the result obtained in each of them is combined in some way. The most used planes return to the RGB space, selecting two of them or using the three [22]. Other planes used are the S and V of the HSV color space [23]. The methods that make use of the vectorial nature of the color in uniform spaces, CIE L*a*b* in [4, 24] and JCh of the color appearance model CIECAM02 in [25], are scarce.

Given the relevance of the disease and the evolution of available algorithms, the number of publications regarding the diagnosis of glaucoma has increased in the last five years. Most of them are based on the calculation of the CDR and its subsequent comparison with a fixed threshold value. Within these methods the chosen color space is mostly RGB, given its simplicity. Moreover, these techniques only use the G [26–28] plane although some use the image in gray scale [29].

Other techniques include some additional measures for distinguishing between healthy and glaucomatous images. In [30] processing is performed in the V plane of the HSV space. Once the OD and the cup are extracted, the CDR and the thickness of the neuroretinal ring are calculated in relation to their location in the OD. If the CDR is greater than 0.3 and the calculated thickness is not distributed as expected, the image is considered glaucomatous.

At the view of this reading we can infer:

(1) They do not take into account the adaptation to human perception nor the advanced theory of color limiting in the information they get from the images.
(2) They mainly use basic processing techniques, especially gray-level approaches.

(3) Despite the fact that high success rates are presented, around 94%, the algorithms are not rigorously validated: the images usually belong to private bases with an unknown degree of difficulty and, therefore, of representativeness. The number of images is very low ranging from 30 to 196. This number is mostly around 40. The image base is usually biased containing a larger number of healthy images. Usually a single database per article is used, which can lead to the method describing well those images and not others.

In this article we propose the use of color information in a perceptually adapted color space CIE L*a*b* along with an advance color distance to compute color variations that characterize OD and cup areas.

## 2  Methods

Regarding the development of the tool, we have selected 200 images from six publicly available databases [31–36]. Some of these retinal fundus images correspond to healthy eyes while others correspond to different ocular pathologies. The images have been chosen because they offer a wide range of appearances, illumination and colors as it can be seen in Fig. 1.

### 2.1  Pre-processing

The initial step involves locating a Region of Interest (ROI) to achieve a considerable reduction in computational resources [37, 38]. This selection was made by initially locating the OD center and retaining a square area centered on it. The total ROI area is 7% of the total eye area. This restriction is usually adopted by other state of the art techniques [1]. Some ROI examples can be seen in Fig. 1.

### 2.2  Perceptually Adapted Color Derivatives

For each pixel on each ROI 25 color directional derivatives are computed, each one corresponding to an angle from 0° to 360° with an interval of 15° of separation. As in [39] a Sobel mask is used in CIE L*a*b* with CIE94 color difference. The derivative can be obtained by subtracting two vectors: the first one containing the positive coefficients of the mask, V+, and the other one with the negatives ones, V− (Fig. 2).

Where a $(x, y) = [L^* (x, y), a^* (x, y), b^* (x, y)]'$ are the color values of a pixel with coordinates (x, y).

The norm of the derivative $v$ is obtained as:

$$v = \|\Delta E_{94}(V^+, V^-)\| \tag{1}$$

with $\Delta E_{94}(V^+, V^-)$ a. the CIE94 color difference between both vectors (Fig. 3).

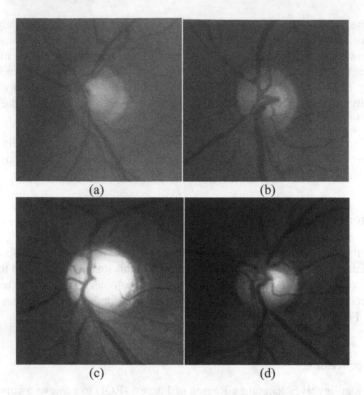

**Fig. 1.** Images (a), (b), (c) and (d) represent some region of interest obtained from the original retinographies. The variability of appearances is a key point for the development of a robust system. (Color figure online)

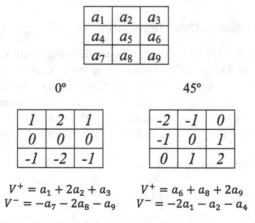

**Fig. 2.** The original image is process in masks and two vectors are computed regarding the orientation selected. The upper matrix represents the image neighbourhood under consideration while the lower matrices are two examples of gradient masks rotated $0°$ and $45°$.

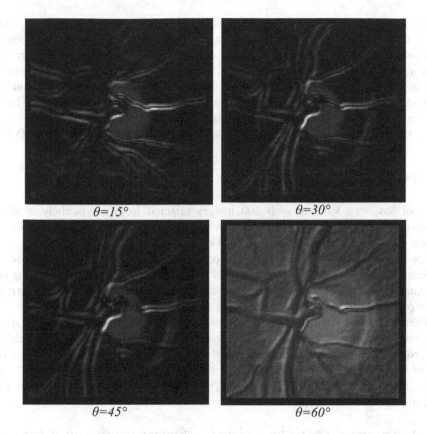

$\theta=15°$     $\theta=30°$

$\theta=45°$     $\theta=60°$

**Fig. 3.** Gradient images corresponding to four different orientations: $\theta = 15°$, $\theta = 30°$, $\theta = 45°$ and $\theta = 60°$

## 2.3 Proposed Features

A feature vector that is mainly compounded by the 25 color gradients previously explained represents each pixel on each image and other characteristics related to its color, its position and its spatial relation to the optic disc center. This vector contains:

(1) 25 color gradients values.
(2) Color values: L*, a* and b*.
(3) 1 value corresponding to distance to the centre.
(4) 1 value corresponding to the angle respect to the centre.
(5) 2 values relating to the position (i, j) of the pixel

## 2.4 Classification

Each pixel, on the base of its feature vector, is assigned to a one of these three possible classes: background, optic disc or cup with a complex tree classifier. Decision tree classifiers are simple yet widely used classifiers [40]. Their structure is similar to a tree with a root, internal nodes and leaves. The root has no incoming edges and zero or more outgoing edges. Internal nodes have one incoming edge and two or more outgoing edges. Moreover, leafs or terminal nodes, have one incoming edge and no outgoing edges [40]. The root and internal nodes are capable to differentiate samples regarding their features while each of the leaves corresponds to a different class. Inside this family of tree classifiers, complex tree is characterized by having many leaves that makes many fine distinctions between classes (maximum number of splits is 100).

## 3   Results and Discussion

The tool has been validate with 200 images selected from six publicly available databases [31–36]. For ground truth generation, the images where manually annotated by two experts that delineated OD and cup areas.

For classification we have selected a complex tree classifier with 10-fold cross validation meaning that the image dataset is split randomly into 90% for training and the remaining 10% for testing, something that is repeated ten times in order to generate unbiased results.

The performance of the proposed system, including accuracy and also the specificity and sensitivity of the classifier for the three categories are presented in Table 1. The proposed technique has achieved a high level of accuracy: 96.63%, a sensitivity (Se) of 98.88%, 79.16% and 79.55%, and a specificity (Sp) of 90.90%, 98.07% and 99.43% respectively for the three classes. The three classes are background (class 1), optic disc (class 2) and cup (class 3).

**Table 1.** Performance of the validated complex tree classifier for the three classes, background (class 1), optic disc (class 2) and cup (class 3).

|             | Background | Optic disc | Cup    |
|-------------|------------|------------|--------|
| Sensitivity | 98.88%     | 79.16%     | 79.55% |
| Specificity | 90.90%     | 98.07%     | 99.43% |
| PPV         | 98.82%     | 77.12%     | 85.04% |
| NPV         | 91.30%     | 98.28%     | 99.18% |

Some examples of the obtained results are shown in Fig. 4. The images used are characterized by a high level of difficulty due to the presence of almost imperceptible cups, abnormal OD sizes, etc. However, the tool is capable to accurate delineate the edges of both areas when results (blue) and ground truth (green) are compared.

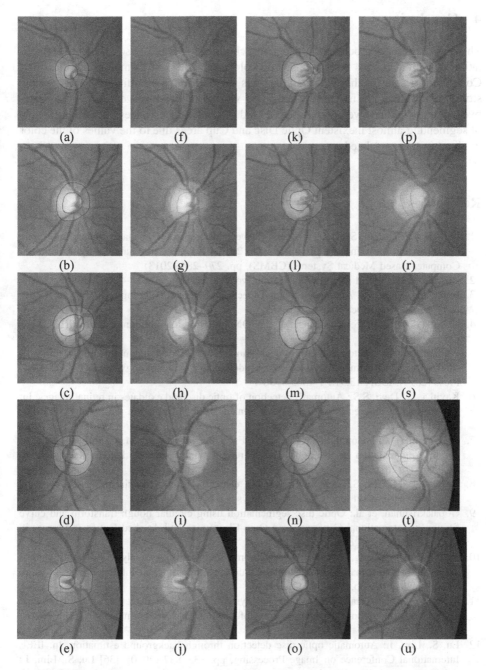

**Fig. 1.** Some examples of the results obtained with the tool. The automatically detected edges of the OD and cup regions are marked in blue over the ROI images (a)–(e) and (l)–(o) while their corresponding manually annotated ground truths are marked in green (f)–(j) and (p)–(u). (Color figure online)

# 4    Conclusion

This paper presents a perception adapted technique for Optic Disc and Cup segmentation on retinal fundus images based on color gradients and Complex tree classifier. Compared to state of the art techniques this method has a significantly more simple structure and does not need to detect blood vessels or quantify vessels bends. Moreover, it not only deals well with different kind of retinal appearances but also is capable of segmenting almost inexistent Optic Disc and Cup areas due to the values of the color gradient in the vessels edges.

# References

1. Fondón, I., Sáez, A., Sánchez, C., van Grinsven, M.: Perceptually adapted method for optic disc detection on retinal fundus images. In: 2013 IEEE 26th International Symposium on Computer-Based Medical Systems (CBMS), pp. 279–284 (2013)
2. Bulletin of the World Health Organization, vol. 82, no. 11, pp. 811–890, November 2004
3. Jelinek, H., Cree, M.: Automated Image Detection of Retinal Pathology. CRC Press, Taylor & Francis Group, Boca Raton (2010)
4. Foster, A., Esnikoff, S.: The impact of VISION 2020 on global blindness. Eye **19**(10), 1133 (2005)
5. Fondón, I., et al.: Automated cup to disc ratio estimation for glaucoma diagnosis in retinal fundus images. In: Image Analysis and Modeling in Ophthalmology, pp. 179–202. CRC Press (2014). ISBN 13:978-1-4665-5930-1
6. Kavitha, D., Devi, S.S.: Automatis detection of optic disc and exudates in retinal images. In: International Conference on Intelligent Sensing and Information Processing (ICISIP), pp. 501–506 (2005)
7. Díaz-Pernil, D., Fondón, I., Peña-Cantillana, F., Gutiérrez-Naranjo, M.A.: Fully automatized parallel segmentation of the optic disc in retinal fundus images. Pattern Recogn. Lett. **83**, 99–107 (2016)
8. Leese, G., Tesfaye, S., Dengler-Harles, M., et al.: Screening for diabetic eye disease by optometrists using slit lamps. J. R. Coll. Phys. Lond. **31**, 65 (1997)
9. Gopalakrishnan, et al.: Optic disc segmentation using circular hough transform and curve fitting. In: International Conference on Opto-Electronics and Applied Optics: Advances in Optical Sciences and Engineering (2015)
10. Ho, C., Pai, T., Chang, H., Chen, H.: An automatic fundus image analysis system for clinical diagnosis of glaucoma. In: International Conference on Complex, Intelligent and Software Intensive Systems, pp. 559–564 (2011)
11. Mubbashar, M., Akram, M.: Automated system for macula detection in digital retinal images. In: International conference on Information and communication Technologies, pp. 1–5 (2011)
12. Lu, S., Lim, J.: Automatic optic disc detection through background estimation. In: IEEE International Conference on Image Processing, pp. 833–837 (2010). [36] Lu, S., Lim, J.: Automatic optic disc detection from retinal images by a line operator. IEEE Trans. on Biomedical Engineering, 88–94 (2011)

13. Abbas, Q., Fondón, I., Jiménez, S., Alemany, P.: Automatic detection of optic disc from retinal fundus images using dynamic programming. In: Campilho, A., Kamel, M. (eds.) ICIAR 2012. LNCS, vol. 7325, pp. 416–423. Springer, Heidelberg (2012). doi:10.1007/978-3-642-31298-4_49

14. Lee, S., Rajeswari, M., Ramachandram, D.: Screening of diabetic retinopathy-automatic segmentation of optic disc in colour fundus images. In: Distributed Frameworks for Multimedia Applications, pp. 1–7 (2006)

15. Marrugo, A., Millan, M.: Retinal image analysis: preprocessing and feature extraction. Reunión Iberoamericana de óptica, Láseres y aplicaciones, 1–8 (2011)

16. Saez, A., Fondon, I., et al.: Optic disc segmentation based on level-set and colour gradients. In: 6th European Conference on Colour in Graphics, Imaging, and Vision 2012, CGIV 2012, pp. 121–125 (2012)

17. Wong, D.W.K., Liu, J., Lim, J.H., et al.: Level-set based automatic cup-to-disc ratio determination using retinal fundus images in ARGALI. In: 30th Annual International Conference of the IEEE Engineering in Medicine and Biology Society, pp. 2266–2269 (2008)

18. Liu, J., Wong, D.W.K., Lim, J.H., et al.: ARGALI: an automatic cup-to-disc ratio measurement system for glaucoma analysis using level-set image processing. In: 13th International Conference on Biomedical Engineering, vol. 2, pp. 559–562 (2009)

19. Hatanaka, Y., Noudo, A.: Automatic measurement of cup to disc ratio based on line profile analysis in retinal images. In: International Conference of the IEEE Engineering in Medicine and Biology Society, pp. 3387–3390 (2011)

20. Almazroa, A., et al.: Optic cup segmentation based on extracting blood vessel kinks and cup thresholding using Type-II fuzzy approach. In: International Conference on Opto-Electronics and Applied Optics: Advances in Optical Sciences and Engineering (2015)

21. Joshi, G.D., Sivaswamy, J., Karan, K., Krishnadas, S.R.: Optic disk and cup boundary detection using regional information. IEEE International Symposium on Biomedical Imaging: From Nano to Macro, pp. 948–951 (2010)

22. Zhang, Z., Liu, J., Cherian, N.S., et al.: Convex hull based neuro-retinal optic cup ellipse optimization in glaucoma diagnosis. In: Annual International Conference of the IEEE Engineering in Medicine and Biology Society, p. 1441 (2009)

23. Zhang, Z., Liu, J., Wong, W.K., et al.: Neuro-retinal optic cup detection in glaucoma diagnosis. In: International Conference on Biomedical Engineering and Informatics, pp. 1–4 (2009)

24. García, A.I., Fondón, I.: Automatic algorithm for optic cup segmentation in retinal fundus images. In: Proceedings of 4th International Symposium on Applied Bioimaging, pp. 40–43 (2015)

25. Fondón, I., Valverde, J.F., Sarmiento, A., Abbas, Q., Jiménez, S., Alemany, P.: Automatic optic cup segmentation algorithm for retinal fundus images based on random forest classifier. In: IEEE Region 8 EuroCon 2015 Conference, Salamanca, Spain, pp. 1–6 (2015)

26. Acharya, U.R., et al.: Automated screening system for retinal health using bi-dimensional empirical mode decomposition and integrated index. Comput. Biol. Med. **75**, 54–62 (2016)

27. Agarwal, A., Gulia, S., et al.: Automatic glaucoma detection using adaptive threshold based technique in fundus image. In: International Conference on Telecommunications and Signal Processing, pp. 416–420 (2015)

28. Dutta, M.K., Mourya, A.K., et al.: Glaucoma detection by segmenting the super pixels from fundus colour retinal images. In: International Conference on Medical Imaging, m-Health and Emerging Communication Systems, pp 86–90 (2015)

29. Aruchamy, S., Bhattacharjee, P., Sanyal, G.: Automated glaucoma screening in retinal fundus images. Int. J. Multimedia Ubiquit. Eng. **10**(9), 129–136 (2015)

30. Ahmad, H., Yamin, A., et al.: Detection of glaucoma using retinal fundus images. In: International Conference on Robotics and Emerging Allied Technologies in Engineering, pp. 321–324 (2014)
31. Hamilton Eye Institute Macular Edema Dataset. http://vibot.u-bourgogne.fr/luca/heimed.php
32. Resolution Fundus (HRF) Image Database. Universität Erlangen-Nürnberg, Pattern Recognition Lab. https://www5.cs.fau.de/research/data/fundus-images/
33. Medical Image Analysis Group. Universidad de La Laguna, España. http://medimrg.webs.ull.es/research/retinal-imaging/
34. Messidor Database. French Ministry of Research and Defense. http://messidor.crihan.fr/index-en.php
35. STructured Analysis of the Retina Database. http://www.ces.clemson.edu/~ahoover/stare/
36. DRIVE Database. http://www.isi.uu.nl/Research/Databases/DRIVE/
37. Liu, J., Wong, D., Lim, J., Li, H., Tan, N., Zhang, Z., Wong, T., Lavanya, R.: ARGALI: an automatic cup-to-disc ratio measurement system for glaucoma analysis using level-set image processing. In: IFMBE Proceedings, vol. 23, pp. 559–562 (2009)
38. Wong, D.W., et al.: Intelligent fusion of cup-to-disc ratio determination methods for glaucoma detection in ARGALI. In: Annual International Conference on IEEE Engineering in Medicine and Biology Society (2009)
39. Sáez, A., Fondón, I., Acha, B., Jiménez, S., Alemany, P., Abbas, Q.: Optic disc segmentation based on level-set and colour gradients. In: Sixth European Conference on Colour in Graphics, Imaging and MCS/10 Vision, pp. 121–125 (2010)
40. Reyna, R.A., Esteve, D., Houzet, D., Albenge, M.: Implementation of the SVM neural network generalization function for image processing. In: Computer Architectures for Machine Perception, pp. 147–151 (2000)

# 2C-SVM Based Radar Detectors in Gaussian and K-Distributed Real Interference

David Mata-Moya[✉], Maria-Pilar Jarabo-Amores, Manuel Rosa-Zurera,
Javier Rosado-Sanz, and Nerea del-Rey-Maestre

Signal Theory and Communications Department, Superior Polytechnic School,
University of Alcalá, Alcalá de Henares, 28805 Madrid, Spain
{david.mata,mpilar.jarabo,manuel.rosa,javier.rosado,nerea.delrey}@uah.es

**Abstract.** This paper tackles the design and evaluation of cost sensitive Support Vector Machine (2C-SVM) based radar detectors in presence of Gaussian and K-Distributed clutter. 2C-SVM based solutions are able to approximate the Neyman-Pearson detector for a specific false alarm rate ($P_{FA}$). Real data acquired in different wind conditions by a coherent, pulsed and X-Band radar were considered. A statistical analysis is carried out to design the 2C-SVM for detecting targets with unknown parameters in Gaussian and non-Gaussian interference. A grid search of the best training parameters to approximate the pair detection probability ($P_D$) and $P_{FA}$ of the NP detector is required. Results prove the capability of the 2C-SVM based detectors to maximize the $P_D$ for a desired $P_{FA}$ independently of the detection problem likelihood functions.

**Keywords:** 2C-SVM · Neyman-Pearson detector · Gaussian clutter · K-Distributed clutter

## 1 Introduction

A traditional security system is based on active radar sensors used for surveillance and monitoring tasks. In Fig. 1, the general structure of a scanning radar is presented. The radar detection problem can be formulated as a binary hypothesis test, where the detector has to decide between target absence (null hypothesis, $H_0$) and target presence (alternative hypothesis, $H_1$). The most extended detector criterion in radar applications is the Neyman- Pearson (NP) detector, which maximizes the Probability of Detection, $P_D$, maintaining the Probability of False Alarm, $P_{FA}$, lower than or equal to a given value [1,2].

If $\tilde{z}$ is the observation vector generated at the output of the synchronous detector and $p(\tilde{z}|H_0)$ and $p(\tilde{z}|H_1)$ are the detection problem likelihood functions, a possible implementation of the NP detector consists in comparing the Likelihood Ratio (LR), $\Lambda(\tilde{z})$, to a threshold selected according to $P_{FA}$ requirements, $\eta_{lr}$ [2], and deciding in favour of $H_1$ when the LR output is higher than the selected threshold, and in favour of $H_0$ when the LR output is lower than the selected threshold (1). This approach requires a complete statistical characterization of the observation vector under both hypotheses, and significant detection

© Springer International Publishing AG 2017
I. Rojas et al. (Eds.): IWANN 2017, Part I, LNCS 10305, pp. 257–268, 2017.
DOI: 10.1007/978-3-319-59153-7_23

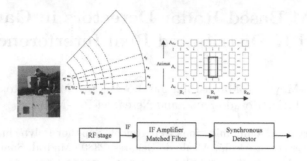

**Fig. 1.** General architecture of a coherent radar receiver

losses are expected when the true likelihood functions are different from those assumed in the LR detector design.

$$\Lambda(\widetilde{\mathbf{z}}) = \frac{p(\widetilde{\mathbf{z}}|H_1)}{p(\widetilde{\mathbf{z}}|H_0)} \underset{H_0}{\overset{H_1}{\gtrless}} \eta_{lr}(P_{FA}) \tag{1}$$

In practice, clutter and target statistics are variable. Although clutter parameters can be estimated from radar measurements, target ones are really difficult to estimate. If the parameters the likelihood functions depend on, $\phi$, are random variables (RVs) and their Probability Density Functions (PDF's) are known, the optimum detector in the NP sense can be implemented by comparing the average likelihood ratio (ALR) to a detection threshold fixed according to $P_{FA}$ requirements [2]. The ALR formulation usually leads to integrals without analytical solution, and suboptimal approaches are proposed: numerical approximations of the ALR, or the Generalized Likelihood Ratio (GLR), which uses the maximum likelihood estimation of the parameters governing the likelihood functions in the LR, as if they were correct [2,3]. Note that GLR test requires infinite number of LRs detector to cover all possible values of $\phi$, so an implementation cannot be carried out. As an alternative the Constrained Generalized Likelihood Ratio (CGLR) is expressed in (2) where $K$ is the finite number of LR detectors designed for equispaced discrete values in the expected variation range of $\phi$.

$$\max_{\varphi_k} \Lambda(\widetilde{\mathbf{z}}, \varphi_k) \underset{H_0}{\overset{H_1}{\gtrless}} \eta_{cglr}(P_{FA}) \quad k = 1, \ldots, K \tag{2}$$

This paper tackles the design of radar detectors based on Support Vector Machines (SVMs) to maximize the detection probability in composite hypothesis testing problems. The possibility of approximating the optimum detector using supervised learning machines trained to minimize a suitable error function been previously studied [4]. Although, the output of a discriminative learning machine permits to obtain an estimate of the posterior probabilities of the (binary) hypotheses if and only if the surrogate cost which is applied for training is a Bregman divergence [5]. SVMs are an intelligent agent that are an approximate implementation of the method of structural risk minimization that can provide good generalization on detection and classification problems without incorporating problem-domain knowledge [6,7]. In [8,9], SVMs, the original SVM formulation (C-SVM) and the cost-sensitive SVM one (2C-SVM), are used to design detectors considering NP and minimax criteria. In [10], a theoretical study about the capabilities of SVMs to approximate the NP detector reveals that only 2C-SVM based solutions are able to approximate the optimum one for any specific $P_{FA}$ value associated with the costs assigned to miss detections and false alarms.

In this paper, the validation of the 2C-SVM based detector with real radar data is considered. Real data acquired by a coherent, pulsed and X-Band radar deployed on Signal Hill by Council for Scientific and Industrial Research (CSIR) [11] were considered. 2C-SVMs, trained in a supervised manner, are designed and evaluated in Gaussian and non-Gaussian radar scenarios. Results confirm the capability of 2C-SVM based solutions to maximize the $P_D$ for a desired $P_{FA}$.

## 2  2C-SVM Based Approximation to the NP Detector

Let's consider a learning machine with one output to classify input vectors $\mathbf{z} = [\Re e(\tilde{z}_1), \Re e(\tilde{z}_2), ..., \Re e(\tilde{z}_P), \Im m(\tilde{z}_1), \Im m(\tilde{z}_2), ..., \Im m(\tilde{z}_P)]^T$ into two hypothesis, $H_0$ and $H_1$. The training set is composed of $N_i$ pre-classified patterns from $H_i$, with desired outputs $t_{H_i}$, $i \in \{0,1\}$, and $N = N_0 + N_1$. The output of the learning machine is denoted by $f(\mathbf{z})$.

In 2C-SVMs, the function implemented by the learning machine ($f(\mathbf{z}) = \mathbf{w}^T \Phi(\mathbf{z}) + b$) is a linear function of the results of mapping the input pattern $\mathbf{z}$ into a higher dimensional space $\mathcal{H}$ with the functions $\Phi(\mathbf{z})$, that are known as *kernel functions* [6]. The parameters of the learning machine are the weights vector $\mathbf{w}$, the bias constant $b$, and the parameters the functions $\Phi(\mathbf{z})$ depend on. The SVM is based on the hyperplane which maximizes the separating margin between the two classes, that can be obtained mathematically by solving the following unconstrained optimization problem defined in expression (3) [9]:

$$\min_{f,\xi,\gamma}\left\{\tfrac{1}{2}\|\mathbf{w}\|^2 + C\gamma \sum_{i\in\mathcal{X}_0} \xi_i + C(1-\gamma) \sum_{i\in\mathcal{X}_1} \xi_i\right\}$$
$$t_{H_i}f(\mathbf{z}_i) \geqslant 1-\xi_i \qquad\qquad i = 1,...,N \qquad (3)$$
$$\xi_i \geq 0 \qquad\qquad i = 1,...,N$$

where $C$ and $\gamma$ control of the cost associated with the two possible errors: $C\gamma$ associated with false alarms and $C(1-\gamma)$ associated with detection losses. $\xi_i$ are slack variables to relax the separability constraints when the training data can not be completely separable by an hyperplane ($\sum_i \xi_i$ is an upper bound on the number of training errors).

The function approximated by a 2C-SVM trained in a supervised manner to minimize (3) when $N \to \infty$ is calculated in [10] and expressed in (4).

$$f_0(\mathbf{z}) = -1 \text{ if } (1-\gamma)P(H_1)f(\mathbf{z}|H_1) < \gamma P(H_0)p(\mathbf{z}|H_0)$$
$$f_0(\mathbf{z}) = 1 \text{ if } (1-\gamma)P(H_1)f(\mathbf{z}|H_1) > \gamma P(H_0)p(\mathbf{z}|H_0) \qquad (4)$$

After training, the 2C-SVM will provide outputs close to $+1$ or $-1$, depending on the conditions expressed in (4). If the output of the 2C-SVM is compared to a threshold $\eta_0 = 0$, the intermediate value between $+1$ and $-1$, the decision rule is equivalent to:

$$f_0(\mathbf{z}) \underset{H_0}{\overset{H_1}{\gtrless}} \eta_0 = 0 \Rightarrow \frac{f(\mathbf{z}|H_1)}{f(\mathbf{z}|H_0)} \underset{H_0}{\overset{H_1}{\gtrless}} \eta_{lr} = \frac{\gamma P(H_0)}{(1-\gamma)P(H_1)} \tag{5}$$

Varying the value of $\gamma$, we can select different thresholds of the likelihood ratio based detector. And varying $\eta_{lr}$, we can implement detectors with pairs $(P_{FA}, P_D)$ corresponding to different points of the Receiver Operating Characteristic (ROC) curve of the NP detector. However, the surrogate cost which is applied for training is not a Bregman divergence, so if the applied threshold is changed, an approximation error to another ROC point has to be assumed. Then, a 2C-SVM based detector can maximize the $P_D$ for a given $P_{FA}$ using the corresponding $\eta_{lr}$ and training the 2C-SVM with the associated $\gamma$. Unfortunately, it is difficult to fix the threshold theoretically, and it must be fixed experimentally. In addition, the training set size is finite and the $C$ parameter will be used to increase the generalization capability. A grid search in $C - \gamma$ space has to be carried out to approximate the desired point of the NP ROC curve.

## 3    Experimental Results

Real radar data acquired by X-Band radar deployed on Signal Hill by Council for Scientific and Industrial Research (CSIR) are used to demonstrate the capability of 2C-SVMs to maximize the $P_D$ for a given $P_{FA}$ value. The datasets used in this study are available to the international radar research community on [12].

Signal Hill location (Fig. 2(a)) provided $140°$ azimuth coverage of which a large sector spanned open sea whilst the remainder looked towards the West Coast coastline from the direction of the open sea. Grazing angles ranging from $10°$ at the coastline to $0.3°$ at the radar instrumented range of 37.28 NM (Nautical Miles) were obtained. The pulse repetition frequency was 2 kHz and the range resolution is 15 m. A collaborative 4.2 m inflatable rubber boat, that can be considered as a point target, was used during some measurements (Fig. 2(b)).

Datasets were recorded with different local wind conditions. The average wind speed varied between 0 knots and 40 knots and the significant wave height ranged between 1 and 4.5 m. Then, in function of the selected dataset and the associated wind conditions, sea echoes can be modelled as Gaussian or non-Gaussian clutter.

The selected files are *Dataset 08-028.TStFA* and *Dataset 10-104.TTrFA*, and main parameters of the acquisitions are summarized in Table 1 [12]. There is no available information about target speed or GPS data. The squared envelopes in logarithmic units of the first pulses of the patterns are presented in Fig. 3.

A CGLR composed of $K = 2P$ LR detectors designed for discrete values of unknown target parameter equally spaced in the variation interval can be

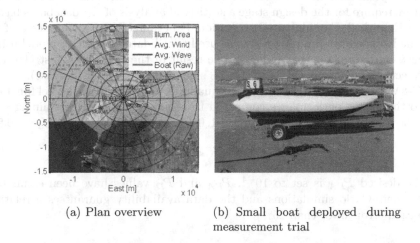

(a) Plan overview

(b) Small boat deployed during measurement trial

**Fig. 2.** Radar environment considered in [12]

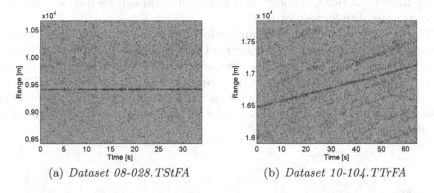

(a) *Dataset 08-028.TStFA*

(b) *Dataset 10-104.TTrFA*

**Fig. 3.** Logarithmic squared amplitude of the first pulses of the patterns for all range cells associated with the selected files

**Table 1.** Acquisition specifications of selected datasets [12]

|  | *Dataset 08-028.TStFA* | *Dataset 10-104.TTrFA* |
|---|---|---|
| Acquisition time | 33.9625 s (1,045 patterns) | 64.48 s (1,984 patterns) |
| Range extend | 2248.4 m (151 gates) | 1,903.7 m (128 gates) |
| Significant wave height | <0.1 m, 246.5 N | 2.48 m, 244.8 N |

used to approximate the NP detector [14,15]. Both CGLR and 2C-SVM based detectors require for the design stage a statistical analysis of the databases based on the Empirical Cumulative Distribution Function (ECDF). Goodness-of-fit test, the Kolmogorov-Smirnov test, was used to validate the clutter model using the *decision* parameter that indicates the result of the hypothesis test (1 if the test rejects the null hypothesis at the 5% significance level, 0 otherwise).

2C-SVMs are designed assuming the quadratic function (6) as kernel function due to the better performance compared to radial basis for large training sets [13]. The training set size is composed of $1,000$ patterns ($P(H_0) = P(H_1) = 0.5$).

$$\Phi(\mathbf{z}_i, \mathbf{z}_j) = (1+ < \mathbf{z}_i, \mathbf{z}_j >)^2 \quad < \mathbf{z}_i, \mathbf{z}_j >= \sum_{l=1}^{L} z_{i,l} z_{j,l} \qquad (6)$$

The desired $P_{FA}$ is set to $10^{-4}$. $P_{FA}$ and $P_D$ values have been estimated using Monte-Carlo simulations and the data availability guarantees a relative estimation error lower than 25%.

### 3.1   Statistical Analysis of Real Radar Data

Attending to significant wave height (Table 1), the sea state is calm (code 1) for *Dataset 08-028.TStFA* and moderate (code 4) for *Dataset 10-104.TTrFA*. Under these conditions, Rayleigh and K-Distribution can be used for modeling sea clutter amplitude, respectively.

The ECDFs of the sea clutter amplitude samples are estimated and compared to theoretical Rayleig and K-Distribution CDF. Samples of $27^{th}$ and $40^{th}$ range gates for *Dataset 08-028.TStFA* and *Dataset 10-104.TTrFA* respectively are considered. The comparisons between the empirical and the theoretical CDFs are presented in Fig. 4. The Visual inspection in Fig. 4 confirms a very good agreement with Gaussian and K-Distributed clutter model respectively. Table 2 details the results provided by the KS-Test.

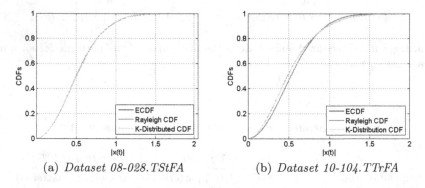

(a) *Dataset 08-028.TStFA*          (b) *Dataset 10-104.TTrFA*

**Fig. 4.** CDFs for the sea clutter amplitude samples of the real radar data

**Table 2.** KS-Test for the sea clutter amplitude samples of the real radar data

*Dataset 08-028.TStFA*

| | Distributions | Parameters | Decision |
|---|---|---|---|
| $\|x\|$ | Rayleigh $(x \geq 0)$ $F_x(x) = 1 - \exp^{-x^2/2\sigma^2}$ | $\sigma = 0.42783$ | KS-Test: 0 |
| $\|x\|$ | K-Distribution $(x \geq 0)$ | $\nu = 1.0040$ | KS-Test: 0 |
| | $F_x(x) = 1 - \left( \frac{2}{2^\nu \cdot \Gamma(\nu)} \left( 2 \cdot x \sqrt{\frac{\nu}{\mu}} \right)^\nu \cdot K_\nu \left( 2 \cdot x \sqrt{\frac{\nu}{\mu}} \right) \right)$ | $\mu = 0.42810$ | |

*Dataset 10-104.TTrFA*

| | Distributions | Parameters | Decision |
|---|---|---|---|
| $\|x\|$ | Rayleigh | $\sigma = 0.4455$ | KS-Test: 1 |
| $\|x\|$ | K-Distribution | $\nu = 3.9357$ | KS-Test: 0 |
| | | $\mu = 0.3969$ | |

The K-distribution is formed by compounding two separate probability distributions, one representing the radar cross-section (RCS) and the other representing speckle. The component representing the RCS is a slowly varying non-negative Gamma process that introduces a power modulation of the local backscatter, consequence of longer wavelength sea waves (texture), $\tau[n]$. The component representing speckle is modeled as a complex Gaussian random process, $g[n]$. As the power modulation is slower than the speckle component, it is possible to approximate the received clutter sequence by the following expression:

$$z_k[n] = \sqrt{\tau[k]}g[n] \qquad n = k, ..., k + L_c - 1 \qquad (7)$$

where $L_c$ is the coherence length of sea texture, the number of time samples for which the texture can be considered constant. Expression (8) can be used to estimate the texture and the hypothetical speckle time sequences, respectively, for each possible $L_c$.

$$\tilde{\tau}[n] = \frac{1}{L_c} \sum_{k=n-\frac{L_c}{2}}^{k=n+\frac{L_c}{2}-1} |z_k[n]|^2 \qquad \tilde{g}[n] = \frac{z_k[n]}{\sqrt{\tilde{\tau}[n]}} \qquad (8)$$

In Fig. 5(a), the KS-Test values obtained by comparing the texture sequence CDF to a gamma distribution for different $L_c$ are presented. The threshold of a 5% significance level is also depicted. It is possible to conclude that the coherence length of sea texture is between 0.08 and 0.43 s. As the 64-pulses-patterns time is 0.0325 s, a constant texture can be considered in each pattern. Additionally, the real and imaginary parts fulfil the Jarque-Bera goodness-of-fit test (it is a goodness-of-fit test of whether sample data have the skewness and kurtosis matching a normal distribution), so the Gaussian distribution fits the real and imaginary parts of speckle. In Fig. 5(b), the speckle amplitude ECDF is depicted confirming the approximation to a Rayleigh distribution.

Proposed detection schemes present an important dependence on the clutter one-lag clutter correlation coefficient ($\rho_c$), so the Autocorrelation Function (ACF) for the pulses associated with the same pattern was studied. The

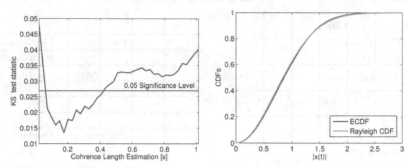

(a) KS-Test statistic of the texture for (b) CDFs for the amplitude of speckle
different $L_c$                          component ($L_c = 0.18s$)

**Fig. 5.** Statistical analysis of the speckle component for $27^{th}$ range cell of *Dataset 10-104.TTrFA*

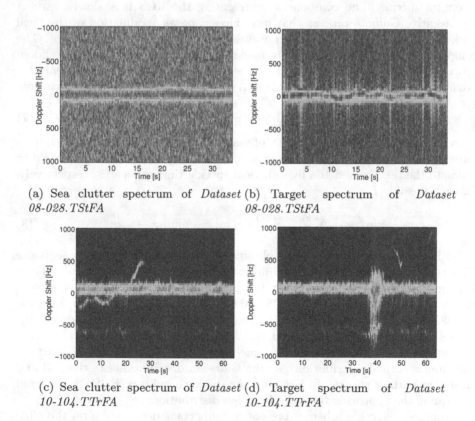

(a) Sea clutter spectrum of *Dataset 08-028.TStFA*    (b) Target spectrum of *Dataset 08-028.TStFA*

(c) Sea clutter spectrum of *Dataset 10-104.TTrFA*    (d) Target spectrum of *Dataset 10-104.TTrFA*

**Fig. 6.** Doppler Shift of range cells of real radar data (RCS $[dBm^2 \cdot Hz^{-1}]$)

estimated $\rho_c$ is equal to 0.8 and 0.966 for *Dataset 08-028.TStFA* and *Dataset 10-104.TTrFA* respectively. In Fig. 6, the spectra for range cells associated with seal clutter and target returns are presented. The spectra corresponding to sea clutter correlated echoes are localized close to the zero Doppler, while spectra of range cells corresponding to target echoes present a variable Doppler shift, $\Omega \in [-0.0491, 0.0982]$ rad for *Dataset 08-028.TStFA* and $\Omega \in [-0.6; 0.6]$ for *Dataset 10-104.TTrFA*.

Clutter and target powers, $p_s$ and $p_c$, were estimated to generate the design CGLR and the 2C-SVM training sets. In Table 3, the estimated power and the mean $SIR = 10 \log(p_s/p_c)$ is detailed.

Table 3. Estimated power levels or real radar data

|  | $p_s$ | $p_c$ | $SIR$ (dB) |
|---|---|---|---|
| *Dataset 08-028.TStFA* | 20.7564 | 0.3510 | 17.72 |
| *Dataset 10-104.TTrFA* | 2.0068 | 0.3969 | 7.04 |

## 3.2   CGLR and 2C-SVM Detection Performance

Attending to statistical analysis, two radar scenarios with different detection problem likelihood functions are considered for designing and testing the considered detectors:

- *Dataset 08-028.TStFA*: CGLR and 2C-SVM are design for detecting targets with unknown Doppler shift in Gaussian interference. Swerling I target to model vessel echoes acquired by marine radars was assumed [16].
- *Dataset 10-104.TTrFA*: detection schemes are designed for detecting targets with unknown Doppler shift in K-distributed interference. Swerling V targets model was selected in order to have the capability of formulating the LR detector analytically [17,18].

In Fig. 7, the grid searchs in the $C$ and $\gamma$ space to select the best 2C-SVM training values in both environments are depicted. The represented logarithmic $P_{FA}$ values were estimated using $\eta_0 = 0$. The detection performances of the 2C-SVMs designed with the suitable pairs $(\gamma, C)$ applied to real radar databases are compared with the $P_D$ provided by the $CGLR$ for $P_{FA} = 10^{-4}$ in Fig. 4. The pair $(\gamma = 0.99, C = 10)$ provides the best approximation to the CGLR performance for $P_{FA} = 10^{-4}$ independently of the case study.

2C-SVM, designed with $(\gamma = 0.99, C = 10)$, based solutions provides similar detection performance to the $CGLR$ detector for the $P_{FA}$ closest to the desired one. Figure 8, show the considered detection schemes outputs for considered radar scenarios. The estimated centroids are depicted and the target trajectories are clearly detected by the $CGLR$ and 2C-SVM detectors with both databases.

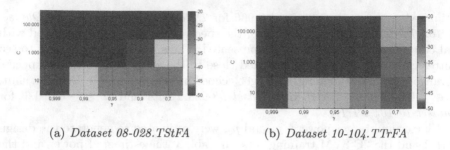

(a) *Dataset 08-028.TStFA*          (b) *Dataset 10-104.TTrFA*

**Fig. 7.** $10 \log(P_{FA})$ resulted from a grid search in $(\gamma, C)$ for 2C-SVM

**Table 4.** 2C-SVM and CGLR detection performances with real radar data

| Detector | | | *Dataset 08-028.TStFA* | | *Dataset 10-104.TTrFA* | |
|---|---|---|---|---|---|---|
| | | | $P_{FA}$ | $P_D$ | $P_{FA}$ | $P_D$ |
| *CGLR* | | | $1.553 \cdot 10^{-4}$ | 0.8016 | $1.0183 \cdot 10^{-4}$ | 0.5101 |
| SVM | $C$ | $\gamma$ | | | | |
| | 10 | 0.99 | $1.0503 \cdot 10^{-4}$ | 0.8008 | $1.1024 \cdot 10^{-4}$ | 0.4879 |
| | 10 | 0.95 | $15.2247 \cdot 10^{-4}$ | 0.8563 | $20.1065 \cdot 10^{-4}$ | 0.8372 |
| | $10^3$ | 0.7 | $0.9145 \cdot 10^{-4}$ | 0.7847 | $23.3584 \cdot 10^{-4}$ | 0.5111 |

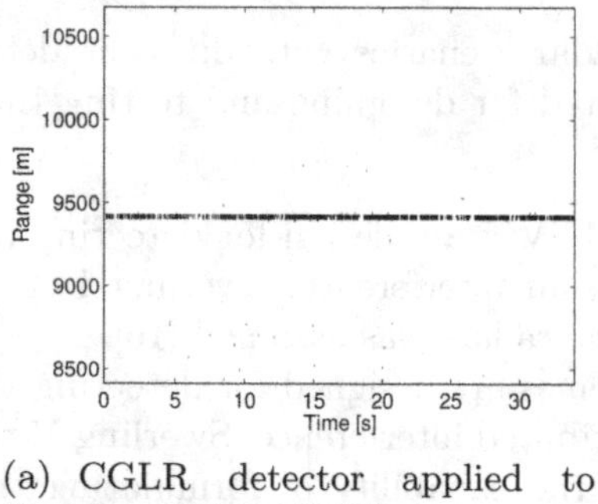

(a) CGLR detector applied to *Dataset 08-028.TStFA*

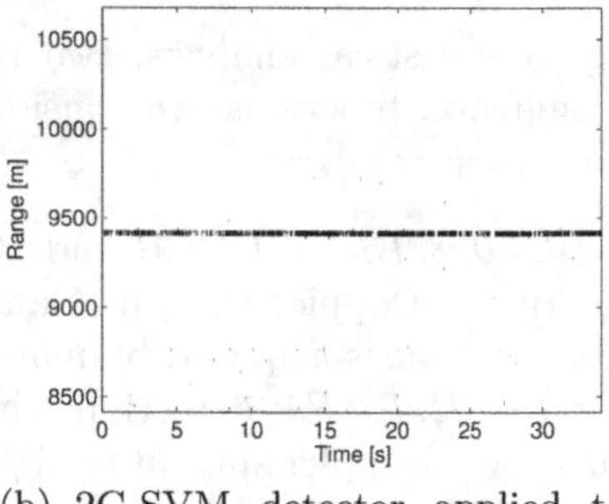

(b) 2C-SVM detector applied to *Dataset 08-028.TStFA*

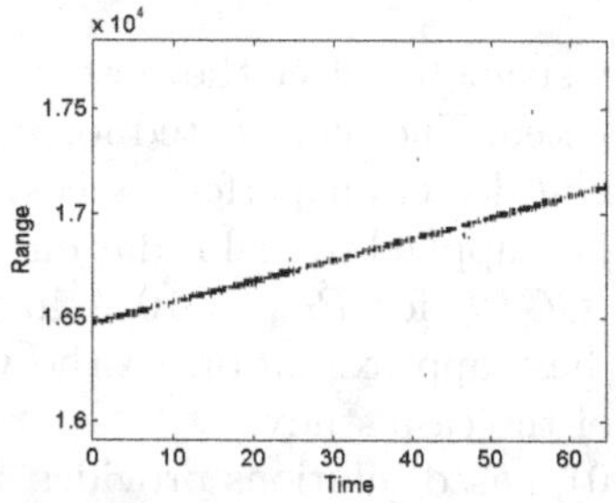

(c) CGLR detector applied to *Dataset 10-104.TTrFA*

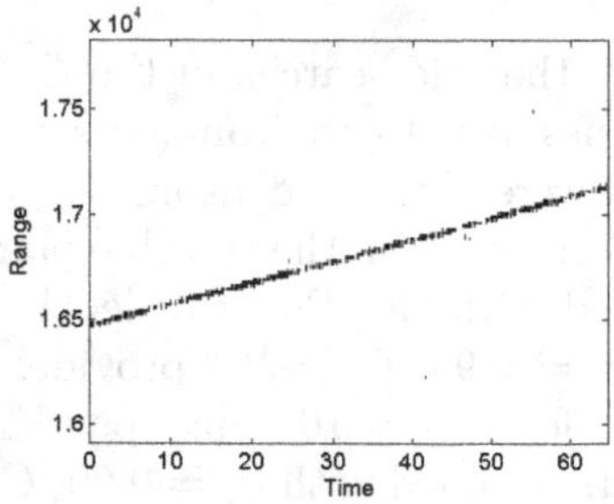

(d) 2C-SVM detector applied to *Dataset 10-104.TTrFA*

**Fig. 8.** Estimated centroids for the real radar matrices

# 4    Conclusions

This paper tackles the problem of designing and validating a 2C-SVM radar detector capable of approximating the optimum Neyman-Pearson detector for given $P_{FA}$ in composite hypothesis testing problems characterized by targets with unknown parameters in Gaussian and K-Distributed clutter.

In [10], a theoretical study to obtain the function the learning machine converges to after training, if a sufficiently large number of training patterns was carried out. The 2C-SVM implements a function with only two values, but they are obtained depending on which is higher $(1 - \gamma)P(H_1)p(\mathbf{z}|H_1)$ or $\gamma P(H_0)p(\mathbf{z}|H_0)$. In this case, the detector that is implemented when the output of the 2C-SVM is compared to a threshold with an intermediate value between the possible outputs, is equivalent to the Neyman-Pearson detector for a fixed pair $(P_{FA}, P_D)$. The values of $P_{FA}$ and $P_D$ varies with the parameter $\gamma$ that can be used to select different points in the ROC curve of the NP detector. As the number of training patterns is finite and the relation between $\gamma$ and $P_{FA}$ is usually unknown, a grid search in $C - \gamma$ space has to be carried out to approximate the desired point of the NP ROC curve.

The evaluation of the 2C-SVM detection performance is also studied using real data available at [12]. Clutter and target statistics are analyzed to design the synthetic training set. Two datasets with different wind conditions are selected to consider two cases study: detecting Swerling I targets with unknown Doppler shift in Gaussian interference and detecting Swerling V targets with unknown Doppler shift in spiky K-Distributed clutter. Sub-optimum approaches to the ALR detector based on the CGLR are considered as reference detectors. Results obtained for $P_{FA} = 10^{-4}$ confirm the similarity of the CGLR and 2C-SVM based solution detection capabilities. 2C-SVMs are quite good systems to approximate the NP detector for a given point of the ROC curve.

**Acknowledgments.** This work has been partially funded by the Spanish "Ministerio de Economía, Industria y Competitividad", under project TEC2015-71148-R, the European FP7 SCOUT Project (Grant Agreement n 607019), and by the University of Alcalá, under project CCG2016/EXP079.

# References

1. Neyman, J., Pearson, E.S.: On the problem of the most efficient test of statistical hypotheses. Philos. Trans. Roy. Soc. Vol A **231**(9), 492–510 (1933)
2. Van Trees, H.L.: Detection, Estimation, and Modulation Theory, vol. 1. Wiley, Hoboken (1968)
3. Sangston, K.J., Gini, F., Greco, M.S.: Coherent radar target detection in heavy-tailed compound-Gaussian clutter. IEEE Trans. Aerospace Electron. Syst. **48**(1), 64–77 (2012)
4. Jarabo-Amores, P., Mata-Moya, D., Gil-Pita, R., Rosa-Zurera, M.: Radar detection with the Neyman-Pearson criterion using supervised-learning-machines trained with the cross- entropy error. EURASIP J. Adv. Sig. Process. **2013**, 1–10 (2013)

5. Bregman, L.M.: The relaxation method of finding the common points of convex sets and its application to the solution of problems in convex programming. USSR Comp. Math. Math. Phys. **7**, 200–217 (1967)
6. Vapnik, V.: The Nature of Statiscal Learning Theory. Springer, Heidelberg (1995)
7. Burges, C.J.C.: A tutorial on support vector machines for pattern recognition. Data Mining Knowl. Discov. **2**, 121–167 (1998)
8. Bach, F., et al.: Considering cost asymmetry in learning classifiers. J. Mach. Learn. Res. **7**, 1713–1741 (2006)
9. Davenport, M.A., Baraniuk, R.G., Scott, C.D.: Tuning support vector machines for minimax and Neyman-Pearson classification. IEEE Trans. Pattern Anal. Mach. Intell. **32**(10), 1888–1898 (2010)
10. Mata-Moya, D., Jarabo-Amores, M.P., Nicolás, J.M., Rosa-Zurera, M.: Approximating the Neyman-Pearson detector with 2C-SVMs. Application to radar detection. Sig. Process. **131**, 364–375 (2016)
11. Herselman, P.L., Baker, C.J., De Wind, H.J.: Analysis of X-Band calibrated sea clutter and small boat reflectivity at medium-to-low grazing angles. Int. J. Navig. Obs. **2008**, 1–14 (2008)
12. CSIR website (2014). http://www.csir.co.za
13. Afifi, A., et al.: Improving the classification accuracy using support vector machines (SVMs) with new kernel. J. Glob. Res. Comput. Sci. **4**, 17 (2013)
14. Nayebi, M.M., et al.: Detection of coherent radar signals with unknown Doppler shift. IEE Proc. Radar, Sonar Navig. **143**(2), 73–86 (1996)
15. Mata-Moya, D., Jarabo-Amores, P., Martin-de-Nicolas, J.: High order neural network based solution for approximating the average likelihood ratio. In: Proceedings of IEEE Statistical Signal Processing Workshop (SSP), pp. 657–660 (2011)
16. Yamaguchi, H., Suganuma, W.: CFAR detection from non-coherent radar echoes using Bayesian theory. EURASIP J. Adv. Sig. Process. **2010**, 19 (2010)
17. Skolnik, M.: Introduction to Radar Systems, 3rd edn. McGraw-Hill, New York (2004)
18. Conte, E., et al.: Radar detection in K-distributed clutter. IEE Proc. Radar, Sonar Navig. **141**(2), 116–118 (1994)

# Uncertainty Analysis of ANN Based Spectral Analysis Using Monte Carlo Method

José Ramón Salinas[1(✉)], Francisco García-Lagos[1],
Javier Díaz de Aguilar[2], Gonzalo Joya[1], and Francisco Sandoval[1]

[1] Grupo ISIS, Dpto. Tecnología Electrónica, ETSI Telecomunicación,
Universidad de Málaga, Campus de Teatinos s/n, 29071 Málaga, Spain
jrsalinasv@gmail.com, lagos@dte.uma.es
[2] Centro Español de Metrología (CEM), Alfar 2, 28760 Tres Cantos, Spain

**Abstract.** Uncertainty analysis of an Artificial Neural Network (ANN) based method for spectral analysis of asynchronously sampled signals is performed. Main uncertainty components contributions, jitter and quantization noise, are considered in order to obtain the signal amplitude and phase uncertainties using Monte Carlo method. The analysis performed identifies also uncertainties main contributions depending on parameters configurations. The analysis is performed simultaneously with the proposed method and two others: Discrete Fourier Transform (DFT) and Multiharmonic Sine Fitting Method (MSFM), in order to compare them in terms of uncertainty. Results show the proposed method has the same uncertainty as DFT for amplitude values and around double uncertainty in phase values.

**Keywords:** Sine-fitting methods · Spectral analysis · ADALINE · ANN · Digital measurement · Uncertainty · Monte-Carlo · DFT

## 1 Introduction

The use of nonlinear electronic components connected to the electricity grid it is becoming more common in our daily life. These components add harmonic and inter-harmonic content to the electric signal, which results in a deterioration of the quality of the power supplied, an increase in losses and a decrease in the reliability of the whole system. In this situation, it is increasingly important to accurately determine the harmonic content of the power signals. For these reasons, several National Metrological Institutes (NMIs) have implemented methods to measure power in non-sinusoidal conditions [1, 2]. These new methods make use of the versatility of the digital techniques, especially considering the possibility of obtaining the spectral analysis of the signals of interest, and are based on the use of traditional algorithms, as DFT. These methods require synchronous sampling in order to be accurate. To overcome this problem, some authors [3, 4] have proposed the use of asynchronous sampling combined with the use of non-synchronous spectral analysis. In particular sine fitting methods [5] have been extended successfully to the multi-harmonic case [6].

Alternatively, a new method based on ANN was presented [7]. The method is based in the implementation of the multi-harmonic sine fitting algorithm [5] by mean of a

I. Rojas et al. (Eds.): IWANN 2017, Part I, LNCS 10305, pp. 269–280, 2017.
DOI: 10.1007/978-3-319-59153-7_24

multilayer perceptron neural network (MLP). The ANN method, as the sine fitting, has the advantage with respect to the traditional spectral analysis methods that it does not require synchronism between generation and sampling, which reduce the complexity of the hardware implementation. In comparison with the conventional multi-harmonic sine fitting method, the ANN method, has a simpler implementation and does not require special adjustments of coefficients for convergence, that depend on the type of signal being analyzed [6]. The propose method has been implemented in the Spanish electrical power primary standard, at Centro Español de Metrología [9, 10]. Additionally, this approach has the advantage to allows, with reduced complexity, modifications of the signal model to be analyzed. In this sense, this work was extended recently to obtain inter-harmonic components in [8].

At a primary level, the traceability of such sampling systems is a complex task due to the complexity of the tests and validations of involved algorithms. The main contribution of this work is the validation for sinusoidal signals of the ANN based spectral analysis method proposed in [7] by means of Monte Carlo method. In order to obtain useful results for real applications, parameters configurations used in laboratory will be used.

Following sections are structured as follows: in Sect. 2 the system to be characterized is briefly presented; in Sect. 3 the Monte Carlo method applied to the characterization of high precision measurement systems is introduced; methodology of the tests carried out is described in Sect. 4 and results are reported in Sect. 5; finally, Sect. 6 exposes the conclusions of this work.

## 2  Network Description

Let us consider a signal of interest, $y(t)$, stationary in the range of analysis, formed by $K$ harmonic frequencies of the fundamental frequency, $f_{ac}$. The signal can be described mathematically in the terms of Eq. (1), where $y(t)$ is formed by the sum of the contributions of the DC component, the fundamental frequency, $f_{ac}$, and its multiples, $f_k = k \cdot f_{ac}$, $k \in [2, K]$.

$$y(t) = C + \sum_{k=1}^{K} [A_k \cdot \cos(2\pi k f_{ac} t) + B_k \cdot \sin(2\pi k f_{ac} t)] \tag{1}$$

The proposed ANN based method [7], MANNFM (Multiharmonic ANN Fitting Method), performs the spectral analysis of a steady state periodical signal composed by harmonics as described in Eq. (1). This is done by means of the multilayer neural network of Fig. 1.

Input data to the ANN are the $N$ time instants - $t[n]$, $n \in [0, N-1]$- at which the samples of $y(t)$, $y[n]$, are taken. The time vector, $t$, is scaled by the factor $2 \cdot \pi$ in order to generate the arguments of cosines and sines of Eq. (1).

The network implements the $K$ order Fourier series synthesis equation. Briefly, Layer 1, implemented by means of an ADALINE neuron without bias and linear transfer function, has the object to implement the scale factor of the fundamental frequency, $f_{ac}$, in the arguments of the cosines and sines of the synthesis equation. The

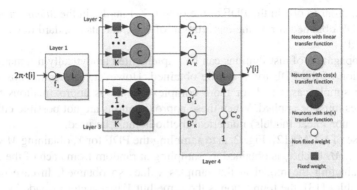

**Fig. 1.** The ANN architecture proposed [7].

output of Layer 1 is connected in parallel to Layers 2 and 3, both of which are implemented by $K$ ADALINE neurons with cosine and sine transfer functions respectively. The position of each neuron in the array of neurons determines its weight value, so that the $k$ scale factor of each harmonic in the cosine and sine arguments is implemented by this weight. As a result, the input to the transfer functions of the neurons of Layers 2 and 3 are the completed arguments of the cosines and sines of the Fourier synthesis equation.

Consequently, the input from the harmonic section to Layer *4* are the arrays $cos(2 \cdot \pi \cdot k \cdot f_{ac} \cdot t)$, $k \in [1, K]$ and $sin(2 \cdot \pi \cdot k \cdot f_{ac} \cdot t)$, $k \in [1, K]$. Layer *4* implements, finally, the sum of the synthesis equation, by means of $2 \cdot K$ neurons with linear transfer functions. The weights of the neurons connected to Layer 2 are identified as the $A_k$ coefficients of Eq. (1), and the weights of the neurons connected to Layer *3* are identified as the $B_k$ coefficients of Eq. (1). In addition, Layer *4* includes the DC component of the Fourier synthesis series, $C$, by means of its bias value, which is available in this layer. Therefore, at this point, the output of Layer *4* is equivalent to the output of the Eq. (1), $y(t)$, in the time instants $t[n]$, $n \in [0, N-1]$.

As mentioned above, this high precision algorithm has been implemented and tested in the measurements system of the Spanish NMI as part of the electrical power primary standard.

## 3   Propagation of Distributions and Simulation Using the Monte Carlo Method

As mentioned previously, due to the complexity of the MANNFM uncertainty evaluation, the MCM is used. In this section the basis of the MCM and its application to this purpose are presented. The Guide to the Expression of Uncertainty in Measurement (GUM) [11] provides the framework for uncertainty evaluation, which has three main stages: formulation, propagation and summarizing. At the *Formulation* stage the $N$ input quantities, $X = (X_1, ..., X_N)$ upon which the output quantity, $Y$, depends, are determined and their probability distribution functions (PDF), are assigned. In the *Propagation* stage, the PDFs for the $X_i$ are propagated through the measurement model,

$Y = f(X)$, in order to obtain the PDF of the output $Y$. Finally, in the *Summarizing* stage, using de PDF obtained for $Y$, the expectation of $Y$, $y$, and its standard deviation, $u(y)$, are obtained[1].

The propagation of distributions can be implemented analytically if a mathematical representation of the PDF for $Y$ can be obtained. However, in practice this option is possible in simple cases and for more complex situations approximations based on Taylor series must be applied. When these approximations are not possible either (e.g., complex or nor linear models) numerical methods must be used.

The base of MCM [12], Fig. 2, is to sampling the PDF for $Y$, obtaining $M$ values $y_r$, for $r = 1, ..., M$. Each $y_r$ is obtained by sampling at random from each of the PDFs for $X_i$, and evaluating the model at the samples values so obtained. In case of several output variables [13], the foundation is the same but $Y$ is a vector provided by a vector function or model $f(X)$.

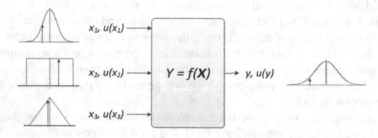

**Fig. 2.** MCM propagation of distributions for 3 independent input quantities

Consequently, in order to apply MCM, it is only necessary that $Y$ can be formed from the values of $X$. Although Monte Carlo method is used as a mean to provide a numerical representation of the PDF for $Y$, it is also a simulation process that provides information about the input/output PDFs relationship of a system modeled as $f(X)$. So, the method can be used not only to obtain output uncertainties of a measurement, but also to simulate a system [14] and evaluate the impact of different parameters as well as the impact of the different uncertainty contributions in the output uncertainties. The second approach will be used in this work.

# 4    Methodology Applied

## 4.1    Simultaneous Comparison of Three Methods

In order to compare MANNFM results with those obtained by other methods, the same tests have been applied to other two methods: DFT [15] and MSFM [6]. DFT can be considered as a reference, given that, in ideal conditions, provides the spectrum of the

---

[1] Although there is a coincidence between some terms used in GUM and this work nomenclatures, since there is no possibility of confusion, GUM nomenclature has been maintained in this section in order to simplify its reading.

signal under analysis without error. When synchronous sampling is not possible, fitting methods are used [3, 5] in order to avoid DFT errors due to spectral leakage. As MANNFM, MSFM is a fitting method, but solved in a conventional way, solving a non-linear equation system iteratively.

As can be observed in Fig. 3, for comparison purposes the same signal, $x(t)$, will be sampled and applied to the three methods. The DFT will process synchronous samples of $x(t)$, $x_s[n]$, while MSFM and MANNFM will process asynchronous samples of $x(t)$, $x_a[n]$, obtained in a set of time instants $t_a[n]$, where $t_a[n] = t[n]\cdot(1 + \xi_a)$, being $\xi_a$ the error of synchronism (relative) between generation and sampling.

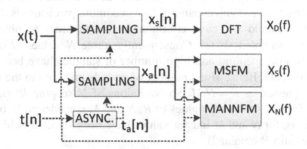

**Fig. 3.** Scheme for the comparison of the three methods

## 4.2    Uncertainties Contributions Considered

Two uncertainty components will be considered in this work, jitter contributing to time uncertainty of each sampled point, the sampling jitter; and quantization contributing to a voltage uncertainty of each sampled point.. Both of them have been theoretically studied for DFT [16, 17].

Regarding jitter, a rectangular distribution has been considered. Taking into account the specifications of usual laboratory instruments, two jitter values (maximum) will be contemplated: $3\cdot10^{-12}$ s and $1\cdot10^{-7}$ s. The first value is the jitter specification for the National Instruments 5922 digitizer, and the second one is the jitter specifications for the Keysight 3458A digital multimeter used in [9, 10].

Concerning the quantization, it has been simulated taken into account three usual laboratory resolutions or number of bits, *Nbits,* values: *16, 18* and *21* bits, which correspond again with the HP 3458A resolution derived from the aperture time ranges configured for high, medium and low sampling frequency respectively. Regarding the full scale parameter, *1.2* V has been used, which is the value for the *1* V range of the HP 3458A in digital operation mode.

In order to evaluate independent and jointly the impact of both contributions to the total uncertainty of the three methods, the scheme presented in Fig. 3 will be simulated with three different sampling situations: (i) samples affected only by quantification; (ii) samples affected only by jitter; and (iii) samples affected by quantification and jitter. Terms *q, j* and *jq* respectively will be used in reference to signals and results related to this three sampling situations.

## 4.3    Parameters Values Considered

Several parameters values (according to real values used in laboratory) of the signal to be sampled, have been considered in order to evaluate their impact to the estimated uncertainty. A sinusoidal signal has been consider, with fundamental frequency, $f_{ac}$, 53 Hz and amplitude $1$ V. Concerning to the phase of the sampled signal, uniformly distributed random values between $\pm \pi$ radians have been generated.

Regarding the ratio $R = f_s/f_{ac}$, being $f_s$ the sampling frequency, three values have been selected, around $24$, $124$ and $261$, related to low, medium and high sampling frequencies, respectively.

Concerning the number of samples, $N$, it has to be considered that (i) the Keysight 3458 internal memory maximum capability (for maximum precision) is $37888$ samples and (ii) $N$ and $M$ should not have common factors in order to obtain maximum information of the sampled signal. Consequently, three $N$ values ($1311$, $7073$ and $15457$), related to low, medium and high number of samples have been selected and used in each ratio value. Being $M$ the number of periods involved in the measurement, $R$ can be also expressed as $R = N/M$. So, variations of $N$ keeping $R$ constant implies also to vary $M$. Table 1 shows the values of $R$, $N$ and $M$ considered. In order to assure the DFT conditions, for $N$ and $M$ integer values have been selected (which imply light variations of the ratio $R$ obtained).

**Table 1.** Sampling frequency parameters values considered to estimate the uncertainty of the system.

| N | | M | R | fs |
|---|------|-----|---------|---|
| H | 15457 | 59 | 261.983 | H |
| M | 7073 | 27 | 261.963 | |
| L | 1311 | 5 | 262.200 | |
| H | 15457 | 125 | 123.656 | M |
| M | 7073 | 57 | 124.088 | |
| L | 1311 | 11 | 119.182 | |
| H | 15457 | 625 | 24.731 | L |
| M | 7073 | 286 | 24.731 | |
| L | 1311 | 53 | 24.736 | |

Finally, about the error of synchronism, $\xi_a$, random values up to $\pm 20\%$, uniformly distributed, have been considered. The Interpolated FFT (IpFFT) [18] algorithm has been used in order to obtain initial estimation of the fundamental frequency for the asynchronous methods. No convergence problems have been detected in any simulation.

## 4.4    Computational Hardware Used

In order to perform the MCM in the different configurations established, large computational resources are required. For this work, resources provided by the University of Málaga's supercomputing center (UMSC) have been used. The computational resources are based in the Cluster Intel E5-2670, with shared memory machines with 2 TB of RAM each and AMD Opteron 6176 and ESX virtualization clusters.

# 5 Results

All tests have been performed in the same way: with the selected parameters and the uncertainty contributions defined in Sect. 4.2, $10^6$ trials of the MCM has been performed over the three methods (as described in Sect. 4.1) in order to obtain a numerical representation of their PDF outputs.

All the obtained PDFs are Gaussian. The values have been expressed in deviations to nominal (absolute) and those deviations have been fitted to a normal PDFs by means of Matlab© *normfit* function, obtaining their mean and standard deviation, $\sigma$. For each output of each method, mean represents the deviation to nominal and standard deviation its uncertainty for a confidence level of 62.8%. In next subsections we will present the most significant results of the tests performed, in terms of mean and standard deviation.

## Methods Comparison

A first test was performed in order to evaluate similarities between methods outputs and, at the same time, evaluate independent and jointly the impact of jitter and quantization to total uncertainty. Maximum jitter was fixed to *100* ns and *Nbits* was set to *16*. The ratio was set to low and *15457* samples were considered (M = 625). Tables 2, 3 and 4 show results considering respectively jitter, quantization and both effects on the three methods, for amplitude and phase of the fundamental frequency.

**Table 2.** Mean ($E$) and standard deviation ($\sigma$) of fundamental frequency deviations to nominal amplitude ($\Delta A_1$) and phase ($\Delta \Phi_1$) for three methods - only jitter considered

|  | $E_j\{\Delta A_1\}$ ($\mu$V) | $\sigma_j (\Delta A_1)$ ($\mu$V) | $E_j\{\Delta \Phi_1\}$ ($\mu$rad) | $\sigma_j (\Delta \Phi_1)$ ($\mu$rad) |
|---|---|---|---|---|
| DFT | 1.06E-04 | 5.50E-02 | 6.69E-04 | 9.55E-02 |
| MPSF | −1.09E-04 | 5.49E-02 | −6.22E-04 | 1.89E-01 |
| MANNFM | −1.11E-04 | 5.49E-02 | −6.21E-04 | 1.89E-01 |

**Table 3.** Mean ($E$) and standard deviation ($\sigma$) of fundamental frequency deviations to nominal amplitude ($\Delta A_1$) and phase ($\Delta \Phi_1$) for three methods - only quantization considered

|  | $E_q\{\Delta A_1\}$ ($\mu$V) | $\sigma_q (\Delta A_1)$ ($\mu$V) | $E_q\{\Delta \Phi_1\}$ ($\mu$rad) | $\sigma_q (\Delta \Phi_1)$ ($\mu$rad) |
|---|---|---|---|---|
| DFT | 6.25E-04 | 1.12E-01 | 1.26E-04 | 1.12E-01 |
| MPSF | 1.16E-03 | 1.12E-01 | −1.28E-03 | 2.41E-01 |
| MANNFM | 1.15E-03 | 1.12E-01 | −1.28E-03 | 2.41E-01 |

**Table 4.** Mean ($E$) and standard deviation ($\sigma$) of fundamental frequency deviations to nominal amplitude ($\Delta A_1$) and phase ($\Delta \Phi_1$) for three methods - jitter and quantization considered

|          | $E_{jq}\{\Delta A_1\}$ ($\mu$V) | $\sigma_{jq}$ ($\Delta A_1$) ($\mu$V) | $E_{jq}\{\Delta \Phi_1\}$ ($\mu$rad) | $\sigma_{jq}$ ($\Delta \Phi_1$) ($\mu$rad) |
|----------|----------|----------|----------|----------|
| DFT      | 7.31E-04 | 1.25E-01 | 7.94E-04  | 1.53E-01 |
| MPSF     | 1.05E-03 | 1.33E-01 | −1.91E-03 | 3.06E-01 |
| MANNFM   | 1.05E-03 | 1.33E-01 | −1.91E-03 | 3.06E-01 |

Several conclusions can be reached with these results. Firstly, the mean deviations to nominal for amplitude and phase are around zero, always lower to $2 \cdot 10^{-9}$ v or $2 \cdot 10^{-9}$ rad. These values have been repeated with the rest of $R$, $N$, jitter and $Nbits$ values. So, from now, we will present only standard deviation values will be presented. Additionally, for the three methods, the phase standard deviation is lightly higher than amplitude standard deviation.

Concerning the fitting methods, results show that both provide almost identical results, which is logical, due both methods solve the same optimization problem in different ways. And comparing DFT and fitting methods, both provides very similar results in amplitude, with practically identical standard deviation, but higher standard deviations (in a factor around 2) is obtained in phase by fitting methods.

Relating to jitter and quantization impact in the three methods results, for the values selected, although both affect to the total standard deviation, quantization has more impact than jitter. On the other hand, results show that their jointly impact is equivalent to the quadratic sum of the independent impacts. That can be explained because although quantization is an uncertainty over the amplitude value and jitter is an uncertainty over the time instant of sampling, jitter is in the last part a noise over the amplitude [1], so both contributions apply, at the end, to the same variable.

### Impact of $R$ and $N$ in Results

In order to evaluate $R$ and $N$ impact on the methods results, for each combination of jitter and number of bit values, the 9 points from Table 1 were simulated. Tables 5 and 6 show standard deviation results obtained for amplitude and phase of the fundamental frequency in the case of jitter value of *100* ns and *16* bits of quantization. In order to unify data as much as possible, both tables simultaneously present values obtained considering only jitter ($j$), only quantization ($q$) and both ($jq$).

Firstly, Tables 5 and 6 confirm in a more general way conclusions obtained in previous sections about methods comparisons. Moreover, from both tables it can be observed that for three methods and three cases $j$, $q$ and $jq$, the $R$ value does not affect the standard deviations of amplitude and phase, while the $N$ value impacts clearly in the results, so that lower $N$ values produces higher standard deviations. The same behaviour can be observed for the rest of jitter and $Nbits$ values combinations.

### Impact of Jitter and Number of Bits on Results

In order to evaluate jitter and resolution impact in the three methods, each combination of *Maxjit* and *Nbits* values defined in Sect. 4.2 were considered for simulation. Regarding $R$, taking into account previous test results, a single value (the intermediate)

**Table 5.** Standard deviation of fundamental frequency amplitude deviations to nominal values ($\mu$V) for three methods in Table 1 points, considering: only jitter ($j$), only quantization ($q$) and both ($jq$)

| | | N | | | | | | | | |
|---|---|---|---|---|---|---|---|---|---|---|
| | | 1311 | | | 7073 | | | 15457 | | |
| | | R | | | R | | | R | | |
| | | 24 | 123 | 262 | 24 | 123 | 262 | 24 | 123 | 262 |
| DFT | q | 4.18E-01 | 4.18E-01 | 4.18E-01 | 1.86E-01 | 1.86E-01 | 1.86E-01 | 1.12E-01 | 1.12E-01 | 1.12E-01 |
| | j | 1.88E-01 | 1.87E-01 | 1.88E-01 | 8.08E-02 | 8.08E-02 | 8.09E-02 | 5.47E-02 | 5.47E-02 | 5.47E-02 |
| | jq | 4.60E-01 | 4.60E-01 | 4.59E-01 | 1.82E-01 | 1.82E-01 | 1.82E-01 | 1.25E-01 | 1.25E-01 | 1.24E-01 |
| MSFM | q | 4.14E-01 | 4.09E-01 | 4.15E-01 | 1.81E-01 | 1.77E-01 | 1.80E-01 | 1.20E-01 | 1.21E-01 | 1.21E-01 |
| | j | 1.88E-01 | 1.88E-01 | 1.89E-01 | 8.08E-02 | 8.08E-02 | 8.08E-02 | 5.47E-02 | 5.46E-02 | 5.46E-02 |
| | jq | 4.55E-01 | 4.53E-01 | 4.56E-01 | 1.97E-01 | 1.95E-01 | 1.96E-01 | 1.33E-01 | 1.32E-01 | 1.32E-01 |
| MANNFM | q | 4.14E-01 | 4.09E-01 | 4.15E-01 | 1.81E-01 | 1.77E-01 | 1.80E-01 | 1.20E-01 | 1.21E-01 | 1.21E-01 |
| | j | 1.88E-01 | 1.88E-01 | 1.89E-01 | 8.08E-02 | 8.08E-02 | 8.08E-02 | 5.47E-02 | 5.46E-02 | 5.46E-02 |
| | jq | 4.55E-01 | 4.53E-01 | 4.56E-01 | 1.97E-01 | 1.95E-01 | 1.96E-01 | 1.33E-01 | 1.32E-01 | 1.32E-01 |

**Table 6.** Standard deviation of fundamental frequency phase deviations to nominals values ($\mu$rad) for three methods in Table 1 points, considering: only jitter (j), only quantization (q) and both (jq)

| | | N | | | | | | | | |
|---|---|---|---|---|---|---|---|---|---|---|
| | | 1311 | | | 7073 | | | 15457 | | |
| | | R | | | R | | | R | | |
| | | 24 | 123 | 262 | 24 | 123 | 262 | 24 | 123 | 262 |
| DFT | q | 4.14E-01 | 4.14E-01 | 4.14E-01 | 2.12E-01 | 2.12E-01 | 2.12E-01 | 1.21E-01 | 1.21E-01 | 1.21E-01 |
| | j | 3.25E-01 | 3.25E-01 | 3.25E-01 | 1.40E-01 | 1.40E-01 | 1.40E-01 | 9.47E-02 | 9.47E-02 | 9.47E-02 |
| | jq | 5.27E-01 | 5.26E-01 | 5.26E-01 | 2.26E-01 | 2.26E-01 | 2.27E-01 | 1.53E-01 | 1.53E-01 | 1.53E-01 |
| MSFM | q | 8.27E-01 | 8.24E-01 | 8.19E-01 | 3.54E-01 | 3.56E-01 | 3.54E-01 | 2.41E-01 | 2.40E-01 | 2.41E-01 |
| | j | 6.50E-01 | 6.51E-01 | 6.55E-01 | 2.80E-01 | 2.80E-01 | 2.80E-01 | 1.89E-01 | 1.89E-01 | 1.89E-01 |
| | jq | 1.05E + 00 | 1.05E + 00 | 1.06E + 00 | 4.53E-01 | 4.53E-01 | 4.53E-01 | 3.06E-01 | 3.06E-01 | 3.06E-01 |
| MANNFM | q | 8.27E-01 | 8.24E-01 | 8.19E-01 | 3.54E-01 | 3.56E-01 | 3.54E-01 | 2.41E-01 | 2.40E-01 | 2.41E-01 |
| | j | 6.50E-01 | 6.51E-01 | 6.55E-01 | 2.80E-01 | 2.80E-01 | 2.80E-01 | 1.89E-01 | 1.89E-01 | 1.89E-01 |
| | jq | 1.05E + 00 | 1.05E + 00 | 1.06E + 00 | 4.53E-01 | 4.53E-01 | 4.53E-01 | 3.06E-01 | 3.06E-01 | 3.06E-01 |

was fixed. Finally, the three $N$ (and $M$) values were additionally considered, in order to observe the $N$ influence.

So, *18* simulation points were performed. In all of them, results obtained consolidate previous tests conclusions: the fitting methods provide identical results which, at the same time respect DFT results, are identical in amplitude and higher by a factor around *around 2* in phase sigma. For this reason and with the aim of synthetize, only MANNFM results will be presented in this section, but all conclusions obtained can be applied to DFT and MSFM.

Results for MANNFM are presented in Tables 7 and 8, which show standard deviations obtained for amplitude and phase of the fundamental frequency at any simulation point. In order to unify data as much as possible, both tables simultaneously present values obtained considering only jitter ($j$), only quantization ($q$) and both ($jq$).

**Table 7.** Standard deviation of fundamental frequency amplitude deviations to nominal values (μV) for *Maxjit*, *Nbits* and *N* values simulated, considering: only jitter (*j*), only quantization (*q*) and both (*jq*)

| | | N | | | | | | | | |
|---|---|---|---|---|---|---|---|---|---|---|
| | | 1311 | | | 7073 | | | 15457 | | |
| | | Nbits | | | Nbits | | | Nbits | | |
| | | 16 | 18 | 21 | 16 | 18 | 21 | 16 | 18 | 21 |
| Maxjit 3 ps | j | 5.68E-06 | 5.68E-06 | 5.68E-06 | 2.42E-06 | 2.42E-06 | 2.42E-06 | 1.64E-06 | 1.64E-06 | 1.64E-06 |
| | q | 3.94E-01 | 1.03E-01 | 1.27E-02 | 1.76E-01 | 4.49E-02 | 5.54E-03 | 1.18E-01 | 2.96E-02 | 3.84E-03 |
| | jq | 3.94E-01 | 1.03E-01 | 1.27E-02 | 1.76E-01 | 4.49E-02 | 5.55E-03 | 1.18E-01 | 2.96E-02 | 3.84E-03 |
| Maxjit 100 ns | j | 1.89E-01 | 1.89E-01 | 1.89E-01 | 8.05E-02 | 8.05E-02 | 8.05E-02 | 5.48E-02 | 5.48E-02 | 5.48E-02 |
| | q | 3.94E-01 | 1.03E-01 | 1.27E-02 | 1.76E-01 | 4.49E-02 | 5.54E-03 | 1.18E-01 | 2.96E-02 | 3.84E-03 |
| | Jq | 4.51E-01 | 2.16E-01 | 1.90E-01 | 1.95E-01 | 9.26E-02 | 8.07E-02 | 1.31E-01 | 6.29E-02 | 5.50E-02 |

**Table 8.** Standard deviation of fundamental frequency phase deviations to nominal values (μrad) for *Maxjit*, *Nbits* and *N* values simulated, considering: only jitter (*j*), only quantization (*q*) and both (*jq*)

| | | N | | | | | | | | |
|---|---|---|---|---|---|---|---|---|---|---|
| | | 1311 | | | 7073 | | | 15457 | | |
| | | Nbits | | | Nbits | | | Nbits | | |
| | | 16 | 18 | 21 | 16 | 18 | 21 | 16 | 18 | 21 |
| Maxjit 3 ps | j | 1.97E-05 | 1.97E-05 | 1.97E-05 | 8.33E-06 | 8.33E-06 | 8.33E-06 | 5.68E-06 | 5.68E-06 | 5.68E-06 |
| | q | 8.68E-01 | 2.04E-01 | 2.63E-02 | 3.53E-01 | 8.82E-02 | 1.10E-02 | 2.36E-01 | 6.05E-02 | 7.42E-03 |
| | jq | 8.68E-01 | 2.03E-01 | 2.63E-02 | 3.53E-01 | 8.83E-02 | 1.10E-02 | 2.36E-01 | 6.04E-02 | 7.43E-03 |
| Maxjit 100 ns | j | 6.56E-01 | 6.56E-01 | 6.56E-01 | 2.78E-01 | 2.78E-01 | 2.78E-01 | 1.89E-01 | 1.89E-01 | 1.89E-01 |
| | q | 8.68E-01 | 2.04E-01 | 2.63E-02 | 3.53E-01 | 8.82E-02 | 1.10E-02 | 2.36E-01 | 6.05E-02 | 7.42E-03 |
| | jq | 1.06E+00 | 6.87E-01 | 6.56E-01 | 4.58E-01 | 2.90E-01 | 2.78E-01 | 3.04E-01 | 1.99E-01 | 1.89E-01 |

First of all, again previous conclusions are confirmed in more general test: as in previous section, higher *N* values results in lower standard deviations for both only jitter and only quantization.

Regarding *Maxjit* effect on *j* results, for the lower *Maxjit* value, *3* ps, we can observe very low standard deviation values, with maxima (for lower *N*) of $5.7 \cdot 10^{-12}$ V and $19.7 \cdot 10^{-12}$ rad. On the other hand, for 100 ns, we can observe maxima standard deviations values of $0.2 \cdot 10^{-6}$ V and $0.6 \cdot 10^{-6}$ rad.

Regarding the quantization effect on points selected, we can observe in Fig. 4 that *N* and *Nbits* variations have comparable effects on the standard deviations values obtained. This can be useful to compensate the effects of one parameter with the other. For example, as we can observe in Fig. 4, the standard deviation obtained for *18* bits and *1311* samples can be nearly obtained with 16 bits if *15457* samples are considered.

Finally, regarding the combined impact of jitter and quantization, it can be observed from Tables 7 and 8 that in the low jitter point, impact of jitter is no significant in front of quantization impact, so that the combined *(jq)* standard deviations are equal to those of the only quantization considered case *(q)*.

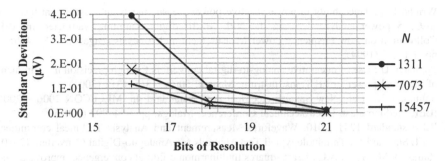

**Fig. 4.** Standard deviation of fundamental frequency amplitude deviations to nominal values ($\mu$V) obtained for *Nbits* and *N* values - only quantization considered

In the case of *Maxjitt* equal 100 ns more similar contributions are obtained, so that both effects contribute to the combined impact. Even so, with the lower *N* and *Nbits* values the main contribution is due to quantization although jitter also contributes. So, if *Nbits* raise, at lower *N*, the quantization contributions decreases; till with 21 bits, the main contribution is due to jitter. When *N* value is increased to it next value, the same behavior is observed, changing the main contribution depending on the *Nbits* value.

## 6  Conclusions and Future Work

Analysis performed has provided practical values of MANNFM standard deviation (amplitude and phase of the fundamental) for its use in practical measurements in laboratory. In the worst case, obtained results show uncertainties - for 95.5% of confidence level ($2\sigma$) - lower than *1.7* $\mu$V and *2.12* $\mu$rad for amplitude and phase respectively. From the methods comparison performed, it has been found that fitting methods, MANNFM and MSFM, provides identical results. Respect DFT, fitting methods are very similar in amplitude but their standard deviation for the phase of the fundamental is higher than that of DFT by a factor around *2*. Besides that, behavior of fitting methods regarding uncertainty contributions (jitter sampling and quantization noise) and sampling parameters (*R* and *N*) follow the same tendencies than DFT.

Further work is also necessary in order to study uncertainties for fundamental and harmonics components when signals with harmonic content are sampled. Other frequencies have to be studied also, in order to identify jitter effect when higher fundamental frequencies are necessary.

**Acknowledgment.** This work was partially supported by the Universidad de Malaga - Campus de Excelencia Andalucia-Tech.

## References

1. Svensson, S.: Power measurement techniques for nonsinusoidal conditions. Doctoral thesis, Chalmers University of Technology, Sweden (1999)

2. Wright, P.S.: Traceability in power/reactive power measurements and assessment tests for 'IEC555 power analyzers', using the NPL Mk.III digital sampling wattmeter. In: IEE Colloquium on Low Frequency Power Measurement and Analysis (Digest No. 1994/203), pp. 1/1–1/6 (1994)

3. Pogliano, U.: Use of integrative analog-to-digital converters for high-precision measurement of electrical power. IEEE Trans. Instrum. Measur. **50**(5), 1315–1318 (2001)

4. Salinas, J.R., et al.: New Spanish electrical power standard. In: MELECON 2006 – 2006 IEEE Mediterranean Electrotechnical Conference, Malaga, pp. 994–997 (2006)

5. IEEE standard 1241, 2010, Waveform Measurement and Analysis Technical committee, IEEE Standard for Terminology and Test Methods for Analog-to-Digital Converters (2010)

6. Ramos, P.M., Serra, A.C.: Least squares multiharmonic fitting: convergence improvements. IEEE Trans. Instrum. Measur. **56**(4), 1412–1418 (2007)

7. Salinas, J.R., García-Lagos, F., Joya, G., Sandoval, F.: Sine fitting multiharmonic algorithms implemented by artificial neural networks. Neurocomputing **72**(16–18), 3640–3648 (2009)

8. Salinas, J.R., García-Lagos, F., de Aguilar, J.D., Joya, G., Lapuh, R., Sandoval, F.: Harmonics and interharmonics spectral analysis by ANN. In: 2016 Conference on Precision Electromagnetic Measurements (CPEM 2016), Ottawa, Canada, pp. 1–2 (2016)

9. Salinas, J.R., et al.: Versatile digital system for high accuracy power measurements. In: CPEM 2006 Digest, pp. 98–99, July 2006

10. Salinas, J.R., de Aguilar, D.J., García-Lagos, F., Joya, G., Sandoval, F., Romero, M.L.: Spectrum analysis of asynchronously sampled signals by means of an ANN method. In: Conference on Precision Electromagnetic Measurements (CPEM) 2014, pp. 422–423, August 2014

11. JGCM 100:2008. Evaluation of measurement data – Guide to the expression of uncertainty in measurement

12. JGCM 101:2008. Evaluation of measurement data – Supplement 1 to the "Guide to the expression of uncertainty in measurement" – Propagation of distributions using a Monte Carlo method

13. JGCM 102:2011. Evaluation of measurement data – Supplement 2 to the "Guide to the expression of uncertainty in measurement" – Extension to any number of output quantities

14. Šíra, M., Mašláň, S.: Uncertainty analysis of non-coherent sampling phase meter with four parameter sine wave fitting by means of Monte Carlo. In: 29th Conference on Precision Electromagnetic Measurements (CPEM 2014), Rio de Janeiro, pp. 334–335 (2014)

15. Oppenheim, A.V., Schafer, R.W.: Discrete-Time Signal Processing. Prentice-Hall, Upper Saddle River (1999)

16. Wadgy, M.F.: Effects of ADC quantization errors on some periodic signal measurements. IEEE Trans. Instrum. Measur. **36**(4), 983–988 (1987)

17. Wagdy, M.F., Awad, S.S.: Effect of sampling jitter on some sine wave measurements. IEEE Trans. Instrum. Measur. **39**(1), 86–89 (1990)

18. Jain, V.K., Collins, W.L., Davis, D.C.: High-accuracy analog measurements via interpolated FFT. IEEE Trans. Instrum. Measur. **28**(2), 113–122 (1979)

# Using Deep Learning for Image Similarity
# in Product Matching

Mario Rivas-Sánchez[1]([✉]), Maria De La Paz Guerrero-Lebrero[2],
Elisa Guerrero[2], Guillermo Bárcena-Gonzalez[2], Jaime Martel[1],
and Pedro L. Galindo[2]

[1] Itelligent Information Technologies, Parque Tecnológico CEEI,
El Puerto de Santa María, Cádiz, Spain
rivassanchez.mario@gmail.com
[2] Department of Computer Science and Engineering,
University of Cádiz, Cádiz, Spain

**Abstract.** Product matching aims at disambiguating descriptions of products
belonging to different websites in order to be able to recognize identical elements and to merge the content from those identical items. Most approaches face
this matter applying various machine learning methods to textual product
descriptions. Recently some authors are including information extracted from an
image associated to a textual description of a product. Modern machine learning
methods, such as content based information retrieval (CBIR) or deep learning,
can be applied to this type of images since they can manage very large data sets
for finding hidden structure within them, and for making accurate predictions.
This information could boost the performance of the traditional textual matching
but at the same time increase the computational complexity of the process. In
this paper we review some CBIR and deep learning models and analyse the
performance of these approaches when they are applied to images for product
matching. The results obtained will help to introduce a combined classifier using
textual and image information.

## 1 Introduction

The growth of the Internet has fuelled the availability of e-commerce marketplaces and
search engines must face with a huge amount of ambiguity and inconsistencies in the
data. The identification and matching of different items as the same product is essential
to obtain accurate and reliable results. The task of Product Matching consists of
identifying, matching and merging records that correspond to the same product from
several data sources [4, 27, 28], allowing the development of tools for product monitoring, product comparison and pricing analysis.

When textual information is used, the matching is based on the application of some
approximate string similarity measure that maps a pair of strings to a real number. From
character-based approaches, through token-based or the use of hybrid methods, there
exists a wide variety of metrics that can be used to obtain some idea of how similar or
different two attribute values are [23, 29, 30].

© Springer International Publishing AG 2017
I. Rojas et al. (Eds.): IWANN 2017, Part I, LNCS 10305, pp. 281–290, 2017.
DOI: 10.1007/978-3-319-59153-7_25

In the case of using image information, Content-Based Image Retrieval (CBIR) [1–4] and deep learning [2, 5–7] techniques can be used in order to link images from different websites.

CBIR is the application of computer vision techniques to measure image similarity. These techniques make use of visual contents to look up images from large scale image databases in agreement with user interests. Different CBIR approaches have been proposed for the purpose of image matching, object detection or image classification. Normally, each CBIR method is specialized in a particular feature type, for example: morphological descriptors, colour descriptors or keypoint descriptors. Frequently, a mashup of different CBIR features are used to reduce the "semantic gap" existing between low and high level pixel images perceived by the human, because in many cases, only one descriptor can not reduce this gap.

In recent years, a new machine learning approach known as deep learning has been applied to many different fields. It is based on algorithms inspired by the structure and function of artificial neural networks but using a deeper architecture in which many layers are connected. In contrast with CBIR techniques, deep learning algorithms are always learning features automatically not requiring human intervention.

In this work, different CBIR and deep learning techniques have been evaluated in the context of linking images for product matching. The dataset to train and validate the models has been built by hand taking images from two different web sites. This study can be useful to append multimedia information in Product Matching processes with regard to obtain better results in conjunction with the textual information.

## 2    Content Based Image Retrieval

CBIR manages the visual contents of an image, such as color, shape, texture and spatial layout to perform and index the image. Normally, these visual contents are described by multi-dimensional feature vectors. To get back images, users hand over the recovery system with examples or sketched figures, then, the system adjusts these examples into internal representation of feature vectors.

Object shapes are significant features used in CBIR, and Fourier descriptors [8] have a special relevance for extracting these features. Some of these characteristics are: low complexity, translation invariance, rotation invariance and scale invariance. All of these properties can be found in different Fourier descriptors implementations [3, 8–11]. In particular Fourier descriptors are necessary for extracting the object shape signature, in this work we applied the centroid distance technique. Figure 1 shows the extraction process, from the original image, first segmentation is applied and then Fourier descriptors are calculated.

Other important CBIR techniques group is called keypoint [1, 12, 13]. These techniques detect the most significant points within a region, working independently of the object shapes, colours and textures. Figure 2 shows the input and output of the keypoint extraction process, which correspond to the original image and the keypoint drawn in green circles, respectively.

In this paper, two different CBIR algorithms based on keypoint techniques are used, SURF (Speeded Up Robust Features) [12] and BRISK (Binary Robust Invariant

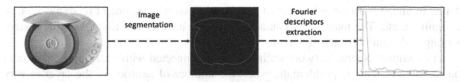

Fig. 1. Fourier descriptors extraction process.

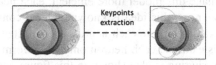

**Fig. 2.** Keypoint extraction process. From the original image (left image) the technique returns the most significant points, green circles in the right image (Color figure online)

Scalable Keypoints) [1]. SURF is considered the most relevant keypoint technique and the most computationally efficient amongst the high-performance methods to date. On the other hand BRISK, shows a very high performance and sometimes lower error ratio and computational cost than SURF.

# 3 Deep Learning

In traditional pattern recognition and machine learning, feature engineering is a labour-intensive task that usually requires a domain expertise to transform the data into a feature vector. Despite the importance of this stage, current learning algorithms lack the ability to extract and manage information from data.

Deep learning allows automatically discovering features from raw data, creating representations of the data that can make easier the extraction of useful information to classification, prediction, etc.

Deep learning methods are formed by the composition of multiple non-linear transformations, that promote the re-use of features, and can potentially lead to progressively more abstract features at higher layers of representations [6]. For example, an image starts with a matrix pixel value. The features learned in the first layer usually represent the presence or absence of edges at particular locations. The second layer detects arrangements of edges. The third layer detects combinations that correspond to parts of objects. Next layers would detect objects. The main characteristic of these methods is that these layers of "features discovering" are not design by human, but automatically learned from raw data [7].

In the last years deep learning has been used in many different fields as image classification [14, 15], natural language processing [16], DNA mutation detections [17, 18], etc.

Convolutional Neural Networks (CNN) are a modification of traditional multilayer perceptrons. These networks are specifically designed to work with inputs of large dimensions or features. Convolutional Neural Network (CNN) can obtain very good results in image processing tasks such as: image classification, object recognition, face recognition, etc. The main CNN characteristics are: local connections, weight sharing, pooling and multiple stacked layers.

In traditional Neural Network each neuron is connected with every other neuron in the same layer. This is problematic when the number of neurons in the layer is very high. For example, when the input is a small image of $200 \times 200$ pixels, the input vector has 120.000 elements ($200 \times 200 \times 3$), consequently each neuron has 120.000 parameters. In order to make this model more efficient, each neuron can be connected only with another few neurons; this idea is known as local connections.

Figure 3 shows an example where each neuron is connected with three neurons of the previous layer; in this example each neuron only has information of three adjacent pixels. Moreover, weight sharing is also shown in this figure, instead of having lots of weights to train, the same weights can be reused at each location, reducing the number of parameters and making networks easier to train.

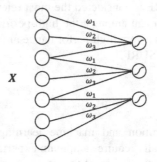

**Fig. 3.** Local connection and weight sharing example. In this configuration each neuron is connected with three neurons of the previous layer, sharing her weights.

CNN are very similar to ordinary Neural Networks, they are made up of neurons that have learnable weights and biases. The typical CNN architecture is composed of several convolutional layers and pooling layers. The convolutional layers make up the core building block of CNN and have the computationally-intensive part of the process; they are usually responsible for detecting local features from the previous layer. Additionally, it is common to periodically insert a pooling layer in-between successive convolutional layers in order to progressively decrease the number of parameters and computation in the network. The CNN model can be very generalizable because the last layer is able to extract high-level features from input data.

The challenging ImageNet dataset [25], have became a benchmark to measure the performance of the different proposals. AlexNet [14], GoogLeNet [21] and ResNet [22] have been the winners in the last few years using CNN-based architectures. In 2012 the model proposed by Alex Krizhevsky et al., AlexNet, achieved a winning top-5 test error rate of 15.3%, compared to 26.2% achieved by the second-best entry. The model consisted of five convolutional layers, some of which followed by a pooling layer and three fully-connected layers with a final 1000-way softmax classification layer.

GoogLeNet [21] reduced the global number of parameters and outperformed AlexNet in 2014. The CNN architecture is formed by 22 convolutional layers and five pooling layers. The final output is a 1000-way softmax classification layer. Finally, In 2015 ResNet [22], composed by 152 layers, won the challenge obtaining a very good generalization performance.

## 4  Design of the Experiment

In this experiment we try to link products that are offered in two different web sites, making use of the corresponding products images. The data set has been generated by hand considering 785 different products and then, taking 785 images of these products from two different web pages.

Figure 4 shows an example of three different products from the data set, in each site the corresponding picture of the object can be similar, but not equal; differences in perspective, orientation, colour and size might exist.

**Fig. 4.** Images from the dataset. Two images of the same product can have similar or very different pictures when they are taken from different web sites.

The classification hit rate has been measured over the validation set, calculating the number of images that have been correctly identified. The matching is carried out using a similarity measure.

Fourier descriptors, BRISK and SURF are the CBIR methods selected to be compared. All of them using the data set that has been split into train (500 images) and validation (285 images) sets.

As similarity measure the number of keypoints in common is used in order to link an image from website 1 with another one from website 2. This number of keypoints can be calculated with or without normalization.

The use of Jaccard index [24] allows the management of normalized data, useful for future data processing. However, this measure uses the number of keypoints in common divided into the total number of keypoints from both sets. The higher the percentage, the more similar the two feature vectors. Although it is easy to interpret, it is extremely sensitive to large sample size and may give erroneous results, especially when the number of keypoints is different between the two vectors under comparison. Then, in this work the absolute number of keypoints (without normalization) is used as similarity measure in CBIR methods. Experiments have been analysed using keypoints ranging from 1 to 500 Keypoints. This upper limit is greater enough in order to cover all the image features.

In practice for deep learning, training an entire CNN from scratch is not an easy task because a dataset of sufficient size is needed. Instead it is usual to use pre-trained models, such as: AlexNet, GoogLeNet and ResNet, thus, only the specific validation set is needed, in our work, 285 images from web 1 and 285 images from web 2.

In addition, these CNN models are usually used for classification purposes, however in this work we are interested in identifying the most similar image, therefore we have considered the last output before the final classification layer as our actual output. In this way we obtain an array of features for each image that is used to measure how similar (cosine similarity [23]) is an image belonging to the validation set of a given web to another image from the validation set of the second web.

## 5  Results

CBIR models have been trained over 500 products and then tested over 285 samples. The results obtained with CBIR-methods do not achieved the 50% of success in any case. The example shown in Fig. 5, can explain the reasons to have such low results. In this figure all the methods failed in the matching, in a case where images from web 1 and web 2 are clearly more similar than the final assigned product. In this example, using SURF method.

**Fig. 5.** First image is the product on web 1, the second image corresponds to the same on web 2 and third image is the image returned by SURF

In order to prevent these classification errors, further experiments have been carried out with SURF. First of all, two different matching methods were used to compare the keypoints set. First method consists of a multiple matching, in which a keypoint from an image can be assigned to various keypoints of the image being compared. The second method consists of a unique matching, where keypoints are assigned just to other keypoint of the image under comparison.

Moreover we have considered different tests varying the number of descriptors. Figure 6 shows the hit rate obtained for both matching approaches, multiple (blue color) and unique (red color), as well as taking into account the number of descriptors, from 1 to 500 keypoints. The use of a number of keypoints lower than 25 is not recommended, since not reliable results can be obtained. For example, in Fig. 6 shows that one keypoint obtains the best result. This is a spurious outcome; when an image with just a keypoint is compared with all dataset images, the number of coincident keypoints is zero, due to this fact, the full dataset is returned as matched images. In general, unique matching presents lower error rates than multiple method, and the use of all keypoints is needed to not to lose information.

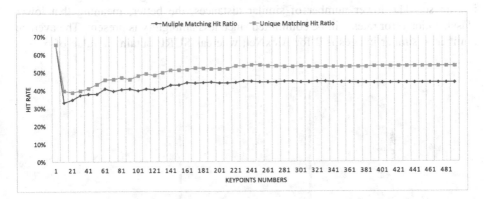

**Fig. 6.** Hit rate obtained per number of descriptors, from 1 to 500, and for both matching approaches of SURF, multiple (blue color) and unique (red color). (Color figure online)

Hit rates for the three CBIR-based methods are shown in Table 1. The best result is obtained by SURF method using unique matching and 500 keypoints, reaching an accuracy equal to 40%.

**Table 1.** Hit Rates obtained with different CBIR methods

| Method | Hit rate |
| --- | --- |
| Fourier descriptors | 1% |
| BRISK | 21% |
| SURF | 40% |

The segmentation stage needed in Fourier descriptors is not suitable for this kind of dataset, since there exists a great variety of background intensity, contrast, brightness, etc. Thus, Fourier descriptors have provided the worst results in classification.

Table 2 shows the results obtained with different deep learning models. The best accuracy is obtained using AlexNet model, although GoogLeNet obtains very similar accuracy, using the cosine distance. These rates are better than SURF hit rate, therefore in general, deep learning can overcome the results obtained by CBIR methods.

**Table 2.** Results obtained with different deep learning models

| Model | Hit rate |
|---|---|
| AlexNet | 60% |
| GooGleNet | 59% |
| ResNet | 48% |

Comparing SURF and AlexNet, the number of images that have obtained the same similarity measure can be another indicator of the precision of the matching. Figure 7 shows the number of similar distances obtained for each image belonging to the validation set. The lower number of similar distances, the better, meaning that lower classification error rates will be committed since less ambiguity is present. The average number of similar images is 2 for AlexNet, whereas SURF obtains a value of 6.

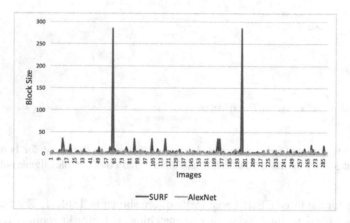

**Fig. 7.** Block size for SURF and AlexNet. Ordinate axis shows the block size and abscissa axis corresponds to each test image

Studying in more detail the results obtained by both methods, SURF and AlexNet, we can observe that some images are so different that no classifier would be able to find the correct image (Fig. 8). If these extreme cases were avoided, the error rate could be reduced in all the methods. In particular AlexNet hit rate would be 67%.

**Fig. 8.** Products from the validation set whose images are totally different in each web page

# 6  Conclusions and Future Work

In this work we have compared different CBIR and deep learning techniques in the context of linking images for product matching. A priori, both approaches are suitable to be applied to this task since they can manage very large data sets for finding hidden structure within them, and for making accurate predictions. Fourier descriptors, SURF and BRISK have been trained and tested using a dataset built by hand, taking images from two different web sites, and pre-trained models (Alexnet, GoogLeNet and ResNet) based on deep learning approaches were validated using the same validation dataset. The matching was carried out using the number of keypoints in common and cosine distance for CBIR and deep learning respectively.

SURF was the best method among the CBIR models, but SURF hit rate did not reach the 50%. We studied the performance over a wide range of keypoints, and we found that a number lower than 25 is not recommended since not reliable results have been found. AlexNet and GoogLeNet have shown better results, obtaining a hit rate around 67%.

In some cases, images from each web page are so different that the matching is not feasible, thus in order to improve the results, future work must addressed the hard task of extent, by hand, the dataset and the generation of a product matching model combining text and images.

# References

1. Leutenegger, S., Chli, M., Siegwart, R.Y.: BRISK: binary robust invariant scalable keypoints. In: 2011 International Conference on Computer Vision, pp. 2548–2555 (2011)
2. Wan, J., et al.: Deep learning for content-based image retrieval: a comprehensive study. In: MM 2014 - Proceedings of the 2014 ACM Conference on Multimedia, pp. 157–166 (2014)
3. Zhang, D., Lu, G.: Study and evaluation of different Fourier methods for image retrieval. Image Vis. Comput. 23(1), 33–49 (2005)
4. Müller, H., Müller, W., Squire, D.M., Marchand-Maillet, S., Pun, T.: Performance evaluation in content-based image retrieval: overview and proposals. Pattern Recogn. Lett. 22(5), 593–601 (2001)
5. Schmidhuber, J.: Deep learning in neural networks: an overview. Neural Netw. 61, 85–117 (2015)
6. Bengio, Y., Courville, A., Vincent, P.: Representation learning: a review and new perspectives. IEEE Trans. Pattern Anal. Mach. Intell. 35(8), 1798–1828 (2013)
7. Lecun, Y., Bengio, Y., Hinton, G.: Deep learning. Nature 521(7553), 436–444 (2015)
8. Zahn, C.T., Roskies, R.Z.: Fourier descriptors for plane closed curves. IEEE Trans. Comput. C–21(3), 269–281 (1972)
9. Zhang, D., Lu, G.: Enhanced generic Fourier descriptors for object-based image retrieval. In: ICASSP, IEEE International Conference on Acoustics, Speech and Signal Processing - Proceedings, vol. 4, pp. IV/3668–IV/3671 (2002)
10. Ahonen, T., Matas, J., He, C., Pietikäinen, M.: Rotation invariant image description with local binary pattern histogram fourier features. In: Salberg, A.-B., Hardeberg, J.Y., Jenssen, R. (eds.) SCIA 2009. LNCS, vol. 5575, pp. 61–70. Springer, Heidelberg (2009). doi:10.1007/978-3-642-02230-2_7

11. He, Q., Ji, Z., Wu, Q.M.J.: Content-based image retrieval using generic fourier descriptor and Gabor filters. In: Proceedings of VISAPP 2008 - 3rd International Conference on Computer Vision Theory and Applications, vol. 1, pp. 525–528 (2008)
12. Bay, H., Ess, A., Tuytelaars, T., Van Gool, L.: Speeded-up robust features (SURF). Comput. Vis. Image Underst. **110**(3), 346–359 (2008)
13. Lowe, D.G.: Distinctive image features from scale-invariant keypoints. Int. J. Comput. Vis. **60**(2), 91–110 (2004)
14. Krizhevsky, A., Sutskever, I., Hinton, G.E.: ImageNet classification with deep convolutional neural networks. Adv. Neural. Inf. Process. Syst. **2**, 1097–1105 (2012)
15. Jia, Y., et al.: Caffe: convolutional architecture for fast feature embedding. In: MM 2014 - Proceedings of the 2014 ACM Conference on Multimedia, pp. 675–678 (2014)
16. Mikolov, T., Sutskever, I., Chen, K., Corrado, G.S., Dean, J.: Distributed representations of words and phrases and their compositionality. In: Advances in Neural Information Processing Systems, pp. 3111–3119 (2013)
17. Leung, M.K.K., Xiong, H.Y., Lee, L.J., Frey, B.J.: Deep learning of the tissue-regulated splicing code. Bioinformatics **30**(12), I121–I129 (2014)
18. Xiong, H.Y.: The human splicing code reveals new insights into the genetic determinants of disease. Science **347**, 6218 (2015)
19. Bengio, Y., Lamblin, P., Popovici, D., Larochelle, H.: Greedy layer-wise training of deep networks. In: Advances in Neural Information Processing Systems, pp. 153–160 (2007)
20. Hinton, G.E., Osindero, S., Teh, Y.-W: A fast learning algorithm for deep belief nets. In: Neural Computation, vol. 18, no. 7, pp. 1527–1554 (2006)
21. Szegedy, C., et al.: Going deeper with convolutions. In: Proceedings of the IEEE Computer Society Conference on Computer Vision and Pattern Recognition, 12 June 2015, vol. 07, pp. 1–9 (2015)
22. He, K., Zhang, X., Ren, S. Sun, J.: Deep residual learning for image recognition. In: Proceedings of the IEEE Computer Society Conference on Computer Vision and Pattern Recognition, 2016, pp. 770–778, January 2016
23. Singhal, A.: Modern information retrieval: a brief overview. IEEE Data Eng. Bull. **24**(4), 35–43 (2001)
24. Leskovec, J., Rajaraman, A., Ullman, J.D.: Mining of Massive Datasets, 2nd edn. Cambridge University Press, Cambridge (2014)
25. Russakovsky, O., et al.: ImageNet large scale visual recognition challenge. Int. J. Comput. Vis. **115**(3), 211–252 (2015)
26. Chisten, P.: Data Matching: Concepts and Techniques for Record Linkage, Entity Resolution and Duplicate Detection. Springer, Heidelberg (2012)
27. Baeza-Yates, R., Ribeiro-Neto, B., et al.: Modern Information Retrieval, vol. 463. ACM press, New York (1999)
28. Winkler, W.E.: Overview of record linkage and current research directions. In: Bureau of the Census (2006)
29. Thor, A.: Toward an adaptive string similarity measure for matching product offers. In: GI Jahrestagung (1), pp. 702–710 (2010)
30. Winkler, W.E.: String Comparator metrics and enhanced decision rules in the Fellegi-Sunter model of record linkage (1990)

# Enhanced Similarity Measure for Sparse Subspace Clustering Method

Sabra Hechmi$^{(\boxtimes)}$, Abir Gallas, and Ezzeddine Zagrouba

LIMTIC Laboratory, Higher Institute of Computer Science of Tunis,
University of Tunis EL Manar, Abou Raihane Bayrouni, 2080 Ariana, Tunisia
hechmi_sabra@yahoo.fr, abir.gallas@yahoo.fr, ezzeddine.zagrouba@fsm.rnu.tn

**Abstract.** Trying to find clusters in high dimensional data is one of the most challenging issues in machine learning. Within this context, subspace clustering methods have showed interesting results especially when applied in computer vision tasks. The key idea of these methods is to uncover groups of data that are embedding in multiple underlying subspaces. In this spirit, numerous subspace clustering algorithms have been proposed. One of them is Sparse Subspace Clustering (SSC) which has presented notable clustering accuracy. In this paper, the problem of similarity measure used in the affinity matrix construction in the SSC method is discussed. Assessment on motion segmentation and face clustering highlights the increase of the clustering accuracy brought by the enhanced SSC compared to other state-of-the-art subspace clustering methods.

**Keywords:** Sparse Subspace Clustering · Similarity measure · Enhanced SSC · Motion segmentation · Face clustering

## 1 Introduction

Clustering is a basic task in machine learning that attempts to segment individuals or objects into meaningful groups. For this reason, several clustering methods have been proposed [1]. However, the choice of the most appropriate method remains an open question that depends essentially on the nature of the data and the area of application. Unfortunately, applied to high dimensional data, conventional clustering methods fail to generate significant results. Indeed, clustering in such spaces is extremely difficult and the calculation of similarity becomes very expensive. Also, in the case of high dimensional spaces, groups of data can be defined only by certain subsets of dimensions and these relevant dimensions may differ from one group to another. Recently, to overcome this challenge, a new technique has been emerged which is the subspace clustering [2]. The purpose of this method is to reveal clusters that exist in multiple underlying subspaces. In fact, subspace clustering can be considered as a generalization of the principal component analysis (PCA) method in which the points do not lie around a single lower dimensional subspace but rather around a union of subspaces. Moreover, this method can be regarded as a special clustering problem where neighbors are

© Springer International Publishing AG 2017
I. Rojas et al. (Eds.): IWANN 2017, Part I, LNCS 10305, pp. 291–301, 2017.
DOI: 10.1007/978-3-319-59153-7_26

not close according to a pre-defined notion of metric but rather belong to the same lower dimensional structure.

Subspace clustering problem arises in many computer vision applications, specially motion segmentation [3] and face recognition [4]. In fact, for these applications, data points in the same cluster i.e. face images of a person under different illumination conditions and feature points of a moving rigid object in a video sequence, lie on a low-dimensional subspace. Thus, clustering a collection of data points is reduced to finding low-dimensional subspaces fitting each group of data. Over the last few years, numerous subspace clustering methods have been proposed. These methods can be broadly divided into four principle categories: algebraic, statistic, iterative and spectral subspace methods [5]. In particular, spectral subspace clustering methods promise to become strong competitors for the rest of methods. In general, the spectral clustering methods aim to construct a similarity matrix $W \in \mathbb{R}^{N \times N}$ by computing the pairwise similarity among all the $N$ data points. Let $G(V, E)$ be an undirected graph where $V$ is the set of $N$ vertices and $E$ is the set of weighted edges $w_{ij}$ between each pair of points $i$ and $j$. Then, the $K$ clusters are obtained by applying the K-means to the subset of $K$ eigenvectors of the laplacian matrix $L \in \mathbb{R}^{N \times N}$ constructed from $W$. In this spirit, spectral subspace clustering methods have the challenge to build the most representative similarity matrix $W$ that captures whether two points belong to the same subspace or not. According to the similarity measure, two main families can be extracted from spectral subspace clustering approaches. Below, each one of them is described:

– **Similarity measure based on principal angles computing** from this category we can cite two methods: Local Subspace Affinity (LSA) [6] and Spectral Local Best Fit Flats (SLBF) [7] which consist as a first step to find the nearest neighbors for each data point and fit a local subspace to this point using PCA. Then, to construct the affinity matrix, a special similarity measure based on the calculation of the $m$ principal angles between each pair of local subspaces is computed. Finally, a spectral technique [8] is applied to have the data segmentation. The advantages of these methods are essentially the conceptual simplicity. However, the neighbors of a point could contain points in different subspaces. Therefore, a fundamental challenge for these methods is to select the most appropriate size of the neighborhood.

– **Similarity measure based on sparse coefficient matrix** the state-of-the-art methods including Sparse Subspace Clustering (SSC) [9,10] and Low Rank Representation (LRR) [11,12] subspace clustering succeed to overcome the problem of neighbors size by directly fit for each point a local subspace neighbors. Indeed, these two methods exploit the fact that every data point in a union of subspaces can be expressed as a linear combination of a few other data points from its own local subspace. This motivation is used to compute the sparse coefficients matrix $C$, which is used then to build a subspace clustering affinity matrix as $W = |C| + |C|^T$. Finally, K-means is applied to the subset of $K$ eigenvectors of the laplacian matrix obtained from $W$.

We are interested here to the SSC method which has showed great clustering results [10]. The SSC has succeeded to provide an exact neighborhood around each point. However, and despite of the robust theoretical bases of the SSC method, the SLBF method succeeded to give also very competitive clustering results [7], due to the reliable estimation of the subspaces and specially to the pairwise similarity measure based on the calculation of the angles between the estimated subspaces.

For this reason, the contribution of this work is to revisit the graph similarity measure in SSC in order to improve the clustering performance. The idea is to exploit the ratio of overlapping membership between local subspaces. This information is added to the pairwise similarity measure $w_{ij}$, for some data points, to consolidate the fact of belonging to the same final subspace or cluster. Thus, we propose a novel similarity measure for the SSC method that will enhance its clustering performance. The rest of the paper is structured as follows. Section 2 reviews the main formulations of the SSC method. In Sect. 3, the proposed method E-SSC is introduced. In Sect. 4, some experiments carried out on 'Feret' and 'Hopkins 155' datasets are produced and finally, a conclusions and ideas for further work are summarized in Sect. 5.

## 2   Related Work

According to [9,10], the SSC is one of the most successful subspace clustering methods. In fact, it is considered as a spectral clustering method where the similarity graph is based on the sparse representation.

Precisely, let the data matrix $Y = [y_1 \, y_2 \, ...y_N]$ that lies in the union of the $l$ subspaces, the key idea behind SSC is what we call the self-expressiveness property of the data. This self-expressiveness means that each data point $y_i$ can be expressed as a linear combination of other data points from the dataset i.e.:

$$y_i = Y c_i, \quad c_{ii} = 0 \tag{1}$$

Where $c_i = [c_{i1} \, c_{i2} \, ... \, c_{iN}]^T$ and $c_{ii} = 0$ eliminates the trivial solution of writing a data point as a linear combination of itself. However, this representation is not unique. Therefore, the main motivation here, that there exist a single solution $c_i$, such that the most of its coefficients are zero. Concerning the non-zero coefficients, they correspond ideally to data points from the same subspace as $y_i$. This solution is named as a subspace-sparse representation and it is obtained by solving the optimization problem defined as:

$$min \, \| \, c_i \, \|_0 \quad s.t. \quad y_i = Y c_i \, \text{ and } \, c_{ii} = 0 \tag{2}$$

Where $\| \, . \, \|_0$ denotes the number of non zero elements of $c_i$. But, since, this problem is NP-hard, $l_0$-norm is replaced by its $l_1$-convex relaxation [13]. Thus, the optimization problem becomes:

$$min \, \| \, c_i \, \|_1 \quad s.t. \quad y_i = Y c_i \, \text{ and } \, c_{ii} = 0 \tag{3}$$

This problem is expressed according to the matrix form as:

$$min \parallel C \parallel_1 \quad s.t. \quad Y = YC \ and \ diag(C) = 0 \qquad (4)$$

Where $C = [c_1 \ c_2 \ ... \ c_N] \in \mathbb{R}^{N \times N}$ is the sparse coefficient matrix.

To deal with real world issues, where data is corrupted by noise, the optimization problem is modified as a LASSO problem which is solved as follows:

$$min \parallel C \parallel_1 + \frac{\lambda_z}{2} \parallel Y - YC \parallel_F^2 \quad s.t \quad diag(C) = 0 \qquad (5)$$

Where $\lambda_z > 0$ is a regularization term in the objective function, $\parallel . \parallel_F$ designs the Frobenius norm selected as the appropriate norm to detect noise.

After finding a set of neighbors for each data point that are estimated to be in the same local subspace as the query point, a similarity graph is constructed based on the sparse representation coefficients between each pair of data such as: $w_{ij} = \mid c_{ij} \mid + \mid c_{ji} \mid$. Eventually, the data segmentation is obtained by applying the K-means to the subset of $K$ eigenvectors of the laplacian matrix $L \in \mathbb{R}^{N \times N}$ constructed from the similarity matrix $W$ where $K$ is the number of clusters.

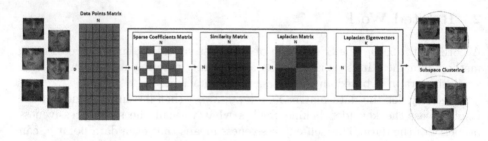

**Fig. 1.** Flowchart presenting the main steps of the SSC method

Figure 1 illustrates a flowchart that summarizes the main steps of the SSC method. Applied to face clustering and motion segmentation [10], SSC appears effective in identifying the underlying subspaces.

## 3    Contribution

The main challenge of the spectral subspace clustering technique is to construct the best similarity matrix which captures whether two points belong to the same subspace or not. According to the similarity measure, two main families can be extracted: the first is the one based on principal angles computing and the second is using the sparse coefficient matrix. The last one is described as: $w_{ij} = \mid c_{ij} \mid + \mid c_{ji} \mid$. In this contribution, we discuss the ability of this measure to describe effectively the similarity between two points. In fact, the weight $w_{ij}$ of the edge between two data points (in the order of 1 in the normalized case)

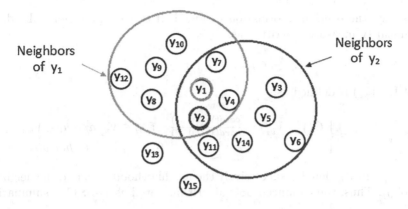

**Fig. 2.** Example of two points $y_1$ and $y_2$ neighborhoods and overlapping

can have an important value. Nevertheless, these points do not really belong to the same final subspace (cluster). Thus, in this case, this value may influence the clustering accuracy. Based on this assumption, we propose an improvement of the existing measure, taking into account the information of the neighborhoods and the overlap between them.

According to the classic SSC, the sparsest representation is obtained when each point is expressed as a linear combination of points in its own subspace. In this way, the idea consists in dividing the data points into two groups. On the one hand, the first group contains the pairs of points that are considered to be highly probable to belong to the same final subspace. Indeed, these points are expressed one by the other, i.e. one is in the neighborhood of the other. On the other hand, the second group incorporates the data points that do not verify the previous condition. For only the first group of data, we will compute the overlapping rate between each pair of local subspaces around two data points. This is done in order to increase the attraction between these points in the similarity graph. For Example, as illustrated in Fig. 2 two points $y_1$ and $y_2$ are written respectively as:

$$y_1 = c_2 y_2 + c_4 y_4 + c_7 y_7 + c_8 y_8 + c_9 y_9 + c_{10} y_{10} + c_{12} y_{12}$$

$$y_2 = c_1 y_1 + c_3 y_3 + c_4 y_4 + c_5 y_5 + c_6 y_6 + c_7 y_7 + c_{11} y_{11} + c_{14} y_{14}$$

Thereby, $y_2$ belongs to the neighborhood of $y_1$ denoted by:

$$V_{y_1} = \{y_2, y_4, y_7, y_8, y_9, y_{10}, y_{12}\}$$

and $y_1$ belongs to the neighborhood of $y_2$ denoted by:

$$V_{y_2} = \{y_1, y_3, y_4, y_5, y_6, y_7, y_{11}, y_{14}\}$$

Afterwards, the degree of overlap between the two neighborhoods $V_{y_1}$ and $V_{y_2}$ is measured as:

$$O(V_{y_1}, V_{y_2}) = \frac{card(V_{y_1} \cap V_{y_2})}{card(V_{y_1} \cup V_{y_2})} \tag{6}$$

In this way, the overlap information is added. If this script is generalized, the new similarity measure is written as:

$$w_{ij} = \mid c_{ij} \mid + \mid c_{ji} \mid + O(V_{y_i}, V_{y_j}) \tag{7}$$

and $O(V_{y_i}, V_{y_j})$ is defined as:

$$O(V_{y_i}, V_{y_j}) = \begin{cases} O(V_{y_i}, V_{y_j}) = \frac{card(V_{y_i} \cap V_{y_j})}{card(V_{y_i} \cup V_{y_j})} & if \ y_i \in V_{y_j} \ and \ y_j \in V_{y_i} \\ 0 & Otherwise \end{cases} \tag{8}$$

Where $V_{y_i}$ and $V_{y_j}$ denotes successively the neighborhood of $y_i$ and the neighborhood of $y_j$. Thus, the enhanced SSC algorithm noted as **E-SSC** is summarized as:

---

**Algorithm 1. E-SSC**

**Input**: A data matrix $Y = [y_1 \ y_2 \ ... \ y_N]$ that lies in the union of $K$ subspaces.

1. Solve the problem (4) for free data or (5) in the case of noise corrupted data.
2. Normalize the columns of $C$ as $c_i = \frac{c_i}{\|c_i\|_\infty}$ [10]
3. Form the graph $W$ with $N$ nodes representing the $N$ data points as:

$$w_{ij} = \mid c_{ij} \mid + \mid c_{ji} \mid + O(V_{y_i}, V_{y_j}) \tag{9}$$

4. Apply spectral clustering to the laplacian matrix.

**Output**: The cluster membership of $Y$.

---

## 4     Experimental Results

In this section, the experiments to verify the effectiveness of the proposed method compared to other subspace clustering methods [5] are assessed. We evaluate the clustering results on two publicly real world datasets, which are 'Hopkins155' [14] for motion segmentation and 'Feret' [15] for face clustering. These datasets are commonly used in testing machine learning algorithms specially for subspace clustering methods [16,17]. To measure the subspace clustering performance, two criteria are adopted:

1. The rate of misclassified points which presents the percentage of the misclassified points in relation to all the classified points. Therefore, it gives an idea of the final clustering quality. This measure is defined as:

$$Error \ \% = \frac{\# of \ misclassified \ points}{\# of \ total \ points} \times 100 \ \% \tag{10}$$

2. The Normalized Mutual Information [18] which is generally used to evaluate the performance of spectral clustering approaches. Let $U$ be the clustering result vector and $T$ be the true label vector. We suppose that $p(U)$ and $p(T)$ are respectively the joint probability mass functions of $U$ and $T$. Then, the NMI measure is defined as:

$$NMI(U,T) = \frac{I(U,T)}{[H(U) + H(T)]/2} \tag{11}$$

Where $H(U)$ and $H(T)$ are the entropies of $p(U)$, $p(T)$ and $I(U,T)$ denotes the mutual information between $p(U)$ and $p(T)$.

### 4.1  Motion Segmentation

Motion segmentation is a fundamental pre-processing step for different computer vision applications including video indexing, video surveillance, traffic monitoring, robotics, etc. The main idea of motion segmentation is to classify a set of tracked feature points into different groups that correspond to different moving objects (Fig. 3).

**Fig. 3.** Samples of images of some video sequences from 'Hopkins155' database with ground truth superimposed.

Indeed, under the affine camera model, the 2-D feature points of a set of 3-D real world data points (from a rigidly moving object) lie to a subspace of dimension at most 4 in $\mathbb{R}^{2F}$ where $F$ is a frame in a video sequence. Therefore, motion segmentation of tracked feature points is reduced to a subspace clustering problem, where each subspace corresponds to a single motion. In this context, the SSC method is known as one the most effective approaches in terms of accuracy [5]. Thus, we aim to prove the performance of our method compared to the SSC and the LRR methods since they use the same similarity measure based on the sparse coefficients. Also, we compare the E-SSC to the SLBF and the LSA methods which use a similarity measure based on the calculation of the angles between subspaces. For these tests, we consider the standard benchmark 'Hopkins 155'. This benchmark can be divided into two main groups: 120 video sequences containing 2 motions and 35 video sequences having 3 motions. For

**Table 1.** Clustering error results on the 'Hopkins 155' dataset

| Algorithms | LRR | LSA | SLBF | SSC | E-SSC |
|---|---|---|---|---|---|
| (a) 2F-dimensional data points | | | | | |
| 2 Motions | | | | | |
| Mean | 4.10 | 4.23 | 1.16 | 1.58 | 1.70 |
| Median | 0.22 | 0.56 | 0.00 | 0.00 | 0.00 |
| 3 Motions | | | | | |
| Mean | 9.89 | 7.02 | 3.63 | 4.40 | 3.90 |
| Median | 6.22 | 1.45 | 0.00 | 0.56 | 0.61 |
| (b) 4K-dimensional data points using PCA | | | | | |
| 2 Motions | | | | | |
| Mean | 4.83 | 3.61 | 1.16 | 1.83 | 1.74 |
| Median | 0.26 | 0.51 | 0.00 | 0.00 | 0.00 |
| 3 Motions | | | | | |
| Mean | 9.89 | 7.65 | 3.63 | 4.40 | 3.90 |
| Median | 6.22 | 1.27 | 0.00 | 0.56 | 0.61 |

both SSC and E-SSC methods, $\lambda_z = \frac{800}{\mu_z}$ where $\mu_z$ is calculated from the data matrix as: $\mu_z \triangleq \min_{} \max_{j \neq i} \mid y_i^T y_j \mid$.

In Table 1, the performance of the subspace clustering methods are compared. According to [10], we use in experiments, two type of data: the original $2F$ dimensional features data points and the $4K$ dimensional projection data points since the feature trajectories of $K$ objects or motions in a video sequence are almost lying in $4K$ dimensional linear subspaces. As shown in the table, the new similarity measure used in the E-SSC method succeeds to give very competitive results specially compared to the SSC method.

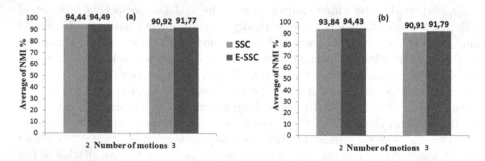

**Fig. 4.** Comparison of the NMI scores on the 'Hopkins 155' dataset. (a): results on $2F$ dimensional data points. (b): results on $4K$ dimensional data points

In Fig. 4, the E-SSC is compared to the SSC method in terms of the average of NMI values recorded for all the video sequences cited above. Thus, we observe that the proposed method out-performs the SSC method for both type of projections $2F$ and $4K$. In fact, the modified similarity measure succeed to provide a better subspace clustering affinities between the nodes of the similarity graph.

## 4.2 Face Clustering

As a second application context, we evaluate the performance of the SSC and the proposed E-SSC methods on the 'b' subset of 'Feret' database for face clustering. This dataset contains 1400 facial images with the size $80 \times 80$ pixels of 200 persons taken under different illumination and pose conditions. In addition, each person has 7 frontal images noted as: 'ba', 'bd', 'be', 'bf', 'bg', 'bj' and 'bk' with different facial expressions (Fig. 5). For the experiment, 40 subjects of this dataset were chosen randomly and divided into 4 groups of data such that each group contains 10 subjects. To see the effect of the number of subjects, we test for all choices of $K \in \{2, 4, 6, 8, 10\}$. Thus, we apply the subspace clustering methods for all the sets of $K$ subjects with $\lambda_z = \frac{60}{\mu_z}$.

**Fig. 5.** Facial images for one person from the Feret database

In Fig. 6(a) the average of misclassification error is presented to prove the performance of the E-SSC compared to the classic SSC. The results recorded by E-SSC are 7.46, 29.71, 38.91, 43.12 and 43.93 for the number of subjects 2, 4, 6,

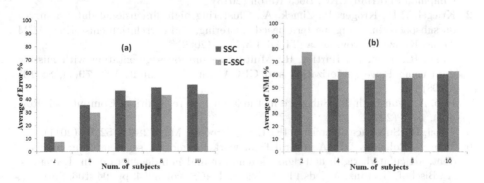

**Fig. 6.** Clustering results on Feret dataset: (a) presents the clustering error rates and (b) presents the NMI scores.

8 and 10 respectively. However, we have 11.45, 35.45, 46.37, 48.91 and 51.07 for the classic SSC. As it can be seen, the proposed E-SSC achieves lower clustering error rates than the SSC method. Moreover, in Fig. 6(b), the average of NMI results is presented. We have recorded 77.75, 62.20, 60.69, 60.86 and 62.83 for all the different choices of subject number (2, 4, 6, 8 and 10 respectively) for E-SSC. Nevertheless, the average NMI scores of the SSC method are: 67.30, 56.05, 55.95, 57.76 and 60.55. Therefore, the E-SSC method gives better results than those given by the SSC method in terms of clustering quality. These results are explained by the effect of the new pairwise similarity measure used by the E-SSC method. In fact, this measure has succeeded in attracting more the data points that are supposed to be in the same final cluster and thus it improves the final clustering accuracy.

## 5    Conclusion

In this paper, we have proposed an enhancement of the SSC method based on a novel pairwise similarity measure. The main motivation is to exploit the idea of overlapping between local subspaces around the data points. Then, we have incorporated this information in the existent measure under some conditions. The experimental results on the 'Hopkins155' database for motion segmentation and on 'Feret' dataset of face clustering, have revealed consistently the performance of the proposed E-SSC compared to the traditional SSC and other state-of-the-art spectral subspace clustering methods. For future work and in the purpose to improve more the clustering accuracy of the E-SSC method, we will focus on transforming our method in the Riemannian manifolds which are known for their ability to detect nonlinear shapes.

## References

1. Aggarwal, C.C., Reddy, C.K. (eds.): Data Clustering: Algorithms and Applications. Chapman and Hall/CRC, Boca Raton (2013)
2. Kriegel, H.P., Kröger, P., Zimek, A.: Clustering high-dimensional data: a survey on subspace clustering, pattern-based clustering, and correlation clustering. ACM Trans. Knowl. Discov. Data (TKDD) **3**(1), 1 (2009)
3. Vidal, R., Tron, R., Hartley, R.: Multiframe motion segmentation with missing data using PowerFactorization and GPCA. Int. J. Comput. Vis. **79**(1), 85–105 (2008)
4. Rao, A., Noushath, S.: Subspace methods for face recognition. Comput. Sci. Rev. **4**(1), 1–17 (2010)
5. Vidal, R.: Subspace clustering. IEEE Sig. Process. Mag. **28**(2), 52–68 (2011)
6. Yan, J., Pollefeys, M.: A general framework for motion segmentation: independent, articulated, rigid, non-rigid, degenerate and non-degenerate. In: Leonardis, A., Bischof, H., Pinz, A. (eds.) ECCV 2006. LNCS, vol. 3954, pp. 94–106. Springer, Heidelberg (2006). doi:10.1007/11744085_8
7. Zhang, T., Szlam, A., Wang, Y., Lerman, G.: Hybrid linear modeling via local best-fit flats. Int. J. Comput. Vis. **100**(3), 217–240 (2012)

8. Ng, A.Y., Jordan, M.I., Weiss, Y.: On spectral clustering: analysis and an algorithm. In: NIPS, vol. 14, no. 2, pp. 849–856 (2001)
9. Elhamifar, E., Vidal, R.: Clustering disjoint subspaces via sparse representation. In: IEEE International Conference on Acoustics Speech and Signal Processing ICASSP 2010, pp. 1926–1929 (2010)
10. Elhamifar, E., Vidal, R.: Sparse subspace clustering: algorithm, theory, and applications. IEEE Trans. Pattern Anal. Mach. Intell. **35**(11), 2765–2781 (2013)
11. Liu, G., Lin, Z., Yu, Y.: Robust subspace segmentation by low-rank representation. In: Proceedings of the 27th International Conference on Machine Learning ICML 2010, pp. 663–670 (2010)
12. Liu, G., Lin, Z., Yan, S., Sun, J., Yu, Y., Ma, Y.: Robust recovery of subspace structures by low-rank representation. IEEE Trans. Pattern Anal. Mach. Intell. **35**(1), 171–184 (2013)
13. Donoho, D.L.: For most large underdetermined systems of linear equations the minimal l1-norm solution is also the sparsest solution. Commun. Pure Appl. Math. **59**(6), 797–829 (2006)
14. Tron, R., Vidal, R.: A benchmark for the comparison of 3-D motion segmentation algorithms. In: IEEE Conference on Computer Vision and Pattern Recognition CVPR 2007, pp. 1–8 (2007)
15. Phillips, P.J., Wechsler, H., Huang, J., Rauss, P.J.: The FERET database and evaluation procedure for face-recognition algorithms. Image Vis. Comput. **16**(5), 295–306 (1998)
16. Chen, G., Lerman, G.: Motion segmentation by SCC on the Hopkins 155 database. In: the 12th International Conference on Computer Vision Workshops (ICCV Workshops), pp. 759–764 (2009)
17. Yin, M., Fang, X., Xie, S.: Semi-supervised sparse subspace clustering on symmetric positive definite manifolds. In: Tan, T., Li, X., Chen, X., Zhou, J., Yang, J., Cheng, H. (eds.) CCPR 2016. CCIS, vol. 662, pp. 601–611. Springer, Singapore (2016). doi:10.1007/978-981-10-3002-4_49
18. Alok, A.K., Saha, S., Ekbal, A.: Development of an external cluster validity index using probabilistic approach and min-max distance. IJCISIM **6**(1), 494–504 (2014)

# Mathematics for Neural Networks

# Neural Network-Based Simultaneous Estimation of Actuator and Sensor Faults

Marcin Pazera$^{(\boxtimes)}$, Marcin Witczak, and Marcin Mrugalski

Institute of Control and Computation Engineering,
University of Zielona Góra, ul. Podgórna 50, 65-246 Zielona Góra, Poland
{M.Pazera,M.Witczak,M.Mrugalski}@issi.uz.zgora.pl

**Abstract.** The paper is devoted to the problem of a neural network-based robust simultaneous actuator and sensor faults estimator design for the purpose of the Fault Diagnosis (FD) of non-linear systems. In particular, the methodology of designing a neural network-based $\mathcal{H}_\infty$ fault estimator is developed. The main novelty of the approach is associated with possibly simultaneous sensor and actuator faults. For this purpose, a Linear Parameter Varying (LPV) description of a Recurrent Neural Network (RNN) is exploited. The proposed approach guaranties a predefined disturbance attenuation level and convergence of the estimator. The final part of the paper presents an illustrative example concerning the application of the proposed approach to the multi-tank system fault diagnosis.

## 1 Introduction

Technological progress leads to increasing the complexity of the processes and systems. Such complexity may result in an increased incidence of their malfunction. It causes the need for developing new and more effective FD methods. In the last few decades, three groups of fault diagnosis methods were developed simultaneously. The first one consists of numerous relatively simple and easy to apply methods which rely on the analysis of the signals from diagnosed system. The second group is represented by qualitative methods demanding deep knowledge about the diagnosed system. The last group is called quantitative methods, which are based on the application of the model of the diagnosed system. The quantitative methods seem to be especially attractive because they can be applied for example in the Fault Tolerant Control (FTC) systems [3,6,7,13,16,18,19,23].

Among the model-based FD methods two main groups can be distinguished: analytical methods and computational approaches [12,14,17,20,24]. The analytical methods are based on the models which are based on the physical description of the diagnosed system. In contrast, the computational approaches rely on the models which only reflect the behaviour of the diagnosed system. Regardless of the kind of fault diagnosis method, it should be reliable for any conditions. In other words, the fault diagnosis method should be robust against the noises, disturbances and model uncertainty [10,14,16,22]. Moreover, the FD method should be universal to be applied for a wide class of systems. Furthermore, the

© Springer International Publishing AG 2017
I. Rojas et al. (Eds.): IWANN 2017, Part I, LNCS 10305, pp. 305–316, 2017.
DOI: 10.1007/978-3-319-59153-7_27

efficient FD should provide a knowledge about fault detection, identification and isolation. Such knowledge is specially required if the method can be applied in the FTC systems.

It should be underlined that the fault diagnosis methods created on the basis of analytical models are mature and widely applied in the industry. Such methods allow simultaneous fault detection, identification and fault estimation. Unfortunately, such methods cannot be applied for some classes of the systems when the analytical model of the diagnosed system is not available. Such drawback is not present in the case of the computational intelligence methods such as Artificial Neural Networks (ANNs) [8]. In this approach, the model can be created in the procedure of system identification on the basis of the measurements from the diagnosed system. It is especially attractive in the case of multidimensional, complex, dynamic and highly non-linear systems. Unfortunately, the ANNs have also some drawbacks, e.g., the training problems or the complex mathematical description which make them difficult to combine with analytical methods. They are rarely available in the form of the state-space neural model with its uncertainty description frequently used for the robust FD [14, 21]. Moreover, the ANNs can be relatively easily applied in the fault detection tasks although it is very difficult to perform the fault isolation and fault identification with their application.

To join the advantages of the neural networks-based models and analytical approaches in the context of their applications in the FD tasks, a novel methodology of designing of Neural Fault Diagnostic Scheme (NFDS) is proposed. The application of the ANNs in the FD system is possible by transformation of the neural model without linearization into a Linear Parameter Varying (LPV) form [1, 2, 4]. It should be underlined that the final NFDS will be described in the ANN-like form. It will facilitate its practical implementation. Moreover, the proposed design procedure boils down to solving a set of Linear Matrix Inequalities (LMIs).

The proposed approach is superior over other FD approaches because it is able to detect and estimate both actuators and sensors fault vectors simultaneously. Moreover, in the developed approach, the robustness of the FD scheme is achieved by minimizing an influence of external disturbances. The resulting methodology guarantees that a prescribed disturbance attenuation level is achieved with respect to the state as well as actuator and sensor fault estimation errors while guaranteeing convergence of the observer.

The paper is organized as follows. Section 2 presents basic information about the RNN model and its LPV representation which can be used in the FD tasks. Subsequently, Sect. 3 describes a novel robust UIO design procedure that can be used for the state and actuators and sensors fault estimation. Section 4 shows an example of the application of the developed approach in the task of the actuators and sensors robust FD of the multi-tank system. The final part of the paper concerns the concluding remarks.

## 2   Problem Statement

Let us consider the following discrete-time system

$$x_{k+1} = f\left(x_k, u_k\right),\tag{1}$$

where $x_k \in \mathbb{R}^n$ and $u_k \in \mathbb{R}^r$ is the state and input vector, respectively, while $f\left(\cdot\right)$ is an unknown non-linear function which are describing the system with respect to the state and input.

As demonstrated in [15], such a system can be efficiently modelled with RNN. Furthermore, by introducing sensor and actuator fault it boils down to the following form:

$$x_{k+1} = A x_k + B\left[u_k + f_{a,k}\right] + A_0\sigma\left(E_1 x_k\right) + B_0\sigma\left(E_2\left[u_k + f_{a,k}\right]\right),\tag{2}$$

$$y_k = C x_k + f_{s,k},\tag{3}$$

where $\sigma\left(\cdot\right)$ is a nonlinear activation function of hidden layers. Matrices $A$, $B$, $A_0$, $B_0$, $E_1$, $E_2$ are the block weight matrices. Moreover, $f_{a,k} \in \mathbb{F}_a \subset \mathbb{R}^r$ is the actuator fault while $f_{s,k} \in \mathbb{F}_s \subset \mathbb{R}^m$ stands for the sensor fault vector.

The goal is to represent (2)–(3) in the LPV-like form:

$$x_{k+1} = A\left(\alpha\right) x_k + B\left(\alpha\right) u_k + B\left(\alpha\right) f_{a,k} + W_1 w_k\tag{4}$$

$$y_k = C x_k + f_{s,k} + W_2 w_k,\tag{5}$$

where $\alpha$ is appropriate scheduling parameter and $w_k$ is an exogenous disturbance vector. Moreover, $W_1$ and $W_2$ denote its distribution matrices. It can be easily shown that $w_k$ can be split in such a way as $w_k = \left[w_{1,k}^T, w_{2,k}^T\right]^T$ where $w_{1,k}$ and $w_{2,k}$ are process and measurement uncertainties, respectively.

It should be noted that the derivation presented here is partially based on [1]. The problem of transforming neural state-space model (2)–(3) into (4)–(5) has the following property

$$\left(A(\alpha), \left(B(\alpha)\right) \in \theta = Co\{\left(A_i, B_i\right), i = 1, \ldots, p\}.\tag{6}$$

Let also define the varying parameter $\alpha_i$ [1]:

$$\alpha_i = \begin{cases} \frac{\sigma\left(E_1^i x_k + E_2^i u_k\right)}{E_1^i x_k + E_2^i u_k}, & E_1^i x_k + E_2^i u_k \neq 0 \\ 1, & E_1^i x_k + E_2^i u_k = 0 \end{cases},\tag{7}$$

where $1 < i < p$ denotes $i$th row of a respective matrix. Then (2), can be rewritten as

$$x_{k+1} = A x_k + D\left[u_k + f_{a,k}\right] + A_0 \Theta E_1 x_k + B_0 \Theta E_2\left[u_k + f_{a,k}\right] + W_1 w_k,\tag{8}$$

with diagonal $\Theta \in \Re^{p \times p}$ in the form

$$\Theta = \text{diag}\left(\alpha_1, \ldots, \alpha_p\right).\tag{9}$$

Using the above results, it is proposed to transform the neural network (2)–(3) into

$$x_{k+1} = Ax_k + B\left[u_k + f_{a,k}\right] + g\left(x_k\right) + h(u_k + f_{a,k}) + W_1 w_k, \qquad (10)$$

$$y_k = Cx_k + f_{s,k} + W_2 w_k, \qquad (11)$$

with $g\left(x_k\right) = \sum_{i=1}^{p} \alpha^i A_0^i E_1^{(i)} x_k$ and $h(u_k) = \sum_{i=1}^{p} \alpha^i B_0^i E_2^{(i)} u_k$.

Furthermore, the neural model can be written in a traditional LPV shape (4)–(5) where $A(\alpha) = A + \sum_{i=1}^{p} \alpha_i A^i$, $B(\alpha) = B + \sum_{i=1}^{p} \alpha_i A^i$. Moreover, as it was shown, no linearization is used for transforming the neural network (2)–(3) into LPV form (4)–(5). Having a general system description, it is possible to develop and estimator which will be able to estimate sensor and actuator fault simultaneously.

## 3   Estimator Design

To handle the above defined problem of simultaneous estimation of the state $x_k$ as well as actuator $f_{a,k}$ and sensor $f_{s,k}$ faults, the following novel observer is proposed:

$$\hat{x}_{k+1} = A\left(\alpha\right)\hat{x}_k + B\left(\alpha\right)u_k + B\left(\alpha\right)\hat{f}_{a,k} + K_x\left(y_k - C\hat{x}_k - \hat{f}_{s,k}\right), \quad (12)$$

$$\hat{f}_{a,k+1} = \hat{f}_{a,k} + K_a\left(y_k - C\hat{x}_k - \hat{f}_{s,k}\right), \qquad (13)$$

$$\hat{f}_{s,k+1} = \hat{f}_{s,k} + K_s\left(y_k - C\hat{x}_k - \hat{f}_{s,k}\right), \qquad (14)$$

where $K_x$, $K_a$, $K_s$ are the gain matrices for the state, actuator and sensor fault, respectively.

Based on (2)–(3) the state estimation error can be described as follows:

$$\begin{aligned} e_{k+1} = x_{k+1} - \hat{x}_{k+1} &= A\left(\alpha\right) + B\left(\alpha\right)u_k + B\left(\alpha\right)f_{a,k} + W_1 w_k \\ &- A\left(\alpha\right)\hat{x}_k - B\left(\alpha\right)u_k - B\left(\alpha\right)f_{a,k} - K_x C x_k - K_x f_{s,k} \\ &- K_x W_2 w_k + K_x C\hat{x}_k + K_x \hat{f}_{s,k} \\ &= \left[A\left(\alpha\right) - K_x C\right]e_k + B\left(\alpha\right)e_{a,k} - K_x e_{s,k} + \left[W_1 - K_x W_2\right]w_k, \end{aligned} \qquad (15)$$

where $e_{a,k}$ and $e_{s,k}$ are the actuator and sensor fault errors, respectively. Furthermore, the dynamics of the actuator fault estimation error is given by:

$$\begin{aligned} e_{a,k+1} = f_{a,k+1} - \hat{f}_{a,k+1} &= f_{a,k+1} + f_{a,k} - f_{a,k} - \hat{f}_{a,k} - K_a C x_k \\ &- K_a f_{s,k} - K_a W_2 w_k + K_a C\hat{x}_k + K_a \hat{f}_{s,k} \\ &= \varepsilon_{a,k} + e_{a,k} - K_a C e_k - K_a e_{s,k} - K_a W_2 w_k, \end{aligned} \qquad (16)$$

with $\varepsilon_{a,k} = f_{a,k+1} - f_{a,k}$ which denotes an error between consecutive samples of the actuator fault. The dynamics of the sensor fault estimation error obeys:

$$\begin{aligned} e_{s,k+1} = f_{s,k+1} - \hat{f}_{s,k+1} &= f_{s,k+1} + f_{s,k} - f_{s,k} - \hat{f}_{s,k} - K_s C x_k \\ &- K_s f_{s,k} - K_s W_2 w_k + K_s C\hat{x}_k + K_s \hat{f}_{s,k} \\ &= \varepsilon_{s,k} + \left[I - K_s\right]e_{s,k} - K_s C e_k - K_s W_2 w_k, \end{aligned} \qquad (17)$$

with $\varepsilon_{s,k} = f_{s,k+1} - f_{s,k}$ which denoting the error between consecutive samples of the sensor fault.

Furthermore, by constructing the super-vectors $\bar{e}_{k+1} = \left[ e_{k+1}^T, e_{a,k+1}^T, e_{s,k+1}^T \right]^T$ and $v_k = \left[ w_k^T, \varepsilon_{a,k}^T, \varepsilon_{s,k}^T \right]^T$, the estimation error of the state and fault can be presented in a compact form

$$\bar{e}_{k+1} = X(\alpha)\bar{e}_k + Zv_k, \tag{18}$$

with:

$$X(\alpha) = \bar{A}(\alpha) - \bar{K}\bar{C}, \tag{19}$$
$$Z = \bar{W} - \bar{K}\bar{V}, \tag{20}$$

where:

$$\bar{A}(\alpha) = \begin{bmatrix} A(\alpha) & B(\alpha) & 0 \\ 0 & I & 0 \\ 0 & 0 & I \end{bmatrix}, \quad \bar{C} = \begin{bmatrix} C & 0 & I \end{bmatrix}, \quad \bar{K} = \begin{bmatrix} K_x \\ K_a \\ K_s \end{bmatrix},$$
$$\bar{W} = \begin{bmatrix} W_1 & 0 & 0 \\ 0 & I & 0 \\ 0 & 0 & I \end{bmatrix}, \quad \bar{V} = \begin{bmatrix} W_2 & 0 & 0 \end{bmatrix}. \tag{21}$$

Based on above results, the following theorem can be defined:

**Theorem 1.** *For a prescribed disturbance attenuation level $\mu > 0$ for the state and fault estimation error (18), the $\mathcal{H}_\infty$ observer design problem for the system (2)–(3) and the observer (12)–(14) is solvable if there exist matrices $P$, $U$ and $N$ such that the following constraints are satisfied:*

$$\begin{bmatrix} I - P & 0 & \bar{A}(\alpha)^T U - \bar{C}^T N^T \\ 0 & -\mu^2 I & \bar{W} U - \bar{V}^T N^T \\ U\bar{A}(\alpha) - N\bar{C} & U\bar{W} - N\bar{V} & P - U - U^T \end{bmatrix} \prec 0. \tag{22}$$

*Proof.* The problem of the designing the $\mathcal{H}_\infty$ observer [11,25] is to obtain matrices $N, U$ and $P$ such that

$$\lim_{k \to \infty} \bar{e}_k = 0 \quad \text{for} \quad v_k = 0, \tag{23}$$

$$\|\bar{e}_k\|_{l_2} \leq \mu \|v_k\|_{l_2} \quad \text{for} \quad v_k \neq 0, \bar{e}_0 = 0. \tag{24}$$

To solve the problem, it is satisfactory to find a Lyapunov function such that

$$\Delta V_k + \bar{e}_k^T \bar{e}_k - \mu^2 v_k^T v_k < 0, \tag{25}$$

where:

$$V_k = \bar{e}_k^T P \bar{e}_k, \quad P \succ 0, \tag{26}$$
$$\Delta V_k = V_{k+1} - V_k. \tag{27}$$

As a consequence by using (18) it is easy to show that

$$
\begin{aligned}
\Delta V_k + \bar{e}_k^T \bar{e}_k - \mu^2 v_k^T v_k = {} & \bar{e}_k^T \left( X \left( \alpha \right)^T P X \left( \alpha \right) + I - P \right) \bar{e}_k \\
& + \bar{e}_k^T \left( X \left( \alpha \right)^T P Z \right) v_k + v_k^T \left( Z^T P X \left( \alpha \right) \right) \bar{e}_k \\
& + v_k^T \left( Z^T P Z - \mu^2 I \right) v_k < 0,
\end{aligned}
\tag{28}
$$

and by introducing

$$
\bar{v}_k = \begin{bmatrix} \bar{e}_k \\ v_k \end{bmatrix},
\tag{29}
$$

it can be shown that (28) can be rewritten to the following form

$$
\bar{v}_k^T \begin{bmatrix} X \left( \alpha \right)^T P X \left( \alpha \right) + I - P & X \left( \alpha \right)^T P Z \\ Z^T P X \left( \alpha \right) & Z^T P Z - \mu^2 I \end{bmatrix} \bar{v}_k \prec 0,
\tag{30}
$$

which is equivalent to

$$
\begin{bmatrix} X \left( \alpha \right)^T \\ Z^T \end{bmatrix} P \left[ X \left( \alpha \right) \ Z \right] + \begin{bmatrix} I - P & 0 \\ 0 & -\mu^2 I \end{bmatrix} \prec 0.
\tag{31}
$$

Now, let us recall the following lemma [5]:

**Lemma 1.** *The following statements are equivalent:*

1. *There exists $X \left( \alpha \right) \succ 0$ such that*

$$
V^T X \left( \alpha \right) V - W \prec 0.
\tag{32}
$$

2. *There exist $X \left( \alpha \right) \succ 0$ such that*

$$
\begin{bmatrix} -W & V^T U^T \\ U V & X \left( \alpha \right) - U - U^T \end{bmatrix} \prec 0.
\tag{33}
$$

Applying Lemma 1 to (30) gives

$$
\begin{bmatrix} I - P & 0 & X \left( \alpha \right)^T U^T \\ 0 & -\mu^2 I & Z^T U^T \\ U X \left( \alpha \right) & U Z & P - U - U^T \end{bmatrix} \prec 0,
\tag{34}
$$

and then substituting:

$$
U X \left( \alpha \right) = U \bar{A} \left( \alpha \right) - U \bar{K} \bar{C} = U \bar{A} \left( \alpha \right) - N \bar{C},
\tag{35}
$$

$$
U Z = U \bar{W} - U \bar{K} \bar{V} = U \bar{W} - N \bar{V},
\tag{36}
$$

completes the proof. □

The final design procedure can be summarized as follows:

1. Find $A_i$, $B_i$ for all $i$ according to $g \left( x_k \right) = \sum_{i=1}^{p} \alpha^i A_0^i E_1^{(i)} x_k$ and $h(u_k) = \sum_{i=1}^{p} \alpha^i B_0^i E_2^{(i)} u_k$,
2. Select $\mu$ and solve LMI (22) to get matrices $N, U$ and $P$,
3. Calculate $\bar{K} = U^{-1} N$,
4. Determine gain matrices $\bar{K} = \left[ K_x^T, K_a^T, K_s^T \right]^T$.

# 4    Illustrative Example

To verify the proposed approach a multi-tank system is employed which is depicted in Fig. 1. Such a system was developed for simulating the real industrial multi-tank system in the laboratory conditions. It can be regularly used to test practically linear and non-linear methodologies used in control, identification and fault diagnosis. The considered system consists of three separate tanks placed in cascade. The tanks are equipped with drain valves as well as electro-valves and level sensors. These sensors based on a hydraulic pressure measurement. Those tanks are differently shaped what implies the nonlinearities of the system. The liquid level in each tank varies in the range from 0 up to $0.35\,[m]$. The lower bottom tank is a water reservoir for the system. The multi-tank system is fed with a DC water pump which is used to fill the upper tank by the liquid. The water outflows from the tanks due to gravity. The multi-tank system exchanges data with the level sensors, it also communicates with valves and a pump with a PC-based digital controller through the dedicated I/O board and the power interface. Real time software is controlled by the I/O board with MATLAB/SIMULINK environment. For more information the reader is referred to [9].

Firstly, following [15], the disturbances influencing the system are distributed through:

$$W_1 = 0.01 I_{3\times 3}, \qquad W_2 = 0.01 I_{3\times 3}, \tag{37}$$

The sampling time of the system is set to $0.01\,[s]$. It should be mentioned that the neural network was trained using Levenberg-Marquardt backpropagation algorithm. Figure 2 presents the real system output (solid line) in all of the tanks and the response of the neural network (dashed line) for a given training input.

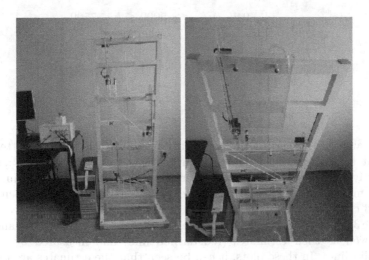

**Fig. 1.** Multi-tank system

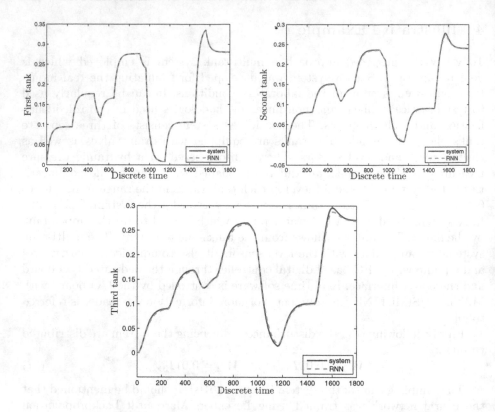

**Fig. 2.** Real water level and its neural network representation

Moreover, let us consider the following fault scenarios:

$$\boldsymbol{f}_{a,k} = \begin{cases} -0.45 \cdot \boldsymbol{u}_k, & 5000 \le k \le 7500, \\ 0, & \text{otherwise,} \end{cases} \tag{38}$$

$$\begin{aligned} \boldsymbol{f}_{s,1,k} &= 0, \\ \boldsymbol{f}_{s,2,k} &= \begin{cases} \boldsymbol{y}_{2,k} + 0.04, & 6000 \le k \le 8000, \\ 0, & \text{otherwise.} \end{cases} \\ \boldsymbol{f}_{s,3,k} &= \begin{cases} \boldsymbol{y}_{3,k} - 0.1, & 7000 \le k \le 9000, \\ 0, & \text{otherwise.} \end{cases} \end{aligned} \tag{39}$$

which means that the actuator fault can be regarded as an intermittent one. Note that temporary sensor faults occurred in two of three sensors, and they were partly at the same time. Moreover, the sensor faults appear during the actuator malfunction. This is a realistic situation that might happen in the industrial conditions.

Figure 3 presents the response of the states for the first, second and third tank as well as their estimates (red dashed line) and measured by a sensor (black solid line). In these plots, it can be seen that the estimates are following the real states irrespective to the fact of fault occurrence. It can be said that the

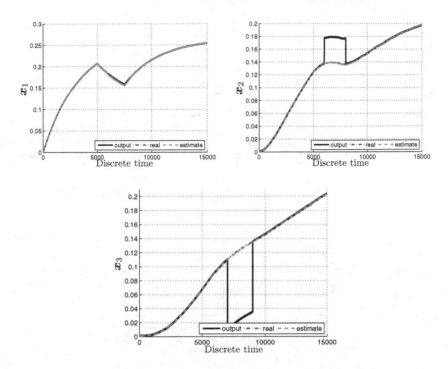

**Fig. 3.** The response of the system (Color figure online)

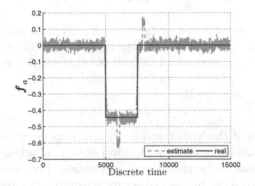

**Fig. 4.** The actuator fault

state estimate is fault-resistant. In Fig. 4, it is shown the actuator fault and its estimate (dashed line). To show the effectiveness of the proposed approach with respect to actuator fault and to simplify the interpretation of the fault signal, it has been scaled to the range from 0 to 1. According to the fault scenario the value $-0.45$ in this graph represent the 45% loss of the effectiveness of the pump. The observer estimates the real fault highly satisfactory, which clearly confirms its effectiveness. Figure 5 presents the sensor faults and their estimates.

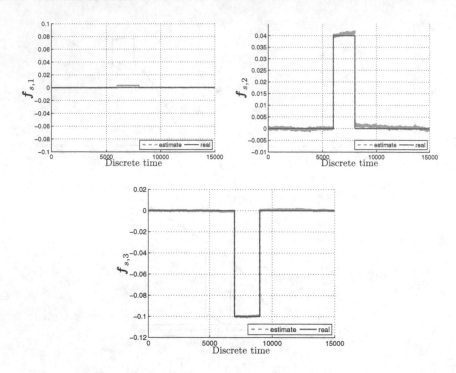

**Fig. 5.** The sensor faults

The observer estimates the sensor faults with very high accuracy. The results obtained with the proposed approach clearly show its quality and recommended its straightforward implementation in fault-tolerant control systems.

## 5    Conclusions

The main objective of this paper was to exploit appealing nonlinear modelling features of neural networks along with novel analytical scheme for the purpose of simultaneous sensor and actuator fault estimation. The usual frameworks presented in the literature can be used either for sensor or actuator fault estimation. This is realised under unrealistic assumption that either all sensors all or actuator all fault free. The proposed approach reduces the above conservatism and the above assumptions are not longer required. The paper presents a full design procedure along with convergence analysis. The final part of the paper shows an illustrative example regarding multi-tank system, which clearly exhibits the efficiency of the proposed approach.

**Acknowledgements.** The work was supported by the National Science Centre of Poland under grant: 2013/11/B/ST7/01110.

# References

1. Abbas, H., Werner, H.: Polytopic quasi-LPV models based on neural state-space models and application to air charge control of a SI engine. In: Proceedings of the 17th World Congress the International Federation of Automatic Control, Seoul, Korea, pp. 6466–6471 (2008)
2. Bendtsen, J.D., Trangbæk, K.: Robust quasi-LPV control based on neural state space models. IEEE Trans. Neural Netw. **13**(2), 355–368 (2002)
3. Blesa, J., Rotondo, D., Puig, V., Nejjari, F.: FDI and FTC of wind turbines using the interval observer approach and virtual actuators/sensors. Control Eng. Pract. **24**, 138–155 (2014)
4. Chen, L., Patton, R., Goupil, P.: Robust fault estimation using an LPV reference model: ADDSAFE benchmark case study. Control Eng. Pract. **49**, 194–203 (2015)
5. de Oliveira, M.C., Bernussou, J., Geromel, J.C.: A new discrete-time robust stability condition. Syst. Control Lett. **37**(4), 261–265 (1999)
6. Ducard, G.: Fault-tolerant Flight Control and Guidance Systems: Practical Methods for Small Unmanned Aerial Vehicles. Springer, Berlin (2009)
7. Gao, Z.-F., Lin, J.-X., Cao, T.: Robust fault tolerant tracking control design for a linearized hypersonic vehicle with sensor fault. Int. J. Control Autom. Syst. **13**(3), 672–679 (2015)
8. Haykin, S.: Neural Networks and Learning Machines. Prentice Hall, New York (2009)
9. INTECO: Multitank System - User's manual (2013). www.inteco.com.pl
10. Isermann, R.: Fault Diagnosis Applications: Model Based Condition Monitoring, Actuators, Drives, Machinery, Plants, Sensors, and Fault-Tolerant Systems. Springer, Berlin (2011)
11. Li, H., Fu, M.: A linear matrix inequality approach to robust $H_\infty$ filtering. IEEE Trans. Sig. Process. **45**(9), 2338–2350 (1997)
12. Li, L., Ding, S.X., Yang, Y., Zhang, Y.: Robust fuzzy observer-based fault detection for nonlinear systems with disturbances. Neurocomputing **174**, Part B, 767–772 (2016)
13. Mahmoud, M., Jiang, J., Zhang, Y.: Active Fault Tolerant Control Systems: Stochastic Analysis and Synthesis. Springer, Berlin (2003)
14. Mrugalski, M.: Advanced Neural Network-based Computational Schemes for Robust Fault Diagnosis. Springer, Heidelberg (2014)
15. Mrugalski, M., Luzar, M., Pazera, M., Witczak, M., Aubrun, C.: Neural network-based robust actuator fault diagnosis for a non-linear multi-tank system. ISA Trans. **61**, 318–328 (2016)
16. Noura, H., Theilliol, D., Ponsart, J.C., Chamseddine, A.: Fault-tolerant Control Systems: Design and Practical Applications. Springer, London (2009)
17. Péni, T., Vanek, B., Szabó, Z., Bokor, J.: Supervisory fault tolerant control of the GTM UAV using LPV methods. Int. J. Appl. Math. Comput. Sci. **25**(1), 117–131 (2015)
18. Rotondo, D., Nejjari, F., Puig, V.: Robust quasi-LPV model reference FTC of a quadrotor uav subject to actuator faults. Int. J. Appl. Math. Comput. Sci. **25**(1), 7–22 (2015)
19. Seybold, L., Witczak, M., Majdzik, P., Stetter, R.: Towards robust predictive fault-tolerant control for a battery assembly system. Int. J. Appl. Math. Comput. Sci. **25**(4), 849–862 (2015)

20. Tayarani-Bathaie, S.S., Vanini, Z.N.S., Khorasani, K.: Dynamic neural network-based fault diagnosis of gas turbine engines. Neurocomputing **125**, 153–165 (2014)
21. Witczak, M.: Toward the training of feed-forward neural networks with the D-optimum input sequence. IEEE Trans. Neural Netw. **17**(2), 357–373 (2006)
22. Witczak, M.: Fault Diagnosis and Fault-Tolerant Control Strategies for Non-Linear Systems: Analytical and Soft Computing approaches. Springer International Publishing, Heidelberg, Germany (2014)
23. Yang, H., Wang, H.: Robust adaptive fault-tolerant control for uncertain nonlinear system with unmodeled dynamics based on fuzzy approximation. Neurocomputing **173**, Part 3, 1660–1670 (2016)
24. Yao, L., Feng, L.: Fault diagnosis and fault tolerant tracking control for the non-Gaussian singular time-delayed stochastic distribution system with PDF approximation error. Neurocomputing **175**, Part A, 538–543 (2016)
25. Zemouche, A., Boutayeb, M., Bara, G.I.: Observers for a class of lipschitz systems with extension to $H_\infty$ performance analysis. Syst. Control Lett. **57**(1), 18–27 (2008)

# Exploring a Mathematical Model of Gain Control via Lateral Inhibition in the Antennal Lobe

Aaron Montero[1](✉), Thiago Mosqueiro[2], Ramon Huerta[1,2],
and Francisco B. Rodriguez[1](✉)

[1] Grupo de Neurocomputación Biológica, Dpto. de Ingeniería Informática,
Escuela Politécnica Superior, Universidad Autónoma de Madrid,
28049 Madrid, Spain
aaron.montero.m@gmail.com, f.rodriguez@uam.es
[2] BioCircuits Institute, University of California,
San Diego, La Jolla, CA 92093-0402, USA

**Abstract.** Bioinspired Neural Networks have in many instances paved the way for significant discoveries in Statistical and Machine Learning. Among the many mechanisms employed by biological systems to implement learning, gain control is a ubiquitous and essential component that guarantees standard representation of patterns for improved performance in pattern recognition tasks. Gain control is particularly important for the identification of different odor molecules, regardless of their concentration. In this paper, we explore the functional impact of a biologically plausible model of the gain control on classification performance by representing the olfactory system of insects with a Single Hidden Layer Network (SHLN). Common to all insects, the primary olfactory pathway starts at the Antennal Lobes (ALs) and, then, odor identity is computed at the output of the Mushroom Bodies (MBs). We show that gain-control based on lateral inhibition in the Antennal Lobe robustly solves the classification of highly-concentrated odors. Furthermore, the proposed mechanism does not depend on learning at the AL level, in agreement with biological literature. Due to its simplicity, this bioinspired mechanism may not only be present in other neural systems but can also be further explored for applications, for instance, involving electronic noses.

**Keywords:** Gain control · Concentration invariance · Neural networks · Olfactory system · Antennal lobe · Pattern recognition

## 1 Introduction

Neuroscience has always intrinsically operated at the intersection of many hard and soft sciences, with a balancing application and fundamental research. Many statistical methods have clear biological inspirations [3,11,13], which are then applied back to relevant scientific problems [1,23,28]. In particular, Neural Networks presented a steep growth of interest due to recent success of deep and wide neural networks [6,14]. Since the recent discovery of the power of using neural

© Springer International Publishing AG 2017
I. Rojas et al. (Eds.): IWANN 2017, Part I, LNCS 10305, pp. 317–326, 2017.
DOI: 10.1007/978-3-319-59153-7_28

networks for pre-processing and feature extraction [7], new exciting developments have been proposed to a wide range of applications in Machine Learning.

Classification of time series recorded from volatiles is one of the applications that have benefited from Neural Networks [3, 4]. Many insects were shown to excel at recognizing patterns based on olfactory information [37], and with response fast response times [20, 36]. The main properties and features of the olfactory pathways of most insects can be modeled as a Single Hidden Layer Network (SHLN) [8, 12], as described in Fig. 1. Odor volatiles trigger waves of excitation in the Olfactory Receptor Neurons (ORNs), which send direct afferents to the Antennal Lobe (AL). In the Antennal Lobe, the Projection Neurons (PNs) and Local Neurons (LNs) operate to extract features [16]. Then, PNs project onto the Kenyon Cells (KCs), in the Mushroom Bodies (MBs). Odor-rewarded information then happens at the Mushroom Body Output Neurons (MBONs). From the point of view of Machine Learning, Hebbian learning takes place between KCs and MBONs [8]. However, one of the key properties employed by the MBs it the sparse code at the KC level [26]. Because part of the LNs are inhibitory, it is possible that one of the operations performed at the AL is to balance the olfactory signal to different levels of concentration [2, 24, 34]. Several studies suggest that this independence over concentration is achieved during the pre-processing phase of olfactory system, by a gain control mechanism [22, 25, 31, 33].

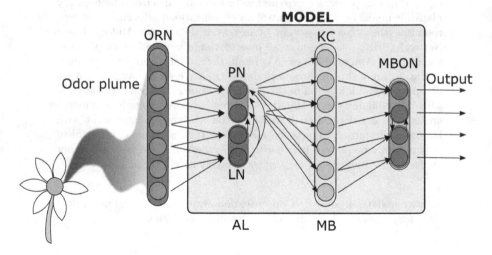

**Fig. 1.** Representation of the model based on a Single Hidden Layer Network (SHLN). Odor volatiles induce currents in the Olfactory Receptor Neurons (ORNs), which relay information into the Antennal Lobe (AL). Our model considers the Projection Neurons (PNs) in the AL as the input layer. Local Neurons (LNs) makes excitatory and inhibitory connections to PNs. Then the information is sent to the Kenyon Cells (KCs), located at the Mushroom Bodies (MBs). Finally, odor identity is encoded at the Mushroom Body Output Neurons (MBONs). Following the biological literature, Hebbian learning is only applied to the connections between KCs and MBONs.

In this paper, we propose an exploratory analysis of a bioinspired mathematical approximation of this gain control mechanism. To model the olfactory system of insects we used a SHLN with the connectivity indicated by in Fig. 1. To stimulate the SHLN, we used Gaussian patterns with variable intensities and variances to model odors with different concentrations. Following experimental observations [26], more concentrated odors created more overlap between different classes. Our gain control mechanism improved significantly the classification performance of the SHLN, increasing the robustness of the neural network to larger concentrations. Because this bioinspired mechanism is only based on the lateral inhibition from local interneurons, it is likely that other neural architectures also employ a similar strategy. Furthermore, because this method only requires the addition of a single layer in an artificial neural network, this mathematical can potentially be developed for applications that depend fast response and low energy, such as devices that record time series from chemical sensors [9].

## 2   Methods

This section is divided into three parts. First, we introduce the proposed gain control mechanism. Next, we describe the computational model that implements this gain control. Finally, we describe the patterns used to simulate different concentration, necessary to test this mechanism.

### 2.1   Lateral Inhibition in the Antennal Lobe

We propose a gain control mechanism based on observations from neuroscience of lateral inhibition from ALs into the PNs. This is assumed to subside excessive activity that may burst from the ORNs, generating a codification of stimulus that is independent of concentration [2]. To develop our gain control mechanism, we assumed that the odor information received by PNs and LNs is proportional, which means that, at the population level,

$$\sum_{j=1}^{N_{PN}} PN_j \approx \alpha \sum_{i=1}^{N_{LN}} LN_i, \qquad (1)$$

where $PN_j$ and $LN_i$ are the neural activities of the $j$-th PN neuron and of the $i$-th LN neuron, respectively. Finally, $\alpha$ is ratio between PN and LN populations ($\sim 2.77$ for the locust, which has $\sim 830$ PNs and $\sim 300$ LNs). This coefficient was used to calculate the total activity of the LNs as a function of the activity of the patterns (PNs), shown in Sect. 2.2. We modeled the lateral inhibition using the following non-linear relation,

$$PN_j = PN_j - PN_j \beta \left( \sum_{i=1}^{N_{LN}} LN_i - \delta \right), \quad j = 1, \ldots, N_{PN}. \qquad (2)$$

In the last equation, $\beta \sim 1/\sum_{i=1}^{N_{LN}} LN_i$ is the weight of the inhibitory connection from LNs to PNs, and $\delta$ is a small threshold. Finally, the non-linear multiplication of $\sum_{i=1}^{N_{LN}} LN_i$ and $PN_j$ assures that PNs with lower activities will remain with lower activities.

## 2.2   Definition of the Input Patterns

We tested the classification performance of the proposed gain control mechanism by using artificial stimulation of odors with different identities and concentrations. We used Gaussians with different centers to encode the identity of the odor (Fig. 2a). Two total number of different odor identities were used, 100 and 1000. Thus, each odor identity elicited the activity of spectrum of PN neurons, in agreement with experiments. To represent different concentrations, we scaled the Gaussian by a sigmoidal function (Fig. 2b). This pattern can be described by the following equation:

$$PN_j = \left[\mathcal{A}_0 + \frac{\mathcal{A}}{1 + \left(\frac{K}{\mathcal{C}}\right)^Q}\right] \exp\left(-\frac{(j - j_k)^2}{2\mathcal{C}}\right), \qquad (3)$$

where $\mathcal{C}$ is the concentration, $j_k$ is the center of the $j$-th odor identity, $\mathcal{A}_0 = 0.05$ is the residual activity, $\mathcal{A} = 0.95$ is a parameter that determines the maximum activity, $K = 10$ defined the concentration at which half of the maximum level was achieved, and $Q = 3.2$ defined the slope of the resulting sigmoid. This

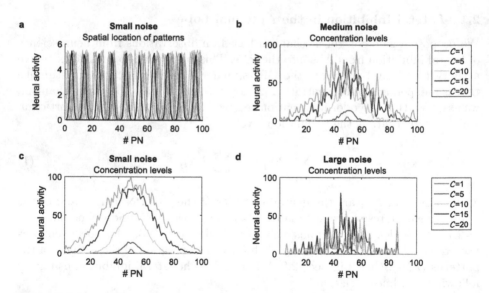

**Fig. 2.** Odor stimulation as represented by Gaussian patterns. (a) Different odor identities are modeled as Gaussians centered at different PN neurons ($\mathcal{C} = 1$). To generate more realistic stimulation, we added noise to Gaussian centers (black) for changing them to another of PNs belonging to the same pattern class (gray). (b) Neural activity variability in function of concentration. (c) Activity in the PN neurons elicited by an odor with small level of noise. (d) Activity in the PN neurons for a larger level of noise.

follows a common psychophysical response of neural populations to different intensities [21].

We used on total five different levels of concentration. Because the sigmoidal function changes the variance of the Gaussian, neighboring PN neurons also become active (in agreement with experiments [29, 30]). Thus, higher concentrations increase the overlap between the activities generated by two different odors. Finally, to simulate the noise experienced in real biological systems, we added a multiplicative, white noise to the Gaussian. We show simulated activity of PN neurons for several values of concentration, for the same odor, with small (Fig. 2c) and large (Fig. 2d) levels of noise.

## 2.3   Neural and Network Model

The computational model is based on a SHLN, following the connectivity described in Fig. 1. We used PNs in the AL as input to the network, KCs as hidden layer, and MBONs the output. The gain control was implemented by a lateral inhibitory connection between LNs and PNs. We selected the number of neurons in each population by matching empirical observations in the MBs of locusts [15].

KCs were implemented as McCulloch-Pitts neurons in all neurons of the hidden and output layers. Let $x_i$ be the state of the $i$-th PN neuron. Then, the state $y_j$ of the $j$-th KC neuron can be written as a function of the PNs,

$$y_j = \varphi \left( \sum_{i=1}^{N_{PN}} c_{ji} x_i - \theta_j \right), \quad j = 1, \ldots, N_{KC}, \tag{4}$$

with $N_{KC} = 5000$ being the number of KCs and $N_{PN} = 100$, the number of PNs. The function $\varphi$ is the Heaviside function. We selected the threshold $\theta_j$ heterogeneously throughout the KCs to improve the classification performance [16, 17] and to maintain the sparsity of KC activity [26]. Finally, $C_{ji}$ was the weight linking the $i$-th PN and $j$-th KC. The connectivity matrix $C$ is determined randomly by independent Bernoulli processes with probability $p_c$ for each existing connection and $1 - p_c$ for each lack of this [5, 8, 10]. The value used for this probability is 0.1, according to [16, 17].

MBONs were trained to identify odors of $N_{class} = 10$ classes, and 5 different concentrations separately. To achieve that, each different class was assigned subgroups of 10 MBONs, where a winner-takes-all mechanism was implemented with lateral inhibition [10, 27, 32]. Thus, on total we simulated $N_{MBON} = 10 \times N_{class} = 100$ MBONs. Thus, the state $z_\ell$ of the $\ell$-th MBON was described as

$$z_\ell = \varphi \left( \sum_{j=1}^{N_{KC}} w_{\ell j} y_j - \frac{1}{N_{class}} \sum_{k=1}^{N_{class}} \sum_{j=1}^{N_{KC}} w_{kj} y_j - \varepsilon_\ell \right), \quad l = 1, \ldots, N_{class}, \tag{5}$$

with $\varepsilon_\ell$ representing the threshold of activation of the $\ell$-th MBON. Because learning only happened between KCs and MBONs, we used a 5-fold cross validation to fit the weights $W_{\ell j}$ using Hebbian learning. Finally, $W$ was initialized as a random matrix with standard normal distribution for the elements, and the updates were controlled by Hebbian probabilities $p_+ = 1$ (strengthening) and $p_- = 0.05$ [8,18].

## 3    Results

The proposed gain control mechanism improved the classification performance, especially for large concentrations (Fig. 3). For lower concentrations, the difference in performance between the network with and without gain control is insignificant. Even for a moderate level of noise (Fig. 3 center), the classification error of the network with gain control is below 10%, while the error of the network without gain control is larger than 20%. If the noise level is further increased (Fig. 3 right), then the performance of both networks drastically drops much. However, the presence of gain control reduces the classification error at least by a fourth (Fig. 3, top right).

We also report that the classification became more robust against noise when the gain control mechanism was applied (Fig. 4). Although the classification error of the network with gain control increases rapidly from medium to large levels of noise, it is always considerable smaller than that of the network without gain

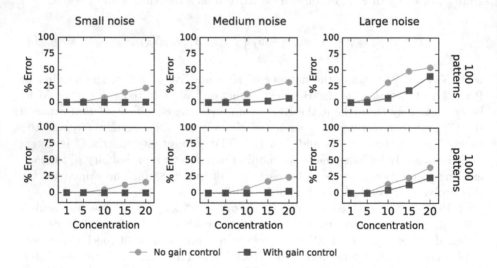

**Fig. 3.** Gain control improves the classification performance, especially for higher concentrations. All results are the average of 10 simulations for the test data set using supervised learning and 5-fold cross-validation. We show the classification performance when either 100 (top) or 1000 (bottom) different patterns were presented to the neural network. Left: With gain control, the classification error rate is insignificant. Center: With moderate levels of noise, the gain control mechanism maintains a reasonable performance. Right: With a high noise, although the performance drops significantly with the concentration, the network that employed gain control is always 10% more accurate than that no gain control.

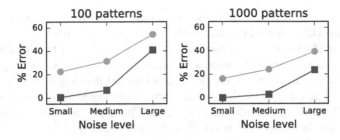

**Fig. 4.** Even for high concentrations, classification using gain control becomes more robust against noise. Both plots show the classification error rate as a function of the level of noise applied to the patterns. To represent the worst case scenario, we used the largest concentration ($\mathcal{C} = 20$).

**Table 1.** Classification errors obtained for higher concentrations and datasets of 1000 patterns. Table with same results from right panel of Fig. 4. First column determines the presence of the gain-control mechanism, and the following columns represent the various levels of noise tested.

| Has gain-control? | Low noise | Medium noise | Large noise |
| --- | --- | --- | --- |
| No | 16.35 | 24.22 | 39.41 |
| Yes | 0.05 | 2.91 | 23.73 |

control. For the largest concentration and with 1000 patterns, the difference between the classification error between using and not using gain control is always greater than 15% (see Table 1 for numerical values). Furthermore, in our simulations with highest level of noise, the classification error of the network with gain control is half the error without gain control.

## 4 Discussions

We studied a mathematical model of gain-control mechanism based on the connectivity presented by the olfactory system of insects. This system is able to recognize the identity of an odorant across a range of concentrations. Our gain control used the known the lateral inhibition that exists in the Antennal lobe (AL) [24], where many researchers believe that a concentration-invariant coding is created [31,33,34]. We showed that our gain control mechanism outperformed another network with the same connectivity but without gain control. Even with large amounts of noise applied to the artificial odor stimuli, the gain control was able to enhance the classification performance by almost 10%.

Our gain control mechanism was able to solve the classification of highly concentrated odors robustly. Because the concentration essentially creates a substantial overlap in the representation of odors from various classes, the performance of classifiers without any gain-control is impaired. Thus, reducing the overlap in the representation of odorants, implementing gain control does not only suppress outbursts of activity from input layers but also robustly improves learning in Mushroom Bodies. To achieve this, we assumed that the population activity of Local Neurons (LNs) scaled linearly with the Projection Neurons (PNs) when the concentration of the odor increased. Furthermore, although the inhibition from LNs must counterbalance the excess of excitation from the Olfactory Receptor Neurons (ORNs), it cannot shut down the communication between these two populations. Thus, evolution must have driven the equilibrium between excitation of PNs by ORNs and inhibition by LNs. There are, however, other possible gain control mechanisms that need to be further studied [35].

Sound and reliable pre-processing bring many direct advantages to biological systems. The existence of inhibitory local interneurons is ubiquitous in many areas of the brain, and in many cases these small populations of neurons may be controlling activity levels [19]. Also, because this gain control condition can be efficiently implemented, by only adding one extra neural population, this solution may also be of interest to applications. One direct application would be arrays of chemical sensors, the resistance of which increases with the concentration of a given chemical [38]. From the point of view of Machine Learning, this methodology only adds two new meta-parameters that needs to be optimized.

**Acknowledgments.** This research was supported by the Spanish Government projects TIN2010-19607 and TIN2014-54580-R, the predoctoral research grant BES-2011-049274, NIH grant R01GM113967 and CNPq grant 234817/2014-3.

# References

1. Brito, J.J., Mosqueiro, T., Ciferri, R.R., de Aguiar Ciferri, C.D.: Faster cloud star joins with reduced disk spill and network communication. Procedia Comput. Sci. **80**, 74–85 (2016)
2. Cleland, T.A., Chen, S.-Y.T., Hozer, K.W., Ukatu, H.N., Wong, K.J., Zheng, F.: Sequential mechanisms underlying concentration invariance in biological olfaction. Front. Neuroeng. **4**, 21 (2011)
3. Diamond, A., Nowotny, T., Schmuker, M.: Comparing neuromorphic solutions in action: implementing a bio-inspired solution to a benchmark classification task on three parallel-computing platforms. Front. Neurosci. **9**, 491 (2015)
4. Fonollosa, J., Sheik, S., Huerta, R., Marco, S.: Reservoir computing compensates slow response of chemosensor arrays exposed to fast varying gas concentrations in continuous monitoring. Sens. Actuators B: Chem. **215**, 618–629 (2015)
5. Garcia-Sanchez, M., Huerta, R.: Design parameters of the fan-out phase of sensory systems. J. Comput. Neurosci. **15**, 5–17 (2003)
6. Haykin, S.: Neural Networks: A Comprehensive Foundation, 2nd edn. Prentice Hall, Upper Saddle River (1999)
7. Hinton, G.E., Osindero, S., Teh, Y.-W.: A fast learning algorithm for deep belief nets. Neural Comput. **18**(7), 1527–1554 (2006)
8. Huerta, R., Nowotny, T., Garcia-Sanchez, M., Abarbanel, H.D.I., Rabinovich, M.I.: Learning classification in the olfactory system of insects. Neural Comput. **16**, 1601–1640 (2004)
9. Huerta, R., Mosqueiro, T., Fonollosa, J., Rulkov, N.F., Rodriguez-Lujan, I.: Online decorrelation of humidity and temperature in chemical sensors for continuous monitoring. Chemometr. Intell. Lab. Syst. **157**, 169–176 (2016)

10. Huerta, R., Nowotny, T.: Fast and robust learning by reinforcement signals: explorations in the insect brain. Neural Comput. **21**, 2123–2151 (2009)
11. Huerta, R., Vembu, S., Amigó, J.M., Nowotny, T., Elkan, C.: Inhibition in multiclass classification. Neural Comput. **24**(9), 2473–2507 (2012)
12. Ito, K., Shinomiya, K., Ito, M., Armstrong, J.D., Boyan, G., Hartenstein, V., Harzsch, S., Heisenberg, M., Homberg, U., Jenett, A., Keshishian, H., Restifo, L.L., Rössler, W., Simpson, J.H., Strausfeld, N.J., Strauss, R., Vosshall, L.B.: A systematic nomenclature for the insect brain. Neuron **81**(4), 755–765 (2014)
13. Dhinesh Babu, L.D., Venkata Krishna, P.: Honey bee behavior inspired load balancing of tasks in cloud computing environments. Appl. Comput. **13**(5), 2292–2303 (2013)
14. LeCun, Y., Bengio, Y., Hinton, G.: Deep learning. Nature **521**(7553), 436–444 (2015)
15. Leitch, B., Laurent, G.: GABAergic synapses in the antennal lobe and mushroom body of the locust olfactory system. J. Comp. Neurol. **372**, 487–514 (1996)
16. Montero, A., Huerta, R., Rodríguez, F.B.: Neuron threshold variability in an olfactory model improves odorant discrimination. In: Ferrández Vicente, J.M., Álvarez Sánchez, J.R., de la Paz López, F., Toledo Moreo, F.J. (eds.) IWINAC 2013. LNCS, vol. 7930, pp. 16–25. Springer, Heidelberg (2013). doi:10.1007/978-3-642-38637-4_3
17. Montero, A., Huerta, R., Rodriguez, F.B.: Regulation of specialists and generalists by neural variability improves pattern recognition performance. Neurocomputing **151**, 69–77 (2015)
18. Montero, A., Huerta, R., Rodriguez, F.B.: Specialist neurons in feature extraction are responsible for pattern recognition process in insect olfaction. In: Ferrández Vicente, J.M., Álvarez-Sánchez, J.R., de la Paz López, F., Toledo-Moreo, F.J., Adeli, H. (eds.) IWINAC 2015. LNCS, vol. 9107, pp. 58–67. Springer, Cham (2015). doi:10.1007/978-3-319-18914-7_7
19. Mosqueiro, T., de Lecea, L., Huerta, R.: Control of sleep-to-wake transitions via fast amino acid and slow neuropeptide transmission. New J. Phys. **16**(11), 115010 (2014)
20. Mosqueiro, T., Strube-Bloss, M., Tuma, R., Pinto, R., Smith, B.H., Huerta, R.: Non-parametric change point detection for spike trains. In: 2016 Annual Conference on Information Science and Systems (CISS), pp. 545–550. IEEE, March 2016
21. Mosqueiro, T.S., Maia, L.P.: Optimal channel efficiency in a sensory network. Phys. Rev. E **88**(1), 12712 (2013)
22. Nowotny, T., Huerta, R., Abarbanel, H.D.I., Rabinovich, M.I.: Self-organization in the olfactory system: rapid odor recognition in insects. Biol. Cyber. **93**, 436–446 (2005)
23. Nowotny, T., Huerta, R.: On the equivalence of Hebbian learning and the SVM formalism. In: 2012 46th Annual Conference on Information Sciences and Systems (CISS), pp. 1–4. IEEE, March 2012
24. Olsen, S.R., Wilson, R.I.: Lateral presynaptic inhibition mediates gain control in an olfactory circuit. Nature **452**(7190), 956–960 (2008)
25. O'Reilly, R.C.: Generalization in interactive networks: the benefits of inhibitory competition and Hebbian learning. Neural Comput. **13**(6), 1199–1241 (2001)
26. Perez-Orive, J., Mazor, O., Turner, G.C., Cassenaer, S., Wilson, R.I., Laurent, G.: Oscillations and sparsening of odor representations in the mushroom body. Science **297**(5580), 359–365 (2002)
27. Rabinovich, M.I., Huerta, R., Volkovskii, A., Abarbanel, H.D., Stopfer, M., Laurent, G.: Dynamical coding of sensory information with competitive networks. J. Physiol. Paris **94**(5–6), 465–471 (2000)

28. Rodriguez-Lujan, I., Hasty, J., Huerta, R.: FBB: a fast Bayesian-bound tool to calibrate RNA-seq aligners. Bioinformatics **33**(2), 210–218 (2017)

29. Rubin, J.E., Katz, L.C.: Optical imaging of odorant representations in the mammalian olfactory bulb. J. Neurophysiol. **23**, 449–511 (1999)

30. Sachse, S., Galizia, C.G.: The coding of odour-intensity in the honeybee antennal lobe: local computation optimizes odour representation. Eur. J. Neurosci. **18**(8), 2119–2132 (2003)

31. Salinas, E., Thier, P.: Gain modulation: a major computational principle of the central nervous system. Neuron **27**, 15–21 (2000)

32. Schürmann, F.W., Frambach, I., Elekes, K.: Gabaergic synaptic connections in mushroom bodies of insect brains. Acta Biol. Hung. **59**, 173–181 (2008)

33. Serrano, E., Nowotny, T., Levi, R., Smith, B.H., Huerta, R.: Gain control network conditions in early sensory coding. PLoS Comput. Biol. **9**(7), e1003133 (2013)

34. Stopfer, M., Jayaraman, V., Laurent, G.: Intensity versus identity coding in an olfactory system. Neuron **39**, 991–1004 (2003)

35. Stopfer, M.: Central processing in the mushroom bodies. Curr. Opin. Insect Sci. **6**, 99–103 (2014). Pests and resistance/Parasites/Parasitoids/Biological control/Neurosciences

36. Strube-Bloss, M.F., Herrera-Valdez, M.A., Smith, B.H.: Ensemble response in mushroom body output neurons of the honey bee outpaces spatiotemporal odor processing two synapses earlier in the antennal lobe. PLoS ONE **7**(11), e50322 (2012)

37. Strube-Bloss, M.F., Nawrot, M.P., Menzel, R.: Mushroom body output neurons encode odor-reward associations. J. Neurosci. Official J. Soc. Neurosci. **31**(8), 3129–3140 (2011)

38. Trincavelli, M., Vergara, A., Rulkov, N., Murguia, J.S., Lilienthal, A., Huerta, R.: Optimizing the operating temperature for an array of mox sensors on an open sampling system. AIP Conf. Proc. **1362**, 225 (2011)

# Optimal Spherical Separability: Artificial Neural Networks

Rama Murthy Garimella[✉], Ganesh Yaparla, and Rhishi Pratap Singh

International Institute of Information Technology, Hyderabad, India
rammurthy@iiit.ac.in, {ganesh.yaparla,
rhishi.pratap}@research.iiit.ac.in

**Abstract.** In this research paper, the concept of hyper-spherical/hyper-ellipsoidal separability is introduced. Method of arriving at the optimal hypersphere (maximizing margin) separating two classes is discussed. By projecting the quantized patterns into higher dimensional space (as in encoders of error correcting code), the patterns are made hyper-spherically separable. Single/multiple layers of spherical/ellipsoidal neurons are proposed for multi-class classification. An associative memory based on hyper-ellipsoidal neuron is proposed.

## 1 Introduction

In an effort to model the biological neural network, perceptron provided an important beginning. Rosenblatt proved the convergence theorem associated with the learning law, when the patterns are linearly separable. The notion of linear separability provided the conceptual basis for statistical learning theory based on support vector machines developed by Vapnick et al. Specifically non-linearly separable patterns are mapped to higher dimension space where they become linearly separable by means of suitable kernel. This approach provided a method of arriving at a feature space where the classification is rendered easy. To progress the investigation, the notion of circular, spherical and hyper-spherical separability concepts are introduced. Using such concept optimal circular/spherical/hyper-spherical separating decision manifolds in 2-class case that maximizes the margin are derived. A novel method of multi-state neuron, called Spherical Neuron is proposed. It is reasoned that such a neuron enables classification of certain type of multiple classes(i.e. structured multi-class classification). Efforts are underway to train single/multi-layer networks of such neurons.

## 2 Circular/Spherical/Hyper-Spherical Separable Patterns: Optimal Separating Circle/Sphere/ Hyper-Sphere

### 2.1 Motivation for Spherical Separability

Pattern's are said to be hyper-spherically separable if there exists a hyper-sphere which separates two classes. While, pattern's are said to be linearly separable if there exists a hyperplane which separates two classes.

© Springer International Publishing AG 2017
I. Rojas et al. (Eds.): IWANN 2017, Part I, LNCS 10305, pp. 327–338, 2017.
DOI: 10.1007/978-3-319-59153-7_29

A necessary condition for linear separability of pattern's belonging to two classes to imply spherical separability is that all the patterns belonging to at-least one of the classes are bounded in $(L)^2$-norm.

**Lemma:** Under the above condition linear separability implies spherical separability.

**Proof:** Consider the patterns belonging to the class which is bounded. The centroid of the pattern is computed using MVEE (minimum volume enclosing ellipsoid).

Since the patterns belonging to two classes are linearly separable, there exists a hyperplane (not necessarily unique) which separates them. Using the above computed center, consider a hypersphere which is tangential to the hyperplane (in the worst case). Such a hypersphere hyper-spherically separates the patterns belonging to 2-classes.

## 2.2  Two Class Classification of Circularly/Spherically/ Hyper-Spherically Separable Patterns in 2/3/N Dimensional Space

Notations used in the section are as following.

- $\omega_1$ and $\omega_2$: classes which are being separated.
- $X$ and $Y$: data points belonging to classes $\omega_1$ and $\omega_2$ respectively. For N dimension case $X, Y \in \mathcal{R}^N$. Similarly for 2 dimension $X, Y \in \mathcal{R}^2$.
- Point $C(c_1, c_2, \ldots, c_N)$: center of the circle/sphere/hyper-sphere which divides the pattern in two classes. For N dimension case $C \in \mathcal{R}^N$.
- $d_{max}, d_{min}$: farthest and closest distances from $C$ of points in $\omega_1$ and $\omega_2$ respectively.
- $C$ is the centre of the MVEE of the class which is bounded. We use Khachiyan's algorithm [6] for the computation of MVEE.

**2-D Separation Case.** Patterns are circularly separable in 2 dimension if there exists a circle which can separate both the classes.

Let $\omega_1, \omega_2$ be circularly separable. There exists a $C \in \mathcal{R}^2$ i.e. $C(c_1, c_2)$. The optimal circle which separates classes is at distance $(d_{max} + d_{min})/2$ from $C$. When a new data point $Z(p, q)$ is given for classification the decision is taken using following function

$$Z(p, q) = \begin{cases} \omega_1 \text{ if } (p - c_1)^2 + (q - c_2)^2 < (d_{max} + d_{min})/2 \\ \omega_2 \text{ otherwise} \end{cases} \tag{1}$$

**3-D Separation Case.** Patterns are spherically separable if there exists a sphere which can separate both the classes.

Let $\omega_1, \omega_2$ be spherically separable. There exists a $C \in \mathcal{R}^3$ i.e. $C(c_1, c_2, c_3)$. The optimal sphere which separates classes is at distance $(d_{max} + d_{min})/2$ from $C$. When a new data point $Z(p, q, r)$ is given for classification the decision is taken using following function

$$Z(p, q, r) = \begin{cases} \omega_1 \text{ if } (p - c_1)^2 + (q - c_2)^2 + (r - c_2)^2 < \\ \quad (d_{max} + d_{min})/2 \\ \omega_2 \text{ otherwise} \end{cases} \quad (2)$$

**N-D Separation Case.** Patterns are hyper-spherically separable if there exists a hyper-sphere which can separate both the classes.

Let $\omega_1, \omega_2$ be hyper-spherically separable. There exists a $C \in \mathcal{R}^N$ i.e. $C(c_1, c_2, \ldots, c_N)$. The optimal hyper-sphere which separates classes is at distance $(d_{max} + d_{min})/2$ from $C$. When a new data point $Z(z_1, z_2, \ldots, z_N)$ is given for classification the decision is taken using following function

$$Z(z_1, z_2, \ldots, z_N) = \begin{cases} \omega_1 \text{ if } (z_1 - c_1)^2 + (z_2 - c_2)^2 + \ldots + \\ \quad (z_N - c_N)^2 < (d_{max} + d_{min})/2 \\ \omega_2 \text{ otherwise} \end{cases} \quad (3)$$

It is clear that if patterns belonging to two classes are linearly separable, they are hyper-spherically separable. But hyper-spherical separability does not imply linear separability (by a hyperplane). For instance, if the patterns belonging to two classes spherically symmetric about the origin, they are clearly not linearly separable.

In the spirit of SVM, we have found the optimal hypersphere separating two classes. Training patterns belonging to classes are presented serially one class after the other, the distance from center (of one class) is varied with every training pattern. Optimal hypersphere is computed for 2-class problem. It readily applies for Multi-class case based on one against rest approach.

### 2.3 Multi Class Classification of Circularly/Spherically/ Hyper-Spherically Separable Patterns in 2/3/N Dimensional Space

Notations for the section are as following.

- $\omega_1, \omega_2, \ldots, \omega_M$: $M$ classes which are being separated. Also $\omega = \omega_1 \cup \omega_2 \cup \ldots \cup \omega_M$.
- $X_i$: data points belonging to classes $i \in (1, M)$. For N dimension case $X \in \mathcal{R}^N$.
- Point $C_i(c_{i1}, c_{i2}, \ldots, c_{iN})$: center of the circle/sphere/hyper-sphere which divides the pattern in $i \in (1, M)$ classes. For N dimension case $C \in \mathcal{R}^N$.
- A class $t_i$ where $i \in (1, M)$ is introduced which contains all the points which lie inside circle/sphere/hyper-sphere by which $\omega_i$ is enclosed. Also $t = t_1 \cup t_2 \cup \ldots \cup t_M$.
- $d_{i1}, d_{i2}$: farthest and closest distances from $C_i$ of points in $t_i$ and $t - t_i$ respectively.

**2-D Case.** Let $\omega_1, \omega_2, \ldots, \omega_M$ be $M$ classes which are circularly separable. We use one vs rest approach to classify the data points. The optimal circle which separates classes is at distance $(d_{i1} + d_{i2})/2$ from the center. When a new data point $Z(a, b)$ is given for classification, then

$$Z(a, b) \in t_i \text{ if } (a - c_{ix})^2 + (b - c_{iy})^2 < (d_{i1} + d_{i2})/2 \quad \forall i \in (1, n) \qquad (4)$$

Z(a,b) belongs to one of $w_x$ in $t_i$, which can be found using distance based algorithms. Similar approach can be followed for 3-D and N-D cases.

It is important to note that, in the theory of error correcting codes, an information word is mapped to the associated codeword using an encoder. Also the coding spheres at hamming distances less than or equal to the minimum distance of the code are disjoint. Using this idea, we project quantized patterns from lower dimension space (in the spirit of SVMs), where they become spherically separable. In the following section we summarize the known results from earlier literature.

Vapnick's idea was to project non-linearly separable patterns into a higher dimensional space to render them linearly separable. This idea motivated the authors to see if certain non-hyper-spherically separable patterns can be rendered as hyper-spherically separable patterns without projecting them into a higher dimensional space. Details are provided in the following section.

## 3   Linear Transformation of Non-hyper-Spherically Separable Patterns to Spherically Separable Patterns: Quadratic Neuron

Traditionally, single artificial neuron called perceptron was based on the concept of linear separability. Rosenblatt proposed a learning law which converges (i.e. the synaptic weights converge), when the patterns are linearly separable. The resulting hyperplane is one among various possible hyperplanes that separate the patterns into two classes.

Vapnick, by introducing the concept of margin, showed that the problem of synthesizing optimal hyperplane (i.e. a hyperplane which maximizes the margin) separating two classes can be formulated as a Quadratic optimization problem.

These two approaches remained as the basis for research related to artificial neural networks (e.g. classification problem).

The authors contemplated on the possibility of combining the logical basis of above two approaches for classification. They succeeded in such an effort by introducing hyper-spherical separability concept. The details are summarized below.

In McCulloh-Pitts neuron, the net contribution is computed using the inner product of weight vector and the vector of the inputs. This net contribution is operated on by signum activation function, to arrive at the neuron output. Such a model of neuron is utilized to classify linearly separable patterns (by a hyperplanes). Generalizing this idea, several researchers proposed a neuron where higher order synaptic operations (e.g. quadratic synaptic operations) are utilized to arrive at the net contribution which is operated on by signum activation function [5].

In such a neuron model, the activation function is retained as signum function. It is thus clear that such models of neuron classify non-linearly separable patterns. Specifically, let $W$ be a symmetric $M \times M$ matrix and $\bar{X}$ be a $M \times 1$ vector of inputs. The output of neurons

$$y = signum\{\bar{X}^T W \bar{X} - T_0\} \tag{5}$$

Here $T_0$ is a threshold value.

**Assumption:** $W$ be a positive symmetric matrix. Hence by cholesky decomposition we have $W = NN^T$, where N is a

Applying it in Eq. (5)

$$\bar{X}^T W \bar{X} = \bar{X}^T N N^T \bar{X} = Z^T Z = \sum_{i=1}^{M} z_i^2 \tag{6}$$

where $Z = N^T \bar{X}$. Thus output of such a neuron is given by

$$y = signum\{\sum_{i=1}^{M} z_i^2 - T_0\} \tag{7}$$

**Claim:** The patterns arrived at by the above linear transformation are hyper-spherically separable.

Note that using above idea, first documented in research monograph [7], NP-hard problem of maximum cut computation is reduced to multi-linear objective function optimization over hypercube [9].

It is well known that homogeneous multivariate polynomial (of degree higher than two) can be expressed in terms of symmetric tensor. Using cholesky type decomposition of symmetric tensor, the results in this section can be generalized.

The approach proposed in this section naturally leads to the idea of transforming the patterns by a non-linear transformation (when they are separable by certain manifold) such that they become spherically separable (without projecting to higher dimensions).

In the view of discussion in Sects. 2 and 3, a natural question arises whether the patterns which are not hyper-spherically separable can be projected to higher dimension where they are hyper-spherically separable (in spirit of SVM design policy). This issue is addressed in the following section. Patterns are preprocessed to ensure that they can be encoded (projected into higher dimension) into suitable codewords of an error correcting code. It is clear that the patterns in various coding spheres are hyper-spherically separable. Such a procedure ensures noise robustness for patterns classification. The following correspondence is identified between pattern classification and coding theory. Here patterns correspond to information words, class centers correspond to codewords and clusters correspond to coding spheres.

In the following section, we briefly refer to some earlier work of Bruck et al. relating associative memories to error correcting codes. Also, we generalize the concept of spherical separability resulting in spherical neuron for multiclass classification.

## 4    Hybrid Neural Networks: Spherical Separability: Spherical Neuron

Bruck and Blaum have shown that hopfield neural network is naturally associated with a graph-theoretical code in the sense that code words are associated with stable states [2]. They generalize the result to linear error correcting codes and non-linear error correcting codes. Effective code words are associated with stable states and vice-versa in the sense that they are the local/global optima of an energy function (associated with the encoder of an error correcting code). Thus effectively a one step associate memory (realized by encoder) performs clustering of data points. In [1], one of the authors proposed hybrid neural networks where encoders are cascaded with multi-layer perceptron, a feedforward network. It is clear that the patterns in different coding spheres are spherically separable. Figure 1 depicts the idea.

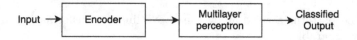

**Fig. 1.** Hybrid neural networks

The elements of pattern vectors can always be quantized such that they belong to the set $\{0, 1, 2, \ldots, p-1\}$ for a suitable chosen prime $p$. Such patterns constitute the information words (as in theory of error correcting codes). They are projected to a higher dimensional space by a suitable encoder to render them hyper-spherically separable. The minimum distance of the block code determines the radius of hamming sphere/euclidean sphere around each code word that renders the codewords hyper-spherically separable. Currently we are investigating the finer details of such approach.

### Spherical Neuron: Multi-class Classification
In the above discussion, in two dimension circularly separable patterns (belonging to two classes) are classified. More generally, hyper spherically separable classes are classified into 2-classes in higher dimensions. Motivated by the idea of ceiling neuron [8], we introduce a novel neuron called Spherical neuron, which performs classification of patterns belonging to multiple classes with certain restrictions/structure. The details of such neuron are presented below.

The inputs belong to $M$ classes. They are clustered around the center $(a_1, a_2, \ldots, a_N) = \bar{A}$. It could be the origin of N-dimensional euclidean space. Let an input vector be $(x_1, x_2, \ldots, x_N)$. The output class $y$ is determined in following manner

$$y = \begin{cases} \text{class 1 if } d(\bar{X}, \bar{A}) \leq r_1 \\ \text{class 2 if } d(\bar{X}, \bar{A}) \leq r_2 \\ \vdots \\ \text{class } M \text{ if } d(\bar{X}, \bar{A}) \geq r_M \end{cases} \qquad (8)$$

Here $d(\bar{X}, \bar{A})$ is the euclidean distance between the vectors $\bar{X}$ and $\bar{A}$. Note here that spherical neuron activation function has resemblance to the neurons utilized in radial basis function networks. It is multi-valued in this case.

Several such neurons are placed in one or more layers to perform fine classification of input patterns.

We can conceive of Ellipsoidal neuron, where the distance $d(\bar{X}, \bar{A})$ is computed in the following manner (with $\bar{A} = \bar{0}$).

$$d(\bar{X}, \bar{0}) = (\frac{x_1}{b_1})^2 + (\frac{x_2}{b_2})^2 + \ldots + (\frac{x_N}{b_N})^2 \qquad (9)$$

Single or multiple layers of ellipsoidal neurons are utilized for fine classification of input data.

Generalizing the notion of spherical neuron for multiclass classification, we propose hyper-ellipsoidal neuron in the following section. Also, associative memory based on such a neuron in proposed.

# 5    Error Correcting Codes: Clustering

The problem of clustering of patterns is well studied in pattern recognition literature and various interesting algorithms such as K-means algorithm are proposed. In most of these algorithms the pattern assume values in euclidean space. Some literature exists on clustering of patterns whose coordinates/components assume integer values (i.e. say vector space over a finite field).

We realized that linear/nonlinear error correcting codes enable clustering of patterns after encoding in a higher dimensional space in the sense of hamming distance. We recognized that such an approach could provide insights into multiclass classification of patterns. To facilitate such an approach the components of patterns must be quantized into integer values which can always be done by choosing the integer value to be a sufficiently large prime number.

Our goal is to ensure that patterns which are not hyper-spherically separable in lower dimensional space be made hyper-spherically separable (in the hamming/euclidean distance) in a higher dimensional space by encoding. We expect that such a goal can always be achieved. There are effectively two approaches.

1. **Covering class region with hyper-spheres and clustering in pattern space:** The class region in pattern space need not be hyper-spherical. But, it

can always be covered with hyper-spheres. Using well known or new clustering algorithms, centroids of clusters/sub-clusters are determined.

These centroids, which form the information words in the coding theory sense, are encoded into codewords/patterns in higher dimensional space. Using the minimum distance of error correcting code, it can always be reasoned that the classes in higher dimensional space are hyper-spherically separable under some conditions.

2. **Pattern encoding and clustering in codeword space:** Patterns in lower dimensional space constitute information words in the coding theory sense. They are encoded into codewords (by projecting into higher dimensional space) by the encoder of a suitable linear/nonlinear error correcting code. The code words are suitably clustered in such a way that the minimum distance increases. Hence, the patterns in higher dimensional space become hyper-spherically separable.

Now, we briefly illustrate the idea of achieving hyper-spherical separability using an error correcting code. Let the pattern vector components be quantised using the alphabet $\{0, 1\}$. For the sake of example, let the pattern vector be the $1 \times 4$ row vector. Also, let the 16 possible such pattern vectors correspond to 16 different classes (i.e. centroids of classes). Now, we employ the encoder of a $(7, 4)$ hamming code to map such pattern vectors (information words) into codewords. It is well known that the minimum distance of $(7, 4)$ hamming code is 3. Hence hamming/euclidean hyper-spheres of radius one centered around the codewords are disjoint. Thus by projecting the patterns 4-dimensional euclidean space to 7-dimensional euclidean space using the encoder we made the patterns hyper-spherically separable.

In general, an $(n, k)$ block code with minimum distance $2t + 1$ over $\{0, 1\}$ alphabet can be utilized to make $k$ dimensional binary pattern vectors (which are not hyper-spherically separable) belonging to different $2^k$ classes to be hyper-spherically separable in 7-dimensional space. The euclidean spheres of radius $\sqrt{t}$(or Hamming spheres of radius $t$) around the codewords are disjoint.

The above approach ensures that the classification is noise tolerant/robust (enabling correction of t errors). We expect that by means of finer quantization, patterns which are not hyper-spherically separable can always be rendered hyper-spherically separable by means of a suitable encoder under some conditions. We are actively investigating this approach for pattern classification using hyper-spherical separability.

# 6    Hyper-Ellipsoidal Neuron: Associative Memory

Hopfield proposed an associative memory [3] based on McCulloch-Pitts neuron (which assumes +1 or −1 values). In literature, associative memories are proposed based on multi state neuron [4]. In this research paper, we propose an associative memory based on hyper-ellipsoidal neuron. Let the state space of such an associative memory be the bounded lattice i.e. each component of the state vector, $\bar{V}(n)$ assumes values in the set $\{0, 1, 2, \ldots, M\}$. Let there be $N$ neurons and let the symmetric synaptic weight matrix be $\bar{W}$. The $i$'th component of the state vector at time $n + 1$ is computed in the following manner

$$V_i(n + 1) = f(net) \tag{10}$$

where

$$net = w_{i1}V_1^2(n) + w_{i2}V_2^2(n) + \ldots + w_{iN}V_N^2(n)$$

$$f(net) = \begin{cases} 0 \text{ if } net < T_1, Threshold \\ 1 \text{ if } T_1 \leq net < T_2 \\ \vdots \\ M - 1 \text{ if } T_{M-1} \leq net < T_M \\ M \text{ if } net \geq T_M \end{cases}$$

As in case of Hopfield associative memory, the neural network is operated in the serial mode (state of only one neuron is updated at a given time i.e. asynchronously) or fully parallel mode (state of all the neurons is updated at any given time i.e. fully synchronously).

It is clear that the network dynamics is periodic (because of choice of $f(.)$). Convergence to stable state (i.e. cycle of length one) or a cycle of certain length is currently being investigated. Energy function based approach to investigate nature of dynamics is being currently pursued.

## 7    Experiments and Results

Following are the results for optimal circular separation case. Two circularly separable classes, having 1000 data points were generated randomly. We have compared our method against SVM with linear, polynomial and rbf kernels and K nearest neighbors algorithm with k = 3. The result is repeated with different noise level (mixing of classes). Figures 2 and 3 show the graphical representation of the results.

**Table 1.** Comparison of classification results among SVM with polynomial, rbf and linear kernels, KNN (n = 3) and our procedure. Numbers represent the accuracy at various noise levels.

| Error | .1 | .15 | .2 | .22 |
|---|---|---|---|---|
| Polynomial | 0.5350 | 0.5749 | 0.4799 | 0.4774 |
| RBF | 1.0 | 0.9825 | 0.9575 | 0.9375 |
| Linear | 0.6474 | 0.4799 | 0.5324 | 0.4774 |
| KNN | 1.0 | 0.9825 | 0.9499 | 0.9425 |
| Our | 1.0 | 0.9840 | 0.9500 | 0.9425 |

Table 1 and Fig. 4 show the comparison among various methods. It is clear that our method provides competitive performance with simple computation.

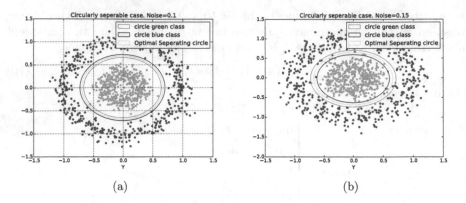

(a)                                                      (b)

**Fig. 2.** Classification with low pattern noise

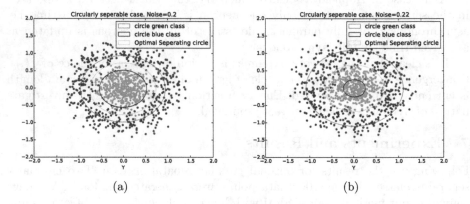

(a)                                                      (b)

**Fig. 3.** Classification with high pattern noise

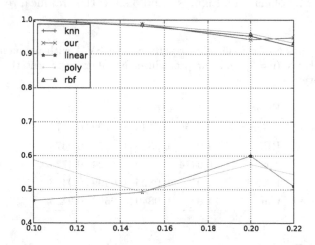

**Fig. 4.** Comparison of various results.

# 8    Conclusion

In this research paper, it is reasoned that the concept of hyper-spherical separability (always implied by linear separability, when one of the classes is bounded) enables efficient classification of multiple classes of linearly separable patterns (one against the rest approach). Also, it is reasoned that certain non-spherically separable patterns can be made spherically separable by a linear transformation without projecting them into higher dimensional space. Using the results from the theory of error correcting codes, it is reasoned that hyper-spherical separability (of quantized patterns) can always be ensured by encoding (i.e. projecting into a higher dimensional space) under some conditions. Finally, the concept of hyper-ellipsoidal neuron and an associative memory based on such a neuron is proposed.

# References

1. Rama Murthy, G.: Hybrid neural networks. In: Proceedings of International Conference of Power System Analysis, Control and Optimization (2008)
2. Bruck, J., Blaum, M.: Neural networks, error correcting codes and polynomials over the binary cube. IEEE Trans. Inf. Theory **35**(5), 976–987 (1989)
3. Hopfield, J.J.: Neural networks and physical systems with emergent collective computational abilities. Proc. Nat. Acad. Sci. U.S.A. **79**, 2554–2558 (1982)
4. Jankowski, S., Lozowski, A., Zureda, J.: Complex-valued multi-state neural associative memory. IEEE Trans. Neural Netw. **7**, 1491–1496 (1996)
5. Hou, Z.G., Song, K.Y., Gupta, M.M., Tan, M.: Neural units with higher-order synaptic operations for robotic image processing applications. Soft Comput.-A Fusion Found. Methodol. Appl. **11**(3), 221–228 (2007)
6. Gács, P., Lovász, L.: Khachiyans algorithm for linear programming. In: König, H., Korte, B., Ritter, K. (eds.) Mathematical Programming at Oberwolfach, pp. 61–68. Springer, Heidelberg (1981)
7. Rama Murthy, G.: Multidimensional Neural Networks: Unified Theory. New Age International (2008)
8. Rama Murthy, G.: Model, Novel Ceiling Neuronal: Artificial Neural Networks. IIIT Technical reports # IIIT/TR/2015/59
9. Rama Murthy, G.: NP Hard Problems: Multi-linear objective optimization problem (manuscript in preparation)
10. Raju, S., Rama Murthy, G., Jha, A., Anil, R.: Dynamics of ordinary and recurrent hopfield networks: novel themes. Accepted for IEEE IACC-2017, January 2017
11. Rama Murthy, G., Anil, R., Dileep, M.: Dynamics of structured complex recurrent hopfield networks. In: Proceedings of International Joint Conference on Neural Networks (IJCNN 2016), Vancouver, Canada, July 2016
12. Rama Murthy, G., Munugoti, S.D., Rayala, A.: Convolutional associative memory: FIR filter model of synapse. In: Arik, S., Huang, T., Lai, W.K., Liu, Q. (eds.) ICONIP 2015. LNCS, vol. 9491, pp. 356–364. Springer, Cham (2015). doi:10.1007/978-3-319-26555-1_40
13. Rama Murthy, G., Gabbouj, M.: Existence and synthesis of complex hopfield type associative memories. In: Rojas, I., Joya, G., Catala, A. (eds.) IWANN 2015. LNCS, vol. 9095, pp. 356–369. Springer, Cham (2015). doi:10.1007/978-3-319-19222-2_30

14. Rama Murthy, G., Gabbouj, M.: On the design of hopfield neural networks: synthesis of hopfield type associative memories. In: Proceedings of IEEE International Joint Conference of Neural Networks (IJCNN 2015), July 2015
15. Rama Murthy, G., Kicanoglu, B., Gabbouj, M.: On the dynamics of a recurrent hopfield network. In: Proceedings of IEEE International Joint Conference on Neural Networks (IJCNN 2015), July 2015
16. Rama Murthy, G., Gabbouj, M.: Linear congruential sequences: feedback and recurrent neural networks. In: Shetty, N.R., Prasad, N.H., Nalini, N. (eds.) Emerging Research in Computing, Information, Communication and Applications, pp. 213–222. Springer, New Delhi (2015). doi:10.1007/978-81-322-2550-8_20
17. Rama Murthy, G., Zolnierek, A., Koszalka, L.: Optimal control, of time varying linear systems: neural networks. In: 2014 International Symposium on Computational and Business Intelligence (ISCBI 2014), New Delhi, India, 6–7 December 2014
18. Rama Murthy, G.: Optimization of quadratic forms: NP hard problems: neural networks. In: 2013 International Symposium on Computational and Business Intelligence (ISCBI 2013), New Delhi, India, 24–26 August 2013
19. Rama Murthy, G.: Distributed signal processing: neural networks. In: IEEE Workshop on Computational Intelligence: Theories, Applications and Future Directions, IIT-Kanpur, 14 July 2013
20. Rama Murthy, G., Nischal, B.: Hopfield-amari neural network: minimization of quadratic forms. In: The 6th International Conference on Soft Computing and Intelligent Systems, Kobe Convention Center, Kobe, Japan, 20–24 November 2012
21. Rama Murthy, G.: Finite impulse response (FIR) filter model of synapses: associated neural networks. In: Proceedings of 4th International Conference on Natural Computation (ICNC 2008), October 2008
22. Rama Murthy, G.: A novel class of generative neural networks. In: Proceedings of 4th International Conference on Natural Computation (ICNC 2008), October 2008
23. Rama Murthy, G.: Hybrid neural networks. In: Proceedings of International Conference on Power System Analysis, Control and Optimization (PSACO 2008), 13–15 March 2008
24. Rama Murthy, G., Praveen, D.: A novel associative memory on the complex hypercubic lattice. In: Proceedings of 16th European Symposium on Artificial Neural Networks (2008)
25. Rama Murthy, G.: Optimal robust filter models of synapse : associated neural networks. In: Proceedings of International Conference on Soft Computing and Intelligent Systems, 27–29 December 2007

# Pre-emphasizing Binarized Ensembles to Improve Classification Performance

Lorena Álvarez-Pérez[✉], Anas Ahachad, and Aníbal R. Figueiras-Vidal

GAMMA-L+/Department Signal Theory and Communications,
University Carlos III of Madrid, Av. de la Universidad 30,
28911 Leganés, Madrid, Spain
{lalvarez,anas,arfv}@tsc.uc3m.es

**Abstract.** Machine ensembles are learning architectures that offer high expressive capacities and, consequently, remarkable performances. This is due to their high number of trainable parameters.

In this paper, we explore and discuss whether binarization techniques are effective to improve standard diversification methods and if a simple additional trick, consisting in weighting the training examples, allows to obtain better results. Experimental results, for three selected classification problems, show that binarization permits that standard direct diversification methods (bagging, in particular) achieve better results, obtaining even more significant performance improvements when pre-emphasizing the training samples. Some research avenues that this finding opens are mentioned in the conclusions.

**Keywords:** Classification · Multi-Layer Perceptron · Ensemble classifiers · Bagging · ECOC

## 1 Introduction

Although the expressive power –the capacity of a machine to establish general input-output correspondences– of one-hidden layer Multi-Layer Perceptrons (MLPs) is theoretically unbounded if an appropriate number of hidden units is included [1,2], the limited number of labeled training examples that are available in practice reduces it.

A way of overcoming this limitation is to build ensembles of MLPs –or other Learning Machines– by diversifying the training of each of them. An ensemble of classifiers is basically composed of a group of machines that try to solve the same problem but under different conditions –diversity– and their outputs are combined aiming at obtaining a system that hopefully is more accurate than any of its members [3]. To achieve this goal, the members of the ensemble, usually known as base learners or simply learners, must be diverse. Put it very simple, two machines are diverse when their errors are not coincident.

In this regard, there are two basic families of ensembles. The first, we are called committees, consists in training the base learners with different examples

© Springer International Publishing AG 2017
I. Rojas et al. (Eds.): IWANN 2017, Part I, LNCS 10305, pp. 339–350, 2017.
DOI: 10.1007/978-3-319-59153-7_30

and their outputs are aggregated, usually by simple procedures (direct averaging or majority vote, for example). Among committees, we can mention two relevant techniques: Bagging [4], in which bootstrap resampling of the samples provides the training sets for the learners, and label switching [5], in which randomly switching examples' labels serves to introduce diversity. The second family of ensembles, which we call consortia, simultaneously train the units and the aggregation. Among consortia, boosting, whose basic forms appeared in [6,7], has proved to be very effective for improving the performance of single learners. The key aspect is to iteratively design and aggregate classification units paying more attention to the examples that present higher classification errors. Mixture of Experts [8], the second most relevant consortia designs, show moderate performance when designed for classification, and the existing modifications to increase it, such as [9], require a huge computational effort.

A different kind of ensembles are those resulting from decomposing a multi-class problem into a number of binary problems whose solutions indicate the class corresponding to each sample. They are called binarization techniques.

The idea of combining binarization methods with one of the above-mentioned techniques (committees or consortia) for solving multi-class problems is not something novel. In [10], it was studied the performance of combining both bagging and binarization techniques over one dataset. The results obtained were not very satisfactory since the overall performance was the same as with only bagging. In [11], bagging and binarized ensembles were evaluated for nine datasets using a bias-variance framework and making comparisons with single neural networks based classifiers. The results obtained outperformed bagging and binarization in some cases, while in other cases gave similar results.

In this work, we take a further step and, we also explore a simple additional alternative that could be applied together with the combination of binarization techniques and standard diversification methods, aiming at increasing the performance of the overall ensemble. This alternative is pre-emphasis, i.e., weighting the training samples according to an auxiliary classifier, taking into account the critical character of each sample, i.e., its proximity to the classification border and its classification error following the ideas of [12,13]. In [14], we proposed flexible enough pre-emphasis approaches that allowed remarkable improvements, requiring a very modest computational cost in the design phase, but not in operation, i.e., to classify unseen samples.

The rest of the paper is organized is follows. In Sect. 2, we present a brief review of ensembles that are produced by binarizing multi-class problems. Section 3 presents the weighting function that is used for pre-emphasis in multi-class problems. Section 4 describes the experimental framework: the selected datasets and the MLP based machines, the ensemble architectures to be studied, the binarization method to be applied, as well as the type of auxiliary classifier that is used to determine the amount of pre-emphasis needed for the training samples. Experimental results and their discussion are also included in this section. The most important conclusions of our work and some open research lines close this contribution in Sect. 5.

## 2    Binarization

There are three popular techniques for reducing a multi-class problem into a series of binary classification problems, namely: One vs. One (OvO), One vs. Rest (OvR) (also known as One vs. All), and Error Correcting Output Codes (ECOC) [15]. ECOC methods come from the area of communications for correcting data errors during transmission. They are based on adding some redundant information to the block to be transmitted, hence obtaining a codeword. In the context of ensembles, the use of ECOC consists in creating base classifiers and training them according to the information obtained from a pre-established code matrix. Experimental work has shown that ECOC offers improvement over OvO and OvR classification methods [15,16]. This is basically the reason why in this work we explore the application of ECOC for binarizing ensembles.

We continue with a short review of ECOC technique. Table 1 shows an example of an ECOC matrix for a 4-class classification problem [16].

**Table 1.** An exhaustive ECOC for a 4-class decision problem [16]. $C_0$ to $C_3$ are the classes, $P_0$ to $P_6$ are the binary problems. The codeword (row) that is nearest to the vector of units' outputs indicates the class to be selected.

| Class | Problem | | | | | | |
|---|---|---|---|---|---|---|---|
| | $P_0$ | $P_1$ | $P_2$ | $P_3$ | $P_4$ | $P_5$ | $P_6$ |
| $C_0$ | 1 | 1 | 1 | 1 | 1 | 1 | 1 |
| $C_1$ | 0 | 0 | 0 | 0 | 1 | 1 | 1 |
| $C_2$ | 0 | 0 | 1 | 1 | 0 | 0 | 1 |
| $C_3$ | 0 | 1 | 0 | 1 | 0 | 1 | 1 |

In the example at hand, each class is assigned a unique binary string of length 7. The string is also called codeword. For example, Class 2 ($C_2$) has the codeword 0011001. During the training process, one binary problem is learned for each column. In this respect, for the first column, we build a binary classifier to separate $\{C_0\}$ from $\{C_1, C_2, C_3\}$. Thus, it seems clear to notice that seven classifiers are trained in this way. To classify a new sample, $\mathbf{x}^{(n)}$, all seven binary classifiers are evaluated to obtain a 7-bit string. Finally, the given sample is classified by computing the similarity between the obtained 7-bit string and the codeword for each class, by using the Hamming distance metric.

## 3    Proposed Pre-emphasis Function

As mentioned in the Introduction, we have considered in this work the training of machines using pre-emphasized samples where the amount of pre-emphasis is determined by an auxiliary classifier.

For binary classifiers, the pre-emphasis used on training sample $\mathbf{x}^{(n)}$ can be written as indicated in the following double-convex combination of three terms:

$$p\left(\mathbf{x}^{(n)}\right) = \alpha + (1-\alpha)\left[\beta\left(e_a^{(n)}\right)^2 + (1-\beta)\left(1 - \left[o_a^{(n)}\right]^2\right)\right] \qquad (1)$$

where $e_a^{(n)}$ is the classification error for input sample $\mathbf{x}^{(n)}$ and $o_a^{(n)}$ is the output of the auxiliary classifier for sample $\mathbf{x}^{(n)}$. It seems clear to notice that the underlying idea of the pre-emphasis function is that the training samples should be weighted based on two measures: How large the error is in the auxiliary classifier and how close its output is to the decision boundary.

The two pre-emphasis parameters, $\alpha$ and $\beta$, have values between zero and one ($0 \le \alpha, \beta \le 1$), and are determined by a process of Cross-Validation (CV) in our experiments.

Extending expression (1) to the case of multi-class formulations is not an easy task since the term $\left(1 - \left[o_a^{(n)}\right]^2\right)$ does not have a direct equivalent. However, if discriminative forms are considered, which are effective for training multi-class machines, it is possible to replace it for $(1 - |o_{ac}^{(n)} - o_{ac'}^{(n)}|)$ or similar forms, where $o_{ac}^{(n)}$ is the softmax output of the auxiliary classifier for the true class and $o_{ac'}^{(n)}$ is the output whose value is the nearest to $o_{ac}^{(n)}$ among the rest.

## 4   Experiments and Their Discussion

### 4.1   Databases

We will limit our presentation of results to a few appropriately selected multi-class databases that are frequently used as benchmark sets for this kind of experiments: the synthetic dataset Firm-Teacher Clave-Direction Classification [17] and two real problems, Satimage [18] and Vehicle [19]. All of them are obtained from the UCI Repository of Machine Learning Databases [20]. Table 2 illustrates the main characteristics of these problems: number of classes, number of dimensions and numbers of training and test samples. When the dataset had no pre-defined train/test partitions (that is, the case of Firm-Teacher Clave-Direction Classification and Vehicle datasets), a random partition has been created with 70/30% for training and test, respectively, keeping the relative proportions of the classes in each subset. From now on, we will denote the databases by their three first letters.

### 4.2   Machines and Their Designs

The architectures under study as well as the single classifier employ MLPs with one-hidden layer as base learners because they are unstable and powerful enough machines. They are trained by the Back-Propagation algorithm to minimize the mean squared error between the desired output and what the network actually

**Table 2.** Characteristics of the benchmark problems.

| Dataset | Notation | # Train samples | # Test samples | Dimension | # Classes |
|---|---|---|---|---|---|
| Firm-Teacher Clave-Direction | Fir | 10800* | - | 16 | 4 |
| Satimage | Sat | 4435 | 2000 | 36 | 6 |
| Vehicle | Veh | 946* | - | 18 | 4 |

*The total number of samples of the datasets without separate train and test sets (Fir and Veh) are listed under the number of training samples. For these datasets, a random partition has been created with 70/30% for training and test, respectively, keeping the relative proportions of the classes in each subset.

outputs, initializing all the weights at random values from a $[-0.2, 0.2]$ uniform distribution. The learning rate for both layers (hidden and output layers) is set to be 0.01, which has been experimentally proven to allow to reach convergence.

As mentioned, the architecture of the MLP based classifiers, explored in the experiments, consists of three layers (input, hidden and output layers). In this architecture, the number of input neurons corresponds to that of the attributes used to characterize the input samples (that is, it is consistent with the dimension of the datasets shown in Table 2), the number of output neurons is related to the classes we are interested in (in a binary classification problem may be enough to have a single output neuron) and the number of hidden neurons (H) depends on the adjustment of the complexity of the MLP. To achieve an optimal ensemble behavior, this appropriate number of hidden neurons has to be fixed. Despite each ensemble can need a different H value, aiming at carrying out a fair comparison in computational terms, we have preferred to fix the same H for all the bagging sampling rates. Then, this parameter is established by means of a 20-run × 5-fold CV.

However, when using the binarization technique, for each binary problem, a different H has to be fixed. In this case, for each dichotomic problem, the value of H is separately obtained also by means of a 20-run × 5-fold CV. This obviously increases the computational cost.

An 80/20 early stopping mode is applied to stop training.

### 4.3   Conventional Diversification

Bagging is carried out by means of conventional bootstrap (sampling with replacement) in our experiments, exploring different sizes B of the diversified training sets: 60, 80, 100, 120 and 140% the size of the original set of training examples. We also examine different number of ensemble units, M = 11, 21, 31, 41, 51, 101 and 201. We remark that these values serve to ensure that performances saturate, as we will see later.

We apply bagging in two different forms:

1. To construct ensembles of different, diversified MLPs;
2. Just at the classification layer, applying the diversification after getting the output values of a single MLP which is trained without diversity.

We will denote these designs by MLP-O (overall) and MLP-T (T-form) architectures, respectively. With this in mind, when we apply bagging to multi-class problems, two different architectures are obtained:

- MLP-BINARIZED-O: As shown in Fig. 1(a), bagging is applied to input data to construct M machines, and the final class is decided by a majority vote of their outputs that are obtained by means of ECOC ensembles that follow each diverse MLP;
- MLP-BINARIZED-T: In this scenario, as illustrated in Fig. 1(b), there is only one MLP structure and bagging is being imposed to its outputs values to construct M diverse ECOC ensembles, whose outputs are also aggregated by a majority vote rule.

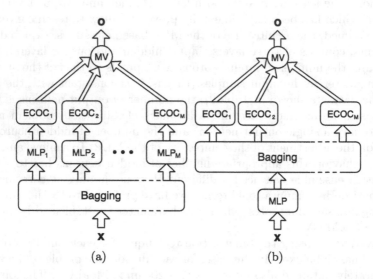

**Fig. 1.** Illustrative architectures showing the ways of combining diversity and binarization techniques in this work for solving multi-class problems: MLP-BINARIZED-O architecture (Fig. 1(a)) and MLP-BINARIZED-T architecture (Fig. 1(b)). MLP is a single Multi-Layer Perceptron; $ECOC_n$ are the diverse ECOC ensembles. MV: majority vote; **x**: input; **o**: output class decision.

It is clear to notice that O structures theoretically require a higher design effort, because M MLP machines have to be trained. To reduce it to affordable computational charges, we apply a frequently used simplification: to design more than M MLP bagging units (450, here) for each value of B, and to select M of them at random. Average performances are obtained for 10 independent selections. Of course, this reduces a little bit the diversity advantage, but we remark that our goal here is just to verify if diversity can improve single shallow MLP performance.

## 4.4   Binarization

In our series of experiments, we will use the C-class exhaustive ECOCs of [16], for the combination between overall and T-form architecture and bagging shown in Fig. 1. According to [16], for classification problems of $3 \leq C \leq 7$ classes, we construct a code length $2^{C-1} - 1$ as follows. Row 1 is all ones. Row 2 consists of $2^{C-2}$ zeros followed by $2^{C-2} - 1$ ones. Row 3 consists of $2^{C-3}$ zeros, followed by $2^{C-3}$ ones, followed by $2^{C-3}$ zeros, followed by $2^{C-3} - 1$ ones. In row $i$, there are alternating runs of $2^{C-i}$ zeros and ones.

## 4.5   Pre-emphasis

In the design of a single classifier, it seems clear that pre-emphasis weighting should be applied to the input training samples. However, for both MLP-BINARIZED-O and MLP-BINARIZED-T architectures previously described, the question arising here is: What is the way to apply the pre-emphasis weighting? In the first scenario, MLP-BINARIZED-O structure, the pre-emphasis is applied to the inputs to each diverse MLP, and in the second situation, MLP-BINARIZED-T architecture, two options have been considered in our experiments: The first is to weight the inputs to the single MLP and the second consists of separately pre-emphasizing each binary problem that appears after applying bagging at the outputs values of the single MLP based classifier. Preliminary results showed that best performances were obtained when weighting the input data to the binarized problems and, consequently, this is the way we have chosen for pre-emphasizing samples in the MLP-BINARIZED-T architecture in our experiments.

To determine the values for the pre-emphasis parameters, $\alpha$ and $\beta$, we used 10-run × 5-fold CV for the three datasets. As mentioned, the values for these parameters that we considered were in the interval $[0, 1]$ at increments of 0.1.

The auxiliary machine used in the experiments that provides the values to compute pre-emphasis weights according to expression (1) is an MLP with one-hidden layer of 50 neurons, because it was experimentally shown that this number of neurons provided the best results over a validation set for the three datasets.

## 4.6   Experimental Results

The results of our experiments (% error rate averages ± standard deviations, averaged for 10 different selections of learners in the case of bagging and for 10 runs when using ECOC ensembles) are shown in Tables 3 (Fir), 4 (Sat), and 5 (Veh) for the test sets. Best results are indicated in boldface (even if statistical differences are not significant). Additionally, we also include the performance of the design which does not apply neither bagging diversity nor ECOC binarization (denoted as MLP), as well as the designs which only apply bagging (indicated as MLP bagging) and the designs which employ ECOC without bagging (marked as MLP-ECOC), aiming at appreciating what is the advantage which can be attributed to the inclusion of the combination of different forms of diversity explored in this work. As mentioned in Sect. 4.3, although different sizes B of the diversified training sets (60, 80, 100, 120 and 140%) have been explored as well as different number of ensemble units (M = 11, 21, 31, 41, 51, 101 and 201), Tables 3, 4 and 5 summarized the most remarkable results.

Having a look at the tree mentioned tables, the first thing to remark is that clear performance saturation effects appear when M or B increase. This indicates that extending the exploration margins is unnecessary. Furthermore, for the three databases under study, all kinds of diversity explored in this work appear to be effective, both O and T bagging: as an example, for Veh dataset, error reduction achieved is about 30%. Obviously, the T-form architecture must be preferred, because it requires lower computational training and operation efforts. Please note that in general, no significant differences appear between separate bagging and ECOC diversifications.

A detailed observation of the mentioned tables leads to the following conclusions:

- For Fir dataset: Directly applying bagging or ECOC achieved success in improving performance. Note that a simple ECOC reduces 18% the error rate, while the best scenario when applying bagging achieves a reduction of about 14%. Results are slightly better when applying jointly ECOC and bagging, M = 31 and B = 120% T-design being the best option in which error rate is approximately reduced up to 22%.
- For Sat dataset: Applying bagging or ECOC separately achieves reducing the error, ECOC ensembles being those which obtain the greater reduction ($\approx$8%), as just happened for the previous dataset. However, when jointly applying ECOC and bagging, results are negative for small values of B, but moderate benefits appear when parameter B increases, M = 21 and B = 140% T-design achieving an error reduction of about 11%.
- For Veh dataset: Simple ECOC approximately reduces 25% the error rate, which is greater than the best error reduction obtained when applying bagging ($\approx$22%). On the other hand, we can observe that jointly applying ECOC and bagging, results are better, especially when M and B parameters increase, M = 51 and B = 120% T-design reaching an error reduction of roughly 31%.
  Please note that if the number of training samples in the datasets is low, the results are more unstable and differences are statistically less significant.

We have taken a further step and we have applied pre-emphasis techniques to the designs that provided the best performances. Table 6 shows the results. For comparative purposes, the results obtained when the pre-emphasis is not applied (see Tables 3, 4 and 5 for further details) are also presented, as well as the results of the single MLP classifier when the input samples are pre-emphasized (MLP PrE) and when they are not pre-emphasized (MLP). For each problem, the values for $\alpha$, $\beta$ obtained by CV are given in parentheses.

It can be observed that the PrE MLP-BINARIZED-T performances are clearly and systematically better than any other results. Error rates decay around 40% in some cases.

**Table 3.** Test (%) error rate averages ± standard deviations for the diversified classifiers applied to problem Firm-Teacher Clave-Direction (Fir).

| MLP | 14.66 ± 0.18 | |
|---|---|---|
| MLP-ECOC | 12.03 ± 0.13 | |
| MLP bagging | B | B |
| M | 120% | 140% |
| 11 | 12.87 ± 0.11 | 12.75 ± 0.15 |
| 21 | 12.73 ± 0.10 | **12.66 ± 0.14** |
| 31 | 12.83 ± 0.09 | 12.71 ± 0.12 |

| MLP-BINARIZED-O bagging | B | B | B |
|---|---|---|---|
| M | 100% | 120% | 140% |
| 11 | 12.75 ± 0.13 | 12.04 ± 0.11 | 12.13 ± 0.10 |
| 21 | 12.64 ± 0.11 | **11.70 ± 0.10** | 12.01 ± 0.11 |
| 31 | 12.58 ± 0.09 | 11.83 ± 0.09 | 11.97 ± 0.09 |
| MLP-BINARIZED-T bagging | B | B | B |
| M | 100% | 120% | 140% |
| 21 | 12.47 ± 0.07 | 11.65 ± 0.08 | 11.97 ± 0.08 |
| 31 | 12.35 ± 0.07 | **11.53 ± 0.07** | 11.86 ± 0.07 |
| 51 | 12.33 ± 0.07 | 11.60 ± 0.07 | 11.75 ± 0.08 |

**Table 4.** Test (%) error rate averages ± standard deviations for the diversified classifiers applied to problem Satimage (Sat).

| MLP | 10.28 ± 0.98 | | |
|---|---|---|---|
| MLP-ECOC | 9.44 ± 0.40 | | |
| MLP bagging | B | B | B |
| M | 100% | 120% | 140% |
| 11 | 10.58 ± 0.84 | 10.04 ± 0.63 | 10.14 ± 0.60 |
| 21 | 10.39 ± 0.71 | **9.88 ± 0.55** | 10.19 ± 0.62 |
| 31 | 10.43 ± 0.69 | 9.94 ± 0.57 | 10.21 ± 0.55 |
| MLP-BINARIZED-O bagging | B | B | B |
| M | 100% | 120% | 140% |
| 11 | 10.34 ± 0.44 | 9.24 ± 0.38 | 9.35 ± 0.39 |
| 21 | 10.23 ± 0.37 | **9.31 ± 0.42** | 9.29 ± 0.37 |

| MLP-BINARIZED-T bagging | B | B |
|---|---|---|
| M | 120% | 140% |
| 11 | 9.31 ± 0.39 | 9.23 ± 0.34 |
| 21 | 9.29 ± 0.36 | **9.18 ± 0.29** |
| 31 | 9.25 ± 0.31 | 9.22 ± 0.28 |

**Table 5.** Test (%) error rate averages ± standard deviations for the diversified classifiers applied to problem Vehicle (Veh).

| MLP | 18.91 ± 3.79 | |
|---|---|---|
| MLP-ECOC | 14.20 ± 3.40 | |
| MLP bagging | B | B |
| M | 120% | 140% |
| 51 | 14.93 ± 3.41 | 14.73 ± 3.16 |
| 101 | 14.80 ± 3.35 | **14.69 ± 3.11** |
| 201 | 14.89 ± 3.27 | 14.75 ± 3.08 |

| MLP-BINARIZED-O bagging | B | B | B |
|---|---|---|---|
| M | 100% | 120% | 140% |
| 31 | 14.72 ± 3.24 | 13.59 ± 3.33 | 13.82 ± 3.20 |
| 51 | 14.68 ± 3.19 | **13.51 ± 3.29** | 13.85 ± 3.21 |
| 101 | 14.51 ± 3.20 | 13.63 ± 3.26 | 13.80 ± 3.18 |

| MLP-BINARIZED-T bagging | B | B | B |
|---|---|---|---|
| M | 100% | 120% | 140% |
| 31 | 14.87 ± 3.29 | 13.19 ± 3.30 | 13.42 ± 3.25 |
| 51 | 14.78 ± 3.19 | **13.04 ± 3.25** | 13.18 ± 3.14 |
| 101 | 14.80 ± 3.15 | 13.08 ± 3.22 | 13.16 ± 3.11 |

**Table 6.** Comparison of test (%) error rate averages ± standard deviations for the benchmark problems considered in our experiments when pre-emphasis is applied.

| Dataset | MLP | | MLP-BINARIZED-T bagging | |
|---|---|---|---|---|
| | No PrE | PrE $(\alpha, \beta)$ | No PrE | PrE $(\alpha, \beta)$ |
| Fir | 14.66 ± 0.18 | 11.87 ± 0.05 (0.3, 0.5) | 11.53 ± 0.07 | 10.34 ± 0.03 (0.4, 0.6) |
| Sat | 10.28 ± 0.98 | 9.26 ± 0.31 (0.4, 0.3) | 9.18 ± 0.29 | 8.13 ± 0.17 (0.4, 0.5) |
| Veh | 18.91 ± 3.79 | 14.05 ± 2.32 (0.6, 0.3) | 13.04 ± 3.25 | 11.59 ± 1.67 (0.5, 0.4) |

All these results seem to indicate that pre-emphasizing the samples of binarized ensembles is extremely useful to obtain high performance MLP-based classifiers and that T-form diversity is fruitful if binarization is applied.

# 5 Conclusions

In this paper, we have not only checked that binarization is effective to improve standard diversification techniques –bagging, in particular–, but we have also seen how a simple technique consisting in weighting the training examples allows to obtain even more important performance improvements. In all the three datasets under study, experimental results show that to combine a flexible enough pre-emphasis function with ECOC binarized and diversified MLP based classifiers, permits an error reduction bigger than their separate application, achieving until 40% error rate reductions in a dataset.

Between the two explored ensemble architectures –O form, corresponding to a full diversification and T-form, in which diversity is applied after designing the MLP based classifier, the second approach reaches lower error rates. Since it is also better from the perspective of the required training and operating computational efforts, it must be preferred for MLP based classifiers. It is also worth mentioning that the saturation performance with respect to the diversification parameters make their selection an easy validation problem.

Needless to say, much more work –considering other problems, other classifiers, and other sources of diversity– is needed to completely appreciate the potential of combining diversity and binarization techniques, including appropriate pre-emphasis sample weighting schemes. Finally, we must remark that to combine this kind of designs with other auxiliary techniques that serve to improve the performance of MLPs classifiers is a promising way to get excellent, or even record performance practical implementations.

**Acknowledgments.** This work has been partly supported by research grants CASI-CAM-CM (S2013/ICE-2845, DGUI-CM and FEDER) and Macro-ADOBE (TEC2015-67719-P, MINECO).

# References

1. Cybenko, G.: Approximation by superpositions of a sigmoidal function. Math. Control Sig. Syst. **2**, 303–314 (1989)
2. Hornik, K., Stinchcombe, M., White, H.: Multilayer feedforward networks are universal approximators. Neural Netw. **2**, 359–366 (1989)
3. Dietterich, T.G.: Ensemble methods in machine learning. In: Kittler, J., Roli, F. (eds.) MCS 2000. LNCS, vol. 1857, pp. 1–15. Springer, Heidelberg (2000). doi:10.1007/3-540-45014-9_1
4. Breiman, L.: Bagging predictors. Mach. Learn. **4**, 123–140 (1996)
5. Hinton, G.E., Osindero, S., Teh, Y.: A fast learning algorithm for deep belief networks. Neural Comput. **18**, 1527–1554 (2006)
6. Freund, Y., Schapire, R.E.: A decision-theoretic generalization of on-line learning and an application to boosting. J. Comput. Syst. Sci. **55**, 119–139 (1997)
7. Schapire, R.E., Singer, Y.: Improved boosting algorithms using confidence-rated predictions. Mach. Learn. **37**, 297–336 (1999)
8. Jacobs, R.A., Jordan, M.I., Nowlan, S.J., Hinton, G.E.: Adaptive mixture of local experts. Neural Comput. **3**, 79–87 (1991)
9. Omari, A., Figueiras-Vidal, A.R.: Feature combiners with gate-generated weights for classification. IEEE Trans. Neural Netw. Learn. Syst. **24**, 158–163 (2013)
10. Leisch, F., Hornik, K.: Combining neural network voting classifiers and error correcting output codes. In: MEASURMENT 1997 (1997)
11. Cemre, Z., Windeatt, T., Yanikogl, B.: Bias-variance analysis of ECOC and bagging using neural nets. In: Okun, O., Valentini, G., Re, M. (eds.) Ensembles in Machine Learning Applications. Studies in Computational Intelligence, vol. 373, pp. 59–73. Springer, Heidelberg (2011)

12. Gómez-Verdejo, V., Ortega-Moral, M., Arenas-García, J., Figueiras-Vidal, A.R.: Boosting by weighting critical and erroneous samples. Neurocomput. **69**, 679–685 (2006)
13. Gómez-Verdejo, V., Arenas-García, J., Figueiras-Vidal, A.R.: A dynamically adjusted mixed emphasis method for building boosting ensembles. IEEE Trans. Neural Netw. **19**, 3–17 (2008)
14. Alvear-Sandoval, R.F., Figueiras-Vidal, A.R.: An experiment in pre-emphasizing diversified deep neural networks. In: 24th European Symposium on Artificial Neural Networks, Computational Intelligence and Machine Learning, pp. 527–532 (2016)
15. Rokach, L.: Pattern Classification Using Ensemble Methods. World Scientific, Singapore (2010)
16. Dietterich, T.G., Bakiri, G.: Solving multi-class learning problems via error-correcting output codes. J. Artif. Intell. Res. **2**, 263–286 (1995)
17. Vurkaç, M.: Clave-direction analysis: a new arena for educational and creative of music technology. J. Music Technol. Educ. **4**, 27–46 (2011)
18. Giannakopoulos, X., Karhunen, J., Oja, E.: An experimental comparison of neural algorithms for independent component analysis and blind separation. Int. J. Neural Syst. **9**, 99–114 (1999)
19. Siebert, J.P.: Vehicle Recognition Using Rule Based Methods. Turing Institute Research Memorandum TIRM-87-018, Glasgow, Scotland (1987)
20. Lichman, M.: UCI Machine Learning Repository. University of California, School of Information and Computer Science, Irvine, CA (2013)

# Dynamics of Quaternionic Hopfield Type Neural Networks

Rama Murthy Garimella[(⊠)] and Rayala Anil

International Institute of Information Technology, Hyderabad, India
rammurthy@iiit.ac.in, anil.rayala@students.iiit.ac.in

**Abstract.** In this research paper, a novel ordinary quaternionic hopfield type network is proposed and the associated convergence theorem is proved. Also, a novel structured quaternionic recurrent hopfield network is proposed. It is proved that in the parallel mode of operation, such a network converges to a cycle of length 4.

## 1   Introduction

Hopfield neural network based on McCulloch-Pitts model of artificial neuron provided an interesting associative memory. The complex valued generalization of such a neural network with suitable complex signum function was proposed by Jankowski et al. [1] and was subject to intensive study by Zurada et al. [2]. To be exact, in [1], the authors utilized phase quantization idea to arrive at a complex Hopfield network that formed the basis for earlier quaternionic neural networks [4]. Also, utilizing Quaternions (a generalization of complex numbers) and suitable definition of complex signum function (as in [1]), an associative memory was studied in [3,4]. It is recently shown in [5] that in such a quaternionic Hopfield network, convergence result conjectured in [3,4] will not hold (i.e. a counter example was provided).

The authors [6] proposed a novel Complex Hopfield network based on magnitude quantization through a different complex Signum function and proved the convergence theorem (with complex hypercube as the State Space). Also, in [7], the authors studied the dynamics of a real valued recurrent Hopfield network. Further in [8], the authors proved convergence theorems associated with structured Complex recurrent Hopfield networks. The generalizations of such neural networks led to the quaternionic Hopfield network discussed in this research paper. This research paper investigates the dynamics of structured quaternionic Hopfield type neural networks. Here we prove the convergence theorems of the network. We expect many applications for our Quaternionic Hopfield Network.

This research paper is organized as follows. In Sect. 2, quaternionic Hopfield network based on quaternionic Hermitian synaptic weight matrix is described and convergence theorem is proved. In Sect. 3, dynamics of interesting structured quaternionic recurrent Hopfield type networks is studied. Experimental results are provided in Sect. 4. Octonionic networks are introduced and their limitations are discussed in Sect. 5. Future work is discussed in Sect. 6 and this paper concludes in Sect. 7.

© Springer International Publishing AG 2017
I. Rojas et al. (Eds.): IWANN 2017, Part I, LNCS 10305, pp. 351–361, 2017.
DOI: 10.1007/978-3-319-59153-7_31

## 2   Dynamics of Ordinary Quaternionic Hopfield Type Network

In research literature, artificial neurons whose output is a complex number were studied. Generalizing this idea, Artificial Neural Networks (ANNS) in which the output is a unit quaternion are proposed and studied. In this research paper we consider artificial neurons whose output belongs to the following set

$$H = \{a + i\ b + j\ c + k\ d \text{ where } a = \pm 1, b = \pm 1, c = \pm 1, d = \pm 1\} \qquad (1)$$

i.e. each artificial neuron is in one of the $2^4 = 16$ states. We consider an artificial neural network with $M$ such neurons. Thus, the state space of such a network is the so called QUATERNIONIC UNIT HYPERCUBE with $16M$ possible elements.

Let the state of $i^{th}$ neuron at time $t$ is denoted by $v_i(t)$. The neurons are connected to each other with the synaptic weights being quaternions. The synaptic weights is a Hermitian quaternions matrix i.e.

$$W = W^* \text{ i.e. } w_{ij} = w_{ji}^*$$

(where $w_{ji}^*$ denotes the conjugate of quaternion $w_{ji}$). The state of $i^{th}$ neuron at time $t + 1$ is computed in the following manner

$$v_i(t+1) = QSIGN\{\sum_{j=1}^{M} w_{ij}\ v_j(t) - T_i\} \qquad (2)$$

where $T_i$ is the threshold at the $i^{th}$ neuron and Quaternionic signum function is defined in the QSIGN

$$QSIGN(a + i\ b + j\ c + k\ d) = sign(a) + i\ sign(b) + j\ sign(c) + k\ sign(d)$$

The Quaternionic Hopfield type neural network operates in the following modes.

- SERIAL MODE: At any given time $t$, state update (to the state at time $t+1$) is performed at only one node/neuron
- FULLY PARALLEL MODE: At any given time $t$, state update is performed simultaneously at all $M$ nodes/neurons
- Other parallel modes: State update is performed at $l$ neurons simultaneously, where $1 < l < M$.

In the state space of quaternionic Hopfield network, there are distinguished states called STABLE STATES. They have a special property such that once the network reached a stable state, it will remain there forever (i.e. no change in the State vector happens after any update.)

The following convergence theorem summarizes the dynamics of a Quaternionic Hopfield type neural network (discussed in this research paper).

**Theorem 1.** *Let $N = (W, T)$ represent a quaternionic Hopfield network. The following dynamics is exhibited by such an artificial neural network.*

- *In the serial mode of operation, the neural network always converges to a stable state.*
- *In the fully parallel mode of operation, the neural network converges to a Stable State or exhibits a cycle of length 2.*

**Proof:** Let $W$ be the quaternionic matrix such that

$$w_{ij} = \bar{w}_{ji} \text{ for } i \neq j \tag{3}$$

$$w_{ij} = 0 \text{ for } i = j \tag{4}$$

Now, let $V(t)$ be the state vector at a given time $t$. Energy at time instant $t$ is

$$E(t) = V(t)^+ \; W \; V(t), \text{ where } + \text{ represents conjugate transpose.} \tag{5}$$

$$= \sum_i \sum_{\substack{j \\ i \neq j}} v_i(t) \; w_{ij} \; v_j(t) \tag{6}$$

Now let us assume that $i^{th}$ neuron is updated. The energy difference is

$$E(t+1) - E(t) = \sum_j (\bar{v}_i(t+1) - \bar{v}_i(t)) \; w_{ij} \; v_j(t) + \sum_j \bar{v}_j(t) \; w_{ji} \; (v_i(t+1) - v_i(t)) \tag{7}$$

$$= (\bar{v}_i(t+1) - \bar{v}_i(t)) \sum_j w_{ij} \; v_j(t) + \left(\sum_j \bar{v}_j(t) \; w_{ji}\right)(v_i(t+1) - v_i(t)) \tag{8}$$

Now,

$$\left(\sum_j \bar{v}_j(t) \; w_{ji}\right)(v_i(t+1) - v_i(t)) = \overline{\left(\sum_j \bar{v}_j(t) \; w_{ji}\right)(v_i(t+1) - v_i(t))}, (\because A = \bar{\bar{A}}) \tag{9}$$

$$= \overline{(\bar{v}_i(t+1) - \bar{v}_i(t))\left(\sum_j \bar{w}_{ji} \; v_j(t)\right)} \tag{10}$$

$$= \overline{(\bar{v}_i(t+1) - \bar{v}_i(t))\left(\sum_j w_{ij} \; v_j(t)\right)}, (\because \bar{w}_{ji} = w_{ij}) \tag{11}$$

Hence,

$$E(t+1) - E(t) = (\bar{v}_i(t+1) - \bar{v}_i(t)) \sum_j w_{ij} v_j(t) + \overline{(\bar{v}_i(t+1) - \bar{v}_i(t))\left(\sum_j w_{ij} \; v_j(t)\right)} \tag{12}$$

$$= 2 * \text{Realpart}\left\{(\bar{v}_i(t+1) - \bar{v}_i(t))\left(\sum_j w_{ij} \; v_j(t)\right)\right\} \tag{13}$$

Now let

$$v_i(t) = a_1 + a_2\ i + a_3\ j + a_4\ k \tag{14}$$

$$v_i(t+1) = b_1 + b_2\ i + b_3\ j + b_4\ k \tag{15}$$

$$\sum_j w_{ij}\ v_j(t) = c_1 + c_2\ i + c_3\ j + c_4\ k \tag{16}$$

Now,

$$\text{Realpart}\{(\bar{v}_i(t+1) - \bar{v}_i(t))(\sum_j w_{ij}\ v_j(t))\} =$$
$$\text{Realpart}\{(b_1 - a_1) + (a_2 - b_2)i + (a_3 - b_3)j + (a_4 - b_4)k \tag{17}$$
$$*(c_1 + c_2 i + c_3 j + c_4 k)\}$$

$$= (b_1 - a_1)c_1 + (b_2 - a_2)c_2 + (b_3 - a_3)c_3 + (b_4 - a_4)c_4 \tag{18}$$

Now, as per our update rule

$$QSIGN(\sum_j w_{ij}\ v_j(t)) = v_i(t+1) \tag{19}$$

So, $b_1$ and $c_1$ are of same sign. $b_2$ and $c_2$ are of same sign. Similarly $b_3, c_3$ and $b_4, c_4$. So Realpart in (18) is $\geq 0$. Hence $E(t+1) - E(t) \geq 0$. So, convergence.

*Note 1.* The above theorem explains the operation of quaternionic Hopfield network as an ASSOCIATIVE MEMORY. The proof for the theorem on parallel mode of operation can be obtained by converting the network to an equivalent model using a bipartite graph and applying the serial case result proved earlier. It is similar to [9].

## 3   Dynamics of Structured Quaternionic Recurrent Hopfield Type Network

In this sector, we consider a quaternionic Hopfield network whose synaptic weight matrix $W$, is not a Hermitian matrix. Specifically we first consider the case where $W$ is Skew Hermitian matrix (whose entries are quaternions) i.e. $W = -W^*$. The following theorem summarizes the dynamics of such a quaternionic recurrent Hopfield network.

**Theorem 2.** *Let $R = (W, T)$ specify a quaternionic recurrent Hopfield network (with $W$ being a Skew Hermitian matrix whose elements are quaternions). Then if the network is operated in a fully parallel mode, a cycle of length 4 in the state Space is observed.*

**Proof:** Let $W$ be the quaternionic matrix such that

$$w_{ij} = -\bar{w}_{ji} \text{ for } i \neq j \tag{20}$$

$$w_{ij} = 0 \text{ for } i = j \tag{21}$$

Let the energy function at time instant $t$ be

$$E(t) = Real\{V(t-1)^+ \; W \; V(t)\} \tag{22}$$

Now,

$$E(t+1) - E(t) = Real\{V(t)^+ \; W \; V(t+1) - V(t-1)^+ \; W \; V(t)\} \tag{23}$$

$$V(t-1)^+ \; W \; V(t) = ((V(t-1)^+ \; W \; V(t))^+)^+$$

$$= ((W \; V(t))^+ V(t-1))^+$$

$$= (V(t)^+ \; W^+ \; V(t-1))^+$$

$$= -(V(t)^+ \; W \; V(t-1))^+ \qquad \because W^+ = -W \tag{24}$$

$$E(t+1) - E(t) = \triangle E = Real\{V(t)^+ \; W \; V(t+1) + (V(t)^+ \; W \; V(t-1))^+\} \tag{25}$$

Let $V(t)^+W = A + B + C + D$ where $A$ is a vector corresponding to real part. $B$ is vector corresponding to $i^{th}$ component. $C$ is vector corresponding to $j^{th}$ component and $D$ is vector corresponding to $k^{th}$ component, i.e. we split the quaternionic matrix into corresponding matrices related to each component.

$$-WV(t) = A^T - B^T - C^T - D^T$$

$$WV(t) = -A^T + B^T + C^T + D^T \tag{26}$$

*Note 2.* Assume that, at any given time $W \; V(t) \neq 0$

Let

$$V(t+1) = V_R(t+1) + V_I(t+1) + V_J(t+1) + V_K(t+1)$$

$$V(t-1) = V_R(t-1) + V_I(t-1) + V_J(t-1) + V_K(t-1)$$

$$\triangle E = Real\{V(t)^+ \; W \; V(t+1)\} + Real\{V(t)^+W \; V(t-1)\}$$

$$= AV_R(t+1) + BV_I(t+1) + CV_J(t+1) + DV_K(t+1) + AV_R(t-1)$$

$$+BV_I(t-1) + CV_J(t-1) + DV_K(t-1)$$

$$= A(V_R(t+1) + V_R(t-1)) + B(V_I(t+1) + V_I(t-1))$$

$$+C(V_J(t+1) + V_J(t-1)) + D(V_K(t+1) + V_K(t-1)) \tag{27}$$

We know $V(t+1) = QSIGN(W \; V(t))$. Now from 26, we can say that elements in $V_R(t+1)$ and $A$ have opposite sign.

$$A(V_R(t+1) + V_R(t-1)) < 0$$

Again from 26, we can say that elements in $V_I(t+1)$ and $B$ have same sign.

$$B(V_I(t+1) + V_I(t-1)) < 0, \because i^2 = -1$$

Similarly, $C(V_J(t+1) + V_J(t-1)) < 0$ and $B(V_K(t+1) + V_K(t-1)) < 0$. So, $\triangle E < 0$. But, we know, in parallel mode, $\triangle E = 0$ during the cycle. So, if we put $\triangle E = 0$ in 27, we get

$$
\begin{aligned}
V_R(t+1) &= -V_R(t-1) \\
V_I(t+1) &= -V_I(t-1) \\
V_J(t+1) &= -V_J(t-1) \\
V_K(t+1) &= -V_K(t-1)
\end{aligned}
\tag{28}
$$

$V(t+1) = -V(t-1)$, implies $V(t+3) = V(t-1)$. We obtain a cycle of length 4.

## 4   Experimental Results

Here we constructed two networks using 10 nodes and 15 nodes respectively. The synaptic weight matrix is an arbitrary Hermitian quaternionic matrix and the initial state is randomly chosen and the network is operated in serial mode. We found that both the networks converged after finite iterations and the resulting energy graphs are as follows (Figs. 1 and 2):

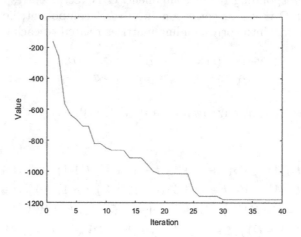

**Fig. 1.** Network with 10 nodes

Note that, the energy values are non-increasing and converged after some finite iterations.

Now, we will look into the parallel mode of operation. A network with 40 nodes is constructed with weight matrix as skew Hermitian quaternionic matrix and initial state is randomly chosen. The network is operated in parallel mode and we found that network reached a cycle of length 4 as predicted by our above proposed theorem. The energy graph is as follows (Fig. 3):

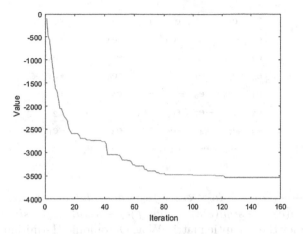

**Fig. 2.** Network with 15 nodes

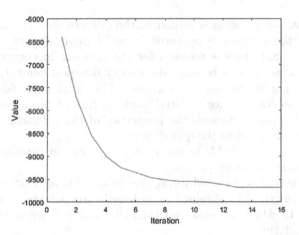

**Fig. 3.** Network with 40 nodes

# 5    Hopfield Network Using Octonions

The natural thought of generalizing the theorems lead to constructing a network based on octonions (8 dimensional numbers) which are extension of quaternions. They are noncommutative and nonassociative. Every octonion $x$ can be denoted as $x = a + be_1 + ce_2 + de_3 + ee_4 + fe_5 + ge_6 + he_7$

where 'a' is the scalar part and remaining are vector parts. The multiplication table of octonions is as follows:

Consider a network of M nodes, where the value of each neuron belongs to the following set $O = \{a + be_1 + ce_2 + de_3 + ee_4 + fe_5 + ge_6 + he_7$ where $a, b, c, d, e, f, g, h = \pm 1\}$ Now, consider the signum function as similar to the quaternion case i.e.

| *     | $e_1$   | $e_2$   | $e_3$   | $e_4$   | $e_5$   | $e_6$   | $e_7$   |
|-------|---------|---------|---------|---------|---------|---------|---------|
| $e_1$ | $-1$    | $e_3$   | $-e_2$  | $e_5$   | $-e_4$  | $-e_7$  | $e_6$   |
| $e_2$ | $-e_3$  | $-1$    | $e_1$   | $e_6$   | $e_7$   | $-e_4$  | $-e_5$  |
| $e_3$ | $e_2$   | $-e_1$  | $-1$    | $e_7$   | $-e_6$  | $e_5$   | $-e_4$  |
| $e_4$ | $-e_5$  | $-e_6$  | $-e_7$  | $-1$    | $e_1$   | $e_2$   | $e_3$   |
| $e_5$ | $e_4$   | $-e_7$  | $e_6$   | $-e_1$  | $1$     | $-e_3$  | $e_2$   |
| $e_6$ | $e_7$   | $e_4$   | $-e_5$  | $-e_2$  | $e_3$   | $-1$    | $-e_1$  |
| $e_7$ | $-e_6$  | $e_5$   | $e_4$   | $-e_3$  | $-e_2$  | $e_1$   | $-1$    |

$OSIGN(a + be_1 + ce_2 + de_3 + ee_4 + fe_5 + ge_6 + he_7) = sign(a) + sign(b)e_1 + sign(c)e_2 + sign(d)e_3 + sign(e)e_4 + sign(f)e_5 + sign(g)e_6 + sign(h)e_7$

Now, Consider the synaptic matrix W as Octonionic Hermitian matrix which means $W = W^*$ i.e. $w_{ij} = w_{ji}^*$. Note that each element of the weight matrix is octonion.

The conjugate of octonion is similar to the quaternion and complex case.

Now, when the network is operated in serial mode(update one node at a time), we found that there is no need for the network to converge unlike the case of quaternions. This is because the energy function defined earlier is neither increasing nor decreasing in this case. This is also verified empirically. Hence there is no convergence in serial mode in case of the above constructed network. This happened because the properties of the octonions are very limited(restricted) compared to the quaternions.

Example: $(AB)^* = B^* A^*$ holds in quaternions but not in octonions. There are many other limitations in properties.

The network is also operated in parallel mode. There is no convergence or cycles of fixed length. The obtained cycle lengths are arbitrary unlike the case of quaternions in which we observed cycles of length 2 when the network is constructed similarly.

Now, we considered the weight matrix to be skew Hermitian octonionic matrix and operated the network in parallel mode which is similar to Sect. 2 but we did not obtain cycles of length 4 unlike the case of quaternions. These lead us to a conclusion that the convergence of a network is very much dependent on the algebra of the numbers involved because it makes the energy function increasing or decreasing or neither of them. So, it is difficult to construct hopfield networks with higher dimensional numbers than quaternions.

## 6   Future Work

Now, we consider the most general quaternionic synaptic weight matrix in the rectangular representation (like complex number based matrix)

$$W = A + i\,B + j\,C + k\,D \tag{29}$$

Note: If C and D are null matrices, then we arrive at the special case of complex valued networks.

Now, the following equality holds true for synaptic weight matrices considered in Sects. 2 and 3.

Hermitian: $A = A^T, B = -B^T, C = -C^T, D = -D^T$

Skew Hermitian: $A = -A^T, B = B^T, C = C^T, D = D^T$

Generalizing, we arrive at 16 possible quaternionic synaptic weight matrices out of which we already studied 2 cases (Hermitian and skew Hermitian). We also want to study the remaining cases. Now, there are another class of matrices which are of the form:

$B = A^T or - A^T, C = A^T or - A^T, D = A^T or - A^T$

where $W = A + i \, B + j \, C + k \, D$.

Those are known as braided matrices which are first introduced in [8]. These type of structured matrices provided good results like cycles of length 8 in [8] for complex valued networks. So, we want to further investigate our proposed quaternionic network using such type of matrices.

Till now, we used the polar representation of the quaternions. Now, we can consider the polar representation and apply various magnitude and phase quantizations as a part of the signum function and look into the dynamics of the network.

In our research efforts, we proved interesting convergence theorems associated with ordinary and recurrent Hopfield networks based on real numbers, complex numbers and quaternions. As in the case of [10], we discuss the existence, synthesis of quaternionic hopfield neural network proposed in this paper as future work. We also want to investigate continuous time quaternionic, complex valued Hopfield neural networks, in the spirit of [11].

Bruck and Blaum [13] proved the connection between hopfield associative memory(with quadratic energy function) as well as Generalized associative memories(with higher degree energy function) and error correcting codes. We expect that quaternionic Hopfield network can be naturally related to generalized error correcting codes. Details are being worked out.

The authors proposed a convolutional associative memory in [12] by modelling the synapse as an FIR filter. The advantage of such a network is that, the state of a neuron is a sequence unlike a single value in case of traditional networks due to which the network is capable of leveraging the multi-model data for an associative memory model. We want to apply it our proposed quaternion network and observe the dynamics of the network as a part of the future work.

We proposed neuronal models in which the state is multiple valued (rather than binary valued). For instance, ceiling neuron, modulo neuron are proposed in our earlier works. Using such multi-state artificial neurons, associative memories can be designed. Starting with an initial state, operation of such novel associative memories either lead to convergence or cycles of various lengths. Specifically complex valued as well as quaternionic associative memories of such type will be investigated for interesting dynamic behaviour.

In various associative memories (real number valued state, complex number valued state, quaternion valued state) proposed by the authors (from [14–20]) and the other researchers, at a single neuron, there is only one threshold value. Thus the state at any neuron is a single scalar value. We propose interesting novel associative memories where the net contribution at any neuron is thresolded using multiple thresold values. Thus, the state at any neuron is a vector. We propose to investigate the dynamic behavious of such associative memories. The reason behind proposal of such associative memories is to tesselate the state space into fine grained regions and increase the state space dimension and size. Thus, it is possible to increase the number of stable states.

### 6.1   Heterogenous Associative Memories

We previously discussed network of associative memories, by connecting a collection of homogenous associative memories in certain topology. But, when we contemplate over the functioning of biological brain, it is clear that audio, video etc. information is stored and retrieved effectively from the memory. Thus, our goal is to emulate such functioning in artificial memories.

We are naturally led to connecting a varied collection of associative memories associated with different types of information in a certain topology. In such a network, the distinct blocks must be coordinated and synchronized for storage and retrieval of information.

## 7   Conclusion

In this research paper, convergence theorem associated with a novel quaternionic hopfield type associative memory is proved. Dynamics of a Structured quaternionic hopfield type network is also discussed when operated in parallel mode. Octonionic networks are introduced and their limitations are discussed. Some interesting ideas are discussed as a part of the future work.

## References

1. Jankowski, S., Lazowksi, A., Zurada, J.: Complex-valued multi-state neural associative memory. IEEE Trans. Neural Netw. **7**, 1491–1496 (1996)
2. Muezzinoglu, M.K., Guzelis, C., Zurada, J.M.: A new design method for the complex-valued multistate Hopfield associative memory. IEEE Trans. Neural Netw. **14**(4), 891–899 (2003)
3. Isokawa, T., Hishimura, H., Saitoh, A., Kmiura, N., Matsui, N.: On the scheme of multistate quaternionic Hopfield neural network. In: Proceedings of Joint 4th International Symposium on Soft Computing Intelligent Systems and 9th International Symposium on advanced Intelligent System (SCIS and ISIS 2008), Nagoya, Japan, pp. 809–813, September 2008
4. Isokawa, T., Nishimura, H., Matsui, N.: Quaternionic neural networks for associative memories. In: Hirose, A. (ed.) Complex-Valued Neural Networks, pp. 103–131. Wiley-IEEE Press, Hoboken (2013)

5. Valle, M.E., de Castro, F.Z.: Theoretical and computational aspects of quaternionic multivalued Hopfield neural networks. In: Proceedings of the IEEE International Joint Conference on Neural Networks 2016 (IJCNN 2016), Vancouver, Canada, July 2016

6. Rama Murthy, G., Praveen, D.: Complex-values neural associative memory on the complex hypercube. In: IEEE Conference on Cybernetics and Intelligent Systems, Singapore, pp. 1–3, December 2004

7. Rama Murthy, G., Kicanaoglu, B., Gabbouj, M.: On the dynamics of a recurrent Hopfield network. In: International Joint Conference on Neural Networks (IJCNN-2015), July 2015

8. Rama Murthy, G., Anil, R., Dileep, M.: Dynamics of structured complex recurrent Hopfield networks. In: Proceedings of IJCNN-2016, Vancouver, July 2016

9. Bruck, J., Goodman, J.W.: A generalized convergence theorem for neural networks. IEEE Trans. Inf. Theory **34**(5), 1089–1092 (1988)

10. Rama Murthy, G., Gabbouj, M.: Existence and synthesis of complex Hopfield type associative memories. In: Rojas, I., Joya, G., Catala, A. (eds.) IWANN 2015. LNCS, vol. 9095, pp. 356–369. Springer, Cham (2015). doi:10.1007/978-3-319-19222-2_30

11. Rama Murthy, G.: Some novel real/complex valued neural network models. In: Advances in Soft Computing, Springer Series, Computational Intelligence, Theory and Applications, Proceedings of 9th Fuzzy days (2006)

12. Rama Murthy, G., Munugoti, S.D., Rayala, A.: Convolutional associative memory: FIR filter model of synapse. In: Arik, S., Huang, T., Lai, W.K., Liu, Q. (eds.) ICONIP 2015. LNCS, vol. 9491, pp. 356–364. Springer, Cham (2015). doi:10.1007/978-3-319-26555-1_40

13. Bruck, J., Blaum, M.: Neural networks, error-correcting codes, and polynomials over the binary n-cube. IEEE Trans. Inf. Theory **35**(5), 976–987 (1989)

14. Raju, S., Rama Murthy, G., Jha, A., Anil, R.: Dynamics of ordinary and recurrent Hopfield networks: novel themes. In: IEEE IACC-2017, January 2017

15. Rama Murthy, G., Gabbouj, M.: On the design of Hopfield neural networks: synthesis of Hopfield type associative memories. In: Proceedings of IEEE International Joint Conference on Neural Networks (IJCNN 2015), July 2015

16. Rama Murthy, G.: Optimization of quadratic forms: NP hard problems: neural networks. In: 2013 International Symposium on Computational and Business Intelligence (ISCBI 2013), New Delhi, India, 24–26 August 2013

17. Rama Murthy, G., Zolnierek, A., Koszalka, L.: Optimal control of time varying linear systems: neural networks. In: 2014 International Symposium on Computational and Business Intelligence (ISCBI 2014), New Delhi, India, 6–7 December 2014

18. Rama Murthy, G.: Distributed signal processing: neural networks. IEEE Workshop on Computational Intelligence: Theories, Applications and Future Directions, IIT-Kanpur, 14 July 2013

19. Rama Murthy, G., Nischal, B.: Network, Hopfield-Amari neural: minimization of quadratic forms. In: The 6th International Conference on Soft Computing and Intelligent Systems, Kobe, Japan, 20–24 November 2012

20. Rama Murthy, G.: Multidimensional Neural Networks Unified Theory. New Age International Publishers, New Delhi (2007)

# Quasi-Newton Learning Methods
# for Quaternion-Valued Neural Networks

Călin-Adrian Popa[✉]

Department of Computer and Software Engineering, Polytechnic University
Timişoara, Blvd. V. Pârvan, No. 2, 300223 Timişoara, Romania
calin.popa@cs.upt.ro

**Abstract.** This paper presents the deduction of the quasi-Newton learning methods for training quaternion-valued feedforward neural networks, using the framework of the HR calculus. Since these algorithms yielded better training results than the gradient descent for the real- and complex-valued cases, an extension to the quaternion-valued case is a natural idea to enhance the performance of quaternion-valued neural networks. Experiments done on four time series prediction applications show a significant improvement over the quaternion gradient descent algorithm.

**Keywords:** Quaternion-valued neural networks · Quasi-Newton algorithms · Time series prediction

## 1 Introduction

Over the last few years, the domain of quaternion-valued neural networks has received an increasing interest. Popular applications of these networks range from chaotic time series prediction [1], color image compression [2], color night vision [3], polarized signal classification [4], to 3D wind forecasting [5–7].

In the 3D and 4D domains, where some signals are naturally expressed in quaternion-valued form, these networks appear as a natural choice for solving problems such as time series prediction. Several methods to increase the efficiency of learning in quaternion-valued neural networks have been proposed, including different network architectures and different learning algorithms, some of which are specially designed for this type of networks, while others are extended from the real-valued case.

One of the most effective from the class of second order methods used to minimize a cost function, is, in theory, the Newton method, see [8]. But because it needs the explicit calculation of the Hessian matrix of the cost function, more precisely its inverse, which is a computationally expensive task, quasi-Newton methods have been developed. They replace the explicit calculation of the Hessian with an approximation of it, which is positive definite by construction, to avoid the convergence problems in the Newton method when the Hessian is not positive definite.

© Springer International Publishing AG 2017
I. Rojas et al. (Eds.): IWANN 2017, Part I, LNCS 10305, pp. 362–374, 2017.
DOI: 10.1007/978-3-319-59153-7_32

First applied, among others, by [9,10] to the training of neural networks, the quasi-Newton learning method has proved to be very efficient in the real-valued and complex-valued [11] cases. Because of these performances, it seems natural to extend this learning algorithm to quaternion-valued neural networks, also.

The rest of the paper is organized as follows: Sect. 2 is an introduction to the $\mathbb{HR}$ calculus, which is a type of calculus used for extending real-valued algorithms to the quaternion-valued domain. Section 3 is concerned with the derivation of the quaternion-valued quasi-Newton algorithms using the $\mathbb{HR}$ calculus. Experimental results of four applications of the proposed algorithms are shown and discussed in Sect. 4, along with a description of each problem. Section 5 is dedicated to presenting the conclusions of this study.

## 2    The $\mathbb{HR}$ Calculus

We will first present the basics of the $\mathbb{HR}$ calculus [12], which will be later used to deduce the quasi-Newton algorithms for a quaternion-valued error function.

Let $\mathbb{H} = \{q_a + \imath q_b + \jmath q_c + \kappa q_d | q_a, q_b, q_c, q_d \in \mathbb{R}\}$ be the algebra of quaternions, where $\imath$, $\jmath$, $k$ are the imaginary units satisfying $\imath^2 = \jmath^2 = \kappa^2 = \imath\jmath\kappa = -1$. For any $\mu \in \mathbb{H}$, we define the operation $q^\mu := \mu q \mu^{-1}$. Then, for any $q = q_a + \imath q_b + \jmath q_c + \kappa q_d \in \mathbb{H}$, we have $q^\imath = \imath q \imath^{-1} = q_a + \imath q_b - \jmath q_c - \kappa q_d$, $q^\jmath = \jmath q \jmath^{-1} = q_a - \imath q_b + \jmath q_c - \kappa q_d$, $q^\kappa = \kappa q \kappa^{-1} = q_a - \imath q_b - \jmath q_c + \kappa q_d$. If $f : \mathbb{H} \to \mathbb{H}$, the $\mathbb{HR}$ derivatives of $f$ can be defined by

$$\begin{pmatrix} \frac{\partial f}{\partial q} \\ \frac{\partial f}{\partial q^\imath} \\ \frac{\partial f}{\partial q^\jmath} \\ \frac{\partial f}{\partial q^\kappa} \end{pmatrix} := \frac{1}{4} \begin{pmatrix} 1 & -\imath & -\jmath & -\kappa \\ 1 & -\imath & \jmath & \kappa \\ 1 & \imath & -\jmath & \kappa \\ 1 & \imath & \jmath & -\kappa \end{pmatrix} \begin{pmatrix} \frac{\partial f}{\partial q_a} \\ \frac{\partial f}{\partial q_b} \\ \frac{\partial f}{\partial q_c} \\ \frac{\partial f}{\partial q_d} \end{pmatrix}.$$

Now, consider a quaternion vector $\mathbf{q} = (q_1, q_2, \ldots, q_N)^T \in \mathbb{H}^N$, which can be written as $\mathbf{q} = \mathbf{q}_a + \imath \mathbf{q}_b + \jmath \mathbf{q}_c + \kappa \mathbf{q}_d \subset \mathbb{H}^N$, where $\mathbf{q}_a, \mathbf{q}_b, \mathbf{q}_c, \mathbf{q}_d \in \mathbb{R}^N$. We have that $\mathbf{q}^\imath = \imath \mathbf{q} \imath^{-1} = \mathbf{q}_a + \imath \mathbf{q}_b - \jmath \mathbf{q}_c - \kappa \mathbf{q}_d$, $\mathbf{q}^\jmath = \jmath \mathbf{q} \jmath^{-1} = \mathbf{q}_a - \imath \mathbf{q}_b + \jmath \mathbf{q}_c - \kappa \mathbf{q}_d$, $\mathbf{q}^\kappa = \kappa \mathbf{q} \kappa^{-1} = \mathbf{q}_a - \imath \mathbf{q}_b - \jmath \mathbf{q}_c + \kappa \mathbf{q}_d \in \mathbb{H}^N$. We denote

$$\overset{\mathcal{H}}{\mathbf{q}} := \begin{pmatrix} \mathbf{q} \\ \mathbf{q}^\imath \\ \mathbf{q}^\jmath \\ \mathbf{q}^\kappa \end{pmatrix} \in \mathbb{H}^{4N}, \ \overset{\mathcal{R}}{\mathbf{q}} := \begin{pmatrix} \mathbf{q}_a \\ \mathbf{q}_b \\ \mathbf{q}_c \\ \mathbf{q}_d \end{pmatrix} \in \mathbb{R}^{4N}, \ \mathbf{J} := \begin{pmatrix} \mathbf{I}_N & \imath\mathbf{I}_N & \jmath\mathbf{I}_N & \kappa\mathbf{I}_N \\ \mathbf{I}_N & \imath\mathbf{I}_N & -\jmath\mathbf{I}_N & -\kappa\mathbf{I}_N \\ \mathbf{I}_N & -\imath\mathbf{I}_N & \jmath\mathbf{I}_N & -\kappa\mathbf{I}_N \\ \mathbf{I}_N & -\imath\mathbf{I}_N & -\jmath\mathbf{I}_N & \kappa\mathbf{I}_N \end{pmatrix},$$

where $\mathbf{I}_N$ is the $N \times N$ identity matrix. With these notations, we can write that

$$\overset{\mathcal{H}}{\mathbf{q}} = \mathbf{J}\overset{\mathcal{R}}{\mathbf{q}}.$$

It can be verified that $\mathbf{J}^H \mathbf{J} = \mathbf{J}\mathbf{J}^H = 4\mathbf{I}_{4N}$, and so we also have that

$$\overset{\mathcal{R}}{\mathbf{q}} = \frac{1}{4}\mathbf{J}^H\overset{\mathcal{H}}{\mathbf{q}}. \tag{1}$$

A function $f : \mathbb{H}^N \to \mathbb{R}$ can now be seen in three equivalent forms

$$f(\mathbf{q}) \Leftrightarrow f(\overset{\mathcal{H}}{\mathbf{q}}) := f(\mathbf{q}, \mathbf{q}^{\imath}, \mathbf{q}^{\jmath}, \mathbf{q}^{\kappa}) \Leftrightarrow f(\overset{\mathcal{R}}{\mathbf{q}}) := f(\mathbf{q}_a, \mathbf{q}_b, \mathbf{q}_c, \mathbf{q}_d).$$

If we define the operators

$$\frac{\partial f}{\partial \mathbf{q}} := \left( \frac{\partial f}{\partial q_1}, \dots, \frac{\partial f}{\partial q_N} \right),$$

$$\frac{\partial f}{\partial \overset{\mathcal{H}}{\mathbf{q}}} := \left( \frac{\partial f}{\partial \mathbf{q}}, \frac{\partial f}{\partial \mathbf{q}^{\imath}}, \frac{\partial f}{\partial \mathbf{q}^{\jmath}}, \frac{\partial f}{\partial \mathbf{q}^{\kappa}} \right),$$

$$\frac{\partial f}{\partial \overset{\mathcal{R}}{\mathbf{q}}} := \left( \frac{\partial f}{\partial \mathbf{q}_a}, \frac{\partial f}{\partial \mathbf{q}_b}, \frac{\partial f}{\partial \mathbf{q}_c}, \frac{\partial f}{\partial \mathbf{q}_d} \right),$$

we have, from the chain rule, that

$$\frac{\partial f}{\partial \overset{\mathcal{H}}{\mathbf{q}}} = \frac{1}{4} \frac{\partial f}{\partial \overset{\mathcal{R}}{\mathbf{q}}} \mathbf{J}^H \Leftrightarrow \frac{\partial f}{\partial \overset{\mathcal{R}}{\mathbf{q}}} = \frac{\partial f}{\partial \overset{\mathcal{H}}{\mathbf{q}}} \mathbf{J}.$$

Now, if we define

$$\nabla_{\mathbf{q}} f := \left( \frac{\partial f}{\partial \mathbf{q}} \right)^H, \quad \nabla_{\overset{\mathcal{H}}{\mathbf{q}}} f := \left( \frac{\partial f}{\partial \overset{\mathcal{H}}{\mathbf{q}}} \right)^H, \quad \nabla_{\overset{\mathcal{R}}{\mathbf{q}}} f := \left( \frac{\partial f}{\partial \overset{\mathcal{R}}{\mathbf{q}}} \right)^T,$$

where $(\cdot)^T$ and $(\cdot)^H$ represent the transpose and the Hermitian transpose, respectively, the previous relations can be written as

$$\nabla_{\overset{\mathcal{H}}{\mathbf{q}}} f = \frac{1}{4} \mathbf{J} \nabla_{\overset{\mathcal{R}}{\mathbf{q}}} f \Leftrightarrow \nabla_{\overset{\mathcal{R}}{\mathbf{q}}} f = \mathbf{J}^H \nabla_{\overset{\mathcal{H}}{\mathbf{q}}} f. \tag{2}$$

Similarly, by defining

$$\nabla_{\mathbf{q}}^2 f := \frac{\partial}{\partial \mathbf{q}} \left( \frac{\partial f}{\partial \mathbf{q}} \right)^H, \quad \nabla_{\overset{\mathcal{H}}{\mathbf{q}}}^2 f := \frac{\partial}{\partial \overset{\mathcal{H}}{\mathbf{q}}} \left( \frac{\partial f}{\partial \overset{\mathcal{H}}{\mathbf{q}}} \right)^H, \quad \nabla_{\overset{\mathcal{R}}{\mathbf{q}}}^2 f := \frac{\partial}{\partial \overset{\mathcal{R}}{\mathbf{q}}} \left( \frac{\partial f}{\partial \overset{\mathcal{R}}{\mathbf{q}}} \right)^T,$$

we obtain that

$$\nabla_{\overset{\mathcal{H}}{\mathbf{q}}}^2 f = \frac{1}{16} \mathbf{J} (\nabla_{\overset{\mathcal{R}}{\mathbf{q}}}^2 f) \mathbf{J}^H \Leftrightarrow \nabla_{\overset{\mathcal{R}}{\mathbf{q}}}^2 f = \mathbf{J}^H (\nabla_{\overset{\mathcal{H}}{\mathbf{q}}}^2 f) \mathbf{J}. \tag{3}$$

## 3   Quasi-Newton Learning

Let's assume that we have a quaternion-valued neural network with an error function denoted by $E : \mathbb{H}^N \to \mathbb{R}$, and an $N$-dimensional weight vector denoted by $\mathbf{w} \in \mathbb{H}^N$. We start with the quasi-Newton algorithm for the real-valued case, in which the function $E(\mathbf{w})$ can be viewed as $E(\overset{\mathcal{R}}{\mathbf{w}})$. The iteration for calculating the value $\overset{\mathcal{R}}{\mathbf{w}}^{*}$, for which the minimum of the function $E(\overset{\mathcal{R}}{\mathbf{w}})$ is attained, is

$$\overset{\mathcal{R}}{\mathbf{w}}_{k+1} = \overset{\mathcal{R}}{\mathbf{w}}_{k} - \alpha_{k}\overset{\mathcal{R}}{\mathbf{H}}_{k}\overset{\mathcal{R}}{\mathbf{g}}_{k}, \tag{4}$$

where $\overset{\mathcal{R}}{\mathbf{g}}_{k} := \nabla_{\overset{\mathcal{R}}{\mathbf{w}}_{k}} E \in \mathbb{R}^{4N}$, and $\overset{\mathcal{R}}{\mathbf{H}}_{k} \in \mathbb{R}^{4N \times 4N}$ is an approximation of the inverse of the Hessian matrix $\nabla^{2}_{\overset{\mathcal{R}}{\mathbf{w}}_{k}} E$. The value of $\alpha_{k} \in \mathbb{R}$ is determined using an inexact line search that minimizes $E(\overset{\mathcal{R}}{\mathbf{w}}_{k+1}) = E(\overset{\mathcal{R}}{\mathbf{w}}_{k} - \alpha_{k}\overset{\mathcal{R}}{\mathbf{H}}_{k}\overset{\mathcal{R}}{\mathbf{g}}_{k})$. In our experiments, we used the golden section search, which is guaranteed to have linear convergence, see [13].

Using (2) and (3), we have that

$$\overset{\mathcal{R}}{\mathbf{g}}_{k} = \mathbf{J}^{H}\overset{\mathcal{H}}{\mathbf{g}}_{k}$$

and

$$\overset{\mathcal{R}}{\mathbf{H}}_{k} = \frac{1}{16}\mathbf{J}^{H}\overset{\mathcal{H}}{\mathbf{H}}_{k}\mathbf{J},$$

where $\overset{\mathcal{H}}{\mathbf{g}}_{k} := \nabla_{\overset{\mathcal{H}}{\mathbf{w}}_{k}} E \in \mathbb{H}^{4N}$, and we took into account the fact that $\overset{\mathcal{R}}{\mathbf{H}}_{k}$ approximates $(\nabla^{2}_{\overset{\mathcal{R}}{\mathbf{w}}_{k}} E)^{-1}$ and $\overset{\mathcal{H}}{\mathbf{H}}_{k}$ approximates $(\nabla^{2}_{\overset{\mathcal{H}}{\mathbf{w}}_{k}} E)^{-1}$. Thus, relation (4) can be written as

$$\frac{1}{4}\mathbf{J}^{H}\overset{\mathcal{H}}{\mathbf{w}}_{k+1} = \frac{1}{4}\mathbf{J}^{H}\overset{\mathcal{H}}{\mathbf{w}}_{k} - \alpha_{k}\frac{1}{16}\mathbf{J}^{H}\overset{\mathcal{H}}{\mathbf{H}}_{k}\mathbf{J}\mathbf{J}^{H}\overset{\mathcal{H}}{\mathbf{g}}_{k},$$

where we also used relation (1), or, equivalently,

$$\overset{\mathcal{H}}{\mathbf{w}}_{k+1} = \overset{\mathcal{H}}{\mathbf{w}}_{k} - \alpha_{k}\overset{\mathcal{H}}{\mathbf{H}}_{k}\overset{\mathcal{H}}{\mathbf{g}}_{k}. \tag{5}$$

Now, the update expression of the inverse Hessian approximation for the *symmetric rank-one (SR1) method* is

$$\overset{\mathcal{R}}{\mathbf{H}}_{k+1} = \overset{\mathcal{R}}{\mathbf{H}}_{k} + \frac{(\overset{\mathcal{R}}{\mathbf{p}}_{k} - \overset{\mathcal{R}}{\mathbf{H}}_{k}\overset{\mathcal{R}}{\mathbf{q}}_{k})(\overset{\mathcal{R}}{\mathbf{p}}_{k} - \overset{\mathcal{R}}{\mathbf{H}}_{k}\overset{\mathcal{R}}{\mathbf{q}}_{k})^{T}}{(\overset{\mathcal{R}}{\mathbf{p}}_{k} - \overset{\mathcal{R}}{\mathbf{H}}_{k}\overset{\mathcal{R}}{\mathbf{q}}_{k})^{T}\overset{\mathcal{R}}{\mathbf{q}}_{k}}, \tag{6}$$

where

$$\overset{\mathcal{R}}{\mathbf{p}}_{k} := \overset{\mathcal{R}}{\mathbf{w}}_{k+1} - \overset{\mathcal{R}}{\mathbf{w}}_{k}, \quad \overset{\mathcal{R}}{\mathbf{q}}_{k} := \overset{\mathcal{R}}{\mathbf{g}}_{k+1} - \overset{\mathcal{R}}{\mathbf{g}}_{k}.$$

Again from (1) and (2), we obtain

$$\overset{\mathcal{H}}{\mathbf{p}}_{k} := \overset{\mathcal{H}}{\mathbf{w}}_{k+1} - \overset{\mathcal{H}}{\mathbf{w}}_{k}, \quad \overset{\mathcal{H}}{\mathbf{q}}_{k} := \overset{\mathcal{H}}{\mathbf{g}}_{k+1} - \overset{\mathcal{H}}{\mathbf{g}}_{k}.$$

Taking the above relations into account, Eq. (6) becomes $\frac{1}{16}\mathbf{J}^{H}\overset{\mathcal{H}}{\mathbf{H}}_{k+1}\mathbf{J} = \frac{1}{16}\mathbf{J}^{H}\overset{\mathcal{H}}{\mathbf{H}}_{k}\mathbf{J} + \frac{(\frac{1}{4}\mathbf{J}^{H}\overset{\mathcal{H}}{\mathbf{p}}_{k} - \frac{1}{16}\mathbf{J}^{H}\overset{\mathcal{H}}{\mathbf{H}}_{k}\mathbf{J}\mathbf{J}^{H}\overset{\mathcal{H}}{\mathbf{q}}_{k})(\frac{1}{4}\mathbf{J}^{H}\overset{\mathcal{H}}{\mathbf{p}}_{k} - \frac{1}{16}\mathbf{J}^{H}\overset{\mathcal{H}}{\mathbf{H}}_{k}\mathbf{J}\mathbf{J}^{H}\overset{\mathcal{H}}{\mathbf{q}}_{k})^{H}}{(\frac{1}{4}\mathbf{J}^{H}\overset{\mathcal{H}}{\mathbf{p}}_{k} - \frac{1}{16}\mathbf{J}^{H}\overset{\mathcal{H}}{\mathbf{H}}_{k}\mathbf{J}\mathbf{J}^{H}\overset{\mathcal{H}}{\mathbf{q}}_{k})^{H}\mathbf{J}^{H}\overset{\mathcal{H}}{\mathbf{q}}_{k}}$, which is equivalent to

$$\overset{\mathcal{H}}{\mathbf{H}}_{k+1} = \overset{\mathcal{H}}{\mathbf{H}}_k + \frac{(\overset{\mathcal{H}}{\mathbf{p}}_k - \overset{\mathcal{H}}{\mathbf{H}}_k\overset{\mathcal{H}}{\mathbf{q}}_k)(\overset{\mathcal{H}}{\mathbf{p}}_k - \overset{\mathcal{H}}{\mathbf{H}}_k\overset{\mathcal{H}}{\mathbf{q}}_k)^H}{(\overset{\mathcal{H}}{\mathbf{p}}_k - \overset{\mathcal{H}}{\mathbf{H}}_k\overset{\mathcal{H}}{\mathbf{q}}_k)^H\overset{\mathcal{H}}{\mathbf{q}}_k}.$$

Up until now we have worked with vectors from $\mathbb{H}^{4N}$. Ideally, we would like to work with vectors directly in $\mathbb{H}^N$. Considering the definition of $\overset{\mathcal{H}}{\mathbf{q}}$ for $\mathbf{q} \in \mathbb{H}^N$, this is done by taking the first $N$ elements of the vector $\overset{\mathcal{H}}{\mathbf{q}}$. For this, we denote

$$\overset{\mathcal{H}}{\mathbf{r}}_k := \overset{\mathcal{H}}{\mathbf{p}}_k - \overset{\mathcal{H}}{\mathbf{H}}_k\overset{\mathcal{H}}{\mathbf{q}}_k,$$

and so

$$\mathbf{r}_k = \mathbf{p}_k - (\mathbf{H}_k^1\mathbf{q}_k + \mathbf{H}_k^2(\mathbf{q}_k)^\imath + \mathbf{H}_k^3(\mathbf{q}_k)^\jmath + \mathbf{H}_k^4(\mathbf{q}_k)^\kappa),$$

where $\mathbf{H}_k^1 := (\overset{\mathcal{H}}{\mathbf{H}}_k)_{11}$, $\mathbf{H}_k^2 := (\overset{\mathcal{H}}{\mathbf{H}}_k)_{12}$, $\mathbf{H}_k^3 := (\overset{\mathcal{H}}{\mathbf{H}}_k)_{13}$, $\mathbf{H}_k^4 := (\overset{\mathcal{H}}{\mathbf{H}}_k)_{14} \in \mathbb{H}^{N\times N}$ are block components of the matrix $\overset{\mathcal{H}}{\mathbf{H}}_k \in \mathbb{H}^{4N\times 4N}$. We can compute

$$(\overset{\mathcal{H}}{\mathbf{p}}_k - \overset{\mathcal{H}}{\mathbf{H}}_k\overset{\mathcal{H}}{\mathbf{q}}_k)^H\overset{\mathcal{H}}{\mathbf{q}}_k = \overset{\mathcal{H}}{\mathbf{r}}_k^H\overset{\mathcal{H}}{\mathbf{q}}_k = 4\mathrm{Re}(\mathbf{r}_k^H\mathbf{q}_k),$$

where $\mathrm{Re}(q)$ represents the real part of the quaternion $q$, i.e. $\mathrm{Re}(q) = q_a$, if $q = q_a + \imath q_b + \jmath q_c + \kappa q_d \in \mathbb{H}$. The iteration (5) becomes

$$\mathbf{w}_{k+1} = \mathbf{w}_k - \alpha_k\left(\mathbf{H}_k^1\mathbf{g}_k + \mathbf{H}_k^2(\mathbf{g}_k)^\imath + \mathbf{H}_k^3(\mathbf{g}_k)^\jmath + \mathbf{H}_k^4(\mathbf{g}_k)^\kappa\right), \tag{7}$$

where the iterations for calculating the matrices $\mathbf{H}_k^1, \ldots, \mathbf{H}_k^4$ are:

$$\mathbf{H}_{k+1}^1 = \mathbf{H}_k^1 + \frac{\mathbf{r}_k\mathbf{r}_k^H}{4\mathrm{Re}(\mathbf{r}_k^H\mathbf{q}_k)},$$

$$\vdots$$

$$\mathbf{H}_{k+1}^4 = \mathbf{H}_k^4 + \frac{\mathbf{r}_k((\mathbf{r}_k)^\kappa)^H}{4\mathrm{Re}(\mathbf{r}_k^H\mathbf{q}_k)}. \tag{8}$$

Relations (7) and (8) now give the *quaternion-valued symmetric rank-one (SR1) method*.

Proceeding in the same manner, we can deduce the update rule of the inverse Hessian approximation for the *Davidon-Fletcher-Powell (DFP) method* (see [14]) in the form

$$\overset{\mathcal{H}}{\mathbf{H}}_{k+1} = \overset{\mathcal{H}}{\mathbf{H}}_k + \frac{\overset{\mathcal{H}}{\mathbf{p}}_k\overset{\mathcal{H}}{\mathbf{p}}_k^H}{\overset{\mathcal{H}}{\mathbf{p}}_k^H\overset{\mathcal{H}}{\mathbf{q}}_k} - \frac{\overset{\mathcal{H}}{\mathbf{H}}_k\overset{\mathcal{H}}{\mathbf{q}}_k\overset{\mathcal{H}}{\mathbf{q}}_k^H\overset{\mathcal{H}}{\mathbf{H}}_k}{\overset{\mathcal{H}}{\mathbf{q}}_k^H\overset{\mathcal{H}}{\mathbf{H}}_k\overset{\mathcal{H}}{\mathbf{q}}_k}.$$

Now, by denoting $\overset{\mathcal{H}}{\mathbf{r}}_k := \overset{\mathcal{H}}{\mathbf{H}}_k\overset{\mathcal{H}}{\mathbf{q}}_k$, we have that

$$\mathbf{r}_k = \mathbf{H}_k^1\mathbf{g}_k + \mathbf{H}_k^2(\mathbf{g}_k)^\imath + \mathbf{H}_k^3(\mathbf{g}_k)^\jmath + \mathbf{H}_k^4(\mathbf{g}_k)^\kappa,$$

and the matrices $\mathbf{H}_k^1, \ldots, \mathbf{H}_k^4$ are computed using the iterations:

$$\mathbf{H}_{k+1}^1 = \mathbf{H}_k^1 + \rho_k \mathbf{p}_k \mathbf{p}_k^H - \sigma_k \mathbf{r}_k \mathbf{r}_k^H$$

$$\vdots$$

$$\mathbf{H}_{k+1}^4 = \mathbf{H}_k^4 + \rho_k \mathbf{p}_k ((\mathbf{p}_k)^\kappa)^H - \sigma_k \mathbf{r}_k ((\mathbf{r}_k)^\kappa)^H, \tag{9}$$

where

$$\rho_k := \frac{1}{4\mathrm{Re}(\mathbf{p}_k^H \mathbf{q}_k)} \text{ and } \sigma_k := \frac{1}{4\mathrm{Re}(\mathbf{q}_k^H \mathbf{r}_k)}.$$

Thus, the *quaternion-valued Davidon-Fletcher-Powell (DFP) method* is given by Eqs. (7) and (9).

The most popular quasi-Newton algorithm is the *Broyden-Fletcher-Goldfarb-Shanno (BFGS) method* (see [15]), for which the following inverse Hessian approximation iteration can be similarly obtained:

$$\overset{\mathcal{H}}{\mathbf{H}}_{k+1} = \overset{\mathcal{H}}{\mathbf{H}}_k - \frac{\overset{\mathcal{H}}{\mathbf{p}}_k \overset{\mathcal{H}}{\mathbf{q}}_k^H \overset{\mathcal{H}}{\mathbf{H}}_k}{\overset{\mathcal{H}}{\mathbf{q}}_k^H \overset{\mathcal{H}}{\mathbf{p}}_k} - \frac{\overset{\mathcal{H}}{\mathbf{H}}_k \overset{\mathcal{H}}{\mathbf{q}}_k \overset{\mathcal{H}}{\mathbf{p}}_k^H}{\overset{\mathcal{H}}{\mathbf{q}}_k^H \overset{\mathcal{H}}{\mathbf{p}}_k} + \frac{\overset{\mathcal{H}}{\mathbf{p}}_k \overset{\mathcal{H}}{\mathbf{q}}_k^H \overset{\mathcal{H}}{\mathbf{q}}_k \overset{\mathcal{H}}{\mathbf{p}}_k^H}{\left(\overset{\mathcal{H}}{\mathbf{q}}_k^H \overset{\mathcal{H}}{\mathbf{p}}_k\right)^2} + \frac{\overset{\mathcal{H}}{\mathbf{p}}_k \overset{\mathcal{H}}{\mathbf{p}}_k^H}{\overset{\mathcal{H}}{\mathbf{q}}_k^H \overset{\mathcal{H}}{\mathbf{p}}_k}.$$

In this case, if we denote $\overset{\mathcal{H}}{\mathbf{r}}_k := \overset{\mathcal{H}}{\mathbf{H}}_k \overset{\mathcal{H}}{\mathbf{q}}_k$ and $\overset{\mathcal{H}}{\mathbf{s}}_k := \overset{\mathcal{H}}{\mathbf{p}}_k \overset{\mathcal{H}}{\mathbf{q}}_k^H$, then

$$\mathbf{r}_k = \mathbf{H}_k^1 \mathbf{g}_k + \mathbf{H}_k^2 (\mathbf{g}_k)^\imath + \mathbf{H}_k^3 (\mathbf{g}_k)^\jmath + \mathbf{H}_k^4 (\mathbf{g}_k)^\kappa$$

and

$$\mathbf{s}_k = \mathbf{p}_k \mathbf{q}_k^H + (\mathbf{p}_k)^\imath ((\mathbf{q}_k)^\imath)^H + (\mathbf{p}_k)^\jmath ((\mathbf{q}_k)^\jmath)^H + (\mathbf{p}_k)^\kappa ((\mathbf{q}_k)^\kappa)^H.$$

Now, the *quaternion-valued Broyden-Fletcher-Goldfarb-Shanno (BFGS) method* is given by relation (7) and the following updates for the matrices $\mathbf{H}_k^1, \ldots, \mathbf{H}_k^4$:

$$\mathbf{H}_{k+1}^1 = \mathbf{H}_k^1 - \rho_k \mathbf{p}_k \mathbf{r}_k^H - \rho_k \mathbf{r}_k \mathbf{p}_k^H$$
$$+ \rho_k^2 \mathbf{s}_k \mathbf{s}_k^H + \rho_k \mathbf{p}_k \mathbf{p}_k^H,$$

$$\vdots$$

$$\mathbf{H}_{k+1}^4 = \mathbf{H}_k^4 - \rho_k \mathbf{p}_k ((\mathbf{r}_k)^\kappa)^H - \rho_k \mathbf{r}_k ((\mathbf{p}_k)^\kappa)^H$$
$$+ \rho_k^2 \mathbf{s}_k ((\mathbf{s}_k)^\kappa)^H + \rho_k \mathbf{p}_k ((\mathbf{p}_k)^\kappa)^H,$$

where

$$\rho_k := \frac{1}{4\mathrm{Re}(\mathbf{q}_k^H \mathbf{p}_k)}.$$

Lastly, of practical importance is the *one step secant (OSS) method* (see [16]), which pertains to the class of quasi-Newton algorithms, although it does not require storing the inverse Hessian approximation, like the BFGS method from which it was derived. The update rule (5) in this case has the form

$$\overset{\mathcal{H}}{\mathbf{w}}_{k+2} - \overset{\mathcal{H}}{\mathbf{w}}_{k+1} = \alpha_{k+1} \left( -\overset{\mathcal{H}}{\mathbf{g}}_{k+1} + A_k \overset{\mathcal{H}}{\mathbf{p}}_k + B_k \overset{\mathcal{H}}{\mathbf{q}}_k \right),$$

where

$$A_k = -\left(\frac{1}{4} + \frac{\mathcal{q}_k^{\mathcal{H}^H\mathcal{H}}\mathcal{q}_k}{\mathcal{q}_k^{\mathcal{H}^H\mathcal{H}}\mathbf{p}_k}\right)\frac{\mathbf{p}_k^{\mathcal{H}^H\mathcal{H}}\mathbf{g}_{k+1}}{\mathcal{q}_k^{\mathcal{H}^H\mathcal{H}}\mathbf{p}_k} + \frac{\mathcal{q}_k^{\mathcal{H}^H\mathcal{H}}\mathbf{g}_{k+1}}{\mathcal{q}_k^{\mathcal{H}^H\mathcal{H}}\mathbf{p}_k},$$

$$B_k = \frac{\mathbf{p}_k^{\mathcal{H}^H\mathcal{H}}\mathbf{g}_{k+1}}{\mathcal{q}_k^{\mathcal{H}^H\mathcal{H}}\mathbf{p}_k}.$$

We can process this to yield the relations that give the *quaternion-valued one step secant (OSS) method*:

$$\mathbf{w}_{k+2} - \mathbf{w}_{k+1} = \alpha_{k+1}\left(-\mathbf{g}_{k+1} + A_k\mathbf{p}_k + B_k\mathbf{q}_k\right),$$

where

$$A_k = -\left(\frac{1}{4} + \frac{\mathbf{q}_k^H\mathbf{q}_k}{\mathrm{Re}(\mathbf{q}_k^H\mathbf{p}_k)}\right)\frac{\mathrm{Re}(\mathbf{p}_k^H\mathbf{g}_{k+1})}{\mathrm{Re}(\mathbf{q}_k^H\mathbf{p}_k)} + \frac{\mathrm{Re}(\mathbf{q}_k^H\mathbf{g}_{k+1})}{\mathrm{Re}(\mathbf{q}_k^H\mathbf{p}_k)},$$

$$B_k = \frac{\mathrm{Re}(\mathbf{p}_k^H\mathbf{g}_{k+1})}{\mathrm{Re}(\mathbf{q}_k^H\mathbf{p}_k)}.$$

In order to apply the quasi-Newton algorithms to quaternion-valued feedforward neural networks, we only need to calculate the gradient $\mathbf{g}_k$ of the error function $E$ at different steps, which we do by using the quaternion-valued back-propagation algorithm.

## 4    Experimental Results

### 4.1    Linear Autoregressive Process with Circular Noise

The prediction of linear signals and chaotic signals represents a popular application of quaternion-valued neural networks. It is a time series regression problem, in which the future value of a signal is estimated based on its past behavior. The following experiments were done in the $M$-step ahead prediction setting, with $M = 1$.

An important benchmark firstly proposed in [17], and used in [18–22] for the complex-valued case, and in [23–26] for the quaternion-valued case, is the prediction of the quaternion-valued circular white noise

$$n(k) = n_a(k) + \imath n_b(k) + \jmath n_c(k) + \kappa n_d(k),$$

where $n_a$, $n_b$, $n_c$, $n_d \sim \mathcal{N}(0,1)$, passed through the stable autoregressive filter given by

$$y(k) = 1.79y(k-1) - 1.85y(k-2) + 1.27y(k-3) - 0.41y(k-4) + n(k).$$

In this and the next experiments, we trained quaternion-valued feedforward neural networks using the general gradient descent algorithm (GD), the quasi-Newton algorithm with symmetric rank-one updates (SR1), the quasi-Newton algorithm with Davidon-Fletcher-Powell updates (DFP), the quasi-Newton algorithm with Broyden-Fletcher-Goldfarb-Shanno updates (BFGS), and the one step secant method (OSS). The learning rate for the GD algorithm was taken to be 0.1, a value which empirically gave the best results.

The tap input of the filter was 4, and, consequently the networks had 4 inputs. A single hidden layer of 4 neurons and an output layer of one neuron complete the architecture of the networks. The activation function for the hidden layer was the fully quaternion hyperbolic tangent function $G^2(q) = \tanh q = \frac{e^q - e^{-q}}{e^q + e^{-q}}$, and the activation function for the output layer was the identity $G^3(q) = q$. Training was done for 5000 epochs with 5000 training samples.

After running each algorithm 50 times, the averaged results are given in Table 1. The table presents a measure of performance called *prediction gain*, defined by $R_p = 10 \log_{10} \frac{\sigma_x^2}{\sigma_e^2}$, where $\sigma_x^2$ represents the variance of the input signal and $\sigma_e^2$ represents the variance of the prediction error. The prediction gain is given in dB, and, because of the way it is defined, a bigger prediction gain means better performance.

In this case, DFP and SR1 gave approximately the same results, with BFGS performing better and OSS worse.

**Table 1.** Experimental results for linear autoregressive process with circular noise

| Algorithm | Prediction gain |
|-----------|-----------------|
| GD        | 4.51            |
| SR1       | 6.73            |
| DFP       | 6.61            |
| **BFGS**  | **7.23**        |
| OSS       | 5.11            |

## 4.2   Linear Autoregressive Process with Noncircular Noise

This experiment involves the prediction of the quaternion-valued noncircular white noise

$$n(k) = n_a(k) + \imath n_b(k) + \jmath n_c(k) + \kappa n_d(k),$$

given by $n_a = \mathcal{N}(0,1)$, $n_b = -0.6 n_a + \mathcal{N}(0,1)$, $n_c = 0.8 n_b + \mathcal{N}(0,1)$, $n_d = 0.8 n_a - 0.4 n_b + \mathcal{N}(0,1)$, passed through the same stable autoregressive filter given by

$$y(k) = 1.79 y(k-1) - 1.85 y(k-2) + 1.27 y(k-3) - 0.41 y(k-4) + n(k).$$

It was also used in [25, 26] to test quaternion-valued learning algorithms.

The tap input of the filter was 4, and so the networks had 4 inputs, 4 hidden neurons, and one output neuron. The only difference between this experiment and the previous one is the noncircular noise $n(k)$. The networks were trained for 5000 epochs with 5000 training samples.

The averaged results after 50 runs of each algorithm are given in Table 2. The measure of performance was the prediction gain, like in the above experiment. The situation is rather different in this case, making a strong point for the extension of all these algorithms to the quaternion-valued case. SR1 performed best in this case, followed by DFP and BFGS with similar results, and finally by OSS. The improvement over the classical gradient descent algorithm is obvious in this experiment, also.

**Table 2.** Experimental results for linear autoregressive process with noncircular noise

| Algorithm | Prediction gain |
|-----------|-----------------|
| GD        | 0.15            |
| **SR1**   | **3.74**        |
| DFP       | 3.59            |
| BFGS      | 3.12            |
| OSS       | 2.61            |

### 4.3  3D Lorenz System

The Lorenz system is given by the ordinary differential equations

$$\frac{\mathrm{d}x}{\mathrm{d}t} = \alpha(y - x)$$
$$\frac{\mathrm{d}y}{\mathrm{d}t} = -xz + \rho x - y$$
$$\frac{\mathrm{d}z}{\mathrm{d}t} = xy - \beta z,$$

where $\alpha = 10$, $\rho = 28$ and $\beta = 2/3$. This chaotic time series prediction problem was used to asses the performance of quaternion-valued neural networks in [1, 26–30].

Like in the above experiments, the tap input of the filter was 4, and so the networks had 4 inputs, 4 hidden neurons, and one output neuron. The networks were trained for 5000 epochs with 1337 training samples, which result from solving the 3D Lorenz system on the interval $[0, 25]$, with initial conditions $(x, y, z) = (1, 2, 3)$.

After running each algorithm 50 times on this problem, the average results are given in Table 3, in the same form as in the above experiments.

In this experiment, DFP and SR1 performed approximately in the same way, OSS slightly better, and the best was BFGS.

**Table 3.** Experimental results for the 3D Lorenz system

| Algorithm | Prediction gain |
|-----------|-----------------|
| GD | 7.56 |
| SR1 | 11.74 |
| DFP | 11.27 |
| **BFGS** | **13.74** |
| OSS | 12.09 |

## 4.4 4D Saito Chaotic Circuit

The last experiment concerns the 4D Saito chaotic circuit given by

$$\begin{bmatrix} \frac{dx_1}{dt} \\ \frac{dy_1}{dt} \end{bmatrix} = \begin{bmatrix} -1 & 1 \\ -\alpha_1 & \alpha_1\beta_1 \end{bmatrix} \begin{bmatrix} x_1 - \eta\rho_1 h(z) \\ y_1 - \eta\frac{\rho_1}{\beta_1}h(z) \end{bmatrix}$$

$$\begin{bmatrix} \frac{dx_2}{dt} \\ \frac{dy_2}{dt} \end{bmatrix} = \begin{bmatrix} -1 & 1 \\ -\alpha_2 & \alpha_2\beta_2 \end{bmatrix} \begin{bmatrix} x_2 - \eta\rho_2 h(z) \\ y_2 - \eta\frac{\rho_2}{\beta_2}h(z) \end{bmatrix},$$

where $h(z) = \begin{cases} 1, & z \geq -1 \\ -1, & z \leq 1 \end{cases}$, is the normalized hysteresis value, and $z = x_1 + x_2$,

$\rho_1 = \frac{\beta_1}{1-\beta_1}$, $\rho_2 = \frac{\beta_2}{1-\beta_2}$. The values of the parameters are $(\alpha_1, \beta_1, \alpha_2, \beta_2, \eta) = (7.5, 0.16, 15, 0.097, 1.3)$. Also a chaotic time series prediction problem, the 4D Saito chaotic circuit was used to benchmark quaternion-valued neural networks in [1, 23, 25, 31–34].

The architectures of the networks were the same as the ones in the previous experiments. We trained the networks for 5000 epochs with 5249 training samples, which result from solving the 4D Saito chaotic circuit on the interval $[0, 10]$, with initial conditions $(x_1, y_1, x_2, y_2) = (1, 0, 1, 0)$.

The prediction gains after 50 runs of each algorithm are given in Table 4.

In this last experiment, OSS had the best performance, followed closely by BFGS, and lastly by SR1 and DFP.

**Table 4.** Experimental results for the 4D Saito chaotic circuit

| Algorithm | Prediction gain |
|-----------|-----------------|
| GD | 5.76 |
| SR1 | 11.71 |
| DFP | 11.10 |
| BFGS | 16.24 |
| **OSS** | **16.94** |

# 5   Conclusions

The deduction of the most known variants of the quasi-Newton algorithm for training quaternion-valued neural networks was presented, starting from the real-valued case and using the framework of $\mathbb{HR}$ calculus to extend these methods to the quaternion-valued case.

Experimental results of four well-known time series prediction problems, each solved using the four variants of the quaternion-valued quasi-Newton algorithm have showed that all the quasi-Newton methods performed better on the proposed problems than the classical gradient descent algorithm, with an improvement in prediction gain of as much as 11 dB in some cases.

The Broyden-Fletcher-Goldfarb-Shanno (BFGS) method had the best performances among the quasi-Newton variants in two of the four applications. The one step secant method (OSS) performed best in one application, the symmetric rank-one (SR1) method in another, and the Davidon-Fletcher-Powell (DFP) method generally performed slightly worse. Thus, no algorithm is always better than all the others, yet another argument for the extension of these learning methods to the quaternion-valued case.

To conclude, their performance in solving different linear and chaotic time series prediction problems prove that quasi-Newton algorithms represent efficient methods for training feedforward quaternion-valued neural networks.

# References

1. Arena, P., Fortuna, L., Muscato, G., Xibilia, M.G.: Neural Networks in Multi-dimensional Domains Fundamentals and New Trends in Modelling and Control. Lecture Notes in Control and Information Sciences, vol. 234. Springer, London (1998)
2. Isokawa, T., Kusakabe, T., Matsui, N., Peper, F.: Quaternion neural network and its application. In: Palade, V., Howlett, R.J., Jain, L. (eds.) KES 2003. LNCS, vol. 2774, pp. 318–324. Springer, Heidelberg (2003). doi:10.1007/978-3-540-45226-3_44
3. Kusamichi, H., Isokawa, T., Matsui, N., Ogawa, Y., Maeda, K.: A new scheme for color night vision by quaternion neural network. In: International Conference on Autonomous Robots and Agents, pp. 101–106, December 2004
4. Buchholz, S., Le Bihan, N.: Polarized signal classification by complex and quaternionic multi-layer perceptrons. Int. J. Neural Syst. **18**(2), 75–85 (2008)
5. Jahanchahi, C., Took, C., Mandic, D.: On HR calculus, quaternion valued stochastic gradient, and adaptive three dimensional wind forecasting. In: International Joint Conference on Neural Networks (IJCNN), pp. 1–5. IEEE, July 2010
6. Took, C., Mandic, D., Aihara, K.: Quaternion-valued short term forecasting of wind profile. In: International Joint Conference on Neural Networks (IJCNN), pp. 1–6. IEEE, July 2010
7. Took, C., Strbac, G., Aihara, K., Mandic, D.: Quaternion-valued short-term joint forecasting of three-dimensional wind and atmospheric parameters. Renew. Energy **36**(6), 1754–1760 (2011)
8. Nocedal, J., Wright, S.: Numerical Optimization. Springer Series in Operations Research, Springer New York (1999)

9. Watrous, R.: Learning algorithms for connectionist networks: applied gradient methods of nonlinear optimization. Technical reports (CIS) MS-CIS-88-62, University of Pennsylvania, July 1988
10. Barnard, E.: Optimization for training neural nets. IEEE Trans. Neural Netw. **3**(2), 232–240 (1992)
11. Popa, C.A.: Quasi-newton learning methods for complex-valued neural networks. In: International Joint Conference on Neural Networks (IJCNN). IEEE, July 2015
12. Xu, D., Xia, Y., Mandic, D.: Optimization in quaternion dynamic systems: gradient, Hessian, and learning algorithms. IEEE Trans. Neural Netw. Learn. Syst. **27**(2), 249–261 (2016)
13. Luenberger, D., Ye, Y.: Linear and Nonlinear Programming. International Series in Operations Research & Management Science, vol. 116. Springer, Heidelberg (2008)
14. Fletcher, R., Powell, M.: A rapidly convergent descent method for minimization. Comput. J. **6**(2), 163–168 (1963)
15. Shanno, D.: Conditioning of quasi-newton methods for function minimization. Math. Comput. **24**(111), 647–656 (1970)
16. Battiti, R.: First and second-order methods for learning between steepest descent and Newton's method. Neural Comput. **4**(2), 141–166 (1992)
17. Mandic, D., Chambers, J.: Recurrent Neural Networks for Prediction: Learning Algorithms, Architectures and Stability. Wiley, New York (2001)
18. Goh, S., Mandic, D.: A complex-valued RTRL algorithm for recurrent neural networks. Neural Comput. **16**(12), 2699–2713 (2004)
19. Goh, S., Mandic, D.: Nonlinear adaptive prediction of complex-valued signals by complex-valued PRNN. IEEE Trans. Signal Process. **53**(5), 1827–1836 (2005)
20. Goh, S., Mandic, D.: Stochastic gradient-adaptive complex-valued nonlinear neural adaptive filters with a gradient-adaptive step size. IEEE Trans. Neural Netw. **18**(5), 1511–1516 (2007)
21. Goh, S., Mandic, D.: An augmented CRTRL for complex-valued recurrent neural networks. Neural Netw. **20**(10), 1061–1066 (2007)
22. Xia, Y., Jelfs, B., Van Hulle, M., Principe, J., Mandic, D.: An augmented echo state network for nonlinear adaptive filtering of complex noncircular signals. IEEE Trans. Neural Netw. **22**(1), 74–83 (2011)
23. Took, C., Mandic, D.: A quaternion widely linear adaptive filter. IEEE Trans. Signal Process. **58**(8), 4427–4431 (2010)
24. Wang, M., Took, C., Mandic, D.: A class of fast quaternion valued variable stepsize stochastic gradient learning algorithms for vector sensor processes. In: International Joint Conference on Neural Networks (IJCNN), pp. 2783–2786. IEEE, August 2011
25. Ujang, C.B., Took, C., Mandic, D.: Quaternion-valued nonlinear adaptive filtering. IEEE Trans. Neural Netw. **22**(8), 1193–1206 (2011)
26. Xia, Y., Jahanchahi, C., Mandic, D.: Quaternion-valued echo state networks. IEEE Trans. Neural Netw. Learn. Syst. **26**(4), 663–673 (2015)
27. Buchholz, S., Sommer, G.: Quaternionic spinor MLP. In: European Symposium on Artificial Neural Networks, pp. 377–382, April 2000
28. Took, C., Mandic, D.: The quaternion lms algorithm for adaptive filtering of hypercomplex processes. IEEE Trans. Signal Process. **57**(4), 1316–1327 (2009)
29. Took, C., Mandic, D., Benesty, J.: Study of the quaternion LMS and four-channel LMS algorithms. In: International Conference on Acoustics, Speech and Signal Processing, pp. 3109–3112. IEEE, April 2009

30. Che Ujang, B., Took, C., Mandic, D.: On quaternion analyticity: enabling quaternion-valued nonlinear adaptive filtering. In: International Conference on Acoustics, Speech and Signal Processing (ICASSP), pp. 2117–2120. IEEE, March 2012
31. Arena, P., Baglio, S., Fortuna, L., Xibilia, M.: Chaotic time series prediction via quaternionic multilayer perceptrons. In: International Conference on Systems, Man and Cybernetics, vol. 2, pp. 1790–1794. IEEE (1995)
32. Arena, P., Fortuna, L., Muscato, G., Xibilia, M.: Multilayer perceptrons to approximate quaternion valued functions. Neural Netw. 10(2), 335–342 (1997)
33. Ujang, C.B., Took, C., Mandic, D.: Split quaternion nonlinear adaptive filtering. Neural Netw. 23(3), 426–434 (2010)
34. Took, C., Mandic, D.: Quaternion-valued stochastic gradient-based adaptive iir filtering. IEEE Trans. Signal Process. 58(7), 3895–3901 (2010)

# Exponential Stability for Delayed Octonion-Valued Recurrent Neural Networks

Călin-Adrian Popa[✉]

Department of Computer and Software Engineering, Polytechnic University
Timişoara, Blvd. V. Pârvan, No. 2, 300223 Timişoara, Romania
calin.popa@cs.upt.ro

**Abstract.** Over the last few years, neural networks with values in multidimensional domains have gained a lot of interest. A non-associative normed division algebra which generalizes the complex and quaternion algebras is represented by the octonion algebra. It does not fall into the category of Clifford algebras, which are associative. Delayed octonion-valued recurrent neural networks are introduced, for which the states and weights are octonions. A sufficient criterion is given in the form of linear matrix inequalities, which assures the global exponential stability of the equilibrium point for the proposed networks. Lastly, a numerical example illustrates the correctness of the theoretical results.

**Keywords:** Octonion-valued neural networks · Global stability · Linear matrix inequality · Time delay

## 1 Introduction

In recent years, neural networks with values in multidimensional domains have been studied with increasing interest. The most popular form of multidimensional neural networks are the complex-valued neural networks. First introduced in the 1970s (see, for example, [1]), they have recently received more attention, especially due to their numerous applications, ranging from those in telecommunications and image processing, to those in complex-valued signal processing (see, for example, [2,3]).

Another type of multidimensional networks, defined on the 4-dimensional quaternion algebra, are the quaternion-valued neural networks. They were first introduced in the 1990s as a generalization of the complex-valued neural networks, see [4,5]. Lately, the have been used in an increasing number of applications, like chaotic time series prediction, the 4-bit parity problem, and, very recently, quaternion-valued signal processing.

A generalization of both the complex and quaternion algebras are the Clifford algebras, which have dimension $2^n$, $n \geq 1$. The numerous applications in physics and engineering of Clifford or geometric algebras made them appealing for use in the field of neural networks, also. Thus, Clifford-valued neural networks were defined in [6,7], and later discussed, for example, in [8]. Because of the Clifford

© Springer International Publishing AG 2017
I. Rojas et al. (Eds.): IWANN 2017, Part I, LNCS 10305, pp. 375–385, 2017.
DOI: 10.1007/978-3-319-59153-7_33

algebras' underlying connection with geometry, possible applications of neural networks with values in these algebras include processing different geometric objects and applying different geometric models to data.

The 8-dimensional algebra of octonions represents a different generalization of the complex and quaternion numbers, which does not fall into the Clifford algebra category. The easiest way to see this is by considering the fact that Clifford algebras are associative, whereas the octonion algebra is not. However, the octonions form a normed division algebra, an important property, especially for applications, which Clifford algebras don't have. This means that a norm and a multiplicative inverse can be defined for them. Moreover, it can be proved that the complex, quaternion, and octonion algebras are the only normed division algebras that can be defined over the field of real numbers.

With many applications in physics and geometry (see [9,10]), octonions have also been successfully applied to signal processing in the very recent years (see [11]). Taking all the above facts into consideration, the definition of octonion-valued neural networks seemed a promising idea, first in the form of feedforward networks [12]. Octonion-valued neural networks may be applied in signal processing and all other areas related to higher-dimensional object processing.

On the other hand, at the beginning of the 1980s, Hopfield introduced an energy function with the purpose of studying the dynamics of fully connected recurrent neural networks, see [13,14]. He also showed that this type of network can be applied to solving combinatorial problems. Since then, Hopfield neural networks have found numerous applications, especially in the synthesis of associative memories, image processing, speech processing, control systems, signal processing, pattern matching, etc.

Over the last few years, generalizations of the Hopfield neural networks to multidimensional domains have appeared. Complex-valued Hopfield networks were proposed in [15–17], quaternion-valued Hopfield networks in [18,19], and Clifford-valued Hopfield networks in [20,21]. As a consequence, taking all the above-discussed facts into account, this paper introduces delayed octonion-valued Hopfield neural networks, which could be applied to solve octonion optimization problems.

The rest of the paper is organized as follows: Sect. 2 gives the definition of delayed octonion-valued Hopfield neural networks, and an assumption and a useful lemma. A sufficient condition for the global exponential stability of the equilibrium point of these networks is given in Sect. 3. The correctness of the theoretical results is proved by a numerical example in Sect. 4. The conclusions are given in Sect. 5.

*Notations*: $\mathbb{R}$ denotes the set of real numbers, $\mathbb{R}^n$ denotes the $n$ dimensional Euclidean space, and $\mathbb{R}^{n \times n}$ the algebra of real square matrices of dimension $n \times n$. $A^T$ denotes the transpose of matrix $A$ and $*$ denotes the symmetric terms in a matrix. $I_n$ denotes the identity matrix of dimension $n$. $|| \cdot ||$ is the vector Euclidean norm or the matrix Frobenius norm. $A > 0$ $(A < 0)$ means that $A$ is a positive definite (negative definite) matrix. $\lambda_{\min}(P)$ is defined as the smallest eigenvalue of positive definite matrix $P$.

## 2    Preliminaries

We start by defining the algebra of octonions and highlighting some of its properties.

The algebra of octonions is defined as

$$\mathbb{O} := \left\{ x = \sum_{p=0}^{7} [x]_p e_p \,\middle|\, [x]_0, [x]_1, \ldots, [x]_7 \in \mathbb{R} \right\},$$

where $e_p$ represent the octonion units, $0 \le p \le 7$. They satisfy the following multiplication table

| × | $e_0$ | $e_1$ | $e_2$ | $e_3$ | $e_4$ | $e_5$ | $e_6$ | $e_7$ |
|---|---|---|---|---|---|---|---|---|
| $e_0$ | $e_0$ | $e_1$ | $e_2$ | $e_3$ | $e_4$ | $e_5$ | $e_6$ | $e_7$ |
| $e_1$ | $e_1$ | $-e_0$ | $e_3$ | $-e_2$ | $e_5$ | $-e_4$ | $-e_7$ | $e_6$ |
| $e_2$ | $e_2$ | $-e_3$ | $-e_0$ | $e_1$ | $e_6$ | $e_7$ | $-e_4$ | $-e_5$ |
| $e_3$ | $e_3$ | $e_2$ | $-e_1$ | $-e_0$ | $e_7$ | $-e_6$ | $e_5$ | $-e_4$ |
| $e_4$ | $e_4$ | $-e_5$ | $-e_6$ | $-e_7$ | $-e_0$ | $e_1$ | $e_2$ | $e_3$ |
| $e_5$ | $e_5$ | $e_4$ | $-e_7$ | $e_6$ | $-e_1$ | $-e_0$ | $-e_3$ | $e_2$ |
| $e_6$ | $e_6$ | $e_7$ | $e_4$ | $-e_5$ | $-e_2$ | $e_3$ | $-e_0$ | $-e_1$ |
| $e_7$ | $e_7$ | $-e_6$ | $e_5$ | $e_4$ | $-e_3$ | $-e_2$ | $e_1$ | $-e_0$ |

The addition of octonions is defined by

$$x + y = \sum_{p=0}^{7} ([x]_p + [y]_p) e_p,$$

and the multiplication is given by the multiplication of the unit octonions shown in the above table. Scalar multiplication is given by

$$\alpha x = \sum_{p=0}^{7} (\alpha [x]_p) e_p,$$

and thus $\mathbb{O}$ is a real algebra. It can be verified using the multiplication table that $e_i e_j = -e_j e_i \ne e_j e_i$, $\forall i \ne j$, $0 < i, j \le 7$, which means that $\mathbb{O}$ is not commutative, and that $(e_i e_j) e_k = -e_i(e_j e_k) \ne e_i(e_j e_k)$, for $i, j, k$ distinct, $0 < i, j, k \le 7$, or $e_i e_j \ne \pm e_k$, which shows that $\mathbb{O}$ is also not associative.

The conjugate of an octonion $x$ is defined by

$$\bar{x} = [x]_0 e_0 - \sum_{p=1}^{7} [x]_p e_p.$$

Using the conjugate, the norm of an octonion can be defined as

$$||x|| = \sqrt{x\overline{x}} = \sqrt{\sum_{p=0}^{7}[x]_p^2},$$

and the inverse of an octonion as $x^{-1} = \frac{\overline{x}}{||x||^2}$. Thus, $\mathbb{O}$ is a normed non-associative division algebra, unlike the 8-dimensional Clifford algebras, which are associative algebras, but not division algebras. In fact, the only three real division algebras that can be defined are the complex, quaternion, and octonion algebras.

We can now introduce octonion-valued Hopfield neural networks, for which the states and weights are from $\mathbb{O}$. The following set of differential equations describes this type of networks:

$$\dot{x}_i(t) = -d_i x_i(t) + \sum_{j=1}^{N} a_{ij} f_j(x_j(t)) + \sum_{j=1}^{N} b_{ij} g_j(x_j(t-\tau)) + u_i, \tag{1}$$

for $i \in \{1, \dots, N\}$, where $x_i(t) \in \mathbb{O}$ is the state of neuron $i$ at time $t$, $d_i \in \mathbb{R}$, $d_i > 0$, is the self-feedback connection weight of neuron $i$, $a_{ij} \in \mathbb{O}$ is the weight connecting neuron $j$ to neuron $i$ without delay, $b_{ij} \in \mathbb{O}$ is the weight connecting neuron $j$ to neuron $i$ with delay, $f_j : \mathbb{O} \to \mathbb{O}$ is the nonlinear octonion-valued activation function of neuron $j$ without delay, $g_j : \mathbb{O} \to \mathbb{O}$ is the nonlinear octonion-valued activation function of neuron $j$ with delay, $\tau \in \mathbb{R}$ is the delay and we assume $\tau > 0$, and $u_i \in \mathbb{O}$ is the external input of neuron $i$, $\forall i, j \in \{1, \dots, N\}$.

The derivative $\frac{dx_i(t)}{dt}$ is defined as the octonion formed by the derivatives of each element $[x_i(t)]_p$ of the octonion $x_i(t)$ with respect to $t$:

$$\dot{x}_i(t) = \frac{dx_i(t)}{dt} := \sum_{p=0}^{7} \frac{d([x_i]_p)}{dt} e_p.$$

Thus, the above set of differential equations has values in $\mathbb{O}$, and the multiplication between the weights and the values of the activation functions is the octonion multiplication.

We need to make an assumption about the activation functions, in order to study the stability of the above defined network.

**Assumption 1.** *The following Lipschitz conditions are satisfied by the octonion-valued activation functions $f_j$ and $g_j$:*

$$||f_j(x) - f_j(x')|| \le l_j^f ||x - x'||, \ \forall x, x' \in \mathbb{O},$$

$$||g_j(x) - g_j(x')|| \le l_j^g ||x - x'||, \ \forall x, x' \in \mathbb{O},$$

*where $l_j^f > 0$ and $l_j^g > 0$ are the Lipschitz constants, $\forall j \in \{1, \dots, N\}$. Moreover, we denote $\overline{L_f} = diag(l_1^f I_8, l_2^f I_8, \dots, l_N^f I_8)$, $\overline{L_g} = diag(l_1^g I_8, l_2^g I_8, \dots, l_N^g I_8)$.*

We will first transform the system of octonion-valued differential Eq. (1) into a real-valued one. For this, we will expand each equation in (1) into 8 real-valued equations:

$$[\dot{x}_i(t)]_p = -d_i[x_i(t)]_p + \sum_{j=1}^{N}\sum_{q=0}^{7}[a_{ij}]_{pq}[f_j(x_j(t))]_q$$

$$+\sum_{j=1}^{N}\sum_{q=0}^{7}[b_{ij}]_{pq}[g_j(x_j(t-\tau))]_q + [u_i]_p, \qquad (2)$$

for $0 \le p \le 7$, $i \in \{1,\ldots,N\}$, where $[x]_{pq}$ represents an entry of the matrix $\mathrm{mat}(x)$, defined by

$$\mathrm{mat}(x) := \begin{bmatrix} [x]_0 & -[x]_1 & -[x]_2 & -[x]_3 & -[x]_4 & -[x]_5 & -[x]_6 & -[x]_7 \\ [x]_1 & [x]_0 & -[x]_3 & [x]_2 & -[x]_5 & [x]_4 & [x]_7 & -[x]_6 \\ [x]_2 & [x]_3 & [x]_0 & -[x]_1 & -[x]_6 & -[x]_7 & [x]_4 & [x]_5 \\ [x]_3 & -[x]_2 & [x]_1 & [x]_0 & -[x]_7 & [x]_6 & -[x]_5 & -[x]_4 \\ [x]_4 & [x]_5 & [x]_6 & [x]_7 & [x]_0 & -[x]_1 & -[x]_2 & -[x]_3 \\ [x]_5 & -[x]_4 & [x]_7 & -[x]_6 & [x]_1 & [x]_0 & [x]_3 & -[x]_2 \\ [x]_6 & -[x]_7 & -[x]_4 & [x]_5 & [x]_2 & -[x]_3 & [x]_0 & [x]_1 \\ [x]_7 & [x]_6 & -[x]_5 & -[x]_4 & [x]_3 & [x]_2 & -[x]_1 & [x]_0 \end{bmatrix}.$$

If we now denote $\mathrm{vec}(x) := ([x]_0, [x]_1, \ldots, [x]_7)^T$, the Eq. (2) can be written as

$$\mathrm{vec}(\dot{x}_i(t)) = -d_i I_8 \mathrm{vec}(x_i(t)) + \sum_{j=1}^{N} \mathrm{mat}(a_{ij})\mathrm{vec}(f_j(x_j(t)))$$

$$+ \sum_{j=1}^{N} \mathrm{mat}(b_{ij})\mathrm{vec}(g_j(x_j(t-\tau))) + \mathrm{vec}(u_i), \qquad (3)$$

for $i \in \{1,\ldots,N\}$. Furthermore, by denoting $y(t) = (\mathrm{vec}(x_1(t))^T, \mathrm{vec}(x_2(t))^T,$ $\ldots, \mathrm{vec}(x_N(t))^T)^T$, $\overline{D} = \mathrm{diag}(d_1 I_8, d_2 I_8, \ldots, d_N I_8)$, $\overline{A} = (\mathrm{mat}(a_{ij}))_{1\le i,j\le N}$, $\overline{B} = (\mathrm{mat}(b_{ij}))_{1<i,j\le N}$, $\overline{f}(y(t)) = (\mathrm{vec}(f_1(x_1(t)))^T, \mathrm{vec}(f_2(x_2(t)))^T, \ldots, \mathrm{vec}$ $(f_N(x_N(t)))^T)^T$, $\overline{g}(y(t-\tau)) = (\mathrm{vec}(g_1(x_1(t-\tau)))^T, \mathrm{vec}(g_2(x_2(t-\tau)))^T, \ldots, \mathrm{vec}$ $(g_N(x_N(t-\tau)))^T)^T$, $\overline{u} = (\mathrm{vec}(u_1)^T, \mathrm{vec}(u_2)^T, \ldots, \mathrm{vec}(u_N)^T)^T$, and also $y = y(t)$, $y^\tau = y(t-\tau)$, system (1) can be written as

$$\dot{y} = -\overline{D}y + \overline{A}\,\overline{f}(y) + \overline{B}\,\overline{g}(y^\tau) + \overline{u}. \qquad (4)$$

The equilibrium point of (4) can now be shifted to the origin, and so the system (4) becomes

$$\dot{\tilde{y}} = -\overline{D}\tilde{y} + \overline{A}\,\tilde{f}(\tilde{y}) + \overline{B}\,\tilde{g}(\tilde{y}^\tau), \qquad (5)$$

where $\tilde{y} = y - \hat{y}$, $\tilde{y}^\tau = y^\tau - \hat{y}$, $\tilde{f}(\tilde{y}) = \overline{f}(\tilde{y}+\hat{y}) - \overline{f}(\hat{y})$, and $\tilde{g}(\tilde{y}^\tau) = \overline{g}(\tilde{y}^\tau + \hat{y}) - \overline{g}(\hat{y})$.

*Remark 1.* Systems (5) and (1) are equivalent, meaning that any property that holds for system (5), will also hold for system (1). For this reason, from now on, we will only study the global exponential stability of the origin of system (5).

We will also need the following lemma:

**Lemma 1 [22].** *For any vector function* $y : [a, b] \rightarrow \mathbb{R}^{8N}$ *and positive definite matrix* $M \in \mathbb{R}^{8N \times 8N}$, *the following linear matrix inequality (LMI) holds:*

$$\left( \int_a^b y(s)ds \right)^T M \left( \int_a^b y(s)ds \right) \leq (b - a) \int_a^b y^T(s)My(s)ds,$$

*where the integrals are well defined.*

## 3    Main Results

We give an LMI-based sufficient condition for the global exponential stability of the origin of (5).

**Theorem 1.** *If Assumption 1 holds, then the origin of system (5) is globally exponentially stable if there exist positive definite matrices* $P$, $Q_1$, $Q_2$, $Q_3$, $S_1$, $S_2$, $S_3$, $S_4$, *positive block-diagonal matrices* $R_1$, $R_2$, $R_3$, $R_4$, *all from* $\mathbb{R}^{8N \times 8N}$, *and* $\varepsilon > 0$, *such that the following linear matrix inequality (LMI) holds*

$$(\Pi)_{9 \times 9} < 0, \tag{6}$$

*where* $\Pi_{1,1} = 2\varepsilon P - P\overline{D} - \overline{D}P + Q_1 + \tau S_2 + \tau^{-1}e^{-2\varepsilon\tau}S_1 + \tau\overline{D}S_1\overline{D} + \overline{L_f}^T R_1\overline{L_f} + \overline{L_g}^T R_3\overline{L_g}$, $\Pi_{1,3} = P\overline{A} - \tau^{-1}e^{-2\varepsilon\tau}S_1 - \tau\overline{D}S_1\overline{A}$, $\Pi_{1,6} = P\overline{B} - \tau\overline{D}S_1\overline{B}$, $\Pi_{2,2} = -e^{-2\varepsilon\tau}Q_1 + \tau^{-1}e^{-2\varepsilon\tau}S_1 + \overline{L_f}^T R_2\overline{L_f} + \overline{L_g}^T R_4\overline{L_g}$, $\Pi_{3,3} = Q_2 + \tau S_3 - R_1 + \tau\overline{A}^T S_1\overline{A}$, $\Pi_{3,6} = \tau\overline{A}^T S_1\overline{B}$, $\Pi_{4,4} = -e^{-2\varepsilon\tau}Q_2 - R_2$, $\Pi_{5,5} = Q_3 + \tau S_4 - R_3$, $\Pi_{6,6} = -e^{-2\varepsilon\tau}Q_1 - R_4 + \tau\overline{B}^T S_1\overline{B}$, $\Pi_{7,7} = -\tau^{-1}e^{-2\varepsilon\tau}S_2$, $\Pi_{8,8} = -\tau^{-1}e^{-2\varepsilon\tau}S_3$, $\Pi_{9,9} = -\tau^{-1}e^{-2\varepsilon\tau}S_4$.

*Proof.* We begin by defining the Lyapunov-Krasovskii functional

$$\begin{aligned}
V(\tilde{y}(t)) = {}& e^{2\varepsilon t}\tilde{y}^T(t)P\tilde{y}(t) \\
& + \int_{t-\tau}^t e^{2\varepsilon s}\tilde{y}^T(s)Q_1\tilde{y}(s)ds \\
& + \int_{t-\tau}^t e^{2\varepsilon s}\tilde{f}^T(\tilde{y}(s))Q_2\tilde{f}(\tilde{y}(s))ds \\
& + \int_{t-\tau}^t e^{2\varepsilon s}\tilde{g}^T(\tilde{y}(s))Q_3 g(\tilde{y}(s))ds \\
& + \int_{-\tau}^0 \int_{t+\theta}^t e^{2\varepsilon s}\dot{\tilde{y}}^T(s)S_1\dot{\tilde{y}}(s)ds d\theta \\
& + \int_{-\tau}^0 \int_{t+\theta}^t e^{2\varepsilon s}\tilde{y}^T(s)S_2\tilde{y}(s)ds d\theta
\end{aligned}$$

$$+ \int_{-\tau}^{0} \int_{t+\theta}^{t} e^{2\varepsilon s} \tilde{f}^T(\tilde{y}(s)) S_3 \tilde{f}(\tilde{y}(s)) ds d\theta$$

$$+ \int_{-\tau}^{0} \int_{t+\theta}^{t} e^{2\varepsilon s} \tilde{g}^T(\tilde{y}(s)) S_4 g(\tilde{y}(s)) ds d\theta.$$

The time derivative of $V$ along the trajectories of system (5) is

$$\dot{V}(\tilde{y}) = e^{2\varepsilon t} \left[ 2\varepsilon \tilde{y}^T P \tilde{y} + \dot{\tilde{y}}^T P \tilde{y} + \tilde{y}^T P \dot{\tilde{y}} + \tilde{y}^T Q_1 \tilde{y} - e^{-2\varepsilon \tau} \tilde{y}^{\tau T} Q_1 \tilde{y}^\tau \right.$$

$$+ \tilde{f}^T(\tilde{y}) Q_2 \tilde{f}(\tilde{y}) - e^{-2\varepsilon \tau} \tilde{f}^T(\tilde{y}^\tau) Q_2 \tilde{f}(\tilde{y}^\tau) + \tilde{g}^T(\tilde{y}) Q_3 \tilde{g}(\tilde{y})$$

$$- e^{-2\varepsilon \tau} \tilde{g}^T(\tilde{y}^\tau) Q_3 \tilde{g}(\tilde{y}^\tau) + \tau \dot{\tilde{y}}^T S_1 \dot{\tilde{y}} - \int_{t-\tau}^{t} e^{2\varepsilon(s-t)} \dot{\tilde{y}}^T(s) S_1 \dot{\tilde{y}}(s) ds$$

$$+ \tau \tilde{y}^T S_2 \tilde{y} - \int_{t-\tau}^{t} e^{2\varepsilon(s-t)} \tilde{y}^T(s) S_2 \tilde{y}(s) ds + \tau \tilde{f}^T(\tilde{y}) S_3 \tilde{f}(\tilde{y})$$

$$- \int_{t-\tau}^{t} e^{2\varepsilon(s-t)} \tilde{f}^T(\tilde{y}(s)) S_3 \tilde{f}(\tilde{y}(s)) ds + \tau \tilde{g}^T(\tilde{y}) S_4 \tilde{g}(\tilde{y})$$

$$\left. - \int_{t-\tau}^{t} e^{2\varepsilon(s-t)} \tilde{g}^T(\tilde{y}(s)) S_4 g(\tilde{y}(s)) ds \right]$$

$$\leq e^{2\varepsilon t} \left[ 2\varepsilon \tilde{y}^T P \tilde{y} + (-\overline{D}\tilde{y} + \overline{A}\tilde{f}(\tilde{y}) + \overline{B}\tilde{g}(\tilde{y}^\tau))^T P \tilde{y} + \tilde{y}^T P(-\overline{D}\tilde{y} \right.$$

$$+ \overline{A}\tilde{f}(\tilde{y}) + \overline{B}\tilde{g}(\tilde{y}^\tau)) + \tilde{y}^T Q_1 \tilde{y} - e^{-2\varepsilon \tau} \tilde{y}^{\tau T} Q_1 \tilde{y}^\tau + \tilde{f}^T(\tilde{y}) Q_2 \tilde{f}(\tilde{y})$$

$$- e^{-2\varepsilon \tau} \tilde{f}^T(\tilde{y}^\tau) Q_2 \tilde{f}(\tilde{y}^\tau) + \tilde{g}^T(\tilde{y}) Q_3 \tilde{g}(\tilde{y}) - e^{-2\varepsilon \tau} \tilde{g}^T(\tilde{y}^\tau) Q_3 \tilde{g}(\tilde{y}^\tau)$$

$$+ \tau \dot{\tilde{y}}^T S_1 \dot{\tilde{y}} - \tau^{-1} e^{-2\varepsilon \tau} \left( \int_{t-\tau}^{t} \dot{\tilde{y}}(s) ds \right)^T S_1 \left( \int_{t-\tau}^{t} \dot{\tilde{y}}(s) ds \right)$$

$$+ \tau \tilde{y}^T S_2 \tilde{y} - \tau^{-1} e^{-2\varepsilon \tau} \left( \int_{t-\tau}^{t} \tilde{y}(s) ds \right)^T S_2 \left( \int_{t-\tau}^{t} \tilde{y}(s) ds \right)$$

$$+ \tau \tilde{f}^T(\tilde{y}) S_3 \tilde{f}(\tilde{y}) - \tau^{-1} e^{-2\varepsilon \tau} \left( \int_{t-\tau}^{t} \tilde{f}(\tilde{y}(s)) ds \right)^T S_3 \left( \int_{t-\tau}^{t} \tilde{f}(\tilde{y}(s)) ds \right)$$

$$\left. + \tau \tilde{g}^T(\tilde{y}) S_4 \tilde{g}(\tilde{y}) - \tau^{-1} e^{-2\varepsilon \tau} \left( \int_{t-\tau}^{t} \tilde{g}(\tilde{y}(s)) ds \right)^T S_4 \left( \int_{t-\tau}^{t} \tilde{g}(\tilde{y}(s)) ds \right) \right],$$

$$(7)$$

where the inequality was deduced using Lemma 1.

The Lipschitz conditions in Assumption 1 are equivalent with

$$||f_j(x) - f_j(x')|| \leq l_j^f ||x - x'||$$

$$\Leftrightarrow ||\text{vec}(f_j(x)) - \text{vec}(f_j(x'))|| \leq l_j^f ||\text{vec}(x) - \text{vec}(x')||,$$

for $j \in \{1, \ldots N\}$, and the analogous ones for the functions $g_j$. Now, from these inequalities we can deduce that there exist positive block-diagonal matrices $R_1 = \text{diag}(r_1^1 I_8, r_2^1 I_8, \ldots, r_N^1 I_8)$, $R_2 = \text{diag}(r_1^2 I_8, r_2^2 I_8, \ldots, r_N^2 I_8)$, $R_3 = \text{diag}(r_1^3 I_8, r_2^3 I_8, \ldots, r_N^3 I_8)$, $R_4 = \text{diag}(r_1^4 I_8, r_2^4 I_8, \ldots, r_N^4 I_8)$, such that

$$0 \le \tilde{y}^T \overline{L_f}^T R_1 \overline{L_f} \tilde{y} - \tilde{f}^T(\tilde{y}) R_1 \tilde{f}(\tilde{y}), \ 0 \le \tilde{y}^{\tau T} \overline{L_f}^T R_2 \overline{L_f} \tilde{y}^\tau - \tilde{f}^T(\tilde{y}^\tau) R_2 \tilde{f}(\tilde{y}^\tau), \quad (8)$$

$$0 \le \tilde{y}^T \overline{L_g}^T R_3 \overline{L_g} \tilde{y} - \tilde{g}^T(\tilde{y}) R_3 \tilde{g}(\tilde{y}), \ 0 \le \tilde{y}^{\tau T} \overline{L_g}^T R_4 \overline{L_g} \tilde{y}^\tau - \tilde{g}^T(\tilde{y}^\tau) R_4 \tilde{g}(\tilde{y}^\tau). \quad (9)$$

Adding inequalities (8) and (9), with inequality (7), yields

$$\dot{V}(\tilde{y}) \le e^{2\varepsilon t} \zeta^T \Pi \zeta, \quad (10)$$

where

$$\zeta = \begin{bmatrix} \tilde{y}^T & \tilde{y}^{\tau T} & \tilde{f}^T(\tilde{y}) & \tilde{f}^T(\tilde{y}^\tau) & \tilde{g}^T(\tilde{y}) \tilde{g}^T(\tilde{y}^\tau) \end{bmatrix}$$
$$\left( \int_{t-\tau}^t \tilde{y}(s) ds \right)^T \left( \int_{t-\tau}^t \tilde{f}(\tilde{y}(s)) ds \right)^T \left( \int_{t-\tau}^t \tilde{g}(\tilde{y}(s)) ds \right)^T \Big]^T,$$

and $\Pi$ is defined by (6). Condition (6) says that $\Pi < 0$, so we can infer from (10) that $\dot{V}(\tilde{y}) < 0$, which means that $V(\tilde{y}(t))$ is strictly decreasing for $t \ge 0$. From the definition of $V(\tilde{y}(t))$, we can further deduce that

$$e^{2\varepsilon t} \lambda_{\min}(P) \|\tilde{y}(t)\|^2 \le e^{2\varepsilon t} \tilde{y}^T(t) P \tilde{y}(t) \le V(t) \le V_0, \ \forall t \ge T, \ T \ge 0,$$

where $V_0 = \max_{0 \le t \le T} V(t)$. Consequently,

$$\|\tilde{y}(t)\|^2 \le \frac{V_0}{e^{2\varepsilon t} \lambda_{\min}(P)} \Leftrightarrow \|\tilde{y}(t)\| \le M e^{-\varepsilon t}, \ \forall t \ge 0,$$

for $M = \sqrt{\frac{V_0}{\lambda_{\min}(P)}}$. Thus, we obtained the global exponential stability for the origin of system (5).

## 4    Numerical Example

We now give a numerical example to prove the correctness of the result derived above.

*Example 1.* Consider the following delayed octonion-valued Hopfield neural network with two neurons:

$$\begin{cases} \dot{x}_1(t) = -d_1 x_1(t) + \sum_{j=1}^2 a_{1j} f_j(x_j(t)) + \sum_{j=1}^2 b_{1j} g_j(x_j(t-\tau)) + u_1, \\ \dot{x}_2(t) = -d_2 x_2(t) + \sum_{j=1}^2 a_{2j} f_j(x_j(t)) + \sum_{j=1}^2 b_{1j} g_j(x_j(t-\tau)) + u_2, \end{cases} \quad (11)$$

where $d_1 = 50$, $d_2 = 40$, and

$$\mathrm{vec}(a_{11}) = (1,1,2,2,1,-1,-1,1)^T, \ \mathrm{vec}(a_{12}) = (2,1,1,-2,2,1,-2,2)^T,$$

$$\mathrm{vec}(a_{21}) = (2,-2,2,1,2,-2,1,2)^T, \ \mathrm{vec}(a_{22}) = (1,2,2,-2,1,1,2,-2)^T,$$

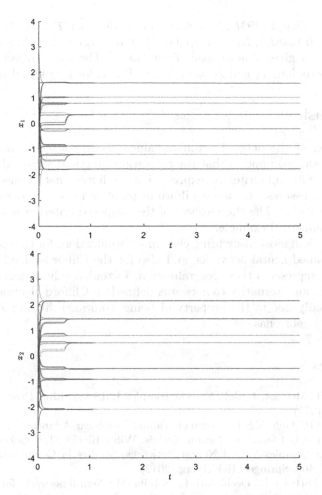

**Fig. 1.** State trajectories of elements of $x_1$ and $x_2$ in Example 1

$$\text{vec}(b_{11}) = (2, 1, 2, 1, -2, 2, -1, 2)^T, \quad \text{vec}(b_{12}) = (-2, 2, -2, 2, 1, 2, -2, 2)^T,$$

$$\text{vec}(b_{21}) = (1, -2, 2, -2, 1, 2, 2, 2)^T, \quad \text{vec}(b_{22}) = (1, 2, 2, 1, 2, -2, -2, 1)^T,$$

$$\text{vec}(u_1) = (10, -20, 30, -40, 50, -70, 80, -90)^T,$$

$$\text{vec}(u_2) = (90, -40, 10, -60, 30, -80, 50, -20)^T,$$

$$f_j\left([x]_p\right) = \frac{1}{1 + e^{-[x]_p}}, \quad g_j\left([x]_p\right) = \frac{1 - e^{-[x]_p}}{1 + e^{-[x]_p}}, \quad p \in \{0, 1, \ldots, 7\}, \quad j \in \{1, 2\}.$$

We have that $l_1^f = l_2^f = \frac{\sqrt{2}}{2}$ and $l_1^g = l_2^g = 2\sqrt{2}$. Also, the constant delay is $\tau = 0.5$.

The global exponential stability of the equilibrium point of system (11) is obtained by solving the LMI in condition (6) in Theorem 1, to get

$\varepsilon = 3$, $R_1 = \text{diag}(3.1392I_8, 3.0388I_8)$, $R_2 = \text{diag}(0.4777I_8, 0.4685I_8)$, $R_3 = \text{diag}(0.2165I_8, 0.1348I_8)$, $R_4 = \text{diag}(0.0017I_8, 0.0016I_8)$. (The values of the other matrices are not given due to space limitations.) The state trajectories of the elements of octonions $x_1$ and $x_2$ are given in Fig. 1, for four initial values.

## 5    Conclusions

The definition of the delayed octonion-valued recurrent neural networks was given. Under the assumption that the activation functions satisfy the Lipschitz condition, a sufficient criterion expressed as a linear matrix inequality was derived, which assures that the equilibrium point of these networks is globally exponentially stable. The effectiveness of the proposed criterion was showed by providing a numerical example.

The future will most likely bring even more applications for the complex- and quaternion-valued neural networks, and also for the Clifford-valued neural networks, which represents their generalization. Octonion-valued neural networks can be used as an alternative to networks defined on Clifford algebras of dimension 8, especially due to the property of being a normed division algebra that the octonion algebra has.

## References

1. Widrow, B., McCool, J., Ball, M.: The complex LMS algorithm. Proc. IEEE **63**(4), 719–720 (1975)
2. Mandic, D.P., Goh, V.S.L.: Complex Valued Nonlinear Adaptive Filters: Noncircularity, Widely Linear and Neural Models. Wiley-Blackwell, Hoboken (2009)
3. Hirose, A.: Complex-Valued Neural Networks. Studies in Computational Intelligence, vol. 400. Springer, Heidelberg (2012)
4. Arena, P., Fortuna, L., Occhipinti, L., Xibilia, M.: Neural networks for quaternion-valued function approximation. In: International Symposium on Circuits and Systems (ISCAS), vol. 6, pp. 307–310. IEEE (1994)
5. Arena, P., Fortuna, L., Muscato, G., Xibilia, M.: Multilayer perceptrons to approximate quaternion valued functions. Neural Netw. **10**(2), 335–342 (1997)
6. Pearson, J., Bisset, D.: Back propagation in a Clifford algebra. Int. Conf. Artif. Neural Netw. **2**, 413–416 (1992)
7. Pearson, J., Bisset, D.: Neural networks in the Clifford domain. In: International Conference on Neural Networks, vol. 3, pp. 1465–1469. IEEE (1994)
8. Buchholz, S., Sommer, G.: On Clifford neurons and Clifford multi-layer perceptrons. Neural Netw. **21**(7), 925–935 (2008)
9. Okubo, S.: Introduction to Octonion and Other Non-associative Algebras in Physics. Cambridge University Press, Cambridge (1995)
10. Dray, T., Manogue, C.: The Geometry of the Octonions. World Scientific, Singapore (2015)
11. Snopek, K.M.: Quaternions and octonions in signal processing - fundamentals and some new results. Przeglad Telekomunikacyjny + Wiadomosci Telekomunikacyjne **6**, 618–622 (2015)

12. Popa, C.-A.: Octonion-valued neural networks. In: Villa, A.E.P., Masulli, P., Pons Rivero, A.J. (eds.) ICANN 2016. LNCS, vol. 9886, pp. 435–443. Springer, Cham (2016). doi:10.1007/978-3-319-44778-0_51

13. Hopfield, J.J.: Neural networks and physical systems with emergent collective computational abilities. Proc. Nat. Acad. Sci. U.S.A. **79**(8), 2554–2558 (1982)

14. Hopfield, J.J.: Neurons with graded response have collective computational properties like those of two-state neurons. Proc. Nat. Acad. Sci. U.S.A. **81**(10), 3088–3092 (1984)

15. Liu, X., Chen, T.: Global exponential stability for complex-valued recurrent neural networks with asynchronous time delays. IEEE Trans. Neural Netw. Learn. Syst. **27**(3), 593–606 (2016)

16. Song, Q., Zhao, Z.: Stability criterion of complex-valued neural networks with both leakage delay and time-varying delays on time scales. Neurocomputing **171**, 179–184 (2016)

17. Song, Q., Yan, H., Zhao, Z., Liu, Y.: Global exponential stability of complex-valued neural networks with both time-varying delays and impulsive effects. Neural Netw. **79**, 108–116 (2016)

18. Valle, M.: A novel continuous-valued quaternionic Hopfield neural network. In: Brazilian Conference on Intelligent Systems (BRACIS), pp. 97–102. IEEE, October 2014

19. Liu, Y., Zhang, D., Lu, J., Cao, J.: Global $\mu$-stability criteria for quaternion-valued neural networks with unbounded time-varying delays. Inf. Sci. **360**, 273–288 (2016)

20. Zhu, J., Sun, J.: Global exponential stability of Clifford-valued recurrent neural networks. Neurocomputing **173**(Part 3), 685–689 (2016)

21. Liu, Y., Xu, P., Lu, J., Liang, J.: Global stability of Clifford-valued recurrent neural networks with time delays. Nonlinear Dyn. **84**(2), 767–777 (2016)

22. Gu, K.: An integral inequality in the stability problem of time-delay systems. In: Proceedings of the 39th IEEE Conference on Decision and Control, pp. 2805–2810 (2000)

# Forward Stagewise Regression
# on Incomplete Datasets

Marcelo B.A. Veras[1], Diego P.P. Mesquita[1], João P.P. Gomes[1(✉)],
Amauri H. Souza Junior[2], and Guilherme A. Barreto[3]

[1] Department of Computer Science, Federal University of Ceará,
Fortaleza, Ceará, Brazil
{marcelobveras,diegoparente,jpaulo}@lia.ufc.br
[2] Department of Computer Science, Federal Institute of Ceará,
Maracanaú, Ceará, Brazil
amauriholanda@ifce.edu.br
[3] Department of Teleinformatics Engineering, Federal University of Ceará,
Fortaleza, Ceará, Brazil
guilherme@deti.ufc.br

**Abstract.** The Forward Stagewise Regression (FSR) algorithm is a popular procedure to generate sparse linear regression models. However, the standard FSR assumes that the data are fully observed. This assumption is often flawed and pre-processing steps are applied to the dataset so that FSR can be used. In this paper, we extend the FSR algorithm to directly handle datasets with partially observed feature vectors, dismissing the need for the data to be pre-processed. Experiments were carried out on real-world datasets and the proposed method reported promising results when compared to the usual strategies for handling incomplete data.

## 1 Introduction

Missing data is a common occurrence in many real-world domains that may have a significant effect on the results of machine learning algorithms. Roughly speaking, in the problem of learning from incomplete datasets, a machine learning algorithm has to learn from input vectors where some of its attributes are unknown. Possible reasons for the absence of these attributes are transmission and storage problems, operator failure, measurement error and etc. [1].

According to Little and Rubin in [2], understanding the missingness mechanism is fundamental to the task of designing solutions to handle the missing data problem. Missing data mechanisms are usually classified into three main groups: Missing Completely at Random (MCAR), Missing at Random (MAR) and Not Missing at Random (NMAR). In MCAR, the missingness of a component is independent of its real value and any value of other components on the dataset. This characterization is often seen as very restrictive and various authors consider that it is very unlikely in real-world applications [3]. A more realistic approach is the MAR mechanism. In MAR, the missingness of a component is independent of the value itself but can be related to the observed values.

© Springer International Publishing AG 2017
I. Rojas et al. (Eds.): IWANN 2017, Part I, LNCS 10305, pp. 386–395, 2017.
DOI: 10.1007/978-3-319-59153-7_34

Finally, MNAR characterizes a whole different situation where the instance is not missing at random. In MNAR the missing probability is related to the value of the missing component and handling such problems usually requires a model of the missingness mechanism. In this work, we consider the case where the probability of a component being missing is not related to its value, hence we adopt the less restrictive option, assuming that the missing data is MAR.

Considering the MAR framework, the simplest strategy to handle missing data is the Listwise Deletion (LD). In this method, only fully observed input vectors are used to build the learning model. Although LD is simple and popular, it may lead to poor modeling as the number of vectors with missing components increases [4]. In such cases, a better solution consists in filling the missing components with likely values. The so-called imputation strategies comprise a variety of methods mostly based on either probabilistic models or regression methods [4]. In the probabilistic approach, the vectors in the dataset are assumed to be i.i.d. random variables and inference is carried out to estimate the missing values. The Conditional Mean Imputation (CMI, [5]) is a widely used statistical imputation method in which the missing components are filled according to their expected values given the observed components of the same vector. In general, one can assume the data follow any distribution, being the multivariate normal distribution the most common use.

It is worth noting that, in the context of machine learning, data imputation based methods consist of pre-processing steps, i.e., the learning process only starts when the missing data vectors are filled or deleted. Recently, [1,6] propose variants of machine learning methods that can handle missing data directly and thus do not require any pre-processing step. In addition to being elegant solutions, those methods also achieved promising results.

The Forward Stagewise Regression (FSR, [8]) algorithm is a linear regression sparse model. According to Hastie et al. [9], there are two main reasons that explain why sparse linear models are preferable to non-sparse ones (e.g., linear models coupled with least-squares estimation). First, sparse models often produce lower variance predictions, and hence good generalization. Second, models with reduced number of nonzero coefficients tend to represent only strong effects of the data, thus eliminating details that may be important to a further analysis. The FSR follows a strategy for constructing a sequence of sparse regression estimates: it starts with all coefficients equal to zero, and iteratively updates the coefficient of the variable that achieves the maximal correlation with the current residual [7].

In this paper we propose a new variant of the FSR algorithm with a built-in mechanism to handle missing data. The proposed model is based on the estimation of the expected correlations between each feature and the vector of residuals at each iteration. To compute the necessary steps, we assume that the data are normally distributed. Results show that our method is able to outperform LD and CMI strategies in various real-world datasets.

The remainder of the paper is organized as follows. Section 2 overviews the FSR algorithm. Section 3 introduces the proposed method to extend the FSR to incomplete data. Section 4 reports the empirical assessment of the proposal, comparing it to the CMI and LD strategies. Conclusions are given in Sect. 5.

## 2    Forward Stagewise Regression

Consider a regression setup in which you are given a set $\mathcal{D} = \{(\mathbf{x}_i, y_i)\}_{i=1}^{N}$ of input/output training examples, such that $\mathbf{x}_1, \cdots, \mathbf{x}_N$ are $p$-dimensional input column vectors and $y_1, \cdots, y_N$ are their respective scalar outputs. Furthermore, define the $N \times p$ matrix $\mathbf{X} = [\mathbf{x}_1, \cdots, \mathbf{x}_N]^T$ and the column vector $\mathbf{y} = [y_1, \cdots, y_N]^T$. We assume a linear relationship between the input and output variables (a linear model) of the form:

$$\mathbf{y} = \mathbf{X}\boldsymbol{\theta} + \mathbf{r}, \tag{1}$$

where $\mathbf{r} \in \mathbb{R}^N$ denotes a column vector of residuals and $\boldsymbol{\theta} = [\theta_1, \cdots \theta_p]^T$ represents the parameters of the linear model.

The goal in sparse linear estimation is to provide an estimate $\hat{\boldsymbol{\theta}}$ of the parameters $\boldsymbol{\theta}$ such that the $l_2$-norm of the residuals is small while having as many as possible entries in $\hat{\boldsymbol{\theta}}$ with values equal to zero. This is usually achieved by the following minimization problem:

$$\hat{\boldsymbol{\theta}} = \arg\min_{\boldsymbol{\theta}'} \|\mathbf{y} - \mathbf{X}\boldsymbol{\theta}'\|_2 + \lambda\|\boldsymbol{\theta}'\|_1, \tag{2}$$

where $\| \cdot \|_2$ and $\| \cdot \|_1$ denote the $l_2$ and $l_1$ norms, respectively, and we use $\boldsymbol{\theta}'$ to distinguish from the actual parameter vector. This formulation leads to a quadratic programming problem and thus many numerical methods can be used to solve it [8]. Among the various methods, the Forward Stagewise Regression algorithm leads to an approximate solution by means of simple iteractive procedure.

The Forward Stagewise Regression algorithm computes $\hat{\boldsymbol{\theta}}$ by iteratively selecting and increasing the value of one of its coefficients $\hat{\theta}_j$ according to the correlation between $\mathbf{X}_j$ and a vector of residuals $\mathbf{r}$. Henceforth, we use $\mathbf{X}_j$ to denote the $j$th column of $\mathbf{X}$, that is, $\mathbf{X}_j = [x_{1j}, x_{2j}, \cdots, x_{Nj}]^T$. In other words, $\mathbf{X}_j$ comprises the values of the $j$th feature of all input points. At the beginning of the FSR, the estimates $\hat{\boldsymbol{\theta}}$ are set to zero so that the vector $\mathbf{r}$ reduces to $\mathbf{y}$. At each iteration, both parameters and residuals are updated. The FSR algorithm is detailed in the following steps:

1. Start with $\hat{\boldsymbol{\theta}}^{(0)} = \mathbf{0}$ and $\mathbf{r}^{(0)} = \mathbf{y}$. In addition, standardize the columns of $\mathbf{X}$ to have zero mean and unit variance.
2. For each iteration $t = 1, 2, \ldots$
3. Find the feature index $j \in \{1, \cdots, p\}$ most correlated with the residual variable at instant $t - 1$[1].

---

[1] We are assuming that the vectors $\{\mathbf{x}_i\}$ are realizations of a $p$-dimensional random variable. Likewise, $\mathbf{r}$ comprises $N$ samples from the residual random variable. We use the method-of-moments estimator for the correlation between $j$th variable and the residual variable, that is, $\frac{1}{N}\mathbf{X}_j^T\mathbf{r}$.

4. Update the parameter estimate according to:

$$\hat{\theta}_j^{(t)} \leftarrow \hat{\theta}_j^{(t-1)} + \delta_j^{(t)}, \quad \text{such that} \quad \delta_j^{(t)} = \begin{cases} \epsilon, & \text{if } \mathbf{X}_j^T \mathbf{r}^{(t-1)} > 0, \\ -\epsilon, & \text{otherwise.} \end{cases} \tag{3}$$

where the step-size $\epsilon > 0$ is a pre-defined constant.

5. Update the vector of residuals as follows:

$$\mathbf{r}^{(t)} \leftarrow \mathbf{r}^{(t-1)} - \delta_j^{(t)} \mathbf{X}_j. \tag{4}$$

6. Go back to step 2 until the residuals are uncorrelated with all the predictors.

## 3   Proposed Method

We now consider the case where some instances of $\mathbf{X}$ have one or more missing entries. We are interested in reformulating the FSR algorithm to handle such case. In this matter, we first need to tackle the problem of estimating the correlation between the $j$-th feature and the residual variable, i.e., the value of $\mathbf{X}_j^T \mathbf{r}$ when some entries of $\mathbf{X}_j$ and/or $\mathbf{r}$ are missing. Under this scenario, we can consider the missing components of $\mathbf{X}$ as random variables. Thus, in the general case where any entry of $\mathbf{X}$ can be missing, the expected value of the desired correlation is given by

$$
\begin{aligned}
\mathbb{E}\left[\mathbf{X}_j^T \mathbf{r}\right] &= \mathbb{E}\left[\mathbf{X}_j^T \mathbf{y} - \mathbf{X}_j^T \mathbf{X}\boldsymbol{\theta}\right] \\
&= \mathbb{E}[\mathbf{X}_j^T \mathbf{y}] - \mathbb{E}[\mathbf{X}_j^T \mathbf{X}\boldsymbol{\theta}] \\
&= \sum_{i=1}^N \left(y_i \mathbb{E}[x_{i,j}]\right) - \sum_{i=1}^N \left(\mathbb{E}[x_{i,j} \mathbf{x}_i^T \boldsymbol{\theta}]\right) \\
&= \sum_{i=1}^N \left(y_i \mathbb{E}[x_{i,j}] - \mathbb{E}[x_{i,j}]\mathbb{E}[\mathbf{x}_i^T \boldsymbol{\theta}] + \text{Cov}[x_{i,j}, \mathbf{x}_i^T \boldsymbol{\theta}]\right) \\
&= \sum_{i=1}^N \left(y_i \mathbb{E}[x_{i,j}] - \left(\mathbb{E}[x_{i,j}] \sum_{k=1}^p \theta_k \mathbb{E}[x_{i,k}] + \sum_{k=1}^p \theta_k \text{Cov}[x_{i,k}, x_{i,j}]\right)\right) \\
&= \sum_{i=1}^N \left(y_i \mathbb{E}[x_{i,j}] - \sum_{k=1}^p \theta_k \left(\mathbb{E}[x_{i,k}]\mathbb{E}[x_{i,j}] + \text{Cov}[x_{i,j}, x_{i,k}]\right)\right) \tag{5}
\end{aligned}
$$

In the missing data scenario, there is uncertainty only on the unobserved/missing entries of $\mathbf{X}$, as the observed values are constants, i.e., $\mathbb{E}[x_{i,j}] = x_{i,j}$ if $x_{i,j}$ is not missing. Likewise, $\text{Cov}[x_{i,j}, x_{i,k}] = 0$ if $x_{i,j}$ or $x_{i,k}$ are not missing.

Equation (5) expresses the expected correlation as a function of the expected values of the inputs and the covariance between different attributes of the same input vector. Let $M_i$ denote the indices of the unobserved entries of $\mathbf{x}_i$. Furthermore, let $O_i = \{1, \ldots, p\} \setminus M_i$. Thus, the vector $\mathbf{x}_i$ can be divided into two parts $[\mathbf{x}_{i,O_i}, \mathbf{x}_{i,M_i}]$.

We are interested in computing the expected value of $\mathbf{X}_j \mathbf{r}$ conditioned on the observed values of $\mathbf{X}$. For that, as shown in Eq. (5), we need to compute the expected value of, and the covariance between, the missing entries of each training point $\mathbf{x}_i$ conditioned on the observed entries of the same vector, compactly written as $\mathbb{E}[\mathbf{x}_{i,M_i}|\mathbf{x}_{i,O_i}]$ and $\mathrm{Cov}[\mathbf{x}_{i,M_i}|\mathbf{x}_{i,O_i}]$.

According to [1], under the assumption that $\mathbf{x}_i \sim \mathcal{N}(\boldsymbol{\mu}, \boldsymbol{\Sigma})$, we can obtain $\mathbb{E}[\mathbf{x}_{i,M_i}|\mathbf{x}_{i,O_i}]$ and $\mathrm{Cov}[\mathbf{x}_{i,M_i}|\mathbf{x}_{i,O_i}]$ as follows:

$$\mathbb{E}[\mathbf{x}_{i,M}|\mathbf{x}_{i,O}] = \boldsymbol{\mu}_M + \boldsymbol{\Sigma}_{MO}\boldsymbol{\Sigma}_{OO}^{-1}(\mathbf{x}_{i,O} - \boldsymbol{\mu}_O), \tag{6}$$

$$\mathrm{Cov}[\mathbf{x}_{i,M}|\mathbf{x}_{i,O}] = \boldsymbol{\Sigma}_{MM} - \boldsymbol{\Sigma}_{MO}\boldsymbol{\Sigma}_{OO}^{-1}\boldsymbol{\Sigma}_{OM}, \tag{7}$$

where we omitted the dependence of $i$ in $M$ and $O$ for simplicity. The subscripts $OO$, $OM$, $MO$ and $MM$ refer to the subsets of the full covariance matrix $\boldsymbol{\Sigma}$ between missing and observed variables of $\mathbf{x}_i$. Additional details can be found in [1].

The FSR for incomplete data can be summarized as follows:

1. Start with $\hat{\boldsymbol{\theta}}^{(0)} = \mathbf{0}$ and $\mathbf{r}^{(0)} = \mathbf{y}$. In addition, standardize the columns of $\mathbf{X}$ to have zero mean and unit variance.
2. For each iteration $t = 1, 2, \ldots$
3. Find the feature index $j$ most correlated with the residual variable at instant $t - 1$:

$$j = \arg\min_{k=1,\ldots,p} \mathbb{E}[\mathbf{X}_k \mathbf{r}^{(t-1)}|\mathbf{X}_O], \tag{8}$$

where $\mathbf{X}_O$ refers to all pairs of indexes $(i, j)$ at which $x_{i,j}$ is observed.
4. Update the parameter estimate according to:

$$\hat{\theta}_j^{(t)} \leftarrow \hat{\theta}_j^{(t-1)} + \delta_j^{(t)}, \quad \text{such that} \quad \delta_j^{(t)} = \begin{cases} \epsilon, & \text{if } \mathbb{E}[\mathbf{X}_j \mathbf{r}^{(t-1)}|\mathbf{X}_O] > 0, \\ -\epsilon, & \text{otherwise.} \end{cases} \tag{9}$$

5. Update the vector of residuals as follows:

$$\mathbf{r}^{(t)} \leftarrow \mathbf{r}^{(t-1)} - \delta_j^{(t)}\mathbb{E}[\mathbf{X}_j|\mathbf{X}_O]. \tag{10}$$

6. Go back to step 2 until the residuals are uncorrelated with all the predictors.

Remarkably, Eq. (5) can be written concisely as:

$$\mathbb{E}[\mathbf{X}_j \mathbf{r}] = \mathbb{E}[\mathbf{X}_j]^T \mathbf{y} - \mathbb{E}[\mathbf{X}_j]^T \mathbb{E}[\mathbf{X}]\boldsymbol{\theta} - \sum_{i=1}^{N} \mathbf{e}_j^T \mathrm{Cov}[\mathbf{x}_i]\boldsymbol{\theta} \tag{11}$$

where $\mathbf{e}_j$ is the $j$th vector of the canonical basis of $\mathbb{R}^p$ and $\mathrm{Cov}[\mathbf{x}_i]$ is the full covariance matrix of the example $\mathbf{x}_i$ (which is the covariance matrix of the missing entries padded with zeros in the components related to the observed entries).

Therefore, one can conclude that the proposed method differs from common imputation strategies as the last term in Eq. (11) takes into account the uncertainty concerning the missing data entries.

# 4    Performance Evaluation

To asses the performance of the proposed method, named Forward Stagewise Regression for Incomplete datasets (FSRI), we carried out a set of experiments

**Table 1.** Datasets description.

|                    | # Features | # Training samples | # Test samples |
| ------------------ | ---------- | ------------------ | -------------- |
| Wine               | 13         | 100                | 78             |
| CPU                | 9          | 139                | 70             |
| Cancer             | 32         | 129                | 65             |
| Automobile price   | 15         | 106                | 53             |
| Forest Fire        | 4          | 344                | 173            |

**Table 2.** Average MSE between the outputs of each linear model and the target outputs. The number of input vectors with missing entries varies from 10% to 50%.

| Wine |            |            |            |            |            |
| ---- | ---------- | ---------- | ---------- | ---------- | ---------- |
|      | 10%        | 20%        | 30%        | 40%        | 50%        |
| FSRI | 6.3404     | 6.5319     | 6.7531     | 7.3600     | 8.1164     |
| CMI  | 6.6190     | 7.5510     | 10.7476    | 22.9362    | 54.9262    |
| LD   | 12.0314    | 23.6968    | 35.8139    | -          | -          |

| CPU  |              |              |              |              |              |
| ---- | ------------ | ------------ | ------------ | ------------ | ------------ |
|      | 10%          | 20%          | 30%          | 40%          | 50%          |
| FSRI | 2.7838e+05   | 2.8849e+05   | 2.9230e+05   | 3.2206e+05   | 3.2950e+05   |
| CMI  | 2.8381e+05   | 3.0676e+05   | 3.4010e+05   | 4.5708e+05   | 6.2148e+05   |
| LD   | 3.3261e+05   | 4.1271e+05   | 7.8900e+05   | 1.3920e+06   | -            |

| Automobile price |              |              |              |              |              |
| ---------------- | ------------ | ------------ | ------------ | ------------ | ------------ |
|                  | 10%          | 20%          | 30%          | 40%          | 50%          |
| FSRI             | 4.2238e+08   | 4.2386e+08   | 4.1773e+08   | 4.2275e+08   | 4.1695e+08   |
| CMI              | 4.4615e+08   | 4.9690e+08   | 7.7811e+08   | 1.8076e+09   | 3.4648e+09   |
| LD               | 1.5076e+09   | 1.1875e+09   | -            | -            | -            |

| Cancer |              |              |              |              |              |
| ------ | ------------ | ------------ | ------------ | ------------ | ------------ |
|        | 10%          | 20%          | 30%          | 40%          | 50%          |
| FSRI   | 8.5119e+04   | 8.2159e+04   | 8.0475e+04   | 7.9577e+04   | 7.7541e+04   |
| CMI    | 9.2285e+04   | 9.4305e+04   | 9.9716e+04   | 1.1450e+05   | 1.6995e+05   |
| LD     | 1.4298e+05   | -            | -            | -            | -            |

| Forest-fire |              |              |              |              |              |
| ----------- | ------------ | ------------ | ------------ | ------------ | ------------ |
|             | 10%          | 20%          | 30%          | 40%          | 50%          |
| FSRI        | 6.4320e+05   | 6.7355e+05   | 6.8613e+05   | 6.9128e+05   | 7.0489e+05   |
| CMI         | 6.4374e+05   | 6.7554e+05   | 6.9017e+05   | 6.9693e+05   | 7.1871e+05   |
| LD          | 6.4787e+05   | 7.2750e+05   | 9.2299e+05   | 1.2284e+06   | 1.6003e+06   |

**Table 3.** Average MSDC between each linear model and the linear model obtained by a FSR on the same dataset. In this experiment we set 0.30 of the maximum norm as the comparison point

| Wine | | | | | |
|------|------|------|------|------|------|
| | 10% | 20% | 30% | 40% | 50% |
| FSRI | 0.0028 | 0.0060 | 0.0114 | 0.0168 | 0.0297 |
| CMI | 0.0031 | 0.0077 | 0.0151 | 0.0232 | 0.0402 |
| LD | 0.0345 | 0.1261 | 0.2264 | 0.2236 | 0.2635 |
| CPU | | | | | |
| | 10% | 20% | 30% | 40% | 50% |
| FSRI | 949.5511 | 1642.1833 | 2362.9053 | 2609.0817 | 2905.4348 |
| CMI | 994.5057 | 1767.8236 | 2581.6698 | 2890.8464 | 3208.9734 |
| LD | 2858.6407 | 4416.8752 | 4713.3837 | 5112.4878 | 5491.9145 |
| Automobile price | | | | | |
| | 10% | 20% | 30% | 40% | 50% |
| FSRI | 4.99136e+05 | 7.89447e+05 | 1.37572e+06 | 1.64525e+06 | 2.34146e+06 |
| CMI | 5.38478e+05 | 9.24256e+05 | 1.65136e+06 | 2.06429e+06 | 3.04505e+06 |
| LD | 6.07145e+06 | 1.43461e+07 | 1.09962e+07 | 1.48738e+07 | 2.57570e+07 |
| Cancer | | | | | |
| | 10% | 20% | 30% | 40% | 50% |
| FSRI | 315.5959 | 689.3359 | 879.4547 | 939.1615 | 941.2634 |
| CMI | 321.0407 | 714.4983 | 984.0727 | 1214.7930 | 1433.9887 |
| LD | 2738.7149 | - | - | - | - |
| Forest-fire | | | | | |
| | 10% | 20% | 30% | 40% | 50% |
| FSRI | 1.3064 | 2.6853 | 3.1192 | 4.2199 | 5.1891 |
| CMI | 1.3349 | 2.9743 | 3.6451 | 5.192 | 6.3540 |
| LD | 1.3788 | 3.7616 | 7.3096 | 10.2142 | 12.6646 |

with 5 arbitrary real-world datasets, available at [10]. We compare FSRI to standard methods used to handle missing data. For each dataset, we varied the amount of inputs with missing variable from 10% to 50%. The description of the datasets is presented in Table 1.

The FSRI was compared to the Listwise Deletion (LD) and the Conditional Mean Imputation (CMI). Both CMI and LD were used as pre-processing steps and the standard FSR was used to generate the linear models. For FSRI and CMI, we estimate the parameters of the data distribution using the Expectation Conditional Maximization (ECM, [5]) algorithm for datasets with missing values.

All methods were compared based on two criteria, the Mean Square Error (MSE) between $\mathbf{y}$ and the results of each model and Mean Squared Difference

**Table 4.** Average MSDC between each linear model and the linear model obtained by a FSR on the same dataset. In this experiment we set 0.45 of the maximum norm as the comparison point

| Wine | | | | | |
|------|------|------|------|------|------|
| | 10% | 20% | 30% | 40% | 50% |
| FSRI | 0.0033 | 0.0065 | 0.0125 | 0.0194 | 0.0348 |
| CMI | 0.0032 | 0.0087 | 0.0179 | 0.0283 | 0.0520 |
| LD | 0.0425 | 0.1681 | 0.3012 | 0.3126 | 0.2575 |

| CPU | | | | | |
|------|------|------|------|------|------|
| | 10% | 20% | 30% | 40% | 50% |
| FSRI | 1277.7359 | 2385.4040 | 3638.0817 | 4105.8572 | 4838.1707 |
| CMI | 1370.3084 | 2707.6797 | 4261.6337 | 4913.7103 | 5758.5741 |
| LD | 4757.6309 | 7546.5486 | 8578.5215 | 9597.8471 | 11329.4762 |

| Automobile price | | | | | |
|------|------|------|------|------|------|
| | 10% | 20% | 30% | 40% | 50% |
| FSRI | 6.39862e+05 | 1.08062e+06 | 1.88349e+06 | 2.19982e+06 | 3.00759e+06 |
| CMI | 6.97030e+05 | 1.26571e+06 | 2.29862e+06 | 2.74333e+06 | 3.96610e+06 |
| LD | 7.66190e+06 | 1.52016e+07 | 2.54035e+07 | 2.24063e+07 | - |

| Cancer | | | | | |
|------|------|------|------|------|------|
| | 10% | 20% | 30% | 40% | 50% |
| FSRI | 815.4529 | 1432.2428 | 1835.1505 | 1571.4583 | 1542.1451 |
| CMI | 859.2785 | 1783.3939 | 2393.9583 | 2772.8872 | 3309.5005 |
| LD | 5932.4867 | - | - | - | - |

| Forest-fire | | | | | |
|------|------|------|------|------|------|
| | 10% | 20% | 30% | 40% | 50% |
| FSRI | 1.8247 | 3.6323 | 4.1453 | 5.2285 | 7.4526 |
| CMI | 1.8937 | 4.2270 | 5.4768 | 8.0466 | 10.3822 |
| LD | 2.3055 | 6.5959 | 13.2102 | 18.6168 | 23.3338 |

between the Coefficients (MSDC) of each linear model and the linear model obtained by a FSR on the same dataset without missing values. All experiments were repeated 500 times. Table 2 shows the average MSE obtained in the experiments.

Beforehand, it is important to clarify that some of the LD results are not filled which indicates that the FSR algorithm was not able to converge due to the significant number of discarded examples. Concerning the other AMSE values, one can see that the average MSE for all methods increase as the number of missing data increases. However, it is noticeable that FSRI had the lowest AMSE for all datasets and missing data percentages. This performance gap is even more significant in the experiments with the highest number of missing data.

**Table 5.** Average MSDC between each linear model and the linear model obtained by a FSR on the same dataset. In this experiment we set 0.60 of the maximum norm as the comparison point

| Wine | | | | | |
|------|------|------|------|------|------|
| | 10% | 20% | 30% | 40% | 50% |
| FSRI | 0.0033 | 0.0084 | 0.0169 | 0.0262 | 0.0426 |
| CMI | 0.0037 | 0.0104 | 0.0221 | 0.0350 | 0.0630 |
| LD | 0.0511 | 0.1926 | 0.3183 | 0.5117 | - |
| CPU | | | | | |
| | 10% | 20% | 30% | 40% | 50% |
| FSRI | 1191.6427 | 2369.9071 | 3796.9097 | 4585.2540 | 5591.5145 |
| CMI | 1313.4053 | 2841.6674 | 4745.0750 | 5810.1135 | 7133.1241 |
| LD | 5174.1832 | 8978.9877 | 11144.0691 | 12598.5600 | 18165.6937 |
| Automobile price | | | | | |
| | 10% | 20% | 30% | 40% | 50% |
| FSRI | 8.20843e+05 | 1.55785e+06 | 2.47574e+06 | 3.08841e+06 | 3.95091e+06 |
| CMI | 8.92548e+05 | 1.80668e+06 | 3.03822e+06 | 3.66187e+06 | 5.14811e+06 |
| LD | 9.34456e+06 | 1.83650e+07 | 2.34479e+07 | - | - |
| Cancer | | | | | |
| | 10% | 20% | 30% | 40% | 50% |
| FSRI | 1471.5130 | 2002.8079 | 2187.5142 | 2280.6016 | 1995.2143 |
| CMI | 1561.4253 | 2983.8447 | 4298.5120 | 4711.2356 | 5891.7460 |
| LD | 7616.6000 | - | - | - | - |
| Forest-fire | | | | | |
| | 10% | 20% | 30% | 40% | 50% |
| FSRI | 2.0451 | 4.1852 | 4.8350 | 5.6707 | 9.1165 |
| CMI | 2.1624 | 5.2306 | 6.8008 | 10.2454 | 14.4056 |
| LD | 3.2613 | 9.3837 | 20.2777 | 28.8248 | 36.5733 |

Along with the MSE, we computed the MSDC metric to quantify the difference between the linear model generated by each method and an ideal linear model obtained by a FSR on a complete (no missing data) dataset. We decided to compare the methods on several instants during the learning process. The instants are defined according to the norm of the weights generated by each method. We considered the norm obtained by the FSR in the complete dataset as the maximum norm and evaluated all method at 3 different ratios of this norm. Such procedure was adopted to provide a fair comparison since different methods show weight vectors with varying norms at each iteration. Tables 3, 4 and 5 show the MSDC values for the ratios 0.3, 0.45 and 0.6.

As can be noticed, the difference between the weight vectors generated by each method and the ideal linear model increased with the number of missing data. Once again FSRI had the best overall performance being less sensible to the presence of missing data.

## 5  Conclusions

In this paper we proposed a variant of the Forward Stagewise Regression algorithm for incomplete datatsets. In the proposed method, named FSRI, we considered the inputs as normally distributed random variables and modified the steps of FSR such that weights are incremented according to the expected correlation of the residuals and each of the features. FSRI was compared to popular strategies to handle missing values and achieved promising results.

It is worth highlighting that the performance of FSRI can be significantly degraded if the normality assumption of the training set does not hold. Hence we are currently working to extend the FSRI formulation for non-Gaussian datasets using nonparametric/semi-parametric models

**Acknowledgments.** The authors acknowledge the support of CNPq (Grant 456837/2014-0 and research fellowship).

## References

1. Eirola, E., Doquire, G., Verleysen, M., Lendasse, A.: Distance estimation in numerical data sets with missing values. Inf. Sci. **240**, 115–128 (2013)
2. Little, R.J.A., Rubin, D.B.: Statistical Analysis with Missing Data, vol. 2. Wiley Interscience, Hoboken (2002)
3. Ding, Y., Simonoff, J.S.: An investigation of missing data methods for classification trees applied to binary response data. J. Mach. Learn. Res. **11**, 131–170 (2010)
4. Garcia-Laencina, P.J., Sancho-Gomez, J.-L., Figueiras-Vidal, A.R.: Pattern classification with missing data: a review. Neural Comput. Appl. **19**, 263–282 (2010)
5. Meng, X.-L., Rubin, D.B.: Maximum likelihood estimation via the ECM algorithm. Biometrika **80**, 267–278 (1993)
6. Belanche, L.A., Kobayashi, V., Aluja, T.: Handling missing values in kernel methods with application to microbiology data. Neurocomputing **141**, 110–116 (2014)
7. Tibshirani, R.J.: A general framework for fast stagewise algorithms. J. Mach. Learn. Res. **16**, 2543–2588 (2015)
8. Hastie, T., Taylor, J., Tibshirani, R., Walther, G.: Forward stagewise regression and the monotone lasso. Electron. J. Stat. **1**, 1–29 (2007)
9. Hastie, T., Tibshirani, R., Friedman, J.: The Elements of Statistical Learning, 2nd edn. Springer New York Inc., New York (2009)
10. Frank, A., Asuncion, A.: UCI Machine Learning Repository University of California, Irvine, School of Information and Computer Sciences (2010)

# Convolutional Neural Networks with the F-transform Kernels

Vojtech Molek$^{(\boxtimes)}$ and Irina Perfilieva

University of Ostrava, Ostrava 701 03, Czech Republic
{vojtech.molek,irina.pefilieva}@osu.cz
http://www.osu.cz/, http://irafm.osu.cz/

**Abstract.** We propose a new convolutional neural network – the FTNet and explain its theoretical background referring to the theory of a higher degree F-transform. The FTNet is parametrized by kernel sizes, on/off activation of weights learning, the choice of strides or pooling, etc. It is trained on the database MNIST and tested on handwritten inputs. The obtained results demonstrate that the FTNet has better recognition accuracy than the automatically trained LENET-5. We have also analyzed the FTNet and LENET-5 rotation invariance.

## 1 Introduction

Deep learning (DL) [1] neural networks have proven themselves as efficient tools for pattern recognition [2–4]. One of the main principles of the DL is based on automatic extraction of "good" features [5] using a general-purpose learning procedure [6,7]. This is opposite to hand designed feature extractors that require a considerable amount of testing time and expert skills [8–10].

In this contribution, we argue with the absolutization of the above given main principle and propose the theoretical background of FTNets – convolutional neural networks (CNN) that use kernels related to a higher degree F-transform [11]. The FTNet is parametrized by kernel sizes, on/off activation of weights learning, the choice of strides or pooling, etc. It is trained on the database MNIST and tested on handwritten inputs.

The obtained results demonstrate that the FTNet has better recognition accuracy than the automatically trained LENET-5 [12]. The efficiency of the proposed FTNet (measured in training time) is higher. Last, but not least, we provide the theoretical justification of a suitability of the FTNet for the problem of recognition.

To confirm our conclusion, we compare the FTNet networks with the LENET-5 (both are trained on the dataset *MNIST*) on various recognition tests. The results are discussed in Sect. 4.2. We have chosen LENET-5, because it was specially designed for dataset *MNIST* whose objects are hand drawn integers from 0 to 9 together with their various transforms. LENET-5 has a reasonable size, good performance accuracy and serves as a prototype for many other convolutional networks. Moreover, LENET-5 and its modifications are included into many modern machine learning frameworks.

© Springer International Publishing AG 2017
I. Rojas et al. (Eds.): IWANN 2017, Part I, LNCS 10305, pp. 396–407, 2017.
DOI: 10.1007/978-3-319-59153-7_35

The structure of the paper is as follows: in Sect. 2 we give a short characteriation of convolutional neural networks; Sect. 3 recalls the main facts about the higher degree F-transform and specifically $F^2$-transform - the technique, which will be used in the proposed FTNet networks; Sect. 4 contains description of tests and discussion of their results.

## 2   Convolutional Neural Networks

Convolutional Neural Networks [13] are hierarchical models capable of learning. The hierarchy consists of layers of units. The layers are connected together in a cascade manner. They can be specified according to their types. One of the types is a convolutional type. Units in convolutional layer are partially connected to units of the previous layer, unlike units in fully connected layers. Each units of a convolutional layer performs operation known as *convolution*[1], thus the layer name. The purpose of a convolutional layer is to extract features. Multiple convolutional layers connected one after another extract features of higher abstractions. Multiple connected convolutional layers are interlarded with *pooling* (*sub-sampling*) layers which should ensure tolerance to translations and distortions (Fig. 1).

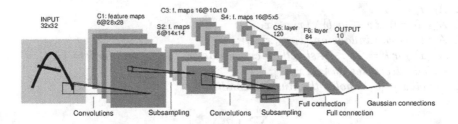

**Fig. 1.** LENET-5 architecture reproduced from [12]

CNN architecture reproduced from paper [13] is considered as perhaps the first that deserves the label *deep* [1]. The difference between deep and shallow networks is not clearly distinguished (more on the topic can be found in the article [1]). Learning deep convolutional neural network (*DNN*) using a learning algorithm (the back-propagation with gradient descent [14]) proved to be computationally heavy. The problem of intense computations was simplified by the advent of programmable GPUs (frameworks *cuDNN, Caffe, Theano, Torch, Tensorflow, etc.*).

CNN is one of the best tools in the task of classification especially image classification. Dataset *MNIST* is an example of benchmark that confirms this claim. The following web page[2] contains error rates of different neural networks

---

[1] Weighted average in case of convolutional layers. Weights are being learned.
[2] http://yann.lecun.com/exdb/mnist/.

sorted into groups by their types. The best neural network displayed on the website has only 0.23% (23 miss-classifications in 10000) error rate. Another example is competition *ILSVRC* (Large Scale Visual Recognition Challenge). In some cases mean average precisions nearly doubled between 2014[3] and 2015[4].

# 3   The F-transform of a Higher Degree ($F^m$-transform)

In this section, we recall the main facts (see [11] for more details) about the higher degree F-transform and specifically $F^2$-transform - the technique, which will be used in the proposed below CNN with the FT kernels (FTNet).

## 3.1   Fuzzy partition

The F-transform is the result of a convolution of an object function (image, signal, etc.) and a generating function of what is regarded as a *fuzzy partition* of a universe.

**Definition 1.** *Let* $n > 2$, $a = x_0 = x_1 < \ldots < x_n = x_{n+1} = b$ *be fixed nodes within* $[a, b] \subseteq \mathbb{R}$. *Fuzzy sets* $A_1, \ldots, A_n : [a, b] \to [0, 1]$, *identified with their membership functions defined on* $[a, b]$, *establish a fuzzy partition of* $[a, b]$, *if they fulfill the following conditions for* $k = 1, \ldots, n$:

1. $A_k(x_k) = 1$;
2. $A_k(x) = 0$ *if* $x \in [a, b] \setminus (x_{k-1}, x_{k+1})$;
3. $A_k(x)$ *is continuous on* $[x_{k-1}, x_{k+1}]$;
4. $A_k(x)$ *for* $k = 2, \ldots, n$ *strictly increases on* $[x_{k-1}, x_k]$ *and for* $k = 1, \ldots, n-1$ *strictly decreases on* $[x_k, x_{k+1}]$;
5. *for all* $x \in [a, b]$ *holds the Ruspini condition*

$$\sum_{k=1}^{n} A_k(x) = 1. \tag{1}$$

The elements of fuzzy partition $\{A_1, \ldots, A_n\}$ are called *basic functions*.

In particular, an $h$-uniform fuzzy partition of $[a, b]$ can be obtained using the so called *generating function*

$$A : [-1, 1] \to [0, 1], \tag{2}$$

which is defined as an even, continuous and positive function everywhere on $[-1, 1]$ except for on boundaries, where it vanishes. Basic functions $A_2, \ldots, A_{n-1}$ of an $h$-uniform fuzzy partition are rescaled and shifted copies of $A$ in the sense that for all $k = 2, \ldots, n-1$;

$$A_k(x) = \begin{cases} A(\frac{x-x_k}{h}), & x \in [x_k - h, x_k + h], \\ 0, & \text{otherwise.} \end{cases}$$

---

[3] http://image-net.org/challenges/LSVRC/2014/results.
[4] http://image-net.org/challenges/LSVRC/2015/results.

Below, we will be working with one particular case of an $h$-uniform fuzzy partition that is generated by the triangular shaped function $A^{tr}$ and its $h$-rescaled version $A_h^{tr}$, where

$$A^{tr}(x) = 1 - |x|, \ x \in [-1, 1], \ \text{and} \ A_h^{tr}(x) = 1 - \frac{|x|}{h}, \ x \in [-h, h].$$

A fuzzy partition generated by the triangular shaped function $A^{tr}$ will be referred to as *triangular shaped*.

## 3.2   Space $L_2(A_k)$

Let us fix $[a, b]$ and its $h$-uniform fuzzy partition $A_1, \ldots, A_n$, where $n \geq 2$ and $h = \frac{b-a}{n-1}$ [5]. Let $k$ be a fixed integer from $\{1, \ldots, n\}$, and let $L_2(A_k)$ be a set of square-integrable functions $f : [x_{k-1}, x_{k+1}] \to \mathbb{R}$. Denote $L_2(A_1, \ldots, A_n)$ a set of functions $f : [a, b] \to \mathbb{R}$ such that for all $k = 1, \ldots, n$, $f|_{[x_{k-1}, x_{k+1}]} \in L_2(A_k)$. In $L_2(A_k)$, we define an *inner product* of $f$ and $g$

$$\langle f, g \rangle_k = \int_{x_{k-1}}^{x_{k+1}} f(x)g(x)d\mu_k = \frac{1}{s_k} \int_{x_{k-1}}^{x_{k+1}} f(x)g(x)A_k(x)dx,$$

where

$$s_k = \int_{x_{k-1}}^{x_{k+1}} A_k(x)dx.$$

The space $(L_2(A_k), \langle f, g \rangle_k)$ is a *Hilbert space*. We apply the Gram-Schmidt process to the linearly independent system of polynomials $\{1, x, x^2, \ldots x^m\}$ restricted to the interval $[x_{k-1}, x_{k+1}]$ and convert it to an orthogonal system in $L_2(A_k)$. The resulting orthogonal polynomials are denoted by $P_k^0, P_k^1, P_k^2, \ldots, P_k^m$.

*Example 1.* Below, we write the first three orthogonal polynomials $P^0, P^1, P^2$ in $L_2(A)$, where $A$ is the generating function of a uniform fuzzy partition, and $\langle \cdot, \cdot \rangle_0$ is the inner product:

$$P^0(x) = 1,$$
$$P^1(x) = x,$$
$$P^2(x) = x^2 - I_2, \ \text{where} \ I_2 = h^2 \int_{-1}^{1} x^2 A(x)dx,$$

If generating function $A^{tr}$ is triangular shaped and $h$-rescaled, then the polynomial $P^2$ can be simplified to the form

$$P^2(x) = x^2 - \frac{h^2}{6}. \tag{3}$$

We denote $L_2^m(A_k)$ a linear subspace of $L_2(A_k)$ with the basis $P_k^0, P_k^1, P_k^2 \ldots, P_k^m$.

---

[5] The text of this and the following subsection is a free version of a certain part of [11] where the theory of a higher degree F-transform was introduced.

## 3.3    $F^m$-transform

In this section, we define the $F^m$-transform, $m \geq 0$, of a function $f$ with polynomial components of degree $m$. Let us fix $[a, b]$ and its fuzzy partition $A_1, \ldots, A_n$, $n \geq 2$.

**Definition 2** [11]. *Let $f : [a, b] \to \mathbb{R}$ be a function from $L_2(A_1, \ldots, A_n)$, and let $m \geq 0$ be a fixed integer. Let $F_k^m$ be the k-th orthogonal projection of $f|_{[x_{k-1}, x_{k+1}]}$ on $L_2^m(A_k)$, $k = 1, \ldots, n$. We say that the n-tuple $(F_1^m, \ldots, F_n^m)$ is an $F^m$-transform of $f$ with respect to $A_1, \ldots, A_n$, or formally,*

$$F^m[f] = (F_1^m, \ldots, F_n^m).$$

*$F_k^m$ is called the $k^{th}$ $F^m$-transform component of $f$.*

Explicitly, each $k^{th}$ component is represented by the $m^{th}$ degree polynomial

$$F_k^m = c_{k,0} P_k^0 + c_{k,1} P_k^1 + \cdots + c_{k,m} P_k^m, \tag{4}$$

where

$$c_{k,i} = \frac{\langle f, P_k^i \rangle_k}{\langle P_k^i, P_k^i \rangle_k} = \frac{\int_a^b f(x) P_k^i(x) A_k(x) dx}{\int_a^b P_k^i(x) P_k^i(x) A_k(x) dx}, \quad i = 0, \ldots, m.$$

**Definition 3.** *Let $F^m[f] = (F_1^m, \ldots, F_n^m)$ be the direct $F^m$-transform of $f$ with respect to $A_1, \ldots, A_n$. Then the function*

$$\hat{f}_n^m(x) = \sum_{k=1}^n F_k^m A_k(x), \quad x \in [a, b], \tag{5}$$

*is called the inverse $F^m$-transform of $f$.*

The following theorem proved in [11] estimates the quality of approximation by the inverse $F^m$-transform in a normed space $L_1$.

**Theorem 1.** *Let $A_1, \ldots, A_n$ be an h-uniform fuzzy partition of $[a, b]$. Moreover, let functions $f$ and $A_k$, $k = 1, \ldots, n$ be four times continuously differentiable on $[a, b]$, and let $\hat{f}_n^m$ be the inverse $F^m$-transform of $f$, where $m \geq 1$. Then*

$$\|f(x) - \hat{f}_n^m(x)\|_{L_1} \leq O(h^2),$$

*where $L_1$ is the Lebesgue space on $[a + h, b - h]$.*

## 3.4    $F^2$-transform in the Convolutional Form

Let us fix $[a, b]$ and its $h$-uniform fuzzy partition $A_1, \ldots, A_n$, $n \geq 2$, generated from $A : [-1, 1] \to [0, 1]$ and its $h$-rescaled version $A_h$, so that $A_k(x) = A(\frac{x - x_k}{h}) = A_h(x - x_k)$, $x \in [x_k - h, x_k + h]$, and $x_k = a + kh$. The $F^2$-transform of a function $f$ from $L_2(A_1, \ldots, A_n)$ has the following representation

$$F^2[f] = (c_{1,0} P_1^0 + c_{1,1} P_1^1 + c_{1,2} P_1^2, \ldots, c_{n,0} P_n^0 + c_{n,1} P_n^1 + c_{n,2} P_n^2), \tag{6}$$

where for all $k = 1, \ldots, n$,

$$P_k^0(x) = 1, \; P_k^1(x) = x - x_k, \; P_k^2(x) = (x - x_k)^2 - I_2, \tag{7}$$

where $I_2 = h^2 \int_{-1}^1 x^2 A(x)dx$, and coefficients are as follows:

$$c_{k,0} = \frac{\int_{-\infty}^{\infty} f(x) A_h(x - x_k)dx}{\int_{-\infty}^{\infty} A_h(x - x_k)dx}, \tag{8}$$

$$c_{k,1} = \frac{\int_{-\infty}^{\infty} f(x)(x - x_k) A_h(x - x_k)dx}{\int_{-\infty}^{\infty} (x - x_k)^2 A_h(x - x_k)dx}, \tag{9}$$

$$c_{k,2} = \frac{\int_{-\infty}^{\infty} f(x)((x - x_k)^2 - I_2) A_h(x - x_k)dx}{\int_{-\infty}^{\infty} ((x - x_k)^2 - I_2)^2 A_h(x - x_k)dx}. \tag{10}$$

In [11,15], it has been proved that

$$c_{k,0} \approx f(x_k), \; c_{k,1} \approx f'(x_k), \; c_{k,2} \approx f''(x_k), \tag{11}$$

where $\approx$ is meant up to $O(h^2)$.

Without going into technical details, we rewrite (8)–(10) into the following discrete representations

$$c_{k,0} = \sum_{j=1}^l f(j)g_0(ks - j), \; c_{k,1} = \sum_{j=1}^l f(j)g_1(ks - j), \; c_{k,2} = \sum_{j=1}^l f(j)g_2(ks - j), \tag{12}$$

where $k = 1, \ldots, n$, $n = \lfloor \frac{l}{s} \rfloor$, $s$ is the so called "stride" and $g_0, g_1, g_2$ are normalized functions that correspond to generating functions $A_h$, $(xA_h)$ and $((x^2 - I_2)A_h)$. It is easy to see that if $s = 1$, then coefficients $c_{k,0}, c_{k,1}, c_{k,2}$ are results of the corresponding discrete convolutions $f \star g_0, f \star g_1, f \star g_2$. Thus, we can rewrite the representation of $F^2$ in (6) in the following vector form:

$$F^2[f] = ((f \star_s g_0)^T \mathbf{P}^0 + (f \star_s g_1)^T \mathbf{P}^1 + (f \star_s g_2)^T \mathbf{P}^2), \tag{13}$$

where $\mathbf{P}^0, \mathbf{P}^1, \mathbf{P}^2$ are vectors of polynomials with components given in (7), and $\star_s$ means that the convolution is performed with the stride $s$, $s \geq 1$.

*Example 2.* We choose the triangular shaped generating function $A^{tr} : [-1, 1] \to [0, 1]$ and consider it on the discrete domain $D = \{-1, -2/3, -1/3, 0, 1/3, 2/3, 1\}$. Below, we show four matrices $G_0, G_{1,1}, G_{1,2} \; G_2$ of $5 \times 5$ kernels[6] that are used for functions of two variables and correspond to the three above considered convolutions with $g_0, g_1, g_2$.

$$G_0 = \begin{pmatrix} 0.000 & 0.000 & 0.000 & 0.000 & 0.000 \\ 0.000 & 0.062 & 0.125 & 0.062 & 0.000 \\ 0.000 & 0.125 & 0.250 & 0.125 & 0.000 \\ 0.000 & 0.062 & 0.125 & 0.062 & 0.000 \\ 0.000 & 0.000 & 0.000 & 0.000 & 0.000 \end{pmatrix}$$

---

[6] The size is determined by only five non-zero values of $A^{tr}$ on $D$.

$$G_{1,1} = \begin{pmatrix} -0.074 & -0.074 & 0. & 0.074 & 0.074 \\ -0.148 & -0.148 & 0. & 0.148 & 0.148 \\ -0.222 & -0.222 & 0. & 0.222 & 0.222 \\ -0.148 & -0.148 & 0. & 0.148 & 0.148 \\ -0.074 & -0.074 & 0. & 0.074 & 0.074 \end{pmatrix}$$

$$G_{1,2} = \begin{pmatrix} -0.074 & -0.148 & -0.222 & -0.148 & -0.074 \\ -0.074 & -0.148 & -0.222 & -0.148 & -0.074 \\ 0.000 & 0.000 & 0.000 & 0.000 & 0.000 \\ 0.074 & 0.148 & 0.222 & 0.148 & 0.074 \\ 0.074 & 0.148 & 0.222 & 0.148 & 0.074 \end{pmatrix}$$

$$G_{2} = \begin{pmatrix} 0.062 & 0.049 & -0.037 & 0.049 & 0.062 \\ 0.049 & -0.049 & -0.148 & -0.049 & 0.049 \\ 0.037 & -0.148 & -0.333 & -0.148 & 0.037 \\ 0.049 & -0.049 & -0.148 & -0.049 & 0.049 \\ 0.062 & 0.049 & -0.037 & 0.049 & 0.062 \end{pmatrix}$$

Let us remark that in the context of convolutional neural networks, matrices $G_0$, $G_{1,1}$, $G_{1,2}$ $G_2$ determine convolution filters. In the context of the F-transform, they depend on the chosen partition of underlying universe and do not depend on the functions they are applied to.

### 3.5   $F^2$-transform in the FTNet Architecture

We propose to modify the LENET-5 [12] and replace convolution-type units in the first and third convolution layers $C_1$ and $C_3$ by the similar units which realize the computation of the $F^2$-transform coefficients according to (12) and adapted to functions of two variables. We use the meaning (11) of the $F^2$ coefficients and specify features in the feature maps of the convolution layer $C_1$ as partial derivatives (positive and negative) of an input function (of two variables) with respect to each single variable up to the second degree. In more details, the six matrices $G_0$, $-G_0$ $G_{1,1}$, $G_{1,2}$, $G_2$, $-G_2$ are convolved with the input 2D image in order to produce the mentioned partial derivatives at uniformly distributed nodes over the image domain. Thus, we have six feature maps in $C_1$. Each feature map of $C_1$ is connected (via subsampling layer $S2$) with each feature map of $C_3$ - thus, we have thirty six feature maps in $C_3$. The meaning of feature in $C_3$ corresponds to all possible mixed partial derivatives up to the third degree.

The features extracted in $C_1$ and $C_3$ are used to classify objects in MNIST. Our justification is based on the Theorem 1 which says that the inverse $F^m$ (particularly, $F^2$) transform approximates any function with sufficient quality. The number 2 of convolutional layers was set up empirically, and this turned out to be sufficient for the recognition purpose from MNIST. Thus, in comparison with the LENET-5 we use less number of convolutional layers. Other layers in the FTNet are of the fully-connected types and serve the same purposes as in the LENET-5.

Let us discuss the learning of convolution filters represented by matrices $G_0$, $-G_0$ $G_{1,1}$, $G_{1,2}$, $G_2$, $-G_2$. These filters can be excluded from the learning procedure on the basis of the mentioned above Theorem regarding the universal approximation. However, if they are learned (and this is confirmed by our tests), then the quality of approximation is adapted to a narrower class of objects (they constitute a certain dataset) and it is better than in the general case (valid for a generic class of objects).

# 4    Experiments and Results

We have used the FTNet (based of the LENET-5 [12]) architecture as the baseline for our experiments. The details of the FTNet architecture are described in Table 1 and Sect. 4.1 where the following notation is used: convolution layers $C_1$ and $C_3$, subsampling layers $S_2$ and $S_4$ and fully connected layers $FC_5$ and $FC_6$. We have examined the impact of the following hyper-parameters: convolution kernel size $\mathcal{D}$, presence and type of the subsampling $\mathcal{S}$, layer weights trainability $\mathcal{T}$, and a form of the layer weights initialization $\mathcal{I}$ on the network performance. We have used all possible combinations of the hyper-parameters of $C_1$, $S_2$, $C_3$ and $S_4$ in the *grid search* with the purpose to select the optimal setting with respect to the quality of recognition (loss function).

**Table 1.** FT-Net architecture.

| Hyper-paramater | $C_1$ | $S_2$ | $C_3$ | $S_4$ | $FC_5$ | $FC_6$ |
|---|---|---|---|---|---|---|
| Kernel size | $5 \times 5$ | - | $5 \times 5$ | - | - | - |
| # feature maps | 6 | - | 36 | - | - | - |
| Stride | $1 \times 1$ | *pooling* | $1 \times 1$ | *pooling* | - | - |
| Pooling size | - | $2 \times 2$ | - | $2 \times 2$ | - | |
| # FC units | - | - | - | - | 500 | 10 |

## 4.1    FTNet Architecture

In this section, we describe details of one particular FTNet architecture where the hyper-parameters are: $\mathcal{D} = 5 \times 5$, $\mathcal{S} = $ max pooling. Below, we characterize other details: layers, connection types, input, intermediate and output objects.

The first layer of the FTNet is convolutional $C_1$, it has 6 feature maps with the size of $28 \times 28$. To ensure the same size of the feature maps and image, padding is used. Each $C_1$ unit has 25 connections to input image. Unit connections are spatially close, forming $5 \times 5$ neighborhood called unit's *receptive field*; the latter overlaps with others unit's receptive fields. The $C_1$ has $25 \cdot 6 + 6$ (trainable) parameters and $784 \cdot 25 \cdot 6 + 6$ connections.

The $C_1$ feature maps are connected to the max pooling layer $S_2$. The $S_2$ units have $2 \times 2$ non-overlapping receptive fields from which they select the maximum. $S_2$ effectively decreases the feature maps size to $14 \times 14$. The $S_2$ has $784 \cdot 6$ connections.

The convolutional layer $C_3$ is connected to $S_2$ outputs. Each $14 \times 14$ output is convolved with all $C_3$ kernels (they are the same as in $C_1$) creating $6^2$ new feature maps with the size of $14 \times 14$. The $C_3$ has $25 \cdot 6^2 + 6^2$ (trainable) parameters and $784/4 \cdot 25 \cdot 6^2 + 6^2$ connections.

The $C_3$ feature maps are connected to the max pooling layer $S_4$ that further reduces their size to $7 \times 7$.

The $S_4$ feature maps are inputs to the fully connected layer $FC_5$. The $FC_5$ has 500 units, each connected to all outputs from $S_4$, therefore the $FC_5$ has $7 \cdot 7 \cdot 6^2 \cdot 500$ (trainable) parameters/connections.

The $FC_5$ output vector is the input to the last fully connected layer $FC_6$ that has 10 units. The $FC_6$ has $500 \cdot 10$ (trainable) parameters/connections. The $FC_6$ output vector goes through the softmax layer. Softmax layer normalizes an input vector to that whose sum of components is equal to 1.

The $C_1$, $C_3$ and $FC_5$ uses Rectified Linear activation function (RELU) [2].

## 4.2   Tests

The proposed network was tested on grayscale images from the database MNIST. The MNIST consists of 70000 $28 \times 28$ images[7]. They were normalized to the size of $20 \times 20$ so that the centering and the color ratio of the original $28 \times 28$ images were preserved.

Two sets of kernels were selected for testing. They determine feature maps in layers $C_1$ and $C_3$. The first set (referred to as "FT2") is composed by the $F^2$-transform kernels represented above by the six matrices $G_0$, $-G_0$ $G_{1,1}$, $G_{1,2}$, $G_2$, $-G_2$. The second set (referred to as "Conventional") is composed by the widely used kernels with the same meaning as the $F^2$-transform ones: they specify partial derivatives (positive and negative) of an image function with respect to each single variable up to the second degree. These kernels are: Gauss, Sobel, Laplace and their derivatives such as -Gauss (multipled by -1), 90Sobel (rotated by 90°) and -Laplace.

Our first test was focused on the choice of an optimal combination with respect to the chosen loss function – the *cross entropy*. With this purpose, we have applied a grid search over all combinations of the hyper-parameters $(\mathcal{D}, \mathcal{S}, \mathcal{T}$ and $\mathcal{I})$.

The results were clustered into 3 groups based on the following values of the loss function: $\{\approx 1.5, \approx 6, \approx 25\}$. We have observed that the clustering (and by this, the quality of recognition) essentially depends on the presence of subsampling. In more details:

**Fig. 2.** Numbers used for testing rotation invariance.

---

[7] 60000 training and 10000 testing.

**Table 2.** Selected optimal networks, their parameters and performance.

| Kernel set | FT2 | FT2 | Conventional | Conventional | Conventional | LENET-5[a] |
|---|---|---|---|---|---|---|
| $C_1$ kernel size | $5 \times 5$ | $5 \times 5$ | $5 \times 5$ | $5 \times 5$ | $5 \times 5$ | $5 \times 5$ |
| $C_1$ train | true | false | true | false | true | true |
| $C_1$ init. | fixed | fixed | fixed | fixed | fixed | random |
| $S_2$ subs. | stride | pooling | pooling | pooling | pooling | pooling |
| $C_3$ kernel size | $5 \times 5$ | $5 \times 5$ | $5 \times 5$ | $3 \times 3$ | $5 \times 5$ | $5 \times 5$ |
| $C_3$ train | true | false | true | false | true | true |
| $C_3$ init. | fixed | random | random | fixed | fixed | random |
| $S_4$ subs. | pooling | pooling | pooling | stride | pooling | pooling |
| Training time | 34.7s | 34.8s | 39.9s | 26.3s | 41.2s | 82.6s |
| Avg. loss | 0.9037 | 0.9487 | 0.9039 | 0.9341 | 0.9008 | 1.56 |
| | none | none | none | none | none | none |
| | [40 - 160]<br>[215 - 345] | [30 - 160]<br>[215 - 345] | [30 - 125]<br>[215 - 345] | [35 - 130]<br>[230 - 345] | [35 - 155]<br>[220 - 345] | [35 - 125]<br>[225 - 350] |
| | [50 - 80]<br>[145 - 275] | [40 - 270] | [50 - 90]<br>[130 - 285] | [45 - 170]<br>[200 - 275] | [45 - 175]<br>[210 - 265] | [50 - 100]<br>[130 - 185]<br>[210 - 275] |
| | [70 - 305] | [65 - 310] | [45 - 305] | [70 - 310] | [70 - 300] | [50 - 300] |
| | [25 - 340] | [35 - 50]<br>[85 - 200]<br>[240 - 340] | [15 - 185]<br>[230 - 335] | [25 - 195]<br>[235 - 340] | [35 - 205]<br>[230 - 335] | [20 - 195]<br>[230 - 340] |
| | [80 - 120]<br>[180 - 305] | [50 - 135]<br>[180 - 245]<br>[285 - 300] | [50 - 115]<br>[190 - 295] | [70 - 255]<br>[275 - 300] | [75 - 120]<br>[180 - 310] | [75 - 115]<br>[180 - 315] |
| | [0 - 5]<br>[85 - 355] | [0 - 30]<br>[80 - 355] | [0 - 25]<br>[95 - 355] | [0 - 15]<br>[90 - 355] | [0 - 10]<br>[85 - 355] | [85 - 350] |
| | [0 - 0]<br>[40 - 355] | [0 - 0]<br>[40 - 355] | [35 - 350] | [0 - 10]<br>[35 - 95]<br>[115 - 355] | [30 - 350] | [35 - 350] |
| | [45 - 140]<br>[235 - 270] | [0 - 140]<br>[245 - 285]<br>[315 - 355] | [50 - 140]<br>[215 - 285]<br>[320 - 345] | [0 - 150]<br>[220 - 285] | [10 - 25]<br>[50 - 140]<br>[240 - 325] | [45 - 145]<br>[230 - 290] |
| | [55 - 320] | [0 - 355] | [30 - 30]<br>[50 - 330] | [0 - 25]<br>[60 - 355] | [45 - 355] | [60 - 325] |

[a] LENET-5 like network with 32 feature maps in $C_1$ and 64 feature maps in $C_3$.

– in the group with the loss value ≈1.5, the subsampling (in the form of the max pooling or stride) accompanies both convolutional layers;
– in the group with the loss value ≈6, the subsampling was applied in combination with exactly one convolutional layer;
– in the group with the loss value ≈25, the subsampling was not applied in combination convolutional layers.

Our second test was focused on the invariance of recognition by the FTNet with respect to rotation. On the basis of the first test, we selected several FTNet configurations for the analysis of the rotation invariance:

1. the one with the (absolute) lowest loss value,
2. the one with the lowest loss value among those with fixed kernels (FT2 or Conventional) in $C_1$ and $C_3$ layers,
3. the one with the lowest loss value among those with trainable kernels (FT2 or Conventional) in $C_1$ and $C_3$ layers.

It is worth noting that the configuration with the absolute lowest loss value (item 1) coincides with the one described in item 3 with the FT2 kernels.

For the purpose of testing, we have created 10 input images (manually) (Fig. 2) and rotated them from 0° up to 355° with the step 5°. All selected FTNets were trained for 10 epochs. In the bottom part of Table 2, we show those angle interval(s) where the network top prediction was not correct.

We have accomplished accuracy of 99.23% on MNIST which is competitive result (see LeCun MNIST web-page). This accuracy was achieved without any distortions on training set.

## 5   Conclusion

In this contribution, we have introduced a new convolutional neural network – the FTNet. In its two convolutional layers, the FTNet uses fixed kernels extracted from the discrete version of the $F^2$-transform. Moreover, it has a certain number of trainable hyper-parameters. The MNIST database was used for the FTNet training. The results were compared with the LENET-5 like network. The tests have shown that the best FTNet configuration performs significantly better recognition (according to the training time and the loss function value) than the LENET-5 like network.

We have also analyzed the FTNet and LENET-5 rotation invariance. We came to the conclusion that all tested networks and their best configurations show similar results. In particular, the rotation invariance was demonstrated only for relatively small angles within the interval [0°, 30°].

Last, but not least, we provided theoretical justification of a suitability of the FTNet for the problem of recognition.

# References

1. Schmidhuber, J.: Neural Netw. **61**, 85 (2015)
2. Krizhevsky, A., Sutskever, I., Hinton, G.E.: Imagenet classification with deep convolutional neural networks, In: Pereira, F., Burges, C.J.C., Bottou, L., Weinberger, K.Q. (eds.) Advances in Neural Information Processing Systems 25, pp. 1097–1105. Curran Associates, Inc. (2012)
3. Szegedy, C., Ioffe, S., Vanhoucke, V.: CoRR abs/1602.07261 (2016)
4. He, K., Zhang, X., Ren, S., Sun, J.: CoRR abs/1512.03385 (2015)
5. Oquab, M., Bottou, L., Laptev, I., Sivic, J.: Learning and transferring mid-level image representations using convolutional neural networks. In: The IEEE Conference on Computer Vision and Pattern Recognition (CVPR), June 2014
6. Werbos, P.J.: Beyond regression: new tools for prediction and analysis in the behavioral sciences, Ph.D. thesis, Harvard University. Department of Applied Mathematics (1974)
7. Rumelhart, D.E., Hinton, G.E., Williams, R.J.: Cogn. Model. **5**, 1 (1988)
8. Lowe, D.G.: Object recognition from local scale-invariant features. In: Proceedings of the Seventh IEEE International Conference on Computer Vision (1999)
9. McConnell, R.: Method of and apparatus for pattern recognition, 28 January 1986. US Patent 4,567,610
10. Maini, R., Aggarwal, H.: Int. J. Image Process. (IJIP) **3**, 1 (2009)
11. Perfilieva, I., Danková, M., Bede, B.: Towards f-transform of a higher degree. In: IFSA/EUSFLAT Conference (2009)
12. LeCun, Y., Bottou, L., Bengio, Y., Haffner, P.: Proc. IEEE **86**, 2278 (1998)
13. Fukushima, K.: Biol. Cybern. **36**, 193 (1980)
14. Williams, D.R.G.H.R., Hinton, G.: Nature **323**, 533 (1986)
15. Perfilieva, I., Kreinovich, V.: Fuzzy Sets Syst. **180**, 55 (2011)

# Class Switching Ensembles for Ordinal Regression

Pedro Antonio Gutiérrez[1]([⊠]), María Pérez-Ortiz[2], and Alberto Suárez[3]

[1] Department of Computer Science and Numerical Analysis,
University of Córdoba, Córdoba, Spain
pagutierrez@uco.es
[2] Department of Quantitative Methods, Universidad Loyola Andalucía,
Córdoba, Spain
mariaperez@uloyola.es
[3] Computer Science Department, Universidad Autónoma de Madrid, Madrid, Spain
alberto.suarez@uam.es

**Abstract.** The term ordinal regression refers to classification tasks in which the categories have a natural ordering. The main premise of this learning paradigm is that the ordering can be exploited to generate more accurate predictors. The goal of this work is to design class switching ensembles that take into account such ordering so that they are more accurate in ordinal regression problems. In standard (nominal) class switching ensembles, diversity among the members of the ensemble is induced by injecting noise in the class labels of the training instances. Assuming that the classes are interchangeable, the labels are modified at random. In ordinal class switching, the ordering between classes is taken into account by reducing the transition probabilities to classes that are further apart. In this manner smaller label perturbations in the ordinal scale are favoured. Two different specifications of these transition probabilities are considered; namely, an arithmetic and a geometric decrease with the absolute difference of the class ranks. These types of ordinal class switching ensembles are compared with an ensemble method that does not consider class-switching, a nominal class-switching ensemble, an ordinal variant of boosting, and two state-of-the-art ordinal classifiers based on support vector machines and Gaussian processes, respectively. These methods are evaluated and compared in a total of 15 datasets, using three different performance metrics. From the results of this evaluation one concludes that ordinal class-switching ensembles are more accurate than standard class-switching ones and than the ordinal ensemble method considered. Furthermore, their performance is comparable to the state-of-the-art ordinal regression methods considered in the analysis. Thus, class switching ensembles with specifically designed transition probabilities, which take into account the relationships between classes, are shown to provide very accurate predictions in ordinal regression problems.

This work has been partially supported by projects TIN2014-54583-C2-1-R, TIN2013-42351-P, TIN2016-76406-P and TIN2015-70308-REDT of the Spanish Ministerial Commission of Science and Technology (MINECO, Spain) and FEDER funds (EU).

© Springer International Publishing AG 2017
I. Rojas et al. (Eds.): IWANN 2017, Part I, LNCS 10305, pp. 408–419, 2017.
DOI: 10.1007/978-3-319-59153-7_36

**Keywords:** Class switching · Ordinal regression · Ensemble learning

# 1  Introduction

Ensemble methods have been successfully employed in numerous machine learning applications, including standard supervised learning problems [2,7,8], clustering [25] and image segmentation [13], among others. The goal in ensemble learning is to build a diverse collection of learners whose predictions are complementary. If the errors of the individual predictors are independent, they can be averaged in a combination step. This results in a global ensemble prediction that is more accurate than the one of individual classifiers. There are a number of strategies that can be used to build ensembles of diverse base learners. One of the most effective approaches to generate variability is to take advantage of the brittle character of the base learners and apply randomization techniques either in the dataset used for induction (e.g. training base learners with different bootstrap samples, as in bagging), or in the learning algorithm itself (e.g. build building random trees by considering splits only within a random subset of features, as in random forest). Another possibility is the injection of noise in the class labels. This technique was first introduced by Breiman under the name of *Output smearing* for regression, or *Output flipping* for classification problems [3]. In regression problems, the values of the dependent variable are contaminated with additive Gaussian noise. For classification, class labels are flipped at random with the restriction that the proportion of instances of the different classes is fixed. A direct extension of this technique, in which the class labels are simply modified at random, without ensuring that the class proportions are maintained, was analysed and seen to be more effective in ensembles of decision trees [19] and neural networks [18].

Despite their usefulness and demonstrated competitive performance, there are some areas of machine learning in which ensembles have been barely used. An example of this is the problem of ordinal regression, where these learning techniques have been applied only recently [9,15,16,21], with promising results. The problem of ordinal regression, also known in the literature as ordinal classification, is a supervised learning task in which the labels to be predicted are discrete, yet present an intrinsic ordering, which is relevant to the prediction problem at hand. For example, those surveys where students evaluate their teachers are usually based on an ordinal scale {*poor, average, good, very good, excellent*}. However, misclassifying *excellent* teachers as *poor* should be far more penalised than misclassifying them as *very good*. While it is possible to simply use standard classification techniques (which ignore the ordering of the labels) or regression techniques (which disregard the discrete nature of the labels and assume a specific distance between them), it is generally advantageous to consider specific methods that take into account the ordinal nature of the problem [12], not only for classification, but also in all the stages of the learning process, such as data preprocessing or performance evaluation. One of the most widely used approaches to ordinal regression are decomposition methods [12], which decompose the original ordinal variable into simpler classification problems [10,21], usually binary tasks, and the predictions of the corresponding classifiers are fused to produce an unique ordinal output. This approach can be seen as an ensemble where the diversity is introduced by the differences found in the classification tasks, and it is very natural in the case of ordinal regression because the classes can be joined according to different strategies (e.g. by a cascade binary utility model [14] or simply mixing neighbouring classes [10]). For example, for a given rank $q$, a direct question could be: "Is the label of pattern $\mathbf{x}$ greater than $q$?". Decomposition methods are popular within the ordinal classification literature, given that any binary classifier can be generally used as a base learner, without the need of reformulating the model to deal with the order of the classes. The main problem with this type of techniques is that the number of learners is relatively low (usually the number of classes minus one) and that fusing the different outputs is not straightforward [4]. Another differentiated group of ensemble-based approaches for ordinal regression are based on the concept of boosting. Some of these strategies rely on the confidence of a binary classifier [11,15,22], which can be used as an ordering preference, while others extend the well-known AdaBoost algorithm [16,23]. Finally, there are other strategies that make use of a base ordinal learner for the ensemble construction and introduce diversity as a term to be optimised during classifier construction [9] or that impose a global constraint to ensure that ordinal requirements are met [24].

In this article, we propose to design class-switching ensembles to address ordinal regression problems. To do so, different techniques are proposed to maintain and exploit the order information, based on a switching probability function that decreases with the rank difference. Our experiments compare seven different approaches in 15 datasets, showing that the ordinal class-switching approach outperforms the standard one and it is competitive with the state-of-the-art methods.

The rest of this paper is structured as follows: Sect. 2 discusses the design principles behind standard class-switching ensembles and describes how they are built. In Sect. 3 the class switching method for ensemble generation is adapted to address ordinal regression problems. The effectiveness of the ensembles generated with the variants of class switching proposed is evaluated in an extensive set of experiments on benchmark ordinal regression problems, whose results are reported and analysed in Sect. 4. Finally, Sect. 5 summarises the contributions of this work and outlines some concluding remarks.

## 2   Class Switching Ensembles

In nominal classification problems, one generally assumes that the classes are independent of each other and their labels interchangeable. When designing machine learning algorithms, these properties are incorporated in the prediction mechanism. For instance, in decision trees the class label assignment is performed at the leaf nodes. Those instances that, as a result of the hierarchy of Boolean queries at the inner nodes of the decision tree, are assigned to a leaf node receive the label of the majority class of the training instances assigned to that node. Similarly, in ensembles, majority voting is used. In the voting process, each predictor is allowed to vote for a single class. In neural networks, 1-of-$K$ encoding is typically used for prediction problems with $K$ classes. In this type of encoding, the $k$th class is represented as a vector of $K$ bits, all of which are 0, except for the $k$th bit, which is equal to 1. The classes can be thought of as lying on the vertices of a regular simplex, and are therefore at equal distance from each other. The noise injection process in nominal class switching also assumes that classes are interchangeable: to build an individual ensemble classifier, a subset of training instances are selected at random. Then, the class label of these instances is modified also in a random fashion, assigning equal probabilities to switching the label to one of the other $K - 1$ classes. Finally, a base learner is generated by applying the same base learning algorithm to the perturbed dataset. Once the ensemble has been completed, the predictions of the individual classifiers are combined by majority voting. Since the realizations are independent, the noise in the class labels is averaged out by the combination process. Furthermore, the variability induced can have a positive effect in the representation capacity of the ensemble.

In particular, when unpruned CART decision trees are used as base learners, class-switching ensembles achieve high prediction accuracy, provided that high switching rates (modifying the class label of a substantial fraction $\approx 0.6(K - 1)/K$ of the training instances) and sufficiently large ensembles (of size $\geq 1000$) are used. Similar improvements can be obtained using neural networks [18]. The goal of this work is to adapt this ensemble construction method to ordinal regression problems, in which class labels are ordered. If this ordering is relevant for prediction, designing a noise injection scheme that takes into account these relations among the class labels should lead to further accuracy improvements for this type of problems.

## 3   Ordinal Class Switching Ensembles

Consider a labelled dataset $\mathcal{D} = \{(\mathbf{x}_n, y_n)\}_{n=1}^{N}$. Assume that the dependent variable takes values in a finite set, $y_n \in \{c_1, c_2, \ldots, c_K\}$, which has an intrinsic ordering; that is, $c_1 < c_2 < \ldots < c_K$, where $<$ is an order relation provided by the nature of the classification problem.

To take advantage of the ordering relation among the class labels, we make the assumption that instances whose class labels are close to each other are more similar than instances whose class labels are further apart. Therefore, when selecting the modified class labels, one should assign higher probability to nearby classes. Specifically, we will choose a set of transition probabilities $\{p_{i \to j}; i, j = 1, \ldots, K\}$ with the following properties: If $|i - j| < |i - k| \Rightarrow p_{i \to j} < p_{i \to k}, \forall i, j, k \in \{1, \ldots, K\}$. Hence, the transition probability matrix should be V-shaped with respect to the class labels $c_k$, in the same way that cost matrices need to be V-shaped in the context of ordinal regression [17].

In this paper, we consider two different possibilities for constructing a V-shaped transition matrix:

- The probability of transition arithmetically decreases when the distance to the original class increases (arithmetic ordinal class switching, AOCS):

$$p_{i \rightarrow j}^* = \begin{cases} p^*, & \text{if } i = j, \\ \dfrac{1 - p^*}{|i - j|}, & \text{if } i \neq j, \end{cases} \qquad (1)$$

where $0 \leq p^* \leq 1$ is the parameter that sets the probability of not transitioning to a different class.

- The probability of transition geometrically decreases when the distance to the original class increases (geometric ordinal class switching, GOCS):

$$p_{i \rightarrow j}^* = \begin{cases} p^*, & \text{if } i = j. \\ \dfrac{1 - p^*}{2^{|i - j|}}, & \text{if } i \neq j. \end{cases} \qquad (2)$$

Given that the transition probabilities must add up to one, these matrices need to be normalised by rows:

$$p_{i \rightarrow j} = \frac{p_{i \rightarrow j}^*}{\sum_{k=1}^{K} p_{i \rightarrow k}^*}. \qquad (3)$$

For example, for an ordinal regression problem with $K = 5$ classes and $p^* = 0.6$, the transition matrices are:

$$\mathbf{P}_{\text{AOCS}} = \begin{pmatrix} 0.54 & 0.18 & 0.12 & 0.09 & 0.07 \\ 0.16 & 0.49 & 0.16 & 0.11 & 0.08 \\ 0.11 & 0.16 & 0.47 & 0.16 & 0.11 \\ 0.08 & 0.11 & 0.16 & 0.49 & 0.16 \\ 0.07 & 0.09 & 0.12 & 0.18 & 0.54 \end{pmatrix}, \mathbf{P}_{\text{GOCS}} = \begin{pmatrix} 0.62 & 0.21 & 0.10 & 0.05 & 0.03 \\ 0.17 & 0.52 & 0.17 & 0.09 & 0.04 \\ 0.08 & 0.17 & 0.50 & 0.17 & 0.08 \\ 0.04 & 0.09 & 0.17 & 0.52 & 0.17 \\ 0.03 & 0.05 & 0.10 & 0.21 & 0.62 \end{pmatrix},$$

Note that, because of the normalization of class probabilities, the probability that a label remains unchanged (elements in the main diagonal of the matrix) decreases for central labels in the ordinal scale. This is sensible in ordinal regression, given that the changes in the extreme labels should be less likely.

## 4   Experiments

This section describes the experiments used to evaluate the performance of the ensembles generated with the variants of ordinal class switching proposed and an analysis of the results of these experiments.

### 4.1   Methods Compared

The experiments are designed to compare the accuracy of ensembles generated with the two ordinal variants of class switching introduced in this paper (AOCS and GOCS, see Sect. 3 with standard (nominal) class switching (NCS, see Sect. 2) and an ensemble that does not make use of a label perturbation strategy for the different members of the ensemble (Orig). All the ensembles generated use classification trees as base learners, as in previous studies on class switching [19]. We used the implementation included in the Python `scikit-learn` machine learning framework [20], in which an optimised version of the CART algorithm is considered for tree induction. This implementation considers heuristic algorithms, where locally optimal decisions are made at each node. This makes the induction process non deterministic, therefore different trees can be obtained depending on the seed used for random number generation. This introduces diversity for the Orig algorithm, where no perturbation of the dataset is performed.

Besides evaluating whether ensembles generated with ordinal class switching methods are more accurate than their nominal counterparts, we compare their performance also with other state-of-the-art ordinal classifiers analysed in [12]; namely, we consider the reduction from ordinal regression to binary support vector machine classifiers (REDSVM) [17], the reformulation of Gaussian processes for ordinal regression (GPOR) [5] including automatic relevance determination, and an ensemble method, the ORBoost method with all margins [15].

## 4.2   Measures of Performance

Different metrics can be used to evaluate ordinal regression classifiers. The most common ones are accuracy ($Acc$) and Mean Absolute Error ($MAE$). $Acc$ is the rate of correctly classified patterns:

$$Acc = \frac{1}{N} \sum_{i=1}^{N} I(y_i^* = y_i),$$

where $y_i$ is the correct label of the $i$th instance, $y_i^*$ is the predicted one, and $I(\cdot)$ is a Boolean test. This measure characterises the global performance in the classification task, without taking into account the ordering of the classes.

The Mean Absolute Error is an average deviation in absolute value of the predicted rank from the true one [1]:

$$MAE = \frac{1}{N} \sum_{i=1}^{N} |\mathcal{O}(y_i) - \mathcal{O}(y_i^*)|,$$

where $\mathcal{O}$, $\mathcal{O}(c_j) = j, 1 \le j \le K$. $MAE$ values range from 0 to $K - 1$ (maximum deviation in number of ranks between two labels). Given that some of the datasets considered are imbalanced, we also consider the average of the $MAE$s across classes ($AMAE$) [1]:

$$AMAE = \frac{1}{K} \sum_{j=1}^{K} MAE_j = \frac{1}{K} \sum_{j=1}^{K} \frac{1}{n_j} \sum_{i=1}^{n_j} |\mathcal{O}(y_i) - \mathcal{O}(y_i^*)|,$$

where $n_j$ is the number of patterns in class $j$. The value of $AMAE$ range from 0 to $K - 1$.

## 4.3   Datasets and Experimental Setup

A battery of 15 ordinal regression datasets is used for evaluation. Their characteristics are summarised in Table 1. The selected datasets are very different in terms of numbers of patterns, attributes, classes and class distribution to ensure that the conclusions of the study cover a sufficiently wide rage of ordinal regression problems. The experimental protocol is similar to the one used in [12]. The results reported are averages over 30 random partitions into training (3/4 of the data) and test sets (1/4 of the data), considering the same partitions than in [12][1]. Consequently, the results for GPOR, ORBoost and REDSVM were directly taken from [12].

**Table 1.** Characteristics of the benchmark datasets used in the experiments.

| Dataset | #Pat. | #Attr. | #Classes | Class distribution |
|---|---|---|---|---|
| ERA | 1000 | 4 | 9 | (92, 142, 181, 172, 158, 118, 88, 31, 18) |
| ESL | 488 | 4 | 9 | (2, 12, 38, 100, 116, 135, 62, 19, 4) |
| LEV | 1000 | 4 | 5 | (93, 280, 403, 197, 27) |
| SWD | 1000 | 10 | 4 | (32, 352, 399, 217) |
| Automobile | 205 | 71 | 6 | (3, 22, 67, 54, 32, 27) |
| Balance-scale | 625 | 4 | 3 | (288, 49, 288) |
| Bondrate | 57 | 37 | 5 | (6, 33, 12, 5, 1) |
| Eucalyptus | 736 | 91 | 5 | (180, 107, 130, 214, 105) |
| Newthyroid | 215 | 5 | 3 | (30, 150, 35) |
| Pasture | 36 | 25 | 3 | (12, 12, 12) |
| Squash-stored | 52 | 51 | 3 | (23, 21, 8) |
| Squash-unstored | 52 | 52 | 3 | (24, 24, 4) |
| Tae | 151 | 54 | 3 | (49, 50, 52) |
| Toy | 300 | 2 | 5 | (35, 87, 79, 68, 31) |
| Winequality-red | 1599 | 11 | 6 | (10, 53, 681, 638, 199, 18) |

In all cases (Orig, NCS, AOCS and GOCS) ensembles of $T = 1001$ predictors are used. The outputs of the ensemble classifiers are combined using a soft voting rule: The global ensemble prediction for a given instance, characterised by the vector of attributes $\mathbf{x}$, is $c_j = \arg\max_{c_i} \sum_{t=1}^{T} p_{ti}(\mathbf{x})$, where $p_{ti}(\mathbf{x})$ is the probability assigned by the $t$th ensemble classifier to the class label $c_i$. These probabilities are approximated as the fraction of samples of the corresponding class in the leaf node to which the instance is assigned. In class switching, after preliminary experiments, $p^*$ is set to 0.6.

The hyperparameters for REDSVM are selected by using a nested five fold cross-validation over the training set. The criterion used to determine the optimal

---

[1] Available at http://www.uco.es/grupos/ayrna/orreview.

hyperparameter values is $MAE$. A Gaussian kernel function is used, and both the value of the cost parameter $C$ and the width of the kernel are determined considering the range $\{10^{-3}, 10^{-2}, \ldots, 10^{3}\}$. In GPOR, the hyperparameters are determined by part of the optimisation process. Following the recommendation given in [15], the ensemble size in ORBoost is $T = 2000$. In this method, normalised sigmoid functions are used as base learners. The smoothness parameter, $\gamma$ is set to 4.

## 4.4 Results

In Table 2 the values of $Acc$, $MAE$ and $AMAE$ obtained by the different methods in the test set are presented. The measures are averages and standard deviations (in a smaller font) over the 30 random training/test partitions.

In Table 3 the average ranks (in terms of the values of $Acc$, $MAE$ and $AMAE$ in the test partitions for the 15 datasets) are presented. Specifically, rank 1 corresponds to the best performance and rank 7 to the worst one. From the results presented in this table, it is apparent that the $Orig$ ensembles have poor performance results. This is probably related to the low diversity among the ensemble classifiers: since the same unperturbed training data are used to build the individual classifiers, the only source of variability is the intrinsic randomness of the base learning algorithm. Both ordinal class switching ensembles (AOCS and GOCS) yield better performance than standard (nominal) class switching (NCS). The geometric decrease (GOCS) leads to better overall results than the arithmetic one (AOCS) when the performance is measured in terms of measures such as $Acc$ and $MAE$. However AOCS outperforms GOCS when the metric is class-specific (e.g. using $AMAE$). In general, the accuracy is competitive with respect to the state-of-the-art methods considered (GPOR and REDSVM). The ordinal class-switching ensembles are clearly more accurate than ORBoost ensembles. The statistical significance of the differences of performance between the different methods is determined using the guidelines given in [6]. Specifically, a non-parametric Friedman's test (at a significance level of $\alpha = 0.10$) has been applied for $Acc$, $MAE$ and $AMAE$ rankings. The confidence interval is, in this case, $C_0 = (0, F_{(\alpha=0.10)} = 1.85)$. The values of the statistic are $F_{Acc}$: $2.53 \notin C_0$, $F_{MAE}$: $3.34 \notin C_0$ and $F_{AMAE}$: $2.17 \notin C_0$. Consequently, the test rejects the null-hypothesis that, as measured by the average rank, the algorithms have similar performance.

Considering AOCS and GOCS as the control methods, we apply the post-hoc Holm's test [6]. The performance of the $i$-th and $j$-th algorithms are compared using the statistic:

$$z = \frac{R_i - R_j}{\sqrt{\frac{J(J+1)}{6N}}},$$

where $J$ is the number of algorithms ($J = 7$, in our case), $N$ is the number of datasets ($N = 15$, in our case) and $R_i$ is the average rank of the $i$-th method (see Table 3). Asymptotically, this statistic is normally distributed, which allows us to quantify the significance of the differences observed at the corresponding

**Table 2.** Average and standard deviation of $Acc$, $MAE$ and $AMAE$ values obtained for the different methods compared

| Dataset | Orig | NCS | AOCS | GOCS | GPOR | ORBoost | REDSVM |
|---|---|---|---|---|---|---|---|
| *Acc* | | | | | | | |
| ERA | $0.257_{0.024}$ | $0.255_{0.025}$ | *$0.259_{0.029}$* | $0.258_{0.028}$ | **$0.288_{0.027}$** | $0.240_{0.021}$ | $0.249_{0.019}$ |
| ESL | $0.642_{0.033}$ | $0.651_{0.035}$ | $0.652_{0.035}$ | $0.653_{0.035}$ | **$0.713_{0.031}$** | $0.677_{0.022}$ | *$0.713_{0.030}$* |
| LEV | $0.614_{0.021}$ | $0.613_{0.024}$ | *$0.621_{0.023}$* | $0.621_{0.022}$ | $0.612_{0.030}$ | $0.609_{0.029}$ | **$0.627_{0.024}$** |
| SWD | $0.540_{0.029}$ | $0.551_{0.026}$ | $0.562_{0.025}$ | $0.562_{0.024}$ | **$0.578_{0.031}$** | $0.561_{0.032}$ | *$0.571_{0.027}$* |
| Automobile | $0.788_{0.061}$ | **$0.820_{0.047}$** | $0.816_{0.049}$ | *$0.817_{0.050}$* | $0.611_{0.073}$ | $0.706_{0.055}$ | $0.683_{0.070}$ |
| Balance-scale | $0.772_{0.019}$ | $0.802_{0.023}$ | $0.775_{0.024}$ | $0.775_{0.024}$ | $0.966_{0.012}$ | *$0.968_{0.016}$* | **$0.999_{0.004}$** |
| Bondrate | $0.463_{0.110}$ | $0.524_{0.074}$ | $0.553_{0.070}$ | $0.559_{0.070}$ | **$0.578_{0.032}$** | $0.542_{0.091}$ | *$0.564_{0.055}$* |
| Eucalyptus | $0.609_{0.030}$ | $0.674_{0.032}$ | *$0.681_{0.032}$* | $0.679_{0.032}$ | **$0.686_{0.034}$** | $0.620_{0.025}$ | $0.638_{0.035}$ |
| Newthyroid | $0.940_{0.035}$ | **$0.969_{0.017}$** | *$0.969_{0.018}$* | *$0.969_{0.018}$* | $0.966_{0.024}$ | $0.958_{0.029}$ | $0.968_{0.023}$ |
| Pasture | *$0.762_{0.105}$* | **$0.781_{0.147}$** | $0.753_{0.138}$ | $0.753_{0.138}$ | $0.522_{0.178}$ | $0.700_{0.121}$ | $0.674_{0.116}$ |
| Squash-stored | $0.624_{0.119}$ | *$0.699_{0.110}$* | **$0.704_{0.101}$** | **$0.704_{0.101}$** | $0.451_{0.100}$ | $0.636_{0.124}$ | $0.621_{0.132}$ |
| Squash-unstored | $0.764_{0.112}$ | *$0.835_{0.085}$* | **$0.839_{0.085}$** | **$0.839_{0.085}$** | $0.644_{0.162}$ | $0.703_{0.098}$ | $0.731_{0.119}$ |
| Tae | $0.582_{0.057}$ | $0.570_{0.071}$ | $0.580_{0.062}$ | $0.580_{0.062}$ | $0.328_{0.041}$ | *$0.597_{0.057}$* | **$0.601_{0.068}$** |
| Toy | $0.885_{0.033}$ | $0.936_{0.025}$ | $0.934_{0.026}$ | $0.933_{0.026}$ | *$0.954_{0.022}$* | $0.948_{0.025}$ | **$0.977_{0.012}$** |
| Winequality-red | $0.615_{0.020}$ | $0.685_{0.017}$ | *$0.685_{0.016}$* | **$0.686_{0.016}$** | $0.606_{0.015}$ | $0.666_{0.021}$ | $0.627_{0.020}$ |
| *MAE* | | | | | | | |
| ERA | $1.383_{0.065}$ | $1.372_{0.065}$ | $1.335_{0.060}$ | $1.334_{0.061}$ | *$1.241_{0.051}$* | $1.250_{0.041}$ | **$1.219_{0.044}$** |
| ESL | $0.385_{0.040}$ | $0.370_{0.042}$ | $0.370_{0.040}$ | $0.367_{0.040}$ | **$0.301_{0.035}$** | $0.340_{0.025}$ | *$0.306_{0.037}$* |
| LEV | $0.425_{0.027}$ | $0.429_{0.031}$ | *$0.417_{0.027}$* | $0.417_{0.026}$ | $0.422_{0.031}$ | $0.434_{0.030}$ | **$0.410_{0.023}$** |
| SWD | $0.499_{0.034}$ | $0.484_{0.031}$ | $0.467_{0.028}$ | $0.466_{0.027}$ | **$0.440_{0.032}$** | $0.463_{0.036}$ | *$0.445_{0.031}$* |
| Automobile | $0.327_{0.108}$ | **$0.263_{0.074}$** | $0.266_{0.074}$ | *$0.265_{0.076}$* | $0.594_{0.131}$ | $0.348_{0.079}$ | $0.403_{0.092}$ |
| Balance-scale | $0.264_{0.028}$ | $0.230_{0.031}$ | $0.250_{0.031}$ | $0.250_{0.031}$ | $0.034_{0.012}$ | *$0.032_{0.017}$* | **$0.001_{0.004}$** |
| Bondrate | $0.739_{0.163}$ | $0.630_{0.104}$ | $0.592_{0.105}$ | *$0.587_{0.103}$* | $0.624_{0.062}$ | **$0.531_{0.110}$** | $0.613_{0.081}$ |
| Eucalyptus | $0.447_{0.035}$ | $0.350_{0.036}$ | *$0.343_{0.036}$* | $0.344_{0.037}$ | **$0.331_{0.038}$** | $0.415_{0.036}$ | $0.395_{0.036}$ |
| Newthyroid | $0.060_{0.038}$ | $0.031_{0.017}$ | *$0.031_{0.018}$* | *$0.031_{0.018}$* | $0.034_{0.024}$ | $0.042_{0.029}$ | **$0.029_{0.022}$** |
| Pasture | *$0.238_{0.105}$* | **$0.219_{0.147}$** | $0.247_{0.138}$ | $0.247_{0.138}$ | $0.489_{0.190}$ | $0.300_{0.121}$ | $0.330_{0.111}$ |
| Squash-stored | $0.424_{0.156}$ | *$0.335_{0.132}$* | **$0.327_{0.122}$** | **$0.327_{0.122}$** | $0.626_{0.148}$ | $0.364_{0.124}$ | $0.346_{0.145}$ |
| Squash-unstored | $0.236_{0.112}$ | *$0.168_{0.086}$* | **$0.161_{0.085}$** | **$0.161_{0.085}$** | $0.356_{0.162}$ | $0.305_{0.106}$ | $0.262_{0.122}$ |
| Tae | $0.521_{0.092}$ | $0.553_{0.104}$ | $0.523_{0.095}$ | $0.523_{0.095}$ | $0.861_{0.155}$ | *$0.504_{0.088}$* | **$0.461_{0.060}$** |
| Toy | $0.117_{0.035}$ | $0.064_{0.025}$ | $0.066_{0.026}$ | $0.067_{0.026}$ | *$0.046_{0.022}$* | $0.052_{0.025}$ | **$0.024_{0.013}$** |
| Winequality-red | $0.457_{0.026}$ | $0.351_{0.019}$ | *$0.349_{0.019}$* | **$0.348_{0.019}$** | $0.425_{0.017}$ | $0.365_{0.021}$ | $0.417_{0.020}$ |
| *AMAE* | | | | | | | |
| ERA | $1.428_{0.122}$ | $1.405_{0.114}$ | **$1.370_{0.099}$** | *$1.372_{0.102}$* | $1.372_{0.108}$ | $1.427_{0.102}$ | $1.413_{0.086}$ |
| ESL | $0.586_{0.086}$ | $0.572_{0.081}$ | $0.569_{0.084}$ | $0.569_{0.083}$ | *$0.477_{0.094}$* | $0.554_{0.096}$ | **$0.459_{0.116}$** |
| LEV | $0.603_{0.061}$ | $0.603_{0.054}$ | **$0.601_{0.051}$** | *$0.602_{0.051}$* | $0.654_{0.050}$ | $0.615_{0.044}$ | $0.620_{0.046}$ |
| SWD | $0.579_{0.052}$ | $0.581_{0.054}$ | *$0.579_{0.052}$* | $0.580_{0.051}$ | $0.589_{0.040}$ | **$0.576_{0.039}$** | $0.608_{0.034}$ |
| Automobile | $0.316_{0.131}$ | $0.320_{0.127}$ | *$0.314_{0.120}$* | **$0.313_{0.120}$** | $0.792_{0.200}$ | $0.400_{0.120}$ | $0.464_{0.135}$ |
| Balance-scale | $0.463_{0.022}$ | $0.440_{0.022}$ | $0.455_{0.023}$ | $0.455_{0.023}$ | $0.051_{0.023}$ | *$0.042_{0.029}$* | **$0.001_{0.003}$** |
| Bondrate | $1.167_{0.320}$ | $1.204_{0.182}$ | $1.161_{0.168}$ | $1.161_{0.163}$ | $1.360_{0.122}$ | **$0.839_{0.260}$** | *$1.144_{0.275}$* |
| Eucalyptus | $0.470_{0.039}$ | $0.378_{0.040}$ | *$0.370_{0.038}$* | $0.372_{0.039}$ | **$0.362_{0.040}$** | $0.434_{0.040}$ | $0.429_{0.039}$ |
| Newthyroid | $0.085_{0.074}$ | *$0.056_{0.035}$* | $0.061_{0.039}$ | $0.061_{0.039}$ | $0.062_{0.049}$ | $0.067_{0.049}$ | **$0.048_{0.040}$** |
| Pasture | *$0.238_{0.105}$* | **$0.219_{0.147}$** | $0.247_{0.138}$ | $0.247_{0.138}$ | $0.489_{0.190}$ | $0.300_{0.121}$ | $0.330_{0.111}$ |
| Squash-stored | $0.452_{0.224}$ | $0.392_{0.197}$ | $0.392_{0.192}$ | $0.392_{0.192}$ | $0.797_{0.234}$ | **$0.368_{0.129}$** | *$0.386_{0.162}$* |
| Squash-unstored | $0.207_{0.135}$ | **$0.170_{0.137}$** | *$0.172_{0.143}$* | *$0.172_{0.143}$* | $0.443_{0.226}$ | $0.341_{0.174}$ | $0.383_{0.184}$ |
| Tae | $0.521_{0.092}$ | $0.553_{0.104}$ | $0.522_{0.095}$ | $0.522_{0.095}$ | $0.863_{0.164}$ | *$0.503_{0.087}$* | **$0.459_{0.059}$** |
| Toy | $0.118_{0.035}$ | $0.063_{0.023}$ | $0.068_{0.026}$ | $0.071_{0.028}$ | *$0.044_{0.025}$* | $0.054_{0.027}$ | **$0.024_{0.016}$** |
| Winequality-red | $0.971_{0.118}$ | $0.958_{0.076}$ | **$0.952_{0.076}$** | *$0.956_{0.078}$* | $1.065_{0.065}$ | $0.974_{0.048}$ | $1.080_{0.083}$ |

For each dataset the best result is marked in **bold** face; the second best in *italics*.

**Table 3.** Average test rankings over the 15 datasets for *Acc*, *MAE* and *AMAE*.

| Method | Orig | NCS | AOCS | GOCS | GPOR | ORBoost | REDSVM |
|--------|------|-----|------|------|------|---------|--------|
| *Acc* | 5.40 | 4.00 | *3.13* | **3.07** | 4.20 | 4.67 | 3.53 |
| *MAE* | 5.80 | 4.20 | 3.40 | *3.20* | 4.33 | 4.07 | **3.00** |
| *AMAE* | 5.00 | 3.93 | **3.00** | *3.40* | 5.13 | 3.73 | 3.80 |

The best result is in bold face and the second one in italics.

**Table 4.** Results of the Holm procedure using AOCS and GOCS as control methods: corrected $\alpha$ values, compared method and $p$-values, ordered by significance.

| $i$ | $\alpha/(7-i)$ | Control method | | | |
|-----|------------|---------|---------|---------|---------|
| | | AOCS | | GOCS | |
| | | Method | $p$-value | Method | $p$-value |
| *Acc* | | | | | |
| 1 | 0.01667 | Orig | 0.00406+ | Orig | 0.00310+ |
| 2 | 0.02000 | ORBoost | 0.05191 | ORBoost | 0.04252 |
| 3 | 0.02500 | GPOR | 0.17630 | GPOR | 0.15079 |
| 4 | 0.03333 | NCS | 0.27190 | NCS | 0.23673 |
| 5 | 0.05000 | REDSVM | 0.61209 | REDSVM | 0.55412 |
| 6 | 0.10000 | GOCS | 0.93265 | AOCS | 0.93265 |
| *MAE* | | | | | |
| 1 | 0.01667 | Orig | 0.00235+ | Orig | 0.00098+ |
| 2 | 0.02000 | GPOR | 0.23673 | GPOR | 0.15079 |
| 3 | 0.02500 | NCS | 0.31049 | NCS | 0.20489 |
| 4 | 0.03333 | ORBoost | 0.39802 | ORBoost | 0.27190 |
| 5 | 0.05000 | REDSVM | 0.61209 | REDSVM | 0.79985 |
| 6 | 0.10000 | GOCS | 0.79985 | AOCS | 0.79985 |
| *AMAE* | | | | | |
| 1 | 0.01667 | GPOR | 0.00684+ | GPOR | 0.02799+ |
| 2 | 0.02000 | Orig | 0.01123+ | Orig | 0.04252 |
| 3 | 0.02500 | NCS | 0.23673 | NCS | 0.49896 |
| 4 | 0.03333 | REDSVM | 0.31049 | AOCS | 0.61209 |
| 5 | 0.05000 | ORBoost | 0.35254 | REDSVM | 0.61209 |
| 6 | 0.10000 | GOCS | 0.61209 | ORBoost | 0.67261 |

+: statistical difference with $\alpha = 0.10$.

significance level ($\alpha$). The value of $\alpha$ is adjusted to take into account that multiple comparisons are made. The adjustment is made by a sequential procedure: the ordered $p$-values are $p_1, p_2, \ldots, p_{J-1}$, so that $p_1 \leq p_2 \leq \ldots \leq p_{J-1}$, and each $p_i$ is compared with $\alpha/(J - i)$. The results of these tests are presented in Table 4. The tests conclude that the differences between AOCS and Orig are statistically significant for $Acc$ and $MAE$, as well as those between GOCS and Orig for the same metrics. In terms of $AMAE$, AOCS is significantly better than GPOR and Orig, while GOCS is significantly better than GPOR.

## 5    Conclusions

In this work, we propose to take into account the ordering between classes to improve the accuracy of class switching ensembles in ordinal regression problems. To this end, the class switching protocol is modified so that, in the noise injection process, a class label that is closer to the original one has a higher probability of being selected. Two variants of the class switching procedure are considered. In the first one (AOCS) the probability of changing the original label to another one decreases arithmetically with the distance between the target label and the original one. In the second one, the decrease with this distance is geometric (GOCS). For global measures of performance, such as accuracy ($Acc$) or the Mean Average Error ($MAE$), the geometric scheme provides the best results. However, when measures that incorporate more information on the relationship among the classes, such as the average of the MAEs across classes (AMAE), are used, the arithmetic scheme performs better than the geometric scheme. Both types of ordinal class switching ensembles are more accurate than standard (nominal) class switching, which assumes that the classes are interchangeable (i.e., it discards the information on the ordering of the classes). Furthermore, their generalization capacity is comparable to, and in some cases better, the state-of-the-art ordinal regression methods.

## References

1. Baccianella, S., Esuli, A., Sebastiani, F.: Evaluation measures for ordinal regression. In: Proceedings of the Ninth International Conference on Intelligent Systems Design and Applications (ISDA 2009), pp. 283–287 (2009)
2. Bauer, E., Kohavi, R.: An empirical comparison of voting classification algorithms: bagging, boosting, and variants. Mach. Learn. **36**(1), 105–139 (1999). http://dx.doi.org/10.1023/A:1007515423169
3. Breiman, L.: Randomizing outputs to increase prediction accuracy. Mach. Learn. **40**(3), 229–242 (2000). http://dx.doi.org/10.1023/A:1007682208299
4. Cardoso, J.S., da Costa, J.F.P.: Learning to classify ordinal data: the data replication method. J. Mach. Learn. Res. **8**, 1393–1429 (2007)
5. Chu, W., Ghahramani, Z.: Gaussian processes for ordinal regression. J. Mach. Learn. Res. **6**, 1019–1041 (2005)
6. Demsar, J.: Statistical comparisons of classifiers over multiple data sets. J. Mach. Learn. Res. **7**, 1–30 (2006)
7. Dietterich, T.G.: Ensemble methods in machine learning. In: Kittler, J., Roli, F. (eds.) MCS 2000. LNCS, vol. 1857, pp. 1–15. Springer, Heidelberg (2000). doi:10. 1007/3-540-45014-9_1
8. Fernández-Delgado, M., Cernadas, E., Barro, S., Amorim, D.: Do we need hundreds of classifiers to solve real world classification problems? J. Mach. Learn. Res. **15**, 3133–3181 (2014). http://jmlr.org/papers/v15/delgado14a.html

9. Fernandez-Navarro, F., Gutiérrez, P.A., Hervás-Martínez, C., Yao, X.: Negative correlation ensemble learning for ordinal regression. IEEE Trans. Neural Netw. Learn. Syst. **24**(11), 1836–1849 (2013). http://dx.doi.org/10.1109/TNNLS.2013. 2268279, jCR (2013): 4.370 (category COMPUTER SCIENCE, THEORY & METHODS, position 2/102 Q1)
10. Frank, E., Hall, M.: A simple approach to ordinal classification. In: Raedt, L., Flach, P. (eds.) ECML 2001. LNCS, vol. 2167, pp. 145–156. Springer, Heidelberg (2001). doi:10.1007/3-540-44795-4_13
11. Freund, Y., Iyer, R., Schapire, R.E., Singer, Y.: An efficient boosting algorithm for combining preferences. J. Mach. Learn. Res. **4**, 933–969 (2003). http://dl.acm.org/ citation.cfm?id=945365.964285
12. Gutiérrez, P.A., Perez-Ortiz, M., Sanchez-Monedero, J., Fernandez-Navarro, F., Hervas-Martinez, C.: Ordinal regression methods: survey and experimental study. IEEE Trans. Knowl. Data Eng. **28**(1), 127–146 (2016)
13. Kim, H., Thiagarajan, J.J., Bremer, P.T.: Image segmentation using consensus from hierarchical segmentation ensembles. In: ICIP, pp. 3272–3276. IEEE (2014)
14. Kwon, Y.S., Han, I., Lee, K.C.: Ordinal pairwise partitioning (OPP) approach to neural networks training in bond rating. Int. Syst. Account. Finan. Manag. **6**(1), 23–40 (1997)
15. Lin, H.-T., Li, L.: Large-Margin thresholded ensembles for ordinal regression: theory and practice. In: Balcázar, J.L., Long, P.M., Stephan, F. (eds.) ALT 2006. LNCS, vol. 4264, pp. 319–333. Springer, Heidelberg (2006). doi:10.1007/ 11894841_26
16. Lin, H.T., Li, L.: Combining ordinal preferences by boosting. In: Proceedings of the European Conference on Machine Learning and Principles and Practice of Knowledge Discovery in Databases (ECML PKDD), pp. 69 83 (2009)
17. Lin, H.T., Li, L.: Reduction from cost-sensitive ordinal ranking to weighted binary classification. Neural Comput. **24**(5), 1329–1367 (2012)
18. Martínez-Muñoz, G., Sánchez-Martínez, A., Hernández-Lobato, D., Suárez, A.: Building ensembles of neural networks with class-switching. In: Kollias, S.D., Stafylopatis, A., Duch, W., Oja, E. (eds.) ICANN 2006. LNCS, vol. 4131, pp. 178–187. Springer, Heidelberg (2006). doi:10.1007/11840817_19
19. Martínez-Muñoz, G., Suárez, A.: Switching class labels to generate classification ensembles. Pattern Recogn. **38**(10), 1483–1494 (2005)
20. Pedregosa, F., Varoquaux, G., Gramfort, A., Michel, V., Thirion, B., Grisel, O., Blondel, M., Prettenhofer, P., Weiss, R., Dubourg, V., et al.: Scikit-learn: machine learning in python. J. Mach. Learn. Res. **12**(Oct), 2825–2830 (2011)
21. Pérez-Ortiz, M., Gutiérrez, P.A., Hervás-Martínez, C.: Projection based ensemble learning for ordinal regression. IEEE Trans. Cybern. **44**(5), 681–694 (2014). http://www.uco.es/grupos/ayrna/elor2013
22. Rennie, J.D.M.: Loss functions for preference levels: regression with discrete ordered labels. In: Proceedings of the IJCAI Multidisciplinary Workshop on Advances in Preference Handling, pp. 180–186 (2005)
23. Riccardi, A., Fernandez-Navarro, F., Carloni, S.: Cost-sensitive AdaBoost algorithm for ordinal regression based on extreme learning machine. IEEE Trans. Cybern. **44**(10), 1898–1909 (2014)
24. Sousa, R., Cardoso, J.S.: Ensemble of decision trees with global constraints for ordinal classification. In: 2011 11th International Conference on Intelligent Systems Design and Applications, pp. 1164–1169, November 2011
25. Strehl, A., Ghosh, J.: Cluster ensembles – a knowledge reuse framework for combining multiple partitions. J. Mach. Learn. Res. **3**, 583–617 (2002)

# Attractor Basin Analysis of the Hopfield Model: The Generalized Quadratic Knapsack Problem

Lucas García[1]($\boxtimes$), Pedro M. Talaván[2], and Javier Yáñez[3]

[1] Facultad de Matemáticas, Universidad Complutense de Madrid, Madrid, Spain
`lucasgarciarodriguez@ucm.es`
[2] Instituto Nacional de Estadística, Madrid, Spain
`pedro.martinez.talavan@ine.es`
[3] Facultad de Matemáticas, Universidad Complutense de Madrid, Madrid, Spain
`jayage@ucm.es`

**Abstract.** The Continuous Hopfield Neural Network (CHN) is a neural network which can be used to solve some optimization problems. The weights of the network are selected based upon a set of parameters which are deduced by mapping the optimization problem to its associated CHN. When the optimization problem is the Traveling Salesman Problem, for instance, this mapping process leaves one free parameter; as this parameter decreases, better solutions are obtained. For the general case, a Generalized Quadratic Knapsack Problem (GQKP), there are some free parameters which can be related to the saddle point of the CHN. Whereas in simple instances of the GQKP, this result guarantees that the global optimum is always obtained, in more complex instances, this is far more complicated. However, it is shown how in the surroundings of the saddle point the attractor basins for the best solutions grow as the free parameter decreases, making saddle point neighbors excellent starting point candidates for the CHN. Some technical results and some computational experiences validate this behavior.

**Keywords:** Artificial neural networks · Optimization · Machine Learning · Hopfield network

## 1 Introduction

Artificial neural networks are frequently used today in Machine Learning to solve all kinds of classification, regression and clustering problems. Moreover, neural networks can also be used to solve optimization problems in areas such as Operations Research, Control Theory, Image Processing, etc. Among such networks, the Hopfield network, the Kohonen network and the Boltzmann machines have been some of the most popular choices of network architectures.

The Hopfield network, also frequently referred to as the Hopfield model, is considered to be one of the milestones [9] for the renaissance of neural networks at the beginning of the 1980s and one of the most influential recurrent networks [1]. Hopfield proposed two models based on the concept of associative memory: the

© Springer International Publishing AG 2017
I. Rojas et al. (Eds.): IWANN 2017, Part I, LNCS 10305, pp. 420–431, 2017.
DOI: 10.1007/978-3-319-59153-7_37

*discrete* model [4], and a generalization that can take all real values in the interval
$[0, 1]$, the *continuous* model [5].

The Continuous Hopfield Network (CHN) is a recurrent neural network with
an associated differential equation, whose states evolve from an initial point to
an equilibrium point by minimizing a Lyapunov function. Since the Lyapunov
function is associated with the objective function of the optimization problem
(i.e. the mapping process), the equilibrium or stable point matches a local optimum of the optimization problem.

The CHN consists of a fully interconnected neural network with $n$ continuous
valued units (neurons) and a smooth sigmoid activation function. The dynamics
of the CHN is described by the differential equation (see Fig. 1):

$$\frac{d\mathbf{u}}{dt} = -\frac{\mathbf{u}}{\tau} + \mathbf{Tv} + \mathbf{i^b}$$

where, $\forall i, j \in \{1, \ldots, n\}$:

> $u_i$  is the current state of neuron $i$
> $v_i$  is the output of neuron $i$
> $T_{i,j}$ is the strength of the connection from neuron $j$ to neuron $i$
> $i_i^b$  is the offset bias of neuron $i$

being the output function $g(u_i)$ a hyperbolic tangent:

$$v_i = g(u_i) = \frac{1}{2}\left(1 + \tanh\left(\frac{u_i}{u_0}\right)\right), \quad u_0 > 0$$

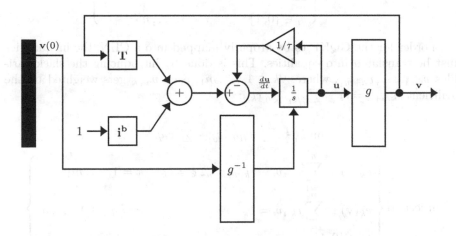

**Fig. 1.** Dynamical system associated to the Continuous Hopfield Network

The existence of an equilibrium point ($\mathbf{u^e}$ such that $\mathbf{u}(t) = \mathbf{u^e}$ $\forall t \geq t_e$ for
some $t_e \geq 0$) is guaranteed [3] if a Lyapunov or energy function exists. As shown

by Hopfield [5], if $\mathbf{T}$ is symmetric, then the following Lyapunov function exists:

$$E(\mathbf{v}) = -\frac{1}{2}\mathbf{v}^t\mathbf{T}\mathbf{v} - (\mathbf{i^b})^t\mathbf{v} + \frac{1}{\tau}\sum_{i=1}^{n}\int_0^{v_i} g^{-1}(x)dx$$

The idea is that the network's Lyapunov function, when $\tau \to \infty$, is associated with the cost function to be minimized in the combinatorial problem. Therefore, the CHN will solve those combinatorial problems which can be expressed as the constrained minimization of:

$$E(\mathbf{v}) = -\frac{1}{2}\mathbf{v}^t\mathbf{T}\mathbf{v} - (\mathbf{i^b})^t\mathbf{v} \tag{1}$$

which has its extremes at the corners of the $n$-dimensional hypercube $[0,1]^n$.

The optimization problem to be mapped in this paper is an extension of the well-known Knapsack Problem.

The Quadratic Assignment Problem (QAP) was introduced by Koopmans and Beckmann in 1957 [7] to solve assignment of industrial plants to locations. In a generalized form, a QAP problem can be stated as a Generalized Quadratic Knapsack Problem (GQKP) in the following way:

$$\min\left\{\frac{1}{2}\sum_{i,j=1}^{n} v_i p_{i,j} v_j + \sum_{i=1}^{n} v_i q_i\right\}$$

$$\text{subject to}\begin{cases}\sum_{i=1}^{n} r_{k,i}v_i \leq b_k & k = 1,\ldots,m_1 \\ \sum_{i=1}^{n} r_{k,i}v_i = b_k & k = m_1+1,\ldots,m \\ v_i \in \{0,1\} & i = 1,\ldots,n\end{cases}$$

In order for the GQKP to be properly mapped into a CHN, the inequalities must be translated into equalities. This is done by introducing the slack variables $v_{n+1},\ldots,v_{n+m_1}$, where, $\forall k \in 1,\ldots,m_1$, each $v_{n+k}$ gets weighted by the coefficient $r_{k,n+k} = b_k - \sum_{j/r_{k,j}<0} r_{k,j}$:

$$\min\left\{\frac{1}{2}\sum_{i,j=1}^{n} v_i p_{i,j} v_j + \sum_{i=1}^{n} v_i q_i\right\}$$

$$\text{subject to}\begin{cases}e_k(\mathbf{v}) \equiv \sum_{i=1}^{n} r_{k,i}v_i + r_{k,n+k}v_{n+k} = b_k & k = 1,\ldots,m_1 \\ e_k(\mathbf{v}) \equiv \sum_{i=1}^{n} r_{k,i}v_i = b_k & k = m_1+1,\ldots,m \\ v_i \in \{0,1\} & i = 1,\ldots,n \\ v_{n+k} \in [0,1] & k = 1,\ldots,m_1\end{cases}$$

The GQKP may be expressed in matrix form in the following way:

$$\min_{\mathbf{v}} \left\{ \frac{1}{2} \mathbf{v}^t \mathbf{P} \mathbf{v} + \mathbf{q}^t \mathbf{v} \right\}$$

$$\text{subject to} \begin{cases} \mathbf{R}\mathbf{v} = \mathbf{b} \\ v_i \in \{0,1\} & i = 1, \ldots, n \\ v_{n+k} \in [0,1] & k = 1, \ldots, m_1 \end{cases}$$

This paper is organized as follows: following the Hopfield neural approach, an illustrative example of a GQKP is introduced in Sect. 2. The mapping of the problem and its parameter setting are respectively obtained in Subsects. 2.1 and 2.2. Then, in Subsect. 2.3, the saddle point of the objective function of this GQKP example is computed. This saddle point can be moved outside of the Hamming hypercube of solutions of the GQKP by varying the free parameter on which it depends; at this point, the entire hypercube is inside the attractor basin of the optimum solution and, consequently, the optimum solution for this example is guaranteed. Some computational experiences (Subsect. 2.4) are provided to validate the previous analytic results. Finally, conclusions and future research are drawn in Sect. 3. Some technical results required in Subsect. 2.2 are detailed in Appendix A.

## 2    Mapping the GQKP into the CHN. An Illustrative Example

The following simple optimization problem is proposed:

$$\min \left\{ \tfrac{1}{2} (4\,v_1{}^2 - 2\,v_2{}^2) \right\}$$
$$\text{subject to } v_1 + v_2 = 1$$

which may be expressed as a GQKP problem with the following values:

$$\mathbf{P} = \begin{bmatrix} 4 & 0 \\ 0 & -2 \end{bmatrix}, \quad \mathbf{q} = \begin{bmatrix} 0 \\ 0 \end{bmatrix}, \quad \mathbf{R} = \begin{bmatrix} 1 & 1 \end{bmatrix}, \quad \mathbf{b} = \begin{bmatrix} 1 \end{bmatrix} \tag{2}$$

Note that no slack variables are required in this example.

### 2.1    Mapping the Problem

Following the procedure by Talaván and Yáñez [12], the mapping between the CHN and the GQKP is carried out.

The feasible solutions set is described by:

$$H_F \equiv \{\mathbf{v} \in H_C : \mathbf{R}\mathbf{v} = \mathbf{b}\} = \begin{cases} e_k(\mathbf{v}) = b_k & k = 1, \ldots, m_1, m_1 + 1, \ldots, m \\ v_i \in \{0,1\} & i = 1, \ldots, n \\ v_{n+k} \in [0,1] & k = 1, \ldots, m_1 \end{cases}$$

$$H_C \equiv \{\mathbf{v} \in H : v_i \in \{0,1\} \quad i = 1, \ldots, n\}$$

and the Hamming hypercube $H \equiv \{\mathbf{v} \in [0,1]^{(n+m_1)}\}$.

Therefore, for the example introduced this section:

$$H_F \equiv \left\{ \begin{matrix} e_1(\mathbf{v}) = v_1 + v_2 = 1 \\ v_1, \ v_2 \in \{0,1\} \end{matrix} \right\}$$

which has two solutions $\left\{ \begin{bmatrix} 1 \\ 0 \end{bmatrix}, \begin{bmatrix} 0 \\ 1 \end{bmatrix} \right\}$, being $\begin{bmatrix} 1 \\ 0 \end{bmatrix}$ the optimal solution.

The optimization problem $min_{\mathbf{v} \in H} E(\mathbf{v})$ is considered to be the *mapping* of the GQKP if

$$E(\mathbf{v}) = E^O(\mathbf{v}) + E^R(\mathbf{v}), \quad \forall \mathbf{v} \in H$$

satisfies

- $E^O(\mathbf{v}) = \alpha(\frac{1}{2}\mathbf{v}^t\mathbf{P}\mathbf{v} + \mathbf{q}^t\mathbf{v})$, being directly proportional to the objective function
- $E^R(\mathbf{v})$ is a quadratic function that penalizes the violated constraints of the problem and guarantees the feasibility of the CHN solution.

Using the energy term proposed by Talaván and Yáñez [12]:

$$E^R(\mathbf{v}) = \frac{1}{2}(\mathbf{Rv})^t \, \Phi \, (\mathbf{Rv}) + \mathbf{v}^t diag\,(\boldsymbol{\gamma})\,(1 - \mathbf{v}) + \boldsymbol{\beta}^t \mathbf{Rv}$$

and the values from Eq. 2, the energy function of the mapped GQKP is:

$$\begin{aligned} E(\mathbf{v}) &= E^O(\mathbf{v}) + E^R(\mathbf{v}) \\ &= \alpha(\tfrac{1}{2}\mathbf{v}^t\mathbf{P}\mathbf{v} + \mathbf{q}^t\mathbf{v}) + \tfrac{1}{2}(\mathbf{Rv})^t \, \Phi \, (\mathbf{Rv}) + \mathbf{v}^t diag\,(\boldsymbol{\gamma})\,(1 - \mathbf{v}) + \boldsymbol{\beta}^t \mathbf{Rv} \\ &= \alpha\,(2\,v_1{}^2 - v_2{}^2) + \tfrac{1}{2}\phi_{1,1}\,(v_1 + v_2)^2 - \gamma_1 v_1\,(v_1 - 1) - \gamma_2 v_2\,(v_2 - 1) \\ &\quad + \beta_1\,(v_1 + v_2) \end{aligned}$$

$$(3)$$

Comparing the energy function obtained in Eq. 3 with the energy function of the CHN, $E(\mathbf{v}) = \frac{1}{2}\mathbf{v}^t\mathbf{T}\mathbf{v} - (\mathbf{i}^b)^t\mathbf{v}$ (Eq. 1), the following values for $\mathbf{T}$ and $\mathbf{i}^b$ are obtained:

$$\mathbf{T} = -\,(\alpha\mathbf{P} + \mathbf{R}^t\mathbf{\Phi}\mathbf{R} - 2 diag\,(\boldsymbol{\gamma})) = -\alpha\begin{bmatrix} 4 & 0 \\ 0 & -2 \end{bmatrix} - \begin{bmatrix} 1 \\ 1 \end{bmatrix}\phi_{1,1}\begin{bmatrix} 1 & 1 \end{bmatrix} + 2\begin{bmatrix} \gamma_1 & 0 \\ 0 & \gamma_2 \end{bmatrix}$$

$$= \begin{bmatrix} -4\alpha - \phi_{1,1} + 2\gamma_1 & -\phi_{1,1} \\ -\phi_{1,1} & 2\alpha - \phi_{1,1} + 2\gamma_2 \end{bmatrix}$$

$$\mathbf{i}^b = -\,(\alpha\mathbf{q} + \mathbf{R}^t\boldsymbol{\beta} + \boldsymbol{\gamma}) = -\alpha\begin{bmatrix} 0 \\ 0 \end{bmatrix} + \begin{bmatrix} 1 \\ 1 \end{bmatrix}\beta_1 - \begin{bmatrix} \gamma_1 \\ \gamma_2 \end{bmatrix} = \begin{bmatrix} -\beta_1 - \gamma_1 \\ -\beta_1 - \gamma_2 \end{bmatrix}$$

## 2.2   Parameter Setting

Once the energy function has been determined, the parameters $(\Phi, \gamma, \beta)$ must be chosen so that any local minimum of the energy function corresponds with a feasible solution of the optimization problem. A stability analysis for the mapped GQKP example is carried out, looking for a parameter setting that guarantees that the CHN is stable in the feasible solutions and unstable in the non-feasible ones. From such analysis (see details in Appendix A), the following set of inequalities is deduced:

$$\begin{cases} -4\alpha - \phi_{1,1} + 2\gamma_1 \geq 0 \\ 2\alpha - \phi_{1,1} + 2\gamma_2 \geq 0 \\ \phi_{1,1} \geq 0 \\ 4\alpha + 2\phi_{1,1} - \gamma_1 + \beta_1 \geq \varepsilon \\ \gamma_2 + \beta_1 \leq -\varepsilon \end{cases}$$

and by solving the system of linear inequalities, the following parameter setting, which leaves $\alpha$ and $\phi_{1,1}$ as free parameters, is obtained:

$$\gamma_1 = 2\alpha + \frac{\phi_{1,1}}{2}, \quad \gamma_2 = \frac{\phi_{1,1}}{2} - \alpha, \quad \beta_1 = -\frac{\alpha}{2} - \phi_{1,1}, \quad \varepsilon = \frac{3\alpha}{2} + \frac{\phi_{1,1}}{2}$$

## 2.3   The Saddle Point

Using the parameter setting obtained in Sect. 2 and the energy function from Eq. 3, the energy function for the GQKP example is:

$$E^O(\mathbf{v}) = \alpha(2v_1^2 - v_2^2), \quad E^R(\mathbf{v}) = \alpha\left(-2v_1^2 + v_2^2 + \frac{3}{2}(v_1 - v_2)\right) - \phi_{1,1}\left(\frac{1}{2}v_1 + \frac{1}{2}v_2 - v_1v_2\right)$$

$$E(\mathbf{v}) = E^O(\mathbf{v}) + E^R(\mathbf{v}) = \frac{3}{2}\alpha(v_1 - v_2) - \frac{1}{2}\phi_{1,1}(v_1 + v_2 - 2v_1v_2)$$

being the partial derivatives of $E(\mathbf{v})$:

$$E_1(\mathbf{v}) = \frac{3\alpha}{2} - \frac{\phi_{1,1}}{2} + v_2\phi_{1,1} \quad E_2(\mathbf{v}) = v_1\phi_{1,1} - \frac{\phi_{1,1}}{2} - \frac{3\alpha}{2}$$

Making the derivatives equal to zero, the saddle point $\mathbf{v}^*$ for $E(\mathbf{v})$ is computed in terms of $\alpha$ and $\phi_{1,1}$:

$$(\mathbf{v}^*)^t = [v_1^*, v_2^*] = \left[\frac{1}{2}\left(1 + \frac{3\alpha}{\phi_{1,1}}\right), \frac{1}{2}\left(1 - \frac{3\alpha}{\phi_{1,1}}\right)\right]$$

Choosing a value for $\phi_{1,1}$ that verifies $\phi_{1,1} \lesssim 3\alpha$, it is guaranteed that the saddle point moves outside of the Hamming hypercube, leaving the entire hypercube inside the attractor basin of the optimum solution, $\left[\begin{smallmatrix}0\\1\end{smallmatrix}\right]$. Figure 2 shows the

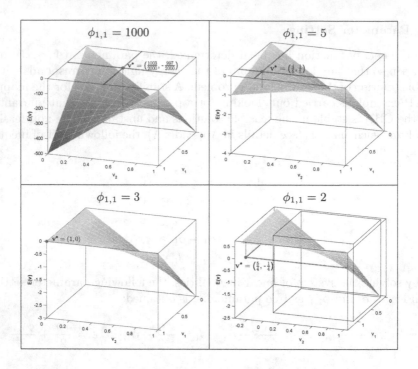

**Fig. 2.** Energy function of the GQKP example with different values for $\phi_{1,1}$

energy function and its saddle point for different values of $\phi_{1,1}$ ($\alpha = 1$). It is shown that choosing values for $\phi$ lower that $3\alpha$ in the energy function leave only one attractor basin inside the Hamming hypercube that leads to the optimum solution.

In a more formal way, it can be noted that for a fixed value of $\alpha$, when $\phi_{1,1}$ tends to $\infty$ the saddle point is:

$$\lim_{\phi_{1,1} \to \infty} (\mathbf{v}^*)^t = \left[\frac{1}{2}, \frac{1}{2}\right]$$

and is outside the Hamming hypercube when $\phi_{1,1}$ tends to 0:

$$\lim_{\phi_{1,1} \to 0} (\mathbf{v}^*)^t = [\infty, -\infty]$$

## 2.4   Computational Experiences

The analytic results from in Sect. 2.3 have also been validated using a CHN simulator developed using MATLAB (The MathWorks Inc., Natick, MA, USA).

Figure 3 shows the attractor basins for the energy functions in Fig. 2, including some intermediate cases. The basins have been obtained by solving multiple

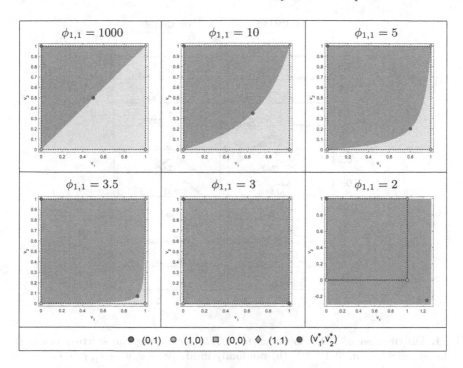

**Fig. 3.** Attractor basins for the energy functions considered in the GQKP example (Color figure online)

CHNs using different starting points (sampling the Hamming hypercube with a $400 \times 400$ array), colored according to the solution that the CHN leads them to.

The darker basin (blue) represents the optimal solution, $\begin{bmatrix} 0 \\ 1 \end{bmatrix}$. The points in the graphs represent the equilibrium points of the CHN: the two solutions $\begin{bmatrix} 0 \\ 1 \end{bmatrix}$ and $\begin{bmatrix} 1 \\ 0 \end{bmatrix}$, the saddle point $\begin{bmatrix} v_1^* \\ v_2^* \end{bmatrix}$ and the corners $\begin{bmatrix} 0 \\ 0 \end{bmatrix}$ and $\begin{bmatrix} 1 \\ 1 \end{bmatrix}$, where the CHN gets trapped if used as starting points. Note how the attractor basin for the best solution gets larger as the value of the free parameter decreases.

Choosing an appropriate value for the free parameter ($\phi_{1,1} \leq 3\alpha$) guarantees that the optimum solution is always obtained independently of the initial point.

In order to measure the impact of the starting point in the CHN in more complex problems such as the TSP, multiple simulations for the TSPLIB [8] problem *CH150* ($N = 150$ cities, i.e. 225000 neurons) are carried out using different values for the free parameter. The quality of the solution of the CHN is measured by computing the performance ratio $\rho = tour\ length/optimum\ tour\ length$.

Figure 4 shows that better solutions are obtained by lowering the value of the free parameter, confirming that the basins for the better solutions get larger as the free parameter decreases.

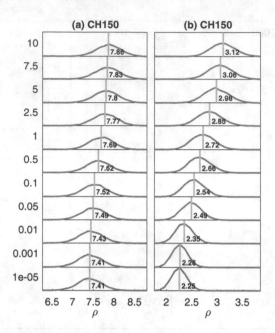

**Fig. 4.** Distribution of CHN solutions for *CH150*, choosing the starting point: (a) randomly anywhere in $[0,1]^{N \times N}$; (b) randomly inside $[\mathbf{v}^* - \varepsilon, \mathbf{v}^* + \varepsilon]$, $\varepsilon > 0$

## 3  Conclusions and Future Research

This paper focuses on the free parameter ( *"global inhibition"*) of the CHN originally proposed by Hopfield applied to the TSP [6]. Historically, researchers in the field have used a small value for the free parameter [10] (as better solutions were obtained) but without explaining the reason behind it. Dealing with simple optimization problems, as in the GQKP illustrative example of Sect. 2, the CHN may guarantee to always obtain the optimum solution. This is thanks to the saddle point being moved outside of the Hamming hypercube given a small enough value of the free parameter, which allows to leave the entire hypercube inside the attractor basin of the optimum solution.

Based on the results shown in this paper about the relationship between the free parameter and the saddle point (Subsect. 2.3), the conjecture of using starting points close to the saddle point is validated empirically using a CHN simulator. Indeed, it is verified that in the surroundings of the saddle point the attractor basins for the best solutions grow as the free parameter decreases.

Whereas for simple GQKP instances the convergence to the optimum solution of the CHN can be assured, this is far more complicated in problems like the TSP. However, as shown in Sect. 2.4 and extending the paper of García et al. [2], it is expected some computational improvement on the solution of the TSP based on the relationship between the free parameter and the saddle point of the CHN, considering it can be found. Research along this line will be continued.

**Acknowledgments.** This research has been partially supported by the Government of Spain, grant TIN2015-66471-P, and by the local Government of Madrid, grant S2013/ICE-2845 (CASI-CAM).

# A    Some Technical Results of the Illustrative Example

Recalling the GQKP problem introduced in Sect. 2, a stability analysis of the valid solutions is carried out, looking for the parameters that guarantee that the CHN is stable in the feasible solutions. The energy function obtained for the simple GQKP problem (see Eq. 3) was:

$$E\left(\mathbf{v}\right) = \alpha \left(2\,{v_1}^2 - {v_2}^2\right) + \tfrac{1}{2}\phi_{1,1}\left(v_1 + v_2\right)^2 - \gamma_1 v_1\left(v_1 - 1\right) - \gamma_2 v_2\left(v_2 - 1\right)$$
$$+ \,\beta_1\left(v_1 + v_2\right)$$

Referencing Talaván and Yáñez [12], the stability of any $\mathbf{v} \in H$ is ensured if:

$$\min_{\mathbf{v}\in H_F} \underline{E}^0(\mathbf{v}) \geq 0 \tag{4}$$

$$\max_{\mathbf{v}\in H_F} \overline{E}^1(\mathbf{v}) \leq 0 \tag{5}$$

$$E_{n+k}(\mathbf{v}) = 0 \quad \forall k \in \{1,\ldots,m_1\}/v_{n+k} \in (0,1) \quad \forall \mathbf{v} \in H_F \tag{6}$$

where

$$E_i(\mathbf{v}) \equiv \frac{\partial E(\mathbf{v})}{\partial v_i} \qquad \underline{E}^0(\mathbf{v}) \equiv \min_{v_i=0} E_i(\mathbf{v}) \qquad \overline{E}^1(\mathbf{v}) \equiv \min_{v_i=1} E_i(\mathbf{v})$$

Although analyzing all feasible solutions would be a very difficult task with a GQKP, this simple example allows rigorous analysis. Thus, the partial derivatives of the energy function will be of the form:

$$E_1\left(\mathbf{v}\right) = \quad 4\alpha v_1 + \phi_{1,1}v_1 + \phi_{1,1}v_2 - 2\gamma_1 v_1 + \gamma_1 + \beta_1$$
$$E_2\left(\mathbf{v}\right) = -2\alpha v_2 + \phi_{1,1}v_1 + \phi_{1,1}v_2 - 2\gamma_2 v_2 + \gamma_2 + \beta_1$$

Condition 4 is satisfied if:

$$\underline{E}^0(\left[\begin{smallmatrix}1\\0\end{smallmatrix}\right]) = \min\{E_2(\left[\begin{smallmatrix}1\\0\end{smallmatrix}\right])\} = \phi_{1,1} + \gamma_2 + \beta_1 \geq 0$$
$$\underline{E}^0(\left[\begin{smallmatrix}0\\1\end{smallmatrix}\right]) = \min\{E_1(\left[\begin{smallmatrix}0\\1\end{smallmatrix}\right])\} = \phi_{1,1} + \gamma_1 + \beta_1 \geq 0$$

and Condition 5 is satisfied if:

$$\overline{E}^1(\left[\begin{smallmatrix}1\\0\end{smallmatrix}\right]) = \max\{E_1(\left[\begin{smallmatrix}1\\0\end{smallmatrix}\right])\} = \quad 4\alpha + \phi_{1,1} - \gamma_1 + \beta_1 \leq 0$$
$$\overline{E}^1(\left[\begin{smallmatrix}0\\1\end{smallmatrix}\right]) = \max\{E_2(\left[\begin{smallmatrix}0\\1\end{smallmatrix}\right])\} = -2\alpha + \phi_{1,1} - \gamma_2 + \beta_1 \leq 0$$

This example does not use Condition 6 as no slack variables are needed.

The instability of any interior point $\mathbf{v} \in H \setminus H_C$ is guaranteed if:

$$T_{i,i} \geq 0 \quad \forall i \in \{1,\ldots,n\}$$

and the stability of valid solutions is obtained if $\Phi$ is positive semidefinite:

$$\phi_{k,l} \geq 0 \quad \forall k, l \in \{1, \dots, m\}$$

The instability of any non-feasible corner is obtained by creating a partition of the set $H_C \setminus H_F$ and forcing the instability conditions for each of the elements in the partition:

$$\vee \begin{cases} \underline{E}^0(\mathbf{v}) \leq -\varepsilon \\ \overline{E}^1(\mathbf{v}) \geq \varepsilon \\ E_{n+k}(\mathbf{v}) \neq 0 \quad \text{for any } v_{n+k} \in (0,1) \end{cases} \quad \text{with } \varepsilon > 0$$

Going forward the partition $H_C \setminus H_F$ is created using the *direct method* introduced by Talaván [11]. In summary, this partition is created considering the different cases which may occur when the constraints for $H_F$ are violated. Thus, given $\mathbf{v} \in H_C \setminus H_F$, the instability will be found from the first unsatisfied constraint, distinguishing inequalities and equations (which may not be satisfied by excess or defect):

$$H_C \setminus H_F = \bigcup_{k=1}^{m_1} W_{k,0} \bigcup_{k=1}^{m} W_{k,1} \bigcup_{k=1}^{m} W_{k,2}$$

where

$$W_{k,0} \equiv \bigcap_{l=1}^{k-1} \{e_l(\mathbf{v}) = b_l\} \cap \{e_k(\mathbf{v}) > b_k\} \cap \{e_k(\mathbf{v}) - r_{k,n+k}\, v_{n+k} \leq b_k\}$$
$$\forall k = 1, \dots, m_1$$

$$W_{k,1} \equiv \bigcap_{l=1}^{k-1} \{e_l(\mathbf{v}) = b_l\} \cap \{e_k(\mathbf{v}) > b_k\} \cap \{e_k(\mathbf{v}) - r_{k,n+k}\, v_{n+k} > b_k\}$$
$$\forall k = 1, \dots, m_1$$

$$W_{k,1} \equiv \bigcap_{l=1}^{k-1} \{e_l(\mathbf{v}) = b_l\} \cap \{e_k(\mathbf{v}) > b_k\} \qquad \forall k = m_1 + 1, \dots, m$$

$$W_{k,2} \equiv \bigcap_{l=1}^{k-1} \{e_l(\mathbf{v}) = b_l\} \cap \{e_k(\mathbf{v}) < b_k\} \qquad \forall k = 1, \dots, m$$

For the GQKP problem being studied, the partition gets reduced to:

$$H_C \setminus H_F = W_{1,1} \cup W_{1,2}$$

where

- $W_{1,1} = \{e_1(\mathbf{v}) > 1\} = \{v_1 + v_2 > 1\}$, which is satisfied if $v_1 = 1$ and $v_2 = 1$:

$$\begin{cases} E_1(\mathbf{v}) > 4\alpha + 2\phi_{1,1} - \gamma_1 + \beta_1 \\ E_2(\mathbf{v}) > -2\alpha + 2\phi_{1,1} - \gamma_2 + \beta_1 \end{cases}$$

and the instability is guaranteed if:

$$\vee \begin{cases} 4\alpha + 2\phi_{1,1} - \gamma_1 + \beta_1 \geq \varepsilon \\ -2\alpha + 2\phi_{1,1} - \gamma_2 + \beta_1 \geq \varepsilon \end{cases}$$

– $W_{1,2} = \{e_1(\mathbf{v}) < 1\} = \{v_1 + v_2 < 1\}$, which is satisfied if $v_1 = 0$ y $v_2 = 0$:

$$\begin{cases} E_1(\mathbf{v}) \leq \gamma_1 + \beta_1 \\ E_2(\mathbf{v}) \leq \gamma_2 + \beta_1 \end{cases}$$

and the instability is guaranteed if:

$$\vee \begin{cases} \gamma_1 + \beta_1 \leq -\varepsilon \\ \gamma_2 + \beta_1 \leq -\varepsilon \end{cases}$$

Therefore, considering the initial conditions and choosing from the recently obtained inequalities, the following set of linear inequalities guarantees the stability of feasible solutions and instability of non-feasible ones:

$$\begin{cases} T_{1,1} = & -4\alpha - \phi_{1,1} + 2\gamma_1 \geq & 0 \\ T_{2,2} = & 2\alpha - \phi_{1,1} + 2\gamma_2 \geq & 0 \\ & \phi_{1,1} \geq & 0 \\ W_{1,1} : 4\alpha + 2\phi_{1,1} - \gamma_1 + \beta_1 \geq & \varepsilon \\ W_{1,2} : & \gamma_2 + \beta_1 \leq -\varepsilon \end{cases}$$

# References

1. Demuth, H.B., Beale, M.H., De Jess, O., Hagan, M.T.: Neural Network Design. PWS Publishing Company, Boston (1996)
2. García, L., Talaván, P.M., Yáñez, J.: Improving the Hopfield model performance when applied to the traveling salesman problem. Soft Comput. 1–15 (2016). doi:10.1007/s00500-016-2039-8
3. Gopal, M.: Modern Control System Theory. New Age International, New Delhi (1993)
4. Hopfield, J.J.: Neural networks and physical systems with emergent collective computational abilities. Proc. Nat. Acad. Sci. **79**(8), 2554–2558 (1982)
5. Hopfield, J.J.: Neurons with graded response have collective computational properties like those of two-state neurons. Proc. Nat. Acad. Sci. **81**(10), 3088–3092 (1984)
6. Hopfield, J.J., Tank, D.W.: "Neural" computation of decisions in optimization problems. Biol. Cybern. **52**(3), 141–152 (1985)
7. Koopmans, T.C., Beckmann, M.: Assignment problems and the location of economic activities. Econom.: J. Econom. Soc. **25**, 53–76 (1957)
8. Reinelt, G.: TSPLIB. A traveling salesman problem library. ORSA J. Comput. **3**(4), 376–384 (1991)
9. Rojas, R.: Neural Networks: A Systematic Introduction. Springer Science & Business Media, Heidelberg (2013)
10. Talaván, P.M., Yáñez, J.: Parameter setting of the Hopfield network applied to TSP. Neural Netw. **15**(3), 363–373 (2002)
11. Talaván, P.M.: Ph.D. dissertation: El modelo de Hopfield aplicado a problemas de optimización combinatoria, Universidad Complutense de Madrid (2003)
12. Talaván, P.M., Yáñez, J.: The generalized quadratic knapsack problem. A neuronal network approach. Neural Netw. **19**(4), 416–428 (2006)

# A Systematic Approach for the Application of Restricted Boltzmann Machines in Network Intrusion Detection

Arnaldo Gouveia and Miguel Correia[✉]

INESC-ID, Instituto Superior Técnico, Universidade de Lisboa, Lisbon, Portugal
miguel.p.correia@tecnico.ulisboa.pt

**Abstract.** A few exploratory works studied Restricted Boltzmann Machines (RBMs) as an approach for network intrusion detection, but did it in a rather empirical way. It is possible to go one step further taking advantage from already mature theoretical work in the area. In this paper, we use RBMs for network intrusion detection showing that it is capable of learning complex datasets. We also illustrate an integrated and systematic way of learning. We analyze learning procedures and applications of RBMs and show experimental results for training RBMs on a standard network intrusion detection dataset.

## 1 Introduction

Deep neural networks have become increasingly popular due to their success in machine learning. Their history goes as far back as 1958 when Rosenblatt published his work on the perceptron concept [18]. Present day forms of deep learning networks include Hopfield Networks, Self-Organizing Maps, Boltzmann Machines, Multi-Layer Perceptrons, Autoencoders or Deep Belief networks.

Most of present day machine learning algorithms are not classifiable as *deep* since they use at most one layer of hidden variables. Bengio and LeCun [2,3], have shown that the internal representations learned by such systems are necessarily simple due to their simple internal structure, being incapable of extracting certain types of complex structure. However this limitation does not apply to specific types of energy-based learning approaches.

Two important classes of Boltzmann Machine (BMs) are the Restricted Boltzmann Machine (RBM) described by a complete bipartite graph, and the Deep RBM that is composed of several layers of RBMs. The BM derived from Hopfield networks and in its initial form was fully node-connected. The concept of RBMs came about due to the difficulty of training a fully connected BM in classification problems. RBMs are designated *restricted* due to the fact that there are no connections among the hidden layer nodes or among the visible layer nodes.

Recently the notion of deep learning gained a lot of attention as a method to model high-level abstractions by composing multiple non-linear layers [14]. Several deep learning network architectures, like deep belief networks [8], deep

© Springer International Publishing AG 2017
I. Rojas et al. (Eds.): IWANN 2017, Part I, LNCS 10305, pp. 432–446, 2017.
DOI: 10.1007/978-3-319-59153-7_38

BMs [19], convolutional neural networks [14], and deep denoising auto-encoders [25], have shown their advantages in specific areas.

Despite their interest, these approaches have barely been applied in Network Intrusion Detection Systems (NIDSs), i.e., for detecting cyber-attacks by inspecting computer network traffic. This paper aims to contribute for closing this gap by introducing a systematic approach for training RBMs for network intrusion detection. The approach considers three important aspects: weight initialization, pre-training, and fine-tuning.

The paper seeks to demonstrate the effectiveness of the approach with an analysis based on a dataset carefully crafted for this purpose: the UNB ISCX intrusion detection evaluation dataset [20]. This dataset is reasonably recent but has been gaining increasing adoption for evaluating NIDSs.

## 2   The Ising Model

The RBM formalism, specifically in terms of synonymous of objective or loss function, is similar to the Ising Model formalism, so it is relevant to address the similarities. Energy-based models are popular in machine learning due to the elegance of their formulation and their relationship to statistical physics. Among these, the Restricted Boltzmann Machine (RBM) is the focus of our work.

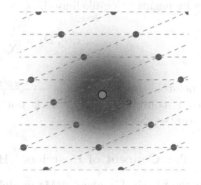

The Ising model was defined by Lenz in 1920 and named after his student Ising, who developed the model in his Ph.D. thesis [13]. Ising solved

**Fig. 1.** Illustrating a 2-dimensional interaction with only the nearest nodes (Color figure online)

the model optimization problem in one dimension. The two-dimensional square lattice Ising model was given an analytic description much later [17]. Such physical models, having two alternate states in an array with mutual interactions, are currently described in physics as spinor Ising models. The mutual interaction among spin units is modeled by an interaction parameter commonly named coupling. Interactions between particles is small and restricted to their neighbourhood. In this model the states align among themselves spontaneously in a way that minimizes a global parameter, e.g., the global energy. In 2D the model topology is usually described as a regular lattice as illustrated in Fig. 1.

## 3   Restricted Boltzmann Machines

### 3.1   Energy-Based Models (EBM)

*Energy-based models* associate an energy figure to each configuration of the state variables. Energy in this context is synonymous of objective or loss function.

Learning, again in this context, corresponds to modifying the energy function as to find minima. In the probabilistic model associated with RBMs the associated probability distribution is described through an energy type function:

$$p(x) = \frac{e^{-E(x)}}{Z} \tag{1}$$

The normalizing factor $Z$ is the *partition function* in the context of physical systems. The normalization is achieved by summing across all available sates and divide.

$$Z = \sum_x e^{-E(x)} \tag{2}$$

An energy-based model can be learnt by performing (stochastic) gradient descent on the empirical negative log-likelihood of the training data. As for logistic regression, we will first define the log-likelihood and then the loss function as being he negative log-likelihood.

$$\mathcal{L}(\theta, \mathcal{D}) = \frac{1}{N} \sum_{x^{(i)} \in \mathcal{D}} \log p(x^{(i)}) \tag{3}$$

The stochastic gradient is $-\frac{\partial \log p(x^{(i)})}{\partial \theta}$ where $\theta$ are the model parameters. However, computationally this is not the best option for this type of energy function (see Sect. 3.4).

### 3.2   The Concept of Restricted Boltzmann Machine

From the physical analogy, BMs model the data with an *Ising model* that is in thermal equilibrium. In this analogy the equivalent to physical spins are called RBM units or nodes. The set of nodes that encode the observed data and the output are called the visible units $\{v_i\}$, whereas the nodes used to model the latent concept and feature space are called the hidden units $\{h_i\}$. For the purpose of explanation, we assume that the visible and hidden units are binary. Alternatives are discussed later.

The RBM concept is similar to the BM concept [1], except that no connections between neurons of the same layer are allowed. Figure 2 depicts the architecture of a RBM, consisting of two layers: the visible layer $\{v_i\}$ and the hidden layer $\{h_i\}$, with $N_v$ and $M_h$ nodes, respectively. Hidden units are used to capture higher level correlations in the data, and the visible units to mirror the data itself. Connections between nodes are restricted so that there are no visible-visible and hidden-hidden connections. Hidden-visible connections are strictly symmetrical. Hence, we have a restricted BM. An RBM is a bi-partite (visible and hidden) BM with full interactivity between visible and hidden units, and no interactivity between units of the same type. RBMs are usually described as energy-based stochastic neural networks composed by two layers of neurons (visible and hidden), in which the learning phase is conducted in an unsupervised fashion. RBMs are a variant of the BM model [1,7]. Here we consider the

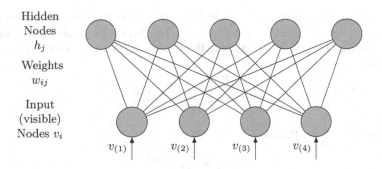

**Fig. 2.** A RBM with 4 inputs and 5 hidden nodes.

case where $h_j$ can only take binary values, and $\sigma_i$ as the standard deviation for each input dimension.

Convergence in BMs can be slow, particularly when the number of units and layers increases. RBMs were introduced to mitigate this issue [22]. The simplification consists of having only one layer of visible and one layer of hidden units with links between units on the same layer erased, allowing for parallel updates of hidden and visible units (Fig. 2).

In Fig. 2, $\mathbf{v} = (v_1, v_2 \cdots v_n)$ and $\mathbf{h} = (h_1, h_2 \cdots h_m)$ are the visible and the hidden vectors, $a_i$ and $b_j$ are their biases, $n$ and $m$ are the dimension of the visible layer and the hidden layer, and $w_{ij}$ is the connection weight matrix between the visible layer and the hidden layer. The visible stochastic binary variables $\mathbf{v} \in \{0,1\}^N$ are connected to hidden stochastic binary variables $\mathbf{h} \in \{0,1\}^M$.

For binary RBMs the energy $E(\mathbf{v}, \mathbf{h})$, which defines the bipartite structure, is given by:

$$E(\mathbf{v}, \mathbf{h}) = - \sum_{i \,\epsilon\, visible} a_i v_k - \sum_{j \,\epsilon\, hidden} b_j h_j - \sum_i \sum_j v_i w_{i,j} h_j \qquad (4)$$

or equivalently:

$$E(\mathbf{v}, \mathbf{h}) = -\mathbf{v}^T \mathbf{a} - \mathbf{b}^T \mathbf{h} - \mathbf{v}^T \mathbf{W} \mathbf{h} \qquad (5)$$

The weight matrix $\mathbf{W}$, the visible bias vector $\mathbf{b}$ and the hidden bias vector $\mathbf{c}$ are the parameters of the model.

RBM satisfies a Boltzmann-Gibbs distribution over all its units. The joint probability of $(\mathbf{v}, \mathbf{h})$ is given by a probability distribution function $P(\mathbf{v}, \mathbf{h})$, where $Z$ is the normalization term to obtain a proper probability distribution function:

$$P(\mathbf{v}, \mathbf{h}) = \frac{1}{Z} e^{-E(\mathbf{v}, \mathbf{h})} \qquad (6)$$

By definition the partition function, which sums over all possible visible and hidden states, is given by:

$$Z = \sum_{\mathbf{v}} \sum_{\mathbf{h}} e^{-E(\mathbf{v}, \mathbf{h})} \qquad (7)$$

The normalizing factor $Z$ is called the partition function by analogy with physical systems. To find $p(v)$, we marginalize over the hidden units. Given a set of training vectors, V, to train a RBM, one aims to maximize the average probability, $p(\mathbf{v})$, $\mathbf{v} \, \epsilon \, V$, where

$$p(\mathbf{v}) = \frac{1}{Z} \sum_{h} e^{-E(\mathbf{v},\mathbf{h})} \tag{8}$$

which can also be written as

$$p(\mathbf{v}) = \frac{1}{Z} e^{-F(\mathbf{v})} \tag{9}$$

where $F(v)$ is the logarithm of the energy function summed over $h$:

$$F(\mathbf{v}) = -log \sum_{h} e^{-E(\mathbf{v},\mathbf{h})} \tag{10}$$

For training efficiency, BMs can be restricted to a bipartite graph with one set of visible neurons and one set of hidden neurons. As shown in Fig. 2 there are only visible-hidden and hidden-visible connections (still symmetric). Therefore hidden units $h_j$ only depend on the visible units $v_j$ and vice-versa, with $b_j$ as the biases for the visible units and $c_j$ for the hidden units:

$$p(h_j = 1|\mathbf{v}) = \sigma(c_j + \sum_{i} w_{ij} v_i) \tag{11}$$

$$p(v_j = 1|\mathbf{h}) = \sigma(b_j + \sum_{i} w_{ij} h_i) \tag{12}$$

### 3.3  Gibbs Sampling

Gibbs sampling is commonly used for obtaining a sequence of observations which are approximated from a specified multivariate probability distribution, when direct sampling is difficult. This sequence can be used to approximate a joint distribution function, e.g., the joint distribution function expressed in Eq. 6 in our case.

In its simplest theoretical description this is how Gibbs sampling would work, with all updates done in parallel, as illustrated in Fig. 3. The most common algorithm for Gibbs sampling is *Contrastive Divergence* (CD), used inside a gradient-descent. A single-step contrastive divergence (CD-1) procedure for a single training example can be summarized as follows [9]:

1. Sample hidden units $\mathbf{h}$ from training example $\mathbf{v}$
2. Sample reconstruction $\mathbf{v}'$ of visible units using $\mathbf{h}$ and then resample $\mathbf{h}'$ from it. (Gibbs sampling step)
3. $w_{ij} \leftarrow w_{ij} - \varepsilon(<v_i h_j>_{data} - <v_i h_j>_{sampled})$

In the previous equation $<.>$ denotes an average. In practice, waiting till $t \rightarrow \infty$ is not practical. However, an alternative consists in modulating the number of iterations for a limited number and extract a final sample:

- After $t$ steps, $(h^{(t)}, v^{(t)})$ is available for sampling
- Sample $(h^{(t)}, v^{(t)})$ assuming a sample accurate enough as an approximation of $P(v, h)$ as $t \to \infty$.

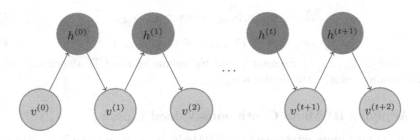

**Fig. 3.** Gibbs sampling in a RBM

As $t \to \infty$, $(v^{(t)}, h^{(t)})$, $v^t$ and $h^t$ will be ever more accurate samples drawn from the RBM's distribution. Nevertheless, one can further speed up the process by using the contrastive divergence (CD) algorithm as explained next.

### 3.4 Contrastive Divergence Algorithm in Detail

It is easy to calculate $<v_i h_j>_{data}$ because there is no direct connections among the hidden units. However, it is difficult to get an unbiased sample of $<v_i h_j>_{model}$. Hinton proposed a faster learning algorithm with contrastive divergence (CD) learning and the change of learning parameter [10]. The partial derivative of the log probability of Eq. 8 with respect to a weight is given by:

$$\frac{\partial \log p(\mathbf{v})}{\partial w_{ij}} = (<v_i h_j>_{data} - <v_i h_j>_{model}) \tag{13}$$

where the angle brackets $<v_i h_j>_{data}$ and $<v_i h_j>_{model}$ are used to denote expectations of the distribution specified by the subscript *data* and *model*. In the log probability, a very simple learning rule for performing stochastic steepest ascent is given by:

$$\Delta w_{ij} = \varepsilon(<v_i h_j>_{data} - <v_i h_j>_{model}) \tag{14}$$

where $\varepsilon$ is a learning rate.

The CD algorithm computes an approximation of the gradient by performing Gibbs sampling for a finite number of steps. This involves initializing the RBM with a training example $v \in V$ and running the RBM for k (often with $k = 1$) steps. CD-k is a generalization $k$ iterations by repeating the sampling process $k$ times. CD makes the following simplifications for computing the gradient:

1. Replace the first term (expectation over all input samples) with a single sample.

2. For the second term, run the chain for fixed $k$ steps:

$$\Delta w_{ij} = \varepsilon(<v_i h_j>_{data} - <v_i h_j>_{sampled}), \tag{15}$$

The bias updates for the visible and hidden layers respectively can be defined by these expressions:

$$\Delta a_i = \varepsilon(<v_i>_{data} - <v_i>_{sampled}), \tag{16}$$

$$\Delta b_j = \varepsilon(<h_j>_{data} - <h_j>_{sampled}) \tag{17}$$

where $<v_i h_j>_{sampled}$, or *reconstructed* by means of the CD algorithm can be computed more efficiently than $<v_i h_j>_{model}$.

### 3.5 Applying RBMs to Continuous-Valued Inputs

With the binary units introduced for RBMs in [8], one can handle continuous-valued inputs by scaling them to the $[0, 1]$ interval and considering each input as the probability for a binary random variable to take the value 1. Previous work on continuous-valued input in RBMs include [4], in which noise is added to sigmoidal units, and the RBM forms a special form of Diffusion Network [15]. The approach followed in this paper starts by acquiring the RBM weights in the pre-training phase by using Rectified Linear Units (for the hidden units) and training with Sigmoid units, which has shown to work very well [6], something that our results confirm.

A continuous RBM (RBM) is a form of RBM that accepts continuous inputs via a different type of contrastive divergence sampling. This allows the CRBM to handle things like image pixels or word-count vectors that are normalized to decimals between zero and one.

## 4   A Systematic Approach for Training RBMs

This section describes an optimized approach for choosing the training parameters of the specific RBMs configured. It takes advantage of Hinton et al.'s results [6]. An optimization approach is important in the context of RBMs due to the complexity of assuring the learning process convergence and for avoiding overfitting. In the end we aim at showing the validity of our approach in the context of network intrusion detection by using a specific dataset. The approach comprehends three major aspects – Weight initialization, Pre-training, and Fine-Tuning – plus a few other parameter optimization choices (see Table 1 and Fig. 4).

### 4.1   The Three Major Aspects

*Pre-training.* As an alternative or in addition to weight initialization techniques, layer-wise unsupervised pre-training can be used to initialize the weights for fine-tuning. For every combination of two adjacent layers, an RBM is trained for a certain number of epochs, whereas an epoch consists of training the network on batches of samples.

*Fine-Tuning.* After pre-training, the instance is fine-tuned using one of several fine-tuning functions (more on those in the next section).

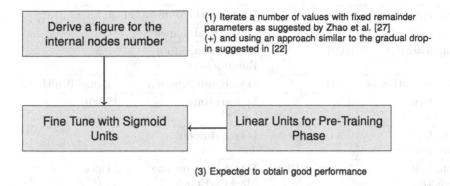

Fig. 4. Approach for training RBMs

*Choosing the Number of Hidden Nodes.* Dimensioning the RBM internals involves testing a set of dimensioning options and extracting the best option using the accuracy maximization criteria.

### 4.2 Using Rectified Linear Units

Hinton et al. have shown that for problems involving real valued data rectified linear units provide better results [16]. Although their research has been over image datasets, there is enough numerical similitude to both problems as the results obtained have been based on a feature based real valued dataset. The present proposal uses Rectified Linear Units for the hidden layer during pre-training. A suggestion over the usage of sigmoid units has been put forward by pre-training the RBM with Rectified Linear Units and fine tuning with Sigmoid Units. The advantages of this approach come in the type of output produced belonging to $\{0, 1\}$ instead of $[0, 1]$.

### 4.3 Parameter Optimization

As stated previously we do not try to optimize the whole set of RBM parameters as it would show to be exhaustive, but instead choose a number of parameters as optimization enablers, based on the literature. The specific values for each parameter used in the experiment is in Table 1

*Number of Training Epochs.* Regularization of neural networks is used to improve generalization by preventing over-fitting. The most straightforward way of preventing over-fitting is early stopping. One of the options for the regularization

**Table 1.** Experimental parameters

| Fine tuning parameters | Value | Fine tuning parameters | Value |
|---|---|---|---|
| rbm.batchSize | 100 | DArch.initialMomentum | 0.5 |
| rbm.lastLayer | True | DArch.finalMomentum | 0.9 |
| rbm.learnRate | 0.001 | DArch.momentum RampLength | 0.5 |
| rbm.weightDecay | 2e-04 | DArch.unitFunction | sigmoidUnitRbm |
| rbm.initial Momentum | 0.5 | bp.learnRate | 0.001 |
| rbm.final Momentum | 0.9 | DArch.dropout | 0 |
| rbm.unit Function | linearUnitRbm | DArch.dropout.one MaskPerEpoch | False |
| rbm.update Function | rbmUpdate | DArch.isClass | True |
| rbm.numCD | 10 | DArch.numEpochs | 50 |
| rbm.num Epochs | 50 | DArch.errorFunction | Mean square error |
| rbm.momentum RampLength | 0.5 | retainData | True |
| DArch.batchSize | 100 | normalizeWeights | True |
| bootstrap | True | preProc.params | method="range" |
| DArch.fine TuneFunction | backpropagation | generateWeights Function | generateWeights GlorotUniform |

approaches used limits the number of iterations is choosing the number of epochs for the training phase to 50. In practice this number has been chosen as a limit value based on empirical criteria since it marked the beginning of an asymptotic behaviour for the error.

*Mini Batches.* Updating the weights using small "mini-batches" of 10 to 100 samples in size has shown to allow better results. In the experiment we used a mini batch size of 100.

*Momentum.* By choosing a momentum value it is possible to modulate the speed of learning in RBM training. At the start of learning, the random initial parameter values may create very large gradients and the system is unlikely to be at a minimum, so it is usually best to start with a low momentum of 0.5. This very conservative momentum typically makes the learning more stable than no momentum at all [6].

*Learning Rate.* A good rule of thumb for setting the learning rate $\varepsilon$ is to look at a histogram of the weight updates and a histogram of the weights [26]. The updates should be about $10^{-3}$ times the weights (to within about an order of magnitude). When a unit has a very large fan-in, the updates should be smaller since many small changes in the same direction can easily reverse the sign of the gradient. Conversely, for biases, the updates can be bigger: $\Delta w_{ij} = \varepsilon(<v_i h_j>_{data} - <v_i h_j>_{sampled})$.

*Weight Decay.* Weight-decay works by adding an extra term to the normal gradient. The extra term is the derivative of a function that penalizes large weights. The simplest penalty function, called L2, is half of the sum of the squared weights times a coefficient which will be called the weight-cost. For an RBM, sensible values for the weight-cost coefficient for L2 weight-decay typically range from 0.01 to 0.00001.

*Weight Initialization.* Weight initialization values represents the starting value of the RBM weights that amplify or mute the input signal coming into each node. Proper weight initialization help training time. In our case a trial has been made with Glorot uniform weight initialization as described in [24], in the perspective of achieving faster convergence.

# 5 Dataset Description

The UNB ISCX Intrusion Detection Evaluation Dataset was developed in order to provide a quality dataset for network intrusion detection research [20]. The approach for defining this dataset involved identifying features that would allow effective detection, while minimizing processing costs. Each record of the dataset is characterized by features that fall into three categories: basic, content, and traffic. These features are described in [5,23].

**Table 2.** Attacks in the UNB ISCX train/test datasets (all attacks from the first exist also in the second)

| Class | Train dataset attacks | Test dataset only attacks |
|---|---|---|
| Probing | portsweep, ipsweep, satan, guesspasswd, spy, nmap | snmpguess, saint, mscan, xsnoop |
| DoS | back, smurf, neptune, land, pod, teardrop, buffer overflow, warezclient, warezmaster | apache2, worm, udpstorm, xterm |
| R2L | imap, phf, multihop | snmpget, httptunnel, xlock, sendmail, ps |
| U2R | loadmodule, ftp write, rootkit | sqlattack, mailbomb, processtable, perl |

The UNB ISCX dataset is composed of sequences of entries in the form of records labeled as either *normal* or *attack*. Each entry contains a set of characteristics of a *flow*, i.e., of a sequence of IP packets starting at a time instant and ending at another, between which data flows between two IP addresses using a transport-layer protocol (TCP, UDP) and an application-layer protocol (HTTP, SMTP, SSH, IMAP, POP3, or FTP). The dataset is fairly balanced with prior class probabilities of 0.466 for the *normal* class and 0.534 for the *anomaly* class.

This dataset is composed of two sub-datasets: a *train dataset*, used for training a NIDS, and a *test dataset*, used for testing. Both have the same structure and contain all four types of attacks. However, the test dataset has more attacks as shown in Table 2, to allow evaluating the ability of algorithms to generalize. The train dataset has around 2.2 GB of data; the test dataset has 0.8 GB.

## 6   The Experiment

We used the Darch R Package package in the experiments [11,12]. This package allows generating deep architecture networks and training them. All parameters not explicitly referred took the default values.

In order to make these features suitable inputs for the visible layer of the RBM, we normalized them to the range $[0, 1]$, and treated them as continuous values. Cross-validation, sometimes called rotation estimation, is a validation technique for assessing how the results of a statistical analysis will generalize to an independent dataset. One round of cross-validation involves partitioning a sample of data into complementary subsets, performing the analysis on one subset (the training set), and validating the analysis on the other subset (the validation or testing set). To reduce variability, multiple rounds of cross-validation are performed using different partitions, and the validation results are averaged over the rounds.

For the starting RBM hidden nodes dimensioning in the experiment, a technique similar to the gradual drop-in of nodes (increasing the number of nodes in each step of the experience) was used with reference values near the $\lceil v/2 \rceil$. The numbers of nodes used in the experience were $16, 17, 18, 19, 20$, plus $25, 30$ for comparison purposes. This has been suggested to be a sufficient approximation for obtaining good results in training RBMs [27]. The most relevant parameters for the RBM as used are depicted in Table 1. This thesis has been validated since the best performance figures are shown to be obtained for an internal node number of 16.

We performed an empirical study to compare the .632+ bootstrap estimator with the repeated 10-fold cross-validation. The results for the pre-training and fine tuning phases were obtained with 125973 samples and 40 predictor variables. The accuracy was used to select the optimal model using the largest value. Pre-processing was used to re-scale the features to the range $[0, 1]$. Bootstrapping was started with 125973 samples and resulted in 79598 unique training samples and 46375 validation samples for this run. Training data was shuffled before each epoch and the final result for each run was taken from the average results of each

**Table 3.** Error estimation (averages for 5 trials)

|                                    | 16     | 17     | 18     | 19     | 20     | 25     | 30     |
|------------------------------------|--------|--------|--------|--------|--------|--------|--------|
| Training MSE                       | 0.137  | 0.120  | 0.122  | 0.123  | 0.109  | 0.115  | 0.106  |
| Training classification error      | 5.55%  | 5.92%  | 6.06%  | 5.79%  | 5.86%  | 5.94%  | 5.75%  |
| Validation MSE                     | 0.137  | 0.120  | 0.122  | 0.123  | 0.110  | 0.115  | 0.107  |
| .632 + MSE                         | 0.137  | 0.120  | 0.122  | 0.123  | 0.110  | 0.115  | 0.107  |
| Validation classification error    | 5.65%  | 5.88%  | 5.99%  | 5.73%  | 5.89%  | 5.94%  | 5.75%  |
| .632 + classification error        | 5.61%  | 5.89%  | 6.02%  | 5.75%  | 5.88%  | 5.95%  | 5.75%  |

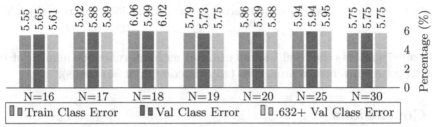

**Fig. 5.** Experimental results for training and validation classification errors (Table 3)

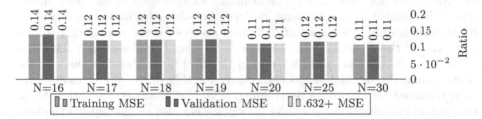

**Fig. 6.** Experimental results for network training and validation errors from Table 3

epoch. The results for the classification and network errors are summarized in Table 3 and put in contrast in Figs. 5 and 6. By looking at the .632+ statistics value we can observe that a negligible amount of bias for the error values is present, since this statistics results similar to the respective values.

In Fig. 5 the training and validation Classification Errors are presented as a graphical example of the results from Table 3. The classification error is the error of the classification given by the RBM for a validation and a training set. In Fig. 6(a) an example of error convergence for the case of an RBM with 16 internal nodes is presented. Similarly, in Fig. 6 the training and validation Mean Square Errors are presented as a graphical example of the results from Table 3. The MSE is the average difference between the expected and the actual output, and it contains both the variance and bias of the estimator. In Fig. 7(b) an example of MSE convergence for the case of a RBM with 16 internal nodes is presented.

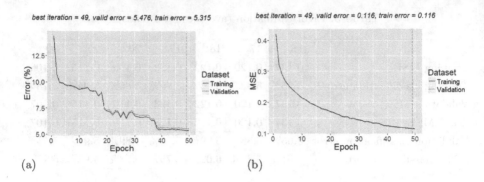

**Fig. 7.** Errors for iterations with 16 internal nodes: (a) classification error; (b) MSE

Figure 7 presents the results for an example with internal node number of 16, showing good convergence properties (with maximum of 50 epochs).

## 7   Conclusions

The application of Restricted Boltzmann Machines in the network intrusion detection field has been gaining momentum. In this work it has been shown that it is possible to surpass some performance obstacles by means of proper RBM optimization given a set of optimized choices for the remaining parameters. Among these the most troublesome is the fact that for using RBMs it is relevant to process real valued features in the $[0, 1]$ range therefore the need for transforming the original values accordingly for this range. Good results have been obtained for the Network Training MSE and Network Validation MSE. In terms of classification the results obtained were also comparable to the best obtained till now in NIDS using RBMs [5] (Table 4).

**Table 4.** Performance metrics

| Metrics | Drop = 0 | Drop = 0.1 | Drop = 0.5 |
|---|---|---|---|
| Accuracy | 0.7609 | 0.7634 | 0.7549 |
| Kappa | 0.5400 | 0.5414 | 0.5258 |
| Specificity | 0.9666 | 0.9368 | 0.9353 |
| Sensitivity | 0.6063 | 0.6331 | 0.6194 |
| Neg pred value | 0.6485 | 0.6574 | 0.6487 |
| Pos pred value | 0.9603 | 0.9302 | 0.9272 |
| Balanced accuracy | 0.7865 | 0.7850 | 0.7773 |

**Acknowledgements.** This work was supported by national funds through Fundação para a Ciência e a Tecnologia (FCT) with reference UID/CEC/50021/2013 (INESC-ID).

# References

1. Ackley, D.H., Hinton, G.E., Sejnowski, T.J.: A learning algorithm for Boltzmann machines. Cogn. Sci. **9**(1), 147–169 (1985)
2. Bengio, Y.: Learning deep architectures for AI. Found. Trends Mach. Learn. **2**(1), 1–127 (2009)
3. Bengio, Y., LeCun, Y.: Scaling learning algorithms towards AI. In: Bottou, L., Chapelle, O., DeCoste, D., Weston, J. (eds.) Large-Scale Kernel Machines. MIT Press, Cambridge (2007)
4. Chen, H., Murray, A.F.: Continuous restricted Boltzmann machine with an implementable training algorithm. IEEE Proc. Vis. Image Sig. Process. **150**(3), 153–158 (2003)
5. Fiore, U., Palmieri, F., Castiglione, A., De Santis, A.: Network anomaly detection with the restricted Boltzmann machine. Neurocomputing **122**, 13–23 (2013)
6. Hinton, G.E.: A practical guide to training restricted Boltzmann machines. In: Montavon, G., Orr, G.B., Müller, K.-R. (eds.) Neural Networks: Tricks of the Trade. LNCS, vol. 7700, pp. 599–619. Springer, Heidelberg (2012). doi:10.1007/978-3-642-35289-8_32
7. Hinton, G., Osindero, S., Teh, Y.-W.: A fast learning algorithm for deep belief nets. Neural Comput. **18**(7), 1527–1554 (2006)
8. Hinton, G., Salakhutdinov, R.: Reducing the dimensionality of data with neural networks. Science **313**(5786), 504–507 (2006)
9. Hinton, G., Srivastava, N., Krizhevsky, A., Sutskever, I., Salakhutdinov, R.: Improving neural networks by preventing co-adaptation of feature detectors, July 2012. arXiv:1207.0580 [cs.NE]
10. Hinton, G.E.: Training products of experts by minimizing contrastive divergence. Neural Comput. **14**(8), 1771–1800 (2002)
11. Hinton, G.E., Osindero, S., Tch, Y.-W.: A fast learning algorithm for deep belief nets. Neural Comput. **18**(7), 1527–1554 (2006)
12. Hinton, G.E., Salakhutdinov, R.R.: Reducing the dimensionality of data with neural networks. Science **313**(5786), 504–507 (2006)
13. Ising, E.: Beitrag zur theorie des ferromagnetismus. Z. Phys. **31**(1), 253–258 (1925)
14. Krizhevsky, A., Sutskever, I., Hinton, G.E.: ImageNet classification with deep convolutional neural networks. In: Pereira, F., Burges, C.J.C., Bottou, L., Weinberger, K.Q. (eds.) Advances in Neural Information Processing Systems, vol. 25, pp. 1097–1105. Curran Associates (2012)
15. Movellan, J.R., Mineiro, P., Williams, R.J.: A Monte Carlo EM approach for partially observable diffusion processes: theory and applications to neural networks. Neural Comput. **14**, 1507–1544 (2002)
16. Nair, V., Hinton, G.E.: Rectified linear units improve restricted Boltzmann machines. In: Frnkranz, J., Joachims, T. (eds.) Proceedings of the 27th International Conference on Machine Learning, pp. 807–814. Omnipress (2010)
17. Onsager, L.: Crystal statistics. I. A two-dimensional model with an order-disorder transition. Phys. Rev. **65**, 117–149 (1944)
18. Rosenblatt, F.: The perceptron: a probabilistic model for information storage and organization in the brain. Psychol. Rev. **65**(6), 386–408 (1958)

19. Salakhutdinov, R.: Learning in Markov random fields using tempered transitions. In: Bengio, Y., Schuurmans, D., Lafferty, J.D., Williams, C.K.I., Culotta, A. (eds.) NIPS, pp. 1598–1606. Curran Associates (2009)

20. Shiravi, A., Shiravi, H., Tavallaee, M., Ghorbani, A.A.: Toward developing a systematic approach to generate benchmark datasets for intrusion detection. Comput. Secur. **31**(3), 357–374 (2012)

21. Smith, L.N., Hand, E.M., Doster, T.: Gradual DropIn of layers to train very deep neural networks. In: Proceedings of the IEEE Conference on Computer Vision and Pattern Recognition, pp. 4763–4771 (2016)

22. Smolensky, P.: Information processing in dynamical systems: foundations of harmony theory. In: Parallel Distributed Processing: Explorations in the Microstructure of Cognition, vol. 1: foundations, pp. 194–281. MIT Press (1986)

23. Tavallaee, M., Bagheri, E., Lu, W., Ghorbani, A.A.: A detailed analysis of the KDD CUP 99 data set. In: Proceedings of the Second IEEE International Conference on Computational Intelligence for Security and Defense Applications, CISDA 2009, pp. 53–58 (2009)

24. Teh, Y.W., Titterington, D.M. (eds.): Proceedings of the 13th International Conference on Artificial Intelligence and Statistics (2010)

25. Vincent, P., Larochelle, H., Lajoie, I., Bengio, Y., Manzagol, P.-A.: Stacked denoising autoencoders: learning useful representations in a deep network with a local denoising criterion. J. Mach. Learn. Res. **11**, 3371–3408 (2010)

26. Welling, M., Teh, Y.W.: Linear response algorithms for approximate inference in graphical models. Neural Comput. **16**, 197–221 (2004)

27. Zhao, X., Hou, Y., Yu, Q., Song, D., Li, W.: Understanding deep learning by revisiting Boltzmann machines: an information geometry approach (2013). CoRR, abs/1302.3931

# Selecting the Coherence Notion in Multi-adjoint Normal Logic Programming

M. Eugenia Cornejo$^{(\boxtimes)}$, David Lobo, and Jesús Medina

Department of Mathematics, University of Cádiz, Cádiz, Spain
{mariaeugenia.cornejo,david.lobo,jesus.medina}@uca.es

**Abstract.** This paper is focused on looking for an appropriate coherence notion which allows us to deal with inconsistent information included in multi-adjoint normal logic programs. Different definitions closely related to the inconsistency concept have been studied and an adaptation of them to our logic programming framework has been included. A detailed reasoning is presented in order to motivate and justify the suitability of the chosen coherence notion.

**Keywords:** Negation operator · Coherence · Multi-adjoint logic programming

## 1 Introduction

Multi-adjoint normal logic programming is a logical theory whose semantic structure is composed by a lattice together with various adjoint pairs and a negation operator. Logic programs of this framework are characterized by using both different implications in their rules and general operators on complete lattices in the bodies of the rules. Recently, the existence of minimal models of multi-adjoint normal logic programs has been studied in [6], where the stable model semantics [14] has been considered.

In spite of inconsistent and contradictory information usually appears in real-life databases, the general tendency of any logical theory consists in either not considering such information and ignoring all the knowledge coming from it or try to repair this information. However, adopting this position is not appropriate due to this conflictive information can be reliable and useful. From this fact arises the importance of handling and measuring the inconsistency [9,17–19]. Madrid and Ojeda-Aciego showed that there are only two possible causes that can lead to obtain an inconsistent residuated logic program: instability and incoherence [20–23]. On the one hand, the instability is associated with the presence of incompatible rules in a logic program which gives rise to the lack of stable models. On the other hand, the incoherence is obtained from the existence of stable models which assign contradictory values to a propositional symbol and to

---

Partially supported by the State Research Agency (AEI) and the European Regional Development Fund (ERDF) project TIN2016-76653-P.

I. Rojas et al. (Eds.): IWANN 2017, Part I, LNCS 10305, pp. 447–457, 2017.
DOI: 10.1007/978-3-319-59153-7_39

its corresponding negation. According to the syntactic structure of multi-adjoint normal logic programs, all the inconsistency causes mentioned above can appear in this framework.

If one program has not stable models, then it is unstable. Hence, the determination of unstable programs is clear, however, when we can say that a program is incoherent, what do we mean by "contradictory values"?, what is the best measure in order to obtain this information?

There exist different notions related to the concept of "coherence" [13, 23–25, 27]. In this paper, we are interested in choosing a suitable notion of coherence for multi-adjoint normal logic programs. Hence, first of all we will provide a survey of the definitions given in the literature which are closely related to the concept of coherence such as self-contradiction in fuzzy sets [27], $x$-consistent interpretation [25], consistency in interlaced bilattices [13] and coherent interpretation [23, 24]. Based on these definitions we will analyze the most suitable definition to be used in the multi-adjoint framework. From this analysis, we will obtain parallel properties. For example, by means of weak involutive negations [26], we will show a natural equivalence condition of the coherence interpretation notion given by Madrid and Ojeda-Aciego [23, 24], and the relationship between the definition of coherence interpretation and the notion of intuitionistic fuzzy set defined on a complete lattice [1] is also introduced.

## 2   Multi-adjoint Normal Logic Programming

First of all, we will introduce a brief summary with essential notions related to the syntax and semantics of multi-adjoint normal logic programming framework.

The algebraic structure considered in this environment is the multi-adjoint normal lattice which is defined as follows.

**Definition 1.** *The tuple* $(L, \preceq, \leftarrow_1, \&_1, \ldots, \leftarrow_n, \&_n, \neg)$ *is a multi-adjoint normal lattice if the following properties are verified:*

1. $(L, \preceq)$ *is bounded lattice, i.e. it has a bottom* $(\bot)$ *and a top* $(\top)$ *element;*
2. $(\&_i, \leftarrow_i)$ *is an adjoint pair in* $(L, \preceq)$*, for* $i \in \{1, \ldots, n\}$*;*
3. $\top \&_i \vartheta = \vartheta \&_i \top = \vartheta$*, for all* $\vartheta \in L$ *and* $i \in \{1, \ldots, n\}$*;*
4. $\neg$ *is a negation operator on* $(L, \preceq)$*, that is,* $\neg$ *is a decreasing mapping satisfying that* $\neg(\bot) = \top$ *and* $\neg(\top) = \bot$*.*

In this paper, we are interested in considering a logic program defined on a multi-adjoint normal lattice $(L, \preceq, \leftarrow_1, \&_1, \ldots, \leftarrow_n, \&_n, \neg)$ together with an additional (symbol of) negation $\sim$. It is important to emphasize that $\neg$ is called "default negation" and $\sim$ is called "strong negation", which should not be confused with the well-known notion of involutive operator. Specifically, the truth value of $\sim\phi$ can straightforwardly be inferred from the program and the value of $\neg\phi$ is obtained from the truth value of $\phi$.

From the algebraic structure described above, the syntax for multi-adjoint normal logic programs is defined as a set of weighted rules of a given language.

The elements appearing in these weighted rules, which can be either (positive) propositional symbols or negated propositional symbols by the strong negation $\sim$, will be called literals. The literals included in $\mathbb{P}$ will be collected in a set denoted by $Lit_{\mathbb{P}}$, whereas the set of propositional symbols appearing in $\mathbb{P}$ will be denoted by $\Pi_{\mathbb{P}}$.

**Definition 2.** *Let* $(L, \preceq, \leftarrow_1, \&_1, \ldots, \leftarrow_n, \&_n, \neg)$ *be a multi-adjoint normal lattice and* $\sim$ *be a strong negation. A* multi-adjoint normal logic program *(MANLP)* $\mathbb{P}$ *is a finite set of weighted rules of the form:*

$$\langle l \leftarrow_i @[l_1, \ldots, l_m, \neg l_{m+1}, \ldots, \neg l_n]; \vartheta \rangle$$

*where* $i \in \{1, \ldots, n\}$, @ *is an aggregator operator,* $\vartheta$ *is an element of* $L$ *and* $l, l_1, \ldots, l_n$ *literals such that* $l_j \neq l_k$, *for all* $j, k \in \{1, \ldots, n\}$, *with* $j \neq k$.

The semantics for MANLPs is given by means of the next definitions.

**Definition 3.** *Given a complete lattice* $(L, \preceq)$, *a mapping* $I \colon Lit_{\mathbb{P}} \rightarrow L$ *which assigns to every literal appearing in* $Lit_{\mathbb{P}}$ *an element of* $L$ *is called* $L$-*interpretation. The set of all* $L$-*interpretations is denoted by* $\mathcal{I}_{\mathfrak{L}}$.

It is worth mentioning that the ordering relation in $(L, \preceq)$ is extended to the set of interpretations in the following way:

$$I_1 \sqsubseteq I_2 \text{ if and only if } I_1(l) \preceq I_2(l), \text{ for all } l \in Lit_{\mathbb{P}} \text{ and } I_1, I_2 \in \mathcal{I}_{\mathfrak{L}}.$$

Before presenting the last concepts associated with the semantics of MANLPs, it is necessary to consider the following notational convention. We will denote as $\dot{\omega}$ the interpretation of a operator symbol $\omega$ under a multi-adjoint normal lattice. In addition, the evaluation of a formula $\mathcal{A}$ under an interpretation $I$ will be denoted as $\hat{I}(\mathcal{A})$. Note that, it will be proceeded inductively as usual, until all propositional symbols in $\mathcal{A}$ are reached and evaluated under $I$.

**Definition 4.** *Given an interpretation* $I \in \mathcal{I}_{\mathfrak{L}}$, *we say that:*

*(1) A weighted rule* $\langle l \leftarrow_i @[l_1, \ldots, l_m, \neg l_{m+1}, \ldots, \neg l_n]; \vartheta \rangle$ *is satisfied by* $I$ *if and only if* $\vartheta \preceq \hat{I}(\langle l \leftarrow_i @[l_1, \ldots, l_m, \neg l_{m+1}, \ldots, \neg l_n])$.
*(2) An* $L$-*interpretation* $I \in \mathcal{I}_{\mathfrak{L}}$ *is a* model *of a MANLP* $\mathbb{P}$ *if and only if all weighted rules in* $\mathbb{P}$ *are satisfied by* $I$.

As we mentioned in the introduction, the existence of stable models has been already studied by the authors in [6]. Now, our goal focuses on studying if the information provided by such stable models is coherent. In order to reach this goal, the first task will be to introduce an appropriate notion of coherence for our logical theory. With this purpose, we will analyze some of the works associated with this research line.

# 3    An Overview on Coherence Notions

To the best of our knowledge, the notions most closely related to the concept of coherence are: self-contradiction in fuzzy sets [27], $x$-consistent interpretation [25], consistency in interlaced bilattices [13] and coherent interpretation [23,24]. In order to compare these notions we need to introduce them using the notation and definitions of the MANLPs framework. Besides this adaptation, different interesting results and remarks will be presented.

## 3.1    Self-contradiction in Fuzzy Sets

Trillas et al. [27] introduced the notion of self-contradiction considering fuzzy sets [31] taking into account a classic principle of propositional logic. A statement $p$ is considered self-contradictory if $p$ violates the principle of contradiction, that is, the statement *"If $p$ is true, then $\neg p$ is true"* has some degree of truth. Considering an implication operator $\leftarrow$ and an involutive negation operator $n$, Trillas et al. established that a fuzzy set $\mu$ is self-contradictory if and only if $n(\mu) \leftarrow \mu = 1$ or, equivalently, $\mu \leq n(\mu)$. Three different definitions of self-contradiction were taken into account in [27], which are rewritten in the multi-adjoint normal logic programming framework next.

**Definition 5.** *Given an interpretation $I \in \mathcal{I}_{\mathfrak{L}}$, we say that:*

(1) *$I$ is strongly self-contradictory if $I(l) \preceq n(I(l))$, for any involutive negation $n$ and for all $l \in Lit_{\mathbb{P}}$.*

(2) *$I$ is weakly self-contradictory if $I(l) \preceq n(I(l))$, for a given involutive negation $n$ and for all $l \in Lit_{\mathbb{P}}$.*

(3) *$I$ is $(n,l)$-locally self-contradictory if $I(l) \preceq n(I(l))$, for a given involutive negation $n$ and a fixed literal $l \in Lit_{\mathbb{P}}$.*

The importance of considering these notions is due to the amount of works carried out on self-contradiction in fuzzy sets. For instance, some fuzzy set theories which avoid self-contradiction were given in [27–29]. Moreover, the self-contradiction notion was also considered in the Atanassov intuitionistic sets framework, providing models for measuring contradiction between this kind of sets [5,8]. Also, a more general notion of self-contradiction can be found in [3,4].

All these different approaches show the wide flexibility of the self-contradiction notion. Thus, it seems reasonable to be considered as a possible definition of consistency in a multi-adjoint normal logic programming framework.[1]

---

[1] In order to proceed with the comparison in Sect. 4 we will consider the definition of weakly self-contradictory.

## 3.2    $x$-consistent Notion

The notion of $x$-consistency was firstly given by Van Nieuwenborgh et al. [25]. The most interesting feature of this definition is that it allows an user to choose, by means of an aggregator operator, how the individual consistencies of the propositional symbols and their negations affect the consistency of an interpretation. Now, we introduce the notion of $x$-consistent interpretation adapted to the multi-adjoint framework.

**Definition 6.** *Let* $\&\colon L \times L \to L$ *be a t-norm,* $A_c\colon L \times L \to L$ *be an aggregator operator, and interpretation* $I \in \mathcal{I}_{\mathfrak{L}}$ *and* $x \in L$. *Given the mapping* $I_c\colon \Pi_{\mathbb{P}} \to L$ *defined as* $I_c(p) = \neg(I(p)\,\&\,I(\sim p))$ *for each* $p \in \Pi_{\mathbb{P}}$, *we say that* $I$ *is* $x$-consistent *if and only if* $x \preceq A_c(\Pi_{\mathbb{P}}, I_c)$.

Note that, the use of an aggregator operator allows us to ignore certain inconsistencies or that certain literals to be more inconsistent than others. It is also important to mention that, from Definition 6, an user can choose a suitable lower bound $x$ from which an interpretation stops being consistent. Although the previously given features in relation to the notion of $x$-consistent interpretation are interesting, the following example shows that this definition is not appropiate for our approach.

*Example 1.* Let $\mathbb{P}$ be a MANLP defined on $([0,1], \preceq, \leftarrow_1, \&_1, \ldots, \leftarrow_n, \&_n, \neg)$ together with the strong negation $\sim$. Considering the Gödel t-norm defined as $x \,\&_G\, y = \min\{x, y\}$, for all $x, y \in [0,1]$, an interpretation $I\colon Lit_{\mathbb{P}} \to [0,1]$ such that $I(l) = 0.5$ for each $l \in Lit_{\mathbb{P}}$, the standard negation $\neg x = 1 - x$, for all $x \in [0,1]$ and the aggregator operator defined as $A(\Pi_{\mathbb{P}}, I_c) = \min\{I_c(p) \mid p \in \Pi_{\mathbb{P}}\}$, we will compute the maximum value $x$ from which the interpretation $I$ is consistent.

For each $p \in \Pi_{\mathbb{P}}$, we obtain $I_c(p) = \neg(I(p)\,\&_G\,I(\sim p)) = 1 - \min\{0.5, 0.5\} = 0.5$ and therefore, $A(\Pi_{\mathbb{P}}, I_c) = 0.5$. As a consequence, the greatest value $x$ such that $I$ is $x$-consistent is $x = 0.5$.                                        □

If we pay attention to the simple definition of the interpretation $I$ in the example above, since the truth value of a propositional $p$ is equal to its negated and it is 0.5, we expect that $I$ be 1-consistent. However, we have obtained that $I$ is at most 0.5-consistent. Hence, the notion of $x$-consistency is not the most suitable option for the framework we are interested in.

## 3.3    Consistency in Interlaced Bilattices

This section includes another fundamental definition in this paper. We need to introduce some previous definitions in order to present the notion of consistency given by Fitting [13].

Bilattices were proposed by Ginsberg [15, 16] as a generalization of the four-valued logic [2, 30]. We are concerned here in a more restricted notion of bilattice, which is known as interlaced bilattice.

**Definition 7** [13]. *The tuple* $(B, \leq_k, \leq_t)$ *composed by a set* $B$ *together with two partial orderings* $\leq_k$ *and* $\leq_t$ *is an* interlaced bilattice *if:*

*(1) each of* $\leq_k$ *and* $\leq_t$ *gives* $B$ *the structure of a complete lattice;*
*(2) the meet and join operations for each partial ordering are monotone with respect to the other ordering.*

Now, we will consider an interlaced bilattice enriched with a special kind of negation operator called conflation.

**Definition 8** [13]. *An interlaced bilattice* $(B, \leq_k, \leq_t)$ *has a* conflation *operator if there is a mapping* $-: B \to B$ *such that:*

*(1) if* $a \leq_k b$ *then* $-a \leq_k -b$;
*(2) if* $a \leq_t b$ *then* $-b \leq_t -a$;
*(3)* $--a = a$.

Unlike what one can expect from a negation operator, the conflation operator of an interlaced bilattice reverses the truth ordering $\leq_t$ and preserves the knowledge ordering $\leq_k$. This fact implies that one knows as much about $-a$ as one knows about $a$.

We are interested in the notion of consistency in interlaced bilattices with conflaction which was introduced in [13] as follows.

**Definition 9** [13]. *Let* $(B, \leq_k, \leq_t)$ *be an interlaced bilattice with a conflation operator* $-$. *An element* $a \in B$ *is* consistent *if and only if* $a \leq_k -a$.

This definition can be straightforwardly extended to the set of interpretations, as it is shown below.

**Definition 10.** *Let* $\mathbb{P}$ *be a MANLP defined on the multi-adjoint normal bilattice* $(B, \leq_k, \leq_t, \leftarrow_1, \&_1, \ldots, \leftarrow_n, \&_n, -)$ *where* $(B, \leq_k, \leq_t)$ *is an interlaced bilattice with a conflation operator* $-$. *An interpretation* $I \in \mathcal{I}_{\mathfrak{B}}$ *is* consistent *if the inequality* $I(l) \leq_k -I(l)$ *holds, for each* $l \in Lit_{\mathbb{P}}$.

Observe that, Definition 10 is closely related to the notion of weak self-contradiction interpretation introduced in Sect. 3.1. In fact, the inequality to be satisfied is the same in both cases. The only difference is the choice of a lattice or an interlaced bilattice as the algebraic structure from which the program is defined. Consequently, we can ensure that the difference between these two notions will affect mainly to the syntax of the program and not the semantics.

Moreover, given an interlaced bilattice $(B, \leq_k, \leq_t)$ with conflaction $-$, it is easy to see that $(B, \leq_k)$ is a complete lattice and $-$ is an involutive decreasing mapping defined on $(B, \leq_k)$. Hence, the notion of weak self-contradiction interpretation can be defined in $(B, \leq_k)$. As a consequence, the notion of consistent interpretation given in this section can be seen as a particular case of weak self-contradiction interpretation.

## 3.4    Coherence Interpretation in Fuzzy Answer Sets

The last notion associated with the coherence concept which we will take into account has been recently considered by Madrid and Ojeda-Aciego in [23,24]. This notion is based on the idea of accepting an interpretation contradicting the next inference rule *"If the truth value of a propositional symbol p is $\vartheta$ then the truth value of $\neg p$ is $n(\vartheta)$"*, where $n$ is a negation operator. This fact involves a possible lack of information but not an excess of information.

**Definition 11** [23]. *Given an interpretation $I \in \mathcal{I}_{\mathfrak{L}}$, we say that $I$ is coherent if and only if the inequality $I(\sim p) \preceq \, \sim I(p)$ holds, for every $p \in \Pi_{\mathbb{P}}$.*

An interesting feature of this notion is that every interpretation lower than a coherent interpretation is also coherent.

**Proposition 1** [23]. *Let $I$ and $J$ be two interpretations satisfying $I \sqsubseteq J$. If $J$ is coherent, then $I$ is coherent as well.*

The contrapositive of Proposition 1 ensures that if an interpretation $I$ is not coherent, then the only possibility is having an excess of information, since every interpretation $J$ being $I \sqsubseteq J$ is also incoherent. This is one of the three main reasons given by Madrid and Ojeda-Aciego [23,24] to guarantee that this coherence notion is a good generalization of the consistency concept. Another reason is the easy implementation of the condition of coherence, due to it only depends on the negation operator. Finally, they emphasize the allowance of lack of information. For instance, the interpretation $I_\perp$, which represents no information, is always coherent.

Hence, this definition together with the notion of self-contradiction are the two more suitable definitions we can consider in order to formalize the meaning of coherence in the multi-adjoint framework.

# 4    Selecting a Suitable Coherence Notion for MANLPs

In the previous section, the notions of self-contradiction, $x$-consistency, consistency in bilattices and coherence interpretation were presented. We have seen that $x$-consistency is not suitable in the logic programming framework which we are interested in. In addition, we have also proven that the notion of consistency in bilattices is just involved in the notion of self-contradiction. Hence, we have two possible selections of the definition of consistency interpretation of a multi-adjoint normal logic program. In this section, we provide two important reasons in order to choose the most suitable consistency notion between them for the multi-adjoint normal framework.

## 4.1    Weak Involutive Negations

The weak involutive negations are a particular case of negation operators, which have been intensively studied in different papers [10–12,26]. They are interesting negation operators which, in particular, generalize the residuated negations as it was shown in [7].

**Definition 12.** *Let* $n: L \to L$ *be a negation operator. We say that* $n$ *is a* weak involutive negation *if and only if* $x \preceq n(n(x))$ *for each* $x \in L$.

Hence, a suitable coherence definition in the multi-adjoint framework should allow the consideration of this kind of negations.

If we analyze the coherence notion given by Madrid and Ojeda-Aciego, we have that we can use weak involutive negations perfectly. Indeed, this kind of negations provides an interesting property.

Madrid and Ojeda-Aciego explained in [23] that an alternative definition to the one given initially by them $(I(\sim p) \preceq \dot{\sim} I(p))$, is the dual inequality $I(p) \preceq \dot{\sim} I(\sim p)$. However, they discard this possibility because the negation operators $\sim$ satisfying $\sim (\sim x) < x$ for some $x \in L$ does not verify the dual inequality.

If we consider the useful weak involutive negations, then the coherence condition $I(\sim p) \preceq \dot{\sim} I(p)$ is equivalent to the dual inequality $I(p) \preceq \dot{\sim} I(\sim p)$, as the following proposition shows.

**Proposition 2.** *Let* $(L, \preceq, \leftarrow_1, \&_1, \ldots, \leftarrow_n, \&_n, \neg)$ *be a multi-adjoint normal lattice,* $\sim$ *a strong negation and* $\mathbb{P}$ *a multi-adjoint normal logic program defined on this lattice. If* $n$ *is a weak involutive negation such that* $n = \dot{\sim}$, *then* $I \in \mathcal{I}_{\mathfrak{L}}$ *is a coherent interpretation if and only if* $I(p) \leq \dot{\sim} I(\sim p)$ *for each* $p \in \Pi_{\mathbb{P}}$.

Therefore, the coherence notion used by Madrid and Ojeda-Aciego works with the weak involutive negations. Nevertheless, from the notion of (weakly) self-contradiction a strong condition arises. If we consider the weak involutive negation $n$ defined as

$$n(x) = \begin{cases} 0 & \text{if } x \neq 0 \\ 1 & \text{if } x = 0 \end{cases}$$

which is the residuated negation associated with the product and Gödel t-norms, we have not any self-contradictory interpretation which is not a good property.

The main problem is that the definition of self-contradiction does not consider both literals, the propositional symbol $p$ and its negated $\sim p$ at the same time.

## 4.2  Intuitionistic Fuzzy Sets

An interesting discussion about the idea of the coherence notion can be made from the point of view of intuitionistic fuzzy sets, which were firstly introduced by Atanassov in [1]. This kind of sets are defined in a complete lattice as follows:

**Definition 13.** *Given a complete lattice* $(L, \preceq)$, *an involutive negation operator* $n: L \to L$ *and a non-empty set* $E$. *An* intuitionistic $L$-fuzzy set $A$ *in* $E$ *is defined as the set:*

$$A = \{\langle x, \mu_A(x), \nu_A(x) \rangle \mid x \in E\}$$

*where the functions* $\mu_A: E \to L$ *and* $\nu_A: E \to L$ *define the degree of membership and the degree of non-membership, respectively, to* $A$ *of the elements* $x \in E$, *and for every* $x \in E$:

$$\mu_A(x) \preceq n(\nu_A(x))$$

After recalling the notion of intuitionistic $L$-fuzzy set, we will present the relationship established between the definition of coherent interpretation given by Madrid and Ojeda-Aciego and the previous one. In order to get this goal, we need to consider two mappings $\mu_A, \nu_A \colon \Pi_\mathbb{P} \to L$ defined as follows $\mu_A(p) = I(p)$ and $\nu_A(p) = I(\sim p)$, for all $p \in \Pi_\mathbb{P}$. Notice that, $I(p)$ represents the degree of truth and $I(\sim p)$ represents the degree of non-truth, for each $p \in \Pi_\mathbb{P}$. Hence, the definition of the mappings $\mu_A$ and $\nu_A$ is reasonable.

Considering an involutive negation $n$ as the operator corresponding to the symbol $\sim$, it can be proved that the inequalities $\mu_A(p) \preceq n(\nu_A(p))$ and $I(\sim p) \preceq \,\dot\sim I(p)$ are equivalent, from which the following result arises.

**Proposition 3.** *Let $\mathbb{P}$ be a MANLP defined on a multi-adjoint normal lattice $(L, \preceq, \leftarrow_1, \&_1, \ldots, \leftarrow_n, \&_n, \neg)$ and whose strong negation is denoted as $\sim$. Consider an interpretation $I \in \mathcal{I}_\mathfrak{L}$ and two mappings $\mu_A, \nu_A \colon \Pi_\mathbb{P} \to L$ defined as $\mu_A(p) = I(p)$ and $\nu_A(p) = I(\sim p)$, respectively. If the negation operator $\dot\sim$ is involutive, then the set*

$$A = \{\langle p, \mu_A(p), \nu_A(p)\rangle \mid p \in \Pi_\mathbb{P}\}$$

*is an intuitionistic $L$-fuzzy set if and only if the interpretation $I$ is coherent.*

As a consequence, the coherence notion used by Madrid and Ojeda-Aciego involves the philosophy of the Atanassov intuitionistic fuzzy sets, that is, if an interpretation can be rewritten as an Atanassov intuitionistic fuzzy set as in Proposition 3, then it is coherent.

This relationship provides another argument in order to choose the suitable notion of coherence in the multi-adjoint framework.

The definition of self-contradiction cannot be related in this sense to this kind of fuzzy sets since it does not consider the propositional symbol $p$ and its negated $\sim p$ at the same time. There are studies which associate intuitionistic fuzzy sets with the definition of self-contradiction [5,8], but they are focused on another problem and they do not provide any added value to the notion of self-contradiction interpretation studied in Sect. 3.1.

According to the discussion carried out throughout the paper, we can conclude that the best option to model the consistency concept in multi-adjoint normal logic programming is the one considered by Madrid and Ojeda-Aciego.

# 5    Conclusions and Future Work

This paper has presented an overview on the most interesting concepts related to the consistency notion. The definitions corresponding to self-contradiction, $x$-consistency, consistency in bilattices and coherence interpretation have been interpreted in the multi-adjoint normal logic programming framework in order to choose an appropriate coherence notion for this framework. In order to make this choice, we have presented a detailed analysis on these notions which has disposed the winner notion based on its affinity with the Atanassov intuisionistic fuzzy sets and the weak involutive negations.

In the future, we will study different operators in order to measure the coherence of multi-adjoint normal logic programs and diverse tools for correcting the possible noise in that programs.

# References

1. Atanassov, K.: Intuitionistic Fuzzy Sets. Springer Physsica-Verlag, Berlin (1999)
2. Belnap, N.D.: A useful four-valued logic. In: Dunn, J.M., Epstein, G. (eds.) Modern Uses of Multiple-Valued Logic, pp. 5–37. Springer Nature, Dordrecht (1977)
3. Bustince, H., Madrid, N., Ojeda-Aciego, M.: A measure of contradiction based on the notion of n-weak-contradiction. In: 2013 IEEE International Conference on Fuzzy Systems (FUZZ-IEEE). Institute of Electrical and Electronics Engineers (IEEE), July 2013
4. Bustince, H., Madrid, N., Ojeda-Aciego, M.: The notion of weak-contradiction: definition and measures. IEEE Trans. Fuzzy Syst. **23**(4), 1057–1069 (2015)
5. Castiñeira, E., Cubillo, S.: Measures of self-contradiction on Atanassov's intuitionistic fuzzy sets: an axiomatic model. Int. J. Intell. Syst. **24**(8), 863–888 (2009)
6. Cornejo, M.E., Lobo, D., Medina, J.: Towards multi-adjoint logic programming with negations. In: Koczy, L., Medina, J. (eds.) 8th European Symposium on Computational Intelligence and Mathematics (ESCIM 2016), pp. 24–29 (2016)
7. Cornejo, M.E., Medina, J., Ramírez-Poussa, E.: Adjoint negations, more than residuated negations. Inf. Sci. **345**, 355–371 (2016)
8. Cubillo, S., Torres, C., Castiñeira, E.: Self-contradiction, contradiction between two Atanassov' s intuitionistic fuzzy sets. Int. J. Uncertain. Fuzziness Knowl.-Based Syst. **16**(03), 283–300 (2008)
9. Dubois, D., Konieczny, S., Prade, H.: Quasi-possibilistic logic and its measures of information and conflict. Fundam. Inf. **57**(2–4), 101–125 (2003)
10. Esteva, F.: Negaciones en retículos completos. Stochastica **I**, 49–66 (1975)
11. Esteva, F., Domingo, X.: Sobre funciones de negación en [0,1]. Stochastica **IV**, 141–166 (1980)
12. Esteva, F., Trillas, E., Domingo, X.: Weak and strong negation functions in fuzzy set theory. In: Proceedings of the XI International Symposium on Multivalued Logic, pp. 23–26 (1981)
13. Fitting, M.: Bilattices and the semantics of logic programming. J. Logic Program. **11**(2), 91–116 (1991)
14. Gelfond, M., Lifschitz, V.: The stable model semantics for logic programming. ICLP/SLP **88**, 1070–1080 (1988)
15. Ginsberg, M.L.: Multi-valued logics. In: AAAI (1986)
16. Ginsberg, M.L.: Multivalued logics: a uniform approach to reasoning in artificial intelligence. Comput. Intell. **4**(3), 265–316 (1988)
17. Grant, J., Hunter, A.: Measuring inconsistency in knowledgebases. J. Intell. Inf. Syst. **27**(2), 159–184 (2006)
18. Jabbour, S., Raddaoui, B.: Measuring Inconsistency through minimal proofs. In: Gaag, L.C. (ed.) ECSQARU 2013. LNCS, vol. 7958, pp. 290–301. Springer, Heidelberg (2013). doi:10.1007/978-3-642-39091-3_25
19. Knight, K.: Measuring inconsistency. J. Philos. Logic **31**(1), 77–98 (2002)
20. Madrid, N., Ojeda-Aciego, M.: On the measure of incoherence in extended residuated logic programs. In: Proceedings of the IEEE International Conference on Fuzzy Systems, FUZZ-IEEE 2009, Jeju, Island, Korea, 20–24 August 2009, pp. 598–603 (2009)

21. Madrid, N., Ojeda-Aciego, M.: Measuring instability in normal residuated logic programs: adding information. In: Proceedings of the IEEE International Conference on Fuzzy Systems, FUZZ-IEEE 2010, Barcelona, Spain, 18–23 July 2010, pp. 1–7 (2010)

22. Madrid, N., Ojeda-Aciego, M.: Measuring instability in normal residuated logic programs: discarding information. In: Hüllermeier, E., Kruse, R., Hoffmann, F. (eds.) IPMU 2010. CCIS, vol. 80, pp. 128–137. Springer, Heidelberg (2010). doi:10.1007/978-3-642-14055-6_14

23. Madrid, N., Ojeda-Aciego, M.: Measuring inconsistency in fuzzy answer set semantics. IEEE Trans. Fuzzy Syst. **19**(4), 605–622 (2011)

24. Madrid, N., Ojeda-Aciego, M.: On the measure of incoherent information in extended multi-adjoint logic programs. In: 2013 IEEE Symposium on Foundations of Computational Intelligence (FOCI). Institute of Electrical and Electronics Engineers (IEEE), April 2013

25. Van Nieuwenborgh, D., De Cock, M., Vermeir, D.: An introduction to fuzzy answer set programming. Ann. Math. Artif. Intell. **50**(3–4), 363–388 (2007)

26. Trillas, E.: Sobre negaciones en la teoría de conjuntos difusos. Stochastica **III**, 47–60 (1979)

27. Trillas, E., Alsina, C., Jacas, J.: On contradiction in fuzzy logic. Soft Comput. - Fusion Found. Methodol. Appl. **3**(4), 197–199 (1999)

28. Trillas, E., Alsina, C., Jacas, J.: On logical connectives for a fuzzy set theory with or without nonempty self-contradictions. Int. J. Intell. Syst. **15**(3), 155–164 (2000)

29. Trillas, E., Renedo, E., Guadarrama, S.: On a new theory of fuzzy sets with just one self-contradiction. In: 10th IEEE International Conference on Fuzzy Systems (Cat. No.01CH37297), vols. 2 and 3, pp. 658–661. Institute of Electrical and Electronics Engineers (IEEE), December 2001

30. Visser, A.: Four valued semantics and the liar. J. Philos. Logic **13**(2), 181–212 (1984)

31. Zadeh, L.: Fuzzy sets. Inf. Control **8**, 338–353 (1965)

# Gaussian Opposite Maps for Reduced-Set Relevance Vector Machines

Lucas Silva de Sousa[✉] and Ajalmar Rêgo da Rocha Neto

Department of Teleinformatics, Federal Institute of Ceará (IFCE),
2081 Treze de Maio Av., Benfica, Fortaleza, Ceará 60040-215, Brazil
lucas.sousa@ppgcc.ifce.edu.br, ajalmar@gmail.com

**Abstract.** The Relevance Vector Machine is a bayesian method. This model represents its decision boundary using a subset of points from the training set, called relevance vectors. The training algorithm of that is time consuming. In this paper we propose a technique for initialize the training process using the points of an *opposite map* in classification problems. This solution approximate the relevance points of the solutions obtained by Support Vector Machines. In order to assess the performance of our proposal, we carried out experiments on well-known datasets against the original RVM and SVM. The GOM-RVM achieved accuracy equivalent or superior than to SVM and RVM with fewer relevance vectors.

**Keywords:** Opposite Maps · Support Vector Machines · Relevance Vector Machine · Support vectors · Relevance vectors

## 1 Introduction

Machine learning researchers seek increasingly efficient and sparse models [9,14,18]. Support Vector Machine(SVM) [6] is a popular sparse model that represents its boundary decision through a subset from the training points, known as support vectors. This support vectors generally are the closest points to the boundary decision. Similarly, the Relevance Vector Machine (RVM) [15] also uses another subset from the training points, known as relevance vectors, to represent its decision function. However such relevance vectors generally are more disperse over the training points than the support vectors, achieving an even more sparse solution.

Two common characteristics between SVM and RVM are: (i) the functional form, although the RVM being a probabilistic model, (ii) and the high training-cost. In order to overcome the high training-cost in SVM, new training approaches were proposed, such as the usage of metaheuristics [4], Divide-and-Conquer solvers [7], and the Opposite Maps [12].

Modifications in the original RVM proposal were performed to modify the training algorithm that started with all training points at once [13]. In [5] the RVM marginal likelihood was analyzed in order to reduce the training cost. Then,

© Springer International Publishing AG 2017
I. Rojas et al. (Eds.): IWANN 2017, Part I, LNCS 10305, pp. 458–468, 2017.
DOI: 10.1007/978-3-319-59153-7_40

in [16], another variation of RVM was proposed with an accelerated training algorithm by starting with an empty set of relevance vectors. Such algorithm was based in a sequential addition and deletion of candidate basis functions. Furthermore, other approaches were applied to improve others aspects of RVM, such as Markov Chain methods [2] and dependent relevance determination [17].

More specifically, the RVM training consists of an iterative process to obtain the posterior distribution parameters. A critical point in this iterative process is the computation of a covariance matrix that implies in an inversion of matrix with order $N$, where $N$ is the number of training points. The Opposite Maps (OM) method is a technique that finds the samples located between classes in a binary classification problem. The OM method was already used to obtain a reduced-set SVM and Least Squares Support Vector Machine (LSSVM) classifiers [12]. In this paper, an OM approach called Gaussian Opposite Maps (GOM) is applied to select the training points to initialize the RVM training procedure, maintaining the efficiency.

The remainder of the paper is organized as follows. In the next section, the Relevance Vector Machine model for classification is described. Then, in Sect. 3, our proposal, the Gaussian Opposite Maps Relevance Vector Machine (GOM-RVM) is presented. Finally the results and discussions are given in Sect. 4.

## 2    Relevance Vector Machines for Classification

The Relevance Vector Machine (RVM) is a bayesian sparse model. The description for such model for classification is based on the work of [1,15].

The simplest representation of a linear discriminant function is obtained by taking a linear function of the input vector so that

$$y(\mathbf{x}, \mathbf{w}) = \mathbf{w}^T \mathbf{x} + w_0, \tag{1}$$

where $\mathbf{w} = (w_1, ..., w_D)$ is called weight vector and $\mathbf{x} = (x_1, ..., x_D)^T$. This is a linear function of the input variables $x_i$ and this imposes significant limitations on the model. Thus, extending the class of models considering linear combinations of fixed nonlinear functions from the input variables, it becomes

$$y(\mathbf{x}, \mathbf{w}) = w_0 + \sum_{j=1}^{M-1} w_i \phi_j(\mathbf{x}), \tag{2}$$

where $\phi_j(\mathbf{x})$ are known as *basis functions*.

The parameter $w_0$ is sometimes called of bias parameter. It is often convenient to define an additional constant basis function value $\phi_0(\mathbf{x}) = 1$ so that

$$y(\mathbf{x}, \mathbf{w}) = \sum_{j=0}^{M-1} w_i \phi_j(\mathbf{x}) = \mathbf{w}^T \boldsymbol{\phi}(\mathbf{x}), \tag{3}$$

where $\mathbf{w} = (w_0, ..., w_{M-1})^T$ and $\boldsymbol{\phi} = (\phi_0, ..., \phi_{M-1})^T$ [1].

Started that, in a two-class classification context, for a given the input $\mathbf{x}$, it is desired to predict the posterior probability of membership of one of the classes. Following the statistical convention and generalize the linear model by applying the logistic sigmoid link function $\sigma(y) = 1/(1 + e^{-y})$ to $y(\mathbf{x}, \mathbf{w})$ and using the Bernoulli distribution for $p(t|\mathbf{w})$. Thus we can write the likelihood as

$$p(\mathbf{t}|\mathbf{w}) = \prod_{n=1}^{N} \sigma\{y(\mathbf{x}_n, \mathbf{w})\}^{t_n} [1 - \sigma\{y(\mathbf{x}_n, \mathbf{w})\}]^{1-t_n}, \tag{4}$$

where, following from the probabilistic specification, the targets $t_n \in \{0, 1\}$ [15].

We can not integrate out the weights analytically and so are denied closed-form expressions for either the weight posterior $p(\mathbf{w}|\mathbf{t}, \boldsymbol{\alpha})$ or the marginal likelihood $p(\mathbf{t}|\boldsymbol{\alpha})$, with $\boldsymbol{\alpha}$ a vector with $N + 1$ *hyperparameters*. For classification problems, we can apply an approximation procedure based on the method of Laplace [11]. This procedure seeks the posterior distribution mode to obtain the weights $\mathbf{w}_{\mathbf{MP}}$ for the current fixed values of $\boldsymbol{\alpha}$. Maximize the log-posterior is equivalent to maximize the posterior, thus the log-posterior is given by

$$\log\{p(\mathbf{t}|\mathbf{w})p(\mathbf{w}|\boldsymbol{\alpha})\} = \sum_{n=1}^{N} [t_n \log(y_n) + (1 - t_n) \log(1 - y_n)] - \frac{1}{2}\mathbf{w}^T \mathbf{A}\mathbf{w}, \tag{5}$$

where $y_n = \sigma\{y(\mathbf{x}_n, \mathbf{w})\}$ and $\mathbf{A} = diag(\alpha_0, \alpha_1, ..., \alpha_N)$. The function (5) is a logistic log-likelihood function and it requires iterative maximization. Second-order Newton methods may be applied effectively, since the Hessian of (5) is explicitly computed.

The method of Laplace is a simple quadratic approximation to the log-posterior around its mode. In this case the function (5) is differentiated twice to give:

$$\nabla_{\mathbf{w}} \nabla_{\mathbf{w}} \log\{p(\mathbf{w}|\mathbf{t}, \boldsymbol{\alpha})\} |_{\mathbf{w}_{\mathbf{MP}}} = -(\boldsymbol{\Phi}^T \mathbf{B} \boldsymbol{\Phi} + \mathbf{A}), \tag{6}$$

where $\mathbf{B} = diag(\beta_1, \beta_2, ..., \beta_N)$ is a diagonal matrix with $\beta_n = \sigma\{y(\mathbf{x}_n)\}[1 - \sigma\{y(\mathbf{x}_n)\}]$ and $\boldsymbol{\Phi}$ is an $N \times M$ matrix, called *design matrix*, whose elements are given by $\boldsymbol{\Phi}_{nj} = \phi(\mathbf{x}_n)$ so that

$$\boldsymbol{\Phi} = \begin{pmatrix} \phi_0(\mathbf{x}_1) & \phi_1(\mathbf{x}_1) & \cdots & \phi_{M-1}(\mathbf{x}_1) \\ \phi_0(\mathbf{x}_2) & \phi_1(\mathbf{x}_2) & \cdots & \phi_{M-1}(\mathbf{x}_2) \\ \vdots & \vdots & \ddots & \vdots \\ \phi_0(\mathbf{x}_N) & \phi_1(\mathbf{x}_N) & \cdots & \phi_{M-1}(\mathbf{x}_N) \end{pmatrix}. \tag{7}$$

In order to obtain the covariance matrix $\boldsymbol{\Sigma}$ for a Gaussian approximation to the posterior over weights centred at $\mathbf{w}_{\mathbf{MP}}$, it is required to negate and invert the Eq. (6). At the mode of $p(\mathbf{w}|\mathbf{t}, \boldsymbol{\alpha})$, using Eq. (6) and knowing the fact that $\nabla_{\mathbf{w}} \log\{p(\mathbf{w}|\mathbf{t}, \boldsymbol{\alpha})\} |_{\mathbf{w}_{\mathbf{MP}}} = 0$, the parameters of posterior are denoted by

$$\boldsymbol{\Sigma} = (\boldsymbol{\Phi}^T \mathbf{B} \boldsymbol{\Phi} + \mathbf{A})^{-1}, \tag{8}$$

$$\mathbf{w}_{\mathbf{MP}} = \boldsymbol{\Sigma} \boldsymbol{\Phi}^T \mathbf{B} \mathbf{t}. \tag{9}$$

Using the statistics $\Sigma$ and $\mathbf{w}_{\mathbf{MP}}$ of the Gaussian approximation, the hyperparameters $\boldsymbol{\alpha}$ are optimized using an iterative re-estimation procedure [10]. The rule for updates the values of $\boldsymbol{\alpha}$ is given by

$$\alpha_i^{new} = \frac{\gamma_i}{\mu_i^2} \tag{10}$$

where $\mu_i$ is the $i$-th posterior mean weight from Eq. (9) and the quantities $\gamma_i$ are given by

$$\gamma_i \equiv 1 - \alpha_i \Sigma_{ii}, \tag{11}$$

with $\Sigma_{ii}$ the $i$-th diagonal element of the posterior covariance from Eq. (8), computed with the current $\boldsymbol{\alpha}$ values.

When the hyperparameters are updated, a significant proportion of them go to infinity and the corresponding weight parameters have posterior distributions that are concentrated at zero. The basis functions associated with such parameters therefore play no role in the predictions made by the model and so are effectively pruned out, resulting in a sparse model [1].

At the convergence of the hyperparameter estimation procedure, we make predictions based on the posterior distribution over the weights, conditioned on the maximizing values $\boldsymbol{\alpha}_{\mathbf{MP}}$ and $\Sigma$. The predictive distribution for a new data $\mathbf{x}_*$ is given by

$$p(t_*|\mathbf{t}, \boldsymbol{\alpha}_{\mathbf{MP}}, \Sigma) = \int p(t_*|\mathbf{w}, \Sigma)p(\mathbf{w}|\mathbf{t}, \boldsymbol{\alpha}_{\mathbf{MP}}, \Sigma)d\mathbf{w}, \tag{12}$$

Since both terms in the integrand are Gaussian, this is readily computed, giving:

$$p(t_*|\mathbf{t}, \boldsymbol{\alpha}_{\mathbf{MP}}, \Sigma) = \mathcal{N}(t_*|y_*, \Sigma_*), \tag{13}$$

with

$$y_* = \mathbf{w}_{\mathbf{MP}}^T \phi(\mathbf{x}_*), \tag{14}$$

$$\Sigma_* = \Sigma_* + \phi(\mathbf{x}_*)^T \Sigma \phi(\mathbf{x}_*). \tag{15}$$

More details about the formulation of RVMs in [1, 15].

## 3 Gaussian Opposite Maps Relevance Vector Machine

In this section, we describe our proposal that implies in the reduction of the training points for RVM training algorithm initialization. In the Subsect. 3.1 the method Gaussian Opposite Maps is presented. This method is used to select the training points. In the Subsect. 3.2 we show how the RVM training algorithm was adopted.

### 3.1 Gaussian Opposite Maps

The Opposite Maps (OM) method is a Self-Organizing Map-based technique used for obtaining reduced-set SVM and LSSVM classifiers [12]. In this paper, the OM method was modified to work with gaussian mixture models instead of Self-Organizing Maps in the issue of select points.

The OM finds points located at the region between two class in a classification problem. Thus this method provides the classifiers such as SVM the support vectors that generally closed to the decision boundary [6].

The procedure used in this paper to find that points is described below:

**INIT.** Are given the values of $\lambda$, that defines the number of points of each class for reduced set, and $\delta$ that is the number of gaussians for the mixture model.

**STEP 1.** Split the available data set $\mathcal{D} = \{(\mathbf{x}_i, t_i)\}_{i=1}^n$ into two subsets:

$$\mathcal{D}^{(1)} = \{(\mathbf{x}_i, t_i)|t_i = 1\}, i = 1, \cdots, n_1 \tag{16}$$
$$\mathcal{D}^{(2)} = \{(\mathbf{x}_i, t_i)|t_i = 0\}, i = 1, \cdots, n_2 \tag{17}$$

where $n_1$ and $n_2$ are the cardinalities of the subsets $\mathcal{D}^{(1)}$ and $\mathcal{D}^{(2)}$, respectively.

**STEP 2.** Execute two Gaussian mixture procedure with $\delta$ Gaussians in order to obtain the probability distribution of subsets $\mathcal{D}^{(1)}$ and $\mathcal{D}^{(2)}$, denoted by $P(\mathcal{D}^{(1)})$ and $P(\mathcal{D}^{(2)})$, respectively.

**STEP 3.** At this step the distributions $P(\mathcal{D}^{(1)})$ and $P(\mathcal{D}^{(2)})$ are used to find the set of *opposite map* prototypes.

**STEP 3.1.** For each $\mathbf{x}_i \in \mathcal{D}^{(1)}$, evaluate the $P(\mathcal{D}^{(2)})$ to create the vector $c^{(2)}$:

$$c_i^{(2)} = P(\mathcal{D}^{(2)}, \mathbf{x}_i), i = 1, \cdots, n_1 \tag{18}$$

where $P(\mathcal{D}^{(2)}, \mathbf{x}_i)$ is the value of probability $P(\mathcal{D}^{(2)})$ at the point $\mathbf{x}_i$.

**STEP 3.2.** For each $\mathbf{x}_i \in \mathcal{D}^{(2)}$, evaluate the $P(\mathcal{D}^{(1)})$ to create the vector $c^{(1)}$:

$$c_i^{(1)} = P(\mathcal{D}^{(1)}, \mathbf{x}_i), i = 1, \cdots, n_2 \tag{19}$$

where $P(\mathcal{D}^{(1)}, \mathbf{x}_i)$ is the value of probability $P(\mathcal{D}^{(1)})$ at the point $\mathbf{x}_i$.

**STEP 3.3.** Let $\mathcal{C}^{(1)} = [c_1^{(1)}, c_2^{(1)}, \cdots, c_{n_1}^{(1)}]$ a decrescent sorted vector with values of $c^{(1)}$ and $\mathcal{B}^{(1)} = [d_1, d_2, \cdots, d_{n_2}]$, so that for all $d_i = (\mathbf{x}_i, t_i) \in \mathcal{D}^{(2)}$, $P(\mathcal{D}^{(1)}, \mathbf{x}_i) = c_i^{(1)}$.

**STEP 3.4.** Let $\mathcal{C}^{(2)} = [c_1^{(2)}, c_2^{(2)}, \cdots, c_{n_1}^{(2)}]$ a decrescent sorted vector with values of $c^{(2)}$ and $\mathcal{B}^{(2)} = [d_1, d_2, \cdots, d_{n_2}]$, so that for all $d_i = (\mathbf{x}_i, t_i) \in \mathcal{D}^{(1)}$, $P(\mathcal{D}^{(2)}, \mathbf{x}_i) = c_i^{(2)}$.

**STEP 4.** At this step the reduced set is formed.

**STEP 4.1.** Let $\mathcal{X}^{(1)}$ be the subset of $\lceil \lambda n_1 \rceil$ first elements in $\mathcal{B}^{(1)}$.

**STEP 4.2.** Let $\mathcal{X}^{(2)}$ be the subset of $\lceil \lambda n_2 \rceil$ first elements in $\mathcal{B}^{(2)}$.

**STEP 4.3.** The reduced set is given by $\mathcal{X}^{(rs)} = \mathcal{X}^{(1)} \cup \mathcal{X}^{(2)}$.

The points yield by OM in artificial datasets are shown in the Fig. 1. Notice that the value of parameter $\lambda$ the quantity of samples selected.

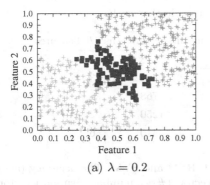

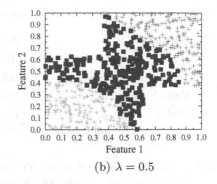

(a) $\lambda = 0.2$                    (b) $\lambda = 0.5$

**Fig. 1.** The points selected by OM for the artificial dataset with overlapping. The gaussian mixture was performed with 5 Gaussians distributions.

## 3.2 Gaussian Opposite Maps for RVM Training

The training of the RVM consists in an iterative process to obtain the posterior distribution parameters given by Eqs. (8) and (9). Notice that the $\Sigma$ value is obtained through of an matrix inversion of order $N$ because in the original RVM training algorithm [15] all elements from the training set are used in the initialization step. However, in each iteration a portion of $\alpha$ values going to infinity meaning that the training samples represented by such values are pruned out. But the training using all samples of training set still is time consuming.

The Gaussian Opposite Maps Relevance Vector Machine (GOM-RVM), replace the complete training set $\mathcal{D}$ by $\mathcal{X}^{(rs)}$, which is obtained after applying the Gaussian Opposite Maps procedure described in the Sect. 3.1. That method is a tentative of obligate that the RVM seek the relevance vectors more closest of the surface decision like in the SVM.

## 4   Simulations and Discussion

We carried out some simulations on five datasets and present some results for GOM-RVM, RVM and SVM in this section. We used some real UCI binary datasets [8] and an artificial one, called Ripley. All datasets are detailed in Table 1 with its full name, abbreviation, total of patterns (#Patterns), and number of features (#Features).

In our simulations 80% of the data examples were randomly selected for training purposes and so the remaining 20% of the examples were used for assessing the classifiers' generalization performance. We carried out 30 executions on each dataset. All experiments were performed with Gaussian kernel for all models and the parameters C of SVM, kernel width of RVM and GOM-RVM, and the $\lambda$ of the GOM-RVM were tuned by applying grid search with 10-fold cross-validation over the training dataset. We have done this in order to follow the common strategy use in RVM, GOM-RVM and SVM for parameter tuning.

**Table 1.** List of datasets used in this work.

| Dataset | Abbreviation | # Patterns | # Features |
|---|---|---|---|
| Breast Cancer Winconsin | BCW | 683 | 9 |
| Haberman's Survival | HAB | 306 | 3 |
| Pima Indians Diabets | PID | 768 | 8 |
| Ripley | RIP | 1250 | 2 |
| Vertebral Column Pathologies | VCP | 310 | 6 |

**Table 2.** Performance metrics for GOM-RVM, RVM and SVM. The accuracy (avg), standard deviation (std), number of relevance vectors (#rv), number of support vectors (#sv) and Friedman test (st), so that (✓) mean equivalent and (✗) mean not equivalent results. The statistical test for RVM and SVM are compared with GOM-RVM performance. The parameter of GOM-RVM $\delta = 5$ in this simulation.

| Dataset | $\lambda$ | GOM-RVM avg | std | #rv | RVM avg | std | #rv | st | SVM avg | std | #sv | st |
|---|---|---|---|---|---|---|---|---|---|---|---|---|
| BCW | 0.4 | 0.95 | 0.01 | 20.9 | 0.96 | 0.00 | 19.7 | ✗ | 0.96 | 0.01 | 53.8 | ✗ |
|  | 0.3 | 0.95 | 0.02 | 19.3 |  |  |  | ✓ |  |  |  | ✗ |
|  | 0.1 | 0.93 | 0.02 | 14.5 |  |  |  | ✓ |  |  |  | ✗ |
| HAB | 0.4 | 0.71 | 0.01 | 1.0 | 0.71 | 0.01 | 3.0 | ✓ | 0.69 | 0.01 | 211.3 | ✗ |
|  | 0.3 | 0.71 | 0.01 | 1.0 |  |  |  | ✓ |  |  |  | ✗ |
|  | 0.1 | 0.71 | 0.01 | 1.0 |  |  |  | ✓ |  |  |  | ✗ |
| PID | 0.4 | 0.71 | 0.02 | 27.0 | 0.74 | 0.03 | 36.3 | ✗ | 0.73 | 0.02 | 363.1 | ✓ |
|  | 0.3 | 0.73 | 0.02 | 19.8 |  |  |  | ✓ |  |  |  | ✗ |
|  | 0.1 | 0.63 | 0.01 | 16.6 |  |  |  | ✓ |  |  |  | ✗ |
| RIP | 0.4 | 0.93 | 0.01 | 11.9 | 0.93 | 0.01 | 18.0 | ✓ | 0.92 | 0.00 | 228.2 | ✗ |
|  | 0.3 | 0.92 | 0.01 | 6.2 |  |  |  | ✗ |  |  |  | ✗ |
|  | 0.1 | 0.51 | 0.07 | 1.0 |  |  |  | ✗ |  |  |  | ✗ |
| VCP | 0.4 | 0.84 | 0.02 | 10.4 | 0.86 | 0.01 | 8.8 | ✗ | 0.80 | 0.03 | 95.0 | ✗ |
|  | 0.3 | 0.84 | 0.02 | 8.5 |  |  |  | ✗ |  |  |  | ✗ |
|  | 0.1 | 0.84 | 0.01 | 2.9 |  |  |  | ✗ |  |  |  | ✗ |

In the Table 2, we report performance metrics in methodology aforementioned $80 - 20\%$ on testing set averaged over 30 independent runs. We also show results of applying the Friedman test [3]. As one we see, the number of relevance vectors in GOM-RVM and RVM are lower than the number of support vectors in SVM. However, the RVM achieved accuracy greater or equivalent than SVM. By the statistical test, only with VCP dataset our proposal did not succeed equivalent values to the RVM, but with $\lambda = 0.1$ and around two relevance vectors, our proposal achieved an accuracy greater than SVM. Another important aspect of the GOM-RVM is the growing of the accuracy alongside $\lambda$ in almost all datasets. Such behavior shows that only the closest points to the decision boundary have

**Table 3.** Time results for GOM-RVM, RVM and SVM. These results were obtained from the mean of 30 independent runs. The results presented are the mean time of cross-validation (cv), mean time for the model fitting (tr) and the mean time of the test procedure (te). For the GOM-RVM all results used $\delta$ equal to 5.

| | GOM-RVM | | | RVM | | | SVM | | |
|---|---|---|---|---|---|---|---|---|---|
| Dataset | $\lambda$ | cv | tr | te | cv | tr | te | cv | tr | te |
| BCW | 0.4 | 195.5346 | 0.6285 | 0.0006 | 40.9026 | 0.0189 | 0.0003 | 91.4337 | 0.0031 | 0.0008 |
| | 0.3 | 96.6354 | 0.0654 | 0.0004 | | | | | | |
| | 0.1 | 151.2689 | 0.2628 | 0.0005 | | | | | | |
| HAB | 0.4 | 80.7062 | 0.0876 | 0.0003 | 20.6726 | 0.0046 | 0.0002 | 63.3218 | 0.0017 | 0.0010 |
| | 0.3 | 49.0774 | 0.0114 | 0.0003 | | | | | | |
| | 0.1 | 74.6451 | 0.0710 | 0.0003 | | | | | | |
| PID | 0.4 | 188.9828 | 0.4533 | 0.0007 | 63.4298 | 0.0401 | 0.0004 | 201.3944 | 0.0770 | 0.0015 |
| | 0.3 | 89.6275 | 0.1227 | 0.0004 | | | | | | |
| | 0.1 | 129.5278 | 0.2336 | 0.0005 | | | | | | |
| RIP | 0.4 | 182.2381 | 0.1972 | 0.0006 | 101.6293 | 0.0545 | 0.0003 | 259.7003 | 0.0344 | 0.0017 |
| | 0.3 | 111.1888 | 0.0373 | 0.0003 | | | | | | |
| | 0.1 | 128.6005 | 0.1376 | 0.0004 | | | | | | |
| VCP | 0.4 | 79.0554 | 0.1385 | 0.0005 | 20.8471 | 0.0126 | 0.0002 | 62.5749 | 0.0024 | 0.0007 |
| | 0.3 | 46.8261 | 0.0646 | 0.0003 | | | | | | |
| | 0.1 | 65.0809 | 0.1966 | 0.0003 | | | | | | |

not all necessary information to the model fitting. We also notice that the number of relevance vectors of the GOM-RVM is generally less than RVM.

In the Table 3, we notice that the time of the RVM is substantially less than the SVM and the GOM-RVM, but the GOM-RVM achieved results similar to the SVM. We see that the time of the GOM-RVM with $\lambda = 0.1$ is greater than the results with $\lambda = 0.3$, an important observation is that for small values of $\lambda$, the convergence of the RVM training algorithm becomes more difficult.

In the Table 4, we analyze the behaviour of the parameter $\lambda$ in GOM-RVM. We see that the Gaussian Opposite Maps achieved a reduction in the training dataset amoung 50% for a $\lambda = 0.4$. The value of $\lambda \times 100$ is equivalent to the percentual of reduction when we have a balanced dataset.

Figure 2 shows the boundary decision of the GOM-RVM, RVM and SVM. We see that the relevance vectors found by the GOM-RVM is located at more closer of the boundary decision than RVM. The number of support vectors to represent a linear function is substantially greater than the ones in other models. We see also that the decision boundary of the GOM-RVM is similar to the RVM decision boundary, although the models use different relevance vectors.

The behaviour of the models is sustained in the non-linear case presented in the Fig. 3. The GOM-RVM found the same number of relevance vectors that the RVM mantaining the shape of the decision boundary. The SVM generate a decision boundary with shape smoothed for the non-linear case, but used more support vectors.

**Table 4.** Reduction rates of the GOM method for the datasets used in this work.

| Dataset | # Patterns | $\lambda$ | # OM | # Red |
|---------|------------|-----------|------|-------|
| BCW | 683 | 0.1 | 115.1 | 83.2% |
|  |  | 0.3 | 229.6 | 66.4% |
|  |  | 0.4 | 272.5 | 60.1% |
| HAB | 306 | 0.1 | 22.1 | 92.8% |
|  |  | 0.3 | 70.0 | 77.1% |
|  |  | 0.4 | 93.0 | 69.6% |
| PID | 768 | 0.1 | 105.6 | 86.2% |
|  |  | 0.3 | 281.0 | 63.4% |
|  |  | 0.4 | 356.7 | 53.6% |
| RIP | 1254 | 0.1 | 252.4 | 79.9% |
|  |  | 0.3 | 611.1 | 51.3% |
|  |  | 0.4 | 740.2 | 41.0% |
| VCP | 310 | 0.1 | 22.0 | 92.9% |
|  |  | 0.3 | 63.1 | 79.7% |
|  |  | 0.4 | 83.8 | 73.0% |

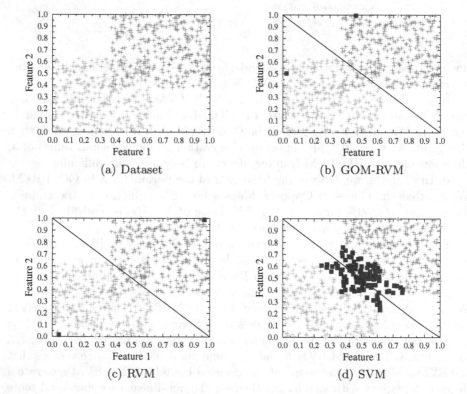

(a) Dataset

(b) GOM-RVM

(c) RVM

(d) SVM

**Fig. 2.** Decision boundary of the models GOM-RVM, RVM and SVM. The Fig. 2(a) is the dataset used for training the classfiers in this simulation.

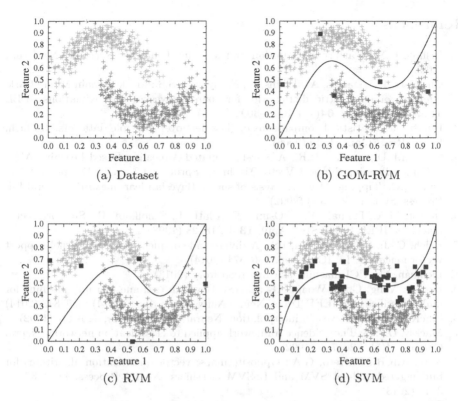

**Fig. 3.** Decision boundary of the models GOM-RVM, RVM and SVM for an artificial binary dataset with overlapping.

## 5  Conclusion

In this work, we proposed a Relevance Vector Machine variation using the Gaussian Opposite Map, called GOM-RVM. The key step in our proposal is to replace the complete training set to the subset yielded by Gaussian Opposite Maps. By employing GOM, we intend to obligate RVM to seek the relevance vectors that lie close to the boundary decision just as the support vectors in SVM.

As one can see in the results, the GOM-RVM achieve accuracy equivalent and in some cases superiors than to SVM and RVM with a less number of relevance vectors. We conclude that all patterns of the training are important to the obtain an efficient model fitting. This is justified by the growing of the accuracy together with the $\lambda$ value. Small values of $\lambda$ are a problem for the convergence of the RVM training algorithm, thus, this affect the time training and cause a falling of the accuracy. Furthermore The points found by the Gaussian Opposite Maps was exactly the points in the overlay region like the original proposed Opposite Maps.

Nowadays, we are working to extend the Gaussian Opposite Maps to work in the kernel space, thus we wish find the opposite map points in a linear problem.

# References

1. Bishop, C.M.: Pattern Recognition and Machine Learning. Springer, New York (2006)
2. Chaspari, T., Tsiartas, A., Tsilifis, P., Narayanan, S.S.: Markov chain Monte Carlo inference of parametric dictionaries for sparse Bayesian approximations. IEEE Trans. Sig. Process. **64**(12), 3077–3092 (2016)
3. Demšar, J.: Statistical comparisons of classifiers over multiple data sets. J. Mach. Learn. Res. **7**(Jan), 1–30 (2006)
4. Dias, M.L.D., Neto, A.R.R.: A Novel Simulated Annealing-Based Learning Algorithm for Training Support Vector Machines. Springer, Cham (2017). pp. 341–351
5. Faul, A.C., Tipping, M.E.: Analysis of sparse Bayesian learning. Adv. Neural Inf. Process. Syst. **1**, 383–390 (2002)
6. Hearst, M.A., Dumais, S.T., Osuna, E., Platt, J., Scholkopf, B.: Support vector machines. IEEE Intell. Syst. Appl. **13**(4), 18–28 (1998)
7. Hsieh, C.-J., Si, S., Dhillon, I.S.: A divide-and-conquer solver for kernel support vector machines. In: ICML, pp. 566–574 (2014)
8. Lichman, M.: UCI machine learning repository (2013)
9. Luo, J., Vong, C.M., Wong, P.K.: Sparse Bayesian extreme learning machine for multi-classification. IEEE Trans. Neural Netw. Learn. Syst. **25**(4), 836–843 (2014)
10. MacKay, D.J.C.: Bayesian interpolation. Neural Comput. **4**(3), 415–447 (1992)
11. Mackay, D.J.C.: The evidence framework applied to classification networks. Neural Comput. **4**(5), 720–736 (1992)
12. Neto, A.R.R., Barreto, G.A.: Opposite maps: vector quantization algorithms for building reduced-set SVM and LSSVM classifiers. Neural Process. Lett. **37**(1), 3–19 (2013)
13. Rasmussen, C.E.: Gaussian Processes for Machine Learning. MIT Press, Cambridge (2006)
14. Sun, B., Ng, W.W.Y., Chan, P.P.K.: Improved sparse LSSVMS based on the localized generalization error model. Int. J. Mach. Learn. Cybern. **8**, 1–9 (2016)
15. Tipping, M.E.: Sparse Bayesian learning and the relevance vector machine. J. Mach. Learn. Res. **1**(3), 211–244 (2001)
16. Tipping, M.E., Anita Faul, J.J., Thomson Avenue, J.J., Avenue, T.: Fast marginal likelihood maximisation for sparse Bayesian models. In: Proceedings of the Ninth International Workshop on Artificial Intelligence and Statistics, pp. 3–6 (2003)
17. Anqi, W., Park, M., Koyejo, O.O., Pillow, J.W.: Sparse Bayesian structure learning with "dependent relevance determination" priors. In: Ghahramani, Z., Welling, M., Cortes, C., Lawrence, N.D., Weinberger, K.Q. (eds.) Advances in Neural Information Processing Systems 27, pp. 1628–1636. Curran Associates Inc., Red Hook (2014)
18. Zhou, S.: Sparse lssvm in primal using cholesky factorization for large-scale problems. IEEE Trans. Neural Netw. Learn. Syst. **27**(4), 783–795 (2016)

# Self-organizing Networks

# Massive Parallel Self-organizing Map and 2-Opt on GPU to Large Scale TSP

Wen-bao Qiao$^{(\boxtimes)}$ and Jean-charles Créput

Le2i FRE2005, CNRS, Arts et Métiers, Univ. Bourgogne Franche-Comté, Besançon,
France
{wenbao.qiao,jean-charles.creput}@utbm.fr,
http://www.multiagent.fr/People:Qiao_wenbao,
http://www.multiagent.fr/People:Creput_jean-charles

**Abstract.** This paper proposes a platform both for parallelism of self-organizing map (SOM) and the 2-opt algorithm to large scale 2-Dimensional Euclidean traveling salesman problems. This platform makes these two algorithms working in a massively parallel way on graphical processing unit (GPU). Advantages of this platform include its flexibly topology preserving network, its fine parallel granularity and it allows maximum ($N/3$) 2-opt optimization moves to be executed with $O(N)$ complexity within one tour orientation and does not cut the integral tour. The parallel technique follows data decomposition and decentralized control. We test this optimization method on large TSPLIB instances, experiments show that the acceleration factor we obtained makes the proposed method competitive, and allows for further increasing for very large TSP instances along with the quantity increase of physical cores in GPU systems.

**Keywords:** Irregular topology · Massive parallel 2-opt · SOM · Doubly linked network · Local spiral search · TSP

## 1 Introduction

The Traveling Salesman Problem (TSP) is well-known NP-hard, which indicates a permutation tour that allows a salesman to travel each city once and return to his starting city. Kohonen Self-organizing Map [1–5] and 2-opt local search have been proved before that they can work together to get an intermediate Euclidean TSP solution considering trade-off between quality and execution time [6]. Also, for further acceleration, parallel strategies both for SOM [7–13] and 2-opt [14–17] have been separately studied for different applications base on various multi-threads devices (SP2, GPU) in past two decades. However, it still exists some problems in these previous works in terms of types of topology, level of parallelism and memory occupancy.

Considering the general SOM working on predefined topological map, its computation time can be divided into following three parts: a time required to determine the closest winner node to the present input; a time required to

© Springer International Publishing AG 2017
I. Rojas et al. (Eds.): IWANN 2017, Part I, LNCS 10305, pp. 471–482, 2017.
DOI: 10.1007/978-3-319-59153-7_41

determine winner node's neighborhood and a time required for updating weights (map updating) [8]. Here, the first operation can be accelerated by using massively parallel local spiral search to simultaneously find the closest winner node for each input data [13,18]. The second and third operations are influenced by network architecture of the topological maps that provide access to neighboring nodes during training process of SOM. Topological maps are usually constructed with unidirectional link network or buffer grid like those SOM applications in [6,11,13,18]. However, training processes on the unidirectional network can only go in one direction. Though algorithms using 2D grid can access arbitrary neighboring node within needed radius centering one winner node, the topological map is limited to regular topology structures, like these work in [6,21]. Besides, method to access neighboring nodes on different topological map plays an important role for applications that require preservation of initial topological relationship between nodes during training process of SOM.

Previous parallel 2-opt implementations can not be defined as massively parallel method for the reason of tour ordering requirement. Like tailored data parallelism that assigns one thread to treat one partial tour' optimization [14], function parallelism that one edge' optimization is executed in parallel [16,19], or geometrically parallelism that assigns one thread to search optimization in divided sub-areas [20].

To solve these problems, in this paper, we propose a platform to implement training process of SOM on irregular topological maps while preserving topological relationship between nodes and a modified 2-opt method to make 2-opt happen in a massive parallel way. We test this platform on GPU taking advantage of its multi-threads' operation on global memory.

The following paper is organized as this: Sect. 2 presents related work about parallel SOM and 2-opt; Sect. 3 presents the proposed platform including parallel SOM working on topological maps and modified 2-opt framework with one concrete implementation. And the last section includes the experiments and discussion.

## 2    Related Work

For parallel implementations of SOM, two common parallel techniques of network partitioning and data partitioning have been used both on CPU- and GPU-based system, which has been discussed and compared in various literatures [7–13]. The main difference between these two techniques is whether the step of node update (map updating) happens independently in different part of the network or simultaneously for all the input data after the training step. Network partitioning technique breaks up the map into parts and each part has a thread to update neurons separately. While using data decomposition technique, map updating does not occur until all the input data has found its winner node and trained its neighbor nodes [12]. For TSP applications, map updating step indicates the relocation of each neuron on Euclidean space.

For parallel SOM specified to TSP applications, Wang et al. [21] proposes parallel SOM implementations based on GPU cellular matrix model proposed by Zhang et al. [18], in which he use buffer grid as neural network structure and assigns one thread to treat all input data in one cell sequentially. His training step of SOM works on the integral buffer grid on global memory [21]. Different from his work, the network in this paper is memorized by doubly linked network both for SOM and 2-opt instead of intermediately by cellular matrix, the cellular matrix model is only used for parallel local spiral search operator, and we assign one thread for one city instead of for one cell.

For parallel 2-opt algorithms, Johnson et al. [15,19] discussed parallel schemes like "geometric partitioning and tour-based partitioning". One geometric partitioning scheme proposed by Karp [20] is based on a recursive subdivision of the overall region containing the cities into rectangles [19]. Verhoeven et al. [14] distinguished parallel 2-opt algorithms between data and function parallelism [14] in which he proposed a tour repartitioning scheme that guarantees their algorithm will not halt until it has found a minimum for the complete problem [14]. Van Luong et al. [17] and Rocki and Sudha [16] adopt parallel strategies similar to "function parallelism" which means one sequential 2-opt is executed in parallel, as Rocki and Sudha [16] distributes the calculation for one edge's exhaustive 2-opt optimization between threads, but only the first edge's optimization has finished, the second edge begins its parallel 2-opt optimization.

## 3   Proposed Methods

Outline of the proposed parallel platform for SOM and 2-opt to large scale TSP applications is shown in Algorithm 1. It mainly includes three parts, first one for initialization step, second one for parallel SOM and last one for parallel 2-opt. Both the two algorithms work base on topological maps constructed by using doubly linked network shown in Figs. 2(b) and 3. "Doubly linked" means that if node $A$ connects (buffers) node $B$, node $B$ should necessarily connect (buffer) node $A$, and every node only buffers its directly connected nodes.

---

**Algorithm 1.** Outline of the proposed platform.

---

1: Initialize topological maps, prepare cellualr matrix for local spiral search and data transmission;
2: **for** iterations **do**
3:    refresh cellular matrix for local spiral search;
4:    kernels <<< ... >>> Parallel SOM processes with one thread for one city;
5: **end for**
6: kernels <<< ... >>> Project result of SOM to be an initial TSP solution;
7: **while** TSP tour can be improved **do**
8:    Refresh TSP tour ordering from random starting point;
9:    kernels <<< ... >>> Simultaneously check each edge's 2-opt optimization in one same tour orientation according to a certain neighborhood edge searching rule;
10:   Serially execute massive non-interacted 2-exchanges;
11: **end while**

---

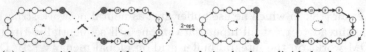

(a) 2-opt without considering tour ordering leads to divided sub-tours.

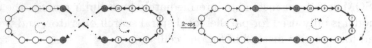

(b) 2-opt considering the tour ordering changes original tour order.

**Fig. 1.** Tour ordering plays an important role for 2-opt optimization.

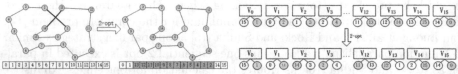

(a) Tour order represented by using buffer memory order.

(b) Tour order represented by using doubly linked network.

**Fig. 2.** Comparison of necessary operation after one same 2-exchange using different TSP tour order representations. (a): The algorithm needs extra temporary memory to invert tour ordering for each 2-exchange. (b): The algorithm just needs to change links of the related four cities and can go easily in two opposite directions from current edge to get possible local optimization.

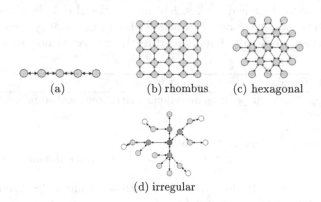

(a)          (b) rhombus     (c) hexagonal

(d) irregular

**Fig. 3.** Topological maps represented by doubly linked network. (a) doubly linked list. (b, c, d) Topologies of rhombus, hexagonal and irregular that respect needed topological properties. They share the same proposed method to access neighboring nodes for one training step of SOM that preserves topological relationship between nodes.

### 3.1   Initialization

The initialization step mainly includes the preparation of SOM maps with initialized neurons, cellular matrix for local spiral search and necessary data transmission from CPU to GPU side.

The initial SOM maps can be predefined with properties that these topologically close nodes correspond to close pixels in images as shown in Fig. 3 or without properties, for example, just add two neighboring nodes to each neuron to be an input TSP solution, as shown in Fig. 2(b). For a TSP instance with $N$ cities ("trainer"), the artificial doubly linked neural network ("learner") is initialized with $2 \times N$ neurons [5]. Coordinate of these initial neurons are initialized according to input cities by setting a small random difference.

One basic building block in this paper is local spiral search operator [22] on 2-dimension Euclidean space. This operator works on cellular matrix [18,23] to find the closest neighboring nodes for the proposed methods. The Euclidean area that contains all the input $N$ points is partitioned into cellular matrix where each cell has its coordinate and buffers nodes lying on corresponding partitioned area. For spiral search operator used by SOM, the algorithm should prepare two cellular matrix for input data ("trainer") and neurons ("learner") separately.

After those steps, the input data, topological maps and the two cellular matrix are copied to GPU side for parallel computing.

### 3.2   Parallel Self-organizing Map

Our parallel implementation of SOM contains following three steps in one epoch (iteration, epo), as shown in Fig. 4.

1. Search winner node : spiral search each input point's closest neuron;
2. Train neighborhood : iteratively access the directly connected neighbor nodes according to topological distance and apply Kohonen's learning low;
3. Refresh cellular matrix : refresh position of each neuron on cellular matrix.

The *search winner node* step mainly indicates massively parallel local spiral search operation working on cellular matrix with one thread for one city. For each input city, the algorithm gets cell coordinate of the cell where this city lies on the "trainer's" cellular matrix, and tries to find this city's closest neuron on "learner's" cellular matrix beginning with the same cell coordinate. If this cell on "learner's cellular does not have neurons, the algorithm searches its neighborhood cells one by one in a spiral manner centering the beginning cell. Every searching operation stops at radius $(lssr + 1)$ or maximum searching range in cellular matrix $(LSSR)$, $lssr$ is the searching radius where the operator encounters the first closest neuron. It has been proved that a single spiral search operation on a bounded data distribution or a uniformed data distribution only takes average $O(1)$ complexity for finding the closest point [22].

The *train neighborhood* step in one epoch follows Kohonen leaning low that is present in Eq. 1. Considering current *epo'* iteration, a winner node $p^*$ has been

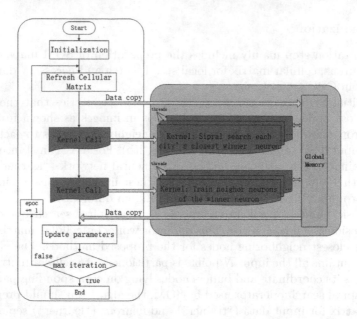

**Fig. 4.** Kernel functions of the parallel implementation of SOM, we assign one thread for one node.

found at previous step for one input city $p$, the Kononen learning low is applied to $p*$ and to neurons within a finite neighborhood of $p^*$ of topological radius $r$. Topological distance from the winner node $d_G$ and learning rate $\alpha(epo)$ affect the learning force of each neuron, $p$ indicates the Euclidean position of input city, $w_k(epo)$ represents Euclidean position of neurons and $k$ indicates different neighborhood neurons. The proposed method applies this learning low on doubly linked network instead of 2D grid, as shown in Figs. 3 and 5(b). After training step of one iteration, the algorithm updates SOM parameters for next iteration by decreasing learning rate $\alpha(epo)$ and training radius $r$.

$$w_k(epo + 1) = w_k(epo) + \alpha(epo) \times exp(-d_G(p^*, k)^2/r^2) \times (p - w_k(epo)) \quad (1)$$

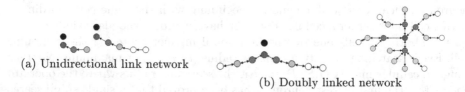

(a) Unidirectional link network

(b) Doubly linked network

**Fig. 5.** SOM' training procedure on two different network architectures. A black node is one input point (a pixel or one city), red node is its winner node. (Color figure online)

We should emphasize that the *train neighborhood* operation accesses all needed neighboring neurons in an iterative manner of one "circle" after another according to their topological distance from winner node. As shown in Fig. 5(b), only all green neurons have been trained, the training procedure begins to train these blue neurons. This iterative operation ensures that neurons with larger topological distance $d_G$ would not move closer to the winner node than neurons with less $d_G$ during one training process, which preserves the predefined topological relationship between nodes.

**Apply SOM to TSP.** When applying SOM to specifically solve TSP applications, every initial city should have a chance to be a "trainer" to train the same network. And it is easy to satisfy this requirement by assigning one thread for one city. Once SOM has stopped, projecting one city to a non-occupied closest neuron to generate an initial TSP solution for further optimization. One TSP solution of SOM is shown in Fig. 6.

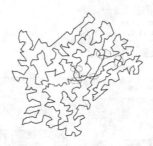

**Fig. 6.** A result of the proposed SOM implementation for lu980.tsp from TSPLIB.

### 3.3 Massively Parallel 2-opt with Data Decomposition

To optimize the TSP solution produced by SOM, various 2-opt strategies can be used, while the nature attributes of 2-opt make its massively parallel implementation become more complex because of tour ordering requirement. Reasons are following: first, one 2-opt move needs to consider tour ordering to avoid cutting the tour, as shown in Fig. 1; second, massive correct 2-opt moves simultaneously found in one same tour orientation may also cut the tour as shown in Fig. 7(b,c). However, these massively *non-interacted 2-exchanges* shown in Fig. 7(a) can be executed in parallel without cutting the tour.

Here, trying to get massively parallel 2-opt optimization while respecting those nature attributes of 2-opt, one simple choice is to adopt a strategy of massively parallel 2-opt evaluation with sequential execution, simplified as *"parallel evaluation but sequential execution"*. Its principle idea is straightforward: the algorithm begins with simultaneously searching each edge's 2-opt optimization in one same tour orientation according to a certain neighborhood edge searching rule; but only those non-interacted 2-exchanges shown in Fig. 7(a) can be sequentially detected and executed on CPU side in one iteration.

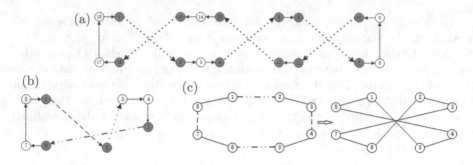

**Fig. 7.** Massively parallel 2-opt framework: *parallel evaluation but sequential execution.* (a) Case of multiple 2-exchanges that do not influence with each other. (b, c) Cases of multiple 2-exchanges interacting with each other: multiple 2-exchanges share one same edge in (b); execution of the two 2-opt moves in (C) will cut the original integral tour.

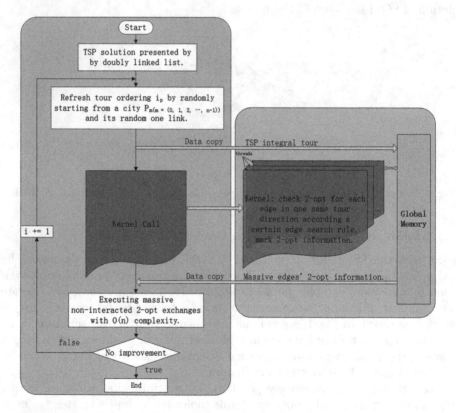

**Fig. 8.** Massively parallel 2-opt framework with a certain edge searching rule.

Overall outline of this massively parallel 2-opt framework is shown in Fig. 8. The algorithm begins with a TSP solution represented by doubly linked list where each city $P_m (m = 0, 1, 2...N - 1)$ has its two and only two links that are used to compose a TSP tour ring. In one iteration, the first step needs to follow current tour solution and assign each city a unique increasing tour order. We use this tour order to detect non-interactive 2-opt moves and indicate current tour orientation in later operations. After this step of "refresh tour order", the algorithm copies necessary data from CPU to GPU side and launches kernel functions to check each edge's 2-opt optimization according to a certain neighboring edge searching rule and store each edge's 2-opt information to this edge's starting city according to current tour orientation. At last, the algorithm copies each edge's optimizing information from GPU to CPU side to sequentially detect and execute these massive non-interacted 2-opt exchanges.

Under this straightforward parallel scheme, various concrete implementations can be applied. For example, the algorithm can simultaneously search 2-opt optimization for each edge along the integral tour or in a local search range. Here, we apply a concrete edge searching rule through searching each edge's 2-opt optimization among its neighborhood edges by using local spiral search operator, namely 2-opt LSS.

## 4    Experiments and Analysis

We implement this combinatorial method on GPU taking advantage of GPU's parallel read/write operation on global memory with atomic control [24]. For neighborhood 2-opt optimization using local spiral search (2-opt LSS), its optimization capability compared with traditionally sequentially exhaustive 2-opt along integral tour (2-opt EAT) is interesting. We test these two strategies base on same TSP solution of SOM. Both the two optimization methods adopt an evaluation strategy of "first optimized first accept". As both these two methods can not further optimize the TSP tour after fixed number of iterations, we set a stop criterion to judge whether the tour has been optimized or not in current iteration.

Parameter setting in Eq. 1 influences result quality and running time of SOM. we apply same parameters shown in Table 1 for different TSP instances. $\alpha_{ini}$ and $\alpha_{final}$ are learning rate at the starting and final epoch; $r_{ini}$ and $r_{final}$ are neighborhood topological radius from the winner node at the starting and final epoch; $epo$ sets the number of iteration; $LSSR$ sets the searching range on cellular matrix for the spiral search operator.

**Table 1.** Parameter setting for SOM.

| $\alpha_{ini}$ | $\alpha_{final}$ | $r_{ini}$ | $r_{final}$ | $epo$ | $LSSR$ |
|------|------|-----|-----|-----|---------|
| 1 | 0.01 | 100 | 0.5 | 100 | Maximum |

Average results of ten tests for each TSP instance are shown in Tables 2 and 3. In each table, 2-opt works based on same results of SOM. For each method

**Table 2.** Test results of sequential implementations.

| TSP instances | Optimum | SOM (sequential) | | | 2-opt EAT (sequential) | | | | 2-opt LSS (sequential) | | | |
|---|---|---|---|---|---|---|---|---|---|---|---|---|
| | | t(s) | %PDM | %PDB | epo | t(s) | %PDM | %PDB | epo | t(s) | %PDM | %PDB |
| uy734 | 79114 | 10.44 | 16.83 | 16.83 | 3.36 | 0.54 | 9.02 | 8.46 | 104.3 | 0.48 | 8.74 | 8.53 |
| zi929 | 95345 | 14.86 | 17.53 | 17.53 | 6 | 0.83 | 7.08 | 6.27 | 399.7 | 0.31 | 7.27 | 6.87 |
| lu980 | 11340 | 13.61 | 8.15 | 8.15 | 2.1 | 0.98 | 7.00 | 6.96 | 28.4 | 0.97 | 6.93 | 6.85 |
| rw1621 | 26051 | 25.14 | 6.54 | 6.54 | 2.1 | 2.78 | 6.10 | 5.99 | 20.1 | 2.74 | 6.25 | 6.18 |
| nu3496 | 96132 | 74.78 | 13.15 | 13.15 | 3.8 | 13.84 | 7.51 | 7.51 | 33.5 | 12.86 | 12.42 | 11.83 |

**Table 3.** Test results of parallel implementations.

| TSP instances | Optimum | SOM (parallel) | | | 2-opt EAT (sequential) | | | | 2-opt LSS (parallel) | | | |
|---|---|---|---|---|---|---|---|---|---|---|---|---|
| | | t(s) | %PDM | %PDB | epo | t(s) | %PDM | %PDB | epo | t(s) | %PDM | %PDB |
| uy734 | 79114 | 1.21 | 18.79 | 11.86 | 4.4 | 0.57 | 8.61 | 7.17 | 117.7 | 0.014 | 8.65 | 7.26 |
| zi929 | 95345 | 2.74 | 25.99 | 21.38 | 4.7 | 0.91 | 10.38 | 8.64 | 274.1 | 0.013 | 11.31 | 8.13 |
| lu980 | 11340 | 1.58 | 10.66 | 8.74 | 2.5 | 1.02 | 7.87 | 6.77 | 32.9 | 0.022 | 8.05 | 6.51 |
| rw1621 | 26051 | 2.46 | 10.35 | 7.68 | 2.2 | 2.77 | 8.26 | 6.43 | 25.4 | 0.039 | 9.01 | 6.47 |
| nu3496 | 96132 | 5.76 | 16.99 | 9.30 | 3.9 | 14.15 | 8.29 | 7.19 | 99.1 | 0.153 | 10.32 | 7.65 |

*2-opt EAT : Exhaustive 2-opt Along the Tour;
*2-opt LSS : 2-opt Local Spiral Search.

listed in these two tables, "t(s)" is the average time taken in one test, including necessary time for generating random data, refreshing TSP tour order and copying data from GPU to CPU; "%PDM" is the percentage deviation between the mean solution and the optimum solution; "%PDB" is the percentage deviation between the best solution and the optimum solution; "epo" indicates the average quantity of iterations in one test.

Comparing running time of the two tables, acceleration of the proposed parallel implementations is obvious. And we think the acceleration factor would be more obvious for larger TSP instances, as the proposed parallel platform both for SOM and 2-opt takes O(N) complexity either for the memory size or for GPU threads.

Comparing solution quality, we should mention that the proposed model deals with a procedure similar to original standard SOM and we did not try to find the best parameters to get best performance of SOM. However, maximum number of iterations, running time and the result quality of the two 2-opt methods are influenced by initial TSP results of SOM.

Visual results of one test in Table 3 are shown in Fig. 9. Our experiments work on the laptop with CPU Intel(R) Core(TM) i7-4710HQ, 2.5 GHz and GPU card GeForce GTX 850M.

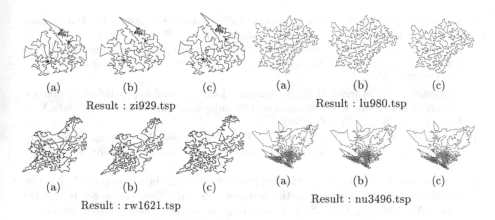

(a)        (b)        (c)          (a)          (b)          (c)

Result : zi929.tsp                    Result : lu980.tsp

(a)        (b)        (c)          (a)          (b)          (c)

Result : rw1621.tsp                   Result : nu3496.tsp

**Fig. 9.** Visual results of one test in Table 3. Columns from left to right for each TSP instance: (a) Results of parallel SOM; (b) Results of sequentially exhaustive 2-opt along tour; (c) Results of massively parallel 2-opt with local spiral search.

## 5   Conclusion

In this paper, we propose an alternative parallel computing platform both for SOM and 2-opt. We test the proposed methods with GPU parallel computation and present the obtained acceleration factor. We believe the acceleration factor will increase for very large scale instances as capacity of parallel devices is growing. Further jobs would concentrate on tests for very large size TSP instances and comparison with more different combinatorial optimization methods.

**Acknowledgments .** This paper is together sponsored by China Scholarship Council (CSC) and LE2I UBFC.

## References

1. Kohonen, T.: The self-organizing map. Proc. IEEE **78**(9), 1464–1480 (1990)
2. Angeniol, B., Vaubois, G.D.L.C., Le Texier, J.-Y.: Self-organizing feature maps and the travelling salesman problem. Neural Netw. **1**(4), 289–293 (1988)
3. Modares, A., Somhom, S., Enkawa, T.: A self-organizing neural network approach for multiple traveling salesman and vehicle routing problems. Int. Trans. Oper. Res. **6**(6), 591–606 (1999)
4. Bai, Y., Zhang, W., Jin, Z.: An new self-organizing maps strategy for solving the traveling salesman problem. Chaos, Solitons Fractals **28**(4), 1082–1089 (2006)
5. Créput, J.-C., Koukam, A.: A memetic neural network for the Euclidean traveling salesman problem. Neurocomputing **72**(4), 1250–1264 (2009)
6. Brocki, Ł., Koržinek, D.: Kohonen self-organizing map for the traveling salesperson problem. In: Jabłoński, R., Turkowski, M., Szewczyk, R. (eds.) Recent Advances in Mechatronics, pp. 116–119. Springer, Heidelberg (2007)
7. Mann, R. Haykin, S.: A parallel implementation of Kohonen's feature maps on the warp systolic computer (1990)

8. Wu, C.-H., Hodges, R.E., Wang, C.-J.: Parallelizing the self-organizing feature map on multiprocessor systems. Parallel Comput. **17**(6), 821–832 (1991)

9. Ienne, P., Thiran, P., Vassilas, N.: Modified self-organizing feature map algorithms for efficient digital hardware implementation. IEEE Trans. Neural Netw. **8**(2), 315–330 (1997)

10. Lawrence, R.D., Almasi, G.S., Rushmeier, H.E.: A scalable parallel algorithm for self-organizing maps with applications to sparse data mining problems. Data Min. Knowl. Disc. **3**(2), 171–195 (1999)

11. Valova, I., Szer, D., Gueorguieva, N., Buer, A.: A parallel growing architecture for self-organizing maps with unsupervised learning. Neurocomputing **68**, 177–195 (2005)

12. Richardson, T., Winer, E.: Extending parallelization of the self-organizing map by combining data and network partitioned methods. Adv. Eng. Softw. **88**, 1–7 (2015)

13. Qiao, W., Créput, J.-C.: Stereo matching by using self-distributed segmentation and massively parallel GPU computing. In: Rutkowski, L., Korytkowski, M., Scherer, R., Tadeusiewicz, R., Zadeh, L.A., Zurada, J.M. (eds.) ICAISC 2016. LNCS, vol. 9693, pp. 723–733. Springer, Cham (2016). doi:10.1007/978-3-319-39384-1_64

14. Verhoeven, M., Aarts, E.H., Swinkels, P.: A parallel 2-opt algorithm for the traveling salesman problem. Future Gener. Comput. Syst. **11**(2), 175–182 (1995)

15. Johnson, D.S., McGeoch, L.A.: Experimental analysis of heuristics for the STSP. In: Gutin, G., Punnen, A.P. (eds.) The Traveling Salesman Problem and its Variations, pp. 369–443. Springer, Heidelberg (2007)

16. Rocki, K., Suda, R.: Accelerating 2-opt and 3-opt local search using GPU in the travelling salesman problem. In: 2012 International Conference on High Performance Computing and Simulation (HPCS), pp. 489–495. IEEE (2012)

17. Van Luong, T., Melab, N., Talbi, E.-G.: GPU computing for parallel local search metaheuristic algorithms. IEEE Trans. Comput. **62**(1), 173–185 (2013)

18. Zhang, N., Wang, H., Creput, J.-C., Moreau, J., Ruichek, Y.: Cellular GPU model for structured mesh generation and its application to the stereo-matching disparity map. In: 2013 IEEE International Symposium on Multimedia (ISM), pp. 53–60. IEEE (2013)

19. Johnson, D.S., McGeoch, L.A.: The traveling salesman problem: a case study in local optimization. Local Search Comb. Optim. **1**, 215–310 (1997)

20. Karp, R.M.: Probabilistic analysis of partitioning algorithms for the traveling-salesman problem in the plane. Math. Oper. Res. **2**(3), 209–224 (1977)

21. Wang, H., Zhang, N., Créput, J.-C.: A massive parallel cellular GPU implementation of neural network to large scale Euclidean TSP. In: Castro, F., Gelbukh, A., González, M. (eds.) MICAI 2013. LNCS, vol. 8266, pp. 118–129. Springer, Heidelberg (2013). doi:10.1007/978-3-642-45111-9_10

22. Bentley, J.L., Weide, B.W., Yao, A.C.: Optimal expected-time algorithms for closest point problems. ACM Trans. Math. Softw. (TOMS) **6**(4), 563–580 (1980)

23. Rajasekaran, S.: On the Euclidean minimum spanning tree problem. Comput. Lett. **1**(1) (2004)

24. CUDA C Programming Guide: CUDA toolkit documentation

# Finding Self-organized Criticality
# in Collaborative Work via Repository Mining

J.J. Merelo[1(✉)], Pedro A. Castillo[1], and Mario García-Valdez[2]

[1] Geneura Team and CITIC, University of Granada, Granada, Spain
{jmerelo,pacv}@ugr.es
[2] Department of Graduate Studies, Instituto Tecnologico de Tijuana,
Tijuana, Mexico
mario@tectijuana.edu.mx

**Abstract.** In order to improve team productivity and the team interaction itself, as well as the willingness of occasional volunteers, it is interesting to study the dynamics underlying collaboration in a repository-mediated project and their mechanisms, because the mechanisms producing those dynamics are not explicit or organized from the top, which allows self organization to emerge from the collaboration and the way it is done. This is why finding if self-organization takes place and under which conditions will yield some insights on this process, and, from this, we can deduce some hints on how to improve it. In this paper we will focus on the former, examining repositories where collaborative writing of scientific papers by our research team is taking place show the characteristics of a critical state, which can be measured by the existence of a scale-free structure, long-distance correlations and *pink* noise when analyzing the size of changes and its time series. This critical state is reached via self-organization, which is why it is called self-organized criticality. Our intention is to prove that, although with different characteristics, most repositories independently of the number of collaborators and their real nature, self-organize, which implies that it is the nature of the interactions, and not the object of the interaction, which takes the project to a critical state. This critical state has already been established in a number of repositories with different types of projects, such as software or even literary works; we will also find if there is any essential difference between the macro measures of the states reached by these and the object of this paper.

## 1 Introduction

The existence of a self-organized critical state [1] in software repositories has been well established [6,7,13,22] and attributed to an stigmergy process [20] in which collaborators interact through the code itself and through messages in other communication media, such as Slack or an IRC chat application, task assignment systems or mailing lists. In this critical state there are specific dynamic behaviors, like small changes provoking *avalanches* of other changes and long-distance correlations that make a particular change in the codebase cause further changes down the line. The dynamics of self-organized criticality is sometimes compared

I. Rojas et al. (Eds.): IWANN 2017, Part I, LNCS 10305, pp. 483–496, 2017.
DOI: 10.1007/978-3-319-59153-7_42

to that of a sand pile [18], in the sense that the actual shape tends to reach a *critical* state, represented in the sand pile by a critical slope, and a single grain of sand creates avalanches unrelated to the frequency of grains falling. This pile of sand is also a simple model of a self-organized system that captures many of its main characteristics, but its behavior is connected to the experience of software developers and paper writers that experience certain periods of stasis followed by *avalanches* of work, new code or new paragraphs without an apparent origin.

This anecdotal experience supports that software teams analyzed through the repositories that support their work might also find themselves in this self-organized critical state. Furthermore, the case for this critical state is supported by several macro measures that certify the non-existence of a particular scale in the size of changes [7,13,22], but in some cases they also exhibit long-range correlations and a *pink noise* [21] in the power spectral density, with *noise* or variations with respect to the *normal* frequency changing in a way that is inversely proportional to it, higher frequency changes getting a smaller spectral density [13].

The state in which the team is has obviously an influence in its productivity, with some authors finding this state favors evolvability of the underlying software system [3], as opposed from the lack of this quality in software created by a top-down organization process. That is why this quality has been mainly studied in open source software systems which follow a more open model of development; however, it might happen that, in the same way it happens in neural systems [10], the self-organized state might be essential to the software development process, as long as it is done in an application that allows collaboration such as a repository managed by a source control system such as git. In fact, some explanations have been offered via conservation laws [8] and other usual complex network mechanisms such as preferential attachment [11].

After some initial exploration of the subject and developing the tools needed to mine repositories in GitHub [15], in this paper we are interested in finding out whether these mechanisms are exclusive to software teams or if, indeed, self-organized criticality can be found also in other repositories. In our research team we are committed to open source and open science, developing all our work in open repositories hosted in GitHub. The repository is open since the first moment of writing a paper, and the repository itself hosts also data and, like in the case of this particular paper, the code used to extract data. We interact throughout writing the paper via comments in the paper and issues, that is, work orders where you can comment and that can also be *closed* once the *issue* has been cleared or fixed. *Developing* a paper using a repository is a good practice that allows an easy distribution of tasks, attribution, and, combined with the use of *literate* programming tools such as Knitr [23], that allow the embedding of code within the text itself, provide a closer relationship between data and report and, of course, easier reproductibility.

That is why, after examining and establishing the existence of this state in the software repository for the Moose Perl library [13] and books written mainly by a single person [14], in this report we are going to work on a repository for several papers in which our research group has been working for different amounts of time, from a few months to more than a year. In particular, one of the papers, which was already the object of a previous report [15] has been chosen since it has been a work in progress for more than one a year until it was eventually published [2]. The other three papers chosen are in one case an evolution of a paper which was initially published in a conference and that is now a work in progress [19], another that contains several papers published in diverse venues and that are evolution of this one [16], and finally a paper that has been in progress for about a year, but has not been finished yet. These papers have been chosen because they had a certain length, with more than 50 commits (or changes). Besides, they were available and we knew their circumstances. Whether they are or not representative of a larger corpus remains to be seen, and we will try to examine this possibility in the conclusions. The papers are plain text with LaTeX commands, and, in some cases, also R commands in those papers using Knitr. The inclusion of lines and commands written in computer languages would indicate that these papers are halfway between a book, which is mainly text, and an application or library, which is mainly code. We will see if this *hybrid* nature translates to the measures taken over the repository and its dynamics.

After presenting a brief state of the art next, followed by the methodology, obtained results will be presented and eventually we will expose our conclusions.

## 2   State of the Art

As far as we know, there has not been a continuing line of research on self-organized criticality in teamwork. Researchers have thoroughly proved that software repositories seem to be in a SOC state, [7,22], including our own reports [13–15] where we examine and establish the existence of repositories in a critical state to which they have arrived via self-organization; the fact that these repositories have different characteristics in terms of the number of users, age and type of information they hold implies that self-organization, as should be expected, is achieved with relative ease. In fact, this state of self-organized criticality quantitatively proves what has been already established via qualitative analysis, the fact that in many successful software projects, developers self-organize [4], which is the preferred way of working in distributed and volunteer teams [5]. In fact, this way of organization matches our own experience in development of open source projects such as [12,17], which are developed mainly by one or a few coders, helped sporadically by other coders that find an error or adapt the code to particular situations. In fact, this self-organization has also been observed in similar projects such as Wikipedia.

This self-organization, eventually, might produce a critical state given the necessary conditions. However, there has been no work going further and proving this even in the case that work is done by a few persons and on repositories that are not devoted to software development.

In this paper we will examine different repositories with the same purpose, all devoted to the collaborative writing of scientific papers, but each with a different age, in order to try and find out if self-organization arrives simply with age and, if so, what seems to be this critical age.

# 3    Methodology

In this paper we will work with the size of changes to a particular set of files
in the repository; since repositories include other artifacts such as images or
style files we, via a wildcard, select only the file or files we are interested in. To
extract information about changes to these files in the repositories, we analyze
the repository using a Perl script that runs over the git log and notes the size of
the changes that have been made to all files.

Since changes include both the insertion and deletion of lines within those
files, the largest of these values is taken; in particular, this means that the

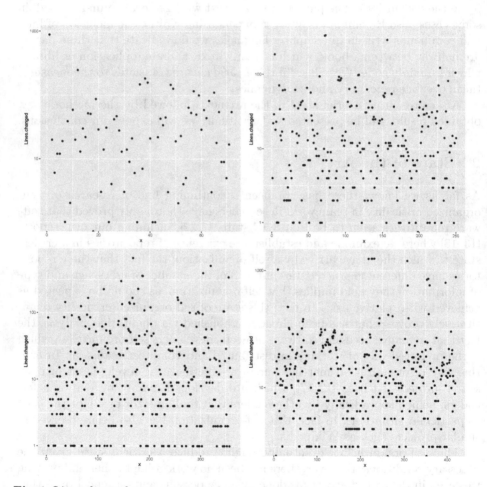

**Fig. 1.** Lines changed per commit in a log-y scale, with $x$ scale corresponding to commit
number. The four charts correspond, starting with the top right, DCAI, PPSN, book
prediction, volunteer computing at the lower bottom. This order is going to be the
same across all figures from now on. This order has been chosen, as it can be observed,
to follow the size of the log, with the paper with the smaller number of commits first
and the one with the largest number of commits last.

addition of all changes will not be equal to the sum of the sizes of all files. A change in two lines will appear in a diff as "2 insertions, 2 deletions", adding up to 0; that is why we consider the larger of these two values; the main reason for doing so is also that in fact, the algorithm that computes changes in the repository examines similitude in lines and counts changes in two lines as two insertions and two deletions. There is no way to find out whether there have been actually two lines added somewhere and two deleted somewhere else, so in absence of that, we opt for the heuristic of using the largest of these two values as change size.

The script generates a .csv file with a single column with the sequence of changes of size in the files of interest in each repository. These files, as well as the repositories where they have been measured, are available with a free license in the repository that also hosts this paper. The sequence of changes for the 4 files is shown in Fig. 1.

The $x$ axes for these timelines does not correspond to physical time, but simply to sequence. In this sense, there is an important difference between our research methodology, which considers discrete changes, to papers such as [9], which take into account *daily* changes. We think that examining discrete changes does not impose a particular rhythm, namely, daily, on the changes, but lets the repository expose its own rhythm; it also allows us to examine slow-changing repositories such as these ones, that can be static for a long time to experience a burst of changes all of a sudden; precisely these changes can indicate an *avalanche* that is a symptom of the underlying self-organized criticality state.

Once the information from the repositories has been extracted, we proceed to analyze it in the next section.

## 4   Results

A summary of the statistical characteristics of the size of the commits, in number of lines, is shown in Table 1. This table shows that, at least from a macro point of view, median and averages are remarkably similar to the ones found in other studies [13,14], with the median between 9 and 22 lines and the average between 24 and 54. The fact that the average is so separated from the median is already a hint that this is a skewed distribution. The book analyzed in [14] had a median

**Table 1.** Summary of statistical measures for the four papers we have been analyzing here; $SD$ stands for "Standard Deviation"

| Name | Mean | Median | SD |
|------|------|--------|-----|
| 2016-DCAI_ALL_ALL | 51.54167 | 21.5 | 110.44531 |
| 2016-ea-languages-PPSN_ea-languages | 24.18800 | 11.0 | 41.21518 |
| 2015_books_ALL | 32.05263 | 9.0 | 67.89211 |
| modeling-volunteer-computing_ALL_ALL | 54.68810 | 13.0 | 193.09572 |

of 10 lines, but a mean of 150 lines changed, in a distribution that is different, much more skewed towards larger sizes, while the software library analyzed in [13] had a median of 9 and a mean of close to 32, which is remarkably similar to one of the papers analyzed here. This implies that the concept of *session*, or size of changes committed together, might be very similar no matter what is the thing that is actually written.

The timeline of the commit sizes is represented in a line chart in Fig. 2 with logarithmic or decimal $y$ scale and smoothing over several commits, either 10 or 20, depending on the color. The $x$ axis is simply the temporal sequence of commits, while the $y$ axis is the absolute size of the commit in number of lines. The serrated characteristic is the same, as well as the big changes in scale, with some periods where small changes happen and other that alternate big with small changes. A certain *rhythm* can be observed, which hints at large-scale correlations, that is influence of changes happening now over changes that occur several, or many, steps afterwards, in the future.

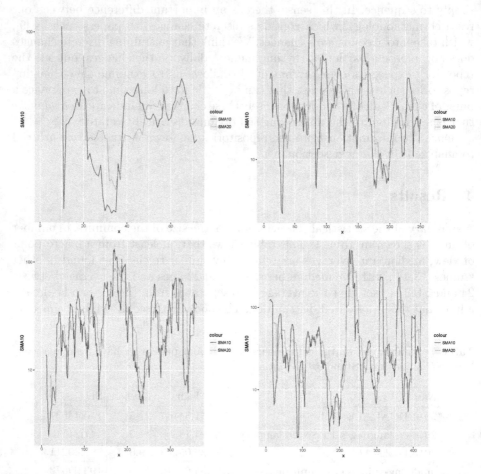

**Fig. 2.** Timeline of changes for the four papers, with lines smoothed over 20 and 10 changes, shown in different colors. (Color figure online)

Besides, these changes in scale might mean that commit sizes are distributed along a Pareto distribution. We will examine this next, representing the number of changes of a particular size in a log-log scale, with linear smoothing to show the trend in Fig. 3.

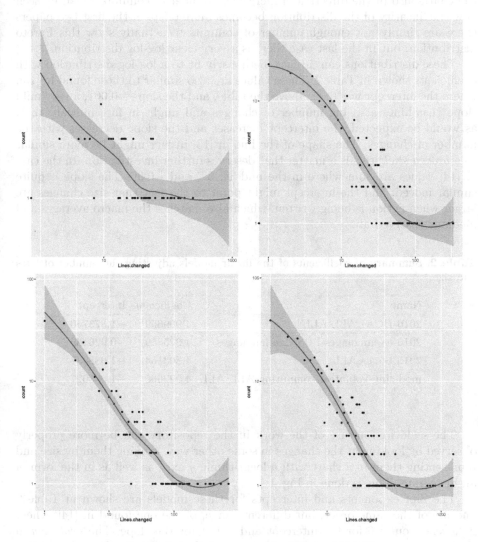

**Fig. 3.** Number of changes vs size in a log-log scale.

This chart show what seems to be a Zipf distribution, with the commit sizes ranked in descending order and plotted with a logarithmic $y$ axis. This distribution shows, in all cases, a *tail* corresponding to big changes. This might be simply a consequence of different practices by different authors, with some preferring atomic changes to single lines or paragraphs and others writing down whole sections; in some cases, it corresponds also to reuse of common parts of papers (authors, acknowledgements, description of a method) to create the initial versions of the paper; finally, in some cases comments are deleted before the final version is submitted, so these *tails* are not really unexpected. If you follow the charts also in the direction of increasing number of commits it can be seen how the linearity of the distribution becomes more crisp; in the first two papers there are simply not enough number of commits to actually show this Pareto distribution, but in the last case there is a very clear log-log distribution.

These distributions can, in fact, be linearly fit to a log-log distribution with coefficients shown in Table 2. These values are also similar to those found in [13], where the intercept was 6.02, above the table, and the slope −0.001, a very mild slope that hints at a big number of changes and might in fact indicate that, as would be expected, the intercept increases and the slope decreases with the number of changes. The shape of the line in [13] in fact might be more similar to a *broken stick*; this is a matter that deserves further investigation. In the case of [14] values are somewhere in the middle, 5.7 and −0.96. The slope is quite similar indeed, and the intercept might point to the fact that size changes are larger when fiction is being written, which also matches the macro averages and medians observed above.

**Table 2.** Summary of coefficients of the linear models adjusting the number of lines and size.

| Name | Coefficient | Intercept |
|---|---|---|
| 2016-DCAI_ALL_ALL | 4.009659 | −1.873740 |
| 2016-ea-languages-PPSN_ea-languages | 4.225859 | −0.926569 |
| 2015_books_ALL | 4.803154 | −1.088837 |
| modeling-volunteer-computing_ALL_ALL | 4.922665 | −1.065027 |

The scale free nature of the work in the repository can be more properly observed by looking at the changes in some other way, ranking them by size and representing them in a chart with a logarithmic $y$ axis, as well as in the form of an histogram. This is done in Fig. 4.

The Zipf exponents and intercepts for these models are shown in Table 3, and are of the same order, but different range, of the one found in [14], where it hovers around 6 for the intercept and −0.01 for the slope. The *evolution* in the nature of the distribution can be observed, from a more or less straight line in the first cases, to something more similar to a broken stick model in the last one, although it can still be linearly fit to a log scale and there is a regime of size changes that is still logarithmic in scale. Whatever the actual distribution, there is no doubt that changes do not organize themselves along a central value and that there is scale-free nature in them, which is, besides, independent of the *age* or total number of changes of the paper, as has been shown above.

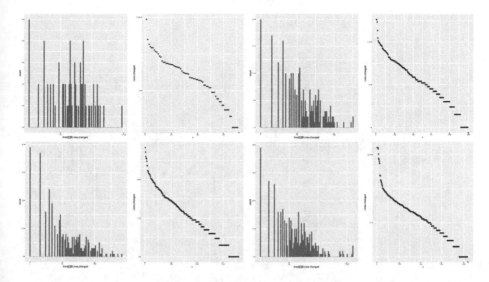

**Fig. 4.** Changes, ordered by size, and represented in a logarithmic $y$ axis. Side by side, the histogram and Zipf chart for the four papers analyzed.

**Table 3.** Summary of Zipf coefficients of the linear models adjusting the number of lines and size.

| Name | Coefficient | Intercept |
| --- | --- | --- |
| 2016-DCAI_ALL_ALL | 5.502401 | −0.0700053 |
| 2016-ea-languages-PPSN_ea-languages | 4.605605 | −0.0177415 |
| 2015_books_ALL | 4.906922 | −0.0141869 |
| modeling-volunteer-computing_ALL_ALL | 5.122338 | −0.0123126 |

Finally, these scale distributions hints at the possibility of long-scale correlations, but in order to find this out, we will have to plot the partial autocorrelation of the sequence, that is, the relationship between the size of a change and the rest of the changes in the sequence. This is computed and plotted in Fig. 5. Autocorrelation is significant only if the lines go over the average plotted as a dashed line. The long distance correlations, already found in [13], are present here. In that case, there was positive autocorrelation in the 21 commit period; in this case, it appears at 25 and 15. It shows, anyway, that the size of a commit has a clear influence further down writing history, with high autocorrelations around 20 commits. In these repositories of increasing age, we find that actual long-distance autocorrelation only happens when they age, with no long distance significant autocorrelation in the first two repositories, and a significant one in the two bottom repositories. In both cases, correlation happens at the distance of 13–25 commits, exactly as it happened before. However, autocorrelation seems to disappear in the older repositories, at least for that long distances. This might

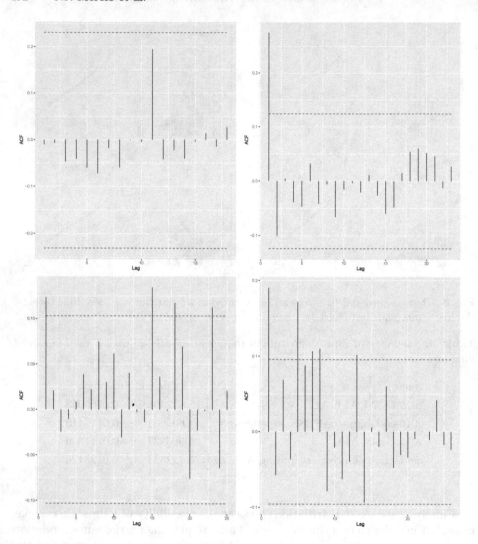

**Fig. 5.** Autocorrelation plot. The order of the papers is the same as in the rest of the figures, from top left to bottom right, DCAI, PPSN, book prediction and volunteer computing.

indicate significant differences for other types of work, but it will need further research to find out, in a more precise way, the ranges of distances where auto-correlation is significant.

Once two of the three features of self-organized criticality have been proved, at least in some of the repositories, we will focus on the third, the presence of *pink* noise, as measured by the power spectral density. This is shown in Fig. 6, where the power spectral density is shown for the four papers. A *pink* noise would be characterized by a spectrum with a negative slope, with decreasing power the higher the frequency.

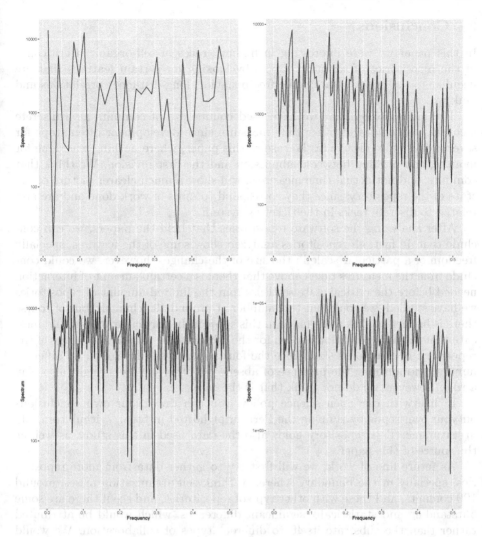

**Fig. 6.** Spectral density of changes. The repos are in the same order as above.

In this case, we see that this *trend* appears with more or less clarity in two of the four papers, the two on the right, although the lower-left paper (book-prediction) also exhibits it to a certain point. In fact, it is much clearer in the second paper, PPSN, which, on the other hand, does not exhibit long-scale autocorrelation.

Once the three main features of systems in self-organized state have been measured for the papers under study, we will present in the next Section our conclusions.

## 5   Conclusions

In this paper we were interesting in finding traces of self-organized criticality in the repositories of scientific papers by looking for certain features that are peculiar to the critical state: scale-free behavior, long-distance correlations and pink we.

The methodology that we have used counts size of commits as a discrete measure, not *dailies* or other time measure, since development often stops for several days more clearly in the case of this paper, where nothing was done for months, and nothing between submission and the first revision. We think that commits, and not actual time measures, will show a much clearer picture of the state of the repository, since they correspond to units of work done and are also related to discrete tasks in the ticketing system.

After analyzing the software repositories that hold the papers, we can conclude that, in fact, all repositories analyzed show some of the features, specially freedom of a particular scale in the size of the changes; however, we could conclude from the measures taken above that there is a certain amount of interaction needed before the critical state settles. From the limited amount of repositories we have studied, we could put this number at around 100 changes, but of course there is further studies to be made in this subject. In particular, this would indicate that the only condition needed for the critical state to arise is the age of the repository, or maybe its size. Since the four papers were developed by different number and authors, the presence of absence of other artifacts might also play a role. However, we do not think that is the case.

In line with our open science policy, you can draw your own conclusions on your own repos by running the Perl script hosted in http://github.com/JJ/literaturame. This repository holds also the data used in this study, as well as the source of this paper.

As future line of work, we will first try to gather data from more repositories, specially in the boundary where we think self-organization arises, around 100 commits, and these with other type of repositories, and see if there are some outstanding and statistically significant differences, which would be attributed rather than the substrate itself, to different types of collaboration. We would also like to make more precise models of the ranked change sizes, as well as the relation between number of changes and its size. A study of particular circumstances of every repository will also help us to understand what self-organization means and, finally, as was our initial objective, if this fact can be used to create methodologies that improve productivity in work teams.

**Acknowledgements.** This work has been supported in part by: de Ministerio español de Economía y Competitividad under project TIN2014-56494-C4-3-P (UGR-EPHEMECH).

## References

1. Bak, P., Tang, C., Wiesenfeld, K.: Self-organized criticality. Phys. Rev. A **38**(1), 364 (1988)

2. Castillo, P.A., Mora, A.M., Faris, H., Merelo, J., García-Sánchez, P., Fernández-Ares, A.J., las Cuevas, P.D., García-Arenas, M.I.: Applying computational intelligence methods for predicting the sales of newly published books in a real editorial business management environment. Knowl.-Based Syst. **115**, 133–151 (2017). http://www.sciencedirect.com/science/article/pii/S0950705116304026

3. Cook, S., Harrison, R., Wernick, P.: A simulation model of self-organising evolvability in software systems. In: IEEE International Workshop on Software Evolvability (Software-Evolvability 2005), pp. 17–22, September 2005

4. Crowston, K., Li, Q., Wei, K., Eseryel, U.Y., Howison, J.: Self-organization of teams for free/libre open source software development. Inf. Softw. Technol. **49**(6), 564–575 (2007). https://www.sciencedirect.com/science/article/pii/S0950584907000080. Qualitative Software Engineering Research

5. Crowston, K., Wei, K., Howison, J., Wiggins, A.: Free/libre open-source software development: what we know and what we do not know. ACM Comput. Surv. (CSUR) **44**(2), 7 (2012)

6. Gao, Y., Zheng, Z., Qin, F.: Analysis of linux kernel as a complex network. Chaos Solitons Fractals **69**, 246–252 (2014)

7. Gorshenev, A., Pis'mak, Y.M.: Punctuated equilibrium in software evolution. Phys. Rev. E **70**(6), 067103 (2004)

8. Hatton, L.: Conservation of information: software's hidden clockwork? IEEE Trans. Softw. Eng. **40**(5), 450–460 (2014)

9. Herraiz, I.: A statistical examination of the evolution and properties of libre software (2009). www.lulu.com

10. Hesse, J., Gross, T.: Self-organized criticality as a fundamental property of neural systems. Front. Syst. Neurosci. **8**, 166 (2014). http://journal.frontiersin.org/article/10.3389/fnsys.2014.00166

11. Lin, Z., Whitehead, J.: Why power laws? An explanation from fine-grained code changes. In: 2015 IEEE/ACM 12th Working Conference on Mining Software Repositories (MSR), pp. 68–75. IEEE (2015)

12. Merelo, J.J., García-Valdez, M., Castillo, P.A., García-Sánchez, P., de las Cuevas, P., Rico, N.: NodIO, a JavaScript framework for volunteer-based evolutionary algorithms: first results. arXiv e-prints, January 2016. http://arxiv.org/abs/1601.01607

13. Merelo, J.J.: Quantifying activity through repository mining: the case of Moose. Technical report, GeNeura/UGR/CITIC (2016). https://doi.org/10.6084/m9.figshare.3756768.v3

14. Merelo-Guervós, J.J.: Self-organized criticality in fiction: the case of #slash, a novel. Technical report 2016-7, GeNeura group, University of Granada (2016). https://www.researchgate.net/publication/307598023_Self-organized_criticality_in_fiction_the_case_of_Slash_a_novel

15. Merelo-Guervós, J.J.: Self-organized criticality in repository-mediated projects. Technical report, GeNeura group, University of Granada, August 2016. https://www.academia.edu/28083421/Self-organized_criticality_in_repository-mediated_projects

16. Merelo-Guervós, J.J., García-Sánchez, P.: Modeling browser-based distributed evolutionary computation systems, CoRR abs/1503.06424 (2015). http://arxiv.org/abs/1503.06424

17. Merelo-Guervós, J.J., Castillo, P.A., Alba, E.: `Algorithm::evolutionary`, a flexible Perl module for evolutionary computation. Soft Comput. **14**(10), 1091–1109 (2010). http://l.ugr.es/000K

18. Paczuski, M., Maslov, S., Bak, P.: Avalanche dynamics in evolution, growth, and depinning models. Phys. Rev. E **53**(1), 414 (1996)

19. Rivas, V.M., Parras-Gutierrez, E., Merelo, J.J., Arenas, M.G., García-Fernández, P.: Web browser-based forecasting of economic time-series. In: Bucciarelli, E., Silvestri, M., Rodríguez-González, S. (eds.) Decision Economics, In Commemoration of the Birth Centennial of Herbert A. Simon 1916–2016 (Nobel Prize in Economics 1978) - Distributed Computing and Artificial Intelligence, 13th International Conference. Advances in Intelligent Systems and Computing, vol. 475, pp. 35–42. Springer, Cham (2016). doi:10.1007/978-3-319-40111-9_5

20. Robles, G., Merelo-Guervós, J.J., Gonzlez-Barahona, J.M.: Self-organized development in libre software projects: a model based on the stigmergy concept. In: Proceedings of the 6th International Workshop on Software Process Simulation and Modeling (ProSim 2005), St. Louis, MO, USA, May 2005, in press

21. Szendro, P., Vincze, G., Szasz, A.: Pink-noise behaviour of biosystems. Eur. Biophys. J. **30**(3), 227–231 (2001)

22. Wu, J., Holt, R.C., Hassan, A.E.: Empirical evidence for SOC dynamics in software evolution. In: 2007 IEEE International Conference on Software Maintenance, pp. 244–254. IEEE (2007)

23. Xie, Y.: Dynamic Documents with R and Knitr, vol. 29. CRC Press, Boca Raton (2015)

# Capacity and Retrieval of a Modular Set of Diluted Attractor Networks with Respect to the Global Number of Neurons

Mario González[1,4](✉), David Dominguez[3], Ángel Sánchez[2],
and Francisco B. Rodríguez[3](✉)

[1] FICA, Universidad de las Américas, Quito, Ecuador
mario.gonzalez.rodriguez@udla.edu.ec
[2] ETSII, Universidad Rey Juan Carlos, 28933 Madrid, Spain
angel.sanchez@urjc.es
[3] Escuela Politécnica Superior, Universidad Autónoma de Madrid,
28049 Madrid, Spain
{david.dominguez,f.rodriguez}@uam.es
[4] FACI, Universidad Estatal de Milagro, Milagro, Ecuador

**Abstract.** The modularity and hierarchical structures in associative networks can replicate parallel pattern retrieval and multitasking abilities found in complex neural systems. These properties can be exhibited in an ensemble of diluted Attractor Neural Networks for pattern retrieval. It has been shown in a previous work that this modular structure increases the single attractor storage capacity using a divide-and-conquer approach of subnetwork diluted modules. Each diluted module in the ensemble learns disjoint subsets of unbiased binary patterns. The present article deals with an ensemble of diluted Attractor Neural Networks which is studied for different values of the global number of network units, and their performance is compared with a single fully connected network keeping the same cost (total number of connections). The ensemble system more than doubles the maximal capacity of the single network with the same wiring cost. The presented approach can be useful for engineering applications to limited memory systems such as embedded systems or smartphones.

**Keywords:** Hopfield network · Ensembled learning · Storage capacity · Network size · Divide-and-conquer approach · Diluted connectivity

## 1 Introduction

One of the main trends of research in Machine Learning, along with "deep learning", is the study of "networks of networks". One central question of interest is comparing a net of nets with overall $N$ neurons and $K$ degree with only one single net with the same size (constant $N \times K$) and check their different performances in terms of storage/load, computational complexity, etc. In this sense

© Springer International Publishing AG 2017
I. Rojas et al. (Eds.): IWANN 2017, Part I, LNCS 10305, pp. 497–506, 2017.
DOI: 10.1007/978-3-319-59153-7_43

recent literature has focused on how the modularity and hierarchical structures in associative networks allow to replicate parallel pattern retrieval and multi-tasking abilities exhibited by complex neural and immune systems as well as the proposal for both theoretical and practical applications [1–4,17].

In this way, the motivation of this work is to increase the information processing (storage and retrieval) capacities using an ensemble Attractor Neural Network (ANN) with modular diluted components. These diluted modules have the advantage of increasing the storage capacity per connection [6], as well as to decrease the wiring/computational cost of the network.

In a recent work [14], a proposed Ensemble of Attractor Neural Network increases the maximal capacity when compared to a single Attractor Network of same cost given by the total number of connections $(N \times K)$, which was kept constant for both the ensemble and the single network systems. Following this finding we compare different ensembles for different number of Network units $N$, keeping the network dilution $\gamma = N/K$, constant. The ensemble and single network systems are studied for non-pathological dilution levels. Diluted networks on different connectivity topologies have been extensively studied [7–9,11], and they proved to overcome both wiring and computational costs. Thus, an ensemble of Attractor Neural Networks (ANN) diluted modules is used in order to improve the processing (storage and retrieval) of unbiased binary random patterns. The main idea is to use a divide-and-conquer approach. A single fully connected ANN is divided into diluted subnetworks, each of which being specialized in learning a disjoint pattern subset. By adding all the ANN components, the storage capacity of the ensemble is increased. It is shown that the modularized ANN model shows an improved patterns' storage capacity, compared to a single ANN model, at similar computational and wiring costs.

In González et al. [14], we have studied the ensemble system keeping the number of units $N$ constant, and comparing the performance of the systems for different values of modules number $n$ and modules dilution $\gamma$. In the aforementioned work we used a fixed value of $N$ as larger as possible for comparison purpose with the statistical mechanics literature. Now we are interested in studying the ensemble systems for different values of network units $N$. The motivation behind the analysis of the network size in terms of number of units is that for small ANN systems (low $N$), the storage of $P$ patterns is limited by the upper value of $P = \alpha_c \times N, \alpha_c \sim 0.14$ for the fully connected network [16], and by $P = \alpha_c \times N, \alpha_c \sim 0.64$ for extremely diluted networks [5]. For large $N$ the network manage to store a reasonable number of patterns. However, this is an issue for small values of $N$. Using an ensemble of ANN the storage of the system is increased [14], which can be useful for system with small number of units $N$. Real world data can be found where the number of patterns is of the order of the number of features $O(P) = N$, in such cases the proposed ensemble can be valuable to deal with the problem.

In the following Sect. 2 we describe the neural and network model. In Sect. 3 the modularized system is described. In Sect. 4 we present the main results and finally, Sect. 5 concludes the paper discussing the implications of our findings.

## 2   Neural and Network Model

In this section the proposed ANN ensemble model is rigorously described, starting with the neural coding, the network topology, and both learning and retrieval dynamics. A schematic representation of the modularized ANN system is illustrated. Finally, the information measures of the network and ensemble system are defined.

### 2.1   Coding, Topology and Dynamics

At any given time $t$, the network state is defined by a set of binary neurons $\sigma^t = \{\sigma_i^t \in \{-1, +1\}, i = 1, \ldots, N\}$. The purpose of the network is to recover a set of independent patterns $\{\xi^\mu, \mu = 1, \ldots, P\}$ that have been stored by a learning process. Each pattern, $\xi^\mu = \{\xi_i^\mu \in \{-1, +1\}, i = 1, \ldots, N\}$, is a set of site-independent unbiased binary random variables, $p(\xi_i^\mu = \{-1, +1\}) = 1/2$.

The synaptic couplings between the neurons $i$ and $j$ are given by the adjacency matrix $J_{ij} \equiv C_{ij} W_{ij}$, where the topology matrix $\mathbf{C} = \{C_{ij}\}$ describes the connection structure of the neural network and in $\mathbf{W} = \{W_{ij}\}$ are the learning weights. The topology matrix is built by random links connecting each neuron to $K$ others uniformly distributed in the network [10]. The network topology is then characterized by the *connectivity* ratio defined by $\gamma = K/N$. An extremely diluted network is obtained as $\gamma \to 0$, and the storage cost of this network is $\|J\| = N \times K$ if the matrix $\mathbf{J}$ is implemented as an adjacency list of $K$ neighbors. The matrix $\mathbf{J}$ is considered to be symmetrical, i.e. $J_{ij} = J_{ji}$.

The task of the network is to retrieve a pattern, $\xi \equiv \xi^\mu$, starting from a neuron state $\sigma^0$ which is close to it. This is achieved through the neuron dynamics

$$\sigma_i^{t+1} = \text{sign}(h_i^t), \tag{1}$$

$$h_i^t \equiv \frac{1}{K} \sum_j J_{ij} \sigma_j^t, \ i = 1, \ldots, N, \tag{2}$$

where $h_i^t$ denotes the local field of neuron $i$ at time $t$.

The learning algorithm updates the weight matrix $\mathbf{W}$ according to the Hebb's rule,

$$W_{ij}^\mu = W_{ij}^{\mu-1} + \xi_i^\mu \xi_j^\mu. \tag{3}$$

Weights start at $W_{ij}^0 = 0$ and after $P$ learning steps, they reach the value $W_{ij} = \sum_\mu^P \xi_i^\mu \xi_j^\mu$. The learning stage displays slow dynamics, being stationary within the time scale of the faster retrieval stage in Eq. (1).

### 2.2   Retrieval Overlap Measures

In order to evaluate the network retrieval performance, two measures are considered: the global overlap and the load ratio. The overlap is used as a measure of information, which is adequate to describe instantaneously the network's ability to retrieve each pattern. In this case, the overlap $m^\mu$ between the stationary neural state $\sigma^*$ and the corresponding pattern $\xi^\mu$ is:

$$m^\mu \equiv \frac{1}{N} \sum_i^N \xi_i^\mu \sigma_i^*, \tag{4}$$

which is the normalized statistical correlation between the learned pattern $\xi^\mu$ and the stationary neural state $\sigma^*$ after a sufficient enough long time.

One lets the network evolve according to Eqs. (1) and (2), and measures the overlap between the network states and the patterns. When the overlap between a given pattern and the corresponding neural states of the network is $m = 1$, the network has retrieved the pattern without noise. When the global overlap $m = 0$, the network carries no macroscopic order. In this case, the corresponding pattern cannot be retrieved. For intermediate values of $m$, where $0 < m < 1$, the pattern has been recovered with a given level of noise $(1 - m)$.

One is also interested in the load ratio $\alpha \equiv P/K$, that accounts for the storage capacity of the network. When the number of stored patterns increases, the noise due to interference between patterns also increases and the network is not able to retrieve them. Thus, the overlap $m$ goes to zero. A good trade-off between a negligible noise (i.e. $1 - m \sim 0$) and the storage of a large pattern set (i.e. a high value of $\alpha$) is desirable for any practical purpose model.

## 3   Modular Set of Diluted Attractor Neural Network

A schematic representation of the single ANN is presented in Fig. 1-left. The connectivity ratio $\gamma$ is diluted with $K < N$. A set of $P$ patterns $\xi$ is presented to the network in a learning phase, represented with the red dashed arrow. Then, this set of patterns is presented in a retrieval phase in order to test the recall abilities of the network in terms of the retrieved patterns load $\alpha$, and the quality of the retrieval $m$. This is represented with the solid black arrow.

In Fig. 1-right, a schematic representation of an ensemble of ANN modules with a number of $n$ components is presented. The connectivity in each $ANN_b$ module $b$ is highly diluted with $K_b \ll N$, $b \in \{1, \ldots, n\}$. Note that, in order to keep the computational cost of the single ANN and the ANN ensemble the same, one uses $K = K_b \times n$. The set of patterns is divided into disjoint subsets of uniform size $P_b = P/n$, and each pattern subset is learned by its corresponding $ANN_b$ module as represented with the red dashed arrows. E.g. $\{\xi^\mu, \mu = 1, \ldots, P/n\}$ for the first module $ANN_b, b = 1$, as shown in Fig. 1-right. The solid black arrows in Fig. 1-right, represent the retrieval stage, in which all the pattern subsets are presented to all ANN modules in order to test the discrimination among them. The target patterns are considered as retrieved by the ANN module with the higher overlap value over the retrieval threshold $\theta_r$, i.e. $max(m_b^\mu) > \theta_r$. For comparison purpose, $\theta_r$ is assumed to take the same value for each component in the ensemble, as well as, for the single ANN system.

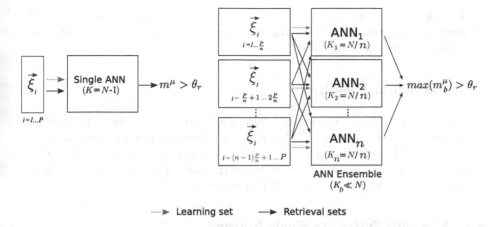

····▸ Learning set     ——▸ Retrieval sets

**Fig. 1.** Schematic representation of a single ANN (left) vs. an ANN ensemble (right) with $n$ components. Red dashed lines represent the learning patterns set flow, black solid lines represent the retrieval patterns sets flow. Overall connectivity in both cases is the same with $K = n \times K_b$, $K_b \ll N$. Retrieval threshold $\theta_r$ is the same for both cases. Color online.

## 3.1 Retrieval Measures of the ANN Ensemble

In order to evaluate the ensemble performance, one may define the retrieval efficiency $R$, as the number of learned patterns that are successfully retrieved $R = \frac{P_r}{P_l}$, where $P_r$ is the overall number of retrieved patterns that satisfy $m^\mu > \theta_r$, and $P_l$ is the overall number of patterns presented to the network during the learning phase. One has that $P_l \geq P_r$. When the super-index $b$ is used, $P_r^b, P_l^b$ refer to the $ANN_b$ module $b$ in the ensemble. Here $\theta_r$ is the retrieval threshold. The mean retrieval overlap $M$ is calculated as the mean retrieval overlap over all patterns subset $\mu \in 1, 2, \ldots, P_l$, $M = \langle m^\mu \rangle = 1/P_l \sum_{\mu=1}^{P_l} m^\mu$. It is worth noting that in the case of the ANN ensemble, the retrieval pattern load is calculated as $\alpha_R = \frac{P_r}{K_b \times n}$, where $n$ is the number of subnetworks. Thus, we use $K_b \times n = K$ constant for all network ensembles studied, where $K$ is the connectivity of the single "dense" network. Also, it is of worth to define the pattern gain $G$ of the ANN ensemble by taking the single ANN system retrieval performance in terms of recovered patterns ($P_r^s$) as baseline, and it is given by $G = P_r^e/P_r^s$. Here $P_r^e$ stands for the number of total recovered patterns by the ANN ensemble and $P_r^s$ stands for the patterns recovered by the single network at the maximum retrieval pattern load $max(\alpha_R)$.

For a highly diluted connectivity the pattern storage is moderate. Although, if one combines several ANN modules in an ensemble learning process, one can increase the overall number of retrieved patterns. The statement of this work is that one can increase the stored number of patterns $\alpha$, with a good quality of retrieval $m$ using an ensemble of ANN modules, with similar computational costs, when compared with a single less-diluted ANN.

In order to measure the performance for comparison purpose, both the ANN ensemble and the single ANN, should have comparable computational costs. This is detailed in the following subsection.

## 4 Simulation Results

The single and ensemble components are studied for different values of $N$ network units, namely $N = \{1, 2, 3, 4, 5\} \times 10^3$. The threshold is kept constant at the value $\theta_R = 0.5$ for all systems. Note that the wiring cost of the system $N \times K \times n$ is the same for all systems under study for the sake of comparison. That is we use $n = \{20, 10, 1\}$ for $\gamma = \{0.05, 0.1, 1.0\}$ respectively.

### 4.1 Ensemble System vs Single Network

In Fig. 2 is shown the performance of three single systems vs. six ensembles systems. In the top panels of Fig. 2 are shown single ($n = 1$) module systems of fully connected networks $\gamma = 1.0$ with $N = 5000, 3000, 1000$. One can appreciate that the larger system manages to store a larger number of patterns $P_r$. For $N = K = 1000$, the network manages to recover $P_r = 135$, for $N = K = 3000$ it recovers $P_r = 395$ and $N = K = 5000$ the recovered patterns are $P_r = 675$. This is expected as the critical load for a fully connected network is $\alpha_c \sim 0.139$ and the recovered patterns are $P_r = \alpha_c K$.

In Fig. 2 middle panels can be appreciated that the performance of ensemble systems with $n = 10$, $\gamma = 0.1$ with same wiring cost ($N \times K \times n$) as the fully connected systems in top panels. The ensemble system modules have degrees $K = 500, 300, 100$ for the left, center and right panels respectively. The corresponding recovered are $P_r = 1300, 780, 260$ for each of these systems. This is equivalent to a pattern gain $G = P_r^e / P_r^s = 1.92$, that is the ensemble systems almost double the performance of the single network systems with the same wiring cost. Again the larger the system in terms of network units $N$, the larger the recovered patterns $P_r$.

Figure 2 bottom panels show the performance of ensemble systems with $n = 20$, $\gamma = 0.05$. The modules in each ensemble system have degree $K = 250, 150, 50$ for the left, center and right panels respectively. The corresponding recovered are $P_r = 1600, 900, 300$ for each of this systems. For the larger system $N = 5000$, $K = 250$, $n = 20$, the ensemble pattern gain is $G \sim 2.4$. Note again that all systems in Fig. 2 have same wiring cost. The ensemble systems outperforms the single system managing to recover more than the double of patterns as presented in Fig. 2. As expected, the fully connected single systems transition from retrieval to non-retrieval is discontinuous for the order parameters $M$ and $R$, while the transition is continuous for the ensemble systems with diluted modules.

A review of this findings is presented in Fig. 3 for the different systems under study. The curves in the figure are built for the values of $P_r$ at the maximum retrieval pattern load $\alpha_R(max)$ depicted in Fig. 2.

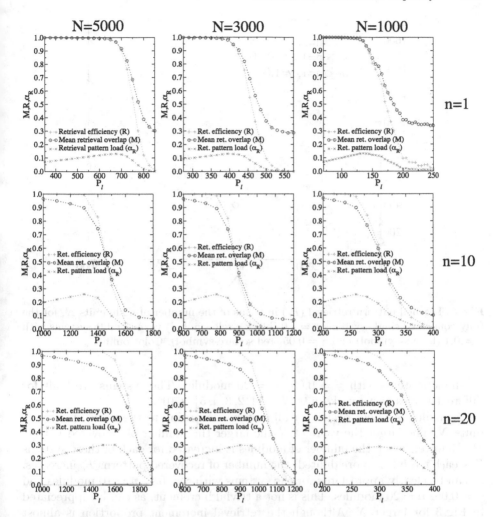

**Fig. 2.** Performance of the systems in terms of the Retrieval efficiency $R$, Mean retrieval overlap $M$ and the Retrieved pattern load $\alpha_R$. **Top panels:** Single fully connected networks $\gamma = 1$, with $N = 5000, 3000, 1000$ in left, center and right panels respectively. Number of modules $n = 1$. **Middle panels:** Ensemble systems with $\gamma = 0.1$, $K = 500, 300, 100$ in left, center and right panels respectively. Number of modules $n = 10$. **Bottom panels:** Ensemble systems with $\gamma = 0.05$, $K = 250, 150, 50$ in left, center and right panels respectively. Number of modules $n = 20$. Color online.

## 4.2    Ensemble Systems Phase Diagrams

Figure 3 summarizes the results for the different performance measures of the ANN ensemble system. In Fig. 3 shows the behavior of the systems under study depicting their performance in terms of $P_r$ for the fully connected single networks $\gamma = 1$, $n = 1$, the ensemble systems with $\gamma = 0.1$, $n = 10$ modules. and the

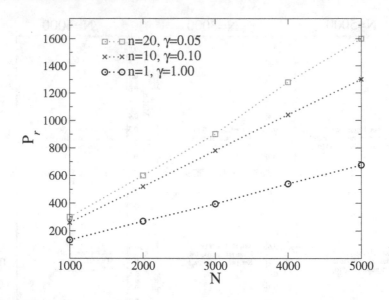

**Fig. 3.** Learned pattern retrieval ($P_r$) in terms of the number of node units $N$, for the fully connected single systems $\gamma = 1.0$ (black circle-symbol) vs. ensemble systems with $\gamma = 0.1$ (blue x-symbol) and $\gamma = 0.05$ (red square-symbol). Color online.

ensemble systems with $\gamma = 0.05$, $n = 20$ modules. The systems are built for different values of network units $N = \{1, 2, 3, 4, 5\} \times 10^3$.

A similar behavior occurs for all systems according to the value of network units $N$. The larger the value of $N$ the larger the number of recovered patters $P_r$. However, when the number of modules is increased for each of these systems (i.e. each module is more diluted) the number of recovered patterns $P_r$ increases, as can be clearly appreciated for the systems with $\gamma = 0.1$, $n = 10$ modules and $\gamma = 0.05$, $n = 20$ modules. This is not a trivial increment, as can be appreciated in Fig. 3 for larger $N$. Although the retrieval increment proportion is almost constant, the lines are not parallel, indicating a better performance for larger systems (larger $N$). The reader should note that the fully connected network (black-circled curve) manages to retrieve $P_r = 675$ patterns, that is a load of $\alpha_c = 0.13$ as expected. The ensemble systems with $\gamma = 0.05, n = 20$ manages to recover $P_r = 1600$ for the same wiring cost $N \times K \times n$.

As fair remark, one can appreciate that the ensemble systems will approach a boundary on their retrieval capacities, given that the increase with the dilution will reach a practical limit where increasing the number of modules in the ensemble will not yield a larger pattern retrieval.

## 5   Conclusions

We have studied ANN ensemble systems with diluted modules for different values of network units $N$ and compared their performances with a with a single attractor network with the same wiring cost. All systems increase the number of stored patterns with the increase of network units $N$, but the ensemble systems achieve a gain over the single systems managing to store more than the double of patterns. Keeping same wiring cost for the ensemble systems one can increase the number of stored patterns using the proposed divide-and-conquer approach. For the systems presented in this paper the increment gain for $N = 5000, \gamma = 0.05, n = 20$ is as large as $G = 2.4$ the number patterns when compared with a similar single network system. The larger the system in terms of $N$, the larger the number of ensemble modules $n$ we can use at non pathological dilution. Our system with $n = 20$, is constrained for comparison purpose by the smaller system $N = 1000$, for which a dilution of $\gamma = 0.05$, gives a degree of $N \times \gamma = 50$. For $N = 5000$ we have a degree of $N \times \gamma = 250$, we could dilute more each module and get more subnetworks. A more detailed study on this issue warrants further investigation.

As commented before, real world applications of ANN face the issue that in order to recover a reasonable number of patterns the number of network units must be larger than the number of patterns $N \gg P$. This issue can be overcome using an ensemble system as the one studied here for different values of $N$. Using this approach would allow the ANN ensemble to deal with real-world data where information is structured and correlated such as fingerprint recognition problems [12, 15]. Also, through the divide-and-conquer parallelizing approach one can apply the modularized ANN to computational costly problems such as video traffic analysis [13].

**Acknowledgments.** This work was funded by the Spanish projects of Ministerio de Economía y Competitividad TIN-2010-19607, TIN2014-54580-R, TIN2014-57458-R (http://www.mineco.gob.es/); and UNEMI-2016-CONV-P-01-01 and DITC-UDLA, Ecuador. The funders had no role in the study design, data collection and analysis, decision to publish, or preparation of the manuscript.

# References

1. Agliari, E., Barra, A., Galluzzi, A., Guerra, F., Moauro, F.: Multitasking associative networks. Phys. Rev. Lett. **109**(26), 268101 (2012)
2. Agliari, E., Barra, A., Galluzzi, A., Guerra, F., Tantari, D., Tavani, F.: Hierarchical neural networks perform both serial and parallel processing. Neural Netw. **66**, 22–35 (2015)
3. Agliari, E., Barra, A., Galluzzi, A., Guerra, F., Tantari, D., Tavani, F.: Metastable states in the hierarchical Dyson model drive parallel processing in the hierarchical Hopfield network. J. Phys. A: Math. Theoret. **48**(1), 015001 (2015). http://stacks.iop.org/1751-8121/48/i=1/a=015001
4. Agliari, F., Barra, A., Galluzzi, A., Guerra, F., Tantari, D., Tavani, F.: Retrieval capabilities of hierarchical networks: from Dyson to Hopfield. Phys. Rev. Lett. **114**(2), 028103 (2015)
5. Amit, D.J., Gutfreund, H., Sompolinsky, H.: Information storage in neural networks with low levels of activity. Phys. Rev. A **35**, 2293–2303 (1987)

6. Derrida, B., Gardner, E., Zippelius, A.: An exactly solvable asymmetric neural network model. Europhys. Lett. **4**, 167–173 (1987)

7. Dominguez, D., Koroutchev, K., Serrano, E., Rodríguez, F.B.: Information and topology in attractor neural network. Neural Comput. **19**(4), 956–973 (2007)

8. Dominguez, D., González, M., Rodríguez, F.B., Serrano, E., Erichsen, R., Theumann, W.K.: Structured information in sparse-code metric neural networks. Phys. A: Stat. Mech. Appl. **391**(3), 799–808 (2012). http://www.sciencedirect.com/science/article/pii/S0378437111007187

9. Dominguez, D., González, M., Serrano, E., Rodríguez, F.B.: Structured information in small-world neural networks. Phys. Rev. E **79**(2), 021909 (2009)

10. Erdös, P., Rényi, A.: On random graphs. I. Publicationes Math. (Debrecen) **6**, 290–297 (1959). http://www.renyi.hu/~p_erdos/Erdos.html#1959-11

11. González, M., Dominguez, D., Rodríguez, F.B.: Block attractor in spatially organized neural networks. Neurocomputing **72**, 3795–3801 (2009)

12. Gonzalez, M., Dominguez, D., Rodriguez, F.B., Sanchez, A.: Retrieval of noisy fingerprint patterns using metric attractor networks. Int. J. Neural Syst. **24**(07), 1450025 (2014)

13. González, M., Dominguez, D., Sánchez, Á.: Learning sequences of sparse correlated patterns using small-world attractor neural networks: an application to traffic videos. Neurocomputing **74**(14–15), 2361–2367 (2011). http://www.sciencedirect.com/science/article/pii/S092523121100155X

14. González, M., Dominguez, D., Sánchez, A., Rodríguez, F.B.: Increase attractor capacity using an ensembled neural network. Expert Syst. Appl. **71**, 206–215 (2017). http://www.sciencedirect.com/science/article/pii/S0957417416306704

15. González, M., del Mar Alonso-Almeida, M., Avila, C., Dominguez, D.: Modeling sustainability report scoring sequences using an attractor network. Neurocomputing **168**, 1181–1187 (2015). http://www.sciencedirect.com/science/article/pii/S0925231215006219

16. Hopfield, J.J.: Neural networks and physical systems with emergent collective computational abilities. Proc. Natl. Acad. Sci. USA **79**(8), 2554–2558 (1982). http://www.pnas.org/cgi/content/abstract/79/8/2554

17. Sollich, P., Tantari, D., Annibale, A., Barra, A.: Extensive parallel processing on scale-free networks. Phys. Rev. Lett. **113**(23), 238106 (2014)

# Opposite-to-Noise ARTMAP Neural Network

Alan Matias[1], Ajalmar Rocha Neto[1(✉)], and Atslands Rocha[2]

[1] Federal Institute of Ceará, Fortaleza, CE, Brazil
alan.matias@ppgcc.ifce.edu.br, ajalmar@ifce.edu.br
[2] Federal University of Ceará, Fortaleza, CE, Brazil
atslands@ufc.br

**Abstract.** Fuzzy ARTMAP (FAM) aims to solve the stability-plasticity dilemma by the adaptive resonance theory (ART). Despite this advantage, category proliferation is an important drawback in Fuzzy ARTMAP due mostly to the overlapping region (noise) between classes. In such a region, the match tracking mechanism is often triggered by raising the vigilance parameter value to avoid future learning errors. In order to overcome this drawback, we propose a Fuzzy ARTMAP-based architecture robust to noise, named OnARTMAP. Our proposal has a two-stage learning process. The first stage requires two new modules, the overlapping region detection module (ORDM) and another one very similar to $ART_a$, called $ART_c$. The ORDM finds the overlapping region between categories and the second one ($ART_c$) computes and stores special categories for overlapping areas (overlapping categories). In the second stage, the weights for conventional categories are estimated from data outside the overlapping area. Consequently, by not considering noise data, the number of categories drops considerably. We can infer from achievements that our proposal in general outperformed Fuzzy ARTMAP, ART-EMAP, $\mu$ARTMAP, and BARTMAP and achieved good data generalization with fewer categories and robustness on noise.

**Keywords:** Neural networks · Adaptive Resonance Theory · Fuzzy ARTMAP · Category proliferation

## 1 Introduction

The Adaptive Resonance Theory develop by Grossberg [13–15] has inspired new architectures of artificial neural networks [3,6,9–11]. As examples, we can highlight ART 1 [4], ART 2 [5] and the well-known Fuzzy ART [8] for unsupervised learning tasks. In a nutshell, ART 1 deals only with binary data, while ART 2 handles both binary and analog data. On the other hand, Fuzzy ART is an architecture that relies on fuzzy set operators for analog data. The supervised learning approach based on Grossberg's theory was introduced in the ARTMAP architecture [7]. To do so, the ARTMAP uses two ART modules: the A-side module ($ART_a$) and the B-side module ($ART_b$) for binary data. As an improvement, Fuzzy ARTMAP was proposed to analog data [6].

© Springer International Publishing AG 2017
I. Rojas et al. (Eds.): IWANN 2017, Part I, LNCS 10305, pp. 507–519, 2017.
DOI: 10.1007/978-3-319-59153-7_44

The Fuzzy ARTMAP can map an arbitrary multidimensional dataset by creating hyperboxes for both input pattern, in the A-side module ($ART_a$), and input label, in the B-side module ($ART_b$). Some Fuzzy ARTMAP advantages are remarkable, such as fast and stable learning, the need of few epochs to achieve stability, the ability to learn quickly and stably new data without catastrophically forgetting past data, and so on. Despite such advantages, Fuzzy ARTMAP is very sensitive to input pattern presentation order as well as to noise data. This is an important drawback concerning the learning process known as category proliferation, since a large number of categories should be included in the module $ART_a$ to represent the input space and its relations to the output space.

The main reason for the category proliferation is the correction of predictive error (in the module $ART_b$) performed by the match tracking mechanism. Due to this correction process, smaller categories should be created inside larger ones. As a result, the learning process of Fuzzy ARTMAP should create too many small and specialized categories. In fact, the larger the overlapping area between classes in a classification task, the larger is the number of small categories within this region. Therefore, one can see that the category proliferation problem is intensified with the degree of class overlapping.

Several works can be found in the literature to handle the category proliferation problem in Fuzzy ARTMAP. Such works are mainly based on changing the Fuzzy ARTMAP architecture [12, 16, 19] or even on changing the geometry by not using hyperrectangles [1, 2, 17, 18]. The models that changes the Fuzzy ARTMAP architecture includes the Boosted ARTMAP [16], which allows non-zero training error in order to improve overall generalization; and $\mu$ARTMAP [12], which uses a probabilistic setting to optimize the categories sizes. The models that changes the category geometry includes the Gaussian ARTMAP [18], a synthesis between a gaussian classifier and the Adaptive Resonance Theory; Ellipsoid ARTMAP [2], which uses hyperellipsoids for data generalization; and Polytope ARTMAP [1], which categories are irregular polytopes.

Taking a closer look at the previous attempts to solve the category proliferation problem, we have not find one that aims, at first, to achieve the overlapping area between classes under the ART framework. In this context, this work focuses on solving the category proliferation problem by detecting the overlapping area (noise) and then creating categories placed far from the noise data. To do so, we propose a novel Fuzzy ARTMAP-based architecture that can identify the overlapping region between classes, if there exists, and exclude this noise data from the training data set for preventing the creation of unnecessary categories. This is accomplished by the addition of a new Fuzzy ARTMAP module for overlapping region detection henceforth called overlapping region detection module (ORDM). Our proposal is named opposite-to-noise ARTMAP (OnARTMAP), since it is able to learn the data outside the noisy region.

Our paper is organized as follows. In the next section, Fuzzy ART and Fuzzy ARTMAP architectures are briefly described. In Sect. 3, our proposal is presented in detail. After that, Sect. 4 presents the simulation achievements carried out for OnARTMAP as well as for other Fuzzy ARTMAP models found in the literature.

In this section, results for artificial and real problems are also included. At last, Sect. 5 presents the conclusions and future works.

## 2   Fuzzy ARTMAP

The Fuzzy ARTMAP architecture has two Fuzzy ART modules, and an additional module to link them, named inter-ART. The Fuzzy ART modules are described below.

### 2.1   Fuzzy ART

The Fuzzy ART module comprises three layers: $F_0$, $F_1$, and $F_2$. The layer $F_0$ stands for the current input vector; the layer $F_1$, that receives both bottom-up input from $F_0$ (i.e., the output of $F_0$) and the top-down input from $F_2$. The layer $F_2$ represents the active category. The activation (output) for $F_0$ is the current input vector $\mathbf{I} \in \mathbb{R}^M$ described by $\mathbf{I} = (\mathbf{a}) = (a_1, \dots, a_i, \dots, a_M)$, so that $a_i \in [0,1]$ and the norm $|\mathbf{I}| = \sum_{i=1}^{M} |a_i|$. An alternative representation for the input vector $\mathbf{I} \in \mathbb{R}^{2M}$ is given by the complement coding $\mathbf{I} = (\mathbf{a}, \mathbf{a}^c)$, where $\mathbf{a}^c = 1 - \mathbf{a}$. Note that the norm

$$|\mathbf{I}| = |(\mathbf{a}, \mathbf{a}^c)| = \sum_{i=1}^{M} a_i + \left(M - \sum_{i=1}^{M} a_i\right) = M. \tag{1}$$

The outputs for $F_1$ and $F_2$ are $\mathbf{x} = (x_1, \dots, x_M)$ and $\mathbf{y} = (y_1, \dots, y_N)$, respectively. Moreover, the $j$-th node of $F_2$ with an adaptive weight vector $\mathbf{w}_j = (w_1, \dots, w_i, \dots, w_M)$ is a category representing a training patterns subset. For input vectors with complement coding, the weight vectors $\mathbf{w}_j = (\mathbf{u}_j, \mathbf{v}_j^c)$ are 2M-dimensional. At the beginning of the training process, each component $w_i$ equals one and, during the process, is monotonically nonincreasing, which let the learning be stable. As for parameters, the Fuzzy ART has the vigilance parameter $\rho \in [0,1]$, the choice parameter $\alpha > 0$, and the learning rate $\beta \in [0,1]$.

For a certain input vector $\mathbf{I}$, the choice function is defined by

$$T_j(\mathbf{I}) = T_j = \frac{|\mathbf{I} \wedge \mathbf{w}_j|}{\alpha + |\mathbf{w}_j|}, \tag{2}$$

where the fuzzy operator AND ($\wedge$) is defined by $\mathbf{x} \wedge \mathbf{y} = min(x_i, y_i)$. Indeed, the **category choice** is given by

$$T_J = \max\{T_j\}_{j=1}^{N}. \tag{3}$$

where $J$ is the index of the chosen category. If more than one $T_j$ is maximal, then the $j$-th category with the lowest index is chosen. In such a situation, $y_J = 1$ and $y_j = 0$, whenever $j \neq J$.

Resonance occurs if the match function, $|\mathbf{I} \wedge \mathbf{w}_J|/|\mathbf{I}|$, of the choice category $J$ meets the **vigilance criterion**

$$\frac{|\mathbf{I} \wedge \mathbf{w}_J|}{|\mathbf{I}|} \geq \rho. \tag{4}$$

In this context, if the vigilance criterion complies with Eq. (4), the choice category $\mathbf{w}_J$ matches and the **updating rule** must be performed, according to

$$\mathbf{w}_J^{new} = \beta(\mathbf{I} \wedge \mathbf{w}_J^{old}) + (1 - \beta)\mathbf{w}_J^{old}. \tag{5}$$

However, if Eq. (4) is not satisfied then category $J$ is no more selected (at least for the current input vector). In this situation, a new category is $(i)$ obtained by Eq. (3) and $(ii)$ evaluated in terms of the vigilance criterion by Eq. (4). The steps $(i)$ and $(ii)$ are performed until the vigilance criterion is satisfied or a new category is created.

## 2.2 Fuzzy ARTMAP Modules

The input vectors of $ART_a$ and $ART_b$ modules are $\mathbf{I} = (\mathbf{a}, \mathbf{a}^c)$ for attributes and $\mathbf{I}^b = (\mathbf{b}, \mathbf{b}^c)$ for labels, respectively. Besides that, for the module $ART_a$, $\mathbf{x}^a = (x_1^a, \ldots, x_{2M_a}^a)$ stands for the layer $F_1^a$, $\mathbf{y}^a = (y_1^a, \ldots, y_{N_a}^a)$ for the layer $F_2^a$, as well as $\mathbf{w}_j^a = (w_{j1}^a, \ldots, w_{j2M_a}^a)$ are the categories (nodes) in the layer $F_2^a$. Similarly, for the module $ART_b$, the $\mathbf{x}^b = (x_1^b, \ldots, x_{2M_b}^b)$ describes the output of the layer $F_1^b$, $\mathbf{y}^b = (y_1^b, \ldots, y_{N_b}^b)$ the output for the layer $F_2^b$, and $\mathbf{w}_k^b = (w_{k1}^b, \ldots, w_{k2M_b}^b)$ the nodes. Moreover, the set $\mathbf{W}^a = \{\mathbf{w}_j^a : \forall j\}$ and $\mathbf{W}^b = \{\mathbf{w}_k^b : \forall k\}$.

As stated before, Fuzzy ARTMAP has a map field in a linking module between $ART_a$ and $ART_b$, called inter-ART. Such a module has a layer $F^{ab}$ whose output is denoted by $\mathbf{x}^{ab} = (x_1^{ab}, \ldots, x_{N_b}^{ab})$ and the categories in the layer $F^{ab}$ are denoted by $\mathbf{w}_j^{ab} = (w_{j1}^{ab}, \ldots, w_{jN_b}^{ab})$, where $j = 1, \ldots, N_a$. The following vectors $\mathbf{x}^a$, $\mathbf{y}^a$, $\mathbf{x}^b$, $\mathbf{y}^b$ and $\mathbf{x}^{ab}$ are set to zero between different input presentations.

The **map field activation** is computed as follows.

$$\mathbf{x}^{ab} = \begin{cases} \mathbf{y}^b \wedge \mathbf{w}_J^{ab} & \text{if the } Jth\ F_2^a \text{ node is active and } F_2^b \text{ is active,} \\ \mathbf{w}_J^{ab} & \text{if the } Jth\ F_2^a \text{ node is active and } F_2^b \text{ is inactive,} \\ \mathbf{y}^b & \text{if } F_2^a \text{ is inactive and } F_2^b \text{ is active, and} \\ \mathbf{0} & \text{if } F_2^a \text{ is inactive and } F_2^b \text{ is inactive.} \end{cases} \tag{6}$$

As one can see that $F^{ab}$ is activated by either or both $F_2^a$ and $F_2^b$ category fields and $\mathbf{x}^{ab} = \mathbf{0}$ if the prediction $\mathbf{w}_j^{ab}$ is unconfirmed by $\mathbf{y}^b$, see Eq. (6). In case of such a mismatch, the match tracking mechanism is triggered to search for a better category.

We highlight that at each input vector presentation, the vigilance parameter $\rho_a$ for $ART_a$ equals a baseline vigilance $\overline{\rho_a}$ that was set up before starting the training process. Moreover, if $|\mathbf{x}^{ab}| < \rho_{ab}|\mathbf{y}^b|$, so that $\rho_{ab}$ is the map field vigilance parameter, then $\rho_a$ is temporarily raised to $|\mathbf{I}^a \wedge \mathbf{w}_J^a|/|\mathbf{I}^a|+\varepsilon$, where $0 < \varepsilon < 1$. This is accomplished to have an other active node $J$ in the layer $F_2^a$, so that $|\mathbf{I}^a \wedge \mathbf{w}_J^a| \geq \rho_a |\mathbf{I}^a|$. This process is performed until a node from $F_2^a$ correctly predicts the activation of the layer $F_2^b$; otherwise, a new node is committed in $ART_a$.

Concerning the map field learning, we initially have $\mathbf{w}_{jk}^{ab}(0) = 1$, $\forall j = 1, \ldots, N_a, \forall k = 1, \ldots, N_b$. When resonance occurs, in which the $J$-th node in $F_2^a$ becomes active, $\mathbf{w}_J^{ab}$ approaches the map field vector $\mathbf{x}^{ab}$. Considering a fast learning approach (i.e., with $\beta = 1$), $\mathbf{w}_J^{ab}$ learns to predict the $ART_b$ category $K$ (i.e., label).

# 3   The Opposite-to-Noise ARTMAP (Proposal)

In this section, our proposal to handle the category proliferation problem is presented. The architecture for the Opposite-to-noise ARTMAP (OnARTMAP) has two additional modules when compared with the conventional Fuzzy ARTMAP. The first module, henceforth named ORDM (Overlapping Region Detection Module), was designed to obtain the overlapping area between classes and the second one ($ART_c$) to store specialized categories (overlapping categories) concerning such an overlapping region. The main idea is to detect, in the first stage of learning, the overlapping area by the ORDM and then to exclude the training patterns inside this categories from the training set. In this stage, the information about the overlapping categories is stored in the module $ART_c$. As expected, the module $ART_c$ is very similar to its counterparts $ART_a$ and $ART_b$. In the second stage, the conventional categories are placed far from this critical zone as usual.

## 3.1   Overlapping Region Detection Module (ORDM)

The ORDM plays an important role in the first stage by detecting the overlapping region and in the second one by checking if there exists an intersection between categories. In the first stage, ORDM receives as input the outputs $\mathbf{y}_a$ and $\mathbf{y}_b$ (in fact, the categories $J$ and $K$) from $ART_a$ and $ART_b$, respectively; and obtains the class associated with the category $J$; i.e., $\mathbf{w}_K^b$. After that, ORDM finds the categories $\mathbf{w}_j^a$ that not belong to the class $\mathbf{w}_K^b$. This can also be expressed as finding $\mathbf{w}_j^{ab} \neq \mathbf{w}_J^{ab}$. Thus, we can describe the weight subset formally as

$$\widetilde{\mathbf{W}}^a = \{\mathbf{w}_{j'}^a : \mathbf{w}_j^{ab} \neq \mathbf{w}_J^{ab}, \; \forall \, j \neq J\}. \tag{7}$$

After having the categories of interest, the ORDM checks the existence of an intersection between the category $\mathbf{w}_J^u$ and the elements $\mathbf{w}_{j'}^u$ of the set $\widetilde{\mathbf{W}}^u$. Let us consider the weight $\mathbf{w}_J^a$ in a geometric point of view as an hyper-rectangle, such that $\mathbf{w}_J^a = (\mathbf{u}_J, \mathbf{v}_J^c)$. The actual geometric representation for $\mathbf{w}_J^a$ is denoted by $\mathbf{w}_J^{a*} = (\mathbf{u}_J, (\mathbf{v}_J^c)^c) = (\mathbf{u}_J, \mathbf{v}_J)$. Similarly, each category as actual hyper-rectangle is $\mathbf{w}_{j'}^a$ in $\widetilde{\mathbf{W}}^a$ and, thus, $\mathbf{w}_{j'}^{a*} = (\mathbf{u}_{j'}, (\mathbf{v}_{j'}^c)^c) = (\mathbf{u}_{j'}, \mathbf{v}_{j'})$.

To compute the existence of an intersection between hyper-rectangles, we can assume that there is no intersection ($\Psi = 0$) and then checking if there is an intersection at each dimension $m = 1, \ldots, M_a$. We can perform that by initially supposing that $\theta = 0$ (an auxiliary variable) and assessing the following conditions

$$\textbf{if } u_{Jm} \leq u_{j'm} \leq v_{Jm} \leq v_{j'm} \textbf{ or}$$
$$u_{j'm} \leq u_{Jm} \leq v_{j'm} \leq v_{Jm} \textbf{ or}$$
$$u_{Jm} \leq u_{j'm} \leq v_{j'm} \leq v_{Jm} \textbf{ or} \tag{8}$$
$$u_{j'm} \leq u_{Jm} \leq v_{Jm} \leq v_{j'm} \textbf{ then}$$
$$\theta = \theta + 1$$

By doing so, if at the end of previous assessment the value of $\theta = M_a$ then the hyper-rectangle $\mathbf{w}_J^{a*}$ and $\mathbf{w}_{j'}^{a*}$ intersect each other (i.e., $\Psi = 1$). Otherwise, there is no intersection ($\Psi = 0$). If there is an intersection between $\mathbf{w}_J^{a*}$ and $\mathbf{w}_{j'}^{a*}$, one can compute the intersection between hyper-rectangles by

$$\mathbf{q}_h = \mathbf{w}_{j'}^a \vee \mathbf{w}_J^a, \tag{9}$$

where the fuzzy set operator OR ($\vee$) is denoted by $\mathbf{x} \vee \mathbf{y} \equiv \max(x_i, y_i)$.

The ORDM output is a set $\mathbf{C}$ of $M_a$-dimensional points $\mathbf{c}$ when intersections occur. As expected, such a set is empty when no intersection occurs. Consider the obtained intersections $\mathbf{q}_h$ and its hyper-rectangle representation $\mathbf{q}_h^* = (\mathbf{u}_h, \mathbf{v}_h)$, in which the lower point ($\mathbf{u}_h$) and upper point ($\mathbf{v}_h$) are described by $\mathbf{c}_h^1 = \mathbf{u}_h$ and $\mathbf{c}_h^2 = \mathbf{v}_h$.

In our proposal, each $\mathbf{q}_h$ has as result two samples $\mathbf{c}_h^1$ and $\mathbf{c}_h^2$. The pattern $\mathbf{c}_h^1$ is given as input to the module $ART_c$ and then the pattern $\mathbf{c}_h^2$. Nevertheless, note that each sample $\mathbf{c}_h$ (i.e., the general term for either $\mathbf{c}_h^1$ or $\mathbf{c}_h^2$) is in a complement coding representation described by $\mathbf{I}^c = (\mathbf{c}_h, 1 - \mathbf{c}_h)$.

## 3.2   The $ART_c$ Module

$ART_c$ is a conventional Fuzzy ART module and, as stated, its input is the $\mathbf{I}^c$ obtained through the ORDM. The overlapping categories $\mathbf{w}_l^c$ are stored in the layer $F_2^c$, such that the weight vector $\mathbf{w}_l^c$ is $2M_a$-dimensional. Initially, $\mathbf{w}_l^c = \mathbf{1}$ and for a certain input $\mathbf{I}^c$, (i) the choice function is computed by Eq. (2) and the maximal activation $T_L^c$ is selected by Eq. (3); (ii) the vigilance parameter $\rho_c$ is used to perform the vigilance criterion described in Eq. (4); and (iii) if the vigilance criterion is satisfied, then the weight vector $\mathbf{w}_L^c$ is updated as described in Eq. (5). As a result, at the end of the first stage, the nodes in the layer $F_2^c$ are the overlapping categories.

## 3.3   Learning Process Overview for OnARTMAP

In this subsection, the two-stage learning for OnARTMAP is detailed. The first stage aims to find and store the overlapping categories, and the second one aims to learn the conventional categories weights. In the first stage, the $ART_a$ module uses a distinct vigilance parameter $\rho_o$, while in the second stage the vigilance parameter is the usual $\rho_a$. In the first stage, the match tracking mechanism is disabled, which means not to raise the vigilance parameter $\rho_o$ when a predictive error occurs. The duration of the first stage is only an epoch long (i.e., each training pattern is presented once). By disabling the match tracking mechanism, the OnARTMAP is able to find the overlapping categories fast and easier.

We present in the Fig. 1 the learning process overview for our proposal. The pictures presented in the Fig. 1(a) would represent an application of OnARTMAP when $\rho_o = 0$ (i.e., $\rho_a = 0$ in firt stage) and $\rho_c = 0$. One can note that, at the end of the first learning stage, the OnARTMAP categories should look like as depicted in the Fig. 1(b), where the blue category is the resulting overlapping category. To do so, when a category $\mathbf{w}_j^a$ achieves the resonance state, the OnARTMAP triggers both ORDM and $ART_c$ modules. The ORDM is in charge of finding the overlapping area between the category $\mathbf{w}_j^a$ and categories $\mathbf{w}_{j'}^a$, not related to the class $\mathbf{w}_K^b$ by the inter-ART module. The module $ART_c$ is responsible for computing these overlapping categories.

When the first learning stage is complete, the stored categories in the layer $F_2^a$ are dropped (see Fig. 1(c)) and the patterns inside the overlapping categories are no more used for training (see Fig. 1(d)). A certain pattern is removed and

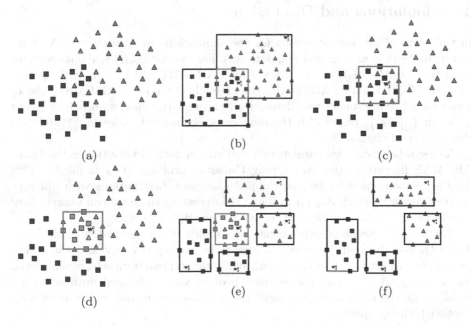

**Fig. 1.** An overview for OnARTMAP: (a) a two classes noise data distribution; (b) OnARTMAP with $\rho_o = 0$ and $\rho_c = 0$ after the first learning stage (in blue, the overlapping category); (c) the categories created in the first stage are dropped; (d) the patterns inside the overlapping category are not used in the second learning stage; (e) no categories are created over the overlapping region; (f) the resulting categories $\mathbf{w}_j^a$. (Color figure online)

not used in the second stage if at least one category $\mathbf{w}_l^c$ contains the input $\mathbf{I}_i^a$. We can check if the $l$-th category contains the $i$-th pattern $\mathbf{I}_i^a$ by

$$\frac{|\mathbf{I}_i^a \wedge \mathbf{w}_l^c|}{|\mathbf{w}_l^c|} = 1. \tag{10}$$

In the second stage, the OnARTMAP then proceeds to learn the resulting training patterns. As one can note, the OnARTMAP does not allow the creation of categories over the overlapping area (see Fig. 1(e)).

For this purpose, the ORDM is called after updating the category $\mathbf{w}_J^a$ to verify if the new weight vector for the category $\mathbf{w}_J^a$ intersects any overlapping category $\mathbf{w}_l^c$ (see Eq. (8)). If there is an intersection then the weight updating is undone. It is worth noticing that, in the second stage, patterns are no more removed. After the second stage, few nodes are generated since categories are forbidden from being created in the overlapping area. An example for conventional categories $\mathbf{w}_J^a$ is depicted in the Fig. 1(f).

Finally, we highlight that the OnARTMAP prediction is performed as the conventional Fuzzy ARTMAP. Thus, for a unseen pattern $\mathbf{I}^a$, we find the category $\mathbf{w}_J^a$ and then the category $\mathbf{w}_K^b$ (label) by the inter-ART association.

## 4    Simulations and Discussion

In this section, we present results for the simulations we carried out. We have results for both synthetic and real data sets. For synthetic and real data sets, we compared OnARTMAP with Fuzzy ARTMAP, ART-EMAP [11] with $Q$-max rule, BARTMAP and $\mu$ARTMAP. We highlight that for ART-EMAP, the $Q$ value for all experiments was determined through rule-of-thumb, which obeys $Q = \min\left\{\frac{C}{2L}, 30\right\}$, where $C$ is the number of committed nodes in $F_2^a$ and $L$ is the number of classes [9].

The synthetic data set consists of a well known data distribution in the Fuzzy ARTMAP literature: the overlapping Gaussian problem (OG problem). This data set is composed by two overlapped Gaussian distributions with different degrees of overlapping. Such a problem is interesting since we can observe how the neural networks stability is affected by the additive noise.

For the OG problem experiment, we employed a conservative limit ($\alpha = 0.00001$). For the real data sets, we used the voting strategy and the choice parameter was set to $\alpha = 0.1$. For both synthetic and real data sets, we employed fast learning ($\beta = 1$) and the learning process was performed until achieving stability for the categories (i.e., until the weights are unchangeable, even with another training epoch).

### 4.1    Artificial Datasets

As stated, overlapping area between classes is the leading cause of the category proliferation problem in Fuzzy ARTMAP since the match tracking mechanism is often triggered to fix learning mistakes. Thus, our experiments are used to demonstrate the performance of the OnARTMAP as the overlap degree increases. We generated three artificial data sets comprised of 2-dimensional Gaussian data. Each data set has two different classes with means $\mu = (0.3, 0.3)$ and $\mu = (0.7, 0.7)$, both with prior probability $1/2$. For the first data set (OG1), we set $\sigma = (0.1, 0.1)$. For the second data set (OG2), we set $\sigma = (0.2, 0.2)$; and finally, for the third data set (OG3), we set $\sigma = (0.3, 0.3)$. We generated 1000 training points and 10000 test points for each dataset. In these experiments, we used $\rho_a = 0$ for FAM, ART-EMAP, BARTMAP and OnARTMAP. For OnARTMAP, we used $\rho_o = 0$, $\rho_c = 0$. The $\mu$ARTMAP parameters $h_{max}$ and $H_{max}$, and the BARTMAP parameter $\epsilon_d$ were relaxed as the Gaussian overlap degree was increased. We show the results for the aforementioned neural networks (NN) in Table 1, which has the accuracy, standard deviation, number of created categories (Nodes) and number of overlapping categories (OC).

**Table 1.** Accuracy, conventional and overlapping categories for the OG Problem.

| Dataset | NN | Accuracy (%) | Nodes | OC |
|---------|-----|--------------|-------|-----|
| OG1 | FAM | $99.38 \pm 0.16$ | 5.0 | - |
|     | ART-EMAP | $99.38 \pm 0.16$ | 5.0 | - |
|     | $\mu$ARTMAP | $99.39 \pm 0.14$ | 2.8 | - |
|     | BARTMAP | $99.15 \pm 0.27$ | **2.0** | - |
|     | OnARTMAP | **$99.57 \pm 0.34$** | 3.0 | 1 |
| OG2 | FAM | $86.68 \pm 1.12$ | 67.2 | - |
|     | ART-EMAP | $87.36 \pm 3.07$ | 67.2 | - |
|     | $\mu$ARTMAP | $88.54 \pm 2.62$ | 13.0 | - |
|     | BARTMAP | $87.12 \pm 1.57$ | 21.1 | - |
|     | OnARTMAP | **$90.62 \pm 1.85$** | **2.9** | 1 |
| OG3 | FAM | $73.67 \pm 0.82$ | 139.0 | - |
|     | ART-EMAP | $76.60 \pm 4.82$ | 139.0 | - |
|     | $\mu$ARTMAP | $77.48 \pm 3.73$ | 13.5 | - |
|     | BARTMAP | $77.33 \pm 1.17$ | 47.5 | - |
|     | OnARTMAP | **$81.10 \pm 1.23$** | **2.9** | 1 |

By analyzing the Table 1, one can infer that the OnARTMAP classification outperforms the other classifiers in all degrees of noise concerning the accuracy. Moreover, the number of OnARTMAP categories created is stable as the overlap degree increases, while the number of categories for Fuzzy ARTMAP, $\mu$ARTMAP and BARTMAP increases as the overlap degree increases. The OnARTMAP stability, concerning the number of categories created, is because of its ability to detect and eliminate overlapping data. As we used $\rho_o = 0$ and $\rho_c = 0$, only one overlapping category was found in all experiments, and this one corresponds to the entire overlapping region in each learning data set.

Figure 2 shows how the OnARTMAP categories copes on the OG problem as the noise degree increases (see the blue rectangle; i.e., the overlapping category). In such a situation, by excluding the complete overlapping information, the OnARTMAP will handle a new problem (in fact, a linear one). By doing so, OnARTMAP will generalize by creating categories, which lay in regions outside (opposite) the overlapping region, preventing over-fitting and improve its generalization capability.

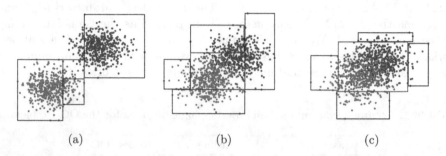

(a)                          (b)                          (c)

**Fig. 2.** OnARTMAP categories for (a) OG1; (b) OG2; and (c) OG3 problems. (Color figure online)

## 4.2   Real Datasets

In these experiments, we used five benchmark data sets for medical diagnosis from the UCI Machine Learning Repository: Vertebral Column Pathology (VCP), Breast Cancer Winsconsin (BCW), Heart Disease (HDE), Pima Indian Diabetes (PID), and Hepatitis (HEP). The information about the data sets (number of data samples, number of classes and number of features) can be seen in Table 2. In our simulations, 80% and 20% of patterns were selected from the entire data set, in each realization, for training and testing purposes, respectively. We carried out 30 realizations of this splitting process for computing the performance metrics.

The Fuzzy ARTMAP neural network and its variation are very sensitive concerning the pattern presentation order, and in some cases, its performance is very

**Table 2.** Real benchmarks used in this work.

| NN | Number of data samples | Number of input features | Number of classes |
|---|---|---|---|
| VCP | 310 | 6 | 2 |
| BCW | 688 | 9 | 2 |
| HDE | 270 | 13 | 2 |
| PID | 768 | 8 | 2 |
| HEP | 80 | 19 | 2 |

**Table 3.** Results for FAM, ART-EMAP, $\mu$ARTMAP, BARTMAP and OnARTMAP on real data sets with voting strategy (T = 5).

| Dataset | NN | Parameters | Epochs | Acc. (%) | Nodes | OC |
|---|---|---|---|---|---|---|
| VCP | FAM | - | 6.5 | 80.38 | 20.01 | - |
| | ART-EMAP | $Q = 5.37$ | 6.5 | 81.61 | 20.01 | - |
| | $\mu$ARTMAP | $h_{max} = 0.15,\ H_{max} = 0.30$ | 43.92 | 81.83 | 38.01 | - |
| | BARTMAP | $\epsilon = 0.1$ | 8.94 | 80.32 | 17.55 | - |
| | OnARTMAP | $\rho_c = 0.95,\ \rho_o = 0.95$ | 3.47 | **83.55** | **11.41** | 7.26 |
| BCW | FAM | - | 4.7 | 96.52 | 13.55 | - |
| | ART-EMAP | $Q = 3.8$ | 4.7 | **97.08** | 13.55 | - |
| | $\mu$ARTMAP | $h_{max} = 0,\ H_{max} = 0.15$ | 15.32 | 92.78 | **4.87** | - |
| | BARTMAP | $\epsilon = 0.1$ | 4.07 | 94.76 | 5.46 | - |
| | OnARTMAP | $\rho_c = 0.85,\ \rho_o = 0.00$ | 4.79 | 96.38 | 13.59 | 3.75 |
| HDE | FAM | - | 5.5 | 65.93 | 23.7 | - |
| | ART-EMAP | $Q = 6.3$ | 5.5 | 68.46 | 23.7 | - |
| | $\mu$ARTMAP | $h_{max} = 0.25,\ H_{max} = 0.50$ | 23.35 | 67.10 | 55.02 | - |
| | BARTMAP | $\epsilon = 0.1$ | 8.17 | 66.67 | 22.05 | - |
| | OnARTMAP | $\rho_c = 0.95,\ \rho_o = 0.00$ | 2.8 | **72.90** | **8.45** | 2.91 |
| HEP | FAM | - | 2.9 | 63.54 | 9.45 | - |
| | ART-EMAP | $Q = 2.72$ | 2.9 | 57.08 | 9.45 | - |
| | $\mu$ARTMAP | $h_{max} = 0.25,\ H_{max} = 0.50$ | 13.08 | 62.92 | 16.65 | - |
| | BARTMAP | $\epsilon = 0.1$ | 3.39 | 61.04 | 8.85 | - |
| | OnARTMAP | $\rho_c = 0.0,\ \rho_o = 0.0$ | 1.73 | **64.17** | **6.15** | 0.6 |
| PID | FAM | - | 7.9 | 69.78 | 71.03 | - |
| | ART-EMAP | $Q = 18.12$ | 7.9 | 70.00 | 71.03 | - |
| | $\mu$ARTMAP | $h_{max} = 0.25,\ H_{max} = 0.50$ | 99.01 | **73.31** | 117.99 | - |
| | BARTMAP | $\epsilon = 0.1$ | 16.85 | 70.65 | 65.29 | - |
| | OnARTMAP | $\rho_c = 0.95,\ \rho_o = 0.00$ | 7.48 | 69.72 | **58.54** | 3.89 |

degraded. In a view to achieve the better approximation about the real performance of the Fuzzy ARTMAP over some data distribution, we employed a voting strategy. Basically, the voting strategy is achieved through a voting committee of size $T$, where each committee member is a trained Fuzzy ARTMAP neural network, and each member is trained with different orders of input presentation.

We compared our proposal OnARTMAP with Fuzzy ARTMAP, ART-EMAP with $Q$-max rule, $\mu$ARTMAP and BARTMAP. The vigilance parameter value was set to $\rho_a = 0$ for all neural networks to achieve max compression in data.

The results by applying the voting strategy in the training stage are described in Table 3. In this table, we have the parameters (Parameters) for each neural network, the number of epochs needed to reach stability (Epochs), the prediction accuracy (Acc.), the mean number of categories created by the $T$ members of the voting committee (Nodes), and the number of overlapping categories (OC) for the OnARTMAP.

In Table 3, the accuracy for OnARTMAP was at least 1.72% higher than the other classifiers on data set VCP and, indeed, the number of needed categories was the smallest. Moreover, the $\mu$ARTMAP achieved the second place in terms of accuracy. As for BCW data set, Fuzzy ARTMAP, ART-EMAP and OnARTMAP achieved equivalent accuracies (96–97%); but $\mu$ARTMAP and BARTMAP achieved accuracies between 92.78% and 94.76%. Concerning the HDE data set, the accuracy for OnARTMAP was about 4–7% higher than the other neural networks, as well as created few categories.

We can also infer on the basis of the Table 3 that the OnARTMAP and Fuzzy ARTMAP performances are similar (in fact, slightly higher) to each other with respect to the data set HEP. Finally, for PDI data set, the accuracy for OnARTMAP is very close to the accuracies for Fuzzy ARTMAP, ART-EMAP and BARTMAP.

## 5   Conclusion

A novel Fuzzy ARTMAP-based architecture called OnARTMAP has been introduced. It implements a new learning approach that avoids to learn the overlapping data. By avoiding the overlapping data, we can prevent over-fitting and improve the accuracy. This can be possible because of its main contribution: the ORDM, responsible for finding the overlapping information, and the $ART_c$ module, responsible for storing the overlapping information. In problems with any similarity with the overlapping Gaussians problem treated is this paper, the OnARTMAP can maintain its stability concerning the number of created categories. Indeed, it preserves its generalization capabilities. This is because that the $\rho_o$ and $\rho_c$ parameters can be easily set, once the data distribution is known.

## References

1. Amorin, D., Delgado, M., Ameneiro, S.: Polytope ARTMAP: pattern classification without vigilance based on general geometry categories. IEEE Trans. Neural Netw. **18**, 1306–1325 (2007)
2. Anagnostopoulos, G., Georgiopoulos., M.: Ellipsoid ART and ARTMAP for incremental clustering and classification. In: IEEE International Joint Conference on Neural Networks, vol. 2, pp. 1221–1226 (2001)
3. Carpenter, G., Gaddam, S.: Biased ART: a neural architecture that shifts attention toward previously disregarded features following an incorrect prediction. Neural Netw. **23**, 435–451 (2010)
4. Carpenter, G., Grossberg, S.: A massively parallel architecture for a self-organizing neural pattern recognition machine. Comput. Vis. Graph. Image Process. **37**, 54–115 (1987)
5. Carpenter, G., Grossberg, S.: ART 2: self-organization of stable category recognition codes for analog input patterns. Appl. Opt. **3**, 129–152 (1990)
6. Carpenter, G., Grossberg, S., Markuzon, N., Reynolds, J., Rosen, D.: Fuzzy ARTMAP: a neural network architecture for incremental supervised learning of analog multidimesional maps. IEEE Trans. Neural Netw. **3**, 698–713 (1992)

7. Carpenter, G., Grossberg, S., Reynolds, J.: ARTMAP: supervised real-time learning and classification of nonstationary data by a self-organizing neural network. Neural Netw. **4**, 565–588 (1991)
8. Carpenter, G., Grossberg, S., Rosen, D.: Fuzzy ART: Fast stable learning and categorization of analog patterns by an adaptive resonance system. Neural Netw. **4**, 759–771 (1991)
9. Carpenter, G., Markuzon, N.: ARTMAP-IC and medical diagnosis: instance counting and inconsistent cases. Neural Netw. **11**, 323–336 (1998)
10. Carpenter, G., Milenova, B., Noeske, B.: Distributed ARTMAP: a neural network for fast distributed supervised learning. Neural Netw. **11**, 793–813 (1998)
11. Carpenter, G., Ross, W.: ART-EMAP: a neural network architecture for object recognition by evidence accumulation. IEEE Trans. Neural Netw. **6**, 805–818 (1995)
12. Gómez-Sánchez, E., Dimitriadis, Y., Cano-Izquierdo, J., Lópes-Coronado, J.: $\mu$ARTMAP: use of mutual information for category reduction in fuzzy ARTMAP. IEEE Trans. Neural Netw. **13**, 58–69 (2002)
13. Grossberg, S.: Adaptive pattern classification and universal recoding: II. Feedback, expectation, olfaction, illusion. Biol. Cybern. **23**, 187–202 (1976)
14. Grossberg, S.: Competitive learning: from interactive activation to adaptive resonance. Cogn. Sci. **11**, 23–63 (1987)
15. Grossberg, S.: Adaptive resonance theory: how a brain learns to consciously attend, learn, and recognize a changing world. Neural Netw. **37**, 1–47 (2013)
16. Verzi, S., Heileman, G., Georgiopoulos, M.: Boosted ARTMAP: modifications to fuzzy artmap motivated by boosting theory. Neural Netw. **19**, 446–468 (2006)
17. Vidgor, B., Lerner, B.: The bayesian ARTMAP. IEEE Trans. Neural Netw. **18**, 1628–1644 (2007)
18. Williamson, J.: Gaussian ARTMAP: a neural network for fast incremental learning of noisy multidimensional maps. Neural Netw. **9**, 881–897 (1996)
19. Zhang, Y., Ji, H., Zhang, W.: TPPFAM: use of threshold and posterior probability for category reduction in fuzzy ARTMAP. Neurocomputing **124**, 63–71 (2014)

# Accuracy Improvement of Neural Networks Through Self-Organizing-Maps over Training Datasets

Daniel Gutierrez-Galan[✉], Juan Pedro Dominguez-Morales,
Ricardo Tapiador-Morales, Antonio Rios-Navarro,
Manuel Jesus Dominguez-Morales, Angel Jimenez-Fernandez,
and Alejandro Linares-Barranco

Robotic and Technology of Computers Lab.,
Department of Architecture and Technology of Computers,
University of Seville, Av. Reina Mercedes S/n, 41012 Sevilla, Spain
dgutierrez@atc.us.es
http://www.rtc.us.es

**Abstract.** Although it is not a novel topic, pattern recognition has become very popular and relevant in the last years. Different classification systems like neural networks, support vector machines or even complex statistical methods have been used for this purpose. Several works have used these systems to classify animal behavior, mainly in an offline way. Their main problem is usually the data pre-processing step, because the better input data are, the higher may be the accuracy of the classification system. In previous papers by the authors an embedded implementation of a neural network was deployed on a portable device that was placed on animals. This approach allows the classification to be done online and in real time. This is one of the aims of the research project MINERVA, which is focused on monitoring wildlife in Doñana National Park using low power devices. Many difficulties were faced when pre-processing methods quality needed to be evaluated. In this work, a novel pre-processing evaluation system based on self-organizing maps (SOM) to measure the quality of the neural network training dataset is presented. The paper is focused on a three different horse gaits classification study. Preliminary results show that a better SOM output map matches with the embedded ANN classification hit improvement.

**Keywords:** Self-organizing map · Artificial neural network · Feedforward neural network · Pattern recognition · Locomotion gaits

## 1 The MINERVA Project

In the last years, the monitoring of wildlife has become a very relevant topic thanks to concepts like the Internet of Things (IoT) and technologies like wireless sensor networks (WSN). Several studies have focused on investigating the

© Springer International Publishing AG 2017
I. Rojas et al. (Eds.): IWANN 2017, Part I, LNCS 10305, pp. 520–531, 2017.
DOI: 10.1007/978-3-319-59153-7_45

best way to gather information about animal patterns using embedded devices that are placed on animals [1–5]. This task is very important when it comes to understand things like the interaction between animals, their survival or even their nutrition habits. Changes in weather, flora or the introduction of non-native species could also affect these activities, making the monitoring of animal motion patterns a very interesting task.

A 2.4-GHz ZigBee-based mobile ad hoc wireless sensor network is presented in [6] to collect motion information from sheep and send it to a base station, which will later be classified into five different behaviors (grazing, lying down, walking, standing and others) using a multilayer perceptron (MLP) artificial neural network (ANN). The accuracy rate of the network is 76.2% without any applied preprocessing method.

MINERVA is a research project whose aim is to study and classify wildlife behavior inside Doñana National Park. The tracking and classification systems that are being used nowadays in the park obtain positional information between two and five times a day (to reduce power consumption) using a GPS and transmit it via GSM. However, biologists need more information to be able to recognize animal patterns. In previous work by the authors, this problem is solved by doing the classification step inside of the collar that is placed on the animals using an embedded implementation of an Artificial Neural Network (ANN) [4] instead of sending the raw sensor information to the database that is later studied by the biologists. This way, several sensors monitoring are carried out, but only the classification result is sent to a base station which later uploads it to a remote database. Hence, less transmissions are needed, which is the activity that consumes most battery power (more than 80% as presented in [7]). Previous studies have used this approach to classify between three horse gaits (standing, walking and trotting) [8], which used different preprocessing techniques applied to the raw data to obtain a better classification result in the embedded ANN. In [4,5] Kalman filter is applied to the input data, obtaining a 81.01% accuracy result. Moreover, in [3], Overall Dynamic Body Acceleration (ODBA) and variance is applied to the same data using different window lengths, achieving up to a 90.3% accuracy. However, to test which preprocessing would have a better accuracy result of the ANN, the whole trial and error method needs to be done. In our case, this task is hard and expensive (in time), so a tool or mechanism to test how good are the preprocessing methods is needed.

In this work, the authors present a novel NN-based mechanism to test the quality of the preprocessed information before having to test it using it as input to the classifier. Self-organizing Maps (SOM), which are a type of ANN, are used to visually show how good the input data is, and how the sensor data differs between each of the classes that want to be classified. This way, if the preprocessing is able to properly sparse the data between each of the classes, the ANN would then have it easier to classify the input information, achieving a better accuracy result. SOMs are usually used for classifying samples which have a features set with different values. The result is a map where samples with similar values are close, and samples with different values are separated, thus appearing sample clusters. The most popular example using SOM is the Fisher's Iris data set [9] problem, where three species of Iris flower have to be classified taking into account some features like sepal length, sepal width, petal length and petal width. In [10], authors using the SOM for processing the characterization of movement patterns of athletes, taking several training session parameters. And in [11], an unsupervised acoustic classification of bird species was done extracting first some features by spectral analysis and using them to classify the species using a SOM. In both cases, several parameters had been extracted in order to be used as SOM input applying complex preprocessing methods. The rest of the paper is structured as follows: Sect. 2 describes the collar device used to gather information from animals, and how this data is obtained. Then, Sect. 3 presents different preprocessing techniques to improve the information that can be extracted from the sensors. Section 4 describes the experiments that have been carried out in this work, as well as the results obtained. At the end, Sect. 5 presents the conclusions of this work.

## 2    Collecting Sensory Information Using a Portable Collar Device

The collar (Fig. 1) collects information from the animal that carries it by using different sensors. It has a MinIMU9V2 inertial measurement unit (IMU), which consists of an LSM303DLHC 3-axis accelerometer, an L3GD20 3-axis gyroscope and a 3-axis magnetometer. Each of these sensors have 12-bits resolution for a more precise data acquisition. Along with the IMU, a GPS is also used, which provides location and time information. The collar has a 2.4 GHz ZigBee-based radio module, which is an open global standard of the IEEE 802.15.4 MAC/PHY [12], to send the obtained information. The collar carries a MicroSD card to store the sensor's information when the animal is outside of the coverage range of the WSN.

The data that is used on this article has been collected from semi-wild horses and different seasons. A total of 30000 samples were obtained during visits to Doñana's National Park from different horses. This data corresponds to three gaits: standing, walking and trotting. Several methods have been used in the literature to classify this kind of locomotion information: Convolutional Neural Networks, Support Vector Machines, statistical methods, etc. High accuracy results have been achieved, as it was presented in Sect. 1. However, these kind of algorithms have a high computational cost, which leads to a high power consumption.

The main aim of the collar is to classify the animal behavior (between three different gait patterns in this work) using the information obtained from the IMU as an input to a feed-forward Artificial Neural Network (ANN) implemented on the collar's microcontroller unit (MCU). To implement an ANN in the collar an open source neural network library called Fast Artificial Neural Network (FANN) [13] has been used. This library allows to implement multilayer ANNs in C programming language in an easy and quick way.

As in MINERVA project the application needs to be focused on low-power consumption devices (capturing an animal to replace its collar is very expensive

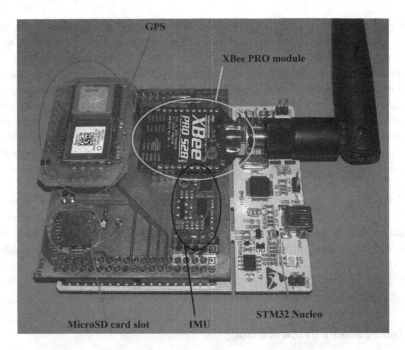

**Fig. 1.** Collar device prototype. Several sensors are necessary to monitor the activity and position of the animal. The XBee module allows to send the collected information to the base station that is placed on the park.

and difficult), an MCU with no Floating Point Unit (FPU) is used. Hence, for this purpose, the FANN library was modified to use fixed point numbers. Processing the sensor's information in the collar allow us to only send the classification result to the base station instead of all the raw data gathered from them, reducing the number of transmissions (as it was studied in [14], transmissions have a very high power consumption compared to the ANN operations) and the length of the packets transmitted, and letting us to know the animal behavior in real time [4].

## 3   Collected Information Processing and Analysis

Processing the sensor information after it has been collected is a common task. This way, the noise that data may have can be reduced, obtaining a better signal, or even extracting information that could be hidden in the raw data.

Data signals provided by the accelerometer vary between $-2g$ and $2g$ in this case. The horse gaits information is included in these signals, and the main challenge is to get these features by performing some math operations to transform the samples, in order to provide a better input to the ANN implemented in the collar.

For this purpose, the authors have carried out several experiments in which different data processing methods were applied. Kalman filter, variance, among others, were calculated and tested using them as an input to a feedforward ANN. However, we did not have any parameter or index that showed us how good the preprocessing step was. Until now, the quality of the processing performed to the input dataset (understanding quality as how good this data is) is evaluated by analyzing the output of the ANN and checking the confusion matrix to see how good the accuracy result is. In this work, we considered that it would be interesting to have a tool which helped us in this task, giving information about the quality of the preprocessing and the input dataset before testing it on the ANN.

Self-Organizing Maps (SOM) [15] is a type of ANN used to cluster input data into groups of similar patterns. Input patterns are compared to each cluster, and associated with the cluster it matches the best. The comparison is usually based on the square of the minimum Euclidean distance. When the best match is found, the associated cluster gets its weights and its neighboring units updated. Preprocessing methods used and carried out tests using SOM will be explained in detail in order to clarify our use of SOM.

### 3.1   Information Preprocessing

While the collar is working, it is continuously collecting data from the IMU sensors. This raw data generally has a lot of noise, but sometimes it can be used as it is, without any previous processing. The fact is that ANNs achieve a better classification result for a specific category when the information from it is distant from the input information of the rest of the classes. In this case, the problem was that, in our dataset, sensor values from different gaits are overlapped because of the accelerometer's range and the nature of the animal movement, as we can see in Fig. 2. So, even though the ANN output was acceptable (around 80.0%), the device needs to have more accuracy, since it will be active only a few hours a day to reduce power consumption. The first solution taken by authors was to implement the Kalman filter [16] in the collar. This method is commonly used by planes and drones, which provides information about the orientation of an object in a 3-dimensional space. Kalman filter uses raw data as input and returns three values: roll, pitch and yaw. Using this parameters a 95.0% of accuracy was achieved by the ANN, but power consumption was increased considerably, reducing the battery life of the collar. In addition, floating point operations (which require a FPU module) are needed to perform these calculations and, as it was presented in previous sections, low power consumption MCUs without FPU are needed in this project, making it a not viable solution.

In search of a simple solution, after several methods were studied, the variance of the raw data was calculated. In this case, since the device has to work in real time, this operation was performed using temporal windows of 1.3 s approximately (40 samples, having a 30.3 Hz sample frequency). When the buffer is full, the variance is calculated and used as input to the ANN. This way the MCU collects data during the enough amount of time to let us know the gait that the

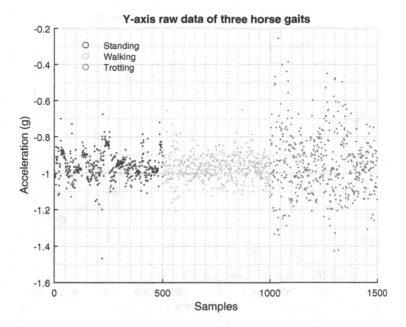

**Fig. 2.** Raw data subset from Y-axis of accelerometer. Three gaits are showed consecutively using different colors (5000 samples per gait). Most samples are located in the same range of values. (Color figure online)

horse is performing. Around 90.0% accuracy was achieved when calculating the variance of the input sensor information [3].

When variance data was shown against time (Fig. 3), it could be seen that some peaks from different gaits were overlapped between them. Therefore, in these cases, the ANN could probably give a wrong result when trying to classify the information. In order to avoid this situation, a window-length-based hull of maximum values was performed. This operation consists on detecting the maximum value of the samples contained in the window and maintaining that value until a greater peak is found or until the end of the window. This way, the ANN input data will be always the peaks of the signal, which are the best ones that represent the gait performed by the horse, except some cases in which an isolated peak is produced by an unusual movement of the horse.

## 3.2 Data Analysis Using Self-Organizing Maps

As it was presented previously, data from different gaits is frequently overlapped. This situation could lead the ANN to a wrong classification. Hence, it would be good to analyze the input dataset and know how the spacial distribution of samples is, to check if it is possible to differentiate the three gaits. In the optimum case, the information of each gait should be well separated in three different clusters. But the real case is that these signals are usually closer to each

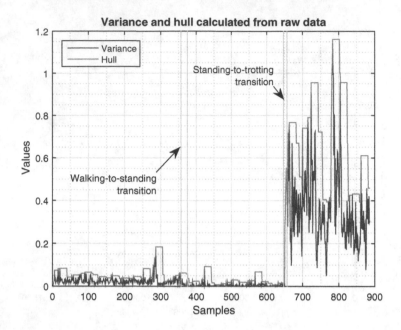

**Fig. 3.** Variance of the raw data from the Y-axis and hull of variance values. Two transition areas can be identified, which are delimited by vertical green lines. In this case, gaits are sorted as walking, standing and trotting. The number on samples was reduced due to window-based preprocessing method. (Color figure online)

other, even overlapped (due to the transitions between gaits and the nature of the horse movement).

For this reason, SOMs were used to analyze the dataset, since they are able to show how good the samples are distributed and point out the existing data clusters, which in this case they will correspond to the gaits performed by horse. Thus, the better differentiated the data is, the better the SOM will be able to represent a map where the three gaits are perfectly distinguished. However, a bad SOM's output does not implies a bad result of the ANN, but it means that samples from a specific category are not well separated from the rest of them.

In addition, a SOM may help us understand the results obtained when using a classification system like a feed-forward neural network, statistical methods with thresholds, etc. The aim is to perform an appropriate sample processing in order to obtain a good SOM's output, and thus, a good accuracy on the classification with the ANN.

## 4  Offline Tests and Results

Several tests were performed using SOMs and applying the different preprocessing methods that were presented in previous sections. These tests were carried out offline, due to the fact that it makes no sense to implement a SOM in the collar's MCU and use it in real time. MATLAB has several toolboxes that allow to train and test different kinds of ANNs. Among them, Neural Network Clustering and Neural Network Pattern Recognition are mostly used to work with SOMs and feedforward neural networks, respectively.

For the Neural Network Clustering Toolbox, the input vector length was different on each case of study, depending on the dataset used (9-samples input vector for the raw data or 3-samples input vector for the variance and the hull). A two-dimensional map with size 8 (i.e. $8 \times 8$ neurons) was trained using this application. The default values of training parameters, like number of epochs and training algorithm, were used. The ANN architecture used to test the dataset consists of a hidden layer with 30 neurons and an output layer with 3 neurons (one per gait). The activation functions used were the sigmoid transfer function in the hidden layer and the softmax transfer function in the output layer. The NN was trained using the backpropagation algorithm, and it was used the same architecture for all performed tests.

Three different tests were carried out, using three datasets (raw, variance and hull) in which the data was processed using the methods explained in previous sections. Both confusion matrix and SOM's output show the results obtained in each test, which can be seen below:

## 4.1   Raw Data

The raw data has the information from the 3 axes of the accelerometer, gyroscope and magnetometer. Therefore, nine neurons are needed in the input layer of the NN and the input vector of the SOM. Much noise is found in the signals, so it is hard to recognize patterns with a high accuracy.

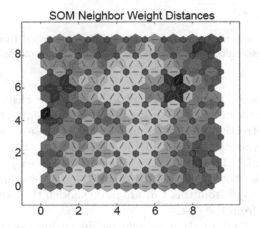
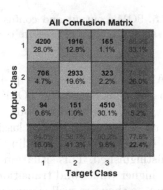

**Fig. 4.** SOM's output (left) and confusion matrix (right) using raw data. Target classes are corresponded with: 1-walking, 2-standing, 3-trotting.

Figure 4 left shows the SOM's output, where lighter colors respresents neurons with similar values and darker colors represent neurons with different values. In the other hand, right side of the picture shows the confusion matrix, where bottom row and right columns are the average per clases and botom-right cell is the hit averate. The hit average obtained by the ANN is 77.6.

## 4.2   Variance

To calculate the variance, only the three axes of the accelerometer were taken into account, because it is the sensor which provides more information about the animal locomotion patterns [5]. The variance was calculated using a 1.3 s length (40 samples) window size, which is approximately the time that the horse takes to perform a full period of any of the gaits studied in this work.

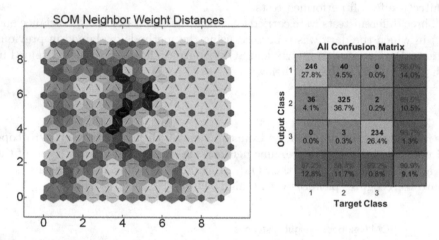

**Fig. 5.** SOM's output and confusion matrix using the variance data calculated from raw samples. Target classes are corresponded with: 1-walking, 2-standing, 3-trotting.

After the samples were processed, the data obtained seems to be clearer than raw data since the SOM's output (Fig. 5) shows two well differentiated areas, which could correspond to trotting and standing/walking because those last two gaits are hard to classify, as can be seen in Fig. 2 in Sect. 3.1. In the area corresponding to standing and walking gaits, the difference is not easy to be distinguished. This situation does not happen with trotting, because samples have higher values and transitions to or from this gait are more sudden than the others, so they are well separated from the rest.

This improvement seen in the SOM's output is reflected in an increase of the hit average achieved by the ANN, where a 90.9% of accuracy was obtained using this process. Furthermore, an improvement in the hit average between both standing and walking was obtained.

## 4.3    Hull

Using this approach, the authors tried to avoid the problem of samples similarity. The input data was the same variance values that were calculated previously, which have now been processed with a hull algorithm using a slice of 20 samples.

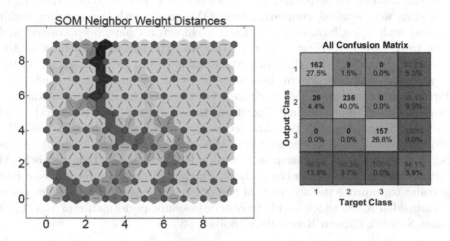

**Fig. 6.** SOM's output and confusion matrix using the hull data calculated from variance values. Target classes are corresponded with: 1-walking, 2-standing, 3-trotting.

Now, the output from the SOM shows three different areas clearly, although there is an area that is not perfectly differentiated as the others (Fig. 6). This situation corresponds to standing and walking gaits, since the movement of the horse's head (where the collar is placed) is almost the same in both cases. The improvement showed in the SOM's output was directly confirmed when the ANN was tested using the calculated values. The hit accuracy obtained with the ANN was 94.1%.

The hull algorithm can be done in real time, due to the fact that the computational cost needed to calculate it is very low. This solution increases the previous one by a 3%, taking only the maximum value in each time slice.

An improvement in walking prediction is a hard task, since more complex operations are needed to extract more information from the samples. However, with this preprocessing technique, we can consider that the collar device is reliable enough to provide information about animal gaits.

## 5    Conclusions

In this paper, the authors have presented a novel mechanism based on SOMs to measure the quality of the input dataset before training and feeding a NN with it. This way, the user is able to know how good the information from the different classes that are contained in the dataset is, and how much they differ from each other. The more separated the information from different classes is, the better accuracy will be achieved by the ANN when classifying samples. Hence, SOM is a useful tool for predicting how good the classification results will be before testing the ANN, to the fact that this process is hard and very expensive in terms of time and money.

In this context, three different experiments have been carried out where three horse gaits were studied, comparing the SOM output with the accuracy result achieved with a feedforward ANN. Both architectures have been trained and tested using MATLAB Neural Network Toolbox. The first experiment consisted on testing the ANN with the raw data obtained from IMU sensor. A 77.9% was achieved with this system; however, the SOM was not able to cluster the information in the three different classes. On the other hand, when the dataset was preprocessed using the hull algorithm, three different areas (clusters) could be seen or distinguished in the obtained map, which was not possible with the previous preprocessing methods. Using this processed dataset as input to the ANN improved the classification, achieving a 94.1% accuracy. Hence, this SOM application became useful for the authors. The hull method can be deployed into the collar to improve the accuracy of the horse gait classification system using the embedded ANN, which is able to obtain the same performance of MATLAB Neural Network Pattern Recognition toolbox [4].

**Acknowledgements.** This work is supported by the excellence project from Andalusia Council MINERVA (P12-TIC-1300) and also by the Spanish government grant (with support from the European Regional Development Fund) COFNET (TEC2016-77785-P). The authors would like to thank Ramon C. Soriguer, Francisco Carro, Francisco Quirós and the EBD-CSIC for their support on the tests that were done in Doñana National Park.

# References

1. Nadimi, E.S., Jørgensen, R.N., Blanes-Vidal, V., Christensen, S.: Monitoring and classifying animal behavior using ZigBee-based mobile ad hoc wireless sensor networks and artificial neural networks. Comput. Electron. Agric. **82**, 44–54 (2012)
2. Rios-Navarro, A., Dominguez-Morales, J.P., Tapiador-Morales, R., Dominguez-Morales, M., Jimenez-Fernandez, A., Linares-Barranco, A.: A sensor fusion horse gait classification by a spiking neural network on SpiNNaker. In: Villa, A.E.P., Masulli, P., Pons Rivero, A.J. (eds.) ICANN 2016. LNCS, vol. 9886, pp. 36–44. Springer, Cham (2016). doi:10.1007/978-3-319-44778-0_5
3. Gutierrez-Galan, D., Dominguez-Morales, J.P., Cerezuela-Escudero, E., Miro-Amarante, L., Gomez-Rodriguez, F., Dominguez-Morales, M.J., Rivas-Perez, M., Jimenez-Fernandez, A., Linares-Barranco, A.: Semi-wildlife gait patterns classification using statistical methods and artificial neural networks. In: 2016 International Joint Conference on Neural Networks (IJCNN). IEEE (2017, accepted for publication)
4. Gutierrez-Galan, D., Dominguez-Morales, J.P., Cerezuela-Escudero, E., Rios-Navarro, A., Tapiador-Morales, R., Rivas-Perez, M., Dominguez-Morales, M.J., Jimenez-Fernandez, A., Linares-Barranco, A.: Embedded neural network for real-time animal behavior classification. Neurocomputing Under review

5. Cerezuela-Escudero, E., Rios-Navarro, A., Dominguez-Morales, J.P., Tapiador-Morales, R., Gutierrez-Galan, D., Martín-Cañal, C., Linares-Barranco, A.: Performance evaluation of neural networks for animal behaviors classification: horse gaits case study. In: Omatu, S., et al. (eds.) Distributed Computing and Artificial Intelligence, 13th International Conference. AISC, vol. 474, pp. 377–385. Springer, Cham (2016)

6. Nadimi, E., et al.: Monitoring and classifying animal behavior using ZigBee-based mobile ad hoc wireless sensor networks and artificial neural networks. Comput. Electron. Agric. **82**, 44–54 (2012)

7. Dominguez-Morales, J.P., et al.: Wireless sensor network for wildlife animals tracking and behavior classifying in Donana. IEEE Commun. Lett. **20**, 2534–2537 (2016). http://ieeexplore.ieee.org/abstract/document/7574341/

8. Harris, S.E.: Horse Gaits, Balance and Movement. Howell Book House, New York (1993). ISBN 0-87605-955-8

9. Vesanto, J., Himberg, J., Alhoniemi, E., Parhankangas, J., et al.: Self-organizing map in Matlab: the SOM toolbox. Proc. Matlab DSP Conf. **99**, 16–17 (1999)

10. Bauer, H.U., Schöllhorn, W.: Self-organizing maps for the analysis of complex movement patterns. Neural Process. Lett. **5**(3), 193–199 (1997)

11. Vallejo, E.E., Cody, M.L., Taylor, C.E.: Unsupervised acoustic classification of bird species using hierarchical self-organizing maps. In: Randall, M., Abbass, H.A., Wiles, J. (eds.) ACAL 2007. LNCS, vol. 4828, pp. 212–221. Springer, Heidelberg (2007). doi:10.1007/978-3-540-76931-6_19

12. IEEE 802 Working Group, et al.: IEEE standard for local and metropolitan area networks-part 15.4: Low-rate wireless personal area networks (LR-WPANS). IEEE Std **802**, 4–2011 (2011)

13. Nissen, S.: Implementation of a fast artificial neural network library (FANN). Report, Department of Computer Science University of Copenhagen (DIKU) 31 29 (2003)

14. Dominguez-Morales, J.P., Rios-Navarro, A., Dominguez-Morales, M., Tapiador-Morales, R., Gutierrez-Galan, D., Cascado-Caballero, D., Jimenez-Fernandez, A., Linares-Barranco, A.: Wireless sensor network for wildlife tracking and behavior classification of animals in doñana. IEEE Commun. Lett. **20**(12), 2534–2537 (2016)

15. Kohonen, T.: Self-Organizing Maps. Springer Series in Information Sciences, vol. 30. Springer, Heidelberg (1995)

16. Welch, G., Bishop, G.: An introduction to the kalman filter (1995)

# Spiking Neurons

# Computing with Biophysical and Hardware-Efficient Neural Models

Konstantin Selyunin(✉), Ramin M. Hasani, Denise Ratasich,
Ezio Bartocci, and Radu Grosu

Vienna University of Technology, Vienna, Austria
{konstantin.selyunin,ramin.hasani,denise.ratasich,ezio.bartocci,
radu.grosu}@tuwien.ac.at

**Abstract.** In this paper we evaluate how seminal biophysical Hodgkin Huxley model and hardware-efficient TrueNorth model of spiking neurons can be used to perform computations on spike rates in frequency domain. This side-by-side evaluation allows us to draw connections how fundamental arithmetic operations can be realized by means of spiking neurons and what assumptions should be made on input to guarantee the correctness of the computed result. We validated our approach in simulation and consider this work as a first step towards FPGA hardware implementation of neuromorphic accelerators based on spiking models.

**Keywords:** TrueNorth model · Hodgkin-Huxley model · Rate encoding · Arithmetic operations · Simulations

## 1 Introduction

Neuroscience and computer engineering are fundamentally different: while, in general, the purpose of a neuroscientist is to understand a nervous system and to develop models capable of explaining its function (from the physical world to models), the purpose of a computer engineer to realize a hardware system that would satisfy initial requirements (from models to the physical world). Despite the differences, from the papers of McCulloch and Pitts [1] neural networks are influencing the development of computer hardware, which result in dedicated chip architectures [2–4]. To design efficient computing systems it is vital to understand the brain and employ principles from nature in hardware design.

We believe that neuromorphic hardware accelerators [5] that co-exist together with the traditional CPU infrastructure will allow to extend the functionality of hardware systems, add flexibility to existing designs, retain established design flow, and reduce overall costs when implemented on a custom-of-the-shelf (COTS) general purpose hardware. Our goal is to develop an open-source implementation of neural models capable of performing computations on FPGAs, to allow inherently parallel, reconfigurable, and easily accessible on the market solutions, which leverage existing design tools.

© Springer International Publishing AG 2017
I. Rojas et al. (Eds.): IWANN 2017, Part I, LNCS 10305, pp. 535–547, 2017.
DOI: 10.1007/978-3-319-59153-7_46

To achieve our goal and build efficient hardware implementation, it is essential to understand how to perform fundamental arithmetic operations, as those set the basis for the higher-level complex processing. In order to analyze the similarities and fundamental differences of computing in biological and hardware systems we study two spiking neural models: biophysically-accurate yet computationally plausible Hodgkin and Huxley [6] neural model, extended with the synapse model from [7], and hardware-efficient digital TrueNorth [4] model. Apart from traditional models of artificial neurons (e.g. perceptron with sigmoid activation function) the two models studied use a spike as a main mechanism to communicate between neurons.

The contributions of the paper can be summarized as follows:

- we formally define the problem of computing a function over spike rates in a neural network;
- we discuss how arithmetic operations can be implemented both using biophysical and digital neural models;
- we elaborate on the role of assumptions on inputs to obtain the correct computation results for both models.

The rest of the paper is organized as follows: Sect. 2 discusses the related work, and Sect. 3 provides a short description of the spiking models under study. Section 4 formalizes the task of computing a function with neural models. Section 5 elaborates on performing computations with Hodgkin-Huxley and TrueNorth neural models. Section 6 offers outlook and our concluding remarks.

## 2    Related Work

Modelling a neuron at a biophysical level is a challenging task: each neuron has on the order of $10^4$ of synapses [8]; state-of-the art models account about 20 ionic channels, 150 state variables and 500 parameters [9]. Moreover, since the synapses are structurally and functionally plastic devices, the dendritic spines of the neuron [10] and the efficacy of synapses change during the operation (e.g. spike-time dependent plasticity [11]). We are aware that taking into account geometrical topology allows to increase the expressiveness and e.g. perform orientation selectivity in the dendritic inputs [12], though to remain computationally efficient we consider only the single-compartment neural models [13].

In the seminal paper [6] Hodgkin and Huxley presented a conductance-based spiking neural model that describes the dynamics of generating an action potential, the role and function of sodium and potassium ionic channels. The model of Hodgkin and Huxley is biophysically accurate [14], and has been refined with other type of ionic channels [9]. A neuron is modelled as an active RC-circuit, in which the opening of ion channels follows in response to influx of external current stimulus. The membrane potential, inward (sodium), and outward (potassium) currents are modelled as a set of differential equations. Although numerous software and hardware implementations (e.g. [15, 16] and [17–20] respectively) of the Hodgkin and Huxley model are available, to the best of our knowledge we are not aware of the works that study computations with spike rates and explicitly compare the biophysical model with hardware-optimized spiking models.

Cassidy et al. [4,21,22] in a series of papers introduced the TrueNorth hardware architecture, which is based on a versatile spiking neuron model. The TrueNorth model [4] is digital with all the parameters being either integers or boolean values, since floating point computations are expensive in hardware. At each time step the digital neuron performs three computational steps: (1) synaptic integration, (2) leak integration, (3) threshold-fire-reset. Although the authors implemented the architecture on the dedicated hardware chip, it is not available on the market. We aim, on the contrary, to develop neuromorphic hardware accelerators on FPGAs and ZYNQ architecture in particular, which are widely available, have an established design flow, and allow AXI-style communication between the processing system and the programmable logic. In this work we implemented a python open-source simulation of both biophysical and digital models and consider this as a first step towards target hardware implementation of neuromorphic accelerators on a ZYNQ FPGA processing system.

In our previous paper [23] we showed how to use the TrueNorth model to monitor MTL specifications on FPGA over discrete time. Although at every time step the neural circuit computed the verdict of a temporal logic specification $\varphi$, the result of this computation is a all-or-none boolean output. In this work we interpret the spike-rate as an integer number, hence allowing both qualitative and quantitative analysis within the framework.

## 3   Neuron and Synapse Modeling

In this section we succinctly recap neural and synapse models that we study.

### 3.1   Modeling Neurons

**The Hodgkin-Huxley Neuron model** qualitatively describes the dynamics of the membrane potential as a function of activation and deactivation of ionic channels such as sodium and potassium together with the leak channel [6]. The model comprises a set of four ordinary differential equations (ODE)s describing the properties of an excitable neuron as follows:

$$C_m \frac{dV_m}{dt} = -(\bar{g}_K n^4 (V_m - E_K) + \bar{g}_{Na} m^3 h(V_m - E_{Na}) + \bar{g}_l(V_m - E_l)) + I_{in}, \quad (1)$$

where $C_m$ and $V_m$ are the membrane capacitance and potential; $\bar{g}_K$, $\bar{g}_{Na}$ and $\bar{g}_l$ are the conductances of the potassium, sodium and leak channels, respectively; $E_K$, $E_{Na}$ and $E_l$ represent the reversal potential of the channels; $n$, $m$ and $h$ are voltage-dependent gating variables for the potassium channel activation, sodium channel activation and sodium channel inactivation, respectively. For the detailed description the reader is referred to [6].

**The TrueNorth Neuron Model** is proposed by IBM [4] and extends leaky-integrate-and-fire model. We review the deterministic part of the TrueNorth model below. For an extended explanation, the reader is referred to [4].

*Synaptic Integration* is the first computational step where every neuron sums up the products of its inputs $A_i(t)$ and weights $s_{ij}$. Every input is enabled by a flag $w_{ij}$. The result is added to its previous membrane potential $V_j(t-1)$. Although in the original model the maximum number of inputs bounded by 255, we drop this restriction and assume that every neuron has a configurable $N \in \mathbb{N}$ number of inputs (the original assumption comes from the chip restrictions):

$$V_j(t) = V_j(t-1) + \sum_{i=0}^{N} A_i(t)\, w_{ij}\, s_{ij} \qquad (2)$$

*Leak Integration* accounts for energy dissipation, self-stimulation, and convergence to an equilibrium in the absence of input. A TrueNorth neuron $n_j$ can exhibit negative, zero or positive leak $\lambda_j$. To express divergent and convergent leak behaviors the leak reverse flag $\epsilon_j$ can be set: in this case the leak changes its sign with the membrane potential's sign (i.e., when the signs are different, the leak forces $V_j$ converge to zero).

$$\Omega_j = (1 - \epsilon_j) + \epsilon_j \mathrm{sgn}(V_j(t)) \qquad (3)$$

$$V_j(t) = V_j(t) + \Omega_j \lambda_j \qquad (4)$$

*Threshold, Fire, Reset* is computed at each time step to generate the binary "all-or-none" output (spike or no spike). A neuron $n_j$ possesses a positive threshold $\alpha_j$ and a negative threshold $\beta_j$. When the membrane potential $V_j$ exceeds $\alpha_j$, the spike is generated, and the membrane potential is reset. The TrueNorth model is extended with three reset modes $\gamma_j$: (0) normal, (1) linear, or (2) non-reset. When $V_j$ falls below the negative threshold $\beta_j$, no spike is generated, although the membrane potential is updated depending on the reset mode $\gamma_j$ and the saturation flag $\kappa_j$.

$$\begin{aligned}
&\texttt{if} \qquad V_j(t) \geq \alpha_j && (5)\\
&\qquad \textsc{Spike} && (6)\\
&\qquad \gamma_j = 0 : V_j(t) = R_j &&\\
&\qquad \gamma_j = 1 : V_j(t) = V_j(t) - \alpha_j &&\\
&\qquad \gamma_j = 2 : V_j(t) = V_j(t) && (7)\\
&\texttt{elseif} \quad V_j(t) < -\beta_j && (8)\\
&\qquad \texttt{if} \quad \kappa_j = 1 &&\\
&\qquad\qquad V_j(t) = -\beta_j &&\\
&\qquad \texttt{else} &&
\end{aligned}$$

$$\gamma_j = 0 : V_j(t) = -R_j$$
$$\gamma_j = 1 : V_j(t) = V_j(t) + \beta_j$$
$$\gamma_j = 2 : V_j(t) = V_j(t) \tag{9}$$

## 3.2   Modeling Synapses

In order to model the current flow between neurons in the Hodgkin-Huxley model (as it accounts for external current stimulus), we implemented three models of synaptic conductance $g_{\text{syn}}$ from [7] and assume that $I_{\text{syn}} \propto g_{\text{syn}}$. The first model (*exponential decay*) assumes that ionic channels open instantaneously upon an arrival of a presynaptic action potential and then $g_{\text{syn}}$ decays exponentially:

$$g_{\text{syn}}(t) = \bar{g}_{\text{syn}} e^{-(t-t_0)/\tau}. \tag{10}$$

The *alpha function* [24] takes into account that the opening of ionic channels is not instantaneous without introducing additional parameters into the model:

$$g_{\text{syn}}(t) = \bar{g}_{\text{syn}} \frac{t - t_0}{\tau} e^{1-(t-t_0)/\tau}. \tag{11}$$

A more comprehensive representation of the dynamics of synaptic conductance can be modeled by the *difference of exponentials* where the rise and decay times are explicitly introduced [7]:

$$g_{\text{syn}}(t) = \bar{g}_{\text{syn}} (e^{-(t-t_0)/\tau_{\text{decay}}} - e^{-(t-t_0)/\tau_{\text{rise}}}). \tag{12}$$

We implement the above models of synaptic conductance (Eqs. 10–12) and employ them in the design of arithmetic operations using the Hodgkin-Huxley model. Figure 1 depicts the normalized excitatory post-synaptic current (EPSC) for synapse models in response to the presynaptic action potential; Table 1 lists the parameters of the models studied in this work.

**Table 1.** Parameters of neuron and synapse model

| Model | Parameters |
|---|---|
| Hodgking-Huxley model | $C_m, \bar{g}_K, \bar{g}_{Na}, \bar{g}_l, E_K, E_{Na}, E_l, \alpha_{\{K,Na\}}, \beta_{\{K,Na\}}$ |
| TrueNorth model | $A_j, w_j, s_j, \lambda_j, \epsilon_j, \gamma_j, \alpha_j, \beta_j, \kappa_j$ |
| Exponential Decay | $\bar{g}_{\text{syn}}, \tau$ |
| Alpha function | |
| Double-exp Synapse | $\bar{g}_s^{max}, \tau_{rise}, \tau_{decay}$ |

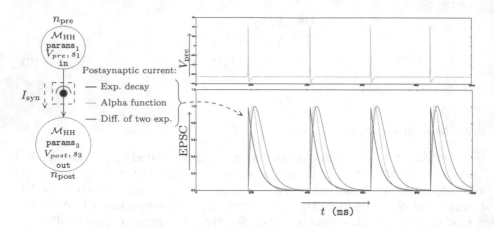

**Fig. 1.** Normalized EPSC in response to pre-synaptic action potentials ($V_{\mathrm{pre}}$)

## 4   Problem Formulation

As we study computation for both biophysical and digital neural models, we assume that: (i) biophysical model operates over real time and real-value domain (it is simulated with a pre-defined rational-value integration step $\Delta t = \frac{q}{r}$, where $q, r \in \mathbb{N}$), (ii) to be efficiently hardware-realizable, the digital neural model operates over discrete time and finite-value domain. A *trial* is an execution of a neural circuit for a time interval $[0, T]$. Each neuron $n_i = (\mathcal{M}, \mathbf{params}_i, V_i, s_i, l_i)$ is characterised by (i) an underlying model $\mathcal{M}$, (ii) a set of parameters $\mathbf{params}_i$, (iii) its membrane potential $V_i$, (iv) a binary spike output $s_i$, and (v) a label $l_i \in \{\mathtt{in}, \mathtt{interm}, \mathtt{out}\}$ of a neuron. For a neural network $N = \{n_1, \cdots, n_m\}$ of $m$ neurons we define its computation $C$ over a time interval in the following way: $C : [0, T] \mapsto V^m \cup S^m$, where $V^m = \{V_1, \cdots, V_m\}$ and $S^m = \{s_1, \cdots, s_m\}$ are membrane potentials and spikes of neurons in the network $N$ respectively.

A spiking activity of a neuron $n_i$ over a trial is defined as a mapping $[0, T] \mapsto s_i$. We assume that a neuron $n_i$ encodes numbers in a spike-count rate [25], and measure spiking activity over a time window $w \subseteq [0, T]$ of the length $\|w\|$:

$$r_i = \frac{\sum_{t \in w} s_i[t]}{\|w\|}. \tag{13}$$

The task of computing a function $f$ in a neural network $N$ then can be formulated as follows: for a given neural model $\mathcal{M}$ find number of neurons with labels $\mathtt{interm}, \mathtt{out}$ and their corresponding parameters $\mathbf{params}$ such that $r_{\{\mathtt{out}_1, \cdots \mathtt{out}_n\}} = f(r_{\{\mathtt{in}_1, \cdots, \mathtt{in}_m\}})$.

## 5   Computations with Neural Models

In this section we describe computations with the Hodgkin-Huxley and the TrueNorth neural models. The circuit topology for the two-argument operations

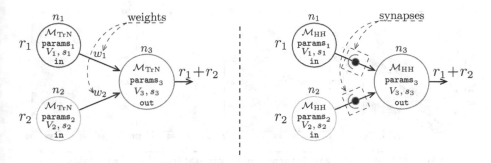

**Fig. 2.** Neural models for computations on spike rates: the TrueNorth model (left) assumes that neurons are connected with discrete weights. For Hodgkin & Huxley model (right) we introduce the synaptic connections between neurons. (Color figure online)

is shown in Fig. 2. Input neurons (blue and green) provide spikes with rates $r_1$ and $r_2$ to the computing neuron $n_3$, which outputs the result $f(r_1, r_2)$.

## 5.1  Computing Addition

**Addition using the TrueNorth model** is realized as follows: we configure the output neuron $n_3$ in the linear reset mode ($\gamma = 1$), the input weights and the positive threshold are set to one. This allows to: (i) generate a spike in the **out** neuron whenever an action potential is generated by the **in** neurons; (ii) memorize in the membrane $V_3$ if two spikes happened at the same time instant, and temporally separate output spikes over the adjacent time steps.

Without any assumptions on input, if the membrane potential is empty at the end of the trial, then the result of addition is correct. If, however, the both inputs arrive at the end of the trial, the output neuron may not be able to generate the correct result when the number of the remaining time steps in the trial is less the value of the membrane potential. This can be mitigated by extending the length of the trial for the output neuron, which we aim to avoid to keep the hierarchical composition simple. In this work we assume random arrival of input, which is implemented in the TrueNorth model as the "rate-store" function [4]; Fig. 3 shows the corresponding simulation results.

**Addition using the Hodgkin-Huxley model** crucially depends on the underlying synapse model. Unlike the TrueNorth, the biophysical Hodgkin-Huxley model is not able to memorize the occurrences of two simultaneous spikes from both inputs in the membrane potential $V_m$. Furthermore, one needs to account for the refractory period, in which no action potential can be initiated. To obtain correct results it is vital to distribute the synaptic current over the time, such that after the refractory period the output neuron still receives enough stimulation. The *alpha function* and the *difference of two exponential* can

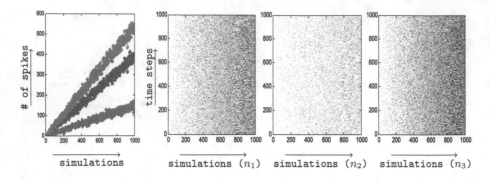

**Fig. 3.** Addition in the spike rates with TrueNorth model: blue + green = red ($n_1$ – blue; $n_2$ – green; $n_3$ – red). We performed 1000 simulation trails for 1000 time steps each. The leftmost plot shows the dependence between input and output spike rates for all the trials. Subsequent 1000 × 1000 plots present the spiking activity of input and output neurons over time during all the trials: a black pixel with coordinates $(i, j)$ denotes a spike in trial $i$ (horizontal axes) at a time step $j$ (vertical axes). (Color figure online)

be used to perform the addition with Hodgkin-Huxley model. Figure 4 shows the simulation result of a trial, where the synapse are modelled as the alpha function.

The fact that at the end of the trial the membrane potential $V_m$ stabilizes at the resting value for a time $T_{\text{stable}} \sim 10\,\text{ms}$ is necessary but not sufficient requirement for producing the correct results.

## 5.2    Computing Constant Multiplication

**Constant multiplication using the TrueNorth model** is implemented as follows: for each occurrence of the input spike, the output neuron generates $C$ spikes, hence the strength of the connection (i.e. its weight) is proportional to $C$. The output neuron $n_3$ is set to the non-reset mode (i.e. $\gamma = 2$) to be able to store all the spikes seen so far. The negative leak $\lambda$ and the saturate flag $\kappa$ ensure that the membrane potential will converge to zero in the absence of input. In the case of constant factor division, we need to output one spike for each $C$ spikes seen so far. To do so, we set a positive threshold $\alpha$ proportional to $C$ and weight $s_0$ to one. We also set the leak to zero and linear reset mode ($\gamma = 2$); see Fig. 5 for the simulation results.

**Constant multiplication using the Hodgkin-Huxley model** can only be performed if the synaptic current and the input spike rate satisfy the following requirements: (i) the length of the synaptic current pulse must be proportional to the multiplication constant $C$; (ii) the input arrival rate is low enough to allow synaptic current attenuate to its resting value before the arrival of the next spike from the pre-synaptic neuron. To satisfy the first requirement it is necessary to control both the amplitude and the width of the synaptic current,

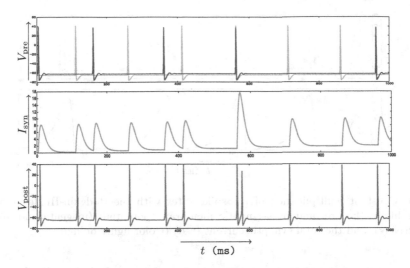

**Fig. 4.** Addition of spike rates with the Hodgkin-Huxley model: the profile of the pre-synaptic voltage profile for the input neurons (blue and cyan), the superposed synaptic current (green), the post-synaptic voltage profile (red) (Color figure online)

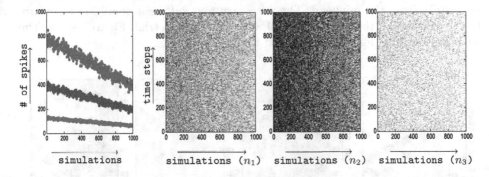

**Fig. 5.** Constant multiplication and division of the spike rates with the TrueNorth model: $2 \cdot$ blue = green, $\frac{1}{3} \cdot$ blue = red ($n_1$ – blue; $n_2$ – green; $n_3$ – red). Refer to the caption of Fig. 3 for interpretation of the spike activity (Color figure onine)

as the "difference of two exponentials" allows to adjust both rise and decay times of the synaptic current, this model shows the best results. Figure 6 shows the simulation trial of performing the multiplication of the input rate by four.

## 5.3   Computing Subtraction

**Subtraction using the TrueNorth model** is realized analogously to addition: the subtrahend though receives the weight of $-1$. Such implementation is inherently sensitive to the input timing: if the spikes from the subtrahend neuron happen before the spikes of the minuend neuron, the circuit computes $max(0, r_1 - r_2)$, i.e. if the actual difference is negative, no spikes are outputted.

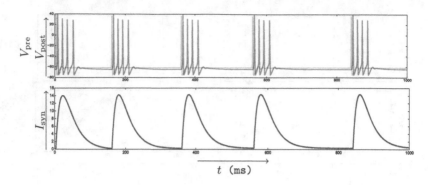

**Fig. 6.** Constant multiplication of the spike rates with the Hodgkin-Huxley model: the profile of the pre- and post-synaptic membrane potentials (magenta and green, respectively), and the total synaptic current (red) (Color figure online)

Conversely, if at a time step $t_i$ the output neuron receives the spike from the minuend, it needs to compute the running result and the correct way would be also to generate an action potential, although a spike from the subtrahend after an arbitrary silence interval would make the running result incorrect until the next spike from the minuend. The necessary and sufficient condition to ensure the correctness of the result is $V = 0$ at the end of the trial. Figure 7 shows the simulation results for the randomized input.

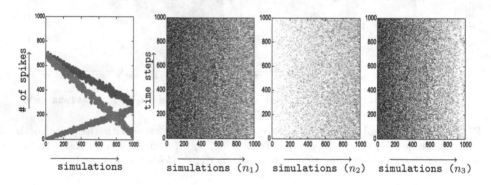

**Fig. 7.** Subtraction in the spike rates with the TrueNorth model: blue − green = red ($n_1$ − blue; $n_2$ − green; $n_3$ − red). Refer to the caption of Fig. 3 for interpretation of the spike activity (Color figure online)

**Subtraction using the Hodgkin-Huxley model** can only be performed when the following assumptions on the inputs are met: since the model does not have a mechanism to memorize the occurrences of spikes from the input neurons, all action potentials of the subtrahend neuron should coincide (up to the small time difference) with the action potential of the minuend neuron.

**Minimum/Maximum using the TrueNorth model** is based on the fact, that the subtraction actually computes $max(0, r_1 - r_2)$. We now can construct the minimum and maximum operators compositionally as follows: $min(r_1, r_2) = r_1 - max(0, r_1 - r_2)$, and $max(r_1, r_2) = r_2 + max(0, r_1 - r_2)$. Figure 8 shows the simulation results of performing these computations.

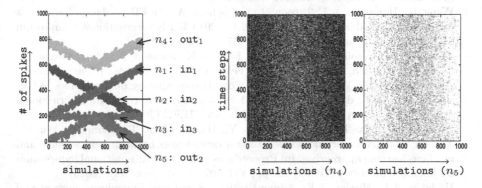

**Fig. 8.** Computing $min/max$ using the TrueNorth model: $r_4 = max(r_1, r_2) + r_3$, and $r_5 = min(r_1, r_2) - r_3$. The neuron $n_3$, which generates an offset of 200 spikes per trial on average, is added to separate the results from the inputs.

# 6   Conclusion and Outlooks

In this paper we showed how inherently different models of spiking neurons that come from neuroscience and computer engineering can be configured to perform the computations on the spike rates. This work steps towards our ultimate goal of developing a unified framework for operating both qualitatively and quantitatively on signals using spiking neurons.

The correctness of the computational results for both models depends on the operation being performed and the spike profile of the inputs. According to the results, the assumption on inputs for obtaining the correct computation results with the TrueNorth model are less restrictive then for the Hodgkin-Huxley model. The model of the synaptic transmission is crucial to perform the computation and obtain correct results using the biophysical neural model.

**Acknowledgment.** This research is supported by the project HARMONIA (845631), funded by a national Austrian grant from FFG (Österreichische Forschungsförderungsgesellschaft) under the program IKT der Zukunft and the EU ICT COST Action IC1402 on Runtime Verification beyond Monitoring (ARVI).

# References

1. McCulloch, W.S., Pitts, W.: A logical calculus of the ideas immanent in nervous activity. Bull. Math. Biophy. **5**(4), 115–133 (1943)
2. da Silva, R.M., de Macedo Mourelle, L., Nedjah, N.: Compact yet efficient hardware architecture for multilayer-perceptron neural networks. SBA: Controle Automacao Soc. Bras. de Automatica **22**, 647–663 (2011)
3. Wang, R., Hamilton, T.J., Tapson, J., van Schaik, A.: An FPGA design framework for large-scale spiking neural networks. In: 2014 IEEE International Symposium on Circuits and Systems (ISCAS), pp. 457–460, June 2014
4. Cassidy, A.S., Merolla, P., Arthur, J.V., Esser, S.K., Jackson, B., Alvarez-icaza, R., Datta, P., Sawada, J., Wong, T.M., Feldman, V., Amir, A., Rubin, D.B.-D., Mcquinn, E., Risk, W.P., Modha, D.S.: Cognitive computing building block: a versatile and efficient digital neuron model for neurosynaptic cores. In: International Joint Conference on Neural Networks (IJCNN). IEEE (2013)
5. Du, Z., Ben-Dayan Rubin, D.D., Chen, Y., He, L., Chen, T., Zhang, L., Wu, C., Temam, O.: Neuromorphic accelerators: a comparison between neuroscience and machine-learning approaches. In: Proceedings of the 48th International Symposium on Microarchitecture, MICRO-48, pp. 494–507. ACM, New York (2015)
6. Hodgkin, A.L., Huxley, A.F.: A quantitative description of membrane current and its application to conduction and excitation in nerve. J. Physiol. **117**(4), 500–544 (1952)
7. Roth, A., van Rossum, M.C.W.: Modeling synapses. Comput. Model. Methods Neurosci. **6**, 139–160 (2009)
8. Schüz, A., Palm, G.: Density of neurons and synapses in the cerebral cortex of the mouse. J. Comp. Neurol. **286**(4), 442–455 (1989)
9. Drion, G., Massotte, L., Sepulchre, R., Seutin, V.: How modeling can reconcile apparently discrepant experimental results: the case of pacemaking in dopaminergic neurons. PLoS Comput. Biol. **7**(5), e1002050 (2011)
10. Trachtenberg, J.T., Chen, B.E., Knott, G.W., Feng, G., Sanes, J.R., Welker, E., Svoboda, K.: Long-term in vivo imaging of experience-dependent synaptic plasticity in adult cortex. Nature **420**(6917), 788–794 (2002)
11. Dan, Y., Poo, M.-M.: Spike timing-dependent plasticity: from synapse to perception. Physiol. Rev. **86**(3), 1033–1048 (2006)
12. Jia, H., Rochefort, N.L., Chen, X., Konnerth, A.: Dendritic organization of sensory input to cortical neurons in vivo. Nature **464**(7293), 1307–1312 (2010)
13. Brette, R.: What is the most realistic single-compartment model of spike initiation? PLOS Comput. Biol. **11**(4), 1–13 (2015)
14. Izhikevich, E.M.: Which model to use for cortical spiking neurons. IEEE Trans. Neural Netw. **15**, 1063–1070 (2004)
15. Hines, M.: NEURON a program for simulation of nerve equations. In: Eeckman, F.H. (ed.) Neural Systems: Analysis and Modeling, pp. 127–136. Springer, Heidelberg (1993)
16. Gewaltig, M.-O., Diesmann, M.: Nest (neural simulation tool). Scholarpedia **2**(4), 1430 (2007)
17. Sarpeshkar, R., Watts, L., Mead, C.: Refractory Neuron Circuits. Caltech Authors, Pasadena (1992)
18. Graas, E.L., Brown, E.A., Lee, R.H.: An FPGA-based approach to high-speed simulation of conductance-based neuron models. Neuroinformatics **2**(4), 417–435 (2004)

19. Indiveri, G., Linares-Barranco, B., Hamilton, T.J., Van Schaik, A., Etienne-Cummings, R., Delbruck, T., Liu, S.-C., Dudek, P., Hafliger, P., Renaud, S., et al.: Neuromorphic silicon neuron circuits. Front. Neurosci. **5**, 73 (2011)
20. Hasani, R.M., Ferrari, G., Yamamoto, H., Kono, S., Ishihara, K., Fujimori, S., Tanii, T., Prati, E.: Control of the correlation of spontaneous neuron activity in biological and noise-activated CMOS artificial neural microcircuits. arXiv preprint arXiv:1702.07426 (2017)
21. Esser, S.K., Andreopoulos, A., Appuswamy, R., Datta, P., Barch, D., Amir, A., Arthur, J.V., Cassidy, A., Flickner, M., Merolla, P., Chandra, S., Basilico, N., Carpin, S., Zimmerman, T., Zee, F., Alvarez-Icaza, R., Kusnitz, J.A., Wong, T.M., Risk, W.P., McQuinn, E., Nayak, T.K., Singh, R., Modha, D.S.: Cognitive computing systems: algorithms and applications for networks of neurosynaptic cores. In: IJCNN, pp. 1–10. IEEE (2013)
22. Amir, A., Datta, P., Risk, W.P., Cassidy, A.S., Kusnitz, J.A., Esser, S.K., Andreopoulos, E., Wong, T.M., Flickner, M., Alvarez-icaza, R., Mcquinn, E., Shaw, B., Pass, N., Modha, D.S.: Cognitive computing programming paradigm: a corelet language for composing networks of neurosynaptic cores. In: International Joint Conference on Neural Networks (IJCNN). IEEE (2013)
23. Selyunin, K., Nguyen, T., Bartocci, E., Nickovic, D., Grosu, R.: Monitoring of MTL specifications with IBM's spiking-neuron model. In: Proceedings of the 19th Design, Automation and Test in Europe Conference and Exhibition, DATE 2016, Dresden, Germany, 14–18 March 2016 (2016)
24. Van Vreeswijk, C., Abbott, L.F., Bard Ermentrout, G.: When inhibition not excitation synchronizes neural firing. J. Comput. Neurosci. **1**(4), 313–321 (1994)
25. Dayan, P., Abbott, L.F.: Theoretical Neuroscience: Computational and Mathematical Modeling of Neural Systems. The MIT Press, Cambridge (2005)

# A SpiNNaker Application: Design, Implementation and Validation of SCPGs

Brayan Cuevas-Arteaga[1], Juan Pedro Dominguez-Morales[2(✉)],
Horacio Rostro-Gonzalez[1], Andres Espinal[3], Angel F. Jimenez-Fernandez[2],
Francisco Gomez-Rodriguez[2], and Alejandro Linares-Barranco[2]

[1] Department of Electronics, DICIS-University of Guanajuato,
Carr. Salamanca-Valle de Santiago Km 3.5+1.8 Palo Blanco,
36885 Salamanca, Guanajuato, Mexico
brayan.cuevas@ugto.mx
[2] Robotic and Technology of Computers Lab., University of Seville,
Av. Reina Mercedes s/n, 41012 Sevilla, Spain
jpdominguez@atc.us.es
[3] Department of Organizational Studies, DCEA-University of Guanajuato,
Fraccionamiento 1, Col. El Establo S/N, 36250 Guanajuato, Guanajuato, Mexico

**Abstract.** In this paper, we present the numerical results of the implementation of a Spiking Central Pattern Generator (SCPG) on a SpiNNaker board. The SCPG is a network of current-based leaky integrate-and-fire (LIF) neurons, which generates periodic spike trains that correspond to different locomotion gaits (i.e. walk, trot, run). To generate such patterns, the SCPG has been configured with different topologies, and its parameters have been experimentally estimated. To validate our designs, we have implemented them on the SpiNNaker board using PyNN and we have embedded it on a hexapod robot. The system includes a Dynamic Vision Sensor system able to command a pattern to the robot depending on the frequency of the events fired. The more activity the DVS produces, the faster that the pattern that is commanded will be.

**Keywords:** SCPGs · Legged robots locomotion · SpiNNaker · Spiking neurons · Hardware based implementations

## 1 Introduction

Robotic locomotion is a highly active research field in artificial intelligence. Nowadays, several methods have been proposed to achieve locomotion in a variety of robots (i.e. wheeled robots, legged robots, swimming robots, flying robots, etc.); particularly for non-wheeled robots, bioinspired locomotion systems may be implemented [1]. These systems, commonly known as Central Pattern Generators (CPGs) imitate behaviors of biological neural mechanisms and they are capable of endogenously produce periodically rhythmic patterns to contribute in locomotion of living beings among other rhythmic activities such as digestion, swallowing, etc. [2]. However, they do not work isolatedly, since they interact

© Springer International Publishing AG 2017
I. Rojas et al. (Eds.): IWANN 2017, Part I, LNCS 10305, pp. 548–559, 2017.
DOI: 10.1007/978-3-319-59153-7_47

with other parts of the central nervous system [3] and afferent sensory information may shape the CPG's outputs [4].

CPG-based locomotion systems have several advantages over non-bioinspired ones, such as: *rhytmicity, stability, adaptability* and *variety* (see [5] for a detailed explanation of these features). These systems have been designed and implemented for biped [6,7], quadruped [8] and hexapod, [9,10] among other kinds of non-wheeled robots (see [1,2,5] for comprehensive reviews). The implementation of CPGs for either software or hardware applications involves previous phases of modeling, analysis and modulation, which comprises the kind of neuron model to use, the coupling and the structure of the connections in a network. The last deals with parameter tunning and gait transitions [5].

As mentioned before, an important aspect on the implementation of CPGs is the selection of a neuron model. In this regard, there are several neuron models with different degrees of plausibility. To date, spiking neurons are considered as the most plausible neuron models, they form Spiking Neural Networks (SNNs) which are considered the third generation of Artificial Neural Networks [11]; CPGs built as SNNs are known as Spiking Central Pattern Generators (SCPGs).

In this paper a fully CPG-based locomotion system for the locomotion of a hexapod robot is proposed. Our proposal covers all phases suggested in [5]: modeling and analysis, modulation and implementation. The CPG-based locomotion system is conceptually based on works published in [8–10]; however we have changed the spiking neuron model and implementation platform by a more plausible one and a brain-like hardware platform respectively, i.e., the current-based integrate-and-fire model and a SpiNNaker board. Three different gaits are generated by the locomotion system, and simulated and implemented on the SpiNNaker board and validated on a real hexapod robot. Moreover, our proposal incorporates gait transitions according to the activity sensed by a DVS camera.

The rest of the paper is organized as follows: Sect. 2.1 provides information concerning the theoretical background around Spiking Central Pattern Generators. Section 2.2 provides information about the hardware used in this work, specifically on the SpiNNaker board. In Sect. 3, the design, implementation and validation of the system is explained in detail. Section 4 presents numerical results of the simulation. In Sect. 5, we present a perspective of the work and we conclude in Sect. 6.

## 2    Materials and Methods

### 2.1    Spiking Central Pattern Generators

SCPGs are a variation of the well-known and widely studied CPGs, which are specialized neural networks capable of endogenously produce rhythmic patterns. The CPGs contribute to living beings to perform actions without consent-effort such as walking among others [2,6]. Also, they have served as the basis of locomotion systems for non-wheeled robots with remarkable advantages over non-bioinspired locomotion systems [5].

SCPGs naturally handle spatiotemporal information as locomotion requires [10], which means that they receive and send information over time. In [8–10] are developed, implemented and tested SCPGs based on a biological study of insect locomotion [12] for legged robots by means of BMS neuron models [13]. The locomotion of legged robots is achieved in these works by means of spike-time-activity where each servo-motor receives a spike train; the presence or absence of spikes indicates the state of a servo-motor at the current time.

The SCPGs use the current-based leaky integrate-and-fire neuron model [14] as processing unit. Such model is one of the standard models in PyNN [15]. Equation (1) shows the equation of the model with fixed threshold and decaying-exponential post-synaptic current, excitatory injection in Eq. (2) and inhibitory injection given by Eq. (3).

$$\frac{dv}{dt} = \frac{ie + ii + i\_offset + i\_inj}{c\_m} + \frac{v\_rest - v}{tau\_m} \tag{1}$$

$$\frac{die}{dt} = -\frac{ie}{tau\_syn\_E} \tag{2}$$

$$\frac{dii}{dt} = -\frac{ii}{tau\_syn\_I} \tag{3}$$

where $v$ stands for the current of membrane potential. The excitatory and inhibitory current injections are expressed by $ie$ and $ii$ respectively. $i\_offset$ represents a base input current added each timestep. $i\_inj$ is an external current injection but in this case it is equal to zero. $c\_m$ is the capacitance of the leaky integrate-and-fire neuron in nano-Farads. $v\_reset$ is the voltage to set the neuron at immediately after a spike. $tau\_m$ means the time-constant of the RC circuit, in milliseconds. $tau\_syn\_E$ and $tau\_syn\_I$ are the excitatory and inhibitory input current decay time constant respectively. The neuron model uses a $tau\_refrac$ value for representing the refractory period, in milliseconds and finally, $v\_thresh$ stands for the threshold voltage at which the neuron will spike.

In PyNN the model is described as **if_curr_exp** and the code reads:

```
eqs = brian.Equations('''
dv/dt = (ie + ii + i_offset + i_inj)/c_m + (v_rest - v)/tau_m : mV
die/dt = -ie/tau_syn_E : nA
dii/dt = -ii/tau_syn_I : nA
tau_syn_E : ms
tau_syn_I : ms
tau_m : ms
c_m : nF
v_rest : mV
i_offset : nA
i_inj : nA
'''
)
```

## 2.2   Spiking Neural Network Architecture (SpiNNaker)

To implement and validate our SCPGs we used a SpiNNaker board [16]. SpiN-Naker is a massively-parallel multicore computing system designed for modeling very large spiking neural networks in real time. Both the system architecture and the design of the SpiNNaker chip have been developed by the Advanced Processor Technologies Research Group (APT) [17], which is based on the School of Computer Science at the University of Manchester. Each SpiNNaker chip consists of 18,200 MHz general-purpose ARM968 cores. The communication between them is done via packets carried by a custom interconnect fabric. The transmission of these packets is brokered entirely by hardware, giving the overall engine and extremely high bisection bandwidth.

In this work, a SpiNNaker 102 machine is used. This board comprises 4 SpiNNaker chips and, hence, it has 72 ARM processor cores deployed as 4 monitor processors, 64 application cores and 4 spare cores. A 100 Mbps Ethernet connection is used as control and I/O interface between the computer and the SpiNNaker board. 5V-1A supply is required for this machine. This platform has been used in previous works by the authors [18,19], proving its robustness and versatility (Fig. 1).

**Fig. 1.** SpiNNaker 102 machine.

## 2.3   Dynamic Vision Sensor

The AER DVS128 retina chip (silicon retina) [20] consists of an array of autonomous pixels that respond to relative changes in light intensity in real-time by placing the address of that specific pixel in an arbitrary asynchronous bus. Only pixels that are stimulated by any change of lightning transmit their addresses (events are produced). Hence, scenarios with no motion do not generate output events. These addresses are called Address Event (AE) and contains the x and y coordinates of the pixel that produced the event.

In this work, an AER DVS128 sensor was used to switch among the three different gaits, which has an array of 128 × 128 pixels. 7 bits are needed to encode each dimension of the array of pixels in this case. This Dynamic Vision Sensor also generates a polarity bit that represents the contrast change, where positive means a light increment and, negative, a light decrement. The DVS128 sensor is placed on the PAER interface that allows parallel AER through the CAVIAR connector [21].

## 3   Design, Implementation and Validation

### 3.1   Design

Three Spiking Central Pattern Generators (SCPGs) have been implemented into SpiNNaker. Each of them represents a different locomotion gait of the hexapod: walk, trot and run. These SCPGs were deployed in the machine using the PyNN library, which allows to easily create populations of neurons, connect them and assign weights to the connections.

SCPGs need an initial potential stimulus to start running. In this case, the resting potential and the threshold of the neurons are set to −65 mV and −50 mV, respectively. Hence, an initial potential of −49 mV is set on the first population to make the SCPG start running at the beginning of the simulation. With this stimulus, the SCPG is able to run infinitely while providing the same output spike pattern.

The current-based Leaky Integrate and Fire neuron model defined in Eq. (1) has been used in this work, and its configuration parameters are presented in Table 1.

**Table 1.** Configuration parameters of the current-based LIF neuron model

| Parameter | Value |
| --- | --- |
| cm | 0.25 |
| tau_m | 20.0 |
| tau_refrac | 2.0 |
| v_reset | −68.0 |
| v_rest | −65.0 |
| v_thresh | −50.0 |
| tau_syn_E | 5.0 |
| tau_syn_I | 5.0 |
| i_offset | 0.0 |

The SNN used to design each of the three locomotion sequences has basically the same architecture. The difference between them are the number of populations used, weights and delays for each of the sequences. It is necessary to mention that each of these sequences is implemented as a different SCPG.

In Fig. 2, we show the spiking neural network topology. Here, red neurons represent the direct stimulus towards the servomotors of the hexapod robot, and the neurons in blue and yellow are used to balance the generation of the correct patterns in each case. As can be seen in the figure, blue connections represent excitatory activity and, on the other hand, yellow connections represent inhibitory activity.

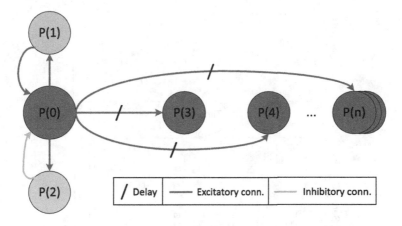

**Fig. 2.** SCPGs neural network architecture. (Color figure online)

To create the complete design of a SCPG it was necessary to find the weight and delay values for each of the connections between populations of neurons. This was carried out using a exhaustive search method.

At the end of the design of the SCPGs, a matrix is obtained, where each element of a column represents a servomotor and each row is an instant of time. Then, the absence or existence of a spike in this matrix will be translated to the movement of a specific servomotor on the hexapod robot. Figure 3 shows the spike trains (SCPG's output) representing the three different locomotion patterns that will perform the hexapod robot.

## 3.2 Implementation

The implementation of the SCPG in SpiNNaker was done using the PyNN toolchain. The results are the three types of spike patterns required for the locomotion. These sequences are shown in Fig. 3. In each figure, X-axis represents the simulation time and Y-axis shows the twelve neurons required for the locomotion of a hexapod robot; one for each servo-motor. The neural activity of each neuron is transduced into electrical signals for moving its associated servo-motor.

Specifically, locomotion corresponds to horizontal (coxa) and vertical (femur) movements, thus six neurons controls the horizontal movement and the other six the vertical one.

The results were corroborated with the state of the art [9]. After verifying that the results were correct in simulation, we proceeded to implement a complete system to validate the implementation of the SCPGs, which is described in the following subsection.

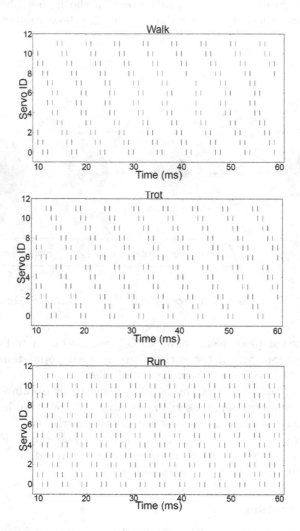

**Fig. 3.** Spike output patterns for each locomotion gait of the hexapod.

### 3.3  Validation

In Fig. 4, a block diagram of our complete validation system can be observed. It consists of a DVS camera, which receives events according to a change of intensity either positive or negative. The information is sent as an Address-Event-Representation (AER) format to an FPGA, which is used to synchronize and encode the incoming information in the specific format used by SpiNNaker (40 bits). The SpiNNaker is connected via Ethernet to a PC. For the mechanical validation of the SCPGs for this last, a hexapod robot was used, which consists of 12 servomotors: 6 to give movement to the extremities horizontally and 6 vertically. Each of the servomotors was connected to a PWM provided by an Arduino Mega board.

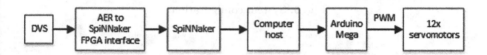

**Fig. 4.** Block diagram of the complete validation system.

The experiment consisted in connecting the complete system and, according to the frequency of events produced by the DVS, a specific locomotion sequence is executed. Different thresholds of event frequencies were set for this purpose. For example, if the DVS produces a few events, the robot would not perform any gait. However, if the DVS produces a moderate event frequency the hexapod would start walking, and so with trotting and running with higher event frequencies. All locomotion sequences were included in the same PyNN script. This scenario corroborates the operation of the complete system including both software and hardware.

## 4  Numerical Results

In Fig. 5, we show the whole system running a locomotion pattern in real time. In such figure, we can easily identify the main elements of the system: a SpiNNaker board configured with a SCPG performing the trot pattern, which can be observed on the oscilloscope. Due to the fact that our oscilloscope can only register four analog signals simultaneously, we decided to register three signals at the same time, i.e. three for each of the movements: horizontal (coxa) and vertical (femur), making thus a total of 12 signals registered for each gait.

To show the effectiveness on replicating locomotion patterns such as those found in vertebrates we performed numerical simulations for the three different (walk, trot and run) gaits using the SpiNNaker board and they are presented in Fig. 6. In such figure, the gaits are presented as follows: in the left side we can find the signals corresponding to the horizontal movements and in the right side for the vertical movements, both for the right legs performing the three gaits. The hexapod robot has a symmetric design. For this reason we only present the results for one of the hexapod sides (the legs on the right).

**Fig. 5.** System configuration

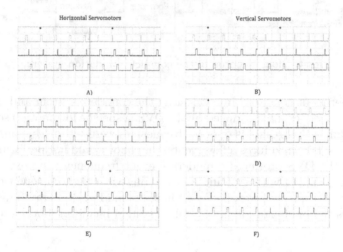

**Fig. 6.** Real time simulation

## 5    Stand-Alone Cognitive Locomotion System

The work presented in this paper is part of a more ambitious project in which the non-wheeled locomotion robot is commanded by a SNN in real-time using neuromorphic sensors and cognition to react to different stimulus. For example, to make the robot to follow an object that produces higher event activity on the DVS sensor it can easily be discriminated not only the frequency of the events, but also by the position on the visual field (left, center or right) to command a pattern that makes the robot turn left or right [22].

The hexapod robot used in the previous section will host an FPGA-based platform that will command the actuators of the robot according to SpiNNaker output stream of events. The circuit on the FPGA will be collecting AER events from the DVS, converting them to 40-bit format in one part. Furthermore, it will host a second circuit that receives 40-bit output from SpiNNaker and it will convert them into the according duty-cycles for the 12 Servo-motors of the hexapod to reproduce the right pattern. By the use of both hardware reconfigurable systems (FPGA and SpiNNaker) in the same robotic platform in a stand-alone capability, any research work could be implemented by properly dividing the functionality between both of them. DVS postprocessing [23] can be done in the FPGA in order to make lighter the algorithms running in SpiNNaker.

## 6 Conclusions

In this work, the authors have presented the design, implementation and validation of Spiking Central Pattern Generators for three locomotion gaits (walk, trot and run). The design was carried out in PyNN by using the current-based Leaky Integrate-and-Fire neuron model. Both implementation and validation were performed on a SpiNNaker board, which controls the locomotion of a real hexapod robot (See Fig. 5) through the generation of periodic spike trains (gaits) sent to the servomotors of the robot. Also, a DVS sensor can be incorporated to provide of sensory feedback and motor response through the locomotion of the hexapod robot. The results obtained are satisfactory comparing it with previous works, due to the fact that the implementation was done in a massively-parallel multi-core computing system, improving power consumption and temporal efficiency.

This opens the way to consider future work such as including the complete system within an hexapod, quadruped or bipedal robot, where the robot is completely stand-alone based on the perceptions obtained through neuromorphic sensors (e.g. DVS and Neuromorphic Cochlea) and processing this information in real-time.

**Acknowledgements.** This work is partially supported by the Spanish government grant (with support from the European Regional Development Fund) COFNET (TEC2016-77785-P). Also, this work has been supported by the Mexican government through the CONACYT project "Aplicación de la Neurociencia Computacional en el Desarrollo de Sistemas Roboticos Biologicamente Inspirados" (269798). The work of Juan P. Dominguez-Morales was supported by a Formación de Personal Universitario Scholarship from the Spanish Ministry of Education, Culture and Sport.

## References

1. Wu, Q., Liu, C., Zhang, J., Chen, Q.: Survey of locomotion control of legged robots inspired by biological concept. Sci. China Ser. F: Inf. Sci. **52**(10), 1715–1729 (2009)
2. Ijspeert, A.J.: Central pattern generators for locomotion control in animals and robots. a review. Neural Netw. **21**(4), 642–653 (2008)
3. Arena, P.: The central pattern generator: a paradigm for artificial locomotion. Soft. Comput. **4**(4), 251–266 (2000)
4. MacKay-Lyons, M.: Central pattern generation of locomotion: a review of the evidence. Phys. Ther. **82**(1), 69–83 (2002)

5. Yu, J., Tan, M., Chen, J., Zhang, J.: A survey on CPG-inspired control models and system implementation. IEEE Trans. Neural Netw. Learn. Syst. **25**(3), 441–456 (2014)
6. Lewis, M.A., Tenore, F., Etienne-Cummings, R.: CPG design using inhibitory networks. In: Proceedings of the 2005 IEEE International Conference on Robotics and Automation, ICRA 2005, pp. 3682–3687. IEEE (2005)
7. Russell, A., Orchard, G., Etienne-Cummings, R.: Configuring of spiking central pattern generator networks for bipedal walking using genetic algorthms. In: IEEE International Symposium on Circuits and Systems, ISCAS 2007, pp. 1525–1528. IEEE (2007)
8. Espinal, A., Rostro-Gonzalez, H., Carpio, M., et al.: Quadrupedal robot locomotion: a biologically inspired approach and its hardware implementation. Comput. Intell. Neurosci. **2016**, Article ID 5615618, 13 p. (2016). doi:10.1155/2016/5615618
9. Rostro-Gonzalez, H., Cerna-Garcia, P.A., Trejo-Caballero, G., Garcia-Capulin, C.H., Ibarra-Manzano, M.A., Avina-Cervantes, J.G., Torres-Huitzil, C.: A CPG system based on spiking neurons for hexapod robot locomotion. Neurocomputing **170**, 47–54 (2015)
10. Espinal, A., Rostro-Gonzalez, H., Carpio, M., Guerra-Hernandez, E.I., Ornelas-Rodriguez, M., Sotelo-Figueroa, M.: Design of spiking central pattern generators for multiple locomotion gaits in hexapod robots by Christiansen grammar evolution. Front. Neurorobotics **10**, 6 (2016). doi:10.3389/fnbot.2016.00006
11. Maass, W.: Networks of spiking neurons: the third generation of neural network models. Neural Netw. **10**(9), 1659–1671 (1997)
12. Grabowska, M., Godlewska, E., Schmidt, J., Daun-Gruhn, S.: Quadrupedal gaits in hexapod animals inter-leg coordination in free-walking adult stick insects. J. Exp. Biol. **215**(24), 4255–4266 (2012). https://www.ncbi.nlm.nih.gov/pubmed/22972892
13. Soula, H., Beslon, G., Mazet, O.: Spontaneous dynamics of asymmetric random recurrent spiking neural networks. Neural Comput. **18**(1), 60–79 (2006)
14. Abbott, L.F.: Lapicques introduction of the integrate-and-fire model neuron (1907). Brain Res. Bull. **50**(5), 303–304 (1999)
15. Davison, A., Brderle, D., Eppler, J., Kremkow, J., Muller, E., Pecevski, D., Perrinet, L., Yger, P.: PyNN: a common interface for neuronal network simulators. Front. Neuroinf. **2**, 11 (2009)
16. Painkras, E., Plana, L.A., Garside, J., Temple, S., Galluppi, F., Patterson, C., Lester, D.R., Brown, A.D., Furber, S.B.: SpiNNaker: a 1-w 18-core system-on-chip for massively-parallel neural network simulation. IEEE J. Solid-State Circ. **48**(8), 1943–1953 (2013)
17. Advanced Processor Technologies Research Group: Spinnaker home page. http://apt.cs.manchester.ac.uk/projects/SpiNNaker. Accessed 22 Jan 2016
18. Dominguez-Morales, J.P., Jimenez-Fernandez, A., Rios-Navarro, A., Cerezuela-Escudero, E., Gutierrez-Galan, D., Dominguez-Morales, M.J., Jimenez-Moreno, G.: Multilayer spiking neural network for audio samples classification using SpiNNaker. In: Villa, A.E.P., Masulli, P., Pons Rivero, A.J. (eds.) ICANN 2016. LNCS, vol. 9886, pp. 45–53. Springer, Cham (2016). doi:10.1007/978-3-319-44778-0_6
19. Rios-Navarro, A., Dominguez-Morales, J.P., Tapiador-Morales, R., Dominguez-Morales, M., Jimenez-Fernandez, A., Linares-Barranco, A.: A sensor fusion horse gait classification by a spiking neural network on SpiNNaker. In: Villa, A.E.P., Masulli, P., Pons Rivero, A.J. (eds.) ICANN 2016. LNCS, vol. 9886, pp. 36–44. Springer, Cham (2016). doi:10.1007/978-3-319-44778-0_5

20. Lichtsteiner, P., Posch, C., Delbruck, T.: A 128 × 128 120 dB 15 μs latency asynchronous temporal contrast vision sensor. IEEE J. Solid-State Circ. **43**(2), 566–576 (2008)
21. Serrano-Gotarredona, R., Oster, M., Lichtsteiner, P., Linares-Barranco, A., Paz-Vicente, R., Gómez-Rodríguez, F., Camuñas-Mesa, L., Berner, R., Rivas-Pérez, M., Delbruck, T., et al.: CAVIAR: A 45k neuron, 5M synapse, 12G connects/s AER hardware sensory-processing-learning-actuating system for high-speed visual object recognition and tracking. IEEE Trans. Neural Netw. **20**(9), 1417–1438 (2009)
22. Jiménez-Fernandez, A., Fuentes-del Bosh, J.L., Paz-Vicente, R., Linares-Barranco, A., Jiménez, G.: Neuro-inspired system for real-time vision sensor tilt correction. In: Proceedings of 2010 IEEE International Symposium on Circuits and Systems (ISCAS), pp. 1394–1397. IEEE (2010)
23. Linares-Barranco, A., Gómez-Rodríguez, F., Villanueva, V., Longinotti, L., Delbrück, T.: A USB3. 0 FPGA event-based filtering and tracking framework for dynamic vision sensors. In: 2015 IEEE International Symposium on Circuits and Systems (ISCAS), pp. 2417–2420. IEEE (2015)

# Smart Hardware Implementation of Spiking Neural Networks

Fabio Galán-Prado[✉] and Josep L. Rosselló

Electronics Engineering Group, Physics Department,
Universitat de les Illes Balears, Palma, Spain
{fabio.galan,j.rossello}@uib.es

**Abstract.** During last years a lot of attention have been focused to the hardware implementation of Artificial Neural Networks (ANN) to efficiently exploit the inherent parallelism associated to these systems. From the different types of ANN, the Spiking Neural Networks (SNN) arise as a promising bio-inspired model that is able to emulate the expected neural behavior with a high confidence. Many works are centered in using analog circuitry to reproduce SNN with a high degree of precision, while minimizing the area and the energy costs. Nevertheless, the reliability and flexibility of these systems is lower if compared with digital implementations. In this paper we present a new, low-cost bio-inspired digital neural model for SNN along with an auxiliary Computer Aided Design (CAD) tool for the efficient implementation of high-volume SNN.

**Keywords:** Neuromorphic hardware · Spiking Neural Network · FPGA

## 1 Introduction

Artificial Neural Networks (ANNs) is an emerging research line of high impact in computational intelligence application. The traditional Von Neumann computational architecture is very efficient in deterministic tasks like calculus, data bases management, file storage and so on.

However, it is really awkward at tasks such as pattern recognition [16] or time series prediction [11]. ANNs are usually implemented on this widely spread old architecture whose main drawback is its sequential nature, which prevents us from exploiting the inherent parallelism of ANNs. For the last years the ANN hardware implementation has been attracting attention because these ones can really exploit the ANNs' intrinsic parallelism [7,11,13]. There are lots of different ANNs such as Feed-Froward Networks, Reservoir Computing systems [11] or Convolutional Neural Networks (CNNs) that can be implemented using different neural models as the classical perceptron neuron or the more bioinspired Spiking Neural models [2,9]. Spiking Neural Netwoks (SNNs), which are considered to be the third generation of artificial neural networks [14], is able to reproduce the real neural behavior with high confidence. Its hardware implementation can be coped with two different point of views: analogical or digital. The former is quite

© Springer International Publishing AG 2017
I. Rojas et al. (Eds.): IWANN 2017, Part I, LNCS 10305, pp. 560–568, 2017.
DOI: 10.1007/978-3-319-59153-7_48

often employed because it gives a high level of precision and minimizes the area and energy required for its functionality. Nevertheless, the latter gives the user more control and is more flexible when it comes to make minor modifications. Digital neural models can be configured in Field Programmable Gate Arrays (FPGAs) that has been widely employed to implement digital neural models with really encouraging results [2,8,10–13]. In this paper a new bio-inspired low-cost digital neuron model is presented which will allow us to efficiently implement a high-volume SNN. For the implementation of huge networks we have set up an auxiliary Computer Aided Design (CAD) tool for its automatic generation.

## 2   Materials and Methods

Biological neurons normally present an unpredictable spike pattern [15] with a clear stochastic or random nature. This fact can suggest that information is mainly codified through the spike firing rate [6], which supports the idea of using a probabilistic codification when building ANNs [4]. A fact that could explain the stochastic nature of spike trains is the synaptic transmission's mechanism, since there exist a probability for transmitters to be released from the presynaptic terminal every time an action potential is transferred through the axon. This apparent lack of neural reliability can be understood as a clever way of implementing a weight for each connection.

In recent works, stochastic mechanisms have been introduced to the SNN models leading to Stochastic Spiking Neural models (SSNNs) [11,12]. This kind of networks combines the firing rate of spike trains and also the degree of correlation between neurons [6,8,12]. In the SSNN model, the neurons are correlated or uncorrelated through the use of the threshold voltage of the membrane potential so that two neurons are correlated if they share the same threshold (otherwise, they will be uncorrelated). Depending on this relationship, the network functionality can change drastically. Making use of this probabilistic coding (through the firing rate and correlation), high-speed pattern recognition systems can be implemented [12]. This is in contrast to neural networks using only a firing rate coding that are unable to provide a high-speed information processing. Probabilistic encoding is much more simpler than any other timing codes such as rank order coding [14,16] or spike-time coding [1,9] and provide evident advantages in the learning process due to its simplicity.

For clarity, in the figures depicted in this work all the correlated neurons are drawn sharing the same color. The stochastic behavior of the neuron synapsis is emulated through the entry layer (see Fig. 4) and a variable threshold voltage.

When using probabilistic encoding, the network functionality depends on the correlation between signals [12]. Apart from that, the exact time at which signals fire is not so important due to the fact that information will be only coded on the firing rate and spike correlation (and not on the shape of the spikes). For simplicity and an energy-efficient digital implementation the spike signals are modeled by boolean signals that only can be settled to either at a high or a low value during the smallest time step $t_{min}$. This time can be understood as the response time of the fastest neuron within the network. The boolean value of the neuron output ($x_k$) is on the high state when there is a spike between $t_i$ and $t_i + t_{min}$, otherwise $x_k$ is in the low state (see Fig. 1). Note that the use of this binary representation does not reduce the information contents of signals.

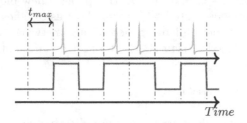

**Fig. 1.** Boolean representation of neural spikes

## 2.1   LIF Digital Neuron Model

The implementation of ANN on FPGAs is a widely extended practice due to the simplicity and short design cycle of the FPGA configuration [2–4,13]. In this paper, we propose a new neuron model which is more efficient in terms of area and energy than a previously-published SSNN neuron model [12] (see Fig. 2). For this purpose we reproduce a Leaky integrate and Fire (LIF) model for the spiking process [5]. Such model is characterized by the fact that the neuron's membrane potential (NMP) decays exponentially to the resting value and by the fact that it remembers how many spikes have taken place in the near past, see Fig. 3. With respect the SSNN model presented in [12], we have implemented two improvements: The exponential decay and a more realistic closed-loop model. As it can be seen in Fig. 4 the former is implemented, on our brand-new design, by a N-bit shift register storing the membrane potential and which is divided by 2 every clock cycle. The latter is implemented by adding a feedback reset from the output signal to the membrane register. As in the previously published neuron model [8,12] there must be also an entry layer which acts as synapses and gives a high or low signal state (where a high state implies the presence of a spike) for every single clock step. The neuron includes a binary comparator to limit the membrane potential to a threshold value provided by a pseudo-random number generator (see Fig. 3).

It can also be seen in Fig. 4 that every neuron has N excitatory entries $x_i$ and M inhibitory entries $x_i'$. Then all these incoming signals are added using an OR or a NOR gate respectively (performing the union function to the input pulses) and joined with an AND gate (performing the intersection function between signals). The final effect is that excitatory signals, $x_i$, activate the neuron by adding a fixed quantity (E) to the membrane potential register contents (see Fig. 4), whereas the inhibitory ones $x_i'$ inhibit such excitatory effect.

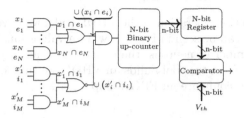

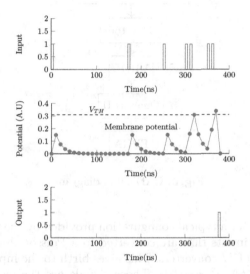

**Fig. 2.** Open-loop stochastic spiking neural model [8,12]

**Fig. 3.** Neuron behavior scheme

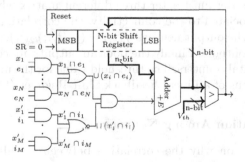

**Fig. 4.** Closed loop neuron's scheme (Color figure online)

## 2.2    The Computer Aided Design (CAD) Tool

In order to speed up the generation of the VHDL code, we have built a CAD tool written in C++. As can be seen in Fig. 5, few data values must be entered in the tool such as the number of neurons, the connectivity among neurons and how they are correlated. Then the tool builds an VHDL file than can be compiled by any synthesis tool to configure an FPGA chip.

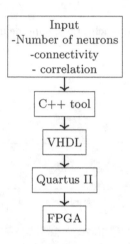

**Fig. 5.** CAD's flux diagram

In Fig. 6 we can see a typical configuration provided by the tool. The network has $M + N$ N-bits inputs that are converted to a Poisson-distributed sequence of spikes by means of $M$ converters. This gives birth to the input boolean vector $\mathbf{u}(\mathbf{t_i}) = (u_1(t_i), u_2(t_i), \ldots, u_M(t_i))$. Those signals are the entry values for the neurons. Simultaneously, the output of the neurons (which are also boolean signals $(x_k)$), can also act as inputs for other neurons, these outputs compose the network state vector $\mathbf{x}(\mathbf{t_i}) = (x_1(t_i), x_2(t_i), \ldots, x_L(t_i))$, where $L$ is the total number of neurons. The user must enter three different matrix which would contain the information associated to the connectivity, the correlation between neurons and the identity of the output neurons $\mathbf{y}(\mathbf{t_i}) = (y_1(t_i), y_2(t_i), \ldots, y_K(t_i))$, where $K$ is the number of outputs. Finally the boolean outputs are converted to binary signals by using digital counters. The threshold value for the membrane potential is generated by means of a Linear Feedback Shift Register (LFSR).

## 2.3    Synchronization Among Neurons

Lets put some light on why the correlation between signals is so important. Depending on the correlation the neuron's functionality can be substantially different [12]. For instance, let us take the results of two input signals with switching activity "$p$" and "$q$" respectively, which have been processed trough

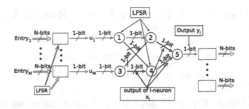

**Fig. 6.** Example of a neural network generated by the CAD tool.

an AND gate of the input circuitry (see Fig. 4). In case the signals are completely correlated, which is denoted here as $p \parallel q$ the AND gate is providing the minimum of both signals as shown in Fig. 7a. Otherwise, if they are completely uncorrelated, (that we denoted as $p \perp q$), the AND gate is performing the product between the inputs (Fig. 7b).

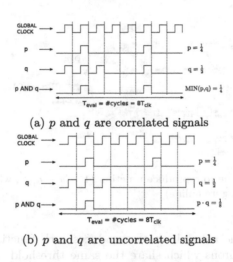

(a) $p$ and $q$ are correlated signals

(b) $p$ and $q$ are uncorrelated signals

**Fig. 7.** Two schemes to show the importance of the correlation between signals

Other than that, signals can also be neither completely correlated nor uncorrelated but a mix of both states. The output activity $r$ when two signals, $p$ and $q$, are evaluated using an OR (or an AND) gate can be obtained from the union (or the intersection for the AND case), which is written as $r = p \cup q$ (or $r = p \cap q$). Thus, we can define the independence factor between two signals as [8]:

$$I(p, q) \equiv \frac{p \cup q - \max(p, q)}{\min(p, q) - pq} \tag{1}$$

Therefore when $p$ and $q$ are completely correlated (uncorrelated at all) $I(p, q)$ will be 0 (1).

It must be noted that the output signals from synchronized (desynchronized) neurons are correlated (uncorrelated).

When two signals are joined or collided by an OR or an AND gate the output signals would follow these expressions:

$$p \cup q = \max(p, q) + (\min(p, q) - pq)I(p, q) \tag{2}$$

$$p \cap q = \min(p, q) - (\min(p, q) - pq)I(p, q) \tag{3}$$

In order to calculate Eq. 3, it has been taken into account the fact that $p \cap q = p + q - p \cup q$

## 3   Results

In order to test the behavior of the new neuron we have designed a network which reproduces the output of a specific mathematical function. The function used is $f(x) = (1 - |x - a|)^{C-1}$, where $a$ is a constant and $C$ is an integer.

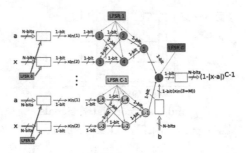

**Fig. 8.** Network's scheme to reproduce function $(1 - |x - a|)^C$, where $C = 11$, $a = 0.5$ and $N = 8bits$ (Color figure online)

An scheme for the network that reproduces such function can be seen on Fig. 8, where the neurons which share the same threshold are indicated using the same color.

As any other network built by our CAD tool, has their entries of N-bits which, are translated into boolean stochastic pulses and then plugged into the network.

The basic unit is composed by five neurons, the first layer's (neurons 1 and 3) duty is synchronizing the two incoming signals, $a$ and $x$. A second layer (neurons 2 and 4) evaluates in parallel both $a - x$ and $x - a$, whose output is connected to a third layer (neuron 5) which performs $(a - x) \cup (x - a)$. Following Eq. 2 such union yields $\min(a - x, x - a)$, which is the absolute value function indeed. This basic unit is replied C-1 times, whose output inhibits the last neuron which forms the fourth layer, $L$. Apart from that, a bias signal, $b$, is connected to the output neuron as an excitatory input, which is set to 1 for the sake of simplicity. Therefore its output is implementing the next expression:

$$y = b \cdot (1 - |x - a|) \cdot (1 - |x - a|) \cdots = (1 - |x - a|)^{C-1} \tag{4}$$

This final output is then converted from the boolean switching signal to a binary number which yields, in fact, the desired function. Such output is shown in Fig. 9, where we can observe that the open-loop and the closed loop neuron behave mostly the same. In addition, by making use of the Quartus II synthesis software it has been checked that the proposed model is more efficient in terms of gate count. This proposed model requires nearly the half number of logical elements to be programmed on the FPGA than the number required when using the previously published open-loop neuron model. However, not only does the new model require less area but it also presents a faster response time than the previous model (10 times faster). The simple network shown in Fig. 9 can be used as a neural comparator where the network is active when both "x" and "a" signals are similar.

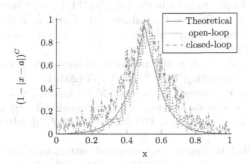

**Fig. 9.** Comparison between the results obtained using a previously published neuron model and the proposed in this paper.

## 4  Conclusion

Software solutions are the main way of implementing Artificial Neural Networks, providing impressive results during the last years. However, hardware implementations arises as an efficient way to further improve the performance of these systems. In this paper we propose an efficient digital circuitry for the implementation of spiking neural networks using a probabilistic coding. Compared to a previously published model [8,11,12], the proposal is up to 100% more efficient in terms of logic elements and is up to 10 times faster in terms of response time with a very faithful functionality.

**Acknowledgment.** This work has been partially supported by the Spanish Ministry of Economy and Competitiveness (MINECO) and the Regional European Development Funds (FEDER) under grant contract TEC2014-56244-R, and fellowship (BES-2015-076161).

# References

1. Bohte, S.M., La Poutré, H., Kok, J.N.: Unsupervised clustering with spiking neurons by sparse temporal coding and multilayer RBF networks. IEEE Trans. Neural Netw. **13**(2), 426–435 (2002)
2. Cassidy, A., Denham, S., Kanold, P., Andreou, A.: FPGA based silicon spiking neural array. In: Biomedical Circuits and Systems Conference BIOCAS 2007, no. 1, pp. 75–78 (2007)
3. Cassidy, A.S., Georgiou, J., Andreou, A.G.: Design of silicon brains in the nano-CMOS era: spiking neurons, learning synapses and neural architecture optimization. Neural Netw. **45**, 4–26 (2013)
4. Kaulmann, T., Dikmen, D., Rückert, U.: A digital framework for pulse coded neural network hardware with bit-serial operation. In: Proceedings - 7th International Conference on Hybrid Intelligent Systems, HIS 2007, pp. 302–307 (2007)
5. Koch, C., Segev, I.: A Bradford book, vol. 2. MIT Press, Cambridge (1998)
6. London, M., Roth, A., Beeren, L., Häusser, M., Latham, P.E.: Sensitivity to perturbations in vivo implies high noise and suggests rate coding in cortex. Nature **466**(7302), 7–123 (2010)
7. Misra, J., Saha, I.: Artificial neural networks in hardware: a survey of two decades of progress. Neurocomputing **74**(1–3), 239–255 (2010)
8. Morro, A., Canals, V., Oliver, A., Alomar, M.L., Galán-Prado, F., Ballester, P.J., Rosselló, J.L.: A stochastic spiking neural network for virtual screening. IEEE Trans. Neural Netw. Learn. Syst. **PP**(99), 1–5 (2017). doi:10.1109/TNNLS.2017. 2657601
9. Natschlager, T., Ruf, B.: Spatial and temporal pattern analysis via spiking neurons. Netw.: Comput. Neural Syst. **9**(731855466), 319–332 (1998)
10. Omondi, A.R., Rajapakse, J.C.: FPGA Implementations of Neural Networks. Springer, Heidelberg (2006)
11. Rossello, J.L., Alomar, M.L., Morro, A., Oliver, A., Canals, V.: High-density liquid-state machine circuitry for time-series forecasting. Int. J. Neural Syst. **26**(5), 1550036 (2016)
12. Rosselló, J.L., Canals, V., Oliver, A., Morro, A.: Studying the role of synchronized and chaotic spiking neural ensembles in neural information processing. Int. J. Neural Syst. **24**(05), 1430003 (2014)
13. Schrauwen, B., D'Haene, M., Verstraeten, D., Campenhout, J.V.: Compact hardware liquid state machines on FPGA for real-time speech recognition. Neural Netw. **21**(2–3), 511–523 (2008)
14. Soltic, S., Kasabov, N.: Knowledge extraction from evolving spiking neural networks with rank order population coding. Int. J. Neural Syst. **20**(06), 437–445 (2010)
15. Steinmetz, P.N., Manwani, A., Koch, C., London, M., Segev, I.: Subthreshold voltage noise due to channel fluctuations in active neuronal membranes. J. Comput. Neurosci. **9**(2), 133–148 (2000)
16. Wysoski, S.G., Benuskova, L., Kasabov, N.: Fast and adaptive network of spiking neurons for multi-view visual pattern recognition. Neurocomputing **71**(13–15), 2563–2575 (2008)

# An Extended Algorithm Using Adaptation of Momentum and Learning Rate for Spiking Neurons Emitting Multiple Spikes

Yuling Luo[1], Qiang Fu[1], Junxiu Liu[1(✉)], Jim Harkin[2],
Liam McDaid[2], and Yi Cao[3]

[1] Guangxi Key Lab of Multi-source Information Mining and Security,
Faculty of Electronic Engineering, Guangxi Normal University,
Guilin 541004, China
{yuling0616, liujunxiu}@mailbox.gxnu.edu.cn
[2] School of Computing and Intelligent Systems, Ulster University,
Derry BT48 7JL, UK
[3] Department of Business Transformation and Sustainable Enterprise,
Surrey Business School, University of Surrey, Surrey GU2 7XH, UK

**Abstract.** This paper presents two methods of using the dynamic momentum and learning rate adaption, to improve learning performance in spiking neural networks where neurons are modelled as spiking multiple times. The optimum value for the momentum factor is obtained from the mean square error with respect to the gradient of synaptic weights in the proposed algorithm. The delta-bar-delta rule is employed as the learning rate adaptation method. The XOR and Wisconsin breast cancer (WBC) classification tasks are used to validate the proposed algorithms. Results demonstrate no error and a minimal error of 0.08 are achieved for the XOR and WBC classification tasks respectively, which are better than the original Booij's algorithm. The minimum number of epochs for XOR and Wisconsin breast cancer tasks are 35 and 26 respectively, which are also faster than the original Booij's algorithm – i.e. 135 (for XOR) and 97 (for WBC). Compared with the original algorithm with static momentum and learning rate, the proposed dynamic algorithms can control the convergence rate and learning performance more effectively.

**Keywords:** Spiking neural networks · Learning rate · Momentum · Self-adaptation

## 1 Introduction

Recently, the third generation of neural networks, namely spiking neural networks (SNNs), was proposed in the approach of [1], where neurons communicate with each other via spikes [2]. In SNNs, the learning algorithm can transfer the information in the timing of spikes and adjust the synapse weights for connected neurons. A supervised learning algorithm, namely SpikeProp, was proposed in the approach of [3], which is based on the back propagation method of the sigmoidal artificial neural networks. It is designed for the SNNs with multiple delayed synapses. To improve the performance of

© Springer International Publishing AG 2017
I. Rojas et al. (Eds.): IWANN 2017, Part I, LNCS 10305, pp. 569–579, 2017.
DOI: 10.1007/978-3-319-59153-7_49

SpikeProp, various training algorithms, e.g. back propagation with momentum [4], QuickProp [5] and heuristic rules [6], have been developed to ensure that the SNN training process can converge to a global minimum. In the approach of [7] researchers tried to speed up SNNs by extending the SpikeProp algorithm, where the delay and the time constant of every connection, and the threshold of the neurons can be trained. Another back-propagating learning model in [8] is a modified version of SpikeProp. It can train SNNs and the neurons can fire multiple times. This type of algorithm is more efficient at learning and has more advantages of the temporal information processing capability of SNNs. It successfully built a lip-reading system, solved Exclusive-OR problem and learned the Parity-n function. However, this algorithm is not fast enough to solve problems. For instance, this algorithm needs more than 100 iterators to solve XOR problem. In addition, the sum of the squared differences easily converge to a local minimum instead of a global minimum. To speed up the algorithm for spiking neurons that emit multiple spikes, a fast learning algorithm via momentum and learning rate self-adaptation is proposed in this paper. The improved algorithm not only can speed up the learning rate but also can prevent the mean squared error from converging to a local minimum.

This paper is organized as follows. Section 2 discusses the spike neuron models and network architectures briefly. The learning rule is derived in Sect. 3. Section 4 presents the proposed self-adaptation method for the momentum factor. Moreover, a dynamic self-adaptation for the learning rate is also introduced. The performance and test results are given Sect. 5 and Sect. 6 concludes the paper.

## 2    The Neuron Model and SNNs Architectures

The neuron model is an important component of SNN. Due to learning algorithms based on gradient descent require computation of partial derivatives, the spike response model (SRM), whose internal state can be expressed intuitively, is widely used in research.

In the framework of the SRM, the state of an output neuron j is described by a single variable $u_j$. When the potential $u_j$ reaches a threshold $\vartheta$, the neuron j fires a spike at time $t_j$. The spike-times of neuron j are described as:

$$F_j = \left\{ t_j^{(f)}; 1 \leq f \leq n \right\} = \left\{ t | u_j(t) = \vartheta \right\} \tag{1}$$

where n denotes the number of spikes and the spike-train $F_j$ is chronologically ordered. If $1 \leq f < g \leq n$, then $t_j^{(f)} < t_i^{(g)}$. The set of presynaptic neuron $i \in \Gamma_j$ is described as:

$$\Gamma_j = \{i | i \text{ is presynaptic to } j\} \tag{2}$$

After the presynaptic neuron fires, the evolution of $u_j$ is given by:

$$u_j(t) = \sum\nolimits_{t_j^{(f)} \in \Gamma_j} \eta \left( t - t_j^{(f)} \right) + \sum\nolimits_{i \in \Gamma_j} \sum\nolimits_{t_i^{(g)} \in F_i} \sum\nolimits_{k=1}^{l} w_{ji}^k \varepsilon(t - t_i^{(g)} - d^k) \tag{3}$$

where $w_{ji}^k$ is the weight of the synapse from input neuron $i$ to output neuron $j$ with a delay of $d^k$. The postsynaptic potential (PSP) induced by one spike is determined by the spike response function $\varepsilon(s)$, which is expressed as:

$$\varepsilon(s) = \frac{s}{\tau} exp(1 - \frac{s}{\tau})H(s) \qquad (4)$$

where $\tau$ is a time constant, H(s) is the Heavy-side step function, and $H(s) = 0$ if $s \leq 0$ and $H(s) = 1$ if $s > 0$. The refractoriness function $\eta(s)$ expresses the internal to depict the relative refractory period, which is expressed as:

$$\eta(s) = -\vartheta exp(-\frac{s}{\tau_r})H(s) \qquad (5)$$

where $\tau_r$ is the time decay constant, which can influence the shape of refractoriness function. The SRM is widely used in many approaches of SNN research [9–12]. The SRM model can give a good approximation of synapse response of a neuron, therefore it is used as the target neuron model for the investigation of fast learning algorithm in this paper.

The SNN is more realistic than traditional artificial neural networks as it emulates internal neuron behavior via firing spikes [13]. As shown in Fig. 1, in general the network architecture consists of a feed forward network of spiking neurons with multiple delayed synaptic terminals, and every neuron in each layer is connected to all the neurons in the next layer; then the network is defined as 'fully connected'. The same neuron network architecture in Fig. 1 is used in this paper.

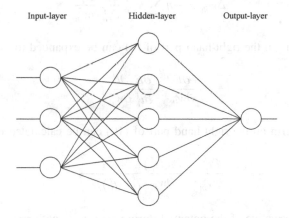

**Fig. 1.** Feed forward spiking neural network [8, 14, 15]

## 3   Error Back Propagation Algorithm

An algorithm based on error back propagation and a fully connected feed forward neural network is derived in this approach. The algorithm can cope with neurons that spike multiple times.

In order to train the network to produce firing times at the output neurons from a set of firing times at the input neurons, the measure function is determined by the difference between the desired output and the actual output for all output neurons. Assume that $t_j^d$ is desired firing time at output neuron $j$ and $t_j^a$ is actual firing time at output neuron $j$. The sum of the squared differences can be described by:

$$E = \frac{1}{2} \sum_{j \in J} \left( t_j^a - t_j^d \right)^2 \tag{6}$$

The mean squared error $(E)$, with respect to one output spike, can be derived by:

$$\frac{\partial E}{\partial t_j^a} = t_j^a - t_j^d \tag{7}$$

To minimize the squared differences E, the weight $w_{ih}^k$ between each separate connection k of the synaptic terminal should be calculated using:

$$\Delta w_{ih}^k = -\eta \frac{\partial E}{\partial w_{ih}^k} \tag{8}$$

where $\eta$ is the learning rate of the network. As neuron $i$ can fire multiple times and all these firing times depend on the weight, this equation can be expended with regard to these spikes:

$$\Delta w_{ih}^k = -\eta \sum_{t_i^f \in F_i} \frac{\partial E}{\partial t_i^f} \frac{\partial t_i^f}{\partial w_{ih}^k} \tag{9}$$

The derivation on the right-hand part of (8) can be expanded to:

$$\frac{\partial E}{\partial w_{ih}^k} = \frac{\partial E}{\partial t_i^f} \frac{\partial t_i^f}{\partial w_{ih}^k} \tag{10}$$

The second term on the right-hand part of (10) can be calculated by:

$$\frac{\partial t_i^f}{\partial w_{ih}^k} = \frac{\partial u_i(t_i^f)}{\partial w_{ih}^k} \frac{-1}{\partial u_i(t_i^f)/\partial t_i^f} \tag{11}$$

The partial derivative of the potential during a spike $t_i^f$ with respect to the weight $(\partial u_i(t_i^f)/\partial w_{ih}^k)$ can be derived by:

$$\frac{\partial u_i(t_i^f)}{\partial w_{ih}^k} = -\sum_{t_i^g \in F_i} \eta' \left( t_i^f - t_i^g \right) \frac{\partial t_i^g}{\partial w_{ih}^k} + \sum_{t_h^l \in F_h} \varepsilon(t_i^f - t_h^l - d_{ih}^k) \tag{12}$$

second term on the right-hand part of (11) (i.e. $\partial u_i(t_i^f)/\partial t_i^f$) can be calculated by:

$$\frac{\partial u_i(t_i^f)}{\partial t_i^f} = -\sum_{t_i^g \in F_i} \eta' \left(t_i^f - t_i^g\right) + \sum_{h \in \Gamma_i} \sum_{t_h^l \in F_h} \sum_k w_{ih}^k \varepsilon' \left(t_i^f - t_h^l - d_{ih}^k\right) \quad (13)$$

The first term on the right-hand side of (10) can be derived by:

$$\frac{\partial E}{\partial t_i^f} = \sum_{j \in \Gamma^i} \sum_{t_i^g \in F_i} \frac{\partial E}{\partial t_j^g} \frac{\partial t_j^g}{\partial t_i^f} \quad (14)$$

where $t_j^g$ is the firing time of a postsynaptic spike and $t_i^f$ is the firing time of a presynaptic spike. The derivative of a postsynaptic spike with respect to a presynaptic can be described as:

$$\frac{\partial t_j^g}{\partial t_i^f} = \frac{\partial u_i(t_j^g)}{\partial t_i^f} \frac{-1}{\partial u_j(t_j^g)/\partial t_j^g} \quad (15)$$

The first term on the right-hand side of (15) can be derived by:

$$\frac{\partial u_j(t_j^g)}{\partial t_i^f} = -\sum_{t_j^l \in F_i} \eta' \left(t_j^g - t_j^l\right) \frac{\partial t_j^l}{\partial t_i^f} - \sum_k w_{ij}^k \varepsilon' \left(t_j^g - t_i^f - d_{ij}^k\right) \quad (16)$$

The weight adaptation rule is derived by combining the Eqs. (14), (11) and (8). The derived algorithm can be applied to networks with more than one hidden layer.

## 4 Momentum and Learning Rate Adaptation

In this section, the methods of adaptation of momentum and learning rate are considered. As the original Booij's algorithm [8] can experience local minima in some optimization problem, a momentum factor is added to the weight update procedure to make the algorithm more robust. In addition, convergence speed is related to the learning rate. The original Booij's algorithm is slow to deal with large datasets, thus the learning rate is needed to be a big value. However, a large learning rate usually lead to low accuracy. Therefore an adaptive learning rate method is proposed to address this problem.

### 4.1 Momentum Adaptation

To enhance the training speed of the SNN, a momentum is added to the weight update procedure as follows:

$$w_{ih}^{k+1} = w_{ih}^k + \eta \frac{\partial E}{\partial w_{ih}^k} + \alpha \Delta w_{ih}^{k-1} \quad (17)$$

where $\alpha$ is the momentum coefficient and the $\Delta w_{ih}^{k-1}$ is the weight correction in the previous iteration. In the approach of [4], $\alpha$ is a static value. The disadvantage is that it cannot guarantee to obtain an optimal value, and the momentum factor is difficult to be tuned. If the momentum is given with small value for the convergence, the improved changes in convergence time would not be expected to be comparable to the time before applying the momentum. In the meantime a large momentum can cause over learning. Thus, as static momentum is equally applied in the entire training, it does not become an effective proposal for choices in reducing the convergence time and improving performance [16]. In the approach of [17], a $mod - DS - \eta\alpha$ rule was proposed to improve the performance. However, this method is restricted to select one of three momentum values. In our proposed algorithm, the optimum value for momentum factor is obtained from the gradient of the error function with respect to the synaptic weight. The weight-change with momentum adaptation is described by:

$$w_{ih}^{k+1} = w_{ih}^k + \eta \frac{\partial E}{\partial w_{ih}^k} + e^{-\delta - \left\| \frac{\partial E}{\partial w_{ih}^k} \right\|} \Delta w_{ih}^{k-1} \tag{18}$$

where $\delta$ is a constant to control the momentum factor. The value of $\delta$ is: $0 < \delta < 1$.

## 4.2  Adaptive Learning Rate Algorithm

The learning rate that has the least validation error, is chosen as the optimum learning rate [18]. The optimum learning rate can guarantee the stable convergence of SNN and improve the convergence speed. If the learning rate is set too high, the convergence speed of the network is fast, but may lead to the network instability, or even the network cannot be operated. If the learning rate is set to a small value, the network convergence speed is slow and consumes long computing time; and in the worst case it probably cannot meet the requirements of practical applications.

To improve the performance of convergence, the momentum can be used in combining with adaptive learning rate [19]. The weight-updating formula can be calculated by:

$$w_{ih}^{k+1} = w_{ih}^k + \eta_{ih}^{k+1} \frac{\partial E}{\partial w_{ih}^k} + e^{-\delta - \left\| \frac{\partial E}{\partial w_{ih}^k} \right\|} \Delta w_{ih}^{k-1} \tag{19}$$

where $\eta_{ih}^{k+1}$ is adaptive learning rate. The dynamic self-adaptation method of learning rate [17, 20] can improve the dynamic properties of the standard algorithm. The $DS - \eta$ learning rule first assigns an initial value to $\eta$, then in each iteration, $\eta_1$ and $\eta_2$ are computed by:

$$\eta_1 = \eta / \rho \tag{20}$$

$$\eta_2 = \eta * \rho \tag{21}$$

where $\eta$ is the learning rate in the previous iteration, and $\rho$ is greater than 1.0. When $\rho > 1.0$, the network can normally work well; but $\rho$ has an optimum value of 1.839 for the network [20]. The disadvantage of this method is that how to select these rates ($\eta_1$ and $\eta_2$). Hence, this method is restricted to select one of these two values [17]. In this paper, the delta-bar-delta rule is used to adjust the learning rate adaptively which was proposed in [6]. The delta-bar-delta rule can be calculated by:

$$\Delta\eta_{ih}^k = \begin{cases} a, & S_{ih}^{k-1}D_{ih}^k > 0 \\ -b\eta_{ih}^k, & S_{ih}^{k-1}D_{ih}^k > 0 \\ 0, & other \end{cases} \tag{22}$$

$$D_{ih}^k = \frac{\partial E}{\partial w_{ih}^k} \tag{23}$$

$$S_{ih}^k = (1 - c)D_{ih}^k + cS_{ih}^{k-1} \tag{24}$$

where a, b, and c are all parameters. The typical values are: $10^{-4} \leq a \leq 0.1$, $0.1 \leq b \leq 0.5$, and $0.1 \leq c \leq 0.7$. The delta-bar-delta method increases learning rate linearly and decreases them exponentially.

## 5    Experimental Results

In order to test the performance of the proposed algorithms, the XOR task and Wisconsin Breast Cancer classification task were selected as exemplar applications. A fully connected feedforward network with multiple delays per connection is employed. Each connection consists of a fixed number of 16 synaptic terminals. These connections have a delay interval of 15 ms and therefore the available synaptic delays are from 1 to 16 ms.

### 5.1    XOR Task

A simple way to encode the XOR dataset by spike times was proposed in [3]. The inputs and outputs for XOR task are binary, and the binary 0's and 1's are directly encoded into firing times of input and the desired firing times of output. For the input variables, an input spike at 1 ms represents logic 0 while a spike at 7 ms represents logic 1. For the output, a spike at 16 ms represents logic 0 while a spike at 10 ms represents logic 1. Such a coding scheme is given by [3]. Therefore the XOR task has four training patterns, including input patterns (1, 1), (1, 7), (7, 1), and (7, 7) with corresponding output patterns 16, 10, 10, and 16 (unit: ms). The network architecture is consisted of three input neurons (two coding neurons and one reference neuron), five hidden neurons and a single output neuron. The parameters $\delta$ and $\tau$ are set to 0.01 and 7, respectively. The static learning rate $\eta$ is set to 1.

Figure 2 and Table 1 show the relationships between epochs and error under different conditions. (a) refers to original Booij's algorithm proposed in [8], it carries out 135 epochs and the error is 1.4. (b) refers to the algorithm with a static momentum, it carries out 42 epochs and the error is 1.7. (c) refers to the algorithm with dynamic momentum, it carries out 35 epochs and the error is 1.2. (d) refers to the algorithm with dynamic momentum and adaptive learning rate, it carries out 51 epochs and the error is 0.7. It can be seen that the adaptive learning method learns within 51 epochs which is much faster than original 135 epochs. And the error is 0.7 in this work, it is 2 times less than original Booij's algorithm. The methods of static momentum factor learns within 42 epochs and adaptive learning rate with static momentum factor learns within 35 epochs. However, the errors of these two methods are higher than the method of adaptation of momentum factor and learning rate.

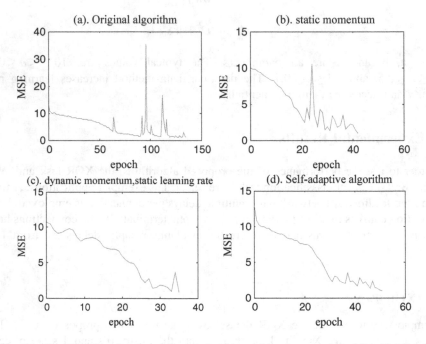

**Fig. 2.** Results of XOR task using different algorithms. (a) refers to original Booij's algorithm [8]. (b), (c), and (d) give the results of the static momentum and learning rate, adaptive momentum and static learning rate, adaptive momentum and adaptive learning rate, respectively.

**Table 1.** The XOR task of different algorithms

| Algorithms | Epochs | Desired output [ms] | Actual output [ms] | Absolute error |
|---|---|---|---|---|
| Original Booij's algorithm [8] | 135 | 16 | 14.6 | 1.4 |
| Static $\alpha$, $\eta = 1$ | 42 | 16 | 17.7 | 1.7 |
| Dynamic $\alpha$, $\eta = 1$ | 35 | 16 | 17.2 | 1.2 |
| This work - Adaptive $\alpha$ and $\eta$ | 51 | 16 | 15.3 | 0.7 |

## 5.2    Wisconsin Breast Cancer Classfication Task

The Wisconsin Breast Cancer (WBC) dataset is used for this testing. It uses breast cytology gained by fine needle aspirations in the University of Wisconsin Hospital and classifies the results into benign or malignant cancer tumors [21]. WBC classification task is a binary classification problem which consists of 669 samples. Each sample consists of 9 attributes and has to be classified as a benign or malignant case of breast cancer. 300 samples are selected to verify the algorithm and they are divided into two groups. One group includes 200 samples for training and another comprises of 100 samples for testing. The 9 attributes in each sample which measure different features of the cytology. And each value can be mapped directly to a linear spike train in the range 1 to 10.

For the input variables, each value can be mapped directly to a linear spike train in range of 1 to 10 ms. For the output variables, the benign and malignant tumours can be coded as an output spike at time 16 ms and 17 ms respectively. Similar to the XOR task, the parameters $\delta$ and $\tau$ are set to 0.01 and 3, respectively. The static learning rate $\eta$ is set to 1. The relationships between epochs and error under different algorithms is shown in Fig. 3 and Table 2. Figure 3(a) refers to original Booij's algorithm, it carries out 97 epochs and the accuracy rate is 79%; (b) refers to the algorithm with a static momentum, it carries out 36 epochs and the accuracy rate is 81%; (c) refers to the

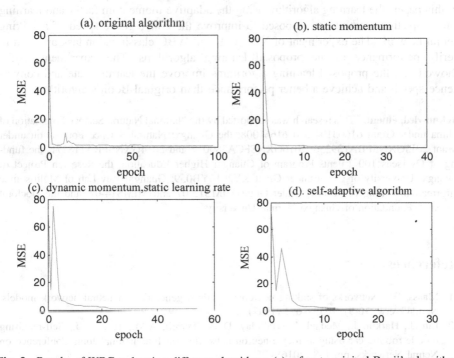

**Fig. 3.** Results of WBC task using different algorithms. (a) refers to original Booij's algorithm [8]. (b), (c), and (d) give the results of the static momentum and learning rate, adaptive momentum and static learning rate, adaptive momentum and adaptive learning rate, respectively.

**Table 2.** The WBC task of different algorithms

| Algorithms | Epochs | Accuracy rate |
|---|---|---|
| Original Booij's algorithm [8] | 97 | 79% |
| Static α, η = 1 | 36 | 81% |
| Dynamic α, η = 1 | 54 | 84% |
| This work - Adaptive α and η | 26 | 92% |

algorithm with dynamic momentum, it carries out 54 epochs and the accuracy rate is 84%; (d) refers to the algorithm with dynamic momentum and adaptive learning rate, it carries out 26 epochs and the accuracy rate is 92%. These experimental results demonstrate that the adaptive learning algorithm with momentum factor and learning rate can learn within 26 epochs and the accuracy is 92%. For most of the iterations sampled, the correct classification results are much higher for the adaptive case than the non-adaptive one. In addition the adaptive learning algorithm learns faster than non-adaptive one. In summary, the error of the proposed algorithm is the smallest and it uses less epochs for the task competition.

## 6   Conclusions

In this paper, the learning algorithm using the adaptive momentum factor and learning rate adaptation method are proposed to improve the convergence speed of a spiking neural network. The experiment of XOR task and WBC classification task are used to verify performance of the proposed learning algorithms. The experiment results showed that the proposed learning algorithms improve the learning rate and convergence speed, and achieve a better performance than original Booij's algorithm.

**Acknowledgement.** This research was supported by the National Natural Science Foundation of China under Grant 61603104 and 61661008, the Guangxi Natural Science Foundation under Grant 2015GXNSFBA139256, 2016GXNSFCA380017 and 2014GXNSFBA118271, the funding of Overseas 100 Talents Program of Guangxi Higher Education, the Research Project of Guangxi University of China under Grant KY2016YB059, Guangxi Key Lab of Multi-source Information Mining & Security under Grant MIMS15-07 and MIMS14-04, and the Doctoral Research Foundation of Guangxi Normal University.

## References

1. Maass, W.: Networks of spiking neurons: the third generation of neural network models. Neural Netw. **10**(9), 1659–1671 (1997)
2. Liu, J., Harkin, J., Mcdaid, L., Halliday, D.M., Tyrrell, A.M., Timmis, J.: Self-repairing mobile robotic car using astrocyte-neuron networks. In: International Joint Conference on Neural Networks, pp. 1–8 (2016)
3. Bohte, S.M., Kok, J.N., La Poutré, H.: Error-backpropagation in temporally encoded networks of spiking neurons. Neurocomputing **48**(1–4), 17–37 (2002)

4. Xin, J., Embrechts, M.J.: Supervised learning with spiking neural networks. In: International Joint Conference on Neural Networks, vol. 3, no. 3, pp. 1772–1777 (2001)
5. McKennoch, S., Liu, D.L.D., Bushnell, L.G.: Fast modifications of the spikeprop algorithm. In: Proceedings of the 2006 IEEE International Joint Conference on Neural Networks, vol. 16, no. 6, pp. 3970–3977 (2006)
6. Jacobs, R.A.: Increased rates of convergence through learning rate adaptation. Neural Netw. 1(4), 295–307 (1988)
7. Schrauwen, B.: Extending spikeprop. In: International Joint Conference on Neural Networks, vol. 1, no. 7, pp. 471–476 (2004)
8. Booij, O., Tat Nguyen, H.: A gradient descent rule for spiking neurons emitting multiple spikes. Inf. Process. Lett. 95(6), 552–558 (2005)
9. Kulkarni, S., Simon, S.P., Sundareswaran, K.: A spiking neural network (SNN) forecast engine for short-term electrical load forecasting. Appl. Soft Comput. J. 13(8), 3628–3635 (2013)
10. Rosado-Muñoz, A., Bataller-Mompeán, M., Guerrero-Martínez, J.: FPGA implementation of spiking neural networks. In: Proceedings of the 1st IFAC Conference on Embedded Systems, Computational Intelligence and Telematics in Control, vol. 45, no. 4, pp. 139–144 (2012)
11. Rosado-Muñoz, A., Bataller-Mompeán, M., Guerrero-Martínez, J.: FPGA implementation of spiking neural networks supported by a software design environment. IFAC Proc. Vol. 45(4), 1934–1939 (2012)
12. Awadalla, M.H.A., Sadek, M.A.: Spiking neural network-based control chart pattern recognition. Alex. Eng. J. 51(1), 27–35 (2012)
13. Dorogyy, Y., Kolisnichenko, V.: Designing spiking neural networks. In: Modern Problems of Radio Engineering, Telecommunications and Computer Science, vol. 6, pp. 124–127 (2016)
14. Ghosh-Dastidar, S., Adeli, H.: A new supervised learning algorithm for multiple spiking neural networks with application in epilepsy and seizure detection. Neural Netw. 22(10), 1419–1431 (2009)
15. Ghosh-Dastidar, S., Adeli, H.: Improved spiking neural networks for EEG classification and epilepsy and seizure detection. Integr. Comput. Aided Eng. 14(4), 187–212 (2007)
16. Kim, E.-M., Park, S.-M., Kim, K.-H., Lee, B.-H.: An effective machine learning algorithm using momentum scheduling. In: Fourth International Conference on Hybrid Intelligent Systems (HIS 2004), pp. 442–443 (2004)
17. Delshad, E., Moallem, P., Monadjemi, S.H.: Spiking neural network learning algorithms: using learning rates adaptation of gradient and momentum steps. In: 2010 5th International Symposium on Telecommunications, no. 1, pp. 944–949 (2010)
18. Chandra, B., Sharma, R.K.: Deep learning with adaptive learning rate using laplacian score. Expert Syst. Appl. 63(5), 1–7 (2016)
19. Huijuan, F., Jiliang, L., Fei, W.: Fast learning in spiking neural networks by learning rate adaptation. Chin. J. Chem. Eng. 20(6), 1219–1224 (2012)
20. Salomon, R., Van Hemmen, J.L.: Accelerating backpropagation through dynamic self-adaptation. Neural Netw. 9(4), 589–601 (1996)
21. Wolberg, W.H., Mangasarian, O.L.: Multisurface method of pattern separation for medical diagnosis applied to breast cytology. Proc. Nat. Acad. Sci. 87(12), 9193–9196 (1990)

# Development of Doped Graphene Oxide Resistive Memories for Applications Based on Neuromorphic Computing

Marina Sparvoli[1(✉)], Mauro F.P. Silva[2], and Mario Gazziro[1]

[1] Universidade Federal do ABC, Santo André, São Paulo, Brazil
marinsparvoli@yahoo.com.br
[2] Faculdades Oswaldo Cruz, São Paulo, São Paulo, Brazil

**Abstract.** Resistive random access memory ReRAM has attracted great attention due to its potential for flash memory replacement in next generation non-volatile memory applications. Among the main characteristics of this type of memory, we have: low energy consumption, high-speed switching, durability, scalability and friendly manufacturing process. This device is based on resistive switching phenomenon for operation, which is reversible and can be played back repeatedly. In this work, eight different devices are developed and fabrication is made as follows: thin films are obtained by dip coating technique. The dip coating apparatus basically consists of a clamp which holds the substrate is dipped in a GO solution (graphene oxide) which containing dopant (copper, iron or silver) or CuO (copper oxide). ITO (indium tin oxide) and aluminum contacts were evaporated. The devices were developed with purpose: intention is record and read information dynamically with appropriate algorithm. There is even the possibility of storing images. With these functions, it would be promising to enter the neuromorphic computing area that is one of the resistive memory applications. ReRAM technology advent represents a paradigm shift for artificial neural networks, being the best candidate for emulation of synaptic plasticity and learning mode.

**Keywords:** Resistive memory · ReRAM · Graphene oxide

## 1 Introduction

Resistive random access memories (ReRAMs) are a class of devices emerging from the new generation of non-volatile memories. Many researchers have made great efforts to understand and develop these new memories because they have simple metal-insulator-metal (MIM) structure [1, 2], ease of recording/reading, high storage density and low power consumption. Resistive switching (RS) is the basic phenomenon for the operation of these memories, in which when a specific electrical voltage is applied in the MIM device, it can undergo switching from its initial insulator resistance state (HRS - high resistance state) to a low resistance state (LRS).

RS has already been observed in several materials such as TiO2, ZnO, NiO, perovskites and some electrolytic solids [1–5], in which two typical behaviors were perceived: unipolar and bipolar. In the unipolar behavior the switching is independent of the polarity applied, whereas in bipolar behavior there is this dependence [4].

© Springer International Publishing AG 2017
I. Rojas et al. (Eds.): IWANN 2017, Part I, LNCS 10305, pp. 580–588, 2017.
DOI: 10.1007/978-3-319-59153-7_50

Among the main characteristics of this type of memory, we have: low energy consumption, high-speed switching, durability, scalability and friendly manufacturing process [1].

There is a strong relationship between the materials used in the composition of these devices and their characteristics. Depending on the composition of the insulator, the voltage and current where the LRS and HRS states will occur can vary greatly; The response time for switching from one state to another or latency is affected and the power consumption for which SET and RESET occur are also influenced. The unipolar or bipolar behavior will depend on the type of material chosen. Contacts also have influence on the phenomenon of resistive switching and filament formation, and may interfere with the mode of memory operation. The geometry of the device is another element that causes impact. This makes it necessary to choose an optimized architecture because of the scalability issue.

Graphene is currently one of the most promising nanomaterials in the world, due to its excellent electrical, thermal and optical properties. Graphene is considered to be the basis of the whole family of carbon materials, with the exception of diamond [6–9]. For its production several methods have been researched [10], exemplified by exfoliation, deposition by CVD (Chemical Vapour Deposition) technique, among others.

Thinking in terms of obtaining the devices, process and waste disposal, we have an optimization, since the process of obtaining the memories is simple, low cost and there is no generation of toxic waste. If we think about the production of materials, some techniques are expensive (such as MBE or molecular beam epitaxy) and others end up generating toxic materials (such as the production of silicon oxide that sometimes uses silane gas). Graphene oxide would be a "green" material which does not pollute and is still derived from carbon.

The production process is simple, efficient and inexpensive. It can be done with low-cost equipment that can be build by the researcher. Moreover, due to the simplicity, the reproducibility of the devices can be made on a large scale.

What is the advantage of using a dopant in material? When the material is doped, in case graphene oxide (GO), its electrical, optical, magnetic and structural characteristics are modified. They can be improved or only directed towards a particular purpose. If we are to think in terms of resistive memories, it is necessary to obtain characteristics that allow a greater miniaturization of the devices, since the flash technology reached its lower limit because the minimum thickness of 15 nm has been reached. In addition, it is expected that there will be a minimum energy consumption with respect to the operation of a ReRAM, with SET and RESET voltage values occurring at low voltages.

Heat dissipation must be considered due to the Joule Effect, so it is important to have control over how the memory will operate in unipolar or bipolar mode. If the ReRAM is unipolar, its SET and RESET will occur for the same polarity; being unipolar, we have that there will be no influence of the electric current, thus avoiding the heating of the device with Joule Effect. In Yoo et al. research [13] devices were developed using pure graphene oxide. Memories presented bipolar behavior and the voltage for both RESET and SET are low, besides the variation in the current values are perceptible. When the GO is doped with some transition metal such as iron, copper or silver, the behavior is changed to unipolar and the SET and RESET occur for smaller voltages, saving energy.

The emergence of portable electronics such as cell phones, MP3 s, digital cameras and netbooks over the last 20 years has led to unbridled demand for better technologies for non-volatile flash memory because of its small cell size and low power consumption. However, scaling of flash memory beyond 15-nm technology is highly problematic due to fundamental limit of cell structure. The cell of a flash memory unit is very similar to the conventional field-effect metal-oxide-semiconductor transistor (MOSFET), except for the additional floating port for storing electrical charges.

Research on non-volatile memories to replace the flash has been very active. Among the most recent and as an alternative that has generated promising results, there is resistive memory (ReRAM) that is based on the resistive switching phenomenon (RS). Basically, non-volatile resistive RAM stores data by creating a resistor in a circuit instead of trapping electrons inside a cell. As a result, while the usual memory read-out latency is hundreds of microseconds, ReRAM reaches 50 ns, a delay time that can fit between main memory and cache memory levels in terms of speed, but at a lower cost.

Resistive Switching refers to the physical phenomena through which the resistance of a dielectric undergoes changes in response to a strong external electric field. It differs from degradation phenomena in dielectrics, which result in a permanent reduction in the resistance so that the change back to the original state is no longer possible [11, 12].

RS process is reversible and can be reproduced countless times. Typically, the change in resistance is non-volatile. Note that these phenomena occur in numerous insulating materials, including oxides, nitrides, chalcogen, organic semiconductor materials. However, the RS phenomenon has been studied more extensively in oxides. ReRAM are based on this phenomenon for operation [13–15].

Basically, memristor is an extremely small (nanometer) component that combines two terminals; when a current flows between them, its resistance increases; when the opposite path is made, the resistance decreases. What is important to consider here is that when current is cut off, the last recorded resistance level is maintained and depending on its value, we can assign a logic level 0 or 1.

RS mechanism is still not very well understood and therefore there are some proposals to explain this process. Predominantly more accepted model in oxide structures is the "conductive filament" model. A conductive path is created inside the insulator when certain electrical voltage is applied, the so-called "forming process" or SET. This creation occurs due to ionic migrations inside the insulator. RESET would be the destruction of this filament. The formation/destruction of filament can occur in two ways: one of these migrations is anionic, in which oxygen atoms migrate towards the anode, leaving behind a path of cations. Another ionic migration that occurs inside the insulator is cationic one, in which one of the electrodes is electrochemically active (such as Ag) and the other being an inert material (Pt, for example), and in addition, the insulator must be conductor of cations.

## 2  Experimental

Copper oxide (CuO) was synthesized using copper (II) acetate as described by Yoo et al. [13]. A colloidal solution of copper acetate in ethanol was used as precursor. 0.3 g copper acetate monohydrate added to 30 mL ethanol was suffered sonication for 1 h. To make CuO thin films, dip coating was used.

Deposition is made as follows: thin films are obtained by dip coating technique. The dip coating apparatus basically consists of a clamp which holds the substrate is dipped in a GO solution (graphene oxide) which containing dopant (cupper, iron or silver) or CuO. ITO (indium tin oxide) and aluminum contacts were evaporated.

The current and voltage relation can be measured by tracer IxV model HP 4140B. A curve is generated by varying the voltage. For the device to be considered a ReRAM resistive memory, the graph of the current in a voltage rising curve must have a format different from the curve obtained for the curve in a downward tension. An important memory device mechanism is the transition from the high resistance state (HRS) to the low resistance state (LRS) under applied voltage variation. It may be more useful for the device to have a fast switching response in the critical voltage where the transition occurs HRS to LRS.

Eight different devices were fabricated:

- ITO/CuO/GO+%1Fe/CuO/Al
- ITO/GO+%1Ag Al
- ITO/GO+%1Cu/Al
- Al/CuO/GO+%1Ag/CuO/Al
- ITO/GO+%0,1Ag/Al
- ITO/CuO/GO+%0,1Ag/CuO/Al
- ITO/CuO/GO+%1Cu/CuO/Al
- ITO/CuO/GO+%1Ag/CuO/Al

## 3  Results

For the eight devices produced there were combinations of aluminum and ITO contacts and graphene oxide (GO) doped with a transition metal (iron, copper or silver) with or without copper oxide (CuO) layer. In Fig. 1 electrical characterization of the resistive memories is observed. HP 4140B was used for current as a function of voltage (IxV) measurements.

It is observed that no memory had a similar behavior. The most interesting and common fact to notice in almost all devices is that there is an abrupt transition from LRS to HRS and in the sequence there is an abrupt transition from HRS to LRS forming a sort of "thorn". It can be noted that there is an abrupt change in current in some tensions, in which Zhang et al. [16], in their work with ITO/ZnO/PCMO/ITO-based RERAMs, associated mobile traps (oxygen vacancies) that were occupied by electrons forming a conducting path, the so-called filament conductor. However, in these voltages where there is occurrence of the "thorn", the weakest point of the conducting filament is destroyed, and according to the authors of this research, results in an electronic properties change.

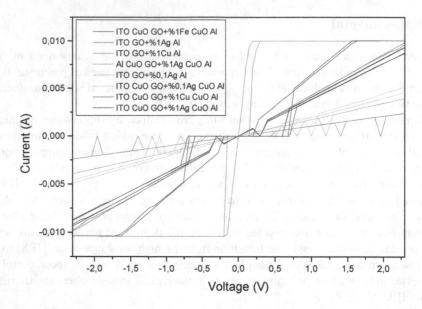

**Fig. 1.** Current as a function of voltage (IxV) measurements.

In graphs below, some comparisons between similar devices were made. The results were grouped for devices with similar structures.

For Fig. 2a, a three layer structure is shown and for Fig. 2b, a five layer structure interspersed with copper oxide. In the graphs of Figs. 2a and b it is possible to observe that devices have unipolar memory behavior since the SET and RESET occur in the same polarity. For Fig. 2b, it is noted that the SET (HRS for LRS) occurs in values around ±0.7 V to ±0.8 V and RESET occurs between ±0.2 V to ±0.3 V, for the same polarity. Being unipolar, there is no influence of the electric current, thus avoiding the joule effect with heating of device. In addition, the SET and RESET occur in a range less than 0.9 V, which implies a lower power consumption for it to operate.

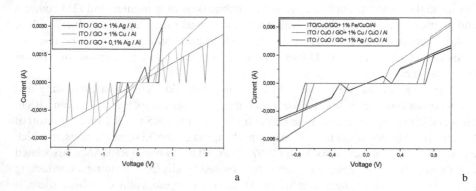

**Fig. 2.** (a) Three layer structure and (b) Five layer structure interspersed with copper oxide.

It can be seen that copper interferes in the behavior of devices with respect to the current as a function of the voltage. Such transition metal interferes with the "thorns" formation. When copper (as GO dopant or oxide form) is added in the ReRAM structure, the tendency is for abrupt change in the resistance state to be more difficult to occur. This element probably mitigates effect that possible oxygen vacancies may cause.

In the case of these devices, as they are formed by GO doped with transition metal (iron or copper or silver), filament will be formed as a function of this metal. Eventually mobile traps will be formed, destroying the filament fragile section and resulting in abrupt changes of resistance state ("thorn"). The behavior of filament is seen in Fig. 3.

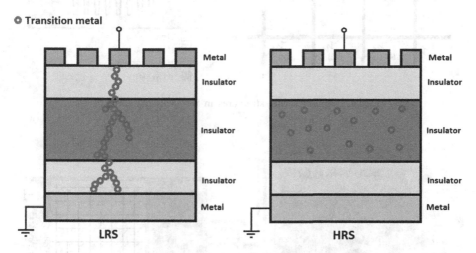

**Fig. 3.** Behavior of filament.

An interesting fact to note is that the used contacts influence the formation of the filament as well. If the graph for the ReRAM based on the Al/CuO/GO+1%Ag/CuO/Al structure is observed, its behavior diverges from the other devices (Fig. 1).

After the individual characterizations of the devices have been completed, memories with structures in 8 × 8 matrix format will be fabricated.

Structures will be fabricated in this way (Fig. 4) in order to develop a circuit whose memory stores 64 bits. The high resistance state HRS will correspond to the low level or 0 and the low resistance state LRS will be the high level or 1.

The algorithm for writing will be based on the code for LED matrix operation, which will be associated with the code to obtain resistance at each position of matrix. In this way, a write/read memory array code will be obtained. The program will be uploaded to the Arduino microcontroller which will manage the operations performed by ReRAMs (Fig. 5)

This memory array will be able to save data. It is possible to record and read information dynamically with appropriate algorithm. There is even the possibility of storing images (referring to original idea of LEDs matrix forming images). With these functions, it would be promising to enter the neuromorphic computing area that is one of the ReRAM applications.

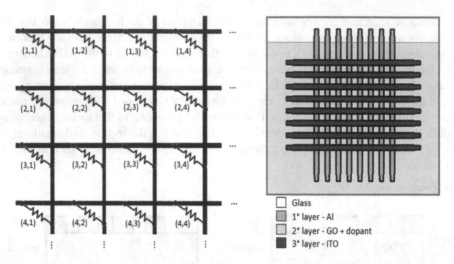

**Fig. 4.** Memories with structures in $8 \times 8$ matrix format.

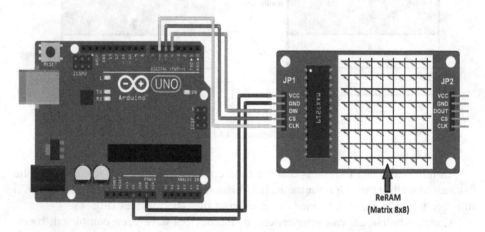

**Fig. 5.** Arduino microcontroller and ReRAMs.

The basic idea in most ReRAM-based neuromorphic approaches is to consider ReRAM devices, or small ReRAM-based circuits, as artificial synapses. According to Indiveri et al. [17], the idea of using ReRAMs as neural simulators comes from Likharev who introduced the "Crossnets" concept, where memory devices serve as interconnectors corresponding to binary synapses.

Human memory is often associated with ability to retain and use information or acquired knowledge, but there is an important function of being related to learning process. In DeSalvo et al. research [18], she mentions that ReRAM technology advent represents a paradigm shift for artificial neural networks, being the best candidate for emulation of synaptic plasticity and learning mode. Thanks to its non-volatility, high switching behavior and reliability, resistive memories can make the operation of machines close to human brain functioning, including mental processes such as visual recognition and problem solving. It would be a great advance if these devices work with genetic algorithms, which would allow a constant evolution.

# 4 Conclusion

Eight different devices were fabricated. It is observed that no memory had a similar behavior. The most interesting and common fact to notice in almost all devices is that there is an abrupt transition from LRS to HRS and in the sequence there is an abrupt transition from HRS to LRS forming a sort of "thorn". This behavior can be associated with mobile traps. Copper interferes in the behavior of devices with respect to the current as a function of the voltage. Such transition metal interferes with the "thorns" formation. When copper (as GO dopant or oxide form) is added in the ReRAM structure, the tendency is for abrupt change in the resistance state to be more difficult to occur. This element probably mitigates effect that possible oxygen vacancies may cause.

This memories will be able to save data. It is possible to record and read information dynamically with appropriate algorithm. There is even the possibility of storing images (referring to original idea of LEDs matrix forming images). With these functions, it would be promising to enter the neuromorphic computing area that is one of the ReRAM applications.

# References

1. Lanza, M.: A review on resistive switching in high-k dielectrics: a nanoscale point of view using conductive atomic force microscope. Materials **7**, 2155–2182 (2014)
2. Marinella, M.J.: Emerging resistive switching memory technologies: overview and current status. IEEE (2014)
3. Akinaga, H., Shima, H.: ReRAM technology; challenges and prospects. IEIEC Electr. Express **9**, 795–807 (2012)
4. Gale, E.M.: TiO2-based Memristors and ReRAM: materials, mechanisms and models (a review). Semicond. Sci. Technol. **29**, 104004 (2014)
5. Choi, H.: Main degradation mechanism in AsTeGeSiN threshold switching devices. Microelectr. Reliabil. **56**, 61–65 (2016)
6. Katsukis, G., Romero-Nieto, C., Malig, J., Ehli, C., Guldi, D.M.: Interfacing nanocarbons with organic and inorganic semiconductors: from nanocrystals/quantum dots to extended tetrathiafulvalenes. Langmuir **28**, 11662–11675 (2012)
7. Kaminska, I., Barras, A., Coffinier, Y., Lisowski, W., et al.: Preparation of a responsive carbohydrate-coated biointerface based on graphene/azido-terminated tetrathiafulvalene nanohybrid material. Appl. Mater. Interfaces **4**, 5386–5393 (2012)
8. Kaminska, I., Das, M.R., Coffinier, Y., Niedziolka-Jonsson, J., et al.: Preparation of graphene/tetrathiafulvalene nanocomposite switchable surfaces. Chem. Commun. **48**, 1221–1223 (2012)
9. Denis, P.A.: Chemical reactivity of electron-doped and hole-doped graphene. J. Phys. Chem. C **117**, 3895–3902 (2013)

10. Lu, W., Soukiassian, P., Boeckl, J.: Graphene: fundamentals and functionalities. MRS Bull. **37**, 1119–1124 (2012)
11. Siemon, A., Menzel, S., Waser, R., Linn, E.: A complementary resistive switch-based crossbar array adder. IEEE J. Emerg. Sel. Top. Circuits Syst. **5**(1), 64–74 (2015)
12. Lee, J.S., Lee, S., Noh, T.W.: Resistive switching phenomena: a review of statistical physics approaches. Appl. Phys. Rev. **2**, 031303 (2015)
13. Yoo, D., Cuong, T.V., Hahn, S.H.: Effect of copper oxide on the resistive switching responses of graphene oxide film. Curr. Appl. Phys **14**, 1301–1303 (2014)
14. Xu, J., Xie, D., Feng, T., et al.: Scaling-down characteristics of nanoscale diamond-like carbon based resistive switching memories. Carbon **75**, 255–261 (2014)
15. Tanaka, H., Kinoshita, K., Yoshihara, M., Kishida, S.: Correlation between filament distribution and resistive switching properties in resistive random access memory consisting of binary transition-metal oxides. AIP Adv. **2**, 022141 (2012)
16. Zhang, R., et al.: Transparent amorphous memory cell: a bipolar resistive switching in ZnO/Pr(0.7)Ca(0.3)MnO(3)/ITO for invisible electronics application. J. Non-Cryst. Solids **406**, 102–106 (2014)
17. Indiveri, G., Linn, E., Ambrogio, S.: ReRAM-based neuromorphic computing. In: Ielmini, D., Waser, R. (eds.) Resistive Switching: From Fundamentals of Nanoionic Redox Processes to Memristive Device Applications. Wiley-VCH Verlag GmbH & Co. KGaA (2016)
18. DeSalvo, B., Vianello, E., Garbin, D., Bichler, O., Perniola, L.: From memory in our brain to emerging resistive memories in neuromorphic systems. IEEE (2015)

# Artificial Neural Networks in Industry
# ANNI'17

# Performance Study of Different Metaheuristics for Diabetes Diagnosis

Fatima Bekaddour[1], Mohamed Ben Rahmoune[2], Chikhi Salim[1], and Ahmed Hafaifa[2(✉)]

[1] SCAL Team, MISC Laboratory, Constantine, Algeria
fatima.bekaddour@gmail.com, slchikhi@yahoo.com
[2] Applied Automation and Industrial Diagnostics Laboratory,
Djelfa University, Djelfa, Algeria
{BR.Mohamed,hafaifa.ahmed.dz}@ieee.org

**Abstract.** The problem of medical data classification involves an optimization phase that may be solved through metaheuristic approaches. In this work, we evaluate the performance in diagnosis of diabetes disease, using Particle Swarm Optimization (PSO), Firefly (FF) and Homogeneity-Based Algorithm (HBA) metaheuristics in conjunction with fuzzy system. Here, the fitness function in the optimization process is the total misclassification cost that is in term of false positive, false negative and unclassifiable rates. The results prove that HBA approach achieves better results than the other metaheuristics. With execution time, FF was faster than the PSO and HBA methods.

**Keywords:** Metaheuristic · PSO · Firefly · HBA · Fuzzy system · PIMA dataset

## 1 Introduction

Diabetes is a complex and chronic disease, which is characterized as an illness that may occur when the pancreas is unable to produce insulin or when this hormone is not effectively used by the human body. Diabetes includes two major classes: type 1, which results in lack of insulin, due to beta-cells destroyment by the human immune system, while type 2 diabetes, results from insulin resistance. Nowadays, diabetes has become a worldwide epidemic that affects a big number of people [14].

Diabetes recognition is an important and difficult task that requires a reliable algorithm in order to reduce the probability of classification error. Recently, there has been several works, interested in the development of automatic tools for healthcare provision. Most of these researches have investigated the development of CAD (Computer Aided-Diagnosis) that provides a second opinion for physicians and helps them in their diagnosis tasks.

Thus, in order to identify the diabetes diagnosis of patient, in this work, we propose the employment of three well-known population-based metaheuristics, namely: PSO (Particle Swarm Optimization), FF (Firefly) and HBA (Homogeneity-Based Algorithm), in conjunction with FIS (Fuzzy Inference System), to evaluate their effectiveness. Note that, in this study, the main goal is to minimize the fitness function value,

© Springer International Publishing AG 2017
I. Rojas et al. (Eds.): IWANN 2017, Part I, LNCS 10305, pp. 591–602, 2017.
DOI: 10.1007/978-3-319-59153-7_51

that represents the total misclassification cost of FIS model, which is in term of FP (false positive), FN (false negative), UC (unclassifiable) errors. The experiments are made on the PIMA diabetes data set, taken from the UCI repository [10]. In the medical field, a (Uc) case means a patient that cannot be diagnosed by the prediction system. This is because of scanty and inadequate information about the patient. In the medical applications, the Uc error is very required, because it may lead to additional examinations that help doctors to make the right diagnosis.

The reminder of this paper is organized as follows: Sect. 2 presents some related works on diabetes diagnosis. In Sect. 3, the PSO, FF and HBA metaheuristics and the FIS model are defined. In Sect. 4, we describe our proposed methodology and we discuss the obtained results, based on PIMA diabetes data set. Finally, in Sect. 5, we conclude the paper.

## 2    Related Works

In the literature, several metaheuristic approaches have been successfully applied in medical diagnosis, such as the PIMA diabetes dataset. In [1], authors developed a Homogeneity-base Algorithm (HBA), for the classification of PIMA diabetes dataset. Authors introduced the concept of Homogeneity Degree (HD), to achieve a simultaneously balance between the fitting and the generalization of the inferred classification models. The HD value calculates the density of learning points in a given homogenous hypersphere. The proposed HBA, compared to standalone data mining approaches such as SVM (Support Vector Machine) and DT (Decision Tree), presented high performance accuracy, but less efficiency due to the HBA's computational time.

In another study [2], Pham and Triantaphyllou proposed a Convexity-Based Algorithm (CBA) approach. In this work, a new concept of convex density (CD) was introduced to optimize the total misclassification cost value, based on defragmenting convex regions. The obtained results, tested on PIMA dataset, present significant improvement compared to other well-known data mining techniques.

Beloufa and Chikh [3] presented a novel approach based on ABC (Artificial Bee Colony) metaheuristic for automatic recognition of diabetes dataset from the UCI repository. In this work, authors modify the original ABC approach by adding a blended genetic operator for better intensification and diversification of the search space. This operator serves mainly to automatically update the Fuzzy Inference System membership functions and rules. Experimental results prove that the proposed ABC metaheuristic, found minimal number of rules, while improving the final classification performance.

Al-Muhaideb and Menai [4] introduced a two-stage metaheuristic optimization method, called HColonies (Hybrid ant-bee colonies), which is a hybrid system between ACO (Ant Colony Optimization) and ABC (Artificial Bee Colony) approaches. The main idea of HColonies is to use ACO approach to create initial population solutions for the ABC metaheuristic. This is done to accelerate the search and obtain good performance results. In the first stage, the Ant-Miner+ is adopted to generate a population of food sources. In the second stage, a modified ABC method, based on new operators is employed, to fit the appropriate problem. Results obtained by the

HColonies method on the PIMA diabetes, illustrate the effectiveness and robustness of the proposed metaheuristic, toward change in its parameters.

## 3  Overview of FIS and the Used Metaheuristics

### 3.1  Particle Swarm Optimization

The first metaheuristic used in this work, is a population-based metaheuristic, called PSO: particle swarm optimization (Kennedy and Eberhart in 1995) [5]. PSO mimics the cooperative behavior concepts of natural organisms such as birds and fish. The PSO algorithm starts with a random swarm that constitutes a set of particles. Each particle 'i' is characterized by its corresponding velocity Vi and its position Xi. Then, at each iteration, particles modify their velocity and their position using formulas (1) and (2).

$$Xi(t) = Xi(t-1) + Vi(t) \tag{1}$$

$$V_i(t) = V_i(t-1) + c_1 r_1 (P_{besti}(t-1) + X_i(t-1)) + c_2 r_2 (G_{besti}(t-1) + X_i(t-1)) \tag{2}$$

where $(c_1, c_2)$ are two factors that represent the cognitive attraction and the social attraction respectively. $(r_1, r_2)$ are two random numbers uniformly distributed between [0, 1]. $(P_{besti}, G_{besti})$ define the best position obtained by a given particle i and the best position ever found in the entire swarm respectively. The different steps of the PSO algorithm are given in Algorithm 1 as follows:

```
Algorithm 1: Particle Swarm Optimization
- Particles initialization.
- While (stopping criteria not met){
      -Evaluate f (Xi):fitness function of each particle i
      -For all particles i do {
          - Compute Velocities Vi using formula(2).
          - Compute the new position Xi using formula(1).
              - If ( f (Xi) <  f (P_besti)) Then P_besti = Xi.
              - If ( f (Xi) <  f (G_besti)) Then G_besti = Xi.
              - Update (Xi, Vi).
                                               }
                                          }
```

### 3.2  Firefly Metaheuristic

The second used metaheuristic is Firefly (FF) algorithm [6]. FF is a nature inspired population based swarm method, that imitates the flashing behaviors of fireflies. Each firefly is considered as a candidate solution in the search space. There are three fundamental key issues, regarding the FF metaheuristic:

- $r_{ij}$: that denote the distance between fireflies i and j.
- Attractiveness ($\beta(r_{ij})$): calculated by the backdrop of the fitness function as follows:

$$\beta(r_{ij}) = \beta_0\, e^{-\gamma r_{ij}^2} \tag{3}$$

Where $\beta_0$ means the attractiveness at r = 0, and $\gamma$ is the absorption coefficient.
- Movement: is determined by the Eq. (4):

$$Xi = Xi + \beta_0\, e^{-\gamma rij2}(Xj - Xi) + \alpha(\text{rand} - 1/2) \tag{4}$$

Where $\alpha$ is a random number, uniformly distributed between [0, 1].

The main steps of FF metaheuristic are summarized in Algorithm 2.

```
Algorithm 2: Firefly metaheuristic
- Fireflies initialization.
- Define the light intensity I.
- Define the absorption coefficient γ.
- Define f(x): fitness function of x=(x1,x2,..,xd).
- While (Max-generations not met){
   - For each firefly i do{
      - For (j=1 to i) do {
         - If ( I (j) >  I (i)) Then {
            Move firefly i toward firefly j using eq(4)
                                  }
            -Attractiveness varies with distance r via e-ᵞʳ²
            -Compute new solutions.
            -Update I(i).
                       }
                    }
   - Rank fireflies and find the best one.

              }
```

### 3.3  Homogeneity-Based Algorithm

In [8], Pham and Triantaphyllou introduced a new metaheuristic called Homogeneity-based Algorithm (HBA). The main objective of HBA metaheuristic, is to achieve a simultaneously balance between the fitting and the generalization, using the concept of homogenous set and homogeneity degree (HD) [1, 7–9]. HBA approach may be applied in conjunction with data mining techniques such as SVM (Support Vector Machine), to minimize the fitness function formula (5):

$$TC = \min(CFP * RateFP + CFN * RateFN + CUC * RateUC) \tag{5}$$

Where: CFP, CFN, CUC are the unit penalty cost for the false positive, false negative and unclassifiable rates respectively. TC represents the total misclassification cost value. From the two inferred models, obtained using a data mining approach, HBA breaks each set into hyperspheres, covering decision regions. Next the Homogeneity Degree (HD) value, corresponding to each hypersphere is calculated, using the Eq. (6):

$$HD(S) = \ln(n_s)/h \qquad (6)$$

Where S is a given homogenous set, $n_s$ is the number of samples in S and $h$ is the minimal most frequent distance in a set S [1, 8, 9]. After that, HBA adopts four thresholds: $(\beta^-, \beta^+, \alpha^-, \alpha^+)$, to expand or break down each homogenous set, using their corresponding HD values. Please note that, $(\alpha^-, \alpha^+)$ are used to expand the negative and the positive homogenous sets respectively. On the other hand, $(\beta^-, \beta^+)$ are used to fragment the negative and the positive homogenous sets respectively [8]. The HBA metaheuristic iterates until all the homogenous sets are processed. Within the scope of HBA, the GA approach is employed, to adjust the $(\beta^-, \beta^+, \alpha^-, \alpha^+)$ factors values.

### 3.4 Fuzzy Inference System

Fuzzy Inference System (FIS) [15] is a well-known artificial intelligence approach, that is based on the theory of fuzzy sets and fuzzy logic to extend the classical crisp sets theory. In the literature, the FIS model has been widely employed in the medical field [16–19].

The basic architecture of the FIS model as shown in Fig. 1, consists of three main phases:

- Fuzzification: transform the crisp input into linguistic variables (fuzzy input).
- Inference Engine (IE): uses the fuzzy input and the rules defined in the knowledge base module, to derive fuzzy sets for each variables.
- Defuzzification: transform the obtained fuzzy output by the IE into crisp output

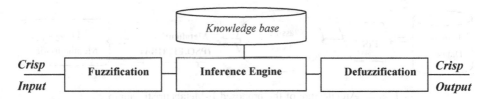

**Fig. 1.** The main structure of the FIS model

The FIS model used in this paper is the Takagi-Sugeno fuzzy model.

# 4 Computational Results

## 4.1 Description of Data

In this study, we employ the PIMA diabetes dataset, obtained from the UCI repository [10]. The main description of PIMA dataset is depicted in Table 1. The PIMA dataset is composed of 768 instances, where 268 samples belong to the positive class, and 500 samples belong to the negative class.

**Table 1.** PIMA diabetes dataset description

| Attribute | Description |
|-----------|-------------|
| Npreg | Number of times pregnants |
| Glu | Plasma glucose concentration |
| Bp | Diastolic blood pressure |
| Skin | Triceps skin fold thickness |
| Insulin | 2-Hour serum insulin |
| BMI | Body mass index |
| PED | Diabetes pedigree function |
| Age | Age of the patient |

## 4.2 Experimental Design

In this work, the procedure of conducting the experiments is based on the fuzzy inference system (FIS) [15] and metaheuristics techniques (PSO, FF, HBA). The individual system was designed and developed with their default parameters values. We developed the FIS approach, using the concept of membership functions and rules, that denote the relationship between input $(x_i)$ and output $(o_i)$. The use of FIS, permit to increase the transparency of the classifier. The main architecture of the proposed F-Metaheuristic is depicted in Fig. 2.

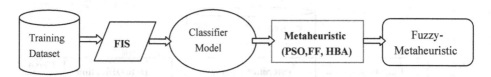

**Fig. 2.** Architecture of the proposed F-Metaheuristic approach

For the development of (PSO, FF, HBA) methods, the objective was to minimize the fitness function values that define the total misclassification cost (TC) of FIS approach. In this work, we propose three different configurations of TC formulas, depicted in Fig. 3.

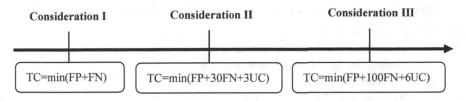

**Fig. 3.** The proposed three scenarios of TC

In the first consideration, we assume that unclassifiable (UC) cost is equal to 0, while FN and FP penalty costs are set to (1, 1) respectively. This is the case on the majority of works in the literature. Therefore, TC can be defined as:

$$TC = \min(1 * RateFP + 1 * RateFN) \tag{7}$$

In the second consideration, we suppose that UC is taken into account, while the FN rate is penalized more than (UC, FP) and the UC is penalized more than the FP rate. Thus, TC is defined as follows:

$$TC = \min(1 * RateFP + 30 * RateFN + 3 * RateUC) \tag{8}$$

In the last consideration, we assume the case where the FN rate is penalized much more than the (FP, UC) costs, while UC still penalized more than FP rate. Therefore, the TC value is described as follows:

$$TC = \min(1 * RateFP + 100 * RateFN + 6 * RateUC) \tag{9}$$

It is surprised that despite the various works reported in the literature, on the PIMA dataset, the majority of researchers still neglects the evaluation of UC cases. It is to be emphasized that the UC error is very essential in medical diagnosis problem. This is because, this may help the physician to make the right decision, based on more examinations. In this work, we compute the UC rates that represent the number of patients which cannot be diagnosed by the classifier. A reason for that is that the information gathered about the patient are scanty or inadequate. The defaults parameters of the three used metaheuristics are given in Table 2.

### 4.3  Measure for Performance Evaluation

In order to conduct a valid experiment, in this work, we propose the use of three different scenarios for the fitness function (TC), as it is described in Sect. 4.2. Please note that TC is in term of three type of errors (FP, FN, UC) that measure mistakes made by the FIS or the F-metaheuristic, during the validation phase. The sensitivity (Se), and the specificity (Sp) are two other well-known metrics that are used in this study. (Se, Sp) measure the true positive and the true negative rates respectively, defined as follows:

**Table 2.** Summary of default parameters used in the implementation of the model

| Approach | Parameters values |
|---|---|
| Fuzzy | Membership number: 2 for each attribute |
| | Membership type: Generalized Bell |
| | FIS type: Takagi-Sugeno |
| | Rules number: 256 |
| PSO | Swarm size: 40 |
| | Generations number: 10–50 |
| | Cognitive attraction (c1) = 0.5 |
| | Social attraction (c2) = 1.25 |
| FF | Population size: 25 |
| | Generations number: 10–100 |
| | Light absorption coefficient = 1 |
| | $\beta 0 = 2$ |
| HBA | Initial ($\beta-$, $\beta+$, $\alpha-$, $\alpha+$): 0 |
| | Final ($\beta-$, $\beta+$, $\alpha-$, $\alpha+$): updated using Genetic Algorithm |
| | Iterations number: 10–50 |
| | $\gamma = 0.01$, is a parameter, used to judge whether a region is homogenous or not |

$$Se = TP/TP + FN \qquad (10)$$

$$Sp = TN/TN + FP \qquad (11)$$

Where (TP, TN) are the true positive and the true negative numbers respectively, while (FP, FN) are false positive and false negative numbers respectively. In this study, we also calculate two other metrics that are:

- Solicitation degree (SD) of each rule, generated by the FIS (F-metaheuristic) approach. SD defines the rules activation degree between 50% and 100%.
- Accuracy: that defines the correct classification rate.

## 4.4    Results and Discussion

We conducted different experiments to juxtapose the three used metaheuristics (PSO, FF, HBA), for the given default parameters, depicted in Table 2. As discussed in Sect. 4.1, we adopt the PIMA diabetes dataset. The experimental methodology ran as presented in Sect. 4.2. The experiments were made in Matlab, version 2012a. As this work presents a comparative study of metaheuristics, the comparative results related to the fitness function value, are presented in Table 3. This table presents the (FP, FN, UC) rates and the obtained TC values. Note that F-PSO means: the FIS system is used in conjunction with PSO metaheuristic. Same definitions are valid for F-FF and F-HBA. The improvement column defines the improvement rate of the FIS-metaheuristic compared to the FIS approach.

**Table 3.** Results obtained under the three proposed considerations

| Nbr. consideration | Approach | FP (%) | FN (%) | UC (%) | TC (%) | Improvement (%) |
|---|---|---|---|---|---|---|
| *Consideration I* | FIS | 67.18 | 0.0 | 32.18 | 67.18 | |
| | F-PSO | 21.35 | 19.8 | 58.85 | 41.15 | 67.18 |
| | F-FF | 16.14 | 19.27 | 11.45 | 35.41 | 47.29 |
| | F-HBA | 0.0 | 8.33 | 55.2 | 8.33 | 87.60 |
| *Consideration II* | FIS | 67.18 | 0.0 | 32.18 | 163.72 | |
| | F-PSO | 21.35 | 19.8 | 58.85 | 791.9 | No.improv |
| | F-FF | 16.14 | 19.27 | 11.45 | 628.6 | No.improv |
| | F-HBA | 0.0 | 8.33 | 55.2 | 415.5 | No.improv |
| *Consideration III* | FIS | 67.18 | 0.0 | 32.18 | 260.26 | |
| | F-PSO | 21.35 | 19.8 | 58.85 | 2354 | No.improv |
| | F-FF | 16.14 | 19.27 | 11.45 | 2011 | No.improv |
| | F-HBA | 0.0 | 8.33 | 55.2 | 1164 | No.improv |

According to Table 3, the average value of TC on PIMA dataset were 28.3%, under the first consideration. This value of TC was less than the TC average value by about 67.35%. In the second consideration, the TC values are equal to (791, 628, 415) for (F-PSO, F-FF, F-HBA) respectively. This value of TC was not optimal than the FIS TC value. The average value of TC is equal to 612%, on the PIMA dataset. The obtained results for the last consideration are also summarized in Table 3. In this case, the obtained average TC value is equal to 1843. The (F-PSO, F-FF, F-HBA) have not yielded optimal TC on the PIMA dataset, compared to the FIS TC value. In addition, the obtained computational results presented in this table, confirm that the three $(C_{FP}, C_{FN}, C_{UC})$ values affect the final TC values.

To make the results clearer, we plotted the (Accuracy, Se, Sp) in Fig. 4. This figure presents also the average improvement of the FIS-metaheuristics, compared to the standards FIS. The experiments illustrate that F-HBA metaheuristic achieved better results in term of (Accuracy, Se, Sp), which were successful for small number of generations. FF was slightly worse than PSO metaheuristic. It clearly appears that the HBA metaheuristic achieves minimal TC error value and better accuracy for small generations number. Also, $(\alpha+, \alpha-, \beta+, \beta-)$ values, played a great role in method performance. However, the F-HBA computational time is higher than (F-PSO, F-FF, Fuzzy) approaches. This is due to the compute of homogenous sets and homogeneity degree. F-FF was slightly faster the F-PSO metaheuristic in term of execution time.

Tables 4 and 5 present the solicitation degree (SD) of some rules, obtained by the F-HBA model, for the TN and UC cases. Rules cited in those tables are:

**R97:**   if (Npreg is sm)& (Glu is bg)& (Bp is bg)& (Skin is sm)& (Insulin is sm)& (BMI is sm)& (PED is sm)& (Age is sm) then Class 97

**R101:**  if (Npreg is sm)& (Glu is bg)& (Bp is bg)& (Skin is sm)& (Insulin is sm)& (BMI is bg)& (PED is sm)& (Age is sm) then Class101

**R109:**  if (Npreg is sm)& (Glu is bg)& (Bp is bg)& (Skin is sm)& (Insulin is bg)& (BMI is bg)& (PED is sm)& (Age is sm) then Class109

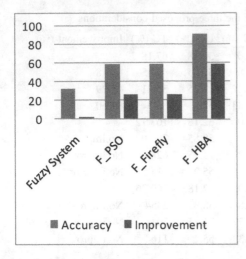 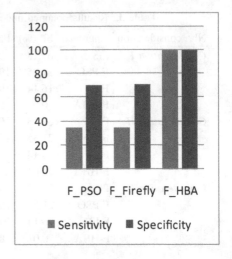

**Fig. 4.** Accuracy, Se, Sp comparisons for PIMA dataset

As described in Table 2, we choose two membership functions for each descriptor (sm: small, bg: big). Figure 5 presents a TN case that activates R97 with SD = 78.22. This example has been misclassified by the FIS, while correctly classified by F-HBA.

**Table 4.** SD of some rules for TN case

| Rule | Solicitation degree |
|------|---------------------|
| [101] | 0.008 |
| [97] | 0.02 |

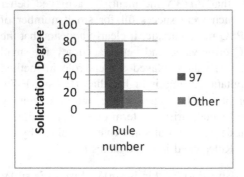

| Parameter | Value |
|-----------|-------|
| **Npreg** | 6.77 |
| **Glu** | 18.84 |
| **BP** | 16.58 |
| **Skin** | 0 |
| **Insulin** | 0 |
| **BMI** | 9.32 |
| **Ped** | 8.99 |
| **Age** | 8.61 |

**Fig. 5.** Example of TN case

For the UC cases, Table 5 illustrates the SD of rules 101 and 97. According to this table, R101 presents the higher SD value. The example depicted in Fig. 6 presents a patient that was correctly classified by the F-HBA approach, while misclassified by the

FIS. In fact, this patient has a Glu value greater than 1.4 g/l and a body mass index greater than 40 kg/m², which indicates a morbid obesity.

**Table 5.** SD of some rules for UC case

| Rule | Solicitation degree |
|------|---------------------|
| [97] | 0.008 |
| [101] | 0.06 |
| [109] | 0.008 |

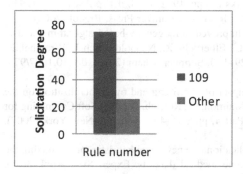

| Parameter | Value |
|-----------|-------|
| Npreg | 0 |
| Glu | 17.77 |
| BP | 16.95 |
| Skin | 20.92 |
| Insulin | 110.96 |
| BMI | 21.28 |
| Ped | 11.78 |
| Age | 9.00 |

**Fig. 6.** Example of UC case

This example has activated R109 with SD equal to 74.45. Experiments proves that the HBA metaheuristic can be of great interest for diabetes disease diagnosis.

## 5 Conclusion

This study investigates the problem of medical data classification that involves an optimization phase, and so the employment of metaheuristics approaches is very recommended. Various works dedicated in the diagnosis of diabetes have been carried out over the last decades. In the present work, three metaheuristic algorithms were developed to identify whether a patient is a subject of diabetes or not. The performance of PSO (Particle Swarm Optimization), FF (Firefly), HBA (Homogeneity-Based Algorithm) has been compared for the minimization of FP (False Positive), FN (False Negative), UC (Unclassifiable) rates of the FIS (Fuzzy Inference System) model. Computational experiments based on PIMA dataset from the UCI repository, illustrate that the HBA metaheuristic obtained better performance among the other used methods. In the near future, we aim to pay more attention to employ other metaheuristics such as Krill Herd [11], Dragonfly [12] and Whale [13] metaheuristics.

# References

1. Pham, H.N.A., Triantaphyllou, E.: Prediction of diabetes by employing a new data mining approach which balances fitting and generalization. In: Yin, L.R. (ed.) Computer and Information Science. SCI, vol. 131, pp. 11–26. Springer, Berlin (2008). Chapter 2
2. Pham, H.N.A., Triantaphyllou, E.: A meta-heuristic approach for improving the accuracy in some classification algorithms. Comput. Oper. Res. **38**, 174–189 (2011)
3. Beloufa, F., Chikh, M.A.: Design of fuzzy classifier for diabetes disease using modified artificial bee colony algorithm. Comput. Methods Programs Biomed. **112**, 92–103 (2013)
4. Al-Muhaideb, S., Menai, M.E.-B.: HColonies: a new hybrid metaheuristic for medical data classification. Appl. Intell. **41**(1), 282–298 (2014)
5. Kennedy, J., Eberhart, R.C.: Particle swarm optimization. In: Proceedings of IEEE International Conference on Neural Networks IV, pp. 1942–1948. IEEE, Piscataway (1995)
6. Yang, X.S.: Nature-Inspired Metaheuristic Algorithms. Luniver Press, Bristol (2008)
7. Bekaddour, F., Amine, C.M.: IHBA: an improved homogeneity-based algorithm for data classification. In: Amine, A., Bellatreche, L., Elberrichi, Z., Neuhold, Erich J., Wrembel, R. (eds.) CIIA 2015. IAICT, vol. 456, pp. 129–140. Springer, Cham (2015). doi:10.1007/978-3-319-19578-0_11
8. Pham, H.N.A., Triantaphyllou, E.: The impact of overfitting and overgeneralization on the classification accuracy in data mining. In: Maimon, O., Rokach, L. (eds.) Soft Computing for Knowledge Discovery and Data Mining, Part 4, pp. 391–431. Springer, New York (2007). Chapter 5
9. Pham, H.N.A., Triantaphyllou, E.: An application of a new meta-heuristic for optimizing the classification accuracy when analyzing some medical datasets. Expert Syst. Appl. **36**(5), 9240–9249 (2009)
10. UCI Repository of Machine Learning Databases, Department of Computer Science, University of California at Irvine. https://archive.ics.uci.edu/ml/datasets/Pima+Indians+Diabetes. Accessed 2017
11. Gandomi, A.H., Alavi, A.H.: Krill herd: a new bio-inspired optimization algorithm. Commun. Non Linear Sci. Numer. Simul. **17**(12), 4831–4845 (2012). doi:10.1016/j.cnsns.2012.05.010
12. Mirjalili, S.: Dragonfly algorithm: a new meta-heuristic optimization technique for solving single-objective, discrete, and multi-objective problems. Neural Comput. Appl. **27**(4), 1053–1073 (2016)
13. Ebrahimi, A., Khamehchi, E.: Sperm whale algorithm: an effective metaheuristic algorithm for production optimization problems. J. Nat. Gas Sci. Eng. **29**, 211–222 (2016)
14. Al-Rubeaan, K., Youssef, A.M., Ibrahim, H.M., Al-Sharqawi, A.H., AlQumaidi, H., AlNaqeb, D., Aburisheh, K.H.: All-cause mortality and its risk factors among type 1 and type 2 diabetes mellitus in a country facing diabetes epidemic. Diabetes Res. Clin. Pract. **118**, 130–139 (2016)
15. Zadeh, L.A.: Fuzzy sets. Inf. Control **8**(3), 338–353 (1965)
16. Miranda, G.H.B., Felipe, J.C.: Computer-aided diagnosis system based on fuzzy logic for breast cancer categorization. Comput. Biol. Med. **64**, 334–346 (2015). http://dx.doi.org/10.1016/j.compbiomed.2014.10.006
17. Nilashi, M., Ibrahim, O., Ahmadi, H., Shahmorad, L.: A knowledge-based system for breast cancer classification using fuzzy logic method. Telemat. Inform. **34**(4), 133–144 (2017)
18. Muthukaruppan, S., Er, M.J.: A hybrid particle swarm optimization based fuzzy expert system for the diagnosis of coronary artery disease. Expert Syst. Appl. **39**, 11657–11665 (2012)
19. Ghazavi, S.N., Liao, T.W.: Medical data mining by fuzzy modeling with selected features. Artif. Intell. Med. **43**(3), 195–206 (2008)

# Randomized Neural Networks for Recursive System Identification in the Presence of Outliers: A Performance Comparison

César Lincoln C. Mattos[1], Guilherme A. Barreto[1(✉)], and Gonzalo Acuña[2]

[1] Department of Teleinformatics Engineering (DETI), Center of Technology,
Federal University of Ceará (UFC), Campus of Pici, Fortaleza, Ceará, Brazil
cesarlincoln@terra.com.br, gbarreto@ufc.br
[2] Departamento de Ingeniería Informática,
Universidad de Santiago de Chile (USACH), Santiago, Chile
gonzalo.acuna@usach.cl

**Abstract.** In this paper, randomized single-hidden layer feedforward networks (SLFNs) are extended to handle outliers sequentially in online system identification tasks involving large-scale datasets. Starting from the description of the original batch learning algorithms of the evaluated randomized SLFNs, we discuss how these neural architectures can be easily adapted to cope with sequential data by means of the famed least mean squares (LMS). In addition, a robust variant of this rule, known as the *least mean M-estimate* (LMM) rule, is used to cope with outliers. Comprehensive performance comparison on benchmarking datasets are carried out in order to assess the validity of the proposed methodology.

**Keywords:** Randomized SLFNs · Online system identification · NARX model · Outliers

## 1 Introduction

A novel class of supervised single-layer feedforward network (SLFN) architectures, generically called Randomized SLFNs, is attracting a great deal of attention from the computational intelligence community in recent years. A few examples of such architectures are the Random Vector Functional Link (RVFL) [15], the Extreme Learning Machine (ELM) [3], and the No-Prop network [14]. All this interest seems to be primarily motivated by the very fast way they are trained, without resorting to a long learning process across several training epochs, as required by the backpropagation algorithm. Even being a valid argument, many real-world applications present challenging features that demand adaptations in the learning algorithms of the aforementioned randomized SLFNs. One of such applications is online dynamical system identification from large scale datasets in the presence of outliers.

© Springer International Publishing AG 2017
I. Rojas et al. (Eds.): IWANN 2017, Part I, LNCS 10305, pp. 603–615, 2017.
DOI: 10.1007/978-3-319-59153-7_52

Dynamical system identification is a regression-like problem where the input and output observations come from time series data [1]. In other words, information about the dynamics (i.e. temporal behavior) of the system of interest must be learned from time series data. Despite the rapidly growing number of successful applications of randomized SLFNs in pattern recognition and regression, their use for nonlinear dynamical system identification has not been fully explored yet, with just a few works available [5,9,12]. In [5,12], for example, the proposed ELM-like models use *batch* learning algorithms based on the ordinary least-squares (OLS) estimation method, which cannot be applied to large scale datasets because it requires a costly matrix inversion and storage of huge data matrices. In [9], a recursive estimation algorithm is proposed aiming at online system identification problems, but despite alleviating the memory storage requirements by using chunks of data samples instead of the whole dataset, the proposed method is still too costly to be applied for large scale datasets.

In what concerns the robustness to outliers, it is widely known that the OLS method is very sensitive to their presence in the estimation data [4]. As a consequence, any randomized SLFN using the OLS method, such as the standard RVFL and ELM networks, will also present a severe degradation in performance when trained with outlier-contaminated data. For online learning in outlier-free scenarios, the No-Prop network [14] becomes a suitable alternative to the standard RVFL and ELM networks because it uses the least mean squares (LMS) rule instead of the OLS method for estimating the output weights. However, like the batch OLS, the adaptive LMS rule is also very sensitive to outliers, an issue that can be resolved by means of an outlier-robust version of it, named the *least-mean M-estimate* (LMM) algorithm [17].

From the exposed, due to the requirements of the applications we are interested in, we pursue randomized nonlinear models capable of fast online learning in large scale datasets *AND* in the presence of outliers. For this purpose, we incorporate into the aforementioned randomized SLFNs, the robust online learning ability of the LMM rule. This strategy is comprehensively evaluated on datasets generated by several benchmarking dynamical systems and shown to be effective. For the sake of completeness, we also introduce the LMM learning rule into the standard backpropagation algorithm in order to carry out a fair performance comparison.

The remainder of the paper are organized as follows. In Sect. 2 we describe all models to be evaluated in this paper, emphasizing the need for adaptive learning rules, such as the LMS rule, for online system identification. In Sect. 3 we show how to replace the original LMS rule with a robust variant by means of concepts from the $M$-estimation framework. A comprehensive performance comparison is presented in Sect. 4. The paper is concluded in Sect. 5.

## 2   Evaluated Models

Let us assume that we have already collected $N$ data pairs $\{(\mathbf{x}_n, d_n)\}_{n=1}^{N}$ for building and evaluating the model, where $\mathbf{x}_n \in \mathbb{R}^p$ is the $n$-th $p$-dimensional input pattern and $d_n \in \mathbb{R}$ is the corresponding target value. Then, let us randomly select $N_1$ ($N_1 < N$) training input-output pairs from the available data pool and arrange the input vectors along the columns of the matrix $\mathbf{X}$ ($p \times N_1$), while the target values are stacked into the column-vector $\mathbf{d}$ ($N_1 \times 1$):

$$\mathbf{X} = [\mathbf{x}_1 \mid \mathbf{x}_2 \mid \cdots \mid \mathbf{x}_{N_1}] \quad \text{and} \quad \mathbf{d} = [d_1 \, d_2 \, \cdots \, d_{N_1}]^T, \tag{1}$$

where the superscript $T$ denotes the transpose of a vector/matrix.

## 2.1   The Random Vector Functional Link Network (RVFL)

The RVFL [15,16] is a randomized SLFN with two pathways for processing information from input units to output neurons. These pathways are then added to form the network's output. The first pathway is a linear one, which directly connects the input units to the output neuron. Mathematically, we get

$$y_n^{(1)} = \mathbf{w}_1^T \mathbf{x}_n, \tag{2}$$

where $\mathbf{w}_1 \in \mathbb{R}^p$ is the corresponding weight vector[1]. The second pathway processes the input vectors through a hidden layer of $q$ ($q \geq 1$) nonlinear neurons; that is,

$$y_n^{(2)} = \mathbf{w}_2^T \mathbf{h}_n, \tag{3}$$

where $\mathbf{w}_2 \in \mathbb{R}^q$ is the corresponding weight vector and $\mathbf{h}_n \in \mathbb{R}^q$ is the hidden activation vector, i.e. the vector containing the outputs of the hidden neurons in response to the current input vector $\mathbf{x}_n$. The vector $\mathbf{h}_n$ is computed as

$$\mathbf{h}_n = \phi(\mathbf{M}\mathbf{x}_n) = [\phi(\mathbf{m}_1^T \mathbf{x}_n + b_1), \ldots, \phi(\mathbf{m}_q^T \mathbf{x}_n + b_q)]^T, \tag{4}$$

where $\phi(\cdot)$ is a nonlinear (e.g. sigmoidal) activation function operating at each component of its argument vector, $\mathbf{M}$ is a $q \times p$ weight matrix, and $b_j, j = 1, \ldots, q$, denotes the bias of the $j$-th hidden neuron. It should be noted that the weight vectors $\mathbf{w}_1$ and $\mathbf{w}_2$ are estimated from data, while the entries of the matrix $\mathbf{M}$ and the biases $b_j$ are randomly sampled either from a uniform or a normal distribution.

If we add the outputs of both pathways, we get

$$y_n = y_n^{(1)} + y_n^{(2)} = \mathbf{w}_1^T \mathbf{x}_n + \mathbf{w}_2^T \mathbf{h}_n = [\mathbf{w}_1^T \mid \mathbf{w}_2^T] \begin{bmatrix} \mathbf{x}_n \\ - \\ \mathbf{h}_n \end{bmatrix} = \mathbf{w}^T \mathbf{z}_n, \tag{5}$$

where $\mathbf{w} = [\mathbf{w}_1^T \mid \mathbf{w}_2^T]^T$ is the $(p+q) \times 1$ vector obtained from the concatenation of the weight vectors $\mathbf{w}_1$ and $\mathbf{w}_2$. By the same token, $\mathbf{z}_n$ is the $(p+q) \times 1$ vector formed from the concatenation of the current input vector $\mathbf{x}_n$ and the current hidden activation vector $\mathbf{h}_n$.

---

[1] We assume that all vectors are column-vectors, unless stated otherwise.

The weight vector $\mathbf{w}$ can be readily estimated via the ordinary least squares (OLS) method by means of the following expression:

$$\mathbf{w} = (\mathbf{Z}\mathbf{Z}^T)^{-1}\mathbf{Z}\mathbf{d}, \tag{6}$$

where $\mathbf{Z} = [\mathbf{z}_1 \,|\, \mathbf{z}_2 \,|\, \cdots \,|\, \mathbf{z}_{N_1}]$ is a $(p+q) \times N_1$ matrix whose $N_1$ columns are the augmented vectors $\mathbf{z}_n = [\mathbf{x}_n^T \,|\, \mathbf{h}_n^T]^T \in \mathbb{R}^{p+q}$, $n = 1, \ldots, N_1$, where $N_1$ is the number of available training input patterns. The vector $\mathbf{d}$ is defined in Eq. (1). To avoid numerical problems, a regularized version of Eq. (6) is commonly used, which is given by

$$\mathbf{w} = (\mathbf{Z}\mathbf{Z}^T + \lambda\mathbf{I})^{-1}\mathbf{Z}\mathbf{d}, \tag{7}$$

where the constant $\lambda > 0$ is the regularization parameter.

## 2.2   The Extreme Learning Machine (ELM)

The ELM network is a recent randomized SLFN introduced by Huang *et al.* [3], whose weights from the inputs to the hidden neurons are randomly chosen, while only the weights from the hidden neurons to the output are analytically determined. Consequently, ELM offers significant advantages such as fast learning speed, ease of implementation, and less human intervention when compared to more traditional SLFNs, such as the Multilayer Perceptrons (MLP) and RBF networks.

From an architectural point of view, the ELM network can be understood as a simplified version of the RVFL in which the direct linear path is removed. Thus, the equations of the ELM are easily obtained as follows:

**Output Computation:** From Eq. (5), once we remove the direct linear pathway, we get

$$y_n = y_n^{(2)} = \mathbf{w}_2^T\mathbf{h}_n, \tag{8}$$

where $\mathbf{h}_n$ is defined as in Eq. (4).

**Estimation of $\mathbf{w}_2$:** In this case, the expression of the OLS estimate in Eq. (6) reduces to

$$\mathbf{w}_2 = (\mathbf{H}\mathbf{H}^T)^{-1}\mathbf{H}\mathbf{d}, \tag{9}$$

where $\mathbf{H} = [\mathbf{h}_1 \,|\, \mathbf{h}_2 \,|\, \cdots \,|\, \mathbf{h}_{N_1}]$ be a $q \times N_1$ matrix whose $N_1$ columns are the hidden activation vectors $\mathbf{h}_n \in \mathbb{R}^q$, $n = 1, \ldots, N_1$, where $N_1$ is the number of available training input patterns.

## 2.3   Sequential Learning Rules for RVFL and ELM

In some applications, such as adaptive channel equalization and online system identification, adaptive learning rules are a better option, where the weight vector $\mathbf{w}$ is updated following the arrival of each input pattern. The input pattern is then discarded after being used for updating the parameters. One of such sequential learning rules is the well-known *least mean squares* (LMS) algorithm, also known as the Widrow-Hoff or Delta rule, which was used recently by Widrow *et al.* [14] to introduce a randomized SLFN architecture, named *No-Propagation* (No-Prop) network. Basically, the No-Prop network is like an ELM network with output weights computed by means of a sequential learning rule.

In order to allow the RVFL network to process sequential data, we replace the standard OLS equation with the LMS rule. For this purpose, let us consider first the instantaneous cost function associated with the output neuron at the presentation of the $n$-th input vector:

$$J\left(\mathbf{w}_n\right) = \frac{1}{2}e_n^2 = \frac{1}{2}(d_n - y_n)^2 = \frac{1}{2}\left(d_n - \mathbf{w}_n^T\mathbf{h}_n\right)^2, \tag{10}$$

where $\mathbf{w}_n \in \mathbb{R}^{p+q}$ is the weight vector of the output neuron at iteration $n$, and $e_n = d_n - y_n$ is the instantaneous error of that neuron at iteration $n$. Then, in order to derive the LMS learning rule, we resort to a stochastic gradient descent recursive formula given by

$$\mathbf{w}_{n+1} = \mathbf{w}_n - \eta\frac{\partial J\left(\mathbf{w}_n\right)}{\partial \mathbf{w}_n} = \mathbf{w}_n + \eta e_n\mathbf{h}_n = \mathbf{w}_n + \eta(d_n - y_n)\mathbf{h}_n, \tag{11}$$

where $0 < \eta \ll 1$ is the learning rate. A widely used variant of the LMS rule, known as the normalized LMS (NLMS) algorithm [2], is given by

$$\mathbf{w}_{n+1} = \mathbf{w}_n + \frac{\eta}{\epsilon + \|\mathbf{h}_n\|^2}e_n\mathbf{h}_n = \mathbf{w}_n + \frac{\eta}{\epsilon + \mathbf{h}_n^T\mathbf{h}_n}e_n\mathbf{h}_n, \tag{12}$$

where $\epsilon$ is a very small positive constant needed to avoid division by zero. The strong points of the LMS and NLMS algorithms are ease of implementation and optimal performance under important practical conditions [13]. For these reasons, the LMS algorithm has enjoyed very widespread application in adaptive filtering and signal processing applications. For instance, it is used in almost every modem for channel equalization and echo canceling.

## 3    A Robust Learning Rule for RVFL and ELM

An important feature of both OLS and LMS rules is that they are derived from cost functions that assign the same importance to all error samples, i.e. all errors contribute the same way to the final solution. Hence, outliers tend to produce large errors and then degrade the parameter estimation process.

Bearing this in mind, a robust variant of the LMS rule, named the Least Mean $M$-Estimate (LMM) algorithm [17], has been introduced for the purpose of better dealing with outliers. The theory behind the LMM rule is provided by an elegant and principled estimation framework, known as $M$-estimation, introduced by Huber and Ronchetti [4]. The letter $M$ stands for "maximum likelihood" type, where robustness is achieved by minimizing a function distinct from the sum of the squared errors.

Based on Huber's theory, the instantaneous cost function to be minimized by the output neuron is now given by

$$J\left(\mathbf{w}_n\right) = \rho(e_n) = \rho(d_n - y_n) = \rho\left(d_n - \mathbf{w}_n^T \mathbf{h}_n\right), \tag{13}$$

where $\mathbf{w}_n \in \mathbb{R}^{p+q}$ is the weight vector of the output neuron at iteration $n$, and $e_n = d_n - y_n$ is the instantaneous error of that neuron at iteration $n$. The function $\rho(\cdot)$ should possess the following properties: $(i)$ $\rho(e_n) \geq 0$; $(ii)$ $\rho(0) = 0$; $(iii)$ $\rho(e_n) = \rho(-e_n)$; and, $(iv)$ $\rho(e_n) \geq \rho(e_{n'})$, for $|e_n| > |e_{n'}|$. For $\rho(e_n) = e_n^2/2$, we get the instantaneous cost function of the standard LMS rule shown in Eq. (10).

Thus, we develop a robust learning rule for the RVFL network as follows:

$$\mathbf{w}_{n+1} = \mathbf{w}_n - \eta \frac{\partial J\left(\mathbf{w}_n\right)}{\partial \mathbf{w}_n} = \mathbf{w}_n - \eta \frac{\partial \rho(e_n)}{\partial \mathbf{w}_n} = \mathbf{w}_n - \eta \frac{\partial \rho(e_n)}{\partial e_n} \frac{\partial e_n}{\partial \mathbf{w}_n}$$

$$= \mathbf{w}_n - \eta \frac{\partial \rho(e_n)}{\partial e_n}(-\mathbf{h}_n) = \mathbf{w}_n + \eta q(e_n) e_n \mathbf{h}_n, \tag{14}$$

where $q(e_n) = \frac{1}{e_n} \frac{\partial \rho(e_n)}{\partial e_n}$ is called the weighting function. The normalized version of the LMM rule is then written as

$$\mathbf{w}_{n+1} = \mathbf{w}_n + \frac{\eta}{\epsilon + \mathbf{h}_n^T \mathbf{h}_n} q(e_n) e_n \mathbf{h}_n, \tag{15}$$

where $\epsilon$ has the same meaning as in Eq. (12). Note that if $\rho(e_n) = e_n^2/2$, then $q(e_n) = 1$, and Eq. (14) reduces to Eq. (11)

In this work, Hampel's three-part function [8] will be used, being defined as

$$\rho(e_n) = \begin{cases} e_n^2/2, & 0 \leq |e_n| < \xi \\ \xi|e_n| - \xi^2/2, & \xi \leq |e_n| < \Delta_1 \\ \frac{\xi}{2}(\Delta_1 + \Delta_2) - \frac{\xi^2}{2} + \frac{\xi}{2}\frac{(|e_n|-\Delta_2)^2}{\Delta_1-\Delta_2}, & \Delta_1 \leq |e_n| < \Delta_2 \\ \frac{\xi}{2}(\Delta_1 + \Delta_2) - \frac{\xi^2}{2}, & \Delta_2 \leq |e_n| \end{cases}, \tag{16}$$

$$q(e_n) = \begin{cases} 1, & 0 \leq |e_n| < \xi \\ \frac{\xi}{e_n}\text{sign}(e_n), & \xi \leq |e_n| < \Delta_1 \\ \frac{\xi}{e_n}\text{sign}(e_n)\frac{|e_n|-\Delta_2}{\Delta_1-\Delta_2}, & \Delta_1 \leq |e_n| < \Delta_2 \\ 0, & \Delta_2 \leq |e_n| \end{cases}, \tag{17}$$

where $\xi, \Delta_1, \Delta_2$ are user-defined thresholds which avoid the influence of inputs with large errors. As in [17], we use $\xi = 1.96\hat{\sigma}_n$, $\Delta_1 = 2.24\hat{\sigma}_n$ and $\Delta_2 = 2.576\hat{\sigma}_n$, where $\hat{\sigma}_n$ is the standard deviation of the output, estimated recursively.

## 4    Simulation and Results

In this section we report the results of the evaluation of four randomized SLFNs, namely: two variants of the RVFL network, named the RVFL-NLMS and the RVFL-NLMM, and two variants of the ELM network, named the ELM-NLMS and ELM-NLMM[2]. We also evaluate the performances of two variants of a single-hidden-layered MLP network trained with the backprop algorithm, using the tanh activation function for the hidden neurons and a linear activation function for the output neuron. The variants of the MLP network differ in the way the weights of the output neuron are adjusted, with one using the LMS rule (MLP-LMS) and the other using the LMM rule (MLP-LMM).

The dynamics of the systems of interest are assumed to be described by a nonlinear autoregressive model with exogenous inputs (NARX) [1]:

$$d_n = f(d_{n-1}, \cdots d_{n-L_y}, u_{n-1}, \cdots, u_{n-L_u}), \qquad (18)$$

where $L_u$ and $L_y$ denote the input and output memory orders, respectively. The target function $f(\cdot) : \mathbb{R}^{L_y + L_u} \to \mathbb{R}$ is unknown and assumed to be nonlinear. Observed data, in the form of an input times series $\{u_n\}$ and an output time series $\{d_n\}_{n=1}^{N}$, are used to build an approximating model $\hat{f}(\cdot)$ for $f(\cdot)$.

Experiments were performed with an artificial and a real-world dataset. The *Artificial* dataset is generated by simulating the following dynamical system [7]:

$$d_n = \frac{d_{n-1}}{1 + d_{n-1}^2} + u_{n-1}^3, \qquad (19)$$

where the training input time series is generated by sampling from an uniform distribution between $-2$ and $2$ (i.e. $u_n \sim U(-2, 2)$), $n = 1, \ldots, 10000$, and the test input time series is given by $u_n = \sin(2\pi i/25) + \sin(2\pi i/10)$, $n = 1, \ldots, 100$. To the resulting output time series $\{d_n\}$, we add zero-mean Gaussian noise with variance equal to 0.65. We use $L_y = 1$ and $L_u = 1$.

The real-world dataset, named *Silver box* [6,10], is an electronic nonlinear feedback laboratory experiment, which simulates a second order mechanical system with a nonlinear spring constant, acting as mass-spring-damper structure. The control input in the mechanical system is the force applied to the mass and its displacement is the output. The electrical circuit is excited with ten different realizations of odd random phase multisine, resulting in 91072 training samples. The test set contains 40000 samples generated with a filtered Gaussian sequence with increasing variance. The regressors' lags are fixed as $L_y = 10$ and $L_u = 10$.

For all the following experiments, time series data are normalized to zero mean and unit variance. All neural models were implemented from scratch using the $R$ software, version 3.3.2, running on Ubuntu 16.04, installed in an Acer notebook, Core i7, 2.40 GHz, 16 GB RAM. We perform experiments with scenarios containing 0%, 5%, 10%, 15%, 20%, 25% and 30% of outliers. The outliers were sampled from $\sigma(d) \times T(0, 2)$, where $\sigma(d)$ is the standard deviation of the original training set and $T(0, 2)$ is a Student-t distribution with zero mean and 2 degrees of freedom.

The models are trained in an online way, where the training samples are presented as they are made available, one after another, within a single full pass

---

[2] The second term in the name of the evaluated randomized SLFN denotes the online learning rule used to estimate the output weights.

of the training data. The figure of merit of the evaluation is the root mean square error (RMSE) values for both one-step ahead (OSA) prediction, where predictions are made using the actual output samples in the regressors:

$$y_n = \hat{d}_n = \hat{f}(d_{n-1}, \cdots d_{n-L_y}, u_{n-1}, \cdots, u_{n-L_u}), \tag{20}$$

and free simulation, where predicted output values are fed back in order to build the output regressor:

$$y_n = \hat{d}_n = \hat{f}(y_{n-1}, \cdots y_{n-L_y}, u_{n-1}, \cdots, u_{n-L_u}),$$
$$= \hat{f}(\hat{d}_{n-1}, \cdots \hat{d}_{n-L_y}, u_{n-1}, \cdots, u_{n-L_u}), \tag{21}$$

where the "hat" symbol $\wedge$ denotes the predicted values.

For each model the number of hidden units and the learning rate were optimized via Bayesian optimization [11] using the mean prediction RMSE within the 10-fold cross-validation performed in the outlier-free training data. We apply a linearly decaying learning rate, i.e., the rate for the $n$-th iteration is given by $\alpha_n = \alpha_1 + (\alpha_{N_1} - \alpha_1)\frac{n-1}{N_1-1}$, where $N_1$ is the number of training samples. Only the final learning rate $\alpha_{N_1}$ is optimized, while its initial value $\alpha_1$ was fixed after preliminary experiments. In Table 1 we summarize the hyperparameters selected for each model. We emphasize that the hyperparameters selection step was executed only once per dataset using outlier-free data, but the adjustment of the models' parameters was performed separately in each contaminated scenario.

**Table 1.** Hyperparameters selected for each evaluated model: the number of hidden units $q$, the initial learning rate $\alpha_1$ and the final learning rate $\alpha_{N_1}$. $\alpha_1$ was fixed after preliminary experiments. The other hyperparameters were determined via Bayesian optimization of the 10-fold cross-validation error in the outlier-free data.

| | Artificial | | | Silver box | | |
|---|---|---|---|---|---|---|
| | $q$ | $\alpha_1$ | $\alpha_{N_1}$ | $q$ | $\alpha_1$ | $\alpha_{N_1}$ |
| ELM-NLMS | 967 | 1 | 2.52e−02 | 981 | 1 | 3.14e−05 |
| ELM-NLMM | 946 | 1 | 3.30e−02 | 997 | 1 | 7.20e−03 |
| RVFL-NLMS | 101 | 1 | 2.19e−03 | 708 | 1 | 3.91e−04 |
| RVFL-NLMM | 319 | 1 | 1.60e−03 | 672 | 1 | 2.84e−02 |
| MLP-LMS | 35 | 0.1 | 3.09e−03 | 31 | 0.01 | 9.98e−03 |
| MLP-LMM | 41 | 0.1 | 2.11e−03 | 43 | 0.01 | 9.88e−03 |

From this table, one can easily note that the RVFL variants require much smaller numbers of hidden neurons than the ELM variants. A possible explanation could rely on the direct linear pathway ($y_n^{(1)}$), which by capturing the linear part of the system dynamics let only the nonlinear (and more difficult) part of the dynamics to the nonlinear pathway ($y_n^{(2)}$).

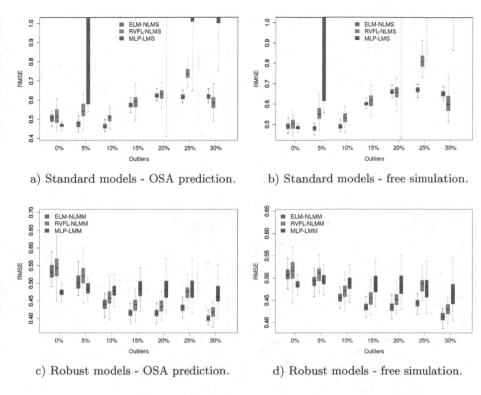

a) Standard models - OSA prediction.

b) Standard models - free simulation.

c) Robust models - OSA prediction.

d) Robust models - free simulation.

**Fig. 1.** Results for the *Artificial* dataset.

The obtained RMSE values and the corresponding variances for the models using the *Artificial* dataset are shown in Fig. 1. Figure 1a stems for the standard algorithms in OSA predictions while Fig. 1b shows mutiple-step-ahead predictions (free simulation). The performances of the robust versions of the algorithms are shown in Figs. 1c (OSA) and d (free simulations). For both kinds of prediction, results are rather consistent and show that no matter the % of contamination by outliers robust algorithms achieve better performances and are rather insensitive to the % of contamination. This is not the case for the standard algorithms that exhibit a certain degree of sensitivity to the % of outliers. In both kinds of prediction MLP-LMS show great variance for some of the % of contamination by outliers and in both cases the introduction of the corresponding robust algorithm greatly reduces these poor performances. In the case of robust algorithms, ELM-NLMM exhibits the best performance followed by RVFL-NLMM and MLP-LMM. There are no great differences in the corresponding performances of standard and robust algorithms depending on the kind of prediction (OSA or free simulation), except for the case of MLP-LMS.

Things are different for the *Silver box* dataset results shown in Fig. 2. In this case robust algorithms (Figs. 2c and d) do not exhibit clear better performances than the standard algorithms as it is the case for the *Artificial* dataset (Fig. 1).

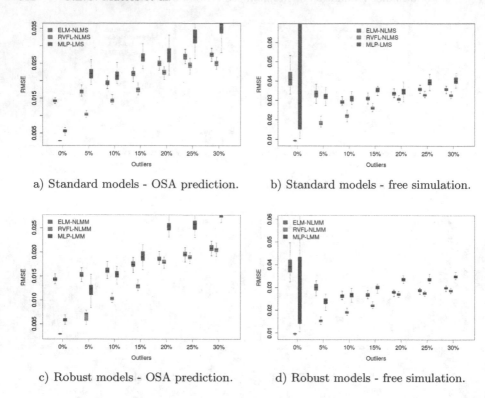

a) Standard models - OSA prediction.

b) Standard models - free simulation.

c) Robust models - OSA prediction.

d) Robust models - free simulation.

**Fig. 2.** Results for the *Silver box* dataset.

A slight reduction of the obtained RMSE and a more consistent and important reduction of the variances are the advantages shown by the robust algorithms. This is more evident in the case of MLP. The RMSE obtained for all models increased significantly when the contamination also increased, specially in the case of OSA predictions. Even robust models are not as insensitive to the % of outliers as in the case shown in Fig. 1.

An additional Wilcoxon signed-rank test for the residuals obtained in the best free simulation with 30% of outliers was performed. Results are reported in Table 2 for the *Artificial* dataset and in Table 3 for the *Silver box* dataset. It can be seen that for the *Silver box* dataset all models perform significantly different from each other and this is due to the small variance they exhibit. This is not the case for the *Artificial* dataset, where some models perform not significantly different from others. This is the case for ELM models which exhibit no significant difference from the MLP-LMS model. RVFL models also exhibit no significative difference with MLP-LMM model.

As a final remark it can be said that even though results are not equally clearly interpretable for both datasets, the robust algorithms achieved consistently better performances. From the point of view of a reduced RMSE and variance for the *Artificial* dataset and from the point of view of a reduced

**Table 2.** $p$-values computed via Wilcoxon signed-rank test for the residuals obtained in the best free simulation on the *Artificial* dataset with 30% of outliers. Red indicates statistically similar results.

|          | ELM-NLMM | RVFL-NLMS | RVFL-NLMM | MLP-LMS | MLP-LMM |
|----------|----------|-----------|-----------|---------|---------|
| ELM-NLMS | 4.608945e−04 | 1.423818e−11 | 2.789759e−06 | 0.3039999415 | 1.054542e-06 |
| ELM-NLMM |          | 3.127407e−09 | 1.738496e−16 | 0.2769369016 | 1.007799e−10 |
| RVFL-NLMS |         |           | 1.314366e−02 | 0.0001062009 | 6.663213e−02 |
| RVFL-NLMM |         |           |           | 0.0086282427 | 7.283775e−01 |
| MLP-LMS  |          |           |           |         | 3.664618e−03 |

**Table 3.** $p$-values computed via Wilcoxon signed-rank test for the residuals obtained in the best free simulation on the *Silver box* dataset with 30% of outliers. All the results were statistically different.

|          | ELM-NLMM | RVFL-NLMS | RVFL-NLMM | MLP-LMS | MLP-LMM |
|----------|----------|-----------|-----------|---------|---------|
| ELM-NLMS | 5.370802e−67 | 7.093561e−75 | 8.371801e−09 | 1.122239e−159 | 8.294698e−86 |
| ELM-NLMM |          | 1.036182e−03 | 9.285395e−100 | 1.466633e−41 | 3.493612e−11 |
| RVFL-NLMS |         |           | 2.397354e−68 | 1.904553e−30 | 9.663732e−02 |
| RVFL-NLMM |         |           |           | 9.958828e−160 | 3.048651e−115 |
| MLP-LMS  |          |           |           |         | 4.956787e−63 |

variance in the case of the experiments performed with the *Silver box* dataset. In which concerns the % of outliers, the behavior of the robust algorithms was clearly insensitive to the increment of the contamination rate for the *Artificial* dataset, although that was not the case for the *Silver box* dataset. Overall, robust versions of ELM and RVFL consistently achieved the best performances.

## 5   Conclusions and Further Work

In this paper we tackled the task of online nonlinear system identification in the presence of outliers with randomized SLFNs. For that purpose, we decided to replace the original *batch* OLS-based learning rules of the RVFL and ELM networks with adaptive LMS-based ones, enabling recursive learning and training from larger datasets. Seeking resilience to outliers, a robust version of the LMS rule, known as LMM rule, was also considered as a learning algorithm.

We performed computational experiments with both an artificial and a real-world datasets using incremental levels of contamination by outliers. The achieved results are promising, with the evaluated robust randomized SLFNs being capable of fast learning of the system dynamics in an online fashion, i.e., without the need of multiple epochs, even in the presence of outliers.

However, despite the appealing results of our evaluation, further experiments are still needed in order to have a clear picture of the pros and cons of the proposed approach. For instance, hyperparameter optimization (i.e. number of hidden units and learning rates) remains an open issue, since it requires costly rounds of cross-validation. Furthermore, we were not able to obtain a strong resilience to outliers in the *Silver box* dataset, when compared to the good results we presented for the *Artificial* dataset.

In this regard, we continue to evaluate the proposed methodology in other benchmarking system identification datasets, as well as experimenting with other adaptive learning rules based, for instance, on the recursive least-squares (RLS) algorithm.

**Acknowledgments.** The first two authors thank the financial support of FUNCAP, CNPq (grant no. 309451/2015-9) and NUTEC. The third author acknowledges partial financial support of Conicyt via Fondef Mineria Grant IT16M100008.

# References

1. Billings, S.A.: Nonlinear System Identification NARMAX Methods in the Time, Frequency and Spatio-Temporal Domains, 1st edn. Wiley, Hoboken (2013)
2. Choi, Y.S., Shin, H.C., Son, W.J.: Robust regularization for normalized LMS algorithms. IEEE Trans. Circ. Syst.-II **53**(8), 627–631 (2006)
3. Huang, G., Huang, G.B., Song, S., You, K.: Trends in extreme learning machines: a review. Neural Netw. **61**(1), 32–48 (2015)
4. Huber, P.J., Ronchetti, E.M.: Robust Statistics. Wiley, Hoboken (2009)
5. Li, M.B., Er, M.J.: Nonlinear system identification using extreme learning machine. In: Proceedings of the 9th International Conference on Control, Automation, Robotics and Vision (ICARCV) 2006, pp. 1–4 (2006)
6. Marconato, A., Sjöberg, J., Suykens, J., Schoukens, J.: Identification of the silverbox benchmark using nonlinear state-space models. IFAC Proc. Vol. **45**(16), 632–637 (2012)
7. Narendra, K.S., Parthasarathy, K.: Identification and control of dynamical systems using neural networks. IEEE Trans. Neural Netw. **1**(1), 4–27 (1990)
8. Rousseeuw, P.J., Leroy, A.M.: Robust Regression and Outlier Detection. Wiley, Hoboken (1987)
9. Salih, D.M., Noor, S.B.M., Merhaban, M.H., Kamil, R.M.: Wavelet network: online sequential extreme learning machine for nonlinear dynamic systems identification. Adv. Artif. Intell. **2015**(184318), 1–10 (2015)
10. Schoukens, J., Nemeth, J.G., Crama, P., Rolain, Y., Pintelon, R.: Fast approximate identification of nonlinear systems. Automatica **39**(7), 1267–1274 (2003)
11. Snoek, J., Larochelle, H., Adams, R.P.: Practical bayesian optimization of machine learning algorithms. In: Advances in Neural Information Processing Systems, pp. 2951–2959 (2012)
12. Tang, Y., Li, Z., Guan, X.: Identification of nonlinear system using extreme learning machine based Hammerstein model. Commun. Nonlinear Sci. Numer. Simul. **19**(9), 3171–3183 (2014)
13. Widrow, B.: Thinking about thinking: the discovery of the LMS algorithm. IEEE Sig. Process. Mag. **22**(1), 100–106 (2005)
14. Widrow, B., Greenblatt, A., Kim, Y., Park, D.: The No-Prop algorithm: a new learning algorithm for multilayer neural networks. Neural Netw. **37**, 182–188 (2013)

15. Pao, Y.-H.: Learning and generalization characteristics of the random vector functional-link net. Neurocomputing **6**, 163–180 (1994)
16. Zhang, L., Suganthan, P.N.: A comprehensive evaluation of random vector functional link networks. Inf. Sci. **367–368**, 1094–1105 (2016)
17. Zou, Y., Chan, S.C., Ng, T.S.: Least mean *M*-estimate algorithms for robust adaptive filtering in impulsive noise. IEEE Trans. Circ. Syst. II **47**(12), 1564–1569 (2000)

# Neural Network Overtopping Predictor Proof of Concept

Alberto Alvarellos[1]([✉]), Enrique Peña[2], Andrés Figuero[2], José Sande[2], and Juan Rabuñal[1]

[1] Department of Information and Communications Technologies, University of A Coruña, Campus Elviña s/n, 15071 A Coruña, Spain
{alberto.alvarellos,juanra}@udc.es
[2] Water and Environmental Engineering Group (GEAMA), University of A Coruña, Campus Elviña s/n, 15071 A Coruña, Spain
{epena,andres.figuero,jose.sande}@udc.es,
http://www.udc.es

**Abstract.** Wave overtopping is a dangerous phenomenon. When it occurs in a commercial port environment, the best case scenario will be the disruption of activities and even this best case scenario has a negative financial repercussion.

Being in disposal of a system that predicts overtopping events would provide valuable information, allowing the minimization of the impact of overtopping: the financial impact, the property damage or even physical harm to port workers.

We designed an overtopping predictor and implemented a proof of concept based on neural networks. To carry out the proof of concept of the system, we created a series of tests in a scaled breakwater physical model, placed on a wave basin. We used a multidirectional wavemaker and video cameras to identify the overtopping events. Using all of the collected data we trained a neural network model that predicts an overtopping based on the simulated sea state.

Once the validity of this approach is determined, we propose the real system design and the resources needed for its implementation.

**Keywords:** Wave overtopping · Overtopping Prediction · Neural Network · Civil Engineering

## 1 Introduction

Spain has an 8000 km coastline, making it the European Union country with the longest coastline. It is also the closest European country to the axis of one of the world's major maritime routes. Its geographical location positions it as a strategic element in international shipping and a logistics platform in southern Europe.

The Spanish Port System includes 46 ports of general interest. The importance of ports, as links in the logistics and transport chains, is supported by the following figures [1]:

© Springer International Publishing AG 2017
I. Rojas et al. (Eds.): IWANN 2017, Part I, LNCS 10305, pp. 616–625, 2017.
DOI: 10.1007/978-3-319-59153-7_53

- They handle nearly 60% of exports and 85% of imports. This accounts for 53% of Spanish foreign trade with the European Union and 96% with third countries
- The State port system's activity contributes with nearly 20% of the transport sector's GDP. This accounts for 1.1% of the Spanish GDP
- It employs more than 35000 workers directly and around 110000 indirectly

There are several weather and sea conditions that can affect the regular activities in a port. One highly disruptive event of port activities is the wave overtopping of the port breakwater.

## 1.1  Wave Overtopping

Wave overtopping is the event that takes place when waves meet a submerged reef or structure. It also happens when waves meet an emerged reef or structure lower than the approximate wave height. The later case is the one that affects a port's breakwater and the one we want to measure.

During an overtopping, two processes, important to the coastal processes, take place: wave transmission and the passing of water over the structure. We studied the passing of water over the structure.

The overtopping phenomenon can occur in three different ways, either independent of each other or combined:

- Green Water, which is defined as the solid step of certain volume above the crown wall of the breakwater due to the rise of the wave (run-up) above the exposed surface of said breakwater
- White Water, that occurs when the wave breaks against the seaside slope. This creates so much turbulence that air is entrained into the water body, forming a bubbly or aerated and unstable current and water springs that reaches the protected area of the structure either by its own impulse or as a result of the wind
- Overtopping as a result of the aerosol generated by the wind passing by the crest of the waves near the breakwater. This is not an especially meaningful even in the case of storms. This case is the less dangerous, its impact on the normal development of port activities is negligible

## 1.2  Overtopping System Goal

The goal of this study was to design an overtopping predictor based on Neural Networks and implement the proof of concept of the design, in order to demonstrate its feasibility and hence its practical potential.

The proof of concept was carried out in the port and coasts laboratory, located in the *Center for Technological Innovation in Construction and Civil Engineering* (CITEEC [2]).

Once we assert the feasibility of the concept, we design the system that would be needed in a real environment and propose its implementation.

## 2    Experiment Design

As we stated before, the proof of concept was performed in port and coasts laboratory of CITEEC. This laboratory has a hydrodynamic experimentation wave basin with a multidirectional wavemaker (see Fig. 1). The hydrodynamic experimentation wave basin is $32 \times 34\,m^2$ and has a depth of 1.1 m. In this basin one can built a reduced-scale model of a breakwater. With this kind of installation we can perform: large scale structural tests on the effects of extreme waves on the model, study the behaviour of dikes, docks and beaches, and test inlets, estuaries, large ports and coastal forms.

**Fig. 1.** Hydrodynamic experimentation wave basin with the scaled port model. Wave-maker in the background

The wave generation system allows the complete control of the generated waves. This wavemaker has an active absorption system. The purpose of the active absorption system is to avoid the reflected waves from structures in the basin to be re-reflected at the paddles and become incident waves. Active absorption is very important for accurate wave generation, especially when testing highly reflective structures occupying a large part of the width of the facility.

Without active absorption re-reflected waves will contaminate the desired incident waves. We want precise control over the test variables, so active absorption is important in our case.

In order to measure the actual incident wave height generated during the tests, and its evolution along the basin, we used several water level probes.

In summary, we have control over the water depth, wave direction and wave height in our test. The variables we can control, available from the wavemaker software and the probes software, are the following (we express them in terms of the simulated real world conditions):

$H_0$ : Sea level, with respect to the zero of the port [3]

$H_s$ : Significant wave height. $H_s$ is the mean of the highest third of the waves in a time-series of waves representing a certain sea state

$T_p$ : The peak wave period is the wave period with the highest energy. The analysis of the distribution of the wave energy as a function of wave frequency for a time-series of individual waves is referred to as a spectral analysis. The peak wave period is extracted from the spectra [4]

$\theta_m$ : Mean wave propagation direction. $\theta_m$ is defined as the mean of all the individual wave directions in a time-series representing a certain sea state

## 2.1   Measuring the Overtopping

The final goal of this proof of concept is to design a sensible system that can predict said event with enough time to take safety measures in a real environment.

In order to identify the overtopping events we installed two video cameras that enabled us to monitor the model breakwater and detect the overtopping events (see Fig. 2).

**Fig. 2.** Frame of an overtopping event as visualized from one of the cameras

The quantification of the volume of water that was overtopping the crown wall of the breakwater in each event using images is a difficult task.

A system that could predict the water volume that is going to surpass the crown wall could enable a more informed decision. This could serve to act differently depending on the severity of the overtopping event, and even minimize its impact.

The goal of this work was to carry out the proof of concept of the system, so we chose not to estimate the volume. One thing we are certain of after the overtopping classification process: an overtopping has occurred. We train the system to predict whether an overtopping event is going to happen or not (delegating the decision of whether to act or not to the operator of the system).

## 2.2 Test Variables

To develop our proof of concept we took measures during stability tests carried out in a scaled physical model of a breakwater (with a 1:55 scale). Those tests reproduced real world sea conditions, ranged from normal to extreme conditions. We tested our system in the end spectrum of the normal conditions and in the extreme conditions of those tests. The overtopping phenomenon is known to happen almost always in this range of conditions, i.e. the sea conditions that produce an overtopping are a subset of this range except some unidentified normal condition. The wind is believed to have an important role in this unidentified, but we cannot introduce it in our model due to infrastructure limitations.

The main goal of the present work is to carry out a proof of concept. As such, it is not complete in its test variables (we did not test all possible values): The range of values tested for each variable are the following (we give the real values simulated, taking the scale into account):

$H_0$ : From 0.5 m to 4.1 m
$H_s$ : From 3 m to 10 m
$T_p$ : From 10 s to $\simeq$20 s
$\theta_m$ : From 290° to 340°

# 3      Predictor Training

We chose a Neural Network as the model for our prediction system. We used the Caret R package [5] to do all the data splitting, model training and validation, and predictions error calculations.

The chosen model is the *nnet* package's Neural Network implementation [6]: this package allows to train feed-forward neural networks with a single hidden layer, and multinomial log-linear models. As we are carrying out a proof of concept, this model is enough for our purpose.

The model training process consisted in the following steps:

1. Load data, clean and join data.
2. Model training:
   (a) choose predictors: calculate predictors correlations
   (b) data slicing (train/test)
   (c) "center" and "scale" the predictors in the train data and store the transformations
   (d) train model on train data
3. estimate model accuracy:
   (a) predict model error on test data
   (b) plot errors and model

We comment the following on some of the steps:

- To split all available data (378 observations) in train and test sets we used the function *createDataPartition*, that allows the creation of balanced splits of the data: if the *y* argument to this function is a factor, the random sampling occurs within each class and preserves the overall class distribution of the data. We used the observation class (a factor) as this argument (*isOvertopping=* "*Y*"| "*N*"). We used the 75% of data for training and 25% for testing.
- In step 2.a we calculated the correlation among the available predictors, in order to discard the highly correlated ones. We can observe (Table 1) that there is no high correlation between any of them, so we used all of them to train the model
- In step 3.a we predict the model error on the test data using the stored transformation for the train data

**Table 1.** Predictors correlation

|         | $H_0$    | $H_s$   | $T_p$   | $\theta_m$ |
|---------|----------|---------|---------|------------|
| $H_0$   | 1        | 0.1478  | −0.0415 | −0.0117    |
| $H_s$   | 0.1478   | 1       | 0.0955  | 0.2689     |
| $T_p$   | −0.0415  | 0.0955  | 1       | −0.0070    |
| $\theta_m$ | −0.0117 | 0.2689  | −0.0070 | 1          |

The training process consists of several steps (see Algorithm 1). The first step is to specify the sets of model parameter values to evaluate. The *nnet* model requires at least 2 parameters:

**size:** number of units in the hidden layer
**decay:** parameter for weight decay. After each update, the weights are multiplied by this parameter. This prevents the weights from growing too large. It can be seen as a gradient descent on a quadratic regularization term

We used a parameters set with 1020 pairs of values of size and decay (known as the parameter grid). This set was created from all the combinations of size from 1 and 20 and decay between 0 and 0.05 in steps of 0.0001.

The second step is to resample the data. We used 10-*fold* Cross-validation for the resampling. This resampling is used to choose among the models that all parameters grid create (each pair size-decay creates one model).

We also set a limit of 10000 iterations for the fitting process (step 3) (instead of the *nnet* package default 100).

## 4  Results

The final model (the one with optimal parameters) has 2 neurons in the hidden layer (see Fig. 3).

**Algorithm 1.** Training process algorithm

```
Define sets of model parameter values to evaluate
for each parameter set do
  for each resampling iteration do
    Hold-out specific samples
    Fit the model on the remainder
    Predict the hold-out samples
  end
  Calculate the average performance across hold-out predictions
end
Determine the optimal parameter set
Fit the final model to all the training data using the optimal parameter
```

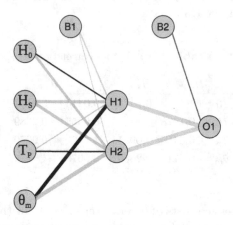

**Fig. 3.** Final Neural Network architecture. The output $O1$ indicates "Y"|"N" when it predicts that there is going to be an overtopping or not

The following values we present are the result of using the final model to predict over the test set.

In the Table 2 we can see the confusion matrix using a threshold of 0.5 for the classification of samples in the test set (i.e. we choose to say that there is going to be an overtopping if $p > 0.5$, where $p$ is the probability of a prediction being of class "Y").

Using this threshold we obtain an accuracy of 0.9362 with $(0.8662, 0.9762)$ as the 95% confidence interval. The balanced accuracy is 0.8741 (our test set has 14 positive classes out of 94).

One important value is the model sensitivity, because we want to correctly classify all the positive cases, i.e. we don't want to miss an overtopping. With a 0.5 threshold we obtain a sensitivity of 0.7857.

Our ideal predictor would be one that predicts all overtopping events, i.e. it should have a high sensitivity. Not predicting an overtopping due to the danger of not predicting an event, i.e. a false negative could result in physical harm

**Table 2.** Final model confusion matrix for a 0.5 threshold

|         |   | Prediction | |
|---------|---|---|----|
|         |   | Y | N  |
| Truth   | Y | 77 | 3 |
|         | N | 3 | 11 |

to a worker. Lowering the threshold we could achieve a sensitivity of 1. The predictor should preferably also have a low false positive rate (low probability of false alarm). A false positive could only cause financial loss, but not physical harm.

The threshold moves these values when we change them, so there is going to be a trade-off between safety and financial loss in choosing the threshold. To test the model performance for different thresholds (instead of just 0.5) we use the ROC curve [7]. The result can be seen in Fig. 4. We can observe that with this predictor we could achieve a sensitivity of 1 changing the threshold, but the probability of false alarm could not be lower than ≃0.12 in such case.

We obtain an AUROC of 0.966, for which we can conclude that our proof of concept resulted in a fairly good model, and that it's worth building a real system and field testing it.

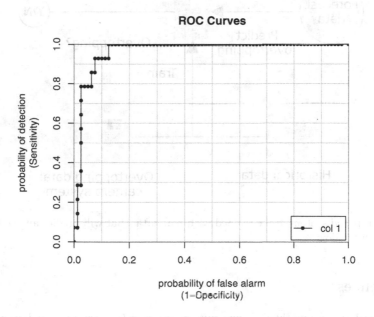

**Fig. 4.** ROC curve for the final model

## 5   Real System Proposal

The ideal Overtopping Prediction System would be one that uses a forecast for the weather and sea state to predict if there is going to be an overtopping event several days in advance. Using a forecast would allow to take the corresponding measures in order to minimize the overtopping impact in time, such as closing access to specific port areas, secure certain infrastructure, remove working material, etc.

In a real environment the wind is known to have an impact on the sea state, so it is going to affect the occurrence of an overtopping event. A real system should use the wind speed and direction as predictors in order to be more accurate.

Taking all of this into account, we propose a system, for a Spanish port, that uses the *Spanish Port Wave Forecast System* [8]. This forecast system predicts the sea state 72 h in advance. The model would be trained with the historical real data stored in the same system (see Fig. 5). Thus, we would be able to use the same variables we used in the proof of concept, plus wind measures.

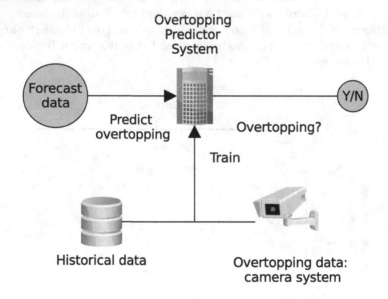

**Fig. 5.** Proposed system, to be trained with real data, that will predict an overtopping with a 72 h window

## References

1. Spanish Port System. http://www.puertos.es/en-us/nosotrospuertos/Pages/Nosotros.aspx. Accessed 05 Apr 2017
2. Center for Technological Innovation in Construction and Civil Engineering (CITEEC). http://www.udc.es/citeec/index-en.html. Accessed 05 Apr 2017

3. Spain Sea Level Reference System. http://portus.puertos.es/Portus/pdf/referenci as/Descripcion_Referencia_NivelDelMar_en.pdf. Accessed 05 Apr 2017
4. Mangor, K.: Shoreline Management Guidelines, 3rd edn. DHI Water & Environment, Horsholm (2004). OCLC: 934930487
5. Kuhn, M.: Building predictive models in R using the caret package. J. Stat. Softw. **28**(1), 1–26 (2008)
6. Venables, W.N., Ripley, B.D., Venables, W.N.: Modern Applied Statistics with S. Statistics and Computing, 4th edn. Springer, New York (2002). OCLC: ocm49312402
7. Hanley, J.A., McNeil, B.J.: The meaning and use of the area under a receiver operating characteristic (ROC) curve. Radiology **143**(1), 29–36 (1982)
8. Spanish Port Wave Forecast System. http://www.puertos.es/en-us/oceanografia/Pages/portus.aspx. Accessed 05 Apr 2017

# Artificial Neural Networks Based Approaches for the Prediction of Mean Flow Stress in Hot Rolling of Steel

Marco Vannucci$^{(\boxtimes)}$, Valentina Colla, and Vincenzo Iannino

TeCIP Institute, Scuola Superiore Sant'Anna,
Via G. Moruzzi, 1, 56124 Pisa, Italy
{mvannucci,colla,v.iannino}@sssup.it

**Abstract.** The problem of the estimation of mean flow stress within a hot rolling mill plant for flat steel products is faced, as the correct estimation of this measure can improve the quality of the final product. Various approaches, from standard empirical methods to advanced architectures based on neural networks, have been tested on industrial data. The results of these tests put into evidence the limit of empirical techniques and the big advantages deriving from the application of neural networks, which are able to efficiently combine process knowledge and data driven models tuning. The best performing approaches reduce the estimation error to one third with respect to standard techniques.

## 1  Introduction

In the steel–making practice, the Hot Rolling (HR) is a process in which the slabs previously produced during the continuous casting process are passed through a sequence of rolls couples (called *stands*), whose aim is the reduction of the thickness of the final product. More in detail, the slabs are heated in a furnace up to a temperature suitable for rolling and subsequently rolled, firstly through the so–called roughing mill and afterwards through the finishing mill, in order to obtainthe desired width and thickness for the product. In this phases the forces applied by the different gauges strongly affect not only the final shape but also the micro–structure of the steel sheet and, as a consequence, its mechanical properties, whose control is of utmost importance for the final product quality. It is thus fundamental to suitably manage the forces applied by the whole set of rolls pairs of the HR Mill (HRM), in order to fulfil the demands for product reliability and efficient rolling. To this aim a quantity named Mean Flow Stress (MFS) must be estimated with good accuracy, since it is the predominant factor of the roll force model [1]. The MFS is the mean value of stress required at each gage in order to allow a continuous plastic deformation of the material. In the framework of the HR process, once a reliable estimate of MFS is provided, suitable application of the forces in the roughing and finishing mill can be pursued. This necessity led, in the last years, to a proficient development of various techniques for the estimation of MFS in modern HRMs.

© Springer International Publishing AG 2017
I. Rojas et al. (Eds.): IWANN 2017, Part I, LNCS 10305, pp. 626–637, 2017.
DOI: 10.1007/978-3-319-59153-7_54

In this paper different approaches are used in order to design a model for the estimation of MFS in the context of an Italian HRM. These approaches include standard methods, Artificial Intelligence (AI)–based methods and a set of approaches based on the use of Artificial Neural Networks (ANN). These techniques are compared in order to put into evidence the benefits gained by the use of advanced ANN–based models that lead to an extremely accurate prediction of MFS. This paper is organized as follows: Sect. 2 presents a review of the standard techniques commonly employed in the industrial plants for the MFS prediction. In Sect. 3 the industrial problem is described and the performance of standard approaches for MFS estimation are shown. In Sect. 4 the ANN-based techniques developed to overcome the criticalities encountered by standard models are described, their performance are evaluated and the main achievements are discussed. Finally, Sect. 5 presents some conclusions and outlines the future perspectives of the proposed approaches.

## 2    Techniques for Mean Flow Stress Prediction

In literature a number of works can be found concerning the modelling of the HR process [2] and the estimation of MFS. An exhaustive analysis of these approaches can be found in [3], where the authors present the main categories of mathematical models based on the exploitation of theoretical knowledge and empirical data and put into evidence the advantages and drawbacks of each of them. The main families of models can be summarized as follows:

**Physical.** These models are based on the actual simulation of the physical processes that occur within the semi–product during the HR. Computational Fluid Dynamics (CFD) and Finite Element Modelling (FEM) techniques are commonly exploited within this category of models, which require a deep and precise knowledge of plant and steel characteristics as well as a considerable computational burden and the use of *ad–hoc* software tools. Despite the noticeable complexity, the predictive performance of these models is often limited, since they cannot perfectly describe all the physical phenomena due to their complexity and to the influence of some non measurable factors, such as friction, yield stress, and system disturbances [4].

**Empirical.** This is the most widely used class of models in the industrial framework. It is able to combine in a straightforward manner the theoretical knowledge of the interaction between input and output variables with the exploitation of real plant data. The most noticeable works correlate the MFS at each rolling stand to the steel chemical composition and the main process parameters, e.g. temperature, strains and strain rates during the rolling. These approaches are often based on statistical analysis and include both simple regressions and more sophisticated methods devoted to the construction of the model [5]. Empirical approaches can embed - at different levels of complexity - theoretical models tuned by means of empirical data. The main advantage of these approaches is their ease of use and adaptability. On the other hand, their generalization capabilities may be affected by the characteristics of the data employed for their tuning (i.e. distribution, presence of noise or outliers). Successful equation–based models for MFS prediction have been proposed independently by Misaka (Eq. 1), Siciliano (Eq. 2) and Poliak (Eq. 3) that embeds and extends Misaka formula [5].

$$MFS = e^{\left(0.126 - 1.75[C] + 0.594[C]^2 + \frac{2851 + 2968[C] - 1120[C]^2}{T}\right)} \cdot \epsilon^\alpha \cdot \epsilon^\beta \tag{1}$$

$$MFS = e^{\frac{2704 + 3345[Nb] + 220[Mn]}{T}} \cdot \epsilon^\alpha \cdot \epsilon^\beta \tag{2}$$

$$MSF = MFS_{Misaka} \cdot (1.09 + 0.056[Mn] \\ + \gamma[Nb] + \delta[Ti] + 0.056[Al] + 0.1[Mo]) \tag{3}$$

These formulae calculate MFS at HR gages (in kgf/mm$^2$) given the steel content of the main chemical elements, such as Carbon and microalloying elements (in wt%), the temperature T (in Kelvin), the applied strain $\epsilon$ and the strain rate $\epsilon_s$. The additional parameters included in the introduced formulae are empirically determined.

Although in Eqs. 1, 2 and 3 the parameters of the formulae are fixed and determined on the basis of the experimental data available to the authors, in [9] it is shown how these parameters can be efficiently tuned on specific plant experimental data through Genetic Algorithms (GA), in order to obtain *customized* models with improved predictive performance.

**Heuristic.** This family includes many AI–based models that are now gaining large acceptance within the steel industry. The main advantage of these models is that, in many cases, a very limited mathematical representation of the physical processes is required, since the output is calculated according to the processing of a series of previous data. ANN–based models are included into this class. Although ANNs are not yet widely used in the context of HRM, an example of exploitation of a simple two–layers Feed Forward Neural Networks (FFNN) can be found in [7,8], where an ANN is used for the prediction of the gages rolling forces achieving satisfactory results.

## 3    Assessment of Standard Approaches on Industrial Data

This work is focused on the practical problem of MFS estimation faced on a Italian HR mill that provided a set of data for model tuning and assessment. These data are related to the rolling of about 5000 coils in the HR mill outlined in Fig. 1 that is composed (from left to right) by four reheating furnaces, a press for the sizing of the slabs (SP), a vertical scale breaker that removes the oxide at the exit of the furnaces (VSB), six stands of the roughing mill (R1–R6), seven stands of the finishing (F1–F7), the run-out-table (ROT), and, finally, the three coilers (C1–C3). The mill is equipped with a complete set of sensors that, together with the plant informative system stored and made available for this

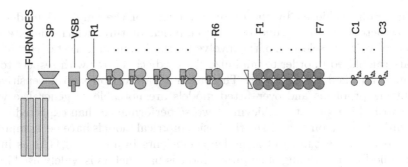

**Fig. 1.** A scheme of the rolling plant handled within this work.

study information concerning the steel chemical composition, the mechanical properties of the product and the rolling parameters for both the roughing and finishing mill.

This work concerns the prediction of MFS for the 7 stages of the finishing mill of this plant by means of approaches of different nature on the basis of the collected data. More in detail, among all the available data the slabs chemical composition and a set of process measures recorded at each of the 7 gages including strains and strain rates are exploited for the tuning of the models introduced in Sect. 1 and their performances are assessed in comparison to the MFS measured on the plant. Models are trained and tested on different datasets in order to fairly evaluate their predictive capabilities on data which have not been previously used for their training, according to the well known *10–fold cross validation* technique. The performance is evaluated at each HRM gage in terms of average percent error $e\%_{MFS}$ on the prediction of MFS through the different samples and cross validation tests. The first tested approach is based on the straightforward application of the original version of the empirical models described by Eqs. 1, 2 and 3 whose performances are depicted in Table 1.

**Table 1.** Performance of empirical models ($e\%_{MFS}$). Each row refers to a gage. In the last row the stand–wise average error is reported.

| Stand | Misaka | Siciliano | Poliak |
| --- | --- | --- | --- |
| 1 | 27.2 | 8.4 | 30.6 |
| 2 | 42.1 | 11.9 | 40.2 |
| 3 | 39.8 | 7.2 | 26.5 |
| 4 | 18.2 | 10.2 | 20.1 |
| 5 | 15.9 | 11.2 | 15.1 |
| 6 | 12.0 | 17.0 | 14.5 |
| 7 | 11.3 | 23.9 | 12.0 |
| Ave. | 23.8 | 12.8 | 22.7 |

The results achieved by the original versions of the empirical models are not satisfactory, probably due to their empirical nature, which, despite of a theoretically driven selection of the involved variables, include a set of coefficients that are calculated in order to minimize the prediction error with respect to the data used for their development. This latter identification step is sensitive to over–fitting problems and over–fitted models are not able to generalize when coping with different data, achieving a worse performance than expected.

In order to overcome this issue the basic empirical models have been improved by means of the re–tuning of formulae coefficients by means of GAs as in [9]. This method for the tuning of empirical models parameters is widely used in the industrial field and was proven to be efficient [6]. This optimization involved for the three models all the coefficients associated to the chemical elements and those ($\alpha$ and $\beta$) associated to strain and strain rate. The GAs search domain for these parameters is in a range between $-30\%$ and $+30\%$ with respect to their original values in order to preserve the original meaning and input/output relations of the standard formulae. The GAs optimization process is set to minimize the average percent error in the MFS prediction through the 7 mill stands.

The results achieved by the improved models are summarized in Table 2 and highlight the great benefit gained through the GAs–driven optimization of the formulae for which the average estimation error is strongly reduced in all the cases. The results put into evidence that the empirical models, despite their high specificity that may limit their performance, can be used in order to obtain optimized versions able to successfully fit different experimental data.

**Table 2.** Performance of empirical models improved by means of GAs ($e\%_{MFS}$).

| Stand | Misaka–GA | Siciliano–GA | Poliak–GA |
|-------|-----------|--------------|-----------|
| 1 | 15.6 | 12.2 | 11.5 |
| 2 | 18.3 | 8.8 | 9.2 |
| 3 | 11.2 | 10.0 | 9.5 |
| 4 | 8.3 | 5.1 | 5.1 |
| 5 | 5.0 | 4.5 | 4.1 |
| 6 | 8.1 | 6.0 | 6.5 |
| 7 | 13.3 | 12.9 | 11.2 |
| Ave. | 11.4 | 8.5 | 8.2 |

# 4    ANNs for Mean Flow Stress Prediction

The noticeable improvement in terms of performance shown by the optimized empirical models encouraged further studies involving more complex and powerful tools, like ANNs, in order to exploit their well acknowledged capabilities of reproducing highly non–linear relationships between inputs and outputs among the data collected on an arbitrary phenomenon. In this section different ANN–based approaches - from the simplest to more complex - are described and tested for the prediction of MFS.

## 4.1   Basic ANN–Based Models

The first approach based on the use of ANNs consists in the straightforward employ of a set of 7 independent Multi–Layer Perceptron (MLP) networks, each one associate to one gage of the finishing mill. Each network has two layers, a number of input neurons equal to the number of input variables and an output neuron. Different values of the number of neurons in the hidden layer $n_h$ have been tested in order to identify the best performing network architecture. In addition, for each network, two variants of the back propagation algorithm have been used as training algorithms: the Levenberg–Marquardt (LM) algorithm and the Bayesian regularization (BR) method.

The set of input variables fed to each network (the same for all the 7 ANNs) has been determined taking into consideration both the experts knowledge and literature survey. The variables mostly influencing MFS according to experts and all those included in Eqs. 1, 2 and 3 have been selected. The final list of variables is composed by: [C], [Mn], [Mo], [Nb], [Ti], T, strain, strain rate. The output for all the networks is the predicted MFS for the associate gage.

A selection of the best results achieved by all the tested network configurations is reported in Table 3, where each column refers to a different configuration of ANN. In header $n_h$ and training method are reported. According to this table, the use of independent ANNs, analogous from the point of view of the input variables included and $n_h$ value, further enhanced the accuracy of MFS prediction, proving its capability to catch the non linear relations among the considered input–output variables.

**Table 3.** Performance ($e\%_{MFS}$) of the set of independent MLP networks.

| Stand | $n_h = 8$, LM | $n_h = 10$, LM | $n_h = 10$ BR |
|-------|------|------|------|
| 1 | 6.5 | 5.6 | 5.2 |
| 2 | 5.4 | 4.6 | 4.4 |
| 3 | 3.3 | 4.0 | 4.1 |
| 4 | 3.5 | 3.1 | 3.6 |
| 5 | 3.8 | 3.9 | 4.0 |
| 6 | 5.4 | 5.3 | 5.7 |
| 7 | 9.2 | 9.5 | 9.2 |
| Ave. | 5.3 | 5.1 | 5.2 |

## 4.2    ANNs Operating on SOM–Clustered Data

Given the interesting performance achieved, the approach described in Sect. 4.1 based on the use of a set of independent ANNs, each one devoted to the estimation of MFS for a particular stand of the mill has been further investigated and enhanced. The idea is to maintain an independent estimator for each stand but, within each of them, to be able to build different ANNs for the different operating conditions of the plant. The main advantage of this approach is the creation of specific models for representing the relation among input and output variables according to various operating conditions that affect such relationship.

For this purpose the available data are clustered by means of the use of a Self Organizing Map (SOM) in order to put into evidence, through the obtained clusters, a set of representative operating conditions. The dataset used for the SOM training includes all the variables related to process conditions (i.e. temperatures, strain and strain rates). Moreover, for the clustering of data, a SOM has been preferred with respect to other methods (i.e. k–means algorithm) in order to obtain a set of clusters that preserves the original data topology and distribution. Once the SOM is trained and data partitions have been created, for each HR mill gage and for each single partition an ANN for MFS prediction for that specific partition and stand is trained. The architecture and the set of input variables are the same as those adopted by the ANNs described in Sect. 4.1. When a new pattern is presented to the estimator, the SOM firstly determines which cluster it belongs to and then activates, for each stand, the corresponding ANN that returns the corresponding estimated MFS. This approach has been tested by varying the number of clusters for data partition $n_c$ and the vale of $n_h$ for the ANNs. The results achieved are summarized, for the best performing configurations, in Table 4, where each column refers to a different configuration of the proposed architecture, described in header, where $n_c$ and $n_h$ for the ANN corresponding to each cluster are reported.

**Table 4.** Performance ($e\%_{MFS}$) of the set of independent MLP networks operating on clustered data

| Stand | $n_c = 3$ Hid: 4, 4, 5 | $n_c = 4$ Hid: 3, 3, 3 | $n_c = 6$ Hid: 3, 3, 2 |
|---|---|---|---|
| 1 | 6.1 | 5.8 | 6.2 |
| 2 | 5.5 | 5.1 | 4.7 |
| 3 | 3.2 | 4.2 | 4.0 |
| 4 | 3.0 | 3.1 | 3.5 |
| 5 | 4.2 | 3.2 | 3.2 |
| 6 | 5.0 | 4.2 | 4.1 |
| 7 | 7.1 | 6.8 | 7.4 |
| Ave. | 4.9 | 4.6 | 4.7 |

Results show that the use of different ANNs in order to estimate the MFS at different gages for various mill operating conditions is fruitful in terms of improvement of estimation accuracy. The percent error reduction is about 0.5%. The adopted data partitioning limited the number of data available for the training of each network and, by consequence, their dimensions in terms of $n_h$. The availability of larger quantities of data will favour the exploitation of more complex structures leading, possibly, to even better performance.

## 4.3    Sequential ANN Predictor

In [10] the authors introduced the use of the so–called sequential MFS predictor in order to improve the estimator accuracy by designing a system able to exploit in an *on–line* manner the information provided by the HR mill sensor through the different phases of the rolling. In [10] in facts it is put into evidence the high correlation between the MFS measured on contiguous stands. In other words the MFS measured for an arbitrary $k^{th}$ stand ($\text{MFS}_k$) of the mill is strongly related to the MFS measured at previous and subsequent ones ($\text{MFS}_{k-1}$ and $\text{MFS}_{k+1}$ respectively). The rolling process is sequential, thus, when the product is going to pass through $k^{th}$ stand, all information concerning the processing at $k - 1^{th}$ stand are available and usable for the prediction of $\text{MFS}_k$.

This concept led to the design of a system directly connected to the HR mill informative system composed by a set of 7 MLP-based ANNs where all networks share the input variables concerning the product properties and general process parameters (the same mentioned in Sect. 4.1) but each arbitrary network $\text{NN}_k$ includes also as input $\text{MFS}_{k-1}$, the MFS measured at previous stand and provided directly by the HR mill control system. In this context, the network devoted to the prediction of first gage MFS does not exploit this latter input variable. The architecture of the sequential model is shown in Fig. 2.

The training of this ANNs set takes places off-line, by using data already available and by adding the respective measured MFS at previous stand to the input variables when training the NN associated to an arbitrary stand. Also in this case different networks configurations in terms of $n_h$ and training algorithms were tested. Table 5 shows the performance of this approach through a selection of the best achieved results. Each column refers to a different configuration of ANN, described in column header, where $n_h$ and training method are reported.

The sequential predictor further lowers the average discrepancy between measured and estimated MFS with respect to the basic predictor composed by 7 independent ANNs. This improvement - apart from the practical advantage of a more reliable estimation of the measure that allows the design of a better mill control - proves the convenience of exploiting the information acquired during the rolling for modelling of the subsequent phases of the process. In this case, among the tested configurations of the system, no one seems to clearly outperform the other ones.

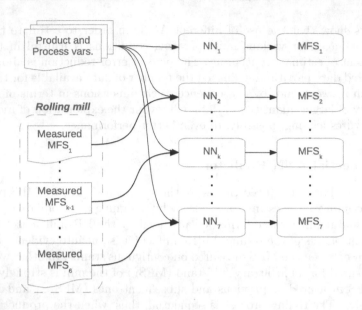

**Fig. 2.** A scheme of the on-line sequential predictor based on ANNs.

**Table 5.** Performance ($e\%_{MFS}$) of the sequential ANN based predictor.

| Stand | $n_h = 7$, LM | $n_h = 8$ BR | $n_h = 10$ BR |
|-------|---------------|--------------|---------------|
| 1 | 4.3 | 4.3 | 4.8 |
| 2 | 4.3 | 4.2 | 3.8 |
| 3 | 3.5 | 3.5 | 3.7 |
| 4 | 3.1 | 3.1 | 3.5 |
| 5 | 3.2 | 3.0 | 3.6 |
| 6 | 4.0 | 4.2 | 5.1 |
| 7 | 8.5 | 8.2 | 8.0 |
| Ave. | 4.4 | 4.4 | 4.6 |

## 4.4   Hybrid Ensemble of ANNs

The problem of MFS estimation has been faced also by means of a novel ensemble approach called *Hybrid Ensemble Method* (HyEM) and proposed in [11]. HyEM is a bagging–inspired ensemble that couples the use of a strong learner (SL) to a set of Weak Learners (WLs). In this work both the SL and all the WLs are two layers MLP networks where $n_h$ is automatically computed on the basis of the available training data. The SL is trained by exploiting all the available training dataset whilst each single WL uses only a subset belonging to a specific region of the input domain determined by means of a SOM–based clustering.

Each learner within the HyEM is coupled to an additional ANN devoted to the estimation of the reliability $R(p)$ of the associated learner when predicting the output for an arbitrary input pattern $p$. The ANN for reliability estimation was introduced and described in detail in [12]. When a new pattern is provided to HyEM, this design enables each learner to provide a candidate output and a measure of its reliability (in terms of error forecast) on its own estimation. In the simulation phase, when a new pattern $p$ is fed, the output of SL and of the WL associated to the domain region the sample belongs to are collected and the HyEM output is *aggregated* on the basis of the relative reliability of the two learners. In [11] the HyEM technique was proven to be extremely efficient on several literature and industrial problems, due to its capability of suitably mixing - through the estimated reliabilities - the robustness of the SL to the specificity of the selected WL, that is able to represent the peculiarities of the input–output relation in the different regions of the domain. In Fig. 3 the simulation procedure of HyEM is graphically depicted. The ANNs devoted to MFS estimation are referred as MainSL and MainWL$_i$, while the associated reliability networks as RelSL and RelWL$_i$. In an analogous manner, their outputs $O$ and $R$ (reliabilities) are specified. All the details on the design of HyEM can be found in [11].

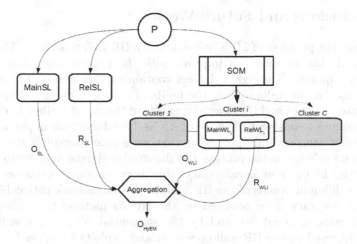

**Fig. 3.** Flow-chart depicting the data flow related to the simulation of the HyEM system when a new pattern $P$ is provided.

In Table 6 the performance of HyEM on MFS prediction are reported. Each column of the table refers to a different architecture of the system characterized by the number of employed WL $N_{WL}$. $N_{WL}$ was varied within the performed tests from a minimum of 4 to a maximum of 12 and Table 6 reports the performance achieved by the best performing ones.

The HyEM approach achieved results comparable to those of the sequential ANNs method. One advantage of HyEM is that, for $N_{WL} = 16$ it was able to keep the accuracy comparable throughout the 7 stands of the finishing mill, which was not achieved by the other approaches due to the criticalities encountered in the

**Table 6.** Performance ($e\%_{MFS}$) of the best performing HyEM configurations.

| Stand | $N_{WL} = 4$ | $N_{WL} = 16$ |
|-------|------------|-------------|
| 1 | 4.9 | 5.0 |
| 2 | 4.2 | 4.7 |
| 3 | 3.8 | 3.9 |
| 4 | 3.3 | 3.7 |
| 5 | 3.9 | 3.7 |
| 6 | 4.4 | 4.4 |
| 7 | 7.5 | 5.1 |
| Ave. | 4.6 | 4.4 |

estimation of last stand MFS. Moreover, due to the limited number of samples composing the training dataset, it was not possible to achieve good results from configurations including higher values of $N_{WL}$ that may have further improved the performance.

## 5    Conclusions and Future Work

In this paper the problem of MFS prediction in a HR mill was faced. The correct estimation of this measure can improve both the process conditions and the final product quality. Through a dataset containing the information related to the production of an Italian plant, the limits of standard empirical approaches were put into evidence and the beneficial effects of the use of artificial intelligence based techniques were shown. The use of ANNs–based methods in particular was analysed by testing several approaches that were able to exploit the available theoretical knowledge on the process and the empirical data in order to simulate efficiently the highly non–linear relationship between input variables and the MFS at the different stands of the HR mill. In this framework noticeable results in terms of accuracy were achieved by the HyEM method that relies on the concept of learners reliability and by the sequential ANNs approach that is directly connected to the HR mill operation and exploits in an on–line manner the data provided by the process itself.

In the future the availability of larger quantity of data will allow a clearer assessment of some of the tested methods. Furthermore, when more plants will be involved in this study, the availability of data coming from different mills will allow the evaluation of the robustness and portability of these approaches.

## References

1. Ginzburg, V.B.: Steel-Rolling Technology: Theory and Practice. Marcel Dekker Inc., New York (1989)
2. Colla, V., Vannucci, M., Valentini, R.: Neural network based prediction of roughing and finishing times in a hot strip mill. Rev. Metal. **46**(1), 15–21 (2010)

3. Dimatteo, A., Vannucci, M., Colla, V.: Prediction of hot deformation resistance during processing of microalloyed steels in plate rolling process. Int. J. Adv. Manuf. Technol. **66**(9–12), 1511–1521 (2013)
4. Kwak, W.J., Kim, Y.H., Park, H.D., Lee, J.H., Hwang, S.M.: FE-based on line model for the prediction of roll force and roll power in hot strip rolling. ISIJ Int. **40**(10), 1013–1018 (2000)
5. Siciliano, F., Leduc, L.L., Hensger, K.: The effect of chemical composition on the hot-deformation resistance during processing of microalloyed steels in thin slab casting/direct rolling process. In: Proceedings of International HSLA 2005 Conference Sanya (CN) (2005)
6. Colla, V., Bioli, G., Vannucci, M.: Model parameters optimisation for an industrial application: a comparison between traditional approaches and genetic algorithms. In: Proceedings - EMS 2008, European Modelling Symposium, 2nd UKSim European Symposium on Computer Modelling and Simulation, pp. 34–39 (2008)
7. Lee, D., Lee, Y.: Application of neural network for improving accuracy of roll force model in hot-rolling mill. Control Eng. Pract. **10**(2), 473–478 (2001)
8. Sungzoon, C., Youngjung, C., Sungchul, Y.: Reliable roll force prediction in cold mill using multiple neural networks. IEEE Trans. Neural Netw. **8**, 874–882 (1997)
9. Di Matteo, A., Vannucci, M., Colla, V.: Prediction of mean flow stress during hot strip rolling using genetic algorithms. ISIJ Int. **54**(1), 171–178 (2014)
10. Vannucci, M., Colla, V., Dimatteo, A.: Improving the estimation of mean flow stress within hot rolling of steel by means of different artificial intelligence techniques. IFAC Proc. Vol. (IFAC-PapersOnLine) **46**(9), 945–950 (2013)
11. Vannucci, M., Colla, V.: Learners reliability estimated through neural networks applied to build a novel hybrid ensemble method. Neural Process. Lett. (accepted for publication)
12. Reyneri, L.M., Colla, V., Sgarbi, M., Vannucci, M.: Self-estimation of data and approximation reliability through neural networks. In: Cabestany, J., Sandoval, F., Prieto, A., Corchado, J.M. (eds.) IWANN 2009. LNCS, vol. 5517, pp. 89–96. Springer, Heidelberg (2009). doi:10.1007/978-3-642-02478-8_12

# Machine Learning for Renewable Energy Applications

# State of Health Estimation of Zinc Air Batteries Using Neural Networks

Andre Loechte[1]([⊠]), Daniel Heming[1], Klaus T. Kallis[2],
and Peter Gloesekoetter[1]

[1] Department of Electrical Engineering and Computer Science,
University of Applied Sciences Muenster, 48565 Steinfurt, Germany
a.loechte@fh-muenster.de
[2] Micro- and Nanotechnologies Group, Faculty of Electrical Engineering and
Information Technology, TU Dortmund University, 44227 Dortmund, Germany

**Abstract.** One major problem of energy storages is degradation. Degradation leads to a loss of capacity and a higher series resistance. One possibility to determine the state of health is the electrochemical impedance spectroscopy. The ac resistance is therefore measured for a set of different frequencies. Previous approaches match the measured impedances with a nonlinear equivalent circuit, which needs a lot of time to solve a nonlinear least squares problem. This paper combines the electrochemical impedance spectroscopy with neural networks to directly model the state of health in order to speed up the estimation. Zinc air batteries are exemplary used as energy storage, as other problems exists, that can be solved by impedance measurements. Optimizing a cost function is used to determine the fastest combination of examined frequencies.

## 1 Introduction

The zinc air battery has a high potential to become one of the leading storage technologies for electrical energy [1]. A big advantage and the main reason of the high energy density is the fact that a zinc air battery only needs one active element in the battery housing (zinc) [2]. The second active element (oxygen) is taken from the ambient air. In addition to the high energy density, zinc air technology has further advantages. For example the low costs. Zinc is the 24th most frequent element. There is more zinc available than copper, plumb and lithium [3]. Furthermore a zinc air battery contains no toxic materials. On the other hand, there are also some disadvantages. Currently the efficiency is about 60% compared to 95% of Li Ion batteries [4]. Another technological disadvantage: zinc air cells can only be used in areas with oxygen.

Two important parameters that need to be known, when a battery is used in a system, are the state of charge and the state of health of the battery. While the state of charge (SoC) represents the remaining capacity of a battery [5], the state of health (SoH) provides information about the degradation of a battery. One method, to measure the state of health is the electrochemical impedance spectroscopy [6]. The ac resistance is therefore measured for a spectrum of different frequencies. Among other things the resulting impedance characteristics

© Springer International Publishing AG 2017
I. Rojas et al. (Eds.): IWANN 2017, Part I, LNCS 10305, pp. 641–647, 2017.
DOI: 10.1007/978-3-319-59153-7_55

differs depending on the grade of aging. This paper combines the electrochemical impedance spectroscopy with a neural network to directly estimate the state of health.

## 2   Problem

There are already publications analyzing the impedance spectra of zinc air batteries. For example Arai et al. examined the impedance spectra of the cathodic part of zinc air batteries by matching the parameter values of an equivalent circuit. This circuit, shown in Fig. 1, contains 2 nonlinear nernst diffusion components, which depend on two parameters diffusion resistance $R_d$ and diffusion factor $K$. The results of the sequence are shown in Table 1. Here an anodic treatment of 6 h correspond to an SoH of 0%. As one can see the values correlate with the state of health.

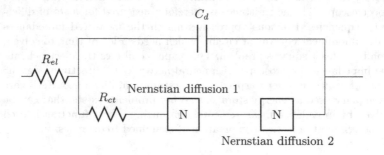

**Fig. 1.** Equivalent circuit of zinc-air cathode

**Table 1.** Parameters depending on SoH

| SoH[%] | $R_{el}[\Omega]$ | $C_d[\text{mF}]$ | $R_{ct}[\Omega]$ | $R_{d1}[\Omega]$ | $K_1[\text{s}^{-1}]$ | $R_{d2}[\Omega]$ | $K_2[\text{s}^{-1}]$ |
|---|---|---|---|---|---|---|---|
| 100 | 0.15 | 2.0 | 0.02 | 0.14 | 20 | 0.25 | 98 |
| 50 | 0.20 | 1.8 | 0.02 | 0.22 | 30 | 0.29 | 110 |
| 0 | 0.25 | 1.5 | 0.03 | 0.25 | 9.1 | 0.55 | 140 |

Examining the time needed for an estimation of the SoH, there are multiple factors to consider. On the one hand there is the measuring time, which does not depend on the evaluation algorithm, but on the chosen frequencies. Lower frequencies have a longer period and need more time to measure.

On the other hand there are factors, that depend on the method. Typically measured impedances are used to calculate the parameters of an equivalent circuit. Because of the nonlinear nernst diffusion components, the parameter values cannot be calculated by solving a linear system of equations, but through a nonlinear least squares algorithm. A Newton-Raphson-like algorithm has to be used in order to fit the values iteratively. This step is obligatory for each evaluation of a new measurement and takes a lot of time and computing capacity. Furthermore this way of evaluation needs to measure at least as much frequencies as existing parameters [9]. This might increases the measuring time, if it would be possible to estimate the SoH with fewer measurements.

There is another step necessary after identifying the parameters. A function, that relate the parameters to the state of health, has to be found.

## 3    Setup

Both parts, the nonlinear least square algorithm and the evaluation of the correlation function results in a long evaluation time that can be solved by using neural networks. While the training of the network still takes some time, the evaluation is fast. Moreover the network can directly model the state of health without calculating the equivalent circuit. In order to check, whether one frequency is enough to determine the SoH, a simple one layer feed forward network ios used. The network is trained by a Levenberg-Marquardt backpropagation algorithm. The equivalent circuit is used as a source for generating training data. As one can see in Table 1, only 3 measurements were taken. In order to get enough training samples, we used a linear interpolation to approximate values in between. This results in the plot of Fig. 2. The figure shows the parameter values, normalized to its starting values. While the resistance of the resistors increases with further degradation, the diffusion capacity sinks.

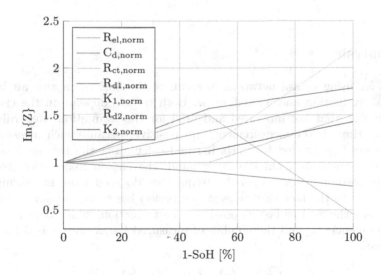

**Fig. 2.** Normalized parameter values depending on SoH

The impedance spectra is calculated for each per mill of aging. Figure 3 shows some sample characteristics in a nyquist plot. Higher degradation results in semicircles with a bigger radius and a light displacement to the right. The datapoints used for the analysis are chosen by its frequencies using 2 values per decade from 100 mHz to 500 Hz. They are combined with normal distributed noise and shown as circles.

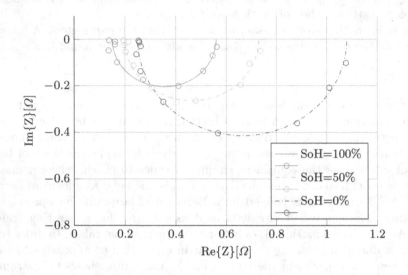

**Fig. 3.** Nyquist plot of impedance depending on SoH

## 4 Analysis

Reasons for using neural networks to estimate the SoH of an zinc air battery are a fast evaluation and a small error. Both targets depend on the choice of frequencies $\omega$, that are measured and used for the estimation. Regarding the evaluation time, fewer measurements are positive, because each measurement takes a time $t_{init}$ needed for initiating and calculating the real an imaginary part of the impedance. On the other hand, the chosen frequencies itself are also an important factor, as smaller frequencies do need more measuring time $t_{measure}$, due to the fact that at least one period has to be measured. In order to determine the SoH as fast as possible a cost function, which depends on the chosen frequencies $\underline{\omega}$ and the number of measured frequencies $n$ is defined and optimized:

$$t = t_{init}(n) + t_{measure}(\underline{\omega}, n).$$

A set-up and evaluation time of 120 ms is used for $t_{init}$:

$$t_{init}(\underline{\omega}) = 120 \text{ ms} \cdot n$$

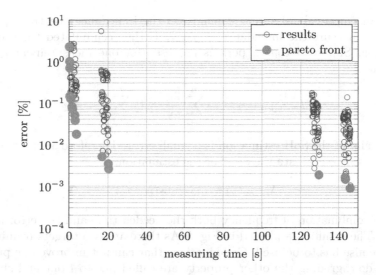

**Fig. 4.** Pareto frontier combining measuring time and error of estimation. (Color figure online)

**Table 2.** Frequency combinations of Pareto front

| $f_1$ [Hz] | $f_2$ [Hz] | $f_3$ [Hz] | $f_4$ [Hz] | $f_5$ [Hz] | $f_6$ [Hz] | $f_7$ [Hz] | $t$ [s] | $\lambda_{error}$ [%] |
|---|---|---|---|---|---|---|---|---|
| 1000 | - | - | - | - | - | - | 0.133 | 2.26 |
| 129 | 359 | - | - | - | - | - | 0.372 | 0.14 |
| 129 | 1000 | - | - | - | - | - | 0.35 | 0.69 |
| 359 | 1000 | - | - | - | - | - | 0.288 | 0.99 |
| 46 | 129 | 359 | 1000 | - | - | - | 0.896 | 0.12 |
| 17 | 46 | 359 | 1000 | - | - | - | 1.55 | 0.08 |
| 17 | 46 | 129 | 1000 | - | - | - | 1.61 | 0.075 |
| 6 | 46 | 129 | 1000 | - | - | - | 2.96 | 0.05 |
| 0.77 | 129 | 359 | 1000 | - | - | - | 16.8 | 0.0051 |
| 17 | 46 | 129 | 359 | 1000 | - | - | 1.77 | 0.07 |
| 6 | 46 | 129 | 359 | 1000 | - | - | 3.11 | 0.037 |
| 6 | 16 | 46 | 359 | 1000 | - | - | 3.77 | 0.017 |
| 0.1 | 0.77 | 17 | 129 | 359 | - | - | 143.4 | 0.0016 |
| 0.77 | 6 | 17 | 46 | 129 | 1000 | - | 20.2 | 0.0034 |
| 0.1 | 6 | 17 | 46 | 129 | 1000 | - | 129.6 | 0.0018 |
| 0.1 | 0.77 | 17 | 129 | 359 | 1000 | - | 143.5 | 0.0014 |
| 0.77 | 6 | 17 | 46 | 129 | 359 | 1000 | 20.3 | 0.0026 |
| 0.1 | 0.77 | 6 | 17 | 46 | 129 | 1000 | 146 | 0.00088 |

This is an empirical value and depends on the impedance spectroscope, that is used to measure the impedance. $t_{measure}$ is the accumulated time for data acquisition. Supposed that 2 periods are used for one measurement $t_{measure}$ results in

$$t_{measure}(\underline{\omega}) = 2 \cdot \sum_{i=1}^{n} \frac{2\pi}{\omega_i} = 4\pi \sum_{i=1}^{n} \frac{1}{\omega_i}.$$

Another aim of the SoH estimation is a small error. Therefore another function $\lambda_{error}$ is used to evaluate the error of the estimation:

$$\lambda_{error}(y) = \max(\text{abs}(y - t))$$

For each combination of frequency both the needed time and the error are calculated. The results are plotted in Fig. 4. As there are two targets to minimize, a compromise has to be used. Combinations that can not improve one property without downgrading the other property are called pareto front and shown in a different color [10]. In order to find a good compromises these pareto optimal combinations are presented in Table 2.

As one can see, using all the frequencies results in the best accuracy, but needs the longest measurement time of all (146 s). The fastest results can be received by using only the highest frequency, which is 1 kHz. The resulting accuracy is 2.26 %. The fastest way to estimate the SoH with a error of 1%, is to use a combination of the frequencies 359 Hz and 1 kHz. This results in a total estimation time of 288 ms.

## 5    Conclusion

This paper showed that a simple neural network is a precise way to determine the state of health of a zinc air battery. There are fewer impedance measurements necessary in contrast of matching parameters of an equivalent circuit which results in a quicker measurement time. The optimization of a cost function leads an optimal combination of frequencies to measure. Further speed up is reached for evaluating the measurements, because instead of a slow nonlinear least squares problem, only a neural network has to be processed.

The combination of electrochemical impedance spectroscopy and neural networks might be used for other problems, too. As the voltage curve of zinc air batteries is very flat, one major problem is to determine the state of charge [2].

## References

1. Pei, P., Wang, K., Ma, Z.: Technologies for extending zinc-air battery's cyclelife: a review. Appl. Energy **128**, 315–324 (2014)
2. Energyzer: Zinc-Air application manual, 12 February 2017. http://data.energizer. com/

3. Greenwood, N.N., Earnshaw, A.: Chemie der Elemente. 1. Auflage, Weinheim (1988). S. 1545
4. Linden, D., Reddy, T.B.: Handbook of Batteries. 3rd edn. McGraw-Hill, New York (2002). Chap. 13/38
5. Chang, W.-Y.: The state of charge estimating methods for battery: a review. ISRN Appl. Math. **2013** (2013). Article id 953792
6. Rezvanizaniani, S.M., Lee, J., Liu, Z., Che, Y.: Review and recent advances in battery health monitoring and prognostics technologies for electric vehicle (EV) safety and mobility. J. Power Sources **256**, 110–124 (2014)
7. Willert-Porada, M., Schmid, M.: Zink-Luft-Batterien für stationäre Energiespeicher, (2013). http://www.lswv.uni-bayreuth.de/de/projekte/aktuell/ziba/index.html
8. Arai, H., Müller, S.: AC impedance analysis of bifunctional air electrodes for metal-air-batteries. J. Electrochem. Soc. **147**, 3584–3591 (2000)
9. Geladi, P., Kowalski, B.R.: Partial least-squares regression - a tutorial. Anal. Chemica Acta **185**, 1–17 (1986)
10. Steuer, R.E.: Multiple Criteria Optimization: Theory Computations and Application. Wiley, Hoboken (1986)

# Bayesian Optimization of a Hybrid Prediction System for Optimal Wave Energy Estimation Problems

Laura Cornejo-Bueno[1], Eduardo C. Garrido-Merchán[2],
Daniel Hernández-Lobato[2], and Sancho Salcedo-Sanz[1(✉)]

[1] Department of Signal Processing and Communications,
Universidad de Alcalá, Alcalá de Henares, Spain
sancho.salcedo@uah.es
[2] Department of Computer Science,
Universidad Autónoma de Madrid, Madrid, Spain

**Abstract.** In the last years, Bayesian optimization (BO) has emerged as a practical tool for high-quality parameter selection in prediction systems. BO methods are useful for optimizing black-box objective functions that either lack an analytical expression, or are very expensive to evaluate. In this paper we show how BO can be used to obtain optimal parameters of a prediction system for a problem of wave energy flux prediction. Specifically, we propose the Bayesian optimization of a hybrid Grouping Genetic Algorithm with an Extreme Learning Machine (GGA-ELM) approach. The system uses data from neighbor stations (usually buoys) in order to predict the wave energy at a goal marine energy facility. The proposed BO methodology has been tested in a real problem involving buoys data in the Western coast of the USA, improving the performance of the GGA-ELM without a BO approach.

**Keywords:** Sea waves energy · Prediction system · Bayesian optimization

## 1 Introduction

Marine Energy is currently one of the most promising sources of renewable energy, due to the huge energy potential of oceans [1], and the well-known benefits of renewable energy resources, such as very low $CO_2$ generation and reduction of oil imports and dependence. However this, the use of marine energy sources is nowadays still minor at a global level, playing only a major role in several offshore islands [2].

There are different technologies within marine energy resources, including ocean wave, tidal and ocean thermal. In this work we focus on wave energy, that uses Wave Energy Converters (WECs) to convert ocean energy into electricity [3]. WECs transform the kinetic energy of wind-generated waves into electricity, by means of either the vertical oscillation of waves or the linear motion of waves,

© Springer International Publishing AG 2017
I. Rojas et al. (Eds.): IWANN 2017, Part I, LNCS 10305, pp. 648–660, 2017.
DOI: 10.1007/978-3-319-59153-7_56

and exhibit some important advantages when compared to alternatives based on tidal converters [4]. On the other hand, wave energy is more difficult to characterize than tidal or ocean thermal, since waves have a highly stochastic nature. As a consequence of this, prediction systems must be used to obtain a proper characterization of waves to mitigate the negative effects of stochastic generation inherent to this technology. In marine wave energy, the two most important wave parameters to be predicted are the significant wave height ($H_{m_0}$) and the wave energy flux ($P$), which characterize the wave energy production from WECs facilities.

The research work on wave energy prediction systems has been intense in the last years, with special incidence in machine learning approaches. One of the first works on this topic was the direct prediction of $H_{m_0}$ using artificial neural networks in [5]. Neural networks have also been applied to other problems of $H_{m_0}$ and $P$ prediction, such as [6], where $H_{m_0}$ and $P$ are inferred from observed wave records using time series neural networks. In [7] a neural network is applied to estimate the wave energy resource in the northern coast of Spain. In [8] a hybrid genetic algorithm-adaptive network-based fuzzy inference system model was developed to forecast $H_{m_0}$ and the peak spectral period at Lake Michigan. In [9,10] different hybrid algorithms mixed with an Extreme Learning Machine neural network were proposed for the estimation of $H_{m_0}$ and $P$, in the context of marine energy applications. Support Vector regression (SVR) has also been applied to marine energy related problems such as in [11]. Similarly, [12,13] proposed to feed SVR approaches with information from radar sources in order to obtain an accurate prediction of $H_{m_0}$. Classification approaches have been applied in [14] to analyze and predict $H_{m_0}$ and $P$ ranges in buoys for marine energy applications. In [15], use of genetic programming for $H_{m_0}$ reconstruction problems was proposed. Finally, in [16] fuzzy logic-based approaches were introduced for $H_{m_0}$ prediction problems.

In this paper we test a BO methodology to improve the performance of a hybrid prediction system for wave energy flux prediction. Specifically, the prediction system was previously presented in [10], and it is formed by a Grouping Genetic Algorithm for feature selection, and an Extreme Learning Machine for carrying out the final energy flux prediction. This hybrid prediction system has a number of parameters that may affect its final performance, and need to be previously specified by the practitioner. Traditionally, these parameters have been manually tuned by a human expert, with experience in both the algorithm and the problem domain. However, it is possible to obtain better results by an automatic fine tuning of the prediction system's parameters. In this case, the parameters of GGA-ELM approach include the probability of mutation in the GGA or the number of neurons in the ELM hidden layer, among others. We propose then to use a Bayesian Optimization (BO) approach to automatically optimize the parameters of the whole prediction system (GGA-ELM), with the aim of improving its performance in wave energy prediction problems. BO has been shown to obtain good results in the task of obtaining good parameter values for prediction systems [22]. In the paper we detail the basic prediction system considered and the BO methodology implemented, along with the improvements obtained in a real problem of wave energy flux prediction in the Western coast of the USA.

# 2    The Hybrid Prediction System Considered

In this paper we will optimize a hybrid prediction system for marine energy applications described in [13]. In this section we describe the main characteristics of this approach, in order to better explain later on the Bayesian optimization carried out on it. The prediction system is a hybrid wrapper approach, formed by a Grouping Genetic Algorithm for feature selection, and an Extreme Learning Machine to carry out the final prediction of $H_{m_0}$ or $P$ from a set of input data.

## 2.1    The Grouping Genetic Algorithm

The grouping genetic algorithm (GGA) is a class of evolutionary algorithm especially modified to tackle grouping problems, i.e., problems in which a number of items must be assigned to a set of predefined groups. It was first proposed by Falkenauer [18]. In the proposed GGA, the encoding, crossover and mutation operators of traditional GAs have been modified to improve the performance of the algorithm in grouping problems. The GGA initially proposed by Falkenauer is a variable-length genetic algorithm. The encoding is carried out by separating each individual in the algorithm into two different parts: the first one is an *assignment* part, which associates each item to a given group. The second one is a *group* part, which defines the number of groups into which the items must be shared. In problems where the number of groups is not previously defined, it is easy to see that the group part varies from one individual to another, so it leads to a variable-length encoding. In the considered hybrid approach for $H_{m_0}$ or $P$ prediction, the GGA's objective is the selection of the best set of features in terms of prediction accuracy. Thus, an individual in the GGA (**c**) has the form $\mathbf{c} = [\mathbf{a}|\mathbf{g}]$. An example of an individual in the proposed GGA for a feature selection problem, with 20 features and 4 groups, is the following:

1 1 2 3 1 4 1 4 3 4 4 1 2 4 4 2 3 1 3 2 | 1 2 3 4

where the group 1 includes features $\{1, 2, 5, 7, 12, 18\}$, group 2 features $\{3, 13, 16, 20\}$, group 3 features $\{4, 9, 17, 19\}$ and finally group 4 includes features $\{6, 8, 10, 11, 14, 15\}$. The dynamics of the algorithm consists of a initialization of the population, at random, and the application of specially tailored operators to come up with the encoding previously described. Specifically, the crossover operator is modified from the standard genetic algorithm (see [13] for details), as follows:

1. Choose two parents from the current population, at random.
2. Randomly select two points for the crossover, from the "Groups" part of parent 1. Then, all the groups between the two cross-points are selected.
3. Insert the selected section of the "Groups" part into the second parent.

4. Modify the "Groups" part of the offspring individual with their corresponding number. Modify also the assignment part accordingly.
5. Remove any empty groups in the offspring individual. In the example considered, it is found that groups 1, 2, 3, and 6 are empty, so we can eliminate these groups' identification number and rearrange the rest. The final offspring is then obtained.

Regarding mutation operator, note that standard mutation usually calls for an alteration of a small percentage of randomly selected parts of the individuals. This type of mutation may be too disruptive in the case of a grouping problem. In our case, a swapping mutation in which two items are interchanged (swapping this way the assignment of features to different groups), is taken into account. This procedure is carried out with a very low probability ($P_m = 0.01$), to avoid increasing of the random search in the process. Finally, a classical tournament selection [19] is applied in order to keep the best individuals in the population as parents of the next generation. This selection mechanism is carried out in terms of a given objective function (fitness value), which in this case is the prediction accuracy obtained by an Extreme Learning Machine neural network.

## 2.2   Fitness Function: The Extreme Learning Machine

An ELM [20] is a fast learning method based on the structure of MLPs with a novel way of training feed-forward neural networks. One of the most important characteristics of the ELM training is the randomness in the process where the network weights are set, obtaining, in this way, a pseudo-inverse of the hidden-layer output matrix. The simplicity of this technique makes the training algorithm extremely fast. Moreover, it must be remarkable its outstanding performance when it is compared to other learning methods, usually better than other established approaches such as classical MLPs or SVRs.

The ELM algorithm can be explained as follows: given a training set

$$\mathbb{T} = (\mathbf{x}_i, \boldsymbol{W}_i) | \mathbf{x}_i \in \mathbb{R}^n, \boldsymbol{W}_i \in \mathbb{R}, i = 1, \cdots, l,$$

an activation function $g(x)$ and number of hidden nodes ($\tilde{N}$),

1. Randomly assign inputs weights $\mathbf{w}_i$ and bias $b_i$, $i = 1, \cdots, \tilde{N}$.
2. Calculate the hidden layer output matrix $\mathbf{H}$, defined as

$$\mathbf{H} = \begin{bmatrix} g(\mathbf{w}_1 \mathbf{x}_1 + b_1) & \cdots & g(\mathbf{w}_{\tilde{N}} \mathbf{x}_1 + b_{\tilde{N}}) \\ \vdots & \cdots & \vdots \\ g(\mathbf{w}_1 \mathbf{x}_l + b_1) & \cdots & g(\mathbf{w}_{\tilde{N}} \mathbf{x}_N + b_{\tilde{N}}) \end{bmatrix}_{l \times N} \quad (1)$$

3. Calculate the output weight vector $\boldsymbol{\beta}$ as

$$\boldsymbol{\beta} = \mathbf{H}^{\dagger} \mathbf{T}, \quad (2)$$

where $\mathbf{H}^\dagger$ stands for the Moore-Penrose inverse of matrix $\mathbf{H}$ [20], and $\mathbf{T}$ is the training output vector, $\mathbf{T} = [\boldsymbol{W}_1, \cdots, \boldsymbol{W}_l]^T$.

The number of hidden nodes ($\tilde{N}$) is a free parameter of the ELM training, and it can be fixed initially, or in a best convenient way, it must be estimated for obtaining good results as a part of a validation set in the learning process. Hence, scanning a range of $\tilde{N}$ values is the solution for this problem.

The Matlab ELM implementation by G.B. Huang, freely available in the Internet [21] has been used in this paper.

## 3   Bayesian Optimization of the Prediction System

Most prediction systems are governed by a set of parameters that control their behavior. For example, in an ELM the number of units in the hidden layer has to be chosen before training; in a SVR system the kernel length-scale and the width of the epsilon insensitive band need to be specified; and in the genetic algorithm described in Sect. 2.1, the probability of mutation and the number of epochs must be chosen initially. Moreover, the final performance of the system may strongly depend on the parameters chosen. Parameter tuning has typically been addressed by human experts, or by a grid search that aims at minimizing the prediction error on a validation set. These solutions have the disadvantage that they may suffer from human bias, and that they do not scale well with dimensionality of the parameter space, respectively.

Bayesian optimization (BO) has emerged as practical tool for parameter selection in prediction systems [22]. More precisely, BO methods are very useful for optimizing black-box objective functions that lack an analytical expression (which means no gradient information), are very expensive to evaluate, and in which the evaluations are potentially noisy [23–25]. The performance of a prediction system on a randomly chosen validation set, when seen as a function of the chosen parameters, has all these characteristics.

Consider a black-box objective $f(\cdot)$ with noisy evaluations of the form $y_i = f(\mathbf{x}_i) + \epsilon_i$, with $\epsilon_i$ some noise term. BO methods are very successful at reducing the number of evaluations of the objective function needed to solve the optimization problem. At each iteration $t = 1, 2, 3, \ldots$ of the optimization process these methods fit a probabilistic model, typically a Gaussian process (GP) [26], to the observations of objective function $\{y_i\}_{i=1}^{t-1}$ collected so far. The uncertainty about the objective function provided by the GP is then used to generate an acquisition function $\alpha(\cdot)$, whose value at each input location indicates the expected utility of evaluating $f(\cdot)$ there. The next point $\mathbf{x}_t$ at which to evaluate the objective $f(\cdot)$ is the one that maximizes $\alpha(\cdot)$. This process is repeated until enough data about the objective has been collected. When this is the case, the GP predictive mean for $f(\cdot)$ can be optimized to find the solution of the optimization problem. This process is described in Algorithm 1.

**for** $t = 1, 2, 3, \ldots, max\_steps$ **do**

  **1:** Find the next point to evaluate by optimizing the acquisition function: $\mathbf{x}_t = \arg\max_{\mathbf{x}} \alpha(\mathbf{x}|\mathcal{D}_{1:t-1})$.

  **2:** Evaluate the black-box objective $f(\cdot)$ at $\mathbf{x}_t$: $y_t = f(\mathbf{x}_t) + \epsilon_t$.

  **3:** Augment the observed data $\mathcal{D}_{1:t} = \mathcal{D}_{1:t-1} \bigcup \{\mathbf{x}_t, y_t\}$.

  **4:** Update the Gaussian process model using $\mathcal{D}_{1:t}$.

**end**

**Result:** Optimize the mean of the Gaussian process to find the solution.

**Algorithm 1.** Bayesian optimization of a black-box objective function.

The key for BO success is that evaluating the acquisition function $\alpha(\cdot)$ is very cheap compared to the evaluation of the actual objective $f(\cdot)$, which in our case requires the re-training of the prediction system. This is so because the acquisition function only depends on the GP predictive distribution for $f(\cdot)$ at a candidate point $\mathbf{x}$. This distribution is a Gaussian characterized by a mean $\mu(\mathbf{x})$ and a variance $\sigma^2(\mathbf{x})$. The GP model provides closed form expressions for these values given the observed data $\mathcal{D}_i = \{(\mathbf{x}_i, y_i)\}_{i=1}^{t-1}$. Thus, the acquisition function can be maximized with little cost. The consequence is that BO methods can employ the acquisition function $\alpha(\cdot)$ to make intelligent decisions about where to evaluate next the objective $f(\cdot)$ with the aim of finding its optimum as quickly as possible. When the actual objective is very expensive, the approach described can save a lot of computational time. Three steps of the BO optimization process are illustrated graphically in Fig. 1 for a toy minimization problem.

An example of an acquisition function is expected improvement (EI) [27]. EI is obtained as the expected value under the GP predictive distribution for $y_i$, of the utility function $u(y_i) = \max(0, \nu - y_i)$, where $\nu = \min(\{y_i\}_{i=1}^{t-1})$ is the best value observed so far. That is, EI measures on average how much we will improve on the current best solution by evaluating the objective at each candidate point. An advantage of EI is that the corresponding acquisition function $\alpha(\cdot)$ can be computed analytically: $\alpha(\mathbf{x}) = \sigma(\mathbf{x})(\gamma(\mathbf{x})\Phi(\gamma(\mathbf{x}) + \phi(\gamma(\mathbf{x}))$, where $\gamma(\mathbf{x}) = (\nu - \mu(\mathbf{x}))/\sigma(\mathbf{x})$ and $\Phi(\cdot)$ and $\phi(\cdot)$ are respectively the c.d.f. and p.d.f. of a standard Gaussian. EI is the acquisition function displayed in Fig. 1.

BO has been recently applied with success in different prediction systems for finding good parameter values. For example, it has been used to find the parameters of topic models based on latent Dirichlet allocation, support vector machines, or deep convolutional neural networks [22]. Finally, BO has been implemented in different software packages. An implementation in python is called Spearmint and is available at https://github.com/HIPS/Spearmint.

## 4   Experiments and Results

This section presents the experiments carried out in order to show the improvement of performance in the system when it is optimized with the Bayesian techniques shown above. We consider a real problem of wave energy flux prediction

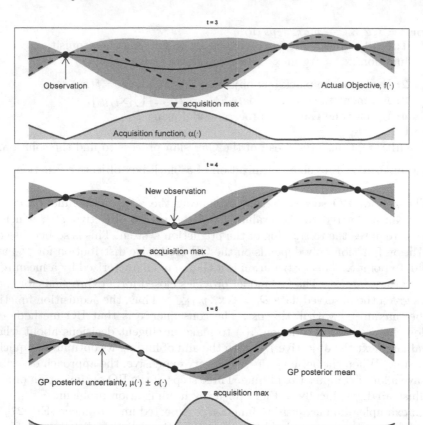

**Fig. 1.** An example of BO on a toy 1D noiseless problem. The figures show a GP estimation of the objective $f(\cdot)$ over three iterations. The acquisition function is shown in the lower part of the plot. The acquisition is high where the GP predicts a low objective and where the uncertainty is high. Those regions in which it is unlikely to find the global minimum of $f(\cdot)$ have low acquisition values and will not be explored.

$(P = 0.49 \cdot H_s^2 \cdot T_e$ kW/m, [17]) from marine buoys. Figure 2 shows the three buoys considered in this study at the Western coast of the USA, whose data bases are obtained from [28]. The objective of the problem is to carry out the reconstruction of buoy 46069 from a number of predictive variables from the other two buoys. Thus, 10 predictive variables measured at each neighbor buoy are considered (a total of 20 predictive variables to carry out the reconstruction). Table 1 shows details of the predictive variables for this problem. Data for two complete years (1st January 2009 to 31st December 2010) are used, since complete data (without missing values in predictive and objective $P$) are available for that period in the three buoys. These data are divided into training set (year 2009) and test set (year 2010) to evaluate the performance of the proposed algorithm.

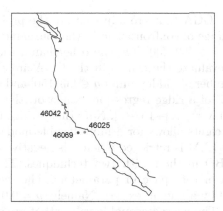

**Fig. 2.** Western USA Buoys considered in this study. In red buoy where the $P$ prediction is carried out from data at blue ones.

**Table 1.** Predictive variables used in the experiments.

| Acronym | Predictive variable | units |
|---------|---------------------|-------|
| WDIR | Wind direction | [degrees] |
| WSPD | Wind speed | [m/s] |
| GST | Gust speed | [m/s] |
| WVHT | Significant wave height | [m] |
| DPD | Dominant wave period | [sec] |
| APD | Average period | [sec] |
| MWD | Direction DPD | [degrees] |
| PRES | Atmospheric pressure | [hPa] |
| ATMP | Air temperature | [Celsius] |
| WTMP | water temperature | [Celsius] |

We have divided this experimental section into two different subsections. First, we show the performance of the BO techniques proposed in the optimization of the specific GGA-ELM prediction algorithm. Second, we will show how the prediction performance is improved when the system is run with the parameters obtained by the BO techniques, i.e. by comparing the performance of the system before and after tuning the parameters with BO.

### 4.1 Bayesian Optimization of the Wave Energy Prediction System Parameters

We evaluate the utility of the BO techniques described in Sect. 3 for finding good parameters for the prediction system described in Sect. 2. More precisely, we try to find the parameters that minimize the RMSE of the best individual found by the GGA on a validation set that contains 33% of the total data available.

The parameters of the GGA that are adjusted are the probability of mutation $p \in [0, 0.3]$, the percentage of confrontation in the tournament $q \in [0.5, 1.0]$, and the number of epochs $e \in [50, 200]$. On the other hand, the parameters of the ELM that is used to evaluate the fitness in the GGA are also adjusted. These parameters are the number of hidden units $n \in [50, 150]$ and the logarithm of the regularization constant of a ridge regression estimator, that is used to find the weights of the output layer $\gamma \in [-15, -3]$. Note that a ridge regression estimator for the output layer weights allows for a more flexible model than the standard ELM, as the standard ELM is retrieved when $\gamma$ is negative and large [29].

We compare the BO method with two techniques. The first technique is a random exploration of the space of parameters. The second technique is a configuration specified by a human expert. Namely, $p = 0.02$, $q = 0.8$, $e = 200$, $n = 150$ and $\gamma = -10$. These are reasonable values that are expected to perform well in the specific application tackled. We set our computational budget to 50 different parameter evaluations for both the BO and the random exploration strategy. After each evaluation, we report the performance of the best solution found. The experiments are repeated for 50 different random seeds and we report average results. All BO experiments are carried out using the acquisition function EI and the software for BO Spearmint.

Figure 3 shows the average results obtained and the corresponding error bars. This figure shows the average RMSE of each method (BO and random exploration) on the validation set as a function of the number of configurations evaluated. The performance of the configuration specified by a human expert is also shown. We observe that the BO strategy performs best. After a few evaluations, it is able to outperform the results of the human expert and it provides results that are similar or better than the ones obtained by the random exploration strategy with a smaller number of evaluations.

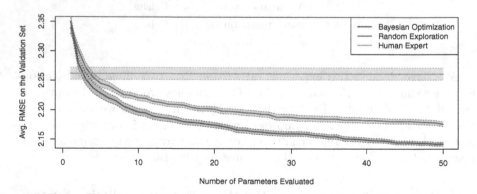

**Fig. 3.** Average results obtained after evaluating the performance of 50 different parameters for the BO technique and a random exploration of the parameter space. The performance of the configuration specified by a human expert is also shown for comparison.

## 4.2  Estimation of the Generalization Performance

In a second round of experiments, we show the performance of the proposed prediction system after its optimization with the BO methodology. Note that after the feature selection process with the GGA-ELM approach, we apply a ELM or a SVR [30,31] to obtain a final prediction of the wave energy flux $P$. Table 2 shows the results obtained for the experiments carried out. We can observe the comparison between ELM and SVR approaches in different scenarios: the prediction obtained with all the features, the prediction obtained with the hybrid algorithm GGA-ELM (without BO methodology), and finally the prediction acquired after the application of the BO process in the GGA-ELM approach. As Table 2 summarizes, we can see how the hybrid GGA-ELM algorithm improves the results obtained by the ELM and SVR approaches (without feature selection). In fact,

**Table 2.** Comparative results of the $P$ estimation by the ELM and SVR approaches after the feature selection by the GGA-ELM in 2010.

| Experiments | RMSE | MAE | $r^2$ |
|---|---|---|---|
| All features-ELM | 3.4183 kW/m | 2.4265 kW/m | 0.6243 |
| All features-SVR | 4.4419 kW/m | 2.8993 kW/m | 0.3129 |
| GGA-ELM-ELM | 2.8739 kW/m | 1.8715 kW/m | 0.7101 |
| GGA-ELM-SVR | 2.6626 kW/m | 1.6941 kW/m | 0.7548 |
| BO-GGA-ELM-ELM | 2.5672 kW/m | 1.7596 kW/m | 0.7722 |
| BO-GGA-ELM-SVR | 2.4892 kW/m | 1.6589 kW/m | 0.7823 |

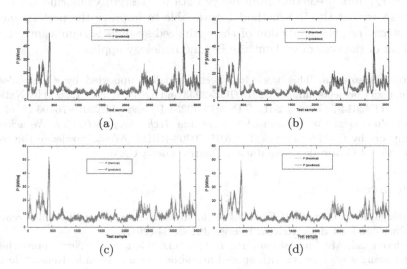

(a)                                    (b)

(c)                                    (d)

**Fig. 4.** $P$ prediction after the feature selection process with the GGA-ELM approach; (a) ELM; (b) SVR; (c) ELM with Bayesian optimization; (d) SVR with Bayesian optimization.

the SVR algorithm improves the most the values of the Pearson's Correlation Coefficient ($r^2$) around 75% in the case of the feature selection method, against the poor 31% when all features are used. Moreover, these results are improved by means of the BO methodology, using ELM and SVR approaches after the GGA-ELM. In the case of the ELM, we get values of the $r^2$ around 77% against the 71% achieved with the GGA-ELM algorithm without the BO improvement. The same behaviour is obtained for the SVR algorithm: we have values around 78% with the application of the BO methodology against the 75% obtained for the GGA-ELM approach when the parameters are fixed by the programmer.

The results of the previous tables can be better visualized in the following graphics. In Fig. 4 the temporary predictions carried out by the ELM and SVR approaches are shown. We can see how the cases (c) and (d) improve the approximation to the real values against the cases (a) and (b) where the BO methodology is not applied.

# 5 Conclusions

In this paper we have shown how a hybrid prediction system for wave energy prediction can be improved by means of Bayesian Optimization (BO) methodology. The prediction system is formed by a grouping genetic algorithm for feature selection, and an Extreme Learning Machine for effective prediction of the target variable, the wave energy flux in this case. After this feature selection process, the final prediction of the wave energy flux is obtained by means of an ELM or a SVR approach. The paper describes in detail the BO methodology, and its specific application in the optimization of the GGA-ELM for a real problem of wave energy flux prediction from buoys data in Western California USA. The results show that the BO methodology is able to improve the performance of the systems, i.e., the prediction of the optimized systems is significantly better than that of the system without the BO methodology applied.

**Acknowledgements.** This work has been partially supported by *Comunidad de Madrid*, under projects number S2013/ICE-2933 and S2013/ICE-2845, and by National projects TIN2014-54583-C2-2-R, TIN2013-42351-P and TIN2016-76406-P of the *Spanish Ministerial Commission of Science and Technology (MICYT)*. We acknowledge support by DAMA network TIN2015-70308-REDT. We acknowledge the use of the facilities of Centro de Computación Científica de la UAM.

# References

1. Arinaga, R.A., Cheung, K.F.: Atlas of global wave energy from 10 years of reanalysis and hindcast data. Renew. Energy **39**, 49–64 (2012)
2. Fadaeenejad, M., Shamsipour, R., Rokni, S.D., Gomes, C.: New approaches in harnessing wave energy: with special attention to small Islands. Renew. Sustain. Energy Rev. **29**, 345–354 (2014)
3. Hong, Y., Waters, R., Boström, C., Eriksson, M., Engström, J., et al.: Review on electrical control strategies for wave energy converting systems. Renew. Sustain. Energy Rev. **31**, 329–342 (2014)

4. Cuadra, L., Salcedo-Sanz, S., Nieto-Borge, J.C., Alexandre, E., Rodríguez, G.: Computational intelligence in wave energy: comprehensive review and case study. Ren. Sustain. Energ. Rev. **58**, 1223–1246 (2016)
5. Deo, M.C., Naidu, C.S.: Real time wave prediction using neural networks. Ocean Eng. **26**(3), 191–203 (1998)
6. Tsai, C.P., Lin, C., Shen, J.N.: Neural network for wave forecasting among multistations. Ocean Eng. **29**(13), 1683–1695 (2002)
7. Castro, A., Carballo, R., Iglesias, G., Rabuñal, J.R.: Performance of artificial neural networks in nearshore wave power prediction. Appl. Soft Comput. **23**, 194–201 (2014)
8. Zanaganeh, M., Jamshid-Mousavi, S., Etemad-Shahidi, A.F.: A hybrid genetic algorithm-adaptive network-based fuzzy inference system in prediction of wave parameters. Eng. Appl. Artif. Intell. **22**(8), 1194–1202 (2009)
9. Alexandre, E., Cuadra, L., Nieto-Borge, J.C., Candil-García, G., del Pino, M., Salcedo-Sanz, S.: A hybrid genetic algorithm - extreme learning machine approach for accurate significant wave height reconstruction. Ocean Model. **92**, 115–123 (2015)
10. Cornejo-Bueno, L., Nieto-Borge, J.C., García-Díaz, P., Rodríguez, G., Salcedo-Sanz, S.: Significant wave height and energy flux prediction for marine energy applications: a grouping genetic algorithm - extreme learning machine approach. Renew. Energy **97**, 380–389 (2016)
11. Mahjoobi, J., Mosabbeb, E.A.: Prediction of significant wave height using regressive support vector machines. Ocean Eng. **36**(5), 339–347 (2009)
12. Salcedo-Sanz, S., Nieto-Borge, J.C., Carro-Calvo, L., Cuadra, L., Hessner, K., Alexandre, E.: Significant wave height estimation using SVR algorithms and shadowing information from simulated and real measured X-band radar images of the sea surface. Ocean Eng. **101**, 244–253 (2015)
13. Cornejo-Bueno, L., Nicto-Borge, J.C., Alexandre, E., Hessner, K., Salcedo-Sanz, S.: Accurate estimation of significant wave height with support vector regression algorithms and marine radar images. Coast. Eng. **114**, 233–243 (2016)
14. Fernández, J.C., Salcedo-Sanz, S., Gutiérrez, P.A., Alexandre, E., Hervás-Martínez, C.: Significant wave height and energy flux range forecast with machine learning classifiers. Eng. Appl. Artif. Intell. **43**, 44–53 (2015)
15. Nitsure, S.P., Londhe, S.N., Khare, K.C.: Wave forecasts using wind information and genetic programming. Ocean Eng. **54**, 61–69 (2012)
16. Özger, M.: Prediction of ocean wave energy from meteorological variables by fuzzy logic modeling. Expert Syst. Appl. **38**(5), 6269–6274 (2011)
17. Goda, Y.: Random Seas and Design of Maritime Structures. World Scientific, Singapore (2010)
18. Falkenauer, E.: The grouping genetic algorithm-widening the scope of the GAs. Belg. J. Oper. Res. Stat. Comput. Sci. **33**, 79–102 (1992)
19. Yao, X., Liu, Y., Lin, G.: Evolutionary programming made faster. IEEE Trans. Evol. Comput. **3**(2), 82–102 (1999)
20. Huang, G.B., Zhu, Q.Y.: Extreme learning machine: theory and applications. Neurocomputing **70**, 489–501 (2006)
21. Huang, G.B.: ELM matlab code. http://www.ntu.edu.sg/home/egbhuang/elm_codes.html
22. Snoek, J., Hugo, L., Adams, R.P.: Practical Bayesian optimization of machine learning algorithms. In: Advances in neural information processing systems (2012)
23. Mockus, J., Tiesis, V., Zilinskas, A.: The application of Bayesian methods for seeking the extremum. Towards Glob. Optim. **2**(117–129), 2 (1978)

24. Brochu, E., Cora, V.M., De Freitas, N.: A tutorial on Bayesian optimization of expensive cost functions, with application to active user modeling and hierarchical reinforcement learning. arXiv preprint arXiv:1012.2599 (2010)
25. Shahriari, B., Swersky, K., Wang, Z., Adams, R.P., de Freitas, N.: Taking the human out of the loop: a review of Bayesian optimization. Proc. IEEE **104**, 148–175 (2016)
26. Rasmussen, C.E.: Gaussian processes for machine learning (2006)
27. Jones, D.R., Schonlau, M., Welch, W.J.: Efficient global optimization of expensive black-box functions. J. Global Optim. **13**(4), 455–492 (1998)
28. NOAA, National Data Buoy Center. http://www.ndbc.noaa.gov/
29. Albert, A.: Regression and the Moore-Penrose pseudoinverse (No. 519.536) (1972)
30. Smola, A.J., Schölkopf, B.: A tutorial on support vector regression. Stat. Comput. **14**, 199–222 (2004)
31. Salcedo-Sanz, S., Rojo, J.L., Martínez-Ramón, M., Camps-Valls, G.: Support vector machines in engineering: an overview. WIREs Data Mining Knowl. Discov. **4**(3), 234–267 (2014)

# Hybrid Model for Large Scale Forecasting
# of Power Consumption

Wael Alkhatib$^{(\boxtimes)}$, Alaa Alhamoud$^{(\boxtimes)}$, Doreen Böhnstedt, and Ralf Steinmetz

Fachgebiet Multimedia Kommunikation, Technische Universität Darmstadt,
S3/20, Rundeturmstr. 10, 64283 Darmstadt, Germany
{wael.alkhatib,alaa.alhamoud,doreen.boehnstedt,
ralf.steinmetz}@kom.tu-darmstadt.de

**Abstract.** After the electricity liberalization in Europe, the electricity market moved to a more competitive supply market with higher efficiency in power production. As a result of this competitiveness, accurate models for forecasting long-term power consumption become essential for electric utilities as they help operating and planning of the utility's facilities including Transmission and Distribution (T&D) equipments. In this paper, we develop a multi-step statistical analysis approach to interpret the correlation between power consumption of residential as well as industrial buildings and its main potential driving factors using the dataset of the Irish Commission for Energy Regulation (CER). In addition we design a hybrid model for forecasting long-term daily power consumption on the scale of portfolio of buildings using the models of conditional inference trees and linear regression. Based on an extensive evaluation study, our model outperforms two robust machine learning algorithms, namely random forests (RF) and conditional inference tree (ctree) algorithms in terms of time efficiency and prediction accuracy for individual buildings as well as for a portfolio of buildings. The proposed model reveals that dividing buildings in homogeneous groups, based on their characteristics and inhabitants demographics, can increase the prediction accuracy and improve the time efficiency.

**Keywords:** Smart grid · Multiple linear regression · Time series models · Random forests · Conditional inference trees

## 1 Introduction

Load forecasting can be defined as the process of estimating the power consumption needs of a specific geographical area in a certain point in time. It plays an essential role in planning the facilities of electric utilities including Transmission and Distribution (T&D) equipments in the demand side management, and in the energy purchases by utilities as well. The accuracy and reliability of forecasting models have a significant impact on electric utilities. On one hand, insufficient power supply due to the underestimation of electricity demand may cause the system to operate in a critical region where a total collapse of the system is

© Springer International Publishing AG 2017
I. Rojas et al. (Eds.): IWANN 2017, Part I, LNCS 10305, pp. 661–672, 2017.
DOI: 10.1007/978-3-319-59153-7_57

possible. On the other hand, the excess power supply due to the overestimation of power consumption leads to high costs for operating too many power supply units and as a result a drop in the investment due to extra energy purchases.

Previously, power utilities could predict the future consumption using statistical metrics regarding economic growth such as the industrial growth index and population statistics such as the growth index of residential buildings. Nowadays, multiple power utilities can operate in the same area in which the customers have different power suppliers to subscribe to and not only one supplier. This makes it difficult for the power utilities to rely on the previously mentioned statistics such as the economic and population growth indexes to predict future consumption.

To assist the future investments of power utilities, we need to provide an estimation of the mean power consumption of current and new constructions based on the historical consumption data and the different factors that affect that consumption. A real-time measurements of residential power consumption can be provided by the installation of smart meters in residential buildings. However, Germany for example will not follow the European Commission program for 80% deployment of smart meters by 2020. Instead, it will adopt a phased approach that will address its specific requirements around energy efficiency and renewable energy integration. This fact triggers the need to design new models which are capable of leveraging the smart metering technology and cope up with the difficulties of integrating smart meters in nowadays networks.

In this work, we propose a new approach to overcome these issues by installing smart meters in a representative subset of the population in a region. This subset should cover the variety of domestic and small and medium enterprises (SME) buildings. Then, by modelling the consumption pattern of the participants in this trial, we can generalize the solution to predict the population's future power consumption. To estimate the long-term power consumption of a population, we integrate the effect of time-independent factors such as building characteristics and demographic features of inhabitants and time-dependent factors such as weather conditions, workdays and holidays.

The paper is organized as follows: Sect. 2 gives an overview of related work in the domain of power consumption forecasting. In Sect. 3, we introduce our concept for the long-term forecasting of power consumption. Sections 4 and 5 focus on the long-term prediction model design while Sect. 6 presents the comparative analysis and evaluation of the proposed model against RF and ctree. Finally, Sect. 7 summarizes the paper and discusses future work.

## 2    Related Work

The problem of modelling and forecasting electrical consumption has been intensively studied in the past decades. Long-term and medium-term forecasting of power consumption are used by the utilities mainly for future planning and maintenance purposes. A wide variety of models have been proposed for the purpose of power consumption forecasting. They can be classified into five categories, namely averaging models [10], regression models [2,5–7,9], time series models [13,17], artificial intelligence models [11,12,16,20], and hybrid models [15,18].

Averaging models are characterized by their simplicity as they make their prediction based on averaging the power consumption of similar points of time horizon such as day, month, and year. They only require the historical consumption information. Aman et al. presented in [1] an empirical comparison between several prediction methodologies for short-term forecasting of power consumption. In their first scenario, they have evaluated the Time of the Week (ToW) averaging model using the 15-min interval load demand in a week calculated as the average over all weeks. This simple model can be used to predict the power consumption in a granularity of 15-min as the kWh value for that interval.

More complex than averaging models, regression tree ($RT$) models build a decision tree to represent the non-linear relationship between the predictors and the response variable. Aman et al. proposed a prediction model based on regression trees to forecast the short-term power consumption of campus micro-grids using indirect indicators [2]. In this work, the authors classify power consumption indicators into direct and indirect. Direct indicators include the historical weather information and the power consumption data from smart meters. Indirect indicators include seasonal patterns such as day of the week, semester and holidays, and academic calendar as well as static knowledge of the building characteristics such as surface area. They provide prediction models at the building and campus levels for daily and 15-min intervals by training a CART regression tree based on the direct and indirect indicators. Also Time series ($TS$) models try to predict future power consumption based on previous historical observations. The commonly used approaches include Moving Average ($MA$), Auto-Regressive Integrated Moving Average ($ARIMA$) and the Pattern Sequence-based Forecasting ($PSF$) [14].

Artificial intelligence techniques such as neural networks, support vector machines, and pattern matching techniques show promising capabilities in forecasting and modelling power consumption. An overview of different AI techniques is provided in [11]. Among all AI-based methods, the technique of artificial neural networks (ANNs) has received substantial attention in forecasting power consumption due to its flexibility in learning load series and modelling the non-linearity between power consumption and the exogenous variables influencing it as well as providing fairly acceptable results. Wan et al. developed an artificial neural network model for modelling the electricity load of campus buildings in [19]. The input data of the network includes building consumption history and the time-depended climate variables such as dew point, rainfall rate, pressure, wind speed, humidity and temperature.

The majority of previous research works for power consumption forecasting focus on homogeneous buildings such as residential or industrial buildings regardless of their differences i.e. demographic data, and building characteristics. Moreover, they consider the prediction of future demand growth of current networks without taking into consideration new or planned constructions. Another limitation of the current research conducted in this field is that it did not take in consideration the difficulties of integrating smart meters in today's networks as well as the geographical structure of the network where each area is monitored independently. In this work, we try to tackle these issues by investigating the possibilities of estimating the long-term daily power consumption for a population out of a representative sample.

## 3   Concept and Dataset

In this work, we follow a multi-step statistical analysis methodology as shown in Fig. 1 in which we use time-dependent predictors such as temperature, business days, and holidays combined with time-independent predictors such as demographic data, and building characteristics to estimate the power consumption of existing and future planned buildings on different scales. In the first step, we build the Building-Performance regression model that correlates the power consumption with time-independent factors by following a stepwise approach for the selection of predictors. This model provides good insights into the average monthly power consumption of individual buildings. Furthermore, it assists the process of excluding the data which belongs to buildings with consumption patterns not representative of the population, in order to reduce the errors in next modelling steps.

In the second step, we investigate the possibility of building a hybrid model which uses conditional inference trees [8] to divide buildings into homogeneous groups using the time-independent factors and then create a multi-linear regression model for each group to estimate the daily power consumption using time-dependent predictors, demographic data, building characteristics and number of available appliances. Later, this model is adapted for the prediction of future power consumption of new buildings by removing the predictors related to

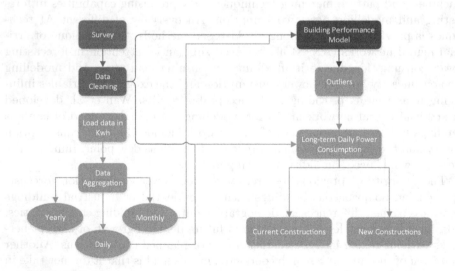

**Fig. 1.** Workflow for long-term daily power consumption forecasting model.

available appliances and part of demographic data. The model will be capable of predicting the daily long-term power consumption for the whole population.

The used dataset in this work is provided by the Commission of Energy Regulation (CER) in Ireland. CER has started a project to collect measurements about power consumption of individual buildings using smart metering technologies. The trials took place over a period of eighteen months during 2009 and 2010. Raw data representing the 30-min power consumption readings in kWh of individual buildings was collected. More than 5000 smart meters were installed in Irish homes and businesses in eight urban areas and three villages [4]. Pre-trial and post-trial surveys were conducted for both residential and business participants. Residential participants, which are considered for the evaluation, provided information about the following aspects in the survey:

- Demographic features of residents such as number of people living in the house, age groups, household income and employment status.
- Physical characteristics of the house such as floor size, house type, number of bedrooms, heating type and insulation.
- Type and number of available electrical appliances in the house.
- Behavioral features of residents such as their usage patterns of electrical appliances as well as their awareness degree of the power each appliance consumes.

## 4   Building-Performance Multiple Regression Model

The building-performance multiple regression model can serve as a reference model for the power usage of the general population by interpreting the effect of different predictors on the average power consumption. The set of predictor variables consists of demographic data, building characteristics, heating sources as well as the number of available appliances. The multiple linear regression model can be expressed in the form:

$$y_i = \beta_1 x_{i1} + \beta_2 x_{i2} + \dots + \beta_p x_{ip} + e_i \tag{1}$$

where $y_i$ is the response variable representing the total power consumption of building $i$ during the trial. $x_1, .., x_p$ refer to the predictor variables where $p$ is the number of predictors. $e_i$ is the estimation error for building $i$ and $\beta_1, .., \beta_p$ are the regression coefficients.

The Multicollinearity due to a potential inter-correlation between different predictors may negatively affect the interpretation of partial regression coefficients and make it difficult to recognize relative importance levels. To avoid any negative effect of multicollinearity, a backward stepwise regression approach is used to select the best model, where iteratively in each iteration a subset of predictors that match best model performance is selected.

This model should represent the performance of the population buildings. Therefore buildings with abnormal power consumption are considered as outliers and their related power measurements are excluded from the dataset and then the model is fitted again. Based on the assumption of normal distribution of the total power consumption, a data point is considered as an outlier if its absolute value of the standardized residual is larger than 2 [9]. By defining the predicted power consumption as $\hat{y}_i$, the standardized residual can take the form:

$$\hat{z}_i = \frac{y_i - \hat{y}_i}{\hat{\sigma}} \tag{2}$$

where $\hat{\sigma}$ is the standard error.

# 5   Hybrid Model for Long-Term Forecasting of Power Consumption

The idea behind developing a hybrid model is to get the benefits of several rigorous modelling techniques in order to achieve a high prediction accuracy. On one hand, modelling the effect of time-independent variables contributes to the prediction of mean power consumption. On the other hand, modelling the effect of time-dependent variables contributes to the modelling of the random error generated by seasonal patterns and temperature changes.

We utilize conditional inference trees (ctree) to group the heterogeneous set of buildings into several homogeneous groups based on time-independent variables, namely building characteristics, demographic data, heating source, and the number of different available appliances. Ctree is a non-parametric class of regression trees embedding tree-structured regression models into a well defined theory of conditional inference procedures [8]. Ctree recursively performs univariate splits of a dataset based on two stages. The first stage is the recursive binary partitioning which proceeds as follows:

1. Using significance test of the global null hypothesis of independence between the predictors and the response variable, the algorithm selects the predictor with the highest association with the response variable based on the p-value corresponding to significance test.
2. Select two subsets of the selected variable to split the observations into two disjoint groups. The splitting point is selected based on another statistical test.
3. Recursively repeat steps 1 and 2.

In the second stage, it fits a constant model in each leaf node of the generated tree. It is important to mention the differences between Ctree and other popular regression tree algorithms such as CART and C4.5. Both CART and C4.5 examine all the possible splits and select the covariate that maximizes the cell purity. Both methods suffer from overfitting and bias towards partitioning based on covariates with multiple splits. The overfitting can be overcome by tree pruning. However, there is no statistical significance analysis that can prove whether there is a significant improvement due to the split or not. On the contrary, Ctree algorithm is statistically more valid, it recursive applies a split based

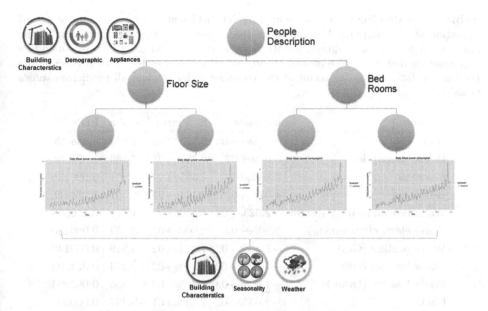

**Fig. 2.** The hybrid model design. Ctree model is used based on time-independent predictors for grouping the buildings into several homogeneous groups. Afterwards, an individual linear regression model is fitted for each group based on building characteristics and time-dependent factors.

on the theory of permutation tests in which partitioning is stopped when there is no significant association between the predictors and the response variable.

After dividing the buildings into several homogeneous groups, a multiple linear regression model is applied on each homogeneous group to model the daily power consumption of that group using time-dependent predictors, namely temperature, holidays, business days, and weekends. Moreover, a subset of time-independent variables including floor size, number of bedrooms, people description, built year and home description is used for the purpose of predicting the base power consumption of different buildings in the same group as shown in Fig. 2. The advantage of using ctree is that the split process tends to apply split on a subset only if a significant improvement can be achieved rather than grouping buildings based on heuristics such as the information gain as is the case in CART algorithm.

## 6   Evaluation

In this section we evaluated the predictive performance of our proposed hybrid model. As a first step, we cleaned the dataset from outliers which are buildings with abnormal power consumption when compared to the majority of buildings with same characteristics. For detecting outliers, we utilized the Building-Performance model explained in Sect. 4. Table 1 shows the main factors contributing to the power consumption of residential buildings. This set mainly

**Table 1.** The Building-performance final model coefficients. Std. error is the standard deviation of the sampling distribution of the estimates of the coefficients under the standard regression assumption. t-statistic is used to test whether the corresponding regression coefficient is different from 0 and $\Pr(> |t|)$ is the p-value of the corresponding t-statistics. Intercept is the mean of the response variable when all predictors values equal 0.

| Coefficient | Estimate | Std. error | t-value | $\Pr(> |t|)$ |
|---|---|---|---|---|
| (Intercept) | $-2.995e+03$ | $3.775e+02$ | $-7.934$ | 8.48e-15 |
| People description | $6.079e+02$ | $8.455e+01$ | $7.189$ | 1.69e-12 |
| Floor size | $2.836e-01$ | $7.365e-02$ | $3.851$ | 0.000128 |
| Bedrooms | $4.378e+02$ | $7.111e+01$ | $6.157$ | 1.25e-09 |
| Water central heating system | $-2.323e+02$ | $1.479e+02$ | $-1.571$ | 0.116663 |
| Water electric(immersion) | $3.509e+02$ | $9.609e+01$ | $3.652$ | 0.000280 |
| Water heating (Gas) | $-5.435e+02$ | $1.423e+02$ | $-3.819$ | 0.000146 |
| Water heating (Oil) | $-2.729e+02$ | $1.200e+02$ | $-2.274$ | 0.023293 |
| Water heating (Other) | $-1.628e+03$ | $8.723e+02$ | $-1.866$ | 0.062394 |
| Cook | $-2.634e+02$ | $7.533e+01$ | $-3.497$ | 0.000501 |
| Tumble dryer | $4.529e+02$ | $1.174e+02$ | $3.857$ | 0.000125 |
| Dishwasher | $4.296e+02$ | $1.281e+02$ | $3.355$ | 0.000837 |
| Electric heater plug in | $1.622e+02$ | $6.928e+01$ | $2.341$ | 0.019511 |
| Stand alone freezer | $3.134e+02$ | $8.788e+01$ | $3.566$ | 0.000388 |
| TV greater 21 | $1.972e+02$ | $5.469e+01$ | $3.605$ | 0.000334 |
| Desktop computers | $5.562e+02$ | $8.141e+01$ | $6.832$ | 1.83e-11 |
| Laptop computers | $3.146e+02$ | $5.755e+01$ | $5.467$ | 6.39e-08 |
| Games consoles | $2.612e+02$ | $6.441e+01$ | $4.056$ | 5.56e-05 |

included the description of people i.e. retired and all over 15 years old, the building characteristics, the number of different appliances, the cooker type, and the water heating source. Thereafter, buildings with abnormal power consumption were excluded from the evaluation. As mentioned before, the iterations of the stepwise regression approach stop when no more improvement of the model accuracy can be achieved and the main features will be fixed then.

After the removal of outliers, we got 892 residential buildings out of 930 used in the Building-Performance model, while the remaining 38 were excluded through the backward stepwise regression approach. Then, we divided the dataset into a training set of 753 residential buildings and another 139 buildings for out of sample accuracy evaluation of the model. This step was done statistically, by classifying the buildings using ctree and selecting 80% of each group for training and the rest for testing. Dividing the buildings into homogeneous insured that the testing sample covers the existing variety in power consumption based on the buildings and the residents characteristics.

After getting a representative sample by excluding buildings with abnormal consumption, we classified the buildings using ctree model into homogeneous groups based on the listed time-independent predictors in Sect. 5. Therefore, ctree model should be configured to produce groups in which buildings are as homogeneous as possible. Ctree uses the argument *mincriterion* as the value 1 - *P-value* corresponding to a significance test of dependency between a singe predictor and the response variable. This value must be exceeded in order to implement a split. In this work we set *mincriterion* to 0.90. The argument *minbucket* defines the minimum sum of weights in a terminal node which, in the default configuration, is equal to the number of data points that belong to a terminal node. These weights of individual buildings can be changed to give different importance levels to different data-points. For our evaluation purposes, we kept the default weights and set *minbucket* to 75, so we have no less than 75 data points for building the multiple linear regression model. After that a separate multi-regression model was designed for each group using time-dependent predictors, and a subset of time-independent predictors Sect. 5.

Figure 3 shows the prediction performance of our hybrid model with ctree's *mincriterion* = 0.90 and *minbucket* = 75. This figure shows the actual aggregated daily power consumption of all buildings compared to the prediction results. The predicted daily total consumption was calculated by predicting the daily power consumption of each individual building for six months in advance using our hybrid model. Thereafter, prediction results of all buildings were aggregated and compared to the sum of actual daily power consumption of all buildings in the dataset.

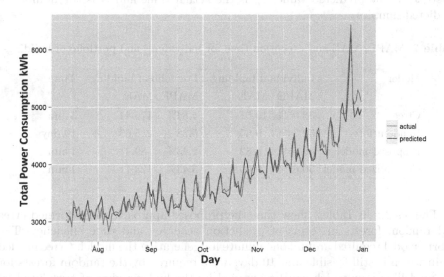

**Fig. 3.** Prediction accuracy of the proposed hybrid model with mincriterion = 0.90 and minbucket = 75.

In order to generalize the proposed model to be capable of predicting the power consumption of new constructions, we removed the factors related to inhabitants such as the number of different appliances as well as how they cook and the demographic data related to the number of people in different age groups.

We compared the performance of our hybrid model against two robust machine learning algorithms, namely conditional inference tree and random forests [3] with the same used datasets for training and testing. For the random forests model, we set the number of bootstrapped trees to grow to ntree = 500. This number should not be too small to insure that each record in the dataset is predicted at least few times. While ctree was used with same configurations as in our model *mincriterion* = 0.90 and *minbucket* = 75.

Table 2 demonstrates a relative comparison between our proposed hybrid model, ctree, random forests and the generalized version of our model in terms of model accuracy and time efficiency. For the accuracy evaluation, the Mean Absolute Percentage Error(MAPE) and the Mean Absolute Error (MAE) were used. MAPE is preferable for reporting since it presents the results as a percentage which makes it more interpretable, while MAE is less sensitive to very large errors in prediction compared to other measures.

$$MAE = \frac{1}{N} \sum_{h=1}^{N} |x_h - \hat{x_h}| \tag{3}$$

$$MAPE = \frac{100\%}{N} \sum_{h=1}^{N} \left| \frac{x_h - \hat{x_h}}{x_h} \right| \tag{4}$$

where $\hat{x_h}$ is the predicted value, $x_h$ is the actual value and N is the number of predicted samples.

**Table 2.** MAPE, MAE and execution time for individuals and portfolio of buildings.

| Model | Individual buildings | | Portfolio of buildings | | Time |
|-------|------|------|------|------|------|
|       | MAPE | MAE  | MAPE | MAE  |      |
| Ctree | 58.65% | 10.51 | 4.84% | 176.44 | 5 min |
| Random forest | 52.34% | 9.65 | 5.38% | 215.63 | 10 days |
| Proposed model | 49.01% | 8.82 | 2.43% | 89.41 | 1 min |
| Generalized model | 50.67% | 9.00 | 3.43% | 123.11 | 1 min |

The results in Table 2 show that the proposed approach outperformed ctree and random forests in terms of prediction accuracy and time efficiency. The hybrid model required around one minute for generating the model. Ctree needed 5 min which is still feasible and 10 days were required by the random forests for the modelling step which can be justified by the high number of trees used by the random forests in order to achieve high accuracy. Moreover, the lowest values of MAPE for individual buildings and portfolio of buildings were also achieved using the proposed hybrid model.

# 7    Discussion and Future Work

In this work, we designed a hybrid model for daily long-term power consumption forecasting on the scale of portfolio of buildings using conditional inference tree and linear regression models. The hybrid model outperformed two robust machine learning algorithms in terms of time efficiency and prediction accuracy. The proposed model showed that, clustering individual buildings into homogeneous groups, based on building's characteristics and their inhabitants demographics, can improve the prediction accuracy and increase time efficiency by reducing the search space. In future work, other modelling techniques will be used instead of the linear regression model to predict individual groups consumption in the hybrid model. Also we are interested in extending this work by designing an ensemble forecasting model by applying multiple modelling techniques on each group of the Ctree leaves. The ensemble model could be a fusion of the predicted values from different models in an equation with different weights for each model.

# References

1. Aman, S., Frincu, M., Chelmis, C., Noor, M.U.: Empirical Comparison of Prediction Methods for Electricity Consumption Forecasting, Department of Computer Science, University of Southern California, Los Angeles, CA, 90089 (2012)
2. Aman, S., Simmhan, Y., Prasanna, V.: Improving energy use forecast for campus micro-grids using indirect indicators, December 2011
3. Breiman, L.: Random forests. Mach. Learn. J. **45**, 5–32 (2001)
4. Commission for Energy Regulation (CER): Electricity Smart Metering Technology Trials Findings Report. ESB Networks, Belgard Square North, Tallaght, Dublin 24 (2011)
5. Fan, S., Hyndman, R.: Short-term load forecasting based on a semi-parametric additive model. IEEE Trans. Power Syst. **27**(1), 134–141 (2012)
6. German, G.: Smoothing and non-parametric regression. Int. J. Syst. Sci. (2001)
7. Hong, T., Gui, M., Baran, M., Willis, H.: Modeling and forecasting hourly electric load by multiple linear regression with interactions. In: 2010 IEEE Power and Energy Society General Meeting, pp. 1–8, July 2010
8. Hothorn, T., Hornik, K., Zeileis, A.: Unbiased recursive partitioning: a conditional inference framework. J. Comput. Graph. Stat. **15**(3), 651–674 (2006)
9. Jiang, H., Lee, Y., Liu, F.: Anomaly Detection, Forecasting and Root Cause Analysis of Energy Consumption for a Portfolio of Buildings Using Multi-step statistical Modeling, US Patent App. 13/098,044 (2012)
10. Coughlin, K., Piette, M.A., Goldman, C., Kiliccote, S.: Estimating demand response load impacts: evaluation of baseline load models for nonresidential buildings in California. Lawrence Berkeley National Lab, Technical report LBNL-63728 (2008)
11. Metaxiotis, K., Kagiannas, A., Askounis, D., Psarras, J.: Artificial intelligence in short-term electric load forecasting: a state-of-the-art survey for the researcher. Energy Convers. Manag. **44**, 1525–1534 (2003)
12. Khotanzad, A., Afkhami-Rohani, R., Lu, T.L., Abaye, A., Davis, M., Maratukulam, D.: Annstlf-a neural-network-based electric load forecasting system. IEEE Trans. Neural Netw. **8**(4), 835–846 (1997)

13. Kohonen, T.: The self-organizing map. Proc. IEEE **78**(9), 1464–1480 (1990)
14. Alvarez, F.M., Troncoso, A., Riquelme, J., Ruiz, J.A.: Energy time series forecasting based on pattern sequence similarity. IEEE Trans. Knowl. Data Eng. **23**(8), 1230–1243 (2011)
15. Mori, H., Takahashi, A.: Hybrid intelligent method of relevant vector machine and regression tree for probabilistic load forecasting. In: 2011 2nd IEEE PES International Conference and Exhibition on Innovative Smart Grid Technologies (ISGT Europe), pp. 1–8, December 2011
16. Rui, Y., El-Keib, A.: A review of ann-based short-term load forecasting models. In: 1995 Proceedings of the Twenty-Seventh Southeastern Symposium on System Theory, pp. 78–82, March 1995
17. Shen, W., Babushkin, V., Aung, Z., Woon, W.: An ensemble model for day-ahead electricity demand time series forecasting. In: Proceedings of the Fourth International Conference on Future Energy Systems, pp. 51–62. ACM, New York (2013)
18. Silipo, R., Winters, P.: Big Data, Smart Energy, and Predictive Analytics Time Series Prediction of Smart Energy Data (2013). http://www.knime.org
19. Wan, S., Yu, X.-H.: Facility power usage modeling and short term prediction with artificial neural networks. In: Zhang, L., Lu, B.-L., Kwok, J. (eds.) ISNN 2010. LNCS, vol. 6064, pp. 548–555. Springer, Heidelberg (2010). doi:10.1007/978-3-642-13318-3_68
20. Yang, X.: Comparison of the LS-SVM based load forecasting models. In: 2011 International Conference on Electronic and Mechanical Engineering and Information Technology (EMEIT), vol. 6, pp. 2942–2945, August 2011

# A Coral Reef Optimization Algorithm for Wave Height Time Series Segmentation Problems

Antonio Manuel Durán-Rosal[1]($\boxtimes$), David Guijo-Rubio[1],
Pedro Antonio Gutiérrez[1], Sancho Salcedo-Sanz[2], and César Hervás-Martínez[1]

[1] Department of Computer Science and Numerical Analysis, University of Córdoba,
Campus de Rabanales, 14071 Córdoba, Spain
{i92duroa,i22gurud,pagutierrez,chervas}@uco.es
[2] Department of Signal Processing and Communications, Universidad de Alcalá,
28805 Alcalá de Henares, Madrid, Spain
sancho.salcedo@uah.es

**Abstract.** Time series segmentation can be approached using meta-heuristics procedures such as genetic algorithms (GAs) methods, with the purpose of automatically finding segments and determine similarities in the time series with the lowest possible clustering error. In this way, segments belonging to the same cluster must have similar properties, and the dissimilarity between segments of different clusters should be the highest possible. In this paper we tackle a specific problem of significant wave height time series segmentation, with application in coastal and ocean engineering. The basic idea in this case is that similarity between segments can be used to characterise those segments with high significant wave heights, and then being able to predict them. A recently meta-heuristic, the Coral Reef Optimization (CRO) algorithm is proposed for this task, and we analyze its performance by comparing it with that of a GA in three wave height time series collected in three real buoys (two of them in the Gulf of Alaska and another one in Puerto Rico). The results show that the CRO performance is better than the GA in this problem of time series segmentation, due to the better exploration of the search space obtained with the CRO.

**Keywords:** Time series segmentation · Coral reef optimization · Genetic algorithms · Significant wave height time series

## 1 Introduction

The effective utilization and management of offshore and coastal resources requires information on ocean waves probability distribution and related events

This work has been partially supported by projects TIN2014-54583-C2-1-R, TIN2014-54583-C2-2-R and the TIN2015-70308-REDT projects of the Spanish Ministerial Commission of Science and Technology (MINECO, Spain) and FEDER funds (EU). Antonio M. Durán-Rosal's research is supported by the FPU Predoctoral Program (Spanish Ministry of Education and Science), grant reference FPU14/03039.

© Springer International Publishing AG 2017
I. Rojas et al. (Eds.): IWANN 2017, Part I, LNCS 10305, pp. 673–684, 2017.
DOI: 10.1007/978-3-319-59153-7_58

[1,2]. Specifically, in order to design effective operational activities and prevent accidents in marine structures, the determination of subsequences of wave height time series (where we could find Extreme Significant Wave Heights (ESWH) values) is a powerful tool. The Significant Wave Height (SWH) is defined as the average (on meters) of the highest one-third of all the wave heights during a 20-minute sampling period [3], and it has been widely researched. Recently, a more specific field, the determination and prediction of Extreme SWHs (ESWHs), has become interesting for many researchers in coastal engineering, marine structures and marine energy facilities. In general, previously proposed methods are based on considering the probability distributions of the Extreme Values (EVs) of SWHs [4]. For example, the work of Muraleedharan et al. [5] proposes the use of quantile regression to model the ESWH distribution, as an alternative to fitting EV distributions based on the tails of data samples. Another popular methodology is the Peaks Over Threshold (POT) [6] (i.e. considering only those values of the time series higher than a predefined threshold), which has been used as a standard approach for ESWH predictions [7]. Neural computation has been recently applied to the prediction of ESWH events [8].

In this paper, we propose an alternative approach to determine ESWHs, considering the corresponding temporal segment where an ESWH is found. This proposal simplifies the time series by a segmentation method, alleviating the difficulty of processing, analysing or mining the series, given their continuous nature. We evaluate two different metaheuristics for this task: a classical genetic algorithm (GA) [9] and a more recent proposal, the coral reef optimization (CRO) [10,11] method. The algorithms autonomously find a set of segments, which are grouped into a clustering step to discover whether one or two of the clusters are representing the extreme event (i.e. they work by grouping ESWH segments). These detection consists on finding segments of the time series that present similar behaviour (similar characteristics) with the objective to determine a cluster which specifically groups the ESWHs in a time series. The segmented time series is then transformed into a sequence of labels (corresponding to each segment), where, ideally, one or two of the labels are related to ESWH segments. This allows determining the nature of ESWHs, and paves the way towards a first prediction approach for this type of extreme events.

Evolutionary computation has been widely used in recent years for segmentation of time series. In this way, segmentation problems are converted into optimization ones, and GAs have proven to be very suitable for solving them [12]. For example, Tseng et al. [13] proposed a GA-based time-series segmentation approach to automatically find appropriate segments and patterns from a time series. Chung et al. [14] proposed a GA-based approach to segment time series, where the user has to specify a set of desired patterns. The CRO is an evolutionary bio-inspired approach, based on the simulation of the processes in a coral reef, such as coral reproduction, growing, fighting for space in the reef or depredation [10]. The CRO simulates the behaviour of a specific natural ecosystem (coral reefs) to tackle optimization problems. In GA and CRO metaheuristics, a random population of individuals is generated, the fitness of these solutions (in our case, segmented time series) is evaluated, and then parents are selected from these individuals. These parents produce offsprings by considering the different exploration processes, specific for each metaheuristic. This process continues until a number of generations is reached or other stop condition if fulfilled. The best individual in the resulting offspring at the end of evolution is returned as the solution of the problem. However, as will be analysed in this paper, there are several important differences between GA and CRO, that produces important differences in performance when facing time-series segmentation problems.

To the best of our knowledge, this is the first work where the CRO is applied to time series segmentation. We compare the results obtained by the CRO and a GA, in different wave height time series. Specifically, the time series considered are collected by two buoys situated in the Gulf of Alaska and in Puerto Rico, respectively, where ESWH are detected every year. The results obtained by the CRO are promising, outperforming the GA in this problem (finding better fitness and segmented time series than the ones obtained by the GA).

The rest of the paper is organized as follows. Section 2 analyzes the problem of time series segmentation and the two methods compared (GA and CRO). Section 3 presents the wave height time series, the experimentation and the results obtained. Finally, Sect. 4 concludes the paper with some concluding remarks.

## 2   Methods

This section describes the most important characteristics of the metaheuristics considered in this paper, the GA and the CRO approaches. Previously, we summarize the problem to be tackled by the proposed algorithms, which is important to fix an encoding of the problem into the algorithms.

### 2.1   Summary of the Problem

Given a time series $Y = \{y_n\}_{n=1}^{N}$, the main objective in segmentation is to divide the $N$ values of the series into $m$ segments, which should present similar characteristics (see next sections). In this way, the time indexes, denoted by $n = 1, \ldots, N$, need to be split into consecutive segments, i.e. $s_1 = \{y_1, \ldots, y_{t_1}\}$, $s_2 = \{y_{t_1}, \ldots, y_{t_2}\}, \ldots, s_m = \{y_{t_{m-1}}, \ldots, y_N\}$, where $t$s are the cut points in ascending order. Note that cut points are the only points which belong to the previous and the next segments, while the rest of the points only belong to one specific segment. The number of segments $m$ and the values of the cut points $t$ have to be discovered by the segmentation algorithm. After segmentation, a clustering process can be applied to the segments for discovering groups of them with similar characteristics. Specifically, in this paper, we consider Significant Wave Height time series, so that we can determine clusters of segments which group extreme wave heights events.

According to this description, each solution/chromosome $(c)$ will be represented by an array of binary values, where each position $c_{t_i}$ stores a binary value

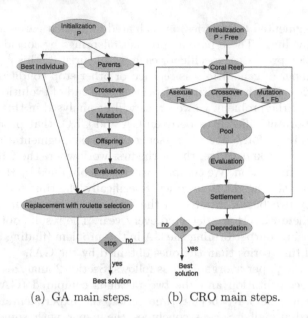

(a) GA main steps.          (b) CRO main steps.

**Fig. 1.** Flowchart comparison of GA and CRO algorithms; (a) GA algorithm; (b) CRO algorithm.

indicating whether the time index $t_i$ of the time series is a cut point of the evaluated solution. The length of the chromosome is the same than the length of the time series ($N$).

In the following sections, we describe the specific elements of the GA and CRO, and also the their common parts. The main steps of the GA are summarised in Fig. 1(a), while Fig. 1(b) shows the main steps of the CRO algorithm. Each of these steps will be further described in the following subsections.

## 2.2   Genetic Algorithm

This section describes those elements which are specific for the GA algorithm (see Fig. 1(a)).

**Initialization.** The population in the GA is a set $P$ chromosomes (binary arrays, as we previously defined), where initially the cut points ($1s$) have been randomly established with a uniform distribution. Two consecutive points are separated by a number of positions in the interval ($s_{min}, s_{max}$), which are the minimum and maximum segment sizes, respectively. It is important to mention that, in this algorithm, the number of elements in the population will be kept constant, being the choice of this parameter very decisive.

**Parents and Reproduction.** In the GA, for the sake of diversity, all chromosomes are selected as parents to perform reproduction (crossover and mutation).

**Crossover.** For each parent, the crossover operator is applied with a given probability, $p_c$. The operator randomly selects the other parent to perform a crossover process. The specific implementation of the crossover operator will be later described, in Sect. 2.4.

**Mutation.** After applying the crossover operator, each individual (crossed or not) is mutated with a given probability, $p_m$. In this paper, we use four different mutation operators (see Sect. 2.4).

**Offspring and Replacement.** The new offspring population, generated by the previous operators, is evaluated by the fitness function (see Sect. 2.4). Parent and offspring populations are then merged, and the replacement process is applied with an elitist roulette wheel selection procedure. It selects $P-1$ individuals for the next generation, each one being represented by a region in a roulette wheel that proportionally corresponds to its fitness function. By repeatedly spinning the roulette wheel, individuals are chosen using stochastic sampling to complete the next population. Finally, to have a population of $P$ individuals, the best solution of the parent population is added to the new population (elitist strategy).

## 2.3   The Coral Reef Optimization Algorithm

This section presents the steps of CRO which are specific for this algorithm (see Fig. 1(b)).

**Initialization.** In this case, the population is also initialised as a set of chromosomes (corals) where 1s are chosen with a uniform distribution $U(s_{min}, s_{max})$. The difference with respect to the GA is that the population has a size of $P$ minus a number of free positions $Free$, predefined by the user. Furthermore, the population size can change during the evolution. This simulates a coral reef, where it is possible to have free locations and occupied positions, and each occupied position represents a coral (solution).

**Asexual Reproduction.** A percentage $(F_a)$ of the best individuals (in terms of fitness function) in the coral reef are selected. Then, random individual (slightly mutated) from this group is chosen to be introduced in a pool of possible solutions.

**Sexual Reproduction.** A percentage $(F_b)$ of the entire population is selected to perform the external reproduction (broadcast spawning and brooding). In this case the broadcast spawning procedure consists in selecting pairs of parents to perform the crossover operator (see Sect. 2.4), though alternative exploration procedures are also possible at this stage. The rest of individuals $(1 - F_b)$ pass to the brooding procedure, i.e. they are mutated using the different operators considered in this paper (see Sect. 2.4). Note that, if an individual is selected to reproduce by broadcast spawning then it will not be reselected for brooding, and vice versa. Note that in the GA previously presented, the same individual can be crossed and then mutated.

**Pool and Settlement in the Reef.** All solutions generated by the asexual and sexual reproductions form a pool of possible solutions. These solutions need to be evaluated by the fitness function (see Sect. 2.4). This pool can be considered as new larvae which will fall over the coral reef. The natural process of the settlement consists on situating the different larvae (possible solutions) into the coral reef. In the CRO, for each solution in the pool, the algorithm chooses a random position in the coral reef. If the location is empty, the new solution is situated in this position. However, if the location is occupied, the new solution will replace the solution in the coral reef only if its fitness function is better. On the contrary, the process is repeated a number of attempts ($N_{att}$) before discarding the solution (considering it as depredated).

**Depredation.** Once the new coral reef is created in each generation, a percentage ($F_d$) of the worst solutions is selected to be depredated. This means that these solutions will be removed from the reef with a $p_d$ probability.

## 2.4   Common Parts of GA and CRO Algorithms

In this section we describe those steps which are shared by both the GA and CRO considered in this paper (see Fig. 1). They are mainly the exploration operators such as crossover and mutation, and also of course the fitness function calculation. Note that the fact that exploration operators are the same for both GA and CRO allows a direct comparison between both algorithms in terms of their searching capability due to the algorithm's structure.

**Crossover Operator.** The crossover operator is applied to two parents. The operator randomly chooses a time index, and then it consists in interchanging the left and right parts of the selected chromosomes, with respect to this time index. In general this type of crossover is known as 1-point crossover operator.

**Mutation Operator.** If an individual is selected to perform the mutation operator, the algorithm chooses, with the same probability (0.25), one of the following four operations: add a number of cut points, remove some cut points, move some cut points to the left, or move some cut points to the right.

**Fitness Evaluation.** A fitness function is needed in order to evaluate whether the cut points are able to lead to compact groups of segments in the time series.

To do this, three steps are sequentially applied in the fitness function calculation: extraction of characteristics, clustering of segments, and final evaluation of the clustering.

1. Extraction of characteristics: when we have to compare two segments for clustering, their lengths can be different. Consequently, we cannot apply a direct comparison of their values. One way of solving this problem is to map the segments into the same dimensional space. In our case, we use a 5-dimensional space, so that each segment $s_i = \{y_{t_{i-1}}, \ldots, y_{t_i}\}$ is mapped by using the function $f : \mathbb{R}^{(1+t_i-t_{i-1})} \rightarrow \mathbb{R}^5$, where $f(s_i) = \left(S_i^2, \gamma_{1i}, \gamma_{2i}, a_i, AC_i\right)$ and the characteristics are defined by:
   - Variance: $S_i^2 = \frac{1}{1+t_i-t_{i-1}} \sum_{n=t_{i-1}}^{t_i} (y_n - \overline{y_i})^2$, where $y_n$ values of the segment, and $\overline{y_i}$ is the average for the $i$-th segment.
   - Skewness: $\gamma_{1i} = \frac{\frac{1}{1+t_i-t_{i-1}} \sum_{n=t_{i-1}}^{t_i} (y_n - \overline{y_i})^3}{S_i^3}$, where $S_i$ is the standard deviation of the $i$-th segment.
   - Kurtosis: $\gamma_{2i} = \frac{\frac{1}{1+t_i-t_{i-1}} \sum_{n=t_{i-1}}^{t_i} (y_n - \overline{y_i})^4}{S_i^4} - 3$.
   - Slope of a linear regression applied to the values of the segment: $a_i = \frac{S_i^{yt}}{\left(S_i^t\right)^2}$, where $S_i^2 = \frac{1}{1+t_i-t_{i-1}} \sum_{n=t_{i-1}}^{t_i} \left(n - \overline{t_i}\right)^2$, and $S_i^{yt} = \frac{1}{1+t_i-t_{i-1}} \sum_{n=t_{i-1}}^{t_i} (n - \overline{t_i}) \cdot (y_n - \overline{y_i})$.
   - Autocorrelation coefficient: $AC_i = \frac{\sum_{n=t_{i-1}}^{t_i} (y_n - \overline{y_i}) \cdot (y_{n+1} - \overline{y_i})}{S_i^2}$.

2. Clustering of segments: after the previous step, the algorithm receives as input a set of mapped segments with five characteristics. A deterministic $k$-means is used then for clustering, where the same clustering should be obtained when the same cut points are evaluated. This means that the centroids have to be initialized always in a deterministic way. The first centroid is chosen as that data point which has the maximum value in the characteristic corresponding to the maximum standard deviation, the second will be the farthest with respect to the first one, the third, the farthest from both, and so on. Note that a scaling in a range $[0,1]$ is necessary for this clustering procedure.

3. Evaluation of the clustering: the last step is the evaluation of the clustering quality. A fitness function which allows determining compact and well separated clusters is the Caliński and Harabasz index, defined as follows:

$$CH = \frac{\text{Tr}(\mathbf{S}_B) \cdot (m - k)}{\text{Tr}(\mathbf{S}_W) \cdot (k - 1)}, \tag{1}$$

where $m$ is the number of patterns (mapped segments in our case), and $\text{Tr}(\mathbf{S}_B)$ and $\text{Tr}(\mathbf{S}_W)$ are the trace of the between and within-class scatter matrices, respectively:

$$\text{Tr}(\mathbf{S}_B) = \sum_{i=1}^{k} n_i \|\bar{\mathbf{c}}_\mathbf{i} - \bar{\mathbf{s}}\|^2, \tag{2}$$

$$\text{Tr}(\mathbf{S}_W) = \sum_{i=1}^{k} \sum_{\mathbf{s} \in c_i} \|\mathbf{s} - \bar{\mathbf{c}}_\mathbf{i}\|^2. \tag{3}$$

where **s** represents the 5-dimensional mapping of a segment, $n_i$ is the number of segments of cluster $i$, $\bar{c_i}$ is the centroid of cluster $i$, $\bar{s}$ is the overall mean of all mapped segments, and $||\mathbf{a} - \mathbf{b}||^2$ is the Euclidean distance between two vectors **a** and **b**. Note that the value of $k$ is fixed in the algorithm.

# 3   Experimental Results and Discussion

This section includes a description of the time series used, the experiments performed and the results obtained, together with the discussion of these experiments and results.

## 3.1   Significant Wave Height Time Series Analyzed

In this paper, the Significant Wave Height time series considered are collected from buoys belonging to the National Buoy Data Center of the USA [3]. Specifically, two buoys situated in the Gulf of Alaska have been chosen, with registration numbers are 46001 and 46075, respectively. The last buoy is located in Puerto Rico with registration number 41043. The resolution of the data correspond to six hours, and data from 1st January 2008 to 31st December 2013 are considered from the buoy 46001, while data from 1st January 2011 to 31st December 2015 are considered for buoys 46075 and 41043. This means a total of 8767, 7303 and 7303 observations, respectively. The complete three wave height time series are shown in Fig. 2.

## 3.2   Experimental Setting

The experimental setting for the time series under study is presented in this subsection. For the GA, the population size is established to $P = 200$; crossover probability is set to $p_c = 0.8$, while the mutation one is $p_m = 0.2$; a 20% of the current number of cut points is modified if a individual is selected to be mutated; the number of generations of the algorithm is $g = 100$. For the CRO algorithm, the reef size is established in $P = 200$, with initial free positions $Free = 20$. The number of generations is also $g = 100$, to test both algorithms in the same conditions. The percentage of asexual reproduction, external reproduction and depredation are established to $F_a = 0.1$, $F_b = 0.8$ and $F_d = 0.2$, respectively. The probability of depredation is set to $p_d = 0.1$. Finally, the number of attempts to replace a solution is $N_{att} = 2$.

For the configuration of the common parts of both algorithms we consider the following parameters: the $k$-means clustering process is allowed a maximum of 20 iterations, the initial minimum and maximum size of the segments are $s_{min} = 20$ and $s_{max} = 120$, respectively; the number of clusters or groups is $k = 4$ for the buoy of Puerto Rico and $k = 5$ for the buoys of Gulf of Alaska. Finally, given the stochastic nature of the GA and CRO processes, the algorithms were run 30 times with different seeds.

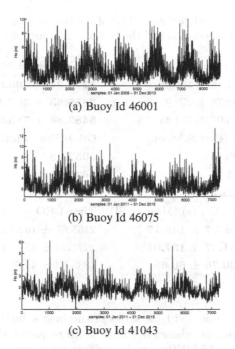

(a) Buoy Id 46001

(b) Buoy Id 46075

(c) Buoy Id 41043

**Fig. 2.** Wave height time series of buoys id 46001 and 46075 in the Gulf of Alaska and buoy id 41043 in Puerto Rico.


### 3.3  Discussion

Table 1 includes the results obtained using CH as the fitness function, both for the GA and the CRO. The results show the mean and the standard deviation of the 30 runs. Furthermore, we include the best solution obtained by each algorithm.

First of all, we compare the performance of both algorithms in terms of the fitness function. As can be seen, the CRO approach obtains the best results in average. For buoy 46001, the CRO improves the GA by a 140.70% of fitness. The same occurs with the rest of the buoys with improvements of 135.16% for buoy 46075 and 131% for buoy 41043. With these results, we can assume that the CRO makes a better exploration in the difficult search space of this problem.

Analyzing the best value obtained by GA and CRO in the 30 repetitions, the results obtained by the CRO for the buoys of the Gulf of Alaska are better than that of the GA, with improvements of 137.39% and 136.09%, respectively. However, for the case of the buoy situated in Puerto Rico (41043), the best solution obtained by the GA is better than the one obtained by the CRO. Nevertheless, if we observe the mean values of the buoy of Alaska (41043), CRO is clearly better, so we can conclude that the best solution obtained by GA is only due to randomness inherent to the algorithm.

**Table 1.** Mean and standard deviation (SD) of the fitness values ($CH$ metric) for the 30 runs of the different algorithms and the best run for each algorithm.

| Buoy id | GA (Mean ± SD) | CRO (Mean ± SD) |
|---------|----------------|-----------------|
| 46001 | 3724.46 ± 421.81 | **5240.24 ± 563.00** |
| 46075 | 2852.74 ± 270.99 | **3855.83 ± 407.10** |
| 41043 | 4185.06 ± 713.43 | **5483.88 ± 785.03** |
| Buoy id | GA (Best Solution) | CRO (Best solution) |
| 46001 | 4405.15 | **6052.28** |
| 46075 | 3413.07 | **4645.01** |
| 41043 | **7025.12** | 6622.19 |
| Buoy id | Time GA (s) (Mean ± SD) | Time CRO (s) (Mean ± SD) |
| 46001 | **2008.87 ± 148.17** | 2168.87 ± 169.12 |
| 46075 | **2051.47 ± 137.01** | 2271.40 ± 157.13 |
| 41043 | **1690.70 ± 60.65** | 1835.63 ± 44.96 |

Graphically, the segmented time series of the best solution obtained by each algorithm for each buoy are shown in Fig. 3. With respect to buoy 46001 and the GA (a), the clustering of ESWHs may be confused, because we cannot detectany

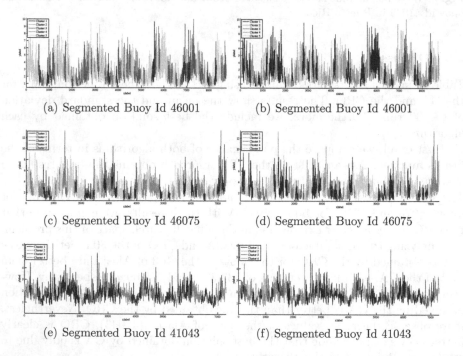

(a) Segmented Buoy Id 46001     (b) Segmented Buoy Id 46001

(c) Segmented Buoy Id 46075     (d) Segmented Buoy Id 46075

(e) Segmented Buoy Id 41043     (f) Segmented Buoy Id 41043

**Fig. 3.** Segmented Wave Height time series of the buoys 46001, 46075 and 41043, for both algorithms (Online version in colour).

colour corresponding with this type of event. However, CRO (b) obtains a segmentation where ESWHs seem to correspond to the red cluster and maybe the pink one. In the case of buoy 46075, it is difficult to detect which cluster groups ESWHs in both the GA and CRO. The GA only detects two wave heights which correspond with the dark blue colour. However, CRO is able to identify extreme wave heights when considering green and pink colours. Finally, for buoy 41043, no ESWH seems to be well separated when using the GA. However, CRO groups ESWHs in the dark blue cluster in combination with the green one.

Finally, another comparison against other four algorithms has been performed. These algorithms correspond to the hybridization of the two algorithms presented (following the same strategy than in [15]): NH and BH are the hybridizations of the GA assuming that data are sampled from a normal or a beta distribution, respectively, while NHCRO and BHCRO are the same for the CRO algorithm. The results, presented in Table 2, show that the hybrid versions are worse than the standard versions, except in the case of the NH with the buoy 46001, which improves the fitness with a value of 5305.51.

**Table 2.** Mean and standard deviation (SD) of the fitness values ($CH$ metric) for the hybrid algorithms.

| Buoy id | NH | BH | NHCRO | BHCRO |
|---|---|---|---|---|
| 46001 | 4709.15 ± 1194.82 | 4729.64 ± 646.88 | 5305.51 ± 596.43 | 4891.06 ± 672.21 |
| 46075 | 3320.25 ± 812.62 | 3330.65 ± 504.72 | 3370.08 ± 791.53 | 3560.81 ± 639.87 |
| 41043 | 3952.11 ± 943.57 | 3375.55 ± 850.00 | 4052.52 ± 1193.55 | 3711.37 ± 1163.01 |

# 4   Conclusions

This paper presents the application of a novel coral reef optimization (CRO) algorithm to time series segmentation, for the detection of extreme values of wave height time series. The algorithm is compared with a standard genetic algorithm (GA), showing their main differences along the generations, and the way to optimize the fitness function. The algorithm is tested on three real wave height time series, two of them are collected from buoys situated at the Gulf of Alaska and the other one from a buoy of Puerto Rico. Experiments show that CRO obtains better results than the GA in the problem. Specifically, the fitness function $CH$ is significantly better in the case of CRO, and the segmented time series show that the clustering of extreme values is also more consistent. Finally, results agree that these algorithms perform a better search than the hybridization assuming normal and beta distribution of the data. We plan to extend this work in different directions: (a) Using more time series for validating better the conclusions obtained; (b) Predicting the occurrence of EWHS events based on the labelled sequences obtained by the CRO segmentation algorithm; and (c) Improving the quality of the segmentation by considering a penalty cost function for guiding the segmentation in a better way.

# References

1. The WAMDI Group: The WAM model—a third generation ocean wave prediction model. J. Phys. Oceanogr. **18**, 1775–1810 (1988)
2. Cornejo-Bueno, L., Nieto-Borge, J.C., Alexandre, E., Hessner, K., Salcedo-Sanz, S.: Accurate estimation of significant wave height with support vector regression algorithms and marine radar images. Coast. Eng. **114**, 233–243 (2016)
3. National Data Buoy Center. National Oceanic and Atmospheric Administration of the USA (NOAA). http://www.ndbc.noaa.gov/. Accessed 29 June 2016
4. Norgaard, J.Q., Andersen, T.L.: Can the rayleigh distribution be used to determine extreme wave heights in non-breaking swell conditions? Coast. Eng. **111**, 50–59 (2016)
5. Muraleedharan, G., Lucas, C., Soares, C.G.: Regression quantile models for estimating trends in extreme significant wave heights. Ocean Eng. **118**, 204–215 (2016)
6. Davison, A.C., Smith, R.L.: Models for exceedances over high thresholds. J. R. Stat. Soc. Ser. B (Methodol.) **52**(3), 393–442 (1990)
7. Caires, S., Sterl, A.: 100-year return value estimates for ocean wind speed and significant wave height from the ERA-40 data. J. Clim. **18**(7), 1032–1048 (2005)
8. Dixit, P., Londhe, S.: Prediction of extreme wave heights using neuro wavelet technique. Appl. Ocean Res. **58**, 241–252 (2016)
9. Eiben, A.E., Smith, J.E.: Introduction to evolutionary computing. Springer, Berlin (2003)
10. Salcedo-Sanz, S., Del Ser, J., Landa-Torres, I., Gil-López, S., Portilla-Figueras, J.: The coral reefs optimization algorithm: a novel metaheuristic for efficiently solving optimization problems. Sci. World J. **2014**, 1–15 (2014)
11. Salcedo-Sanz, S.: A review on the coral reefs optimization algorithm: new development lines and current applications. Prog. Artif. Intell. **6**(1), 1–15 (2017)
12. Chung, F.L., Fu, T.C., Ng, V., Luk, R.W.: An evolutionary approach to pattern-based time series segmentation. IEEE Trans. Evol. Comput. **8**(5), 471–489 (2004)
13. Tseng, V.S., Chen, C.H., Huang, P.C., Hong, T.P.: Cluster-based genetic segmentation of time series with DWT. Pattern Recogn. Lett. **30**(13), 1190–1197 (2009)
14. Chung, F.L., Fu, T.C., Luk, R., Ng, V.: Evolutionary time series segmentation for stock data mining. In: Proceedings of 2002 IEEE International Conference on Data Mining, ICDM 2003, pp. 83–90 (2002)
15. Durán-Rosal, A.M., de la Paz-Marín, M., Gutiérrez, P.A., Hervás-Martínez, C.: Identifying market behaviours using european stock index time series by a hybrid segmentation algorithm. Neural Process. Lett. **1**, 1–24 (2017)

# Satellite Based Nowcasting of PV Energy over Peninsular Spain

Alejandro Catalina, Alberto Torres-Barrán, and José R. Dorronsoro[✉]

Dpto. Ing. Informática, Universidad Autónoma de Madrid, Madrid, Spain
jose.r.dorronsoro@iic.uam.es

**Abstract.** In this work we will study the use of satellite-measured irradiances as well as clear sky radiance estimates as features for the nowcasting of photovoltaic energy productions over Peninsular Spain. We will work with three Machine Learning models (Lasso and linear and Gaussian Support Vector Regression-SVR) plus a simple persistence model. We consider prediction horizons of up to three hours, for which Gaussian SVR is the clear winner, with a quite good performance and whose errors increase slowly with time. Possible ways to further improve these results are also proposed.

**Keywords:** Photovoltaic energy · Nowcasting · EUMETSAT · Support Vector Regression · Lasso · Clear Sky models

## 1 Introduction

Solar energy is possibly the fastest growing renewable energy nowadays and, together with wind, the fastest growing energy source overall. As it is the case with wind, solar energy can come from large, utility-scale photovoltaic (PV) and thermosolar plants, with installed power in the tens of megawatts (mW), but unlike wind and thermosolar, PV energy is also growing in small installations, such as rooftop solar, with much lower capacities (about 20 kilowatts may be typical for a home) but with a much larger number of installations. Moreover, when coupled with the very fast developments in battery storage, solar residential PV energy has an enormous potential grow path, provided that the current trend of fast decreasing prices of solar panels and home batteries keeps going on.

While this accelerated growth offers obvious clean energy advantages, it also has other important side effects, such as the so called "duck curve", present in places with large installed solar energy such as California, which measures the difference between the total load curve and its solar-based component. This curve has its lower values around mid-day, when solar energy reaches its daily peak, but grows markedly at the evening (the neck of the duck) when solar energy drops and other sources have to come on line. While utility-scale solar energy has the greatest effect on the duck curve, large rooftop solar installed power in places like Hawaii causes another effect, the Nessie curve, so called because residential rooftop PV energy production substracts from the energy load, which

I. Rojas et al. (Eds.): IWANN 2017, Part I, LNCS 10305, pp. 685–697, 2017.
DOI: 10.1007/978-3-319-59153-7_59

becomes then markedly lower around noon than earlier or later in the day and, thus, vaguely resembles the shape of the hypothetical Loch Ness monster.

The overall solar energy growth and side effects such as the duck and Nessie curves make it clear that good solar energy forecasting will be increasingly crucial for the management of the electricity grid. In the last years there has been a large research effort on various aspects of solar, and specifically PV, energy prediction; see for instance [2,9,16] for general recent overviews of the area. From a methodological perspective, there are several possibilities to organize and classify different approaches to PV energy forecasting. One often used is to consider the mathematical modeling techniques to be applied, such as standard or exogenous time series methods, machine learning algorithms, physical or engineering approaches or hybrid combinations of all these. Another is to consider the explanatory variables other than, of course, PV energy readings, to be used in models, such as Numerical Weather Prediction (NWP) inputs, sky camera images or either ground- or satellite-based measurements. Finally, there is the desired forecasting horizon: very short (up to one hour), short (up a few hours within the same day) or medium-long (from one to several days ahead).

Each forecasting horizon has its own main goals. For instance, when dealing with day-ahead (or longer) time horizons, one is mostly concerned with energy generation planning in accordance with predicted demand, or with plant maintenance. The main goal in the very short and short horizons is usually to maintain the grid stability. Of course, the concrete forecasting problem to be tackled limits and makes precise the above choices: the very localized point of view of a plant operator will be different from that of a large scale producer or a Transmission System Operator (TSO), most likely to be concerned with energy aggregations over relatively large areas, or from that of a electricity market agent that has to carry out energy purchases or meet delivery contracts.

Our interest here is in short term, same-day PV forecasting over a wide area, namely, that of Peninsular Spain. While this deals with a concrete region in terms of geography and extension, it may have several common aspects with PV forecasting in other areas such as, say, California or Texas in the U.S. or the Mediterranean countries. Besides the time horizon, our problem also largely determines the methods to be used. Physical or engineering approaches are best suited to local PV prediction and a purely time series approach is likely to fail to capture the influence of the overall atmospheric conditions; we are thus basically left with ML-based methods. Moreover, our problem choice also determines the variables to use. Since ground measurements or sky cameras cannot adequately cover the area under study, we are left with PV production readings, NWP forecasts, either as such or transformed to PV energy predictions, and satellite based measurements. We review next these input variable sources.

The most widely used information for day-ahead forecasts comes from NWP systems, such as those of the European Center for Medium Weather Forecasts (ECMWF) or the US Global Forecasting System (GFS). Two main forecasts are run every day starting at UTC 00 and 12 h and are widely available about 6 h later, say at UTC 06 and 18 h. It is clear that only one of these forecasts will be useful for solar energy. For instance, in places like Spain, the NWP run starting at 00 h will be useful for same-day solar energy forecasts, while the 12 h run will not. Many NWP providers offer intermediate runs starting at UTC 06 and 18 h and available again about 6 h later. In such a scenario, the 06 h update would be used for Spain in the afternoon and early evening while morning solar energy forecasts would be derived from the 00 h run. In any case, only two NWP updates will happen daily, which are not enough for short time solar energy forecast updates; other sources are thus needed, among which the most obviously useful is satellite-derived information.

Possibly the best known procedure to estimate solar irradiance from satellite readings is the HELIOSAT method [6]. HELIOSAT is a physical model that relies on the high resolution visible channel satellite measurements to analyze and track cloud behavior and its influence on irradiance values. At an hour $H$ HELIOSAT estimates actual global irradiance by scaling down a model-derived clear-sky irradiance estimate using a dimensionless cloud index derived from the visible channel satellite images at that hour; the model goal is to capture the effects on irradiance of atmospheric transmittance. Irradiance forecasts for hours $H + 1, H + 2, \ldots$, are then obtained in the same way from estimates of future cloud positions derived via a Motion Vector Field model. HELIOSAT was initially proposed for irradiance forecasting but it has been also used to derive short term PV energy forecasts at individual plants and also its aggregation at a regional level (see [10,17]).

In [3] we departed from this use of visible channel satellite images to derive cloud structure and then PV energy estimates to follow a purely Machine Learning (ML) approach in which PV energy is the learning target and the inputs were a subset of the 11 spectral bands in EUMETSAT's Meteosat satellites. These bands range from visible to long-wavelength infrared ones and we selected those having the largest correlation with PV energy. Notice that satellites measure reflected radiance and not the actual incoming irradiance that is transformed into PV energy. However, these reflected radiances were taken as a proxy for incoming irradiation and used to predict PV energy at hour $H$ from channel readings at hours $H - K$, $0 \le K \le 3$, which was done using sparse linear Lasso and Gaussian kernel Support Vector Regression (SVR) models. The present contribution maintains the overall approach in [3] but extending and substantially improving its results, with our main new contributions being:

1. We replace our previous single hourly readings with the more precise averages of the four 15' satellite readings available each hour.
2. We will incorporate into our forecasting variables a clear sky (CS) Global Horizontal Irradiance estimate at each Meteosat grid point, which will result in much better PV forecasts.
3. We build a global CS model of Peninsular Spain by averaging these CS values which, in turn, makes it possible to define a very simple PV energy persistence model that we use as a benchmark.

4. We greatly simplify the $H - K$ models in [3] working with just a single model instead of the three morning, afternoon and evening models used there.
5. We substantially improve the short term forecasts in [3], giving rather accurate nowcasting predictions up to the three hours in advance that we report here but possibly even longer.

The paper is organized as follows. In Sect. 2 we will briefly review the EUMETSAT satellite data system, the satellite readings most relevant to PV energy measures and the concrete reading choices we make for PV nowcasting; we will also discuss in this section the CS estimates we will incorporate to our models. In Sect. 3 we will briefly describe the ML models we will use; besides the Lasso and Gaussian Support Vector Regression (SVR) models considered in [3] we also have used the LIBLINEAR [5] implementation of a linear SVR model. As it is the case with Lasso, this is in principle justified by the very high dimensionality of the independent variables, although the performance of both linear models clearly lags behind that of the Gaussian SVR. We will also describe in Sect. 3 a simple, CS-based persistence model that we use as a benchmark; while such an $H - K$ model is not competitive for $K \geq 3$, it may be hard to beat for $K = 1$ and even $K = 2$, particularly near noon. Numerical results for the PV prediction of Peninsular Spain are given in Sect. 4; we find the results for Gaussian SVR quite remarkable. Finally, in Sect. 5 we briefly recap our contributions and discuss further ways to improve on them.

## 2    Satellite Data and Clear Sky Models

### 2.1    EUMETSAT Satellite Data

The geostationary Meteosat satellites operated by the European Organisation for the Exploitation of Meteorological Satellites (EUMETSAT) cover Europe, Africa and the Atlantic Ocean. The current Meteosat Second Generation (MSG) satellites are equipped with the Spinning Enhanced Visible and Infrared Imager (SEVIRI) technology. They work at different wavelengths and spatial resolutions. Over Europe and parts of Africa, the visible channel measures reflected radiance on the 0.6–0.9 μm visible wavelength range at the finest $1 \times 1$ km resolution. A slightly coarser resolution of about $3 \times 3$ km is used for the remaining 11 channels, with wavelengths that go from 0.6 μm at Channel 1 to the long infrared 13.4 μm at Channel 11. Some of these channels target specific physical variables such as absorption of water-vapor (Channel 5), ozone (Channel 8) or $CO_2$ (Channel 11). Their measures are provided in near real-time every 15 min.

We will use 15' Meteosat readings at UTC hours 0 to 23 for the years 2013, 2014 and 2015 that we have downloaded from the EUMETSAT Data Centre. Global PV readings for Peninsular Spain are kindly provided by Red Eléctrica de España (REE). Given our global PV target and large area, we downsample Meteosat's resolution to that of a 0.125°. As in [3] we will only consider grid locations over Peninsular Spain, which results in 3,391 points. PV energy readings at hour $H$ correspond to the energy produced during the entire hour ending at $H$. Because of this we will average the four 15' readings up to hour $H$ as a proxy of radiances over that hour. We will use the 2013 year for train, 2014 for hyperparameter validation and 2015 for test. We will only consider daylight hours, dropping satellite readings outside the UTC 08 to 20 h range. Notice that for most of the year there will be PV energy in Spain before 08 UTC; we are interested, however, in predicting PV energy up to 3 h ahead and that is the reason we will not consider earlier hours. After dropping the non selected hours we are finally left with 4,638 hours per year, due to some missing data.

As mentioned we will not consider the high resolution visible channel. In the remaining 11 channels a reflectance value is computed as the fraction of their radiance to the maximum solar irradiance. Moreover, radiances are converted into equivalent brightness temperatures through an empirical formula [15] and we have thus a total of 22 measures per grid point. In [3] we computed the correlations between PV energy and the averages of each measure over the grid points considered and selected the radiances of the IR016, IR039 and VIS008 channels plus the brightness temperature of channel IR039 as the ones most relevant. We have repeated this analysis using now the averaged 15' readings and the same correlations are observed, so we use the same variable subset.

## 2.2 Clear Sky Model

Clear Sky (CS) models usually estimate the Direct Normal Irradiance (DNI) and Global Horizontal Irradiance (GHI) at a certain point, taking into account its height as well as a number of physical parameters, such as atmospheric pressure, temperature, air turbidity and others. In order to provide good local modeling they usually require the careful calibration of these parameters at the concrete sites where they are to be used. There have been many proposals in the literature; here we will work with the one introduced by [8] which is implemented in the Python library pvlib [13] We have run this model for each of our 3,391 grid points in Peninsular Spain over the 3 years considered; we derive their altitudes from their geopotentials in the ECMWF's orographical model for Spain with the same 0.125° resolution and leave the other parameters at their pvlib defaults.

When building the $H - K$ models that forecast PV energy at hour $H$ from satellite readings at hour $H - K$, we will add the CS GHI estimates at hours $H$ and $H - K$ as extra variables at each point. For obvious reasons, only the CS value at hour $H$ is used in the same hour $H$ models. Our goal for this is to introduce a time context into the models which enables us to work with a single model over the entire day, avoiding the three separate morning, afternoon and evening intra-day models used in [3]. They were used there to avoid "model confusion" by distinguishing possibly similar PV target readings at morning and evening hours which, however, correspond to very different satellite readings. This was however somewhat cumbersome and the time context that the CS values provide not only simplifies model building but also gives better results. After adding the CS values for each of the grid points, we end up with a yearly data matrix of dimensions $4,638 \times 16,956$ for the $H$ model and $4,638 \times 20,347$ for the rest.

## 3   Machine Learning Models

As mentioned, we will use three well established ML models, Lasso and linear and Gaussian kernel Support Vector Regression (SVR); some of them have already been recently used in literature [11]. Additionally, we will consider a CS-based persistence model as a baseline benchmark for the others. We briefly describe them next.

### 3.1   Persistence

Our baseline CS model simply predicts PV energy at hour $H$ as a scaling of the energy at hour $H - K$, where the scaling factor is given as a ratio of the corresponding averages of CS values over the grid points. Its formulation for an $H - K$ model with a horizon forecast of $K$ hours is thus

$$\widehat{PV}_H = \frac{CS_H}{CS_{H-K}}\, PV_{H-K} \tag{1}$$

where $PV_H, \widehat{PV}_H$ denote the real and predicted PV energy at an hour $H$ and $CS_H$ the averaged value of the clear sky radiances over the area under consideration. While very simple, the baseline CS models will give in general reasonable values and particularly good, hard to beat ones around noon, specially in the mid-year, mostly sunny, months.

### 3.2   The Lasso Model

Given an $N$ pattern sample $\{(x^1, y^1), \ldots, (x^N, y^N)\}$ with $p$-dimensional inputs $\mathbf{X}^p$ and 1-dimensional targets $\mathbf{y}$, the Lasso solution [7] $b^*, \mathbf{w}^*$ minimizes the $\ell_1$ regularized loss

$$\ell_L(w, b) = \frac{1}{2}\|\mathbf{y} - \mathbf{w}^T\mathbf{X} - b\|^2 + \lambda\|\mathbf{w}\|_1. \tag{2}$$

The sparsity introduced by the $\ell_1$ regularization helps to avoid possible singularities in the sample covariance matrix, particularly in cases such as ours where sample sizes of $\approx$4,500 are much smaller than the features dimension of $\approx$20,000. This is achieved because the $\ell_1$ penalty drives many coefficients towards zero; in turn, this allows automatic (i.e., wrapper-based) feature selection and makes possible model interpretation in terms of the specific grid positions of the non-zero coefficients.

### 3.3   Linear and Gaussian Support Vector Regression

The large dimension of our problem motivates our choice for the Linear SVR model, which is reportedly good in problems of high dimensionality [5]. Using the previous notation, the Linear Support Vector Regression (SVR) cost function is

$$\ell_S(w, b) = \sum_p [y^p - w \cdot x^p - b]_\epsilon + \frac{1}{C}\|w\|_2^2 \tag{3}$$

where we use $\ell_2$ regularization and the $\epsilon$-insensitive loss $\ell(y, \hat{y}) = [y - \hat{y}]_\epsilon = \max\{|y - \hat{y}| - \epsilon, 0\}$, that defines an $\epsilon$-wide, penalty-free "error tube" around the model. Notice that this loss-regularization combination is one of the different possibilities in the LIBLINEAR implementation in Scikit-learn [12].

The initial and more standard way to find the optimal $w^*, b^*$ in a SVR model is to rewrite (3) as a constrained minimization problem which is then transformed using Lagrangian theory into a much simpler dual problem, the one actually being solved; see [14]. The optimal $w^*, b^*$ are then obtained from the dual solution through the KKT equations. It turns out that the dual problem only involves patterns through their dot products and a natural extension to improve on a purely linear model is to apply the kernel trick [14]. It replaces the initial dot products $x \cdot x'$ with the values $k(x, x')$ of a positive definite kernel $k$ that can be written as $k(x, x') = \phi(x) \cdot \phi(x')$, where the $x$ are mapped through $\phi(x)$ into a larger, possibly infinite, dimensional Hilbert space $H$. We thus arrive to a non linear model $f(x) = W \cdot \phi(x) + b$ for which the optimal $W^* \in H$ can be written as $W^* = \sum \alpha_p^* \phi(x^p)$. We thus have

$$f(x) = b^* + W^* \cdot \phi(x) = b^* + \sum \alpha_p^* \phi(x^p) \cdot \phi(x) = b^* + \sum_{\alpha_p^* > 0} \alpha_p^* k(x^p, x), \quad (4)$$

where the $x^p$ for which $|\alpha_p^*| > 0$ are the Support Vectors (SVs); the standard kernel choice is the Gaussian one $e^{-\gamma \|x - x'\|^2}$. Notice that in our case, the SVs lend themselves to a temporal interpretation as the most relevant day-hour pairs, given that their radiances define the centers of the model different Gaussians.

## 3.4 Hyper-parameter Tuning

We will work with the Lasso and the LIBLINEAR and LIBSVM implementations of SVR in [12]. All these models require a careful hyper-parameter tuning to find the optimal $\lambda$ for Lasso, $C, \epsilon$ for Linear SVR and $C, \epsilon, \gamma$ for Gaussian SVR. We used for this the Python library *optunity* [4] and its default option *Particle Swarm* algorithm for hyper-parameter search. Given the natural temporal ordering of the data, we will use for this 2013 as a training set and 2014 as a validation set. Table 1 shows the optimal hyper-parameters for our models.

**Table 1.** Hyper-parameters of the Lasso and Linear SVR models.

| Model | Parameter | K | | | |
|---|---|---|---|---|---|
| | | 0 | 1 | 2 | 3 |
| Lasso | $\lambda$ | 0.020 | 0.017 | 0.012 | 0.016 |
| Linear SVR | $C$ $(\times 10^2)$ | 9.410 | 17.742 | 6.349 | 13.511 |
| | $\epsilon$ | 1.690 | 1.670 | 2.800 | 3.000 |
| Gaussian SVR | $C$ $(\times 10^3)$ | 6.820 | 18.389 | 2.559 | 10.229 |
| | $\epsilon$ | 0.011 | 0.022 | 0.018 | 0.036 |
| | $\gamma$ $(\times 10^{-3})$ | 0.241 | 0.238 | 0.231 | 0.244 |

# 4  Results

We recall that our goal is to predict PV energy production at hour $H$ from satellite readings at hour $H - K$, i.e., to work with a $K$ prediction horizon. We will consider the $K = 1, 2, 3$ horizons as well as the $K = 0$ case and denote the resulting models as m0, m1, m2 and m3; while not useful from an operative point of view, the m0 model offers a "best possible" baseline with which we can compare the others. For the Lasso and Linear SVR models we normalize features to 0 mean and 1 standard deviation; for the Gaussian SVR we scale them into a $[-1, 1]$ range to better control the Gaussian kernel behavior when dealing with largely different pattern pairs. For a more homogeneous comparison we will report errors within the time range between 08 and 20 UTC hours. PV energy production after hour 20 is negligible and it is obvious that the results before 08 UTC of, say, the m2 and, more so, the m3 models won't be good, as they have to predict substantial PV energy at 08 UTC from very small readings at 06 UTC and 05 UTC respectively.

After hyper-parameterization, the models were trained over both the 2013 and 2014 years. Table 2 shows a summary of the overall hourly average test errors over 2015 for each model and prediction horizon (we omit the $H$ persistence model for obvious reasons); its rightmost column gives the daily error averages. As it can be seen, the Gaussian (G) SVR results are clearly better than those of Lasso and linear (L) SVR essentially across all hours. Moreover, its errors degrade much more slowly as the prediction horizon increases; in fact, the average error of the Gaussian $H - 3$ model is comparable with those of the $H - 1$ Lasso and

**Table 2.** Average hourly Lasso, Linear SVR, Gaussian SVR and CS Persistence test errors.

| $K$ | Models | Hour | | | | | | | | | | | | Av. |
|---|---|---|---|---|---|---|---|---|---|---|---|---|---|---|
| | | 8 | 9 | 10 | 11 | 12 | 13 | 14 | 15 | 16 | 17 | 18 | 19 | 20 | |
| 0 | Lasso | 2.32 | 2.53 | 2.69 | 2.71 | 2.58 | 2.77 | 2.78 | 2.46 | 2.39 | 2.15 | 0.98 | 0.64 | 0.40 | 2.11 |
| | L SVR | 1.50 | 2.19 | 2.63 | 2.57 | 2.86 | 3.05 | 2.96 | 2.73 | 2.30 | 1.69 | 1.10 | 0.61 | 0.27 | 2.03 |
| | G SVR | 1.02 | 1.77 | 1.94 | 1.94 | 1.91 | 1.90 | 1.81 | 1.88 | 1.72 | 1.43 | 0.82 | 0.46 | 0.17 | 1.44 |
| 1 | CS | 18.25 | 3.99 | 3.09 | 3.04 | 2.26 | 1.34 | 1.84 | 2.66 | 2.66 | 1.74 | 0.54 | 0.23 | 0.04 | 3.21 |
| | Lasso | 2.46 | 3.19 | 3.64 | 3.43 | 3.29 | 2.85 | 2.59 | 2.44 | 2.34 | 2.70 | 1.81 | 0.82 | 0.21 | 2.44 |
| | L SVR | 1.83 | 2.51 | 2.78 | 3.01 | 3.18 | 3.34 | 3.07 | 2.96 | 3.15 | 2.55 | 1.74 | 1.00 | 0.35 | 2.42 |
| | G SVR | 1.25 | 1.80 | 2.04 | 1.85 | 1.92 | 1.82 | 1.86 | 1.86 | 1.67 | 1.45 | 1.10 | 0.54 | 0.24 | 1.49 |
| 2 | CS | 47.86 | 24.71 | 8.91 | 6.14 | 5.49 | 3.31 | 2.56 | 4.01 | 4.31 | 2.93 | 1.01 | 0.30 | 0.05 | 8.58 |
| | Lasso | 3.93 | 4.27 | 4.63 | 4.23 | 4.02 | 3.60 | 3.25 | 2.99 | 3.15 | 3.87 | 2.49 | 1.21 | 0.38 | 3.23 |
| | L SVR | 2.77 | 3.25 | 3.64 | 3.62 | 4.05 | 4.40 | 4.97 | 4.94 | 4.85 | 4.50 | 2.76 | 1.59 | 0.78 | 3.55 |
| | G SVR | 1.98 | 2.60 | 2.78 | 2.59 | 2.40 | 2.21 | 2.05 | 2.20 | 2.08 | 1.71 | 1.23 | 0.59 | 0.36 | 1.91 |
| 3 | CS | 69.06 | 45.09 | 24.72 | 12.42 | 8.43 | 6.39 | 3.45 | 4.38 | 5.13 | 3.69 | 1.41 | 0.35 | 0.05 | 14.20 |
| | Lasso | 4.66 | 4.63 | 5.66 | 5.36 | 5.04 | 4.87 | 4.21 | 3.86 | 3.66 | 4.37 | 2.95 | 1.51 | 0.62 | 3.95 |
| | L SVR | 3.10 | 3.77 | 4.06 | 4.42 | 4.85 | 5.37 | 6.14 | 6.53 | 6.34 | 5.79 | 3.71 | 2.10 | 1.13 | 4.41 |
| | G SVR | 2.75 | 3.27 | 3.69 | 3.29 | 3.19 | 2.80 | 2.61 | 2.49 | 2.37 | 1.87 | 1.30 | 0.65 | 0.30 | 2.35 |

**Table 3.** Gaussian SVR test errors of the m0 models per hour and month.

| Month | Hour | | | | | | | | | | | | | Av. |
|---|---|---|---|---|---|---|---|---|---|---|---|---|---|---|
| | 8 | 9 | 10 | 11 | 12 | 13 | 14 | 15 | 16 | 17 | 18 | 19 | 20 | |
| January | 0.37 | 1.37 | 1.96 | 1.70 | 1.74 | 1.53 | 1.47 | 1.25 | 1.28 | 0.73 | 0.21 | 0.00 | 0.00 | 1.05 |
| February | 0.70 | 1.72 | 2.08 | 1.92 | 2.12 | 2.43 | 2.26 | 2.17 | 1.49 | 1.46 | 0.42 | 0.19 | 0.00 | 1.46 |
| March | 1.07 | 2.82 | 2.58 | 2.56 | 2.27 | 2.21 | 2.11 | 2.05 | 1.86 | 1.54 | 1.05 | 0.35 | 0.00 | 1.73 |
| April | 1.24 | 1.57 | 1.35 | 1.39 | 1.59 | 1.42 | 1.15 | 1.47 | 1.83 | 1.74 | 1.21 | 0.49 | 0.23 | 1.28 |
| May | 1.04 | 1.06 | 1.99 | 1.45 | 1.31 | 1.31 | 1.35 | 1.48 | 1.16 | 1.22 | 1.04 | 0.86 | 0.18 | 1.19 |
| June | 1.13 | 1.48 | 1.62 | 1.33 | 1.36 | 1.44 | 1.91 | 1.87 | 1.76 | 2.17 | 1.67 | 1.33 | 0.60 | 1.51 |
| July | 1.41 | 2.18 | 2.22 | 2.07 | 1.53 | 1.19 | 1.44 | 2.05 | 2.34 | 2.68 | 1.28 | 0.76 | 0.44 | 1.66 |
| August | 1.66 | 2.16 | 2.33 | 2.81 | 2.67 | 2.52 | 2.61 | 2.72 | 2.47 | 1.91 | 1.52 | 0.99 | 0.59 | 2.07 |
| September | 1.02 | 1.47 | 1.49 | 1.65 | 2.05 | 1.86 | 1.88 | 1.59 | 1.39 | 1.49 | 0.73 | 0.55 | 0.01 | 1.32 |
| October | 1.34 | 1.61 | 1.66 | 2.29 | 2.24 | 2.50 | 1.82 | 2.04 | 1.81 | 1.06 | 0.42 | 0.00 | 0.00 | 1.45 |
| November | 0.80 | 1.73 | 1.39 | 1.44 | 1.71 | 2.17 | 1.71 | 1.61 | 1.52 | 0.58 | 0.22 | 0.01 | 0.01 | 1.15 |
| December | 0.47 | 2.05 | 2.67 | 2.64 | 2.28 | 2.22 | 1.98 | 2.34 | 1.76 | 0.53 | 0.01 | 0.00 | 0.00 | 1.46 |
| Average | 1.02 | 1.77 | 1.94 | 1.94 | 1.91 | 1.90 | 1.81 | 1.88 | 1.72 | 1.43 | 0.82 | 0.46 | 0.17 | 1.44 |

**Table 4.** Gaussian SVR test errors of the m1 models per hour and month.

| Month | Hour | | | | | | | | | | | | | Av. |
|---|---|---|---|---|---|---|---|---|---|---|---|---|---|---|
| | 8 | 9 | 10 | 11 | 12 | 13 | 14 | 15 | 16 | 17 | 18 | 19 | 20 | |
| January | 0.61 | 1.58 | 2.26 | 1.66 | 1.91 | 1.59 | 1.53 | 1.60 | 1.22 | 0.83 | 0.50 | 0.00 | 0.00 | 1.18 |
| February | 0.99 | 2.51 | 2.38 | 1.82 | 1.89 | 1.86 | 2.37 | 2.16 | 1.92 | 1.41 | 0.67 | 0.14 | 0.00 | 1.55 |
| March | 1.61 | 1.90 | 3.31 | 2.26 | 2.45 | 2.29 | 2.08 | 2.29 | 1.64 | 1.42 | 1.14 | 0.34 | 0.00 | 1.75 |
| April | 1.73 | 1.28 | 1.71 | 1.56 | 1.41 | 1.77 | 1.66 | 1.50 | 1.39 | 1.33 | 1.34 | 1.01 | 0.36 | 1.39 |
| May | 1.42 | 1.15 | 1.83 | 1.36 | 1.42 | 1.19 | 1.46 | 1.42 | 1.39 | 1.40 | 1.00 | 0.77 | 0.45 | 1.25 |
| June | 1.32 | 1.70 | 1.85 | 1.54 | 1.08 | 1.19 | 1.33 | 1.67 | 1.85 | 2.39 | 2.49 | 1.20 | 0.65 | 1.56 |
| July | 1.38 | 1.98 | 1.96 | 1.65 | 1.17 | 0.94 | 0.94 | 1.27 | 1.82 | 2.40 | 1.99 | 1.05 | 0.76 | 1.49 |
| August | 1.36 | 2.06 | 2.09 | 2.20 | 2.55 | 2.49 | 2.34 | 2.52 | 2.44 | 2.05 | 1.59 | 1.35 | 0.67 | 1.98 |
| September | 1.43 | 1.72 | 1.48 | 1.77 | 1.81 | 1.66 | 1.71 | 1.72 | 1.42 | 1.49 | 1.26 | 0.64 | 0.01 | 1.39 |
| October | 1.76 | 2.32 | 1.75 | 1.95 | 2.47 | 2.09 | 2.26 | 1.56 | 1.43 | 1.28 | 0.80 | 0.00 | 0.00 | 1.51 |
| November | 0.88 | 1.69 | 1.72 | 1.83 | 1.55 | 1.61 | 1.75 | 1.85 | 1.49 | 0.84 | 0.34 | 0.01 | 0.01 | 1.20 |
| December | 0.54 | 1.74 | 2.11 | 2.56 | 3.36 | 3.12 | 2.92 | 2.80 | 2.07 | 0.63 | 0.01 | 0.00 | 0.00 | 1.68 |
| Average | 1.25 | 1.80 | 2.04 | 1.85 | 1.92 | 1.82 | 1.86 | 1.86 | 1.67 | 1.45 | 1.10 | 0.54 | 0.24 | 1.49 |

linear SVR models and quite close to their $H$ models errors. The behavior of the CS persistence model is also remarkable. Its average errors are much larger but this is due to the expectedly bad behavior of the $H - K$ models at the beginning of the day, where they shouldn't be used. On the other hand, its

**Table 5.** Gaussian SVR test errors of the m2 models per hour and month.

| Month | Hour | | | | | | | | | | | | | Av. |
|---|---|---|---|---|---|---|---|---|---|---|---|---|---|---|
| | 8 | 9 | 10 | 11 | 12 | 13 | 14 | 15 | 16 | 17 | 18 | 19 | 20 | |
| January | 0.94 | 1.77 | 3.37 | 2.86 | 2.30 | 2.03 | 1.53 | 1.75 | 1.69 | 0.88 | 0.36 | 0.00 | 0.00 | 1.50 |
| February | 1.06 | 4.22 | 4.50 | 3.43 | 3.04 | 2.85 | 2.53 | 2.85 | 2.49 | 1.86 | 1.22 | 0.11 | 0.00 | 2.32 |
| March | 2.61 | 3.26 | 2.98 | 4.27 | 3.71 | 2.92 | 2.62 | 2.55 | 2.17 | 1.79 | 1.07 | 0.36 | 0.00 | 2.33 |
| April | 2.64 | 2.47 | 1.94 | 2.47 | 1.95 | 1.97 | 2.31 | 2.15 | 1.90 | 1.79 | 1.53 | 0.92 | 0.60 | 1.89 |
| May | 2.86 | 2.11 | 2.13 | 1.22 | 1.60 | 1.44 | 1.28 | 1.66 | 1.33 | 1.71 | 1.38 | 0.73 | 0.45 | 1.53 |
| June | 2.22 | 2.33 | 2.53 | 2.09 | 1.49 | 1.16 | 1.27 | 1.36 | 2.01 | 2.34 | 2.46 | 1.74 | 0.85 | 1.84 |
| July | 2.14 | 2.29 | 2.25 | 2.07 | 1.61 | 1.31 | 1.18 | 1.32 | 1.73 | 2.45 | 2.44 | 1.38 | 1.05 | 1.78 |
| August | 2.16 | 2.22 | 2.19 | 2.39 | 2.07 | 2.64 | 2.37 | 2.69 | 2.95 | 2.95 | 2.05 | 1.13 | 1.31 | 2.24 |
| September | 2.24 | 2.26 | 2.49 | 2.26 | 2.33 | 1.91 | 2.05 | 2.42 | 2.41 | 1.71 | 1.33 | 0.64 | 0.01 | 1.85 |
| October | 2.60 | 3.33 | 3.32 | 2.66 | 3.01 | 2.59 | 2.24 | 2.37 | 2.09 | 1.40 | 0.61 | 0.00 | 0.00 | 2.02 |
| November | 1.33 | 2.57 | 2.53 | 2.58 | 2.44 | 1.91 | 1.66 | 1.75 | 1.47 | 0.72 | 0.23 | 0.01 | 0.01 | 1.48 |
| December | 0.98 | 2.43 | 3.12 | 2.72 | 3.28 | 3.73 | 3.53 | 3.54 | 2.73 | 0.91 | 0.01 | 0.00 | 0.00 | 2.08 |
| Average | 1.98 | 2.60 | 2.78 | 2.59 | 2.40 | 2.21 | 2.05 | 2.20 | 2.08 | 1.71 | 1.23 | 0.59 | 0.36 | 1.91 |

**Table 6.** Gaussian SVR test errors of m3 models per hour and month.

| Month | Hour | | | | | | | | | | | | | Av. |
|---|---|---|---|---|---|---|---|---|---|---|---|---|---|---|
| | 8 | 9 | 10 | 11 | 12 | 13 | 14 | 15 | 16 | 17 | 18 | 19 | 20 | |
| January | 1.26 | 2.39 | 3.57 | 4.14 | 4.24 | 3.02 | 2.24 | 2.06 | 1.81 | 1.07 | 0.34 | 0.00 | 0.00 | 2.01 |
| February | 1.86 | 4.52 | 6.11 | 4.85 | 4.02 | 3.95 | 3.43 | 2.87 | 3.20 | 2.26 | 1.19 | 0.47 | 0.00 | 2.98 |
| March | 2.23 | 4.59 | 4.47 | 4.02 | 5.25 | 4.84 | 3.77 | 3.47 | 2.80 | 2.11 | 1.27 | 0.40 | 0.00 | 3.02 |
| April | 4.45 | 2.87 | 3.23 | 3.24 | 3.30 | 2.53 | 2.85 | 2.88 | 2.66 | 2.16 | 1.47 | 0.79 | 0.70 | 2.55 |
| May | 4.19 | 3.08 | 3.52 | 2.03 | 1.71 | 1.95 | 1.82 | 1.68 | 1.84 | 1.78 | 1.48 | 0.79 | 0.46 | 2.03 |
| June | 3.11 | 2.59 | 3.11 | 2.51 | 1.90 | 1.55 | 1.48 | 1.50 | 1.81 | 2.27 | 2.41 | 1.71 | 0.98 | 2.07 |
| July | 3.18 | 2.70 | 2.23 | 1.82 | 1.61 | 1.42 | 1.55 | 1.55 | 1.68 | 2.14 | 2.35 | 1.53 | 0.70 | 1.88 |
| August | 3.75 | 2.54 | 2.75 | 2.52 | 2.60 | 2.86 | 2.82 | 2.87 | 3.07 | 3.45 | 2.60 | 1.17 | 0.68 | 2.59 |
| September | 3.77 | 3.17 | 3.32 | 3.50 | 3.11 | 2.58 | 2.51 | 2.74 | 2.62 | 1.78 | 1.38 | 0.90 | 0.01 | 2.41 |
| October | 2.80 | 4.80 | 3.97 | 3.68 | 3.61 | 3.11 | 2.99 | 2.69 | 2.28 | 1.52 | 0.72 | 0.00 | 0.00 | 2.48 |
| November | 1.44 | 3.10 | 3.63 | 2.96 | 3.29 | 2.38 | 2.05 | 1.79 | 1.66 | 0.96 | 0.41 | 0.01 | 0.01 | 1.82 |
| December | 1.01 | 2.89 | 4.40 | 4.26 | 3.64 | 3.45 | 3.82 | 3.80 | 2.97 | 0.99 | 0.01 | 0.00 | 0.00 | 2.40 |
| Average | 2.75 | 3.27 | 3.69 | 3.29 | 3.19 | 2.80 | 2.61 | 2.49 | 2.37 | 1.87 | 1.30 | 0.65 | 0.30 | 2.35 |

evening errors tend to be the best ones and its afternoon errors are comparable and sometimes better than those of Lasso and linear SVR. A plausible reason is the many near-clear sky days of the Iberian Peninsula, particularly in areas with large PV productions.

The clearly best average behavior of the G SVR model also holds when we consider its hourly errors on a monthly basis. They are given in Tables 3, 4, 5 and 6 and, as it can be seen, hourly errors also increase quite moderately with the prediction horizons for all months. Moreover monthly average errors are quite stable; the worst months seem to be February and March, most likely because of the atmospheric instability to be expected in them. August also shows large errors, the reason being here its high PV energy production, much larger than the one from, say, November to February. Summing things up, satellite information combined with Gaussian SVR can be used to set up a PV energy nowcasting procedure with a quite good performance.

## 5   Conclusions and Further Work

Satellite-based information has been widely used to nowcast solar irradiance values and PV energy productions, usually from an atmospheric physics perspective. Here we propose the direct exploitation of these readings by feeding them into ML models to nowcast the PV energy production of Peninsular Spain, predicting the energy at hour $H$ from Meteosat satellite data at hour $H - K$, where $K = 1, 2, 3$ for nowcasting purposes and $K = 0$ for model benchmarking and control. We have worked with the radiances of the IR016, IR039 and VIS008 channels and the brightness temperature of channel IR039, which we downsample from Meteosat's initial finer resolution to a coarser $0.125°$ one. Moreover we have added at each point of the resulting grid a theoretical estimation of its Clear Sky (CS) irradiance value and considered three well known ML models, Lasso and linear and Gaussian SVR, plus a simple CS-based persistence approach.

Our results here greatly simplify and improve previous results in [3]; in particular, Gaussian SVR over satellite and CS features gives remarkable results, with rather low mean absolute errors that degrade slowly when the prediction horizon $K$ increases. While also good, the performance of Lasso and linear SVR is clearly below that of Gaussian SVRs. On the other hand, the performance of the CS-based persistence is rather good around noon and at the evenings, although markedly worse, as it was to be expected, in the morning. Another advantage of using CS features is that they allow adding time effects into the models and make possible to use a single model across all day hours instead of the morning, afternoon and evening models used in [3].

The results here are given for hourly energy prediction updates but this can be easily extended to 15' updates, given that satellite readings are available at that frequency. Besides this improvement, there are other areas for further work. A clear one is to consider longer prediction horizons. The large variations of PV energy across long horizons makes this a difficult problem but, on the other hand, the slow increase of the Gaussian SVR errors also points to a possibly solid behavior when applied beyond the $K = 3$ horizon considered here. Another improvement is the addition of other features. A clear option is using day-ahead NWP radiance predictions, that should also particularly help when longer horizons are considered. On the other hand, this will make pattern dimensions even greater than they are now and some form of feature reduction is likely to help. Lasso does precisely that, but as our results show, it doesn't provide by itself the best forecasts. This suggests to explore Lasso models as a kind of feature selectors that are then further exploited by Gaussian SVRs. In this line, and given the natural grouping of features by grid points, it may also be interesting to use for this group versions of Lasso, as those proposed in [1]. Finally, it is also important to consider nowcasting at reduced areas, such as individual plants or islands, which may be isolated form larger grids and where PV energy fluctuations are harder to manage. We are currently studying these issues.

**Acknowledgments.** With partial support from Spain's grants TIN2013-42351-P, TIN2016-76406-P, TIN2015-70308-REDT and S2013/ICE-2845 CASI-CAM-CM. Work supported also by project FACIL–Ayudas Fundación BBVA a Equipos de Investigación Científica 2016, and the UAM–ADIC Chair for Data Science and Machine Learning. The second author is also supported by the FPU–MEC grant AP-2012-5163. We thank Red Eléctrica de España for useful discussions and making available PV energy data and gratefully acknowledge the use of the facilities of Centro de Computación Científica (CCC) at UAM.

# References

1. Alaíz, C.M., Dorronsoro, J.R.: The generalized group lasso. In: International Joint Conference on Neural Networks, IJCNN 2015, Killarney, Ireland, 12–17 July 2015, pp. 1–8 (2015)
2. Antonanzas, J., Osorio, N., Escobar, R., Urraca, R., Martinez de Pison, F.J., Antonanzas-Torres, F.: Review of photovoltaic power forecasting. Sol. Energy **136**, 78–111 (2016)
3. Catalina, A., Torres-Barrán, A., Dorronsoro, J.R.: Machine learning prediction of photovoltaic energy from satellite sources. In: Woon, W.L., Aung, Z., Kramer, O., Madnick, S. (eds.) DARE 2016. LNCS, vol. 10097, pp. 31–42. Springer, Cham (2017). doi:10.1007/978-3-319-50947-1_4
4. Claesen, M., Simm, J., Popovic, D., Moreau, Y., De Moor, B.: Easy hyperparameter search using Optunity (2014). arXiv preprint arXiv:1412.1114
5. Fan, R.-E., Chang, K.-W., Hsieh, C.-J., Wang, X.-R., Lin, C.-J.: LIBLINEAR: a library for large linear classifcation. J. Mach. Learn. Res. **9**(August), 1871–1874 (2008)
6. Hammer, A., Heinemann, D., Hoyer, C., Kuhlemann, R., Lorenz, E., Müller, R., Beyer, H.G.: Solar energy assessment using remote sensing technologies. Remote Sens. Environ. **86**(3), 423–432 (2003)
7. Hastie, T., Tibshirani, R., Friedman, J.: The Elements of Statistical Learning. Springer, New York (2009)
8. Ineichen, P., Perez, R.: A new airmass independent formulation for the linke turbidity coefficient. Sol. Energy **73**(3), 151–157 (2002)
9. Inman, R.H., Pedro, H., Coimbra, C.: Solar forecasting methods for renewable energy integration. Prog. Energy Combust. Sci. **39**(6), 533–576 (2013)
10. Kühnert, J., Lorenz, E., Heinemann, D.: Satellite-based irradiance and power forecasting for the German energy market. In: Kleissl, J. (ed.) Solar Energy Forecasting and Resource Assessment, pp. 267–297. Academic Press, Cambridge (2013)

11. Mohammed, A.A., Yaqub, W., Aung, Z.: Probabilistic forecasting of solar power: an ensemble learning approach. In: Neves-Silva, R., Jain, L.C., Howlett, R.J. (eds.) Intelligent Decision Technologies. SIST, vol. 39, pp. 449–458. Springer, Cham (2015). doi:10.1007/978-3-319-19857-6_38

12. Pedregosa, F., Varoquaux, G., Gramfort, A., Michel, V., Thirion, B., Grisel, O., Blondel, M., Prettenhofer, P., Weiss, R., Dubourg, V., Vanderplas, J., Passos, A., Cournapeau, D., Brucher, M., Perrot, M., Duchesnay, E.: Scikit-learn: machine learning in Python. J. Mach. Learn. Res. **12**, 2825–2830 (2011)

13. Photovoltaic Performance Modeling Collaborative Group: The PVLIB-Python Library. Sandia National Laboratory

14. Schölkopf, B., Smola, A.J.: Learning with Kernels: Support Vector Machines, Regularization, Optimization, and Beyond. MIT Press, Cambridge (2001)

15. Tjemkes, S., Stuhlmann, R., Hewison, T., Müller, J., Gartner, V., Rota, S.: The conversion from effective radiances to equivalent brightness temperatures. Technical report, EUMETSAT 10 (2012)

16. Wan, C., Zhao, J., Song, Y., Zhao, X., Lin, J., Zechun, H.: Photovoltaic and solar power forecasting for smart grid energy management. CSEE J. Power Energy Syst. **1**, 38–46 (2015)

17. Wolff, B., Kühnert, J., Lorenz, E., Kramer, O., Heinemann, D.: Comparing support vector regression for PV power forecasting to a physical modeling approach using measurement, numerical weather prediction, and cloud motion data. Sol. Energy **135**, 197–208 (2016)

# A Study on Feature Selection Methods for Wind Energy Prediction

Rubén Martín-Vázquez$^{(\boxtimes)}$, Ricardo Aler, and Inés M. Galván

EVANNAI, Carlos III University of Madrid,
Avda. Universidad, 30, 28911 Leganés, Spain
ruben.martin@uc3m.es

**Abstract.** This work deals with wind energy prediction using meteorological variables estimated by a Numerical Weather Prediction model in a grid around the wind farm of interest. Two machine learning techniques have been tested, Support Vector Machine and Gradient Boosting Regression, in order to study their performance and compare the results. The use of meteorological variables estimated in a grid generally implies a large number of inputs to the models and the performance of models might decrease. Hence, in this context, the use of feature selection algorithms might be interesting in order to improve the generalization capability of models and/or reduce the number of attributes. We have compared three feature selection techniques based on different paradigms: Principal Components Analysis, ReliefF, and Sequential Forward Selection. Energy production data has been obtained from the Sotavento experimental wind farm. Meteorological variables have been obtained from European Centre for Medium-Range Weather Forecasts, for a $5 \times 5$ grid around Sotavento.

**Keywords:** Wind power prediction · Numerical weather prediction · Support Vector Machine · Gradient Boosting · Feature selection

## 1 Introduction

The prediction of wind energy is becoming an important issue in the context of renewable energy sources. However, its uncertainty makes it difficult to integrate in the electricity grid. Machine learning techniques can be used to improve the quality of wind energy prediction. There is a predominance of articles that use Artificial Neural Networks (ANN) to make predictions using historical data [1,2]. Damousis et al. [3] use fuzzy logic and genetic algorithms to predict wind power using as inputs the wind direction and speed. There are also several articles that use Support Vector Machines (SVM) to make predictions for wind energy. For example, Mohandes et al. [4] predict wind speed using historical wind speed data and later, it is compared with an ANN. Heinermann and Kramer [5], on the other hand, use SVM to predict wind power using wind speed as input. There are other articles that use Numerical Weather Prediction (NWP) to estimate wind power energy. Alonso et al. [6] use Random Forests and Gradient Boosting

© Springer International Publishing AG 2017
I. Rojas et al. (Eds.): IWANN 2017, Part I, LNCS 10305, pp. 698–707, 2017.
DOI: 10.1007/978-3-319-59153-7_60

with 8 meteorological variables for predicting wind energy at Sotavento. In a
similar way Martín et al. [7] use meteorological variables to make solar energy
predictions in the state of Oklahoma.

In this article, we address the problem of wind energy prediction from meteorological variables estimated by a Numerical Weather Prediction (NWP) model.
Two machine learning algorithms have been used to train the model: Support
Vector Machines (SVM) [8,9] and Gradient Tree Boosting (GTB) [10]. The aforementioned comparison of models has been tested on the Sotavento experimental
wind farm, from which electricity production data has been obtained. The inputs
to the models are 22 meteorological variables obtained from the European Centre for Medium-Range Weather Forecasts (ECMWF), for a $5 \times 5$ grid around
Sotavento.

The use of a grid implies a large number of input features for the machine
learning models. Given that not all variables at all locations might be relevant,
and that many of them might contain redundant information, in this article we
explore the application of feature selection algorithms. Feature selection is an
important task when datasets have a large number of variables. Reducing the
number of variables can sometimes improve the performance of models. The aim
is to study whether the number of features can be reduced, while maintaining
or improving the generalization ability of the model. Attribute selection methods may be grouped as filter and wrapper techniques [11]. Filter approaches use
general characteristics of data to evaluate and select attribute subsets, while
wrapper approaches select attributes according to the evaluation provided by a
specific regression algorithm [12]. In this work, we have tried three feature selection algorithms, based on different paradigms: Principal Component Analysis
(PCA), ReliefF, and Sequential Forward Selection (SFS) [11]. PCA is a widely
used unsupervised method for dimensionality reduction and very useful when
redundant variables are present. ReliefF is a filter method, that ranks attributes
according to their relevance, but it is prone to select redundant attributes. SFS is
a wrapper method that greedily adds the most relevant features for a particular
classifier.

The rest of the paper is organized as follows. In Sect. 2 it is explained the
data source and how it is preprocessed. Section 3 presents techniques that are
used in the present work. Results are presented in Sect. 4. Finally, in Sect. 5 some
conclusions are drawn and further work is presented.

## 2   Data

The Sotavento experimental wind farm is located in "A Serra da Loba", concretely at coordinates N: 43.354377 W: 7.8812133. From the website of Sotavento
it is possible to download historical: "wind speed" given in m/s, "direction" given
in degrees and "energy" given in kWh. For the experiment, years from 2005 to
2010 are selected with 10 min measurement interval.

The European Centre for Medium-Range Weather Forecasts, ECMWF, is
an independent intergovernmental organization. It operates one of the largest

supercomputer complexes in Europe and the world's largest archive of numerical weather prediction data. On their website there is a large number of numerical weather prediction models and reanalysis made based on predictions. In this work, to extract input variables the model chosen is ERA-20C because it has a high number of variables and the measurements dates coincide with the Sotavento data. Although some variables are clearly more related to wind than others, we have preferred to include a large amount of variables and let the models combine them in an optimal way, or the feature selection algorithms choose the most relevant ones. The 22 variables selected are shown in Table 1.

**Table 1.** List of variables used.

| Variables | |
|---|---|
| 2 m temperature (K) | 10 m U wind component $(ms^{-1})$ |
| 10 m V wind component $(ms^{-1})$ | 100 m U wind component $(ms^{-1})$ |
| 100 m V wind component $(ms^{-1})$ | Convective available potential energy $(Jkg^{-1})$ |
| Forecast logarithm of surface roughness for heat | Forecast surface roughness (m) |
| Instantaneous eastward turbulent surface stress $(Nm^{-2})$ | Instantaneous northward turbulent surface stress $(Nm^{-2})$ |
| Leaf area index, high vegetation $(m^2m^{-2})$ | Leaf area index, low vegetation $(m^2m^{-2})$ |
| Neutral wind at 10 m u-component $(ms^{-1})$ | Neutral wind at 10 m v-component $(ms^{-1})$ |
| Soil temperature level 1 (K) | Soil temperature level 2 (K) |
| Soil temperature level 3 (K) | Soil temperature level 4 (K) |
| Surface pressure (Pa) | Vertical integral of temperature $(Kkgm^{-2})$ |
| Vertical integral of divergence of kinetic energy flux $(Jm^{-2})$ | Vertical integral of water vapor $(kgm^{-2})$ |

Those variables are obtained in a 5 × 5 grid around Sotavento as shown in Fig. 1 (Sotavento itself is marked with letter A). The distance between the coordinates of the grid is 0.125°.

The years ranging from 2005 to 2010 are selected, both included, as they are the same as in Sotavento data. Although Sotavento provides energy generation data every 10 min, ERA-20C provides complete records only every six hours, therefore the latter is the time period used in this article. The data set has 22 variables and each variable has 25 coordinate measurement per instant of time, that is, 550 columns with input data. For each day, there are 4 predictions. As there are 6 years and one of them is a leap year, there are 8764 instances. Instances where one (or several) of the variables contain NAs (Not Available

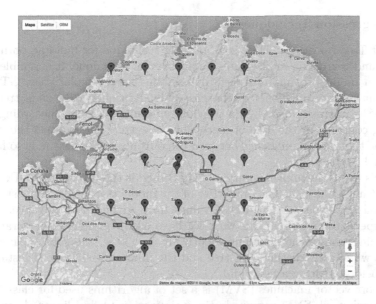

**Fig. 1.** Grid centered at Sotavento

data), are removed from the dataset. Only full instances are kept, resulting in 5778 instances. Table 2 shows the number of instances broken down by year.

**Table 2.** Number of instances by year.

| Year | Instances |
|------|-----------|
| 2005 | 1246 |
| 2006 | 1268 |
| 2007 | 1117 |
| 2008 | 178 |
| 2009 | 913 |
| 2010 | 1056 |

## 3 Techniques

In this section, techniques that are going to be used in this paper will be briefly explained: the two regression techniques (SVM and GTB) and the three feature selection techniques (PCA, ReliefF, and SFS).

Performance of methods has been evaluated using the Mean Absolute Error (MAE), given by:

$$MAE = \frac{1}{n} \sum_{t=1}^{n} |e_t| \qquad (1)$$

where $n$ is the number of patterns and $|e_t|$ is the absolute error for each pattern.

## 3.1  Regression Techniques

Gradient Tree Boosting (GTB) is a boosting technique for building both classification and regression models [10]. As a boosting technique, it ensembles weak prediction models to improve performance. In this particular case, GTB uses regression trees as members of the ensemble. It sequentially adds trees that focus on the mistakes made by the previous ensemble members, according to the boosting concept. In the case of regression, new trees try to learn the pseudo-residuals, that is, the difference between the outputs of the ensemble so far, and the actual outputs.

In this work, the implementation of GTB used is the R package XGBoost created by Chen et al. [13]. The properties of the model trained by Xgboost depend on several hyper-parameters whose values have to be set in advance. Among these hyper-parameters, in this work we have used the number of iterations in the training phase, the maximum depth for each tree, the subsample ratio of columns when constructing each tree and the minimum number of instances in a node before creating a new child.

Support vector machines (SVM) is a set of algorithms used for classification and regression [8]. SVM use quadratic optimization to find linear classification boundaries that maximize the margin between the two classes. Large margins give better generalization capabilities. The non linear case is reduced to the linear one by projecting data into larger dimensionality spaces. This projection is done implicitly by means of kernel functions. SVM can also be used for general regression prediction. In this case, it searches for a function that has a smaller error than a value given as a hyper-parameter and, at the same time, a function as flat as possible.

In this work, the implementation of SVM used is from the R package e1071 created by Meyer et al. [14]. A gaussian kernel is chosen, and the hyper-parameters used in this algorithm are the cost parameter (C) to adjust the error tolerance and gamma to adjust the variance of the model.

## 3.2  Feature Selection Algorithms

Principal Component Analysis (PCA) is a linear transformation of the input attributes of the original dataset into a set of components that try to keep as much variance as possible while reducing linear correlation between attributes. PCA ranks attributes according to the explained variance and can be used for dimensionality reduction [15].

The ReliefF algorithm [16,17] is a robust method to feature interactions. It estimates the quality of attributes according to how well their values distinguish between instances that are near to each other. It evaluates an attribute by repeatedly sampling an instance and considering the value of the given attribute for the nearest instance of the same and different class. ReliefF can also be used in regression problems. In our work, we have used the implementation of CORElearn provided for the R language [18].

Sequential Forward Selection (SFS) is the simplest greedy search algorithm but it is also very effective [11]. The SFS algorithm starts with an empty feature set X. Then, repeatedly, it adds the most significant feature with respect to X. A feature is considered significant with respect to X if a model trained with X and the feature obtains the minimum error, compared to the rest of features. The algorithm stops when no feature improves the result from the last iteration. SFS is a wrapper technique because it uses a machine learning model in order to determine the degree of relevance of a subset of attributes.

## 4    Experimental Results

In order to evaluate the techniques, 6-fold crossvalidation has been used, where each of the folds is a complete year (2005, 2006, 2007, 2008, 2009, and 2010). First, models using the 22 meteorological variables in the $5 \times 5$ grid (550 input features) are trained with XGBoost and SVM. In order to achieve optimal results, their parameters need to be tuned. For this purpose, we have used grid search, a systematic procedure that evaluates all possible combinations of parameter values and selects the best performing one. For XGBoost, the parameters tuned and their explored ranges are: nrounds = {50, 100, 200, 300, 400, 500, 1000}, max_depth = {1, 2, 3, 4, 5}, colsample_bytree = {0, 0.2, 0.4, 0.6, 0.8, 1}, and min_child_weight = {1, 2, 3}. For SVM, two parameters have been adjusted: Gamma = {0, 0.01, 0.1, 0.25, 0.5, 0.75, 1}, and C = {0.1, 1, 10, 100, 500}. Parameter tuning is carried out for each fold. For instance, if 2005 is the test fold, years from 2006 to 2010 are used for grid search. In the case of SVM, grid search yields the same parameter values for all folds: C = 1 and Gamma = 0.01. In the case of XGBoost, parameter values change only slightly depending on the fold. Table 3 show the best parameters for each.

**Table 3.** Best XGBoost parameter values.

| Year | Nrounds | MaxDepth | Colsample bytree | Child weight |
|------|---------|----------|------------------|--------------|
| 2006 | 50      | 3        | 0.6              | 1            |
| 2007 | 100     | 3        | 0.8              | 1            |
| 2008 | 100     | 3        | 0.6              | 1            |
| 2009 | 100     | 3        | 0.6              | 1            |
| 2010 | 150     | 2        | 0.6              | 1            |

Table 4 displays the test Mean Absolute Error (MAE) (Eq. 1) for SVM and XGboost for every fold. The last two rows shows the average MAE and the standard deviation. Results show that SVM performs better than XGboost on average and also for most of the folds. Since SVM reports the best results, the feature selection algorithms will be carried out only for SVM.

**Table 4.** SVM and XGBoost test MAE broken down by fold.

| Test year | SVM | XGBoost |
|---|---|---|
| 2005 | 267.63 | 281.78 |
| 2006 | 240.33 | 257.62 |
| 2007 | 277.17 | 272.67 |
| 2008 | 261.75 | 270.78 |
| 2009 | 275.44 | 291.19 |
| 2010 | 286.04 | 289.25 |
| Average | 268.06 | 277.22 |
| Std | 15.94 | 12.71 |

Feature selection algorithms have been applied by means of the following methodology. For each fold, the train partition is decomposed into a train and validation sets by choosing, every three days, the first two days for training and the last one for validation. Then, each feature selection algorithm is applied on the training set, returning a ranking of the attributes. Finally, the subset of n attributes that minimizes the error on the validation set is selected. This is done by training a model with the n first attributes (according to each method's ranking) and evaluating it on the validation set, for n from 1 to the total number of attributes. Once the optimal n is known, the train and validation sets are joined, and a final model is trained with n features, and then tested on the test set. This strategy is similar to the one used in the Caret R package [19,20] for parameter tuning.[1] It allows using a larger data partition for training the final model, and also, it uses the same partition for training than the all-features model (results shown in Table 4). Hence, the comparison is more fair. In any case, it is important to remark that all models are finally evaluated on a completely independent partition (the test set).

Table 5 shows the results of the three selection methods with SVM. It can be seen that although PCA is the method whose reduction in number of features is largest, the test MAE is always worse than the rest of methods. Also, PCA MAE is much larger than the error obtained with the complete set of initial features (287.33 vs. 268.06). SFS is also able to drastically reduce the number of features (from 550 to 26), while returning an error only slightly worse than the original one (270.49 vs. 268.06). Finally, ReliefF manages to improve the error (263.24 vs. 268.06) while also reducing the number of features, but not to a large extent as the previous techniques (from 550 to 455.83, on average). If results are analyzed per fold, comparing Tables 4 and 5, it can be seen that PCA obtains worse MAEs for all folds, SFS improves MAE in two folds, and ReliefF improves four folds and the other two are very similar.

---

[1] https://topepo.github.io/caret/model-training-and-tuning.html.

**Table 5.** Test MAE for different feature selection algorithms and number of features, broken down by fold.

| Test year | PCA | | ReliefF | | SFS | |
|-----------|-------|-------------|-------|-------------|-------|-------------|
| | MAE | N. Features | MAE | N. Features | MAE | N. Features |
| 2005 | 285.42 | 17 | 268.31 | 550 | 264.70 | 28 |
| 2006 | 259.11 | 17 | 237.27 | 479 | 249.00 | 25 |
| 2007 | 292.71 | 19 | 265.26 | 343 | 278.21 | 27 |
| 2008 | 290.60 | 27 | 249.49 | 360 | 264.08 | 45 |
| 2009 | 280.96 | 17 | 273.86 | 550 | 273.95 | 15 |
| 2010 | 315.16 | 16 | 285.22 | 453 | 293.01 | 16 |
| Average | 287.33 | 18.83 | 263.24 | 455.83 | 270.49 | 26 |
| Std | 18.19 | 4.12 | 17.26 | 89.66 | 14.94 | 10.84 |

# 5 Conclusions

In this article three feature selection methods have been compared for wind energy prediction at the Sotavento experimental farm from ECMWF meteorological variables, estimated for a $5 \times 5$ grid. 22 ECMWF variables have been chosen, resulting in 550 input features, which might contain redundant information. Experimental results have been carried out using data for six years. The cross-validation methodology has been used, using each year as a test fold, in order to validate the methods more extensively and avoid biases from individual years.

Experimental results show that the number of input features can be reduced, while maintaining the generalization capability of models. The reduction of features depends on the feature selection algorithm. In this study, the three feature selection techniques behave differently with respect to error and number of selected features. ReliefF obtains the best performance in terms of error (in fact, ReliefF is able to improve the error over the initial model with all the features). But ReliefF does not reduce the number of features as drastically as the others. SFS has a good balance between error and number of features selected: the error is only slightly worse than the all-features model and it reduces the 550 original features to only 26. PCA provides the worse performance because its associated error is comparatively worse than the original model. This can be due to PCA being an unsupervised method, while both ReliefF and SFS are supervised. The large difference between ReliefF and SFS in the number of selected features might be explained because ReliefF ranks attributes according to relevance, but it does not discard redundant attributes. That is, if several attributes are relevant but correlated, they will all be selected. With SFS, the attribute that add the most information to the current set is included, and therefore, relevant but redundant attributes will be likely ranked last.

As future work, it could be interesting to study the influence of using past values of wind energy as input to machine learning models, together with the

706 R. Martín-Vázquez et al.

most relevant features (meteorological variables and locations in the grid) found in this work. We expect that this information could improve the performance of models. We also intend to study the performance of wind energy forecasting models at different lead times.

**Acknowledgements.** The authors acknowledge financial support granted by the Spanish Ministry of Science under contract ENE2014-56126-C2-2-R.

# References

1. Li, G., Shi, J.: On comparing three artificial neural networks for wind speed forecasting. Appl. Energy **87**(7), 2313–2320 (2010)
2. Cao, Q., Ewing, B.T., Thompson, M.A.: Forecasting wind speed with recurrent neural networks. Eur. J. Oper. Res. **221**(1), 148–154 (2012)
3. Damousis, I.G., Alexiadis, M.C., Theocharis, J.B., Dokopoulos, P.S.: A fuzzy model for wind speed prediction and power generation in wind parks using spatial correlation. IEEE Trans. Energy Convers. **19**(2), 352–361 (2004)
4. Mohandes, M.A., Halawani, T.O., Rehman, S., Hussain, A.A.: Support vector machines for wind speed prediction. Renew. Energy **29**(6), 939–947 (2004)
5. Heinermann, J., Kramer, O.: Precise wind power prediction with SVM ensemble regression. In: Wermter, S., Weber, C., Duch, W., Honkela, T., Koprinkova-Hristova, P., Magg, S., Palm, G., Villa, A.E.P. (eds.) ICANN 2014. LNCS, vol. 8681, pp. 797–804. Springer, Cham (2014). doi:10.1007/978-3-319-11179-7_100
6. Alonso, Á., Torres, A., Dorronsoro, J.R.: Random forests and gradient boosting for wind energy prediction. In: Onieva, E., Santos, I., Osaba, E., Quintián, H., Corchado, E. (eds.) HAIS 2015. LNCS, vol. 9121, pp. 26–37. Springer, Cham (2015). doi:10.1007/978-3-319-19644-2_3
7. Martín, R., Aler, R., Valls, J.M., Galván, I.M.: Machine learning techniques for daily solar energy prediction, interpolation using numerical weather models. Concurr. Comput. Pract. Exp. **28**, 1261–1274 (2016)
8. Cortes, C., Vapnik, V.: Support-vector networks. Mach. Learn. **20**(3), 273–297 (1995)
9. Vapnik, V.N.: The Nature of Statistical Learning Theory. Springer, New York (1995)
10. Friedman, J.H.: Stochastic gradient boosting. Comput. Stat. Data Anal. - Nonlinear Methods Data Min. **38**(4), 367–378 (2002)
11. Guyon, I., Elisseeff, A.: An introduction to variable and feature selection. J. Mach. Learn. Res. **3**, 1157–1182 (2003)
12. Liu, H., Yu, L.: Toward integrating feature selection algorithms for classification and clustering. IEEE Trans. Knowl. Data Eng. **17**(4), 491–502 (2005)
13. Chen, T., He, T., Benesty, M.: XGBoost: eXtreme Gradient Boosting, R package version 0.4-3 (2016)
14. Meyer, D., Dimitriadou, E., Hornik, K., Weingessel, A., Leisch, F.: e1071: Misc Functions of the Department of Statistics, Probability Theory Group (Formerly: E1071), TU Wien, R package version 1.6-7 (2015)
15. Jolliffe, I.: Principal Component Analysis. Wiley Online Library, New York (2002)
16. National Conference on Artificial Intelligence, American Association for Artificial Intelligence (eds.): Proceedings of the Tenth National Conference on Artificial Intelligence, San Jose, California, 12–16 July 1992. AAAI Press [u.a.], Menlo Park (1992). OCLC: 830954541

17. Kononenko, I., Šimec, E., Robnik-Šikonja, M.: Overcoming the myopia of inductive learning algorithms with RELIEFF. Appl. Intell. **7**(1), 39–55 (1997)
18. Robnik-Sikonja, M., Savicky, P.: CORElearn: Classification, Regression and Feature Evaluation, R package version 1.48.0 (2016). (Contributed by John Adeyanju Alao)
19. Kuhn, M.: Caret package. J. Stat. Softw. **28**(5), 1–26 (2008)
20. Kuhn, M., Johnson, K.: Applied Predictive Modeling, vol. 26. Springer, New York (2013)

# Combining Reservoir Computing and Over-Sampling for Ordinal Wind Power Ramp Prediction

Manuel Dorado-Moreno[1]([⊠]), Laura Cornejo-Bueno[2],
Pedro Antonio Gutiérrez[1], Luis Prieto[3], Sancho Salcedo-Sanz[2],
and César Hervás-Martínez[1]

[1] Department of Computer Science and Numerical Analysis,
Universidad de Córdoba, Córdoba, Spain
manuel.dorado@uco.es
[2] Department of Signal Processing and Communications,
Universidad de Alcalá, Alcalá de Henares, Spain
[3] Department of Energy Resource, Iberdrola, Madrid, Spain

**Abstract.** Wind power ramp events (WPREs) are strong increases or decreases of wind speed in a short period of time. Predicting WPREs is of vital importance given that they can damage the turbines in a wind farm. In contrast to previous binary approaches (ramp versus non-ramp), a three-class prediction is proposed in this paper by considering: negative ramp, non-ramp and positive ramp, where the natural order of the events is clear. The independent variables used for prediction include past ramp function values and meteorological data obtained from physical models (reanalysis data). The proposed methodology is based on reservoir computing and an over-sampling process for alleviating the high degree of unbalance of the dataset (non-ramp events are much more frequent than ramps). The reservoir computing model is a modified echo state network composed by: a recurrent neural network layer, a nonlinear kernel mapping and an ordinal logistic regression, in such a way that the order of the classes can be exploited. The standard synthetic minority oversampling technique (SMOTE) is applied to the reservoir activations, given that the direct application over the input variables would damage its temporal structure. The performance of this proposal is compared to the original dataset (without over-sampling) and to nominal logistic regression, and the results obtained with the oversampled dataset and ordinal logistic regression are found to be more robust.

**Keywords:** Wind power ramp events · Reservoir computing · SMOTE · Over-sampling · Ordinal prediction · Kernel mapping

This work has been subsidized by the TIN2014-54583-C2-1-R, TIN2014-54583-C2-2-R and TIN2015-70308-REDT projects of the Spanish Ministerial Commission of Science and Technology (MINECO, Spain) and FEDER funds (EU). Manuel Dorado-Moreno's research has been subsidized by the FPU Predoctoral Program of the Spanish Ministry of Education, Culture and Sport, grant reference FPU15/00647.

© Springer International Publishing AG 2017
I. Rojas et al. (Eds.): IWANN 2017, Part I, LNCS 10305, pp. 708–719, 2017.
DOI: 10.1007/978-3-319-59153-7_61

# 1 Introduction

Wind power ramp events (WPREs) are one of the most dangerous weather events for wind farms, because they can potentially damage the wind turbines if they are not managed properly [10,23], increasing this way the maintenance cost of wind farms. Thus, the prediction of WPREs is extremely important for energy companies [6,9] because it reduces maintenance costs and optimises the power generation of wind farms. These WPREs are originated from complex meteorological processes, such as crossing fronts [11]. The ERA-Interim database (a global atmospheric reanalysis database) is an interesting source of information to predict WPREs [3,11], allowing high quality predictive variables in any location at global level, with a high resolution and accuracy.

Moreover, WPRE prediction can be tackled as a multi-class classification problem, considering three classes of events: negative ramp, non-ramp and positive ramp. The main problem associated with this kind of prediction via multi-class classification is that the number of non-ramps is significantly higher than that of ramp events, and the pattern distribution is even more skewed than for the binary case. In this paper, we approach three-class WRPE prediction as an ordinal regression problem (given that the categories show a clear order) and combine an Echo State Network (ESN) with an over-sampling pre-processing step to tackle the imbalanced nature of the data. The proposed method is an extension of the reservoir computing (RC) architecture presented in [7] for WPRE prediction, which exploits the temporal information of wind power function and includes the reanalysis data in the time instant to be predicted. We modify it by using an ordinal regression classifier and over-sampling the ESN hidden activations.

Ordinal regression refers to those classification problems where there exist a natural ordering among the categories, thresholds models being one of the most widely used and successful techniques [13]. Moreover, many real life problems related with renewable energies have been approached by ordinal regression techniques [12]. RC is a branch of recurrent neural networks (RNNs), which focuses on tackling well-known RNN training difficulties, such as the gradient vanishing, which makes the optimization algorithm convergence very difficult, or the slow transition of inputs through the network, because of the existing cycles among neurons. RC models have also been tested in several renewable energy-related problems [5,17,20], showing good performance. The ESN [16] is one of the possibilities for RC. It is based on the following characteristics: (1) ESNs are randomly created and not modified during the training process. (2) They keep a dynamical memory with non-linear transformations of the inputs. This part of the model is known as *reservoir* in the literature. (3) The network output is calculated as a linear combination of the outputs of the reservoir. In our case, we will apply a nonlinear kernel mapping [24] together with an ordinal threshold model [25]. The training approach for these networks is based on randomly creating the reservoir and adjusting then the rest of network parameters, in a similar way to ELMs [15].

There are several ways of handling imbalanced classification problems [18], such as cost-sensitive learning, over-sampling the minority patterns [4] or under-sampling majority ones [21]. It has been proved that cost-sensitive learning leads to over-fitting [14], while under-sampling leads to poor results because important information is discarded, so that over-sampling is usually preferred, where the synthetic minority over-sampling technique (SMOTE) [4] algorithm is the most popular method. SMOTE is based on the interpolation of randomly selected minority patterns. Directly over-sampling the original time series (in our case, the ramp power function) would destroy the temporal structure of the series. However, given that the temporal structure is only exploited by the first layer of the ESN [26], we propose to feed the reservoir and oversample its activations (outputs), interpolating then the temporal relations of the data. Three different wind farms distributed in the Spanish geography are used in this paper to validate the model and the over-sampling methodology proposed for WPRE prediction, and the results show that the combined use of an ESN, a nonlinear ordinal regression method and a over-sampling procedure leads to better performance than other alternative approaches.

The remainder of the paper is structured in the following way: next section describes the databases considered, which come from three different wind farms in Spain. Section 3 presents the main characteristics of the proposed model architecture. The over-sampling methodology used is described in Sect. 4, whereas Sect. 5 evaluates the performance of the proposed RC system. Finally, Sect. 6 gives some conclusions about the work carried out.

## 2    Database

In this section, we explain the characteristics of the information sources considered in this problem of WPRE prediction, the transformations carried out, and how the data was merged. The first source of information corresponds to wind data, hourly obtained from three wind farms distributed on the Spanish geography, as shown in Fig. 1. We will calculate the WPREs as objective values to be forecast using different predictive variables. The second source of information to obtain the predictive variables is the ERA-Interim project [8], which stores weather data every 6 hours. These data are computed using physical models, not recorded by sensors, so that future estimations can be obtained and used for predicting posterior WPREs.

### 2.1    Wind Farms Data

Let $S_t : \mathbb{R}^k \to \mathbb{R}$ be a ramp function evaluated to decide whether there is a WPRE or not. There are several definitions of $S_t$ [10], all them involving power production ($P_t$) criteria at the wind farm or wind turbine. In this paper, we have used the following one:

$$S_t = P_t - P_{t-\Delta t_r}, \tag{1}$$

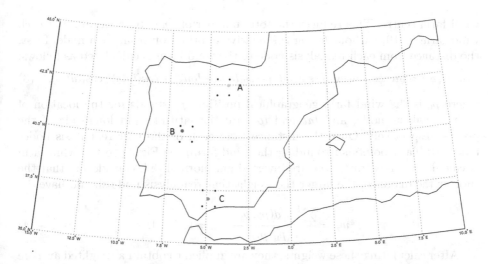

**Fig. 1.** Location of the three wind farms and the reanalysis nodes considered in the study.

where $\Delta t_r$ is the time interval considered (6 h in our case).

Using $S_t$, the classification problem can be stated by defining a threshold value $S_0$, in the following way:

$$y_t = \begin{cases} \mathcal{C}_{\text{NR}}, & \text{if } S_t \leq -S_0, \\ \mathcal{C}_{\text{NoR}}, & \text{if } -S_0 < S_t < S_0, \\ \mathcal{C}_{\text{PR}}, & \text{if } S_t \geq S_0. \end{cases} \tag{2}$$

where $\{\mathcal{C}_{\text{NR}}, \mathcal{C}_{\text{NoR}}, \mathcal{C}_{\text{PR}}\}$ are the different categories of events to be detected, i.e. negative ramps (NRs), non-ramps (NoRs) and positive ramps (PRs).

In this case, we have considered $S_0$ as a percentage of the wind farm rated power (specifically, 50% of the wind farm rated power [10]). The WPRE prediction problem also involves a vector of predictive variables. In our case, we will use meteorological data from reanalysis as input data ($\mathbf{z}$) (which are defined in the next section), together with the past values of the ramp function ($S_{t-1}, S_{t-2}, \ldots$), considering $t$ with 6-hour resolution.

## 2.2   Reanalysis Data

For each of the three wind farms, we have 40 predictors, corresponding to 10 variables per reanalysis node (4 reanalysis node are considered in each wind farm). These variables are computed each 0.75 km all around the globe through a physical model and contain wind speed, pressure and temperature measures at different heights. In order to avoid dealing with so many variables, which in many cases can be highly correlated and introduce noise in the model, we perform a weighted average depending on the distance from each reanalysis node to the wind farm centre. This reduces the total number of reanalysis variables at each wind farm to 10, without losing the relative information from each node. First, the distance from each reanalysis node to the wind farm is calculated as follows:

$$d(p_0, p_j) = \arccos(\sin(lat_0) \cdot \sin(lat_j) \cdot \cos(lon_0 - lon_j) + \cos(lat_0) \cdot \cos(lat_j)), \quad (3)$$

where $p_0$ is the wind farm geographical position, $p_j$ stands for the location of each reanalysis node, and $lat$ and $lon$ are the latitude and longitude of the points, respectively. Once the distance from each of the four reanalysis nodes (the four black points surrounding the wind farm, see Fig. 1) to the wind farm is calculated, these distances are inverted and normalized, considering that the shorter the distance, the larger the weight that reanalysis node should have:

$$w_i = \frac{\sum_{j=0}^{4} d(p_0, p_j)}{d(p_0, p_i)}, \quad i = 1, \ldots, 4. \tag{4}$$

After calculating these weights, they are applied to obtain a weighted average for each of the 10 variables.

## 3    Reservoir Computing Proposal

In this paper we propose a model based on the standard ESN architecture, replacing the regression of the output layer with a combination of a nonlinear kernel mapping and an ordinal logistic regression model.

### 3.1    Architecture and Training Algorithm

We now describe the proposed ESN architecture which solves the WPRE prediction including past ramp function values and ERA-Interim reanalysis data. Figure 2 shows the structure proposed, where some symbols must be explained: $S_t$ and $z_{t+1}$, in the input layer, are the input vectors (ramp function and 10 reanalysis variables in time $t$ and $t+1$, respectively). The use of $z_{t+1}$ in the input layer for predicting $y_{t+1}$ is plausible, as mentioned in Sect. 2, because it is given by meteorological models which can produce reliable estimations for 6 h ahead. In the hidden layer, the reservoir has $M$ neurons randomly interconnected, and the weights from the inputs to the reservoir are also random. Finally, we have a kernel mapping layer which applies an approximated RBF feature map, and an output layer that will predict the class of the different events.

The standard Ordinal Logistic Regression (OLR) [22] is a linear model dividing the patterns into categories by using a set of thresholds. However, the linear nature of the model limits its performance, as shown in previous studies [13]. That is the reason why we consider a kernel mapping applied to the activations of the reservoir and the reanalysis data. Specifically, we use approximated explicit feature maps, given that the standard kernel trick can be very costly. A Random Kitchen Sinks [24] map is considered, which approximates the feature map of a radial basis function (RBF) kernel by Monte Carlo approximation of its Fourier transform.

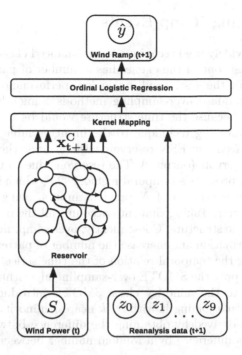

**Fig. 2.** Proposed ESN architecture

The methodology proposed to train the previously presented architecture is the following:

1. Create a reservoir of size $M$, randomly interconnecting the neurons and generating the weights $\mathbf{W}$ by a Gaussian probability distribution. The weights of the connections that link the inputs with the reservoir neurons ($\mathbf{W}^{in}$) are generated using the same distribution. The bias of the reservoir neurons are randomly adjusted using a uniform distribution.
2. Harvest the reservoir states, feeding the reservoir inputs from time $t = 1$ to $t = M$, so that all the network connections would have received a signal, allowing to obtain the full vector $\mathbf{x}_t$.
3. Combine $\mathbf{x}_t$ and $\mathbf{z}_{t+1}$ and apply the kernel mapping.
4. Compute the output weights by training an ordinal logistic regression model, linearly projecting the patterns and establishing thresholds to differentiate each of the three classes ($\mathcal{C}_{NR}$, $\mathcal{C}_{NoR}$ and $\mathcal{C}_{PR}$).

Once the network is trained, it can be used for real-time prediction of ramps, discarding the pattern corresponding to $t = 0$, since there would be no information from an earlier instant.

## 4   Over-Sampling Ramp Events

Over-sampling is a widely used technique for imbalanced classification databases [2,19], where, at least, one of the classes has a number of patterns significantly lower or higher than the rest, resulting in low performance for the minority classes. However, standard over-sampling methods cannot be directly applied to time series data, because the time structure would be destroyed. Instead of applying the over-sampling technique to the input variable $(y_t)$, the proposal of this paper is to feed the ESN reservoir with $y_t$ and, then, oversample the activations of the reservoir (outputs). The reservoir, due to the cycles found in its internal neurons, presents a temporal memory [26], which is able to maintain the information of a maximum of $N$ previous inputs, being $N$ the number of neurons in the reservoir. Taking that into account, all the outputs of the reservoir maintain the time structure. Consequently, over-sampling can be applied to interpolate this information and increase the number of patterns of the minority class, without losing the temporal relation of the time series.

Specifically, we apply the SMOTE over-sampling algorithm [4]. For each pattern corresponding to the minority class, SMOTE introduces synthetic samples along the segments joining it with its $k$ nearest minority neighbours. Each synthetic pattern is derived by obtaining the difference between both patterns and multiplying this difference by a random number between 0 and 1 (random interpolation).

## 5   Experiments

This section describes the experiments considered for validating the proposal presented in Sect. 3.

### 5.1   Evaluation Metrics

Several measures can be used for evaluating ordinal classifiers. The most common ones in machine learning are the Mean Absolute Error $(MAE)$ and the Mean Zero-one Error $(MZE)$ [13], being $MZE = 1 - Acc$, where $Acc$ is the accuracy or correct classification rate. However, these measures may not be the best option, for example, when measuring performance in the presence of class imbalance [1], and/or when the costs of different errors vary markedly. In this way, in order to take into account different aspects of the classifier, six different metrics have been considered:

- The **correct classification rate**, $(CCR)$ is defined by:

$$CCR = \frac{1}{N} \sum_{i=1}^{N} (I(y_i^* = y_i)), \tag{5}$$

where $I(\cdot)$ is the zero-one loss function, $y_i$ is the desired output for pattern $\mathbf{x}_i$, $y_i^*$ is the prediction of the model, and $N$ is the total number of patterns in the dataset. $CCR$ values vary from 0 to 1, and it represents global performance in the classification task. It does not take the category order into account, and it is not recommended for imbalanced datasets.

– The **sensitivities** of each class ($S_{\mathrm{NR}}$, $S_{\mathrm{NoR}}$ and $S_{\mathrm{PR}}$), which represent the model ability to correctly predict each type of event:

$$S_{\mathrm{NR}} = \frac{CC_{\mathrm{NR}}}{N_{\mathrm{NR}}}, \ S_{\mathrm{NoR}} = \frac{CC_{\mathrm{NoR}}}{N_{\mathrm{NoR}}}, \ S_{\mathrm{PR}} = \frac{CC_{\mathrm{PR}}}{N_{\mathrm{PR}}}, \tag{6}$$

where $CC_{\mathrm{NR}}$, $CC_{\mathrm{NoR}}$ and $CC_{\mathrm{PR}}$ are the number of correctly classified $NR$, NoR and $PR$ events, and $N_{\mathrm{NR}}$, $N_{\mathrm{NoR}}$ and $N_{\mathrm{PR}}$ ($N_{\mathrm{NR}} + N_{\mathrm{NoR}} + N_{\mathrm{PR}} = N$) are the total number of NR, NoR and PR events.

– The **geometric mean of the sensitivities** ($GMS$) of each class is a geometric average of the correct classification percentages per class:

$$GMS = \sqrt[3]{S_{\mathrm{NR}} \cdot S_{\mathrm{NoR}} \cdot S_{\mathrm{PR}}}. \tag{7}$$

We include this measure because the prediction problem is highly imbalanced, so that trivial classifiers, which ignore one of the classes, can be easily recognized by $GMS = 0$.

– The **average mean absolute error** ($AMAE$) [1] is the mean of $MAE$ classification errors throughout the classes, where $MAE$ is the average absolute deviation of the predicted class from the true class (in number of categories of the ordinal scale). It is able to mitigate the effect of imbalanced class distributions. It is defined by:

$$AMAE = \frac{1}{J} \sum_{j=1}^{J} MAE_j, \text{with } MAE_j = \frac{1}{N_j} \sum_{i=1}^{N_j} |\mathcal{O}(y_i) - \mathcal{O}(y_i^*)|, \tag{8}$$

where $1 \leq j \leq J$, $\mathcal{O}(\mathcal{C}_{\mathrm{NR}}) = 1, \mathcal{O}(\mathcal{C}_{\mathrm{NoR}}) = 2, \mathcal{O}(\mathcal{C}_{\mathrm{PR}}) = 3$. $MAE$ values range from 0 to $J - 1$, as do those of $AMAE$.

## 5.2  Experimental Design

The three wind farms of Fig. 1 have been used in the experimental validation of this proposal. To evaluate the results, the three datasets have been split in the same way: the last 365 days will be used as the test set, and the rest of the database as the training set (the specific dates for each wind farm are shown in Table 1).

The model presented in Sect. 3 (which will be referred to as kernel ordinal logistic regression, KOLR) has been compared against different alternatives: (1) We compare nominal and ordinal versions of the logistic regression model. The idea is to evaluate whether the order of categories can be exploited to obtain more robust models. In this way, the OLR model presented in Sect. 3 is compared against a nominal logistic regression (LR), i.e. a standard multinomial logistic regression model, where the *softmax* transformation is used for decomposing the multiclass problem. (2) We study if the use of a nonlinear kernel mapping

**Table 1.** Characteristics of the different datasets considered in the experimentation, including the periods used for training and test and the number of WPREs

| Farm | #NR | #NoR | #PR | Train interval | Test interval |
|------|-----|------|-----|----------------|---------------|
| A | 602 | 13852 | 706 | 11/1/2002–29/10/2011 | 30/10/2011–29/10/2012 |
| B | 1077 | 12463 | 1074 | 1/1/2002–17/2/2012 | 18/2/2012–17/2/1013 |
| C | 617 | 13857 | 649 | 2/3/2002–30/6/2012 | 11/7/2012–30/6/2013 |

improves the performance of the models, by comparing the results against the alternative of directly connecting the reservoir to the logistic regression classifier. Consequently, four methods will be compared: ordinal logistic regression with or without kernel mapping (OLR and KOLR, respectively) and nominal logistic regression also with or without kernel (LR and KLR, respectively).

For kernel alternatives (KLR and KOLR), the RBF kernel width ($\gamma$) has been adjusting by using a nested 5-fold cross-validation over the training set, with the following range of values: $\gamma \in \{2^{-12}, 2^{-11}, ..., 2^{-8}\}$. For the four logistic regression models, we need to adjust the regularization parameter ($\alpha$) which encourages parsimony in the model coefficients. In this case, the range is $\alpha \in \{2^{-5}, 2^{-4}, ..., 2^{-1}\}$. In both cases, model selection is based on maximising the minimum sensitivity, i.e. $MS = \min\{S_{\mathrm{NR}}, S_{\mathrm{NoR}}, S_{\mathrm{PR}}\}$.

The parameters of the reservoir have been configured as follows: the number of neurons in the reservoir is $M = 50$, assuming this is a sufficient number of neurons to address the problem, and it does not represent a high computational cost. All the reservoir weights are in the range $[-1, 1]$.

Finally, the SMOTE over-sampling method was applied with $k = 5$ nearest neighbours, as proposed in [4]. For each minority class ($\mathcal{C}_{\mathrm{NR}}$ and $\mathcal{C}_{\mathrm{PR}}$), we generate as many synthetic patterns as needed so that the number of patterns of the class is a 70% of the number of patterns of the majority class ($\mathcal{C}_{\mathrm{NoR}}$). Synthetic patterns are only considered during the training phase and never used for the test set.

## 5.3    Results

This section discusses the results of the different algorithms, which are shown in Table 2, where the four algorithms previously discussed (KOLR, OLR, KLR and LR) are compared when using the two versions of the datasets (the original and the one oversampled using SMOTE). The $GMS$ metric is the one better reflecting the balance of good classification considering the different classes, given that it highly penalises sensitivities close to 0. We can observe how the KOLR algorithm always obtains the best $GMS$ score for two of the wind farms, and the second one for the other.

Although some algorithms lead to high $CCR$ values (e.g., the original versions of the datasets, before applying over-sampling), this is because the classifiers obtained are trivial, classifying all patterns as non-ramp events ($S_{\mathrm{NR}} \approx 0$ and

$S_{PR} \approx 0$). Clearly, the over-sampling process is needed to obtain models able to correctly recognise ramps.

**Table 2.** Results of the different methods evaluated for wind farms A, B and C.

| Wind farm | Version | Algorithm | $GMS$ | $CCR$ | $AMAE$ | $S_{NR}$ | $S_{NoR}$ | $S_{PR}$ |
|---|---|---|---|---|---|---|---|---|
| A | Original | KOLR | 0.0000 | **0.9021** | 0.6667 | 0.0000 | **1.0000** | 0.0000 |
| | | OLR | 0.0000 | *0.9008* | 0.6623 | 0.0153 | *0.9978* | 0.0000 |
| | | KLR | 0.0000 | **0.9021** | 0.6667 | 0.0000 | **1.0000** | 0.0000 |
| | | LR | 0.0000 | **0.9021** | 0.6667 | 0.0000 | **1.0000** | 0.0000 |
| | Oversampled | KOLR | **0.6451** | 0.6316 | **0.3745** | **0.7692** | 0.6297 | 0.5542 |
| | | OLR | 0.6054 | 0.5923 | 0.4041 | 0.6461 | 0.5938 | *0.5783* |
| | | KLR | 0.5784 | 0.6567 | 0.4496 | *0.6769* | 0.6708 | 0.4096 |
| | | LR | *0.6170* | 0.6157 | *0.3928* | 0.6461 | 0.6158 | **0.5903** |
| B | Original | KOLR | 0.0000 | **0.8655** | 0.6667 | 0.0000 | **1.0000** | 0.0000 |
| | | OLR | 0.0000 | *0.8648* | 0.6637 | 0.0103 | *0.9984* | 0.0000 |
| | | KLR | 0.0000 | **0.8655** | 0.6667 | 0.0000 | **1.0000** | 0.0000 |
| | | LR | 0.0000 | **0.8655** | 0.6667 | 0.0000 | **1.0000** | 0.0000 |
| | Oversampled | KOLR | *0.5525* | 0.5404 | *0.4636* | *0.5979* | 0.5372 | *0.5252* |
| | | OLR | 0.5483 | 0.5370 | **0.4613** | 0.5876 | 0.5340 | *0.5252* |
| | | KLR | 0.4743 | 0.5651 | 0.5355 | 0.5154 | 0.5855 | 0.3535 |
| | | LR | **0.5612** | 0.5061 | 0.4754 | **0.6597** | 0.4912 | **0.5454** |
| C | Original | KOLR | 0.0000 | **0.9151** | 0.6667 | 0.0000 | **1.0000** | 0.0000 |
| | | OLR | 0.0000 | **0.9151** | 0.6667 | 0.0000 | **1.0000** | 0.0000 |
| | | KLR | 0.0000 | **0.9151** | 0.6667 | 0.0000 | **1.0000** | 0.0000 |
| | | LR | 0.0000 | **0.9151** | 0.6667 | 0.0000 | **1.0000** | 0.0000 |
| | Oversampled | KOLR | **0.5593** | 0.5182 | *0.4699* | *0.6250* | 0.5119 | **0.5468** |
| | | OLR | *0.5511* | 0.5752 | **0.4430** | **0.6875** | 0.5771 | 0.4218 |
| | | KLR | 0.4452 | *0.6633* | 0.5569 | 0.5468 | *0.6886* | 0.2343 |
| | | LR | 0.5389 | 0.5202 | 0.4995 | *0.6250* | 0.5170 | *0.4843* |

The best result for each wind farm is in bold face and the second one in italics.

The $AMAE$ values can be used to check how the classifiers are behaving with respect to the order of the different categories (i.e., low $AMAE$ values mean that the classifiers do not usually confuse PR events with NR events, and vice-versa). Checking these results in Table 2, ordinal models (OLR and KOLR) generally lead to lower $AMAE$ than nominal ones (LR and KLR), the best $AMAE$ results being obtained by OLR in 2 of the wind farms, and KOLR in the other one. The differences of $AMAE$ between OLR and KOLR are very low, thus, the other metrics have to be analysed to decide which is the best performing one.

The analysis of the individual sensitivities must be cautious, given that a good result in one class usually leads to poor performance in the others. KOLR is the only algorithm able to simultaneously obtain a sensitivity higher than 0.5 for all classes and the three wind farms, also obtaining some of the best results for $GMS$ and $AMAE$.

# 6   Conclusions

This paper proposes and evaluates a modification of an ESN architecture for predicting WPREs in an ordinal scale (negative ramps, non-ramp events and positive ramps, where $C_{NR} < C_{NoR} < C_{PR}$), also considering the high degree of imbalance of the database. The architecture is based on combining past values of the ramp function and reanalysis data. Given that we consider a linear ordinal regression method, a kernel mapping is applied to the input of the classifier, and an over-sampling procedure is considered to avoid trivial predictors. Three wind farms in Spain are used for evaluating the performance of the proposal. The results obtained show that over-sampling is necessary to obtain an acceptable performance for minority classes in this problem, and that the combination of an ordinal classifier with the kernel mapping leads to the best performance. An interesting extension of this work could be considering more than three classes, so that more fine-grain information could be extracted from the predictions.

# References

1. Baccianella, S., Esuli, A., Sebastiani, F.: Evaluation measures for ordinal regression. In: Proceedings of the 9th International Conference on Intelligent Systems Design and Apps, pp. 283–287 (2009)
2. Bach, M., Werner, A., Zywiec, J., Pluskiewicz, W.: The study of under- and over-sampling methods' utility in analysis of highly imbalanced data on osteoporosis. Inf. Sci. **384**, 174–190 (2017)
3. Cannon, D.J., Brayshaw, D.J., Methven, J., Coker, P.J., Lenaghan, D.: Using reanalysis data to quantify extreme wind power generation statistics: a 33 year case study in Great Britain. Renew. Energy **75**, 767–778 (2015)
4. Chawla, N.V., Bowyer, K.W., Hall, L.O., Kegelmeyer, W.P.: SMOTE: synthetic minority over-sampling technique. J. Artif. Intell. Res. **16**, 321–357 (2002)
5. Crisostomi, E., Gallicchio, C., Micheli, A., Raugi, M., Tucci, M.: Prediction of the Italian electricity price for smart grid applications. Neurocomputing **170**, 286–295 (2015)
6. Cui, M., Ke, D., Sun, Y., Gan, D., Zhang, J., Hodge, B.M.: Wind power ramp event forecasting using a stochastic scenario generation method. IEEE Trans. Sustain. Energy **6**(2), 422–433 (2015)
7. Dorado-Moreno, M., Durán-Rosal, A.M., Guijo-Rubio, D., Gutiérrez, P.A., Prieto, L., Salcedo-Sanz, S., Hervás-Martínez, C.: Multiclass prediction of wind power ramp events combining reservoir computing and support vector machines. In: Luaces, O., Gámez, J.A., Barrenechea, E., Troncoso, A., Galar, M., Quintián, H., Corchado, E. (eds.) CAEPIA 2016. LNCS, vol. 9868, pp. 300–309. Springer, Cham (2016). doi:10.1007/978-3-319-44636-3_28

8. Dee, D.P., Uppala, S.M., Simmons, A.J., Berrisford, P., Poli, P., et al.: The ERA-interim reanalysis: configuration and performance of the data assimilation system. Q. J. R. Meteorol. Soc. **137**, 553–597 (2011)
9. Foley, A.M., Leahy, P.G., Marvuglia, A., McKeogh, E.J.: Current methods and advances in forecasting of wind power generation. Renew. Energy **37**, 1–8 (2012)
10. Gallego-Castillo, C., Cuerva-Tejero, A., López-García, O.: A review on the recent history of wind power ramp forecasting. Renew. Sustain. Energy Rev. **52**, 1148–1157 (2015)
11. Gallego-Castillo, C., García-Bustamante, E., Cuerva-Tejero, A., Navarro, J.: Identifying wind power ramp causes from multivariate datasets: a methodological proposal and its application to reanalysis data. IET Renew. Power Gener. **9**(8), 867–875 (2015)
12. Gutiérrez, P.A., Salcedo-Sanz, S., Hervás-Martínez, C., Prieto, L.: Ordinal and nominal classification of wind speed from synoptic pressure patterns. Eng. Appl. Artif. Intell. **26**(3), 1008–1015 (2012)
13. Gutiérrez, P.A., Pérez-Ortiz, M., Sánchez-Monedero, J., Fernández-Navarro, F., Hervás-Martínez, C.: Ordinal regression methods: survey and experimental study. IEEE Trans. Knowl. Data Eng. **28**, 127–146 (2016)
14. Haixiang, G., Yijing, L., Shang, J., Mingyun, G., Yuanyue, H., Bing, G.: Learning from class-imbalanced data: review of methods and applications. Expert Syst. Appl. **73**, 220–239 (2017)
15. Huang, G., Zhu, Q., Siew, C.: Extreme learning machone: theory and applications. Neurocomputing **70**, 489–501 (2006)
16. Jaeger, H.: The "echo state" approach to analysing and training recurrent neural networks. GMD Report 148, German National Research Center for Information Technology, pp. 1–43 (2001)
17. Jayawardene, I., Venayagamoorthy, G.K.: Reservoir based learning network for control of two-area power system with variable renewable generation. Neurocomputing **170**, 428–438 (2015)
18. Kotsiantis, S., Kanellopoulos, D., Pintelas, P.: Handling imbalanced datasets: a review. GESTS Int. Trans. Comput. Sci. Eng. **30**, 25–30 (2006)
19. Li, J., Fong, S., Sung, Y., Cho, K., Wong, R., Wong, K.K.L.: Adaptive swarm cluster-based dynamic multi-objective synthetic minority over-sampling technique algorithm for tackling binary imbalanced datasets in biomedical data classification. BioData Min. **9**, 1–15 (2016)
20. Liu, D., Wang, J., Wang, H.: Short-term wind speed forecasting based on spectral clustering and optimised echo state networks. Renew. Energy **78**, 599–608 (2015)
21. Luengo, J., Fernández, A., García, S., Herrera, F.: Addressing data complexity for imbalanced data sets: analysis of SMOTE-based oversampling and evolutionary undersampling. Soft. Comput. **15**(10), 1909–1936 (2011)
22. McCullagh, P.: Regression models for ordinal data. J. R. Stat. Soc. **42**(2), 109–142 (1980)
23. Ouyang, T., Zha, X., Qin, L.: A survey of wind power ramp forecasting. Energy Power Eng. **5**, 368–372 (2013)
24. Rahimi, A., Recht, B.: Random features for large scale kernel machines. In: NIPS, vol. 3, no. 4, p. 5 (2007)
25. Rennie, J.D.M., Srebro, N.: Loss functions for preference levels: regression with discrete ordered labels. In: Proceedings of the IJCAI Multidisciplinary Workshop on Advances in Preference Handling, pp. 180–186 (2005)
26. Rodan, A., Tiňo, P.: Minimum complexity echo state network. IEEE Trans. Neural Netw. **22**(1), 131–144 (2011)

# Arbitrated Ensemble for Solar Radiation Forecasting

Vítor Cerqueira[1,2(✉)], Luís Torgo[1,2], and Carlos Soares[1,2]

[1] INESC TEC, Porto, Portugal
{vmac,ltorgo}@inesctec.pt, csoares@fe.up.pt
[2] University of Porto, Porto, Portugal

**Abstract.** Utility companies rely on solar radiation forecasting models to control the supply and demand of energy as well as the operability of the grid. They use these predictive models to schedule power plan operations, negotiate prices in the electricity market and improve the performance of solar technologies in general. This paper proposes a novel method for global horizontal irradiance forecasting. The method is based on an ensemble approach, in which individual competing models are arbitrated by a metalearning layer. The goal of arbitrating individual forecasters is to dynamically combine them according to their aptitude in the input data. We validate our proposed model for solar radiation forecasting using data collected by a real-world provider. The results from empirical experiments show that the proposed method is competitive with other methods, including current state-of-the-art methods used for time series forecasting tasks.

**Keywords:** Solar radiation forecasting · Renewable energy · Ensemble methods · Metalearning · Time series

## 1 Introduction

The sun is a fundamental source of energy to our planet. Over the years many technologies have been developed to take advantage of the solar energy. We use this energy to produce electricity, heating water and buildings, lighting and destroying toxic waste.

In this paper we present a novel method for solar radiation forecasting. Specifically, we focus on daily predictions of Global Horizontal Irradiance (GHI). Utility companies use solar radiation forecasting systems to support decision making process in several ways. They use them to predict if the energy produced by a given solar technology can meet the daily electricity demand, to balance electricity market prices and schedule power plant operations. Engineers leverage solar radiation forecasting systems to improve the performance and economics of solar radiation technologies, for example photovoltaic devices. Moreover, solar radiation forecasts enable a dynamic configuration of air-conditioning systems within buildings to optimise their efficiency. In sum, a data-driven solar radiation forecasting system can provide a way to maximise the performance of solar

© Springer International Publishing AG 2017
I. Rojas et al. (Eds.): IWANN 2017, Part I, LNCS 10305, pp. 720–732, 2017.
DOI: 10.1007/978-3-319-59153-7_62

technologies while reducing operating costs. This paper presents an novel time series model for GHI forecasting. The proposed method leverages the predictive power of ensemble methods, combining individual learning models with different inductive bias using a metalearning strategy. We explore ways of combining the predictions of forecasters in a dynamic – online – fashion. In time evolving environments the process causing the underlying data is prone to change over time and the combined model should adapt accordingly.

We use metalearning (e.g. [4]) to analyse the expertise of each individual forecaster across the time series of solar radiation. We can then use this meta knowledge to dynamically weight the predictions of base learners according to their competence in a given observation. If we expect a given forecaster to perform poorly in some subset of the data we render it a low weight in the combination rule. On the other hand, if we are optimistic about some learner in our ensemble we augment its weight with respect to the other learners.

The intuition behind our approach is that different learning models may have different areas of expertise across the input space. That is, in a given test observation, some individual learner may be better than the combined model and different individual learners will be better on different cases. In effect, we can learn about the learning process of each base-learner. In fact, we hypothesise that the underlying process causing the series of solar radiation follows a recurring pattern due to seasonal factors [7]. Consequently, we hypothesise that the metalearning layer enables the combined model to better detect changes between different regimes and quickly adapt itself to the environment.

Our metalearning strategy follows an Arbitrating scheme [13, 22], in which we build a meta-learner for each base-learner comprising the ensemble. Each meta-learner is specifically designed to assess how apt its base counterpart is to make a prediction in a given observation. This is accomplished by analysing how the error incurred by a given learning model relates to the characteristics of the data. At test time, the base-learners are weighted according to their degree of competence in the input observation, estimated by the predictions of the meta-learners.

Our goal is to predict the next value of the series of solar radiation. We use regression models as our base-learners by transforming the solar radiation time series into an Euclidean space using time delay embedding [29]. Furthermore, in order to augment the information about the data we also use external predictors such as weather reports. In summary, the contributions of this paper are the following:

- An arbitrated ensemble for GHI forecasting. The ensemble includes a metalearning layer based on an arbitrating scheme, used to dynamically combine individual models;
- We use the Arbitrating strategy to dynamically weight individual models, whereas typical applications select the most reliable model at each test query.

We start by outlining the related work in Sect. 2. The methodology is addressed in Sect. 3, where we formalise and explain our contributions. The Case Study is briefly described in Sect. 4, along with the pre-processing steps and descriptive statistics. The experiments and respective results are presented and discussed in Sect. 5. Finally, the conclusions are produced in Sect. 6, along with some remarks about future work and reproducibility of the proposed methods.

## 2     Related Work

In this paper we focus on ensembles with self-adapting mechanisms to predict solar radiation, which is a time series with a numerical outcome. Ensemble methods for numerical predictions problems have a vast literature. We refer to the survey written in [19] for a complete overview on ensemble approaches for these tasks.

Building adaptable models is important in dynamic real-world environments in which data is constantly changing over time due to several factors, for example, seasonality. Our proposed method is motivated by the core concepts behind Arbitrating classifiers [13,22]. Arbitrating is an ensemble method used to combine classifiers according to their expertise on the input data. The expertise of a base-learner is learnt with a corresponding meta-model, which learns the loss of its base counterpart according to a set of meta-features. At test time, the classifier with greatest confidence in the input data-point is selected to make a prediction. The authors reason that each meta-model holds information about which part of the data its base counterpart works best and considers when it can make a reliable prediction.

Other forms for dynamically combining models for time series forecasting with numerical outcome were proposed in the literature. In [26], the authors use Zoomed Ranking [28] approach to rank and select time series forecasting models. MetaStream is proposed in [25]. The authors summarise the dynamics of the recent and upcoming observations in a data stream to either select or combine regression models. In [15,32] presented other two approaches that use the characteristics of time series in a meta-level to improve the combination of individual forecasters. They use these characteristics to induce several rules to weight or select between different models.

Our approach is different from the existing literature in the sense that we apply an arbitrating scheme to meta-learn and weight the individual base-learners. To the best of our knowledge, this is the first application of an Arbitrating scheme for time series prediction with numerical outcome, particularly solar radiation forecasting.

### 2.1     Solar Radiation Forecasting

Several solar radiation forecasting models have been proposed in the literature. The most typical approaches rely on regression and time series analysis models (e.g. [3,8,23]). The connectionist approach of Artificial Neural Networks is also of common use, for example in [11,18] or [27]. In our paper we focus on daily forecasts, but the temporal granularity typically ranges from hourly to weekly forecasts.

Many approaches also incorporate external features in their methodology, such as [16] or [1]. These typically include weather information.

## 3   Global Horizontal Irradiance Forecasting

A given solar technological device collects solar radiation in two ways: direct radiation and diffuse radiation. In this paper we aim at predicting the Global Horizontal Irradiance (GHI), which can be derived by summing direct radiation with diffuse radiation and accounting for the sun's position.

GHI forecasting is a particular instance of time series forecasting tasks. We start addressing the methodology by presenting the main notation employed throughout this section:

**Time Series:** A time series is a temporal sequence of values $Y = \{y_1, y_2, \ldots, y_n\}$, where $y_i$ is the value of $Y$ at time $i$ and $n$ is the length of $Y$;

**Embedded Time Series:** $Y^K$ denotes the embedded time series with embedding dimension $K$. We use time delay embedding to represent $Y$ in an Euclidean space with embedding dimension $K$, according to [29]. In effect, we generate the following matrix:

$$Y^K = \begin{bmatrix} y_1 & y_2 & \cdots & y_{K-1} & y_K \\ \vdots & \vdots & \vdots & \vdots & \vdots \\ y_{i-K+1} & y_{i-K+2} & \cdots & y_{i-1} & y_i \\ \vdots & \vdots & \vdots & \vdots & \vdots \\ y_{n-K+1} & y_{n-K+2} & \cdots & y_{n-1} & y_n \end{bmatrix} \tag{1}$$

Each row denotes an embedding vector $v_r, \forall r \in \{1, \ldots, n - K + 1\}$. Our goal is to predict the next point in the series, represented by the last column in Matrix 1;

**External Predictors:** $Y^{ext}$ denotes the set of external predictors computed for each embedding vector $v \in V$. These include external information (e.g. weather data) which helps to model the target concept;

**Base-Learners:** We denote as $M$ the set of $m$ base-learners comprising the ensemble $S$;

**Meta-Learners:** $\overline{M}^j$ is a meta-learner for $M^j$, with $j \in \{1, \ldots, m\}$;

**Base-Learners Loss:** $e_i^j$ represents the absolute loss of $M^j$ in the observation $y_i$;

**Base-Learners weights:** $w_i^j$ denotes the weights assigned to $M^j$ for predicting the value of $y_i$.

Our methodology for GHI forecasting settles on the three main steps: An offline (**i**) training step of $M$ and the online iterative steps: (**ii**) Meta-learning of $\overline{M}$ and (**iii**) prediction of $y_{t+1}$ using $M$ which is dynamically weighted according to $\overline{M}$.

## 3.1 Learning $M$

In the first step we train the learning models $M$ which are then combined to make a prediction. Concretely, each $M^j, \forall\, j \in \{1, \ldots, m\}$ is individually trained using the available $Y_{tr}^K$, the embedded time series combined with the external predictors $Y^{ext}$. $M$ is composed of individual regression models with different inductive bias. Different models (e.g. Gaussian Processes and Neural Networks) hold different assumptions regarding the underlying data. This divergence across base-learners comprising $S$ encourages diversity in the ensemble – a fundamental ingredient in the ensemble recipe [5].

## 3.2 Metalearning $\overline{M}$

The metalearning step of our methodology is an online process run at test time. Our objective in applying this metalearning strategy is to extract the information about the expertise of each individual model in $M$ across the series of water consumption.

We use a metalearning layer for arbitrating among competing individual learners. However, instead of selecting the most reliable model (as in [13, 22]), we use the meta-knowledge to weight the base learners according to their expertise in the input signal.

Formally, each meta-learner $\overline{M}^j, \forall\, j \in \{1, \ldots, m\}$, is trained to build a model for $e^j = f(\overline{X})$, where $f$ denotes the regression function. $\overline{X}$ represent the meta-features, i.e., the set of **features** used in the **meta**-level by the meta-learners in $\overline{M}$. $\overline{X}$ is composed by the primitive features used by $M$ along with some summary statistics. These statistics are computed for each embedding vector and characterise the recent dynamics of the series as well as its structure.

We conduct this meta regression analysis to understand how the loss of a given base-learner relates to the different dynamics of the series. In effect, we can explore forms of capitalising from these relationships. Specifically, we use the information from $e$ to dynamically weight the base-learners $M$.

## 3.3 Predicting $y_{t+1}$

When a new observation $y_{t+1}$ arrives for prediction we combine the predictions of $M$ with the meta information from $\overline{M}$. The arbitrating layer composed by $\overline{M}$ is able to predict how well each base learner in $M$ will perform with respect to each other. If $\overline{M}^j$ predicts that its counterpart $M^j$ will make a large error ($\hat{e}_{t+1}^j$) relative to the other base learners ($\hat{e}_{t+1}^l, \forall\, l \in \{1, \ldots, m\} \backslash \{j\}$) then $M^j$ will be assigned a small relative weight in the final prediction. Conversely, if $\hat{e}_{t+1}^j$ is predicted to be small (also with respect to the loss of other base learners), $M^j$ will be important for the upcoming prediction. Even though the learning models comprising $M$ are trained in a batch way, the models in $\overline{M}$ are updated after every test observation. Moreover, the predictions by $\overline{M}$ are produced for each test observation, rendering an online nature to our method. Formally, we measure the weights of each base-learner using the following equation:

$$w_{t+1}^j = \frac{erfc(\hat{e}_{t+1}^j)}{\sum_{i \in \overline{M}} erfc(\hat{e}_{t+1}^j)} \qquad (2)$$

where $\hat{e}_{t+1}^j$ is the prediction made by $\overline{M}^j$ for the absolute loss that $M^j$ will incur in $y_{t+1}$. The function $erfc$ denotes the complementary Gaussian error function which is formalised as follows:

$$erfc(x) = \frac{4}{\sqrt{\pi}} \int_x^\infty e^{-t^2} dt \qquad (3)$$

The final prediction is a weighted average of the predictions made by the base-learners $\hat{y}_{t+1}^j$ with respect to $w_{t+1}^j$ computed according to Eq. 4:

$$\hat{y}_{t+1} = \sum_{j=1}^m \hat{y}_{t+1}^j \times w_{t+1}^j \qquad (4)$$

The proposed methodology is summarised in Algorithm 1.

## 4 Case Study

Our study was conducted using data collected by the Oak Ridge National Laboratory [17], in Tennessee, USA. The solar radiation data includes global horizontal radiation, direct radiation and diffuse horizontal radiation. These were harvested using a rotating shadow-band radiometer, a low-cost equipment for measuring solar radiation.

The data is collected in an hourly basis. Our sample ranged from 19-01-2009 to 19-01-2017, totalling 70151 observations. Additionally, other external variables were collected: the average air temperature, relative humidity, average wind speed and precipitation levels. These follow the same granularity and temporal scope as the solar radiation data.

### 4.1 Pre-processing

We focused our work on daily forecasts so we aggregated the data by day, reaching a total of 2922 observations across the above-mentioned time-span. The units of the solar radiation levels are in watts per square meter $(W/m^2)$.

Direct radiation and diffuse radiation levels are used as predictor variables in our model. Concretely, we use the information of these attributes from the previous day as well as the mean of the last $K$ days. To augment the information of solar radiation levels we also include the mean and standard deviation of the embedding vectors described in Matrix 1 as predictors. Moreover, from the hourly average air temperature we design two features: max and mean air

---

**Algorithm 1:** Arbitrated Ensemble for Solar Radiation Forecasting

---

**Input:**
– Global Horizontal Irradiance time series $Y$;
– External predictors $Y^{ext}$;
– embedding dimension $K$.

/* Learning Step                                                              */
Embed $Y$ and attach $Y^{ext} \rightarrow Y_{tr}^K$
**foreach** $M^j \in M$ **do**
  |  train $M^j$ using $Y_{tr}^K$
**end**

---

/* Metalearning Step                                                          */
**for** *all available test observations* $Y_{ts}$ **do**
  |  Embed $Y_{ts}$ and attach $Y^{ext} \rightarrow Y_{ts}^K$
  |  $\overline{X} \leftarrow$ Extract Metafeatures from $Y_{ts}^K$
  |  **foreach** $\overline{M}^j \in \overline{M}$ **do**
  |    |  train $\overline{M}^j$ to model: $e^j = f(\overline{X})$
  |  **end**
**end**

---

/* Predicting new observations                                                */
**foreach** *upcoming observation* $y_{t+1}$ **do**
  |  Embed $y_{t+1}$ along with the $K - 1$ most recent observations and attach $y_{t+1}^{ext}$
  |  Compute predictions $\hat{e}_{t+1}^j$ from models $\overline{M}^j \in \overline{M}$
  |  Compute weights $w_{t+1}^j$
  |  Compute predictions $\hat{y}_{t+1}^j$ from models $M^j \in M$
  |  Compute final prediction $\hat{y}_{t+1} = \sum_{j=1}^m \hat{y}_{t+1}^j \cdot w_{t+1}^j$
  |  Add $y_{t+1}$ to $Y_{ts}$ and return to metalearning step
**end**

---

temperature of a given day. From the precipitation levels we create a logical variable that describes if it did or did not rain in a given day.

In Figs. 1 and 2 we present a view to the solar radiation dynamics. Figure 1 shows the mean and respective deviation of solar radiation levels per day of the year. As expected, solar radiation is higher in the warmer seasons of the year. Nonetheless, it also presents a complex structure with several peaks across the days. Figure 2 illustrates the solar radiation per mean temperature, grouped by days that had and had not any rainfall. It also contains LOESS curves that indicate a positive correlation between temperature and solar radiation. Moreover, days without rainfall have considerably higher solar radiation than rainy ones.

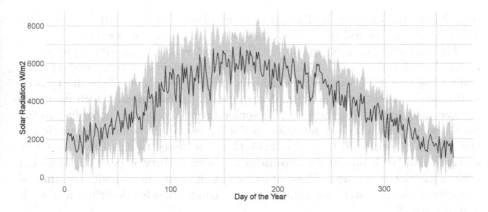

**Fig. 1.** Mean and standard deviation of solar radiation per day of the year (in $W/m^2$)

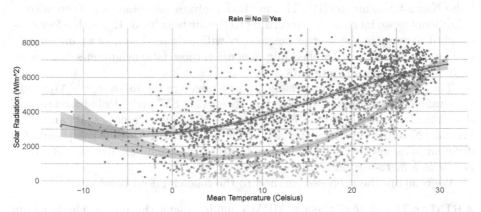

**Fig. 2.** Solar radiation by mean temperature, grouped by days that had rainfall and days that had not any rainfall

## 5    Empirical Experiments

In this section we present the empirical experiments carried out to validate the proposed method for solar radiation forecasting. These address the following research questions:

**Q1:** Is it beneficial to weight individual forecasters according to an Arbitrating scheme for solar radiation forecasting tasks?;

**Q2:** How does the performance of the proposed method relates to the performance of the state-of-the-art methods for solar radiation forecasting tasks?

The experiments were carried out using performance Estimation [30] R package. The methods used in the experiments were evaluated using the Root Mean Squared Error (RMSE) and the Mean Absolute Error (MAE) using a Monte Carlo procedure with **10** repetitions. For each repetition, a random point is picked from the time series. The previous window comprising **40%** of $t$ is used for training and the following window of **20%** of $t$ is used for testing.

The metafeatures used by $\overline{M}$ are the primitive ones, described previously in the Sect. 4.1, together with the following characteristics computed at each embedding vector: (**i**) **kurtosis**, which is a measure of flatness of the data distribution with respect to a Gaussian distribution; (**ii**) **skewness**, which measures the symmetry of the distribution; (**iii**) **series trend**, calculated according to the ratio between the standard deviation of the series and the standard deviation of the differenced series; (**iv**) **serial correlation**, estimated using a Box-Pierce test statistic; and (**v**) **long-range dependence**, using a Hurst exponent estimation with a wavelet transform. These statistics summarise the overall structure of the time series of water consumption. For a comprehensive description of each statistic see [32].

We estimate the optimal embedding dimension $(K)$ using the method of False Nearest Neighbours [12]. This method analyses the behaviour of the nearest neighbours as we increase $K$. According to the authors from [12], with a low suboptimal $K$ many of the nearest neighbours will be false. Then, as we increase $K$ and approach an optimal embedding dimension those false neighbours disappear. $K$ is set to **6** in our experiments.

The base-learners $M$ comprising the ensemble are the following: MARS [20], Generalized Linear Models [6], Random Forest [34], SVM [10], Rule-based regression [14], Generalized Boosted Regression [24], Gaussian Processes [10] and Feed Forward Neural Networks [31]. Each of the individual learners is composed of 6 different parameter settings adding up to **48** learning models. We use a Random Forest as a meta-learner model.

We compare the proposed method to the following four baselines:

**ARIMA:** The state-of-the-art ARIMA model, using the function $auto.arima$ from the forecast R package [9]. This function automatically tunes ARIMA to an **optimal parameter setting**;

**ARIMAX:** Similar to the one above, but augmented with the external features outlined in the case study section;

**NN:** A feed forward neural network with a single hidden layer. The neural network was optimized using a grid search procedure using a total of 56 parameter combinations. The final parameter setting was 7 hidden units and a weight decay of 0.2;

**BT:** Bagged Trees from [21]. This bagging approach is specifically designed for time series forecasting tasks;

**S:** This is a variant of the proposed method, but stripped of the metalearning layer. That is, $M$ is trained in advance and their predictions are simply averaged at run-time using the arithmetic mean;

**Blending:** We use a metalearning technique called Blending to combine the individual learning models. Blending was introduced in [2] in their winning solution for the well known Netflix prize. In practice, it is a variant of Stacking [33] in which out-of-bag predictions are produced with a holdout strategy.

The results from the Monte Carlo experiments are reported in Table 1. Besides the baselines described above, **AE** denotes the proposed method for GHI forecasting tasks.

Relative to **S**, our approach is able to overcome its performance, with similar deviance across Monte Carlo repetitions, both in terms of RMSE and MAE. Overall we conclude that indeed our metalearning approach is beneficial for solar radiation forecasting tasks (**Q1**).

The performance of the proposed method are slightly better than the ARIMAX approach, which shows the competitiveness of our method. Other state-of-the-art baselines such as NN and ARIMA (without external features) perform clearly worse than our method. In effect, we also conclude that our hypothesis **Q2** is valid. Our method was also able to overcome the Blending approach to model combination, which is a widely used technique.

In summary, our experiments validate our hypothesis that our proposed method is able to model the different dynamics of solar radiation with a competitive performance relative to state-of-the-art methods.

**Table 1.** Average results from the methods using RMSE and MAE

| Method | RMSE | MAE |
| --- | --- | --- |
| ARIMA | $1407.561 \pm 88.06$ | $1129.49 \pm 76.32$ |
| ARIMAX | $728.43 \pm 35.46$ | $579.18 \pm 29.03$ |
| BT | $1068.96 \pm 52.90$ | $856.57 \pm 44.85$ |
| NN | $1232.26 \pm 610.20$ | $1000.47 \pm 533.57$ |
| Blending | $731.49 \pm 40.67$ | $566.80 \pm 32.93$ |
| S | $808.82 \pm 46.99$ | $649.36 \pm 39.12$ |
| AE | $\mathbf{721.31 \pm 41.47}$ | $\mathbf{566.22 \pm 35.49}$ |

# 6   Conclusions

In this paper we presented a new method for GHI forecasting tasks. We argued that the planning of operations related to solar radiation is an important topic with economical and social impact. Our proposed method settles on a metalearning scheme called Arbitrating, introduced before by [13,22]. We extend their ideas to GHI forecasting tasks.

We leverage the Arbitrating strategy to dynamically weight individual models in an ensemble. We reasoned that the series of water consumption follows a recurring pattern with different regimes. In effect, our approach allows a fast detection and adaptation to the different regimes causing the data.

Results from numerical experiments suggest that our metalearning is worthwhile. Moreover, we empirically demonstrate that the proposed method is competitive with other state-of-the-art techniques for GHI forecasting tasks, such as Neural Networks and the classical time series model ARIMA.

Future work includes: (**i**) generalise the proposed methodology for other time series forecasting tasks; (**ii**) Compare the proposed method against a time-dependent combining heuristic (e.g. recent performance of individual learners).

In the interest of reproducible research our methods are publicly available as an **R** package called **tsensembler**[1].

**Acknowledgements.** This work is financed by the ERDF - European Regional Development Fund through the Operational Programme for Competitiveness and Internationalisation - COMPETE 2020 Programme within project POCI-01-0145-FEDER-006961, and by National Funds through the FCT - Fundação para a Ciência e a Tecnologia (Portuguese Foundation for Science and Technology) as part of project UID/EEA/50014/2013; Project "NORTE-01-0145-FEDER-000036" is financed by the North Portugal Regional Operational Programme (NORTE 2020), under the PORTUGAL 2020 Partnership Agreement, and through the European Regional Development Fund (ERDF).

# References

1. Bacher, P., Madsen, H., Nielsen, H.A.: Online short-term solar power forecasting. Sol. Energy **83**(10), 1772–1783 (2009)
2. Bell, R.M., Koren, Y., Volinsky, C.: The bellKor 2008 solution to the netflix prize. Statistics Research Department at AT&T Research (2008)
3. Boland, J.: Time series modelling of solar radiation. In: Badescu, V. (ed.) Modeling Solar Radiation at the Earths Surface, pp. 283–312. Springer, Heidelberg (2008)
4. Brazdil, P., Carrier, C.G., Soares, C., Vilalta, R.: Metalearning: Applications to Data Mining. Springer Science & Business Media, Heidelberg (2008)
5. Brown, G.: Ensemble learning. In: Sammut, C., Webb, G.I. (eds.) Encyclopedia of Machine Learning, pp. 312–320. Springer, Boston (2010)
6. Friedman, J., Hastie, T., Tibshirani, R.: Regularization paths for generalized linear models via coordinate descent. J. Stat. Softw. **33**(1), 1–22 (2010)
7. Gama, J., Kosina, P.: Tracking recurring concepts with meta-learners. In: Lopes, L.S., Lau, N., Mariano, P., Rocha, L.M. (eds.) EPIA 2009. LNCS, vol. 5816, pp. 423–434. Springer, Heidelberg (2009). doi:10.1007/978-3-642-04686-5_35
8. Huang, J., Korolkiewicz, M., Agrawal, M., Boland, J.: Forecasting solar radiation on an hourly time scale using a coupled autoregressive and dynamical system (cards) model. Sol. Energy **87**, 136–149 (2013)
9. Hyndman, R.J., Athanasopoulos, G., Razbash, S., Schmidt, D., Zhou, Z., Khan, Y., Bergmeir, C., Wang, E.: Forecast: Forecasting functions for time series and linear models, R package version 5.6 (2014)
10. Karatzoglou, A., Smola, A., Hornik, K., Zeileis, A.: kernlab - an S4 package for kernel methods in R. J. Stat. Softw. **11**(9), 1–20 (2004)
11. Kemmoku, Y., Orita, S., Nakagawa, S., Sakakibara, T.: Daily insolation forecasting using a multi-stage neural network. Sol. Energy **66**(3), 193–199 (1999)

---

[1] https://github.com/vcerqueira/tsensembler.

12. Kennel, M.B., Brown, R., Abarbanel, H.D.: Determining embedding dimension for phase-space reconstruction using a geometrical construction. Phys. Rev. A **45**(6), 3403 (1992)
13. Koppel, M., Engelson, S.P.: Integrating multiple classifiers by finding their areas of expertise. In: AAAI-1996 Workshop On Integrating Multiple Learning Models, pp. 53–58. Citeseer (1996)
14. Kuhn, M., Weston, S., Keefer, C., Quinlan, R.: Cubist: Rule- and Instance-Based Regression Modeling, R package version 0.0.18 (2014)
15. Lemke, C., Gabrys, B.: Meta-learning for time series forecasting and forecast combination. Neurocomputing **73**(10), 2006–2016 (2010)
16. Lorenz, E., Hurka, J., Heinemann, D., Beyer, H.G.: Irradiance forecasting for the power prediction of grid-connected photovoltaic systems. IEEE J. Sel. Top. Appl. Earth Obs. Remote Sens. **2**(1), 2–10 (2009)
17. Maxey, C., Andreas, A.: Oak ridge national laboratory (ORNL); rotating shadowband radiometer (RSR); oak ridge, tennessee (DATA). Technical report, National Renewable Energy Laboratory (NREL), Golden, CO, USA (2007)
18. Mellit, A., Pavan, A.M.: A 24-h forecast of solar irradiance using artificial neural network: application for performance prediction of a grid-connected PV plant at Trieste, Italy. Sol. Energy **84**(5), 807–821 (2010)
19. Mendes-Moreira, J., Soares, C., Jorge, A.M., Sousa, J.F.D.: Ensemble approaches for regression: a survey. ACM Comput. Surv. (CSUR) **45**(1), 10 (2012)
20. Milborrow, S.: Earth: Multivariate Adaptive Regression Spline Models. Derived from mda:mars by Trevor Hastie and Rob Tibshirani (2012)
21. Oliveira, M., Torgo, L.: Ensembles for time series forecasting. In: ACML Proceedings of Asian Conference on Machine Learning. JMLR: Workshop and Conference Proceedings (2014)
22. Ortega, J., Koppel, M., Argamon, S.: Arbitrating among competing classifiers using learned referees. Knowl. Inf. Syst. **3**(4), 470–490 (2001)
23. Reikard, G.: Predicting solar radiation at high resolutions: a comparison of time series forecasts. Sol. Energy **83**(3), 342–349 (2009)
24. Ridgeway, G.: GBM: Generalized Boosted Regression Models, R package version 2.1.1 (2015)
25. Rossi, A.L.D., de Leon Ferreira, A.C.P., Soares, C., De Souza, B.F., et al.: Metastream: a meta-learning based method for periodic algorithm selection in time-changing data. Neurocomputing **127**, 52–64 (2014)
26. dos Santos, P.M., Ludermir, T.B., Prudencio, R.B.C.: Selection of time series forecasting models based on performance information. In: Fourth International Conference on Hybrid Intelligent Systems, HIS 2004, pp. 366–371. IEEE (2004)
27. Sfetsos, A., Coonick, A.: Univariate and multivariate forecasting of hourly solar radiation with artificial intelligence techniques. Sol. Energy **68**(2), 169–178 (2000)
28. Soares, C., Brazdil, P.B.: Zoomed ranking: selection of classification algorithms based on relevant performance information. In: Zighed, D.A., Komorowski, J., Żytkow, J. (eds.) PKDD 2000. LNCS, vol. 1910, pp. 126–135. Springer, Heidelberg (2000). doi:10.1007/3-540-45372-5_13
29. Takens, F.: Detecting strange attractors in turbulence. In: Rand, D., Young, L.-S. (eds.) Dynamical Systems and Turbulence, Warwick 1980. LNM, vol. 898, pp. 366–381. Springer, Heidelberg (1981). doi:10.1007/BFb0091924
30. Torgo, L.: An Infra-Structure for Performance Estimation and Experimental Comparison of Predictive Models, R package version 0.1.1 (2013)
31. Venables, W.N., Ripley, B.D.: Modern Applied Statistics with S, 4th edn. Springer, New York (2002). ISBN 0-387-95457-0

32. Wang, X., Smith-Miles, K., Hyndman, R.: Rule induction for forecasting method selection: meta-learning the characteristics of univariate time series. Neurocomputing **72**(10), 2581–2594 (2009)
33. Wolpert, D.H.: Stacked generalization. Neural Netw. **5**(2), 241–259 (1992)
34. Wright, M.N.: Ranger: A Fast Implementation of Random Forests, R package version 0.3.0 (2015)

# Modeling the Transformation of Olive Tree Biomass into Bioethanol with Reg-CO²RBFN

Francisco Charte Ojeda[1], Inmaculada Romero Pulido[2],
Antonio Jesús Rivera Rivas[1]([✉]), and Eulogio Castro Galiano[2]

[1] Department of Computer Science, University of Jaén, Jaén, Spain
{fcharte,arivera}@ujaen.es
[2] Department of Chemical, Environmental and Materials Engineering,
University of Jaén, Jaén, Spain
{iromero,ecastro}@ujaen.es

**Abstract.** Research in renewable energies is a global trend. One remarkable area is the biomass transformation into biotehanol, a fuel that can replace fossil fuels. A key step in this process is the pretreatment stage, where several variables are involved. The experimentation for determining the optimal values of these variables is expensive, therefore it is necessary to model this process. This paper focus on modeling the production of biotehanol from olive tree biomass by data mining methods. Notably, the authors present Reg-CO²RBFN, an adaptation of a cooperative-competitive designing method for radial basis function networks. One of the main drawbacks in this modeling is the low number of instances in the data sets. To compare the results obtained by Reg-CO²RBFN, other well-known data mining regression methods are used to model the transformation process.

**Keywords:** Regression models · Data mining · Enzymatic hydrolisis · Olive tree biomass

## 1 Introduction

Nowadays, the interest for renewable energy is increasing [1]. Olive tree biomass (OTB) is an abundant organic residual in Mediterranean countries that can be converted in bioethanol. Due to the benefits of olive oil, olive tree cultivation is expanding worldwide. In these places olive trees are periodically pruned to rejuvenate them, as well as to prevent propagation of diseases. The process for obtaining bioethanol from OTB has the following advantages: large availability of OTB, low cost, $CO_2$ emissions reduction when this fuel is used, and a decreasing of dependency on energy imports.

The basic steps in the production of bioethanol from OTB are pretreatment, enzymatic hydrolysis and fermentation. One of the main steps is the pretreatment stage [2], whose performance can be affected by several factors such as processing time, temperature and the use of different salts. Therefore, for the

© Springer International Publishing AG 2017
I. Rojas et al. (Eds.): IWANN 2017, Part I, LNCS 10305, pp. 733–744, 2017.
DOI: 10.1007/978-3-319-59153-7_63

experts involved in the field it is important to adequately model or characterize this process. A key challenge in this modeling is the low number of data samples (instances) available in the experimentation, due to the high costs involved in obtaining them.

Data mining methods have been successfully used for regression tasks [3] even in the renewable energy field [4–7]. Together with Multilayer Perceptron (MLPs) [8], Radial Basis Function Networks (RBFNs) are one the best well-known and important Artificial Neural Network (ANN). The efficacy of RBFNs has been proved in many areas, including regression [9]. The authors have developed $CO^2RBFN$, a cooperative-competitive evolutionary RBFN design method [10] that has been successfully used for regression in the modeling of concentrated photovoltaic modules [5].

In this paper Reg-$CO^2RBFN$, an adaptation of $CO^2RBFN$, is applied to the problem of modeling the bioethanol conversion process from OTB. The main objective of this adaptation is facing the training with a low number of instances per data set. With this aim, a supervised clustering algorithm is introduced in the initialization step of the algorithm. Other recognized data mining methods, such as a MLP trained with the backpropagation technique and a Suppport Vector Machine, have been also used to model the same process. The performance of Reg-$CO^2RBFN$ and these other methods will be compared.

This paper is organized as follows: Sect. 2 depicts the bioethanol production process. The Reg-$CO^2RBFN$ method is described in Sect. 3. In Sect. 4 the experimental framework is outlined and the results obtained are presented. The results produced by the proposed method are compared with that obtained by other data mining models, as explained in this section. Finally, the conclusions appear in Sect. 5.

## 2    Bioethanol Production from Olive Tree Biomass

OTB is a lignocellulose material and it is considered a promising candidate to be transformed into renewable fuels, therefore substituting the well-known fossil fuels [1] at some extent. The pruning of olive trees generates a huge amount of biomass that is dismissed or must be eliminated, for example by burning it. Taking into account environmental considerations, the transformation of OTB into bioethanol implies several advantages. It is a clean way of eliminating organic residuals, and implies a net reduction of $CO_2$ emissions when this fuel is used. In addition, it reduces the dependency from fuel imports.

The process of converting a lignocellulose material into bioethanol includes three steps: pretreatment, enzymatic hydrolysis and fermentation. Pretreatment is a key step in the whole process and has a direct influence over the hydrolysis step [2]. The objective of this first step is to achieve maximal fermentation yields and rates. The use of $FeCl_3$, a metal salt, has demonstrated its efficacy in the pretreatment phase. This salt concentration, along with the settings for temperature and time, define the input variables of the pretreatment. From the pretreatment experiments, liquids and pretreated solids are obtained and separated by filtration. At this time, the content of glucose and hemicellulosic sugars in both fractions can be determined. Then, the pretreated solids are further submitted to enzymatic hydrolysis under standard conditions, and the enzymatic hydrolysis yield for each experiment (grams of glucose in the hydrolysate per gram of glucose in the pretreated material or in the original material) is evaluated.

This research was carried out in Jaén (Spain), collecting the OTB from its olive tree fields. Due to the costs involved in performing empirical pretreatment analysis, only twenty experiments were conducted. The Box-Behnken methodology was used to equally distribute the values of experiments (instances). For a further explanation see [11].

In summary, the objective is to model the process of transforming a lignocellulose material into bioethanol. In this model the input variables are: duration of the pretreatment (Time), its temperature (Temp) and the molar concentration of $FeCl_3$ (FeCl3). The output variables are the Enzymatic Hydrolysis Yields in Raw Material (EHYRM) and Enzymatic Hydrolysis Yields in Pretreated Material (EHYPM). The low number of the conducted experiments implies a challenge for most learning methods.

## 3    Reg-CO²RBFN: Adaptation of CO²RBFN for Regression Tasks

In this paper a hybrid evolutionary cooperative-competitive model for designing RBFNs is proposed. This section starts providing a brief introduction to this type of neural networks, in Sect. 3.1. Then, in Sect. 3.2, the detailed description of the proposed model, Reg-CO²RBFN, is addressed.

### 3.1    Radial Basis Function Networks

RBFNs are an artificial neural network paradigm [12] with remarkable characteristics, such as simple topological structure and universal approximation ability [9]. The topology of an RBFN is composed by three feed-forward connected layers: an input layer with $n$ nodes, a hidden layer with $m$ neurons or RBFs, and an output layer with one node for regression problems.

The neurons in the hidden layer present a radially-symmetric basis activation function, $\phi_i : R^n \rightarrow R$, which can be defined with several shapes, being the Gaussian function (1) the most widely used.

$$\phi_i(x) = e^{-(\|x-c_i\|/d_i)^2} \tag{1}$$

where $c_i \in R^n$ is the center of the basis function $\phi_i$, $d_i \in R$ is the width (radius), and $\|\|$ is typically the Euclidean norm on $R^n$. This expression is the one used in this paper as RBF.

The output neuron implements the weighted sum of RBF outputs in the hidden layer, as can be seen in Eq. 2, where $w_i$ are the weights. Each one of them represents the contribution of one RBF to the output node.

$$f(\boldsymbol{x}) = \sum_{i=1}^{m} w_i \phi_i(\boldsymbol{x}) \tag{2}$$

The main goal in the RBFN design process consists in determining the centers and widths of the neurons (RBFs) in the hidden layer, as well as the linear output weights connecting the RBFs to the output neuron.

An important paradigm for designing RBFNs is Evolutionary Computation (EC) [13–15]. EC uses natural evolution and stochastic searching to design optimization algorithms. Concretely, EC evolves a population of individuals according to operators as mutation, recombination or selection, and each individual in the population receives a measure proportional to its quality, called fitness. Reviews on EC applied to RBFN design can be found in [16,17].

## 3.2   The Proposed Method, Reg-CO$^2$RBFN

As mentioned, the presented method is a cooperative-competitive evolutionary proposal for designing RBFNs. In this methodology each individual of the population corresponds to one RBF, and the entire population implements the whole solution. The individuals cooperate towards a definitive solution, but they must also compete for survival.

In this environment, in which the solution depends on the behavior of many components, the fitness of each individual is known as credit assignment. In order to measure the credit assignment of an individual three factors have been proposed: the RBF contribution to the network output, the error in the basis function radius and the degree of overlapping among RBFs.

The application of the operators is determined by a Fuzzy Rule-Based System (FRBS). The inputs of the FRBS are the three parameters used for credit assignment, and the outputs are the operators' application probability.

---

**Algorithm 1.** CO2RBFN algorithm main steps.

---
1: Initialize RBFN

2: **while**(Stop condition is not met)                              ▷ Training loop
3:      Evaluate RBFs
4:      Apply operators to RBFs
5:      Substitute the eliminated RBFs
6:      Select the best RBFs

---

In the adapted version presented in this work, the network initialization step has been changed. The new initialization is based on supervised clustering [18]. Taking into account the low number of instances, the aim is introducing a previous analysis of the data, before the evolutionary phase, that reinforces the learning process. The main steps of Reg-CO$^2$RBFN, detailed below, are shown in Algorithm 1 in pseudocode.

**RBFN Initialization.** To initialize the RBFN, an adapted version of the k-means algorithm [19] is used. K-means is a clustering that aims to partition the data set into $k$ clusters. Each pattern belongs to a cluster and the cluster prototype is the mean of the patterns that belongs to this cluster. Thus, each RBF center, $c_i$, is assigned to a previously calculated cluster. According to [18] the modification carried out consists in taking into account the input and the output features of the data set during the cluster determination process. The RBF widths, $d_i$, will be set to the average distance between the centers. Finally, the RBF weights, $w_{ij}$, are set to zero.

**RBFN Training.** To adjust the RBF weights the Singular Value Decomposition (SVD) algorithm [20] is used. SVD is a deterministic technique typically used in matrix resolution.

**RBF Evaluation.** In order to evaluate the fitness or credit assignment of each RBF $\phi_i$ in the cooperative-competitive environment, three parameters, $a_i$, $e_i$, $o_i$ are defined:

- The contribution, $a_i$, of the RBF $\phi_i$, $i = 1 \ldots m$, is determined by considering its maximum weight. An RBF with a low weight will have a low contribution and so on.
- The error measure, $e_i$, for each RBF $\phi_i$, is obtained by calculating the Root Mean Square Error (RMSE) (3) inside its width:

$$e_i = \sqrt{\frac{\sum_{t=1}^{n} (f_t - y_t)^2}{n}} \tag{3}$$

  where $f_t$ is the output of the model (2) for a point inside the width of RBF $\phi_i$, $y_t$ is the real output at the same point, and $n$ is the number of points inside the RBF $\phi_i$.
- The overlapping of the RBF $\phi_i$ and the other RBFs is quantified by using the parameter $o_i$. This parameter is computed by taking into account the fitness sharing methodology [21], whose aim is to maintain the diversity in the population. This factor is expressed as indicated in Eq. 4, where $o_{ij}$ measures the overlapping of the RBFs $\phi_i$ and $\phi_j$, $j = 1 \ldots m$.

$$o_i = \sum_{j=1}^{m} o_{ij} \tag{4}$$

$$o_{ij} = \begin{cases} (1 - \|\phi_i - \phi_j\|/d_i) & if \ \|\phi_i - \phi_j\| < d_i \\ 0 & otherwise \end{cases} \tag{5}$$

**Applying Operators to RBFs.** Four operators are defined in order to be applied to the RBFs:

- Operator Remove: eliminates an RBF.
- Operator Random Mutation: the number of coordinates to modify of the selected RBF is randomly generated between 1 and 25% of the total number of coordinates. The chosen coordinates are modified in a percentage between 5% and 25% of the old width. The width is adjusted in a similar way to the coordinates.
- Operator Biased Mutation: modifies the width and some coordinates of the center, between 1 and 25% of the total number of coordinates, using local information of the RBF environment. Concretely, the patterns inside the RBF width are determined and the new coordinates are allocated in the average of the corresponding patterns coordinates. This technique follows the clustering methodology [19]. The width is obtained by changing its old values to a random number (between 5% and 25%).
- Operator Null: in this case any operator is applied.

The probability of applying an operator to a given RBF is determined by means of a Mandani-type FRBS [22] which represents expert knowledge in order to obtain a simple and accurate RBFN.

The inputs of this system are the parameters used to define the credit assignment of the RBF, and the outputs represent the probability of applying Remove, Random Mutation, Biased Mutation and Null operators, respectively. Table 1 shows the rule base used to relate the described antecedents and consequents.

**Table 1.** Fuzzy rule base representing expert knowledge in the design of RBFN

| | Antecedents | | | Consequents | | | | | Antecedents | | | Consequents | | | |
|---|---|---|---|---|---|---|---|---|---|---|---|---|---|---|---|
| | $v_a$ | $v_e$ | $v_o$ | $p_{remove}$ | $p_{rm}$ | $p_{bm}$ | $p_{null}$ | | $v_a$ | $v_e$ | $v_o$ | $p_{remove}$ | $p_{rm}$ | $p_{bm}$ | $p_{null}$ |
| R1 | L | | | M-H | M-H | L | L | R6 | | H | | M-H | M-H | L | L |
| R2 | M | | | M-L | M-H | M-L | M-L | R7 | | | L | L | M-H | M-H | M-H |
| R3 | H | | | L | M-H | M-H | M-H | R8 | | | M | M-L | M-H | M-L | M-L |
| R4 | | L | | L | M-H | M-H | M-H | R9 | | | H | M-H | M-H | L | L |
| R5 | | M | | M-L | M-H | M-L | M-L | | | | | | | | |

For example, the FRBS promotes that an RBF with a poor credit assignment, low contribution, high error and high overlapping is eliminated. Otherwise the RBF is maintained.

**Introduction of New RBFs.** In this step, the removed RBFs are substituted by new ones. Each new RBF is located at the center of the area with maximum error. Its width is set to the average of the RBFs. Finally, the weights are set to zero.

**Replacement Strategy.** In this step the newly generated RBFs are compared with the original ones, and those with the best behavior are incorporated into the new population.

# 4  Experimentation and Results

As mentioned, applying the Box-Behnken methodology twenty chemical pretreatment experiments have been conducted to equally space the values of the data samples (instances) [11], obtaining twenty different data patterns. These make up the working data set, used to carry out the following modeling experimentation.

## 4.1  Experimental Framework

The input variables used to train the models are the following:

- Duration of the pretreatment (Time), measured in minutes and with values that oscillates between 0 and 30.
- Temperature of the pretreatment (Temp), which values go from 120 to 180 centigrade degrees.
- Molar concentration of $FeCl_3$ (FeCl3), whose values are in the range 0.050 to 0.275 M.

The considered output variables are the Enzymatic Hydrolysis Yields in Raw Material (EHYRM) and Enzymatic Hydrolysis Yields in Pretreated Material (EHYPM). All the values are summarized in Table 2.

As the number of available patterns is small (only 20), consequently the leave one out validation is applied. This implies that 20 models haven been developed, each one of them using 19 patterns for training and the remainder one for testing.

Two evaluation metrics have been computed to assess the methods performance: RMSE (*Root Mean Square Error*) and the coefficient of determination, $R^2$. RMSE is defined in Eq. 6, where $n$ is the number of instances, $f_t$ is the output of the model and $y_t$ is the real output for the t-th instance, respectively.

$$RMSE = \sqrt{\frac{\sum_{t=1}^{n}(f_t - y_t)^2}{n}} \tag{6}$$

The quality of the calculated model from the training data is also evaluated with $R^2$, that obtains the fit between the predicted and the real data (7).

$$R^2 = \left(\frac{cov(f, y)}{\sigma_f \sigma_y}\right)^2 \tag{7}$$

where $cov(f, y)$ represents the covariance between the model output and the real output, and $\sigma_f$ and $\sigma_y$ are the standard deviations for the model output and the real output, respectively.

**Table 2.** Values from the OTB pretreatment experiments

| n | Input variables | | | Output variables | |
|---|------|------|-------|-------|-------|
|   | Time | Temp | FeCl3 | EHYRM | EHYPM |
| 1 | 30 | 120 | 0.125 | 10.14 | 19.71 |
| 2 | 0 | 140 | 0.050 | 12.23 | 19.92 |
| 3 | 15 | 160 | 0.050 | 19.57 | 34.61 |
| 4 | 0 | 140 | 0.200 | 11.66 | 19.43 |
| 5 | 30 | 140 | 0.200 | 19.67 | 39.94 |
| 6 | 15 | 140 | 0.125 | 14.62 | 23.17 |
| 7 | 15 | 140 | 0.125 | 15.10 | 27.34 |
| 8 | 30 | 160 | 0.125 | 28.07 | 55.80 |
| 9 | 15 | 120 | 0.050 | 11.85 | 18.85 |
| 10 | 30 | 140 | 0.050 | 11.08 | 19.81 |
| 11 | 15 | 120 | 0.200 | 8.76 | 14.63 |
| 12 | 0 | 160 | 0.125 | 18.80 | 33.69 |
| 13 | 15 | 140 | 0.125 | 14.10 | 25.70 |
| 14 | 15 | 160 | 0.200 | 36.50 | 75.54 |
| 15 | 0 | 120 | 0.125 | 12.62 | 20.07 |
| 16 | 0 | 180 | 0.200 | 25.58 | 93.16 |
| 17 | 0 | 160 | 0.275 | 28.73 | 66.87 |
| 18 | 30 | 180 | 0.200 | 20.17 | 93.55 |
| 19 | 30 | 160 | 0.275 | 38.85 | 88.71 |
| 20 | 30 | 180 | 0.275 | 11.36 | 96.15 |

In order to compare the results obtained for the proposed method, other two data mining methods are used: MLP-BR and NU-SVR. The methods' names and their implementation has been obtained from KEEL [23]. A brief description of this methods follows:

– MLP-BR. This algorithm is a implementation of the well known MLP [8], one of the most popular ANN methods. An MLP consists of multiple layers of interconnected hidden nodes. The hidden nodes receive the inputs from the input layer and calculate outputs which depend on the input and their activation function. The output calculated by the hidden nodes is forwarded to the next layer and so on up to reach the network output layer. As learning technique the model uses back-propagation, the value predicted by the model is compared with the real one, and the committed error is used to adjust the weights connecting the units in each layer. This process continues until a small error is obtained.

MLP-BR [24], the algorithm used in our experiments, is essentially a multi-layer perceptron designed to produce a continuous output as prediction, so it is adapted to regression tasks.

- NU-SVR. This implementation follows the *Support Vectorial Machine* (SVM) [25] technique. The training of SVMs consists on solving a quadratic optimization problem to find the maximum separation margin between pattern categories. And adaptation of the basic algorithm, named NU-SVM [26], changes the $C$ parameter by other called $NU$ which is easier to adjust.

In [27], SVMs can be seen as an universal tool for solving many task such as regression problems. The implementation used in this experimentation, NU-SVR [28], is a regression SVM based on the SMO (*Sequential Minimal Optimization*) [29] algorithm. This algorithm reduces the training process and therefore is more efficient than the traditional learning model.

The values of the configuration parameters for each method are set to the default ones proposed by the respective authors. The number of executions is established to three, as these are non-deterministic methods. The main parameters used for the algorithms are shown in Table 3.

**Table 3.** Parameter specification for the algorithms employed in the experimentation.

| Algorithm | Parameter | Value |
|---|---|---|
| Reg-CO$^2$RBFN | Generations of the main loop | 100 |
| | Number of RBFs | 8 |
| MLP-BP | Hidden_layer | 1 |
| | Hidden_nodes | 8 |
| | Transfer | Htan |
| | Eta | 0.15 |
| | Alpha | 0.10 |
| | Lambda | 0.0 |
| NU-SVM | KERNELtype | RBF |
| | C | 100.0 |
| | Eps | 0.001 |
| | Degree | 1 |
| | Gamma | 0.01 |
| | Coef0 | 0.0 |

## 4.2   Results and Analysis

For each output variable of the hydrolysis process, one model is determined by each data mining method. The provided performance indicators are average values from these 20 runs. The results obtained with the training data sets

**Table 4.** Results obtained from training data sets

| Variable | Reg-CO$^2$RBFN | | MLP-BP | | NU-SVM | |
|---|---|---|---|---|---|---|
| | $R^2$ ↑ | RMSE ↓ | $R^2$ ↑ | RMSE ↓ | $R^2$ ↑ | RMSE ↓ |
| EHYRM | **0.8846** | **2.8440** | 0.6807 | 5.4435 | 0.6622 | 6.6978 |
| EHYPM | **0.9919** | **2.5868** | 0.9072 | 10.7019 | 0.8532 | 18.2879 |

are shown in Table 4. For each method, the RMSE and the $R^2$ coefficient are calculated. The best results are highlighted in bold. For RMSE lower values are better, whereas for $R^2$ higher values are better.

In summary, that the Reg-CO$^2$RBFN method outperforms the remaining methods, both in RMSE error and $R^2$ coefficient for all output variables, can be concluded from Table 4. By observing the RMSE metric, that MLP-BP and NU-SVM are clearly outperformed by Reg-CO$^2$RBFN for both variables can be seen, since MLP-BP and NU-SVM perform quite bad. These results can be explained by the low number of instances in the training data sets. Regarding the $R^2$ coefficient, Reg-CO$^2$RBFN obtain remarkable values, specially for the EHYPM variable.

The RMSE obtained by the models from the tests data sets are shown in Table 5. As before, best results are highlighted in bold.

**Table 5.** Results obtained from test data sets

| Variable | Reg-CO$^2$RBFN | MLP-BP | NU-SVM |
|---|---|---|---|
| EHYRM | **6.5544** | 7.1525 | 7.8550 |
| EHYPM | **7.1723** | 9.8838 | 22.5404 |

For the test data sets, it must be noticed that Reg-CO$^2$RBFN outperforms the other methods, both for the EHYRM and specially for the EHYPM variable, whereas Nu-SVM achieve very bad results, possibly attributable to the low number of instances available in the data sets as noted before.

## 5   Conclusions

The chemical research described in this work, carried out in Jaén, has allowed to obtain bioethanol from olive tree biomass. This study represents another contribution inside the renewable energy field. There are various parameters or variables involved in the process of obtaining bioethanol. In addition, the experimentation is expensive and also time consuming. For this reasons, it is important to automatically obtain a model able to explain the process.

In this paper three data mining methods have been used to modeling the process: Reg-CO$^2$RBFN, an RBFN designing technique developed by the authors, an MLP and an SVM. The main problem for the data mining methods is the low number of experiments (instances) available due to the high cost previously mentioned.

The data mining methods have been trained and have produced a model for each output variable. The results obtained show that Reg-$CO^2$RBFN has a good behavior in spite of the low training data available, and outperforms the remaining data mining methods for the output variables in the study.

As future work a multi-target regression version of Reg-$CO^2$RBFN will be studied.

**Acknowledgments.** This work is partially supported by the Spanish Ministry of Science and Technology under project TIN2015-68454-R.

# References

1. Behera, S., Arora, R., Nandhagopal, N., Kumar, S.: Importance of chemical pretreatment for bioconversion of lignocellulosic biomass. Renew. Sustain. Eng. Rev. **36**, 91–106 (2014). http://dx.doi.org/10.1016/j.rser.2014.04.047
2. Ravindran, R., Jaiswal, A.K.: A comprehensive review on pre-treatment strategy for lignocellulosic food industry waste: challenges and opportunities. Bioresour. Technol. **199**, 92–102 (2016). http://dx.doi.org/10.1016/j.biortech.2015.07.106
3. Maimon, O., Rokach, L.: The Data Mining and Knowledge Discovery Handbook, 2nd edn. Springer, Heidelberg (2010)
4. Saavedra-Moreno, B., Salcedo-Sanz, S., Paniagua-Tineo, A., Prieto, L., Portilla-Figueras, A.: Seeding evolutionary algorithms with heuristics for optimal wind turbines positioning in wind farms. Renew. Eng. **36**(11), 2838–2844 (2011)
5. Rivera, A., Garca-Domingo, B., del Jesus, M., Aguilera, J.: Characterization of concentrating photovoltaic modules by cooperative competitive radial basis function networks. Expert Syst. Appl. **40**(5), 1599–1608 (2013). http://dx.doi.org/10.1016/j.eswa.2012.09.016
6. García-Domingo, B., Carmona, C., Rivera-Rivas, A., Jesus, M.D., Aguilera, J.: A differential evolution proposal for estimating the maximum power delivered by CPV modules under real outdoor conditions. Expert Syst. Appl. **42**(13), 5452–5462 (2015)
7. Kusiak, A., Zheng, H., Song, Z.: Short-term prediction of wind farm power: a data mining approach. IEEE Trans. Eng. Convers. **24**(1), 125–136 (2009)
8. Rumelhart, D.E., Hinton, G.E., Williams, R.J.: Learning internal representations by error propagation. Technical report, DTIC Document (1985)
9. Park, J., Sandberg, I.: Universal approximation using radial-basis function networks. Neural Comput. **3**, 246–257 (1991)
10. Pérez-Godoy, M., Rivera, A., del Jesus, M., Berlanga, F.: $CO^2$RBFN: an evolutionary cooperative-competitive RBFN design algorithm for classification problems. Soft. Comput. **14**(9), 953–971 (2010)
11. López-Linares, J., Romero, I., Moya, M., Cara, C., Ruiz, E., Castro, E.: Pretreatment of olive tree biomass with FeCl3 prior enzymatic hydrolysis. Bioresour. Technol. **128**, 180–187 (2013). doi:10.1016/j.biortech.2012.10.076
12. Broomhead, D., Lowe, D.: Multivariable functional interpolation and adaptive networks. Complex Syst. **2**, 321–355 (1988)

13. Holland, J.: Adaptation in Natural and Artificial Systems. The University of Michigan Press, Ann Arbor (1975)

14. Goldberg, D.: Genetic Algorithms in Search, Optimization, and Machine Learning. Addison-Wesley, Reading (1989)

15. Bäck, T., Hammel, U., Schwefel, H.: Evolutionary computation: comments on the history and current state. IEEE Trans. Evol. Comput. $1(1)$, 3–17 (1997)

16. Harpham, C., Dawson, C., Brown, M.: A review of genetic algorithms applied to training radial basis function networks. Neural Comput. Appl. $13$, 193–201 (2004)

17. Buchtala, O., Klimek, M., Sick, B.: Evolutionary optimization of radial basis function classifiers for data mining applications. IEEE Trans. Syst. Man Cybern. B $35(5)$, 928–947 (2005)

18. Runkler, T.A., Bezdek, J.C.: Alternating cluster estimation: a new tool for clustering and function approximation. IEEE Trans. Fuzzy Syst. $7(4)$, 377–393 (1999). doi:10.1109/91.784198

19. Hartigan, J., Wong, M.: Algorithm as 136: a k-means clustering algorithm. J. R. Stat. Soc. Ser. C (Appl. Stat.) $28(1)$, 100–108 (1979)

20. Widrow, B., Lehr, M.: 30 years of adaptive neural networks: perceptron, madaline and backpropagation. Proc. IEEE $78(9)$, 1415–1442 (1990)

21. Goldberg, D., Richardson, J.: Genetic algorithms with sharing for multimodal function optimization. In: Proceedings of the Second International Conference on Genetic Algorithms, pp. 41–49 (1987)

22. Mandani, E., Assilian, S.: An experiment in linguistic synthesis with a fuzzy logic controller. Int. J. Man Mach. Stud. $7(1)$, 1–13 (1975)

23. Alcalá-Fdez, J., Fernández, A., Luengo, J., Derrac, J., García, S., Sánchez, L., Herrera, F.: Keel data-mining software tool: data set repository, integration of algorithms and experimental analysis framework. J. Mult.-Valued Logic Soft Comput. $17$, 255–287 (2011)

24. Rojas, R.: Neural Networks. Springer, Heidelberg (1996). doi:10.1007/978-3-642-61068-4

25. Boser, B.E., Guyon, I.M., Vapnik, V.N.: A training algorithm for optimal margin classifiers. In: Proceedings of the Fifth Annual Workshop on Computational Learning Theory - COLT 1992, pp. 144–152. ACM Press, New York (1992). doi:10.1145/130385.130401

26. Schölkopf, B., Smola, A.J., Williamson, R.C., Bartlett, P.L.: New support vector algorithms. Neural Comput. $12(5)$, 1207–1245 (2000). doi:10.1162/089976600300015565

27. Vapnik, V., Vapnik, V., Golowich, S.E., Smola, A.: Support vector method for function approximation, regression estimation, and signal processing. Adv. Neural Inf. Process. Syst. $9$, 281–287 (1996)

28. Fan, R., Chen, P., Lin, C.: Working set selection using the second order information for training SVM. J. Mach. Learn. Res. $6$, 1889–1918 (2005)

29. Platt, J.: Sequential Minimal Optimization: A Fast Algorithm for Training Support Vector Machines (1998)

# A Hybrid Neuro-Evolutionary Algorithm for Wind Power Ramp Events Detection

Laura Cornejo-Bueno, Adrián Aybar-Ruiz, Carlos Camacho-Gómez,
Luis Prieto, Alberto Barea-Ropero, and Sancho Salcedo-Sanz[✉]

Department of Signal Processing and Communications, Universidad de Alcalá,
Alcalá de Henares, Madrid, Spain
sancho.salcedo@uah.es

**Abstract.** In this work, a hybrid system for wind power ramps events prediction in wind farms is proposed. The system is based on modelling the prediction problem as a binary classification problem from atmospheric reanalysis data inputs. On the other hand, a hybrid neuro-evolutive algorithm is proposed, which combines Artificial Neuronal Networks such as Extreme Learning Machines, with evolutionary algorithms to optimize the trained models. The phenomenon under study occurs with a very low probability, for this reason the problem is so unbalanced, and it is necessary to resort to techniques focused on obtain good results by means of a reduction of the samples from the majority class, as the SMOTE approach. A feature selection is performed by the evolutionary algorithm in order to choose the best trained model. Finally, this model is evaluated by a test set and its accuracy performance is given. The accuracy obtained in the results is quite good in terms of classification performance.

**Keywords:** Evolutionary algorithms · Extreme Learning Machine · SMOTE · Wind power ramp events

## 1 Introduction

Currently, wind energy is the most important sustainable energy source in the world, in terms of annual growing, penetration in the power system and economic impact [1]. One of the problems of wind energy is that it exhibits intermittent generation (depending on the weather conditions) [2], which makes difficult its integration in the system. Wind power generation forecasting is therefore a key factor to improve this integration [3–7]. Among the different issues in wind power forecasting, one of the most significant is the existence of wind power ramp events (WPREs) in wind farms [8,9]. WPREs consist of large fluctuations of wind power in a short period of time, leading to a significant increasing or decreasing of the power produced in the wind farm. The accurate prediction of WPREs has been reported as an effective method to mitigate the economic impact of these events in wind generation power plants [8,10].

© Springer International Publishing AG 2017
I. Rojas et al. (Eds.): IWANN 2017, Part I, LNCS 10305, pp. 745–756, 2017.
DOI: 10.1007/978-3-319-59153-7_64

Following [10], the prediction of WPREs has been tackled with two major families of techniques: statistical approaches and physical models. Statistical approaches include different methodologies such as autorregressive moving average (ARMA) algorithms, dynamic programming, neural computation techniques or kernel methods, etc. In [11] a hybrid ARMA – Hidden Markov Model approach is proposed to short-term wind speed prediction, including wind ramp events. Experiments in two locations of the USA (one in Pacific Northwest and one in southern Wisconsin) show a good performance of the methodology proposed, using surface wind speed and direction time series to estimate future values of the wind speed. In [12], a neural network approach for switching between three different regimes of WPREs (ramp-up, ramp-down and no-ramp) is proposed. Depending on the WPRE type (evaluated using a gradient time series of the wind speed), a different neural network is trained, with specific structure and training process. Results of the application of this approach in data from Spanish wind farms are reported.

Regarding physical models applied to WPREs detection and prediction, there are a good amount of techniques proposed in the literature. For example, in [13] the evaluation of a conventional wind power forecasting methodology based on the combination of two Numerical Weather Models (NWMs) is carried out. Recently, in [14] data from Global Circulation Models (reanalysis data) are used to identify possible meteorological caused of WPREs. A methodology based on wavelets and PCA is applied to estimate the best set of features (predictive variables) to estimate WPREs.

In spite of this huge previous research both on statistical approaches and physical models, there have been very few works that consider both WPRE prediction paradigms together. In [15] the possibility of using statistical techniques to carry out a down-scaling process with application in WPRE detection is suggested. A similar approach was first proposed in [16] for short-term wind speed prediction, but without direct application to WPREs prediction.

This paper presents a novel hybrid WPRE classification framework that merges parts of physical models with state-of-the-art statistical approaches. Moreover, the problem is so unbalanced, for this reason we have to apply another technique, the SMOTE approach, to improve the results in the classification process. Specifically, the work presents a hybrid system for WPRE classification based on Extreme Learning Machine (ELM) combined with evolutionary algorithms, plus the SMOTE technique, in which predictive variables are obtained from the ERA-Interim reanalysis data. It is proposed the use of direct reanalysis data as predictive variables of machine learning classification technique, (ELM). Results in data from a wind farm in Spain will show the performance of the proposed system.

The remainder of the paper is structured in the following way: next section presents the problem definition that is tackled in this paper, in which the WPRE prediction is formulated as a regression task. Section 3 presents the data and predictive variables involved in the paper. In Sect. 4 the main techniques that are tested to solve the WPRE regression problem are described. In turn, Sect. 5

shows the experimental part of the work, where the results are obtained in three WPRE prediction problems located at wind farm in Spain. Section 6 closes the paper by giving some final concluding remarks on the work carried out.

## 2   Problem Definition

Following [8], a WPRE can be characterized by the following parameters:

- Magnitude ($\Delta P_r$): defined as the variation in power produced in the wind farm or wind turbine during the ramp event.
- Duration ($\Delta t_r$): time period during which the ramp event is produced.
- Ramp rate ($\Delta P_r / \Delta t_r$): defined as the ramp intensity.
- Timing ($t_0$): a time instance related to the ramp occurrence, which can be defined as the starting point or the central point of the wind ramp event.

Let $S_t : \mathbb{R}^k \to \mathbb{R}$ be the so-called *ramp function*, which evaluates if a WPRE occur or not. On the other hand, $S_t$ can be defined by several forms, but the power ($P_t$) produced in the wind farm or wind turbine is included in all of them. The most common ones are:

$$S_t^1 = P_{t+\Delta t_r} - P_t \tag{1}$$

$$S_t^2 = \max([P_t, P_{t+\Delta t_r}]) - \min([P_t, P_{t+\Delta t_r}]) \tag{2}$$

Using any of the previous definitions of $S_t$, and setting a threshold called $S_0$, a binary label for each ramp event can be defined:

$$I_t = \begin{cases} 1 \; if \;\; S_t \geq S_0 \\ 0 \; otherwise \end{cases} \tag{3}$$

Regarding the set of predictive variables, we propose to take into account the meteorological processes in the wind farms, because they are the physical precursors of WPREs. These meteorological variables can be obtained from different meteorological prediction systems, including reanalysis data [14], which will be detailed in the next section. The problem tackled can be defined as follows: Let $\mathbf{X}_t = \{x_{t1}, \ldots, x_{tN}\}$, $t = 1, \ldots, T$ be a time series of $N$ predictive variables, and considering $T$ associated values of the ramp function $I_t$, the problem consist of training a model $\mathcal{M}$ in a subset of $(\mathbf{X}_t, S_t)^{\mathrm{T}}$ (training set), for the purpose of that model $\mathcal{M}$ be applied to a test set $(\mathbf{X}_t, S_t)^{\mathcal{R}}$, where a classification error ($e$) is minimized.

## 3   Data and Predictive Variables

For the development of this work, we use a reanalysis data set, which will training our classification model. A *reanalysis* project consists on combining meteorological observations in the past with current forecast models, in order to obtain numerical weather forecasts in a regular grid that covers the entire planet. Currently, a reanalysis project extend from 1979 to nowadays, offering us the state of the Earth system through twice-daily global numerical weather forecasts, air quality analysis, atmospheric composition monitoring, climate monitoring, ocean circulation analysis, hydrological prediction, etc. Maybe the most important one is the *ERA-Interim reanalysis project*, which is the latest global atmospheric reanalysis produced by the European Centre for Medium-Range Weather Forecasts (ECMWF) [17]. ERA-Interim is a global atmospheric reanalysis from 1979, continuously updated in real time. The data assimilation system used to produce ERA-Interim is based on a 2006 release that includes a 4-dimensional variational analysis (4D-Var) with a 12-h analysis window. The spatial resolution of the data set is approximately 15 km, on 60 vertical levels from the surface up to 0.1 hPa. ERA-Interim provides 6-hourly atmospheric fields on model levels, pressure levels, potential temperature and potential vorticity, and 3-hourly surface fields. At this time, a new project is being developed, ERA-5, which will present more resolution in respect of the covered area and the number of pressure levels, and it will present 1-hourly atmospheric fields also.

In order to solve the WPRE prediction problem in this paper, we consider wind and temperature-related predictive variables from ERA-Interim at 4 specific points in the neighborhood of the area under study, as Fig. 1 shows. The variables considered as predictors (Table 1) are taken at different pressure levels (surface, 850 hPa and 500 hPa). A total of 12 predictive variables per ERA-Interim node are considered at time $t$, i.e. in this problem $\mathbf{X}_t$ is formed by $N = 48$ predictive variables. The ERA-Interim time resolution for the predictive variables (6 h) sets in this case the ramp duration taken into account $(\Delta t_r = 6)$. These predictive variables are obtained for a wind farm located in the southern of Spain as Fig. 1 shows. Moreover, we have samples of the wind speed in that area with which the power $P$ will be calculated.

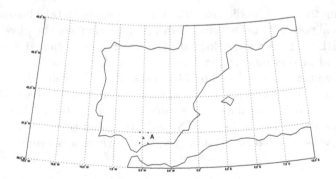

**Fig. 1.** Wind farm considered for the experiments. The four closest nodes from the Era-Interim reanalysis (predictive variables) are also displayed in the picture.

Then, the ramps functions can be written as follows:

$$S_t^1 = P_{t+6} - P_t \tag{4}$$

**Table 1.** Predictive variables considered at each node from the ERA-Interim reanalysis.

| Variable name | ERA-Interim variable |
|---|---|
| skt | Surface temperature |
| sp | Surface pression |
| $u_{10}$ | Zonal wind component ($u$) at 10 m |
| $v_{10}$ | Meridional wind component ($v$) at 10 m |
| temp1 | Temperature at 500 hPa |
| up1 | Zonal wind component ($u$) at 500 hPa |
| vp1 | Meridional wind component ($v$) 500 hPa |
| wp1 | Vertical wind component ($\omega$) at 500 hPa |
| temp2 | Temperature at 850 hPa |
| up2 | Zonal wind component ($u$) at 850 hPa |
| vp2 | Meridional wind component ($v$) at 850 hPa |
| wp2 | Vertical wind component ($\omega$) at 850 hPa |

$$S_t^2 = \max([P_t, P_{t+6}]) - \min([P_t, P_{t+6}]) \tag{5}$$

Thus, each model ($\mathcal{M}$) analyzed in this paper must be trained with the data $(\mathbf{X}_t, S_t^1)^\mathrm{T}$ or $(\mathbf{X}_t, S_t^2)^\mathrm{T}$, where $S_t^1$ and $S_t^2$ are computed using Eqs. (4) and (5), respectively.

We can see more information about the wind farm data used in Table 2.

**Table 2.** Details of the data set obtained from ERA-Interim

| Wind farm | A |
|---|---|
| Number of samples | 15435 |
| Number of ramp samples | 727 |
| % of ramp samples | 4.71% |
| Years of samples | 2002:2013 |

## 4    Computational Methods

A hybrid binary classification system is proposed in this paper, in such a way that an evolutionary algorithm looks for the best set of features for an Extreme Learning Machine. In this section we also describe how to manage the unbalance characteristics of the data base.

### 4.1    Unbalanced Data Processing

The classification of unbalanced data is a very common problem in our days. A data set is unbalanced when its classes are not represented in a equitable way, i.e., the number of samples belonging to a class is much higher than the others. Frequently, the cost of a wrong classification of an unusual sample is higher than in the opposite case. When we want evaluate the performance in a machine learning algorithm, its prediction's accuracy is verified. However, this procedure is not appropriate when the data set is unbalanced, because we can obtain a high percentage of accuracy at the expense of always predict the majority class. For this reason it is necessary to apply alternative techniques in order to evaluate the algorithm used. In this paper we use the main diagonal of the confusion matrix. Our classification system is binary, hence we have only two classes: ramp and not ramp. The confusion matrix's size will be $2 \times 2$ as Table 3 shows.

**Table 3.** Confusion matrix.

|          | Not ramp             | Ramp                |
| -------- | -------------------- | ------------------- |
| Not ramp | True Negatives (TN)  | False Positives (FP) |
| Ramp     | False Negatives (FN) | True Positives (TP)  |

The positive ones (ramp) are the minority class, and the first measure of the main diagonal of the confusion matrix will be: $(TN + TP)/2$.

## 4.2 Synthetic Minority Over-Sampling Technique (SMOTE)

Several studies have demonstrated that the treatment of the unbalanced data set's samples improve the results obtained by the classifier's training. In fact, [18] minimized the number of samples of the majority class maintaining the population of the minority class in order to improve the results in the training process. [19] treated the problem as a combination of the over-sampling of the minority class and the sub-sampling of the majority class. In one of their experiments it was demonstrated that, reducing the number of samples of the majority class, the best results was obtained when the classes was represented equitably. The combination of both techniques provided us major improvements in our results. Synthetic Minority Over-sampling Technique (SMOTE) [20] is an useful over-sampling tool for dealing with unbalanced classification problems. The objective is to increase the number of samples of the minority class by means of the formation of synthetic samples, working on the characteristics space. An over-sampling is applied to the minority class by means of the selection of every sample in this class and introducing synthetic samples as of its the nearest K-neighbors (KNN) of the same class. Depending on the over-sampling applied, neighbors will be selected randomly from their KNN. In our case, a implementation $k = 5$ is used. Let $X = [x_1, ..., x_n]$ be a vector of characteristics and $n$ the number of features, for which its KNNs are calculated, and let $Y$ be one of the its KNN with the same size. The synthetic sample, $Z$, is defined as: $Z = X + (X - Y) \times rand(0, 1)$, where $rand(0, 1)$ causes the selection of an aleatory point in the segment between two particular features. This approach makes that the decision region of the minority class become most general. Finally, the technique SMOTE is merged with an aleatory reduction of the samples from the majority class. This combination makes that the classifier's preference for the majority class is inverted for the minority class.

### 4.3   Proposed Hybrid Approach

Our system is formed by ELM as classifier approach which also provides the fitness function of the evolutionary algorithm, whereas the evolutionary algorithm is focused in the feature selection process. With these tools, the experiments carried out in the wind farm located in Spain will be developed in an independently way. The system proposed consists of two phases: the validation phase, and the training and test phase. The most important one is the validation step, because in this part the main features will be selected for the following phases. Moreover, each phase uses a determined set of characteristics. In our case the split for the different phases is: 20% for the validation set, 60% for the training set and the remaining 20% for the test set. Figure 2 outlines the described method.

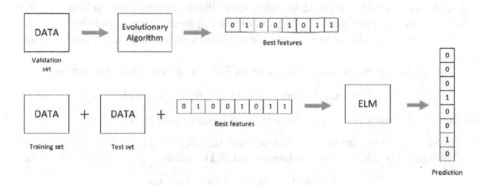

**Fig. 2.** Diagram of the classification system

**Evolutionary Algorithm.** An evolutionary algorithm is a search method that uses the Darwin's Theory in order to solve optimization problems. It is based on a population of solutions which are evolved through successive generations in which a number of genetic operators are applied: selection, crossover and mutation. Selection is the process whereby individuals of the population are selected by means of a ranking process, where only the individuals with the best fitness are kept for the next iterations. The fitness function used in this problem is a measure of the confusion matrix's main diagonal. It is a good measure to evaluate the performance of the classifier proposed, hence the objective will be to maximize the value returned for the fitness function calculation, which is obtained from the output of the ELM algorithm. Once the individuals are selected, the crossover operator is applied. Firstly, the binary strings are coupled at random. Secondly, for each pair of strings, an integer position along the string is selected uniformly at random. Two new strings are composed by swapping all bits between the selected position. Finally the mutation is applied with a very low probability and the offspring is obtained [21]. In this work the evolutionary algorithm play a part in the validation process.

With this procedure the best set of features are selected in order to improve the classification procedure in the following steps: training and test phases.

**Extreme Learning Machine.** A brief explanation of the **Extreme Learning Machine (ELM)** is going to carried out in this Section. The use of this method is currently widespread around the world due to the advantages presents in the algorithm. Specifically, an ELM [22] is a fast learning method based on the structure of MLPs with a novel way of training feed-forward neural networks. One of the most important characteristics of the ELM training is the randomness in the process where the network weights are set, obtaining, in this way, a pseudo-inverse of the hidden-layer output matrix. The simplicity of this technique makes the training algorithm extremely fast. Moreover, it must be remarkable its outstanding performance when is compared to other learning methods, usually better than other established approaches such as classical MLPs or SVRs. Furthermore, the universal approximation capability presents in the ELM network, as well as its classification capability, have been already proven [23].

The ELM algorithm can be explain as follows: given a training set

$$\mathbb{T} = (\mathbf{x}_i, \boldsymbol{W}_i) | \mathbf{x}_i \in \mathbb{R}^n, \boldsymbol{W}_i \in \mathbb{R}, i = 1, \cdots, l,$$

an activation function $g(x)$ and number of hidden nodes $(\tilde{N})$,

1. Randomly assign inputs weights $\mathbf{w}_i$ and bias $b_i$, $i = 1, \cdots, \tilde{N}$.
2. Calculate the hidden layer output matrix $\mathbf{H}$, defined as

$$\mathbf{H} = \begin{bmatrix} g(\mathbf{w}_1\mathbf{x}_1 + b_1) \cdots g(\mathbf{w}_{\tilde{N}}\mathbf{x}_1 + b_{\tilde{N}}) \\ \vdots \quad \cdots \quad \vdots \\ g(\mathbf{w}_1\mathbf{x}_l + b_1) \cdots g(\mathbf{w}_{\tilde{N}}\mathbf{x}_N + b_{\tilde{N}}) \end{bmatrix}_{l \times \tilde{N}} \tag{6}$$

3. Calculate the output weight vector $\beta$ as

$$\beta = \mathbf{H}^\dagger \mathbf{T}, \tag{7}$$

where $\mathbf{H}^\dagger$ stands for the Moore-Penrose inverse of matrix $\mathbf{H}$ [22], and $\mathbf{T}$ is the training output vector, $\mathbf{T} = [\boldsymbol{W}_1, \cdots, \boldsymbol{W}_l]^T$.

The number of hidden nodes $(\tilde{N})$ is a free parameter of the ELM training, and it can be fixed initially, or in a best convenient way, it must be estimated for obtaining good results as a part of a validation set in the learning process. Hence, scanning a range of $\tilde{N}$ values is the solution for this problem.

The Matlab ELM implementation by Huang, freely available in the Internet [24] has been used in this paper.

# 5    Experimental Part

This section presents the experimental evaluation of the hybrid classification method proposed in this paper for a real problem of WPREs prediction. Specifically, a wind farm in Spain is considered, which location is shown in Fig. 1. Note that the wind farm selected covers the South of Spain, characterized by a determined wind regime. The data available for this wind farm ranges from 02/03/2002 to 30/06/2013. A pre-processing step to remove missing and corrupted data was carried out. Note that it is only kept data every 6 h (00 h, 06 h, 12 h and 18 h), to match the predictive variables from the ERA-Interim to the objective variables.

Due to our classification data sets are unbalanced, when a model is trained, whether it is the validation, training or test phase, a SMOTE method is applied to the data set with combinations of over-sampling ($os$) of the 0%, 50%, 100% and 200%, and reduction of the majority class, under-sampling, ($us$) of the 0%, 25%, 50% and 75%, with a total of 16 combinations. Note that the first iteration ($os = 0\%, us = 0\%$) is the case without SMOTE technique.

The following part starts with the validation phase. This process is carried out by means of the evolutionary algorithm. It consists of 10 iterations and a binary codification is used. The population is initialized with 50 individuals, each of one has a total of 48 features. Subsequently, the fitness of the population is estimated by means of the ELM algorithm. As 16 combinations of SMOTE are used, it is necessary to train 16 ELM's and calculating its confusion matrix's averages. The best result and also its corresponding combination of the $os$ and $us$ achieved are saved. The best accuracy is (average diagonal of the confusion matrix) around 95.61% as Table 4 shows.

Table 4. Best validation results of the wind farm A

| Number of hidden neurons | 20 | 40 | 60 | 80 | 100 | 150 | 200 | 300 |
|---|---|---|---|---|---|---|---|---|
| Over-sampling/under-sampling | 50/75 | 200/50 | 100/75 | 200/75 | 200/50 | 200/75 | 200/50 | 200/50 |
| Precision | 91.58% | 90.51% | 90.57% | 92.32% | 90.34% | 92.64% | 95.27% | 94.23% |
| Average diagonal of the confusion matrix | 90.48% | 91.97% | 93.36% | 93.26% | 94.26% | 94.45% | 95.14% | **95.61%** |

With respect to the training and test part, several experiments with different numbers of hidden neurons have been carried out. In this case, this data set is not altered with SMOTE technique because it represents the final samples that have to be classified. Hence, 16 models are trained for each of the numbers of hidden neurons and the best one is selected. Table 5 summarize the results obtained in this part. As in the validation case, it is remarkable the good results obtained for the wind farm A. This is a very unbalanced data set with a 4.71% of samples in the ramp class. Despite of, it is a wind farm with a very low number of ramp samples, it presents high hits of the ramp class, with a value of the average diagonal of the confusion matrix around 89% to the different numbers of hidden neurons (Table 6).

**Table 5.** Best test results of the wind farm A

| Number of hidden neurons | 40 | 60 | 80 | 100 | 150 | 200 | 300 | 500 |
|---|---|---|---|---|---|---|---|---|
| Over-sampling/ under-sampling | 200/75 | 0/50 | 200/50 | 200/0 | 100/25 | 50/50 | 100/0 | 0/25 |
| Precision | 86.03% | 90.37% | 90.05% | 87.40% | 91.22% | 86.49% | 88.62% | 86.29% |
| Average diagonal of the confusion matrix | 87.75% | 88.39% | 89.53% | 89.45% | 88.83% | 88.65% | **90.10%** | 86.58% |

**Table 6.** Confusion matrix of the wind farm A

| | No ramp | Ramp |
|---|---|---|
| No ramp | **88.47%** | 11.52% |
| Ramp | 8.27% | **91.72%** |

## 6 Conclusions

This paper proposes a hybrid classifier system of WPREs in wind farms. The proposed system is formed by an Evolutionary algorithm for feature selection, together with an ELM for providing the final WPRE classification. An SMOTE algorithm is also considered in order to deal with the unbalanced characteristics of the WPRE data. Results in real data from a wind farm located at the south of Spain have show a high accuracy of the system, over 90% in detecting WPRE from reanalysis data.

**Acknowledgements.** This work has been partially supported by *Comunidad de Madrid*, under project number S2013/ICE-2933, and by project TIN2014-54583-C2-2-R of the *Spanish Ministerial Commission of Science and Technology (MICYT)*. The authors acknowledge support by DAMA network TIM2015-70308-REDT.

## References

1. Kumar, Y., Ringenberg, J., Depuru, S.S., Devabhaktuni, V.K., Lee, J.W., Nikolaidis, E., Andersen, B., Afjeh, A.: Wind energy: trends and enabling technologies. Renew. Sustain. Energy Rev. **53**, 209–224 (2016)
2. Yan, J., Liu, Y., Han, S., Wang, Y., Feng, S.: Reviews on uncertainty analysis of wind power forecasting. Renew. Sustain. Energy Rev. **52**, 1322–1330 (2015)
3. Tascikaraoglu, A., Uzunoglu, M.: A review of combined approaches for prediction of short-term wind speed and power. Renew. Sustain. Energy Rev. **34**, 243–254 (2014)

4. Salcedo-Sanz, S., Pastor-Sánchez, A., Prieto, L., Blanco-Aguilera, A., García-Herrera, R.: Feature selection in wind speed prediction systems based on a hybrid coral reefs optimization - extreme learning machine approach. Energy Convers. Manag. **87**, 10–18 (2014)
5. Renani, E.T., Elias, M.F., Rahim, N.A.: Using data-driven approach for wind power prediction: a comparative study. Energy Convers. Manag. **118**, 193–203 (2016)
6. Capellaro, M.: Prediction of site specific wind energy value factors. Renew. Energy **87**(1), 430–436 (2016)
7. Munteanu, I., Besancon, G.: Identification-based prediction of wind park power generation. Renew. Energy **97**, 422–433 (2016)
8. Gallego-Castillo, C., Cuerva-Tejero, A., López-García, O.: A review on the recent history of wind power ramp forecasting. Renew. Sustain. Energy Rev. **52**, 1148–1157 (2015)
9. Ouyang, T., Zha, X., Qin, L.: A survey of wind power ramp forecasting. Energy Power Eng. **5**, 368–372 (2013)
10. Cui, M., Ke, D., Sun, Y., Gan, D., Zhang, J., Hodge, B.M.: Wind power ramp event forecasting using a stochastic scenario generation method. IEEE Trans. Sustain. Energy **6**(2), 422–433 (2015)
11. Barber, C., Bockhorst, J., Roebber, P.: Auto-regressive HMM inference with incomplete data for short-horizon wind forecasting. In: Proceedings of the 24th Annual Conference on Neural Information Processing Systems (NIPS), pp. 1–9 (2010)
12. Gallego-Castillo, C., Costa, A., Cuerva-Tejero, A.: Improving short-term forecasting during ramp events by means of regime-switching artificial neural networks. Adv. Sci. Res. **6**, 55–58 (2011)
13. Cutler, N.J., Kay, M., Jacka, K., Nielsen, T.S.: Detecting, categorizing and forecasting large ramps in wind farm power output using meteorological observations and WPPT. Wind Energy **10**, 453–470 (2007)
14. Gallego-Castillo, C., García-Bustamante, E., Cuerva-Tejero, A., Navarro, J.: Identifying wind power ramp causes from multivariate datasets: a methodological proposal and its application to reanalysis data. IET Renew. Power Gener. **9**(8), 867–875 (2015)
15. Cutler, N.J., Outhred, H.R., MacGill, I.F., Kay, M.J., Kepert, J.D.: Characterizing future large, rapid changes in aggregated wind power using numerical weather prediction spatial fields. Wind Energy **12**(6), 542–555 (2009)
16. Salcedo-Sanz, S., Pérez-Bellido, Á.M., Ortiz-García, E.G., Portilla-Figueras, A., Prieto, L., Paredes, D.: Hybridizing the fifth generation mesoscale model with artificial neural networks for short-term wind speed prediction. Renew. Energy **34**(6), 1451–1457 (2009)
17. Dee, D.P., Uppala, S.M., Simmons, A.J., Berrisford, P., Poli, P., et al.: The ERA-Interim reanalysis: configuration and performance of the data assimilation system. Q. J. R. Meteorol. Soc. **137**, 553–597 (2011)
18. Kubat, M., Matwin, S.: Addressing the curse of imbalanced training sets: one-sided selection. In: Proceedings of the Fourteenth International Conference on Machine Learning, pp. 179–186. Morgan Kaufmann (1997)
19. Ling, C.X., Li, C.: Data mining for direct marketing: problems and solutions. In: Proceedings of the Fourth International Conference on Knowledge Discovery and Data Mining, pp. 73–79. AAAI Press (1998)
20. Chawla, N.V., Bowyer, K.W., Hall, L.O., Kegelmeyer, W.P.: Smote: synthetic minority over-sampling technique. J. Artif. Intell. Res. **16**, 321–357 (2002)

21. Salcedo-Sanz, S., Prado-Cumplido, M., Pérez-Cruz, F., Bousoño-Calzón, C.: Feature selection via genetic optimization. Int. Conf. Artif. Neural Netw. **2415**, 547–552 (2002)
22. Huang, G.B., Zhu, Q.Y.: Extreme learning machine: theory and applications. Neurocomputing **70**, 489–501 (2006)
23. Huang, G.B., Zhou, H., Ding, X., Zhang, R.: Extreme learning machine for regression and multiclass classification. IEEE Trans. Syst. Man Cybern. Part B **42**(2), 513–529 (2012)
24. Huang, G.B.: ELM Matlab code. http://www.ntu.edu.sg/home/egbhuang/elm_codes.html

# Erratum to: A Novel Technique to Estimate Biological Parameters in an Epidemiology Problem

Antone dos Santos Benedito[1] and Fernando Luiz Pio dos Santos[2(✉)]

[1] Institute of Biosciences of Botucatu, São Paulo State University,
Botucatu, Brazil
antone@ibb.unesp.br
[2] Department of Biostatistics, Institute of Biosciences of Botucatu,
São Paulo State University, Distrito Rubião Júnior,
Botucatu, SP 18618-689, Brazil
flpio@ibb.unesp.br
http://www.ibb.unesp.br

**Erratum to:**
**Chapter "A Novel Technique to Estimate Biological**
**Parameters in an Epidemiology Problem" in:**
**I. Rojas et al. (Eds.):**
**Advances in Computational Intelligence (Part I), LNCS 10305,**
**DOI: 10.1007/978-3-319-59153-7_10**

The original version of the paper starting on p. 112 was revised. An acknowledgement has been added, it reads: "The authors would like to thank the Brazilian agencies CAPES for the master's scholarship provided and FAPESP for the financial support received". The original chapter was corrected.

The updated online version of this chapter can be found at
http://dx.doi.org/10.1007/978-3-319-59153-7_10

© Springer International Publishing AG 2017
I. Rojas et al. (Eds.): IWANN 2017, Part I, LNCS 10305, p. E1, 2017.
DOI: 10.1007/978-3-319-59153-7_65

# Author Index